SEDIMENT PROVENANCE

SEDIMENT PROVENANCE

INFLUENCES ON COMPOSITIONAL CHANGE FROM SOURCE TO SINK

Edited by

RAJAT MAZUMDER
Department of Applied Geology, Faculty of Engineering and Science, Curtin University, Sarawak, Malaysia

ELSEVIER

AMSTERDAM • BOSTON • HEIDELBERG • LONDON • NEW YORK • OXFORD
PARIS • SAN DIEGO • SAN FRANCISCO • SINGAPORE • SYDNEY • TOKYO

Elsevier
Radarweg 29, PO Box 211, 1000 AE Amsterdam, Netherlands
The Boulevard, Langford Lane, Kidlington, Oxford OX5 1GB, United Kingdom
50 Hampshire Street, 5th Floor, Cambridge, MA 02139, United States

Library of Congress Cataloging-in-Publication Data
A catalog record for this book is available from the Library of Congress

British Library Cataloguing-in-Publication Data
A catalogue record for this book is available from the British Library

ISBN: 978-0-12-803386-9

For information on all Elsevier publications
visit our website at https://www.elsevier.com/

Coarse to very coarse-grained scoriaceous sandstone (dark colored) interbanded with fine sandstone/siltstone (light colored) and mudstone (brownish), Mio-Pliocene Misaki Formation, Miura Peninsula, Japan. The coarse sandstones are normally graded (turbidites) and were derived from volcanoes. The finer clastics are indigenous background sediments formed in a deep marine sedimentary basin (2000–3000 m deep) in an arc-arc collision zone and thus have different sediment provenance from the coarser clastics. Most of the soft sediment deformation structures preserved within laterally continuous and selective stratigraphic horizons have been interpreted as seismite.

Publisher: Candice Janco
Acquisition Editor: Amy Shapiro
Editorial Project Manager: Tasha Frank
Production Project Manager: Paul Prasad Chandramohan
Designer: Mathew Limbert

Typeset by TNQ Books and Journals

Dedicated to my wife, Sumana Mazumder, for her support and positivity.

Contents

6. Physico-Chemical Characteristics of the Barremian-Aptian Siliciclastic Rocks in the Pondicherry Embryonic Rift Sub-basin, India 85

N. CHAKRABORTY, S. SARKAR, A. MANDAL, W. MEJIAMA, H.A. TAWFIK, R. NAGENDRA, P.K. BOSE, P.G. ERIKSSON

7. Petrological and Geochemical Constraints on Provenance, Paleoweathering, and Tectonic Setting of Clastic Sediments From the Neogene Lambir and Sibuti Formations, Northwest Borneo 123

R. NAGARAJAN, J.S. ARMSTRONG-ALTRIN, F.L. KESSLER, J. JONG

8. What Are the Origins of V-Shaped (Chevron) Dunes in Madagascar? The Case for Their Deposition by a Holocene Megatsunami 155

D. ABBOTT, V. GUSIAKOV, G. RAMBOLAMANANA, D. BREGER, R. MAZUMDER, K. GALINSKAYA

9. The Contourite Problem 183

G. SHANMUGAM

10. Fluvial Systems, Provenance, and Reservoir Development in the Eocene Brennan Basin Member of the Duchesne River Formation, Northern Uinta Basin, Utah 255

T. SATO, M.A. CHAN

Contributors

D. Abbott City College of New York, New York, NY, United States; Lamont-Doherty Earth Observatory of Columbia University, Palisades, NY, United States

P. Acquafredda Università degli Studi di Bari, Bari, Italy

A. Agangi Curtin University, Bentley, WA, Australia

J.S. Armstrong-Altrin Universidad Nacional Autónoma de México, México D.F., México

S. Balakrishnan Pondicherry University, Pondicherry, India

R. Baldacconi Freelancer, Taranto, Italy

A. Basu Indiana University, Bloomington, IN, United States

V.C. Bennett The Australian National University, Canberra, ACT, Australia

P.K. Bose Jadavpur University, Kolkata, India

D. Breger Lamont-Doherty Earth Observatory of Columbia University, Palisades, NY, United States; Micrographic Arts, Saratoga Springs, NY, United States

N. Chakraborty Jadavpur University, Kolkata, India

P.P. Chakraborty University of Delhi, New Delhi, India

M.A. Chan University of Utah, Salt Lake City, UT, United States

J. Chiarenzelli St. Lawrence University, Canton, NY, United States

A.R. Chivas University of Wollongong, Wollongong, NSW, Australia

G. da Costa University of Johannesburg, Auckland Park, South Africa

K. Das Hiroshima University, Higashi-Hiroshima, Japan

P. Dasgupta Durgapur Government College, Durgapur, India

S. De Presidency University, Kolkata, India

W. deLorraine St. Lawrence Zinc Company, Gouverneur, NY, United States

A. Dey Jadavpur University, Kolkata, India

P.G. Eriksson University of Pretoria, Pretoria, South Africa

C.L. Fergusson University of Wollongong, Wollongong, NSW, Australia

V. Festa Università degli Studi di Bari, Bari, Italy

C.R.L. Friend Glendale, Oxon, United Kingdom

K. Galinskaya Brooklyn College, New York, NY, United States

S. Ghosh Presidency University, Kolkata, India

V. Gusiakov Tsunami Laboratory, ICMMG SD RAS, Novosibirsk, Russia

Y. Han China University of Geosciences, Beijing, China

Z. Han Shandong University of Science and Technology, Qingdao, China

R.A. Henderson James Cook University, Townsville, QLD, Australia

A. Hofmann University of Johannesburg, Auckland Park, South Africa

K. Horie National Institute for Polar Research, Tokyo, Japan

M. Ibanez-Mejia Massachusetts Institute of Technology, Cambridge, MA, United States; University of Rochester, Rochester, NY, United States

J. Jong JX Nippon Oil and Gas Exploration (Deepwater Sabah) Limited, Kuala Lumpur, Malaysia

F.L. Kessler Goldbach Geoconsultants O & G, Glattbach, Aschaffenburg, Germany

D. Kratzmann Santa Rosa Junior College, Petaluma, CA, United States

S. Lisco Università degli Studi di Bari, Bari, Italy

D.G.F. Long Laurentian University, Sudbury, ON, Canada

M. Lupulescu New York State Museum, Albany, NY, United States

A. Mandal Jadavpur University, Kolkata, India

G. Mastronuzzi Università degli Studi di Bari, Bari, Italy

R. Mazumder Curtin University, Sarawak, Malaysia

W. Mejiama Osaka City University, Osaka, Japan

M. Moretti Università degli Studi di Bari, Bari, Italy

V. Moretti Regione Puglia – Servizio Ecologia – Ufficio Programmazione, Politiche Energetiche, Bari, Italy

S. Mukherjee Jadavpur University, Kolkata, India

J. Mukhopadhyay Presidency University, Kolkata, India; University of Johannesburg, Auckland Park, South Africa

R. Nagarajan Curtin University, Miri, Sarawak, Malaysia

R. Nagendra Anna University, Chennai, India

A.P. Nutman University of Wollongong, Wollongong, NSW, Australia; Chinese Academy of Geological Sciences, Beijing, China

R. Offler University of Newcastle, Callaghan, NSW, Australia

M. Pisarska-Jamroży Geological Institute, Adam Mickiewicz University, Poznań, Poland

G. Rambolamanana University of Antananarivo, Antananarivo, Madagascar

C.A. Rosiere Federal University of Minas Gerais, Belo Horizonte, Brazil

S. Saha University of Delhi, New Delhi, India

S. Sanyal Jadavpur University, Kolkata, India

S. Sarkar Jadavpur University, Kolkata, India

T. Sato INPEX Corporation, Tokyo, Japan

R. Scotti Freelancer, Taranto, Italy

B. Selleck Colgate University, Hamilton, NY, United States

P. Sengupta Jadavpur University, Kolkata, India

G. Shanmugam The University of Texas at Arlington, Arlington, TX, United States

H.A. Tawfik Tanta University, Tanta, Egypt

M. Tropeano Università degli Studi di Bari, Bari, Italy

Y. Tsutsumi National Science Museum, Tsukuba, Japan

A.J. (Tom) Van Loon Geocom Consultants, Benitachell, Spain

G.M. Young University of Western Ontario, London, ON, Canada

CHAPTER

1

Sediment Provenance: Influence on Compositional Change From Source to Sink

R. Mazumder

Curtin University, Sarawak, Malaysia

OUTLINE

The term "provenance" originates from the Latin word "provenire," meaning to originate. Although commonly used to indicate source or parent rock from which sediments were generated, the term "provenance" actually encompasses *all* factors related to sediment production, with "specific reference to the composition of the parent rocks as well as the physiography and climate of the source area" (Weltje and Eynatten, 2004). Sedimentary provenance data play a critical role in assessing palaeogeographic reconstructions, in constraining lateral displacements in orogens, in characterizing crust that is no longer exposed, in mapping depositional systems, in subsurface correlation, and in predicting reservoir quality (Haughton et al., 1991; Weltje and Eynatten, 2004; Garzanti et al., 2014; Bhattacharya et al., 2016).

The source to sink (S2S) is an approach that connects areas of sediment production with sites of transfer and locations of storage through the quantification of earth processes in a budgetary manner (Walsh et al., 2016; Bhattacharya et al., 2016). Understandably, sediment transport, climate, life, environment, diagenesis/lithification, and contemporaneous tectonism also have significant influences on sediment composition/geochemistry along the way from source to sink. The recent special issue of *Earth Science Reviews* (Walsh et al., 2016) presents several interesting recent to Miocene S2S sediment provenance studies on

Sediment Provenance
http://dx.doi.org/10.1016/B978-0-12-803386-9.00001-0

different continents. One of the critical areas that deserves closer scrutiny by the S2S community is linking the present and the past (Walsh et al., 2016). As pointed out by Walsh et al. (2016), "there continues to be too much community disconnect among 'modern' (process), Quaternary and deep-time researchers." It must be noted that researchers have undertaken provenance analysis of much older (as old as early Archean) sedimentary deposits of the major cratonic blocks of the world, including those of Antarctica and Greenland (see Eriksson et al., 2004 and references therein). In spite of significant technological development and consequent scientific advancement in last 20 years, there is almost no memoir/special publication/book that treats sedimentary rocks from an S2S perspective. This book provides a critical and comprehensive overview as well as new data-based sediment provenance analyses from Precambrian to recent from several continents and will fill in the gap in the knowledge base.

The content of the book has been divided into 19 chapters. The first (Basu) is a critical appraisal of the conceptual evolution and the enhanced scope of inquiries into the provenance of siliciclastic sediments. Van Loon et al. have traced the source of bio/siliciclastic beach sands of the Apulian Coast of Italy. Their analyses reveal a wave-eroded lithified sand source for the beach sands and contribution from a wide variety of organisms. Van Loon and Pisarska-Jamroży have undertaken a detailed heavy mineral study of Pleistocene sandurs, ice-marginal valley and a nearby river in Poland, and have shown that heavy mineral analyses can significantly contribute to the reconstruction of the pathway of sedimentary particles and of the changes in the heavy-mineral spectra from source to sink. The hydraulic conditions prevailing during sediment transportation have the prime control on sediment dispersal patterns, and thus have a significant influence on the changes in sediment composition during the journey from source to sink. Dasgupta has critically reviewed the problematic aspects of paleohydraulic parameter reconstructions from primary sedimentary structures and believes that quantitative methodology for the precise estimation of paleohydraulic parameters from depositional sedimentary structures "is yet to be developed through systematic laboratory and field experiments that can be repeated and empirically verified."

Sedimentological analysis of the Lower Cretaceous siliciclastic rocks (sandstones) of the Pondicherry embryonic rift basin, India by Sarkar et al. clearly reveals cratonic source gaining relative maturity toward the distal depositional setting. Variable degrees of mixing of felsic and mafic components and source-shifting as a consequence of rifting have been established by these authors. Nagarajan et al. have undertaken petrographic and geochemical analyses of Neogene Sibuti and Lambir formations, east Malaysia (Borneo). Their research indicates derivation of sediments from recycled felsic provenance in a predominantly continental to passive margin setting associated with rifting of the proto-South China Sea during the early to middle Miocene. The origin of "V"-shaped elongated dune complexes of Madagascar (Chevron complexes) is disputed; Abbott et al. have argued against the Aeolian origin of these dune complexes. Their sedimentological (grain-size), micropaleontological, and geochronological data from three dune complexes of Madagascar indicate these dune complexes are the depositional product of a Holocene megatsunami possibly related to a Holocene landslide, or bolide impact (Abbott et al.). Many fundamental problems of contourite research have been pointed out by Shanmugam in his detailed and critical review. The contourite domain, according to Shanmugam, is "still in a state of flux after nearly 60 years of research" because of those fundamental problems.

Continental sequences generally record a strong influence of sediment source on depositional facies and provide excellent opportunities for S2S analyses. Sato and Chan have undertaken a detailed sedimentological analysis of the Eocene Duchesne River formation of the Uinta Basin, Utah, USA, and have demonstrated how different source inputs control sedimentary facies development and sandstone petrophysical properties in the sink. Their study reveals the importance of sediment provenance analysis for exploration of fluvial sandstone reservoirs. Van Loon et al. have examined a series of lenses of limestone breccia from the Late Cambrian (Furongian) Chaomidian Formation in Shandong Province, China and interpreted these as a consequence of fragmentation followed by sliding of a breccia layer from the parent layer (the source) to its depositional site (the sink). Long has examined cherts of Upper Jurassic to Lower Cretaceous Tantalus Formation, in south-central Yukon, Canada. His study reveals that a large slab of Cache Creek was obducted over strata of the Yukon–Tanana terrane, and this now eroded slab was the source of chert in the Tantalus piggyback basins.

Late Neoproterozoic to early Mesozoic sedimentary succession of the Tasmanides of eastern Australia developed in an active plate margin setting. Multidisciplinary research undertaken by Fergusson revels provenance switching between the developments of igneous-dominated detritus related to adjoining magmatic arcs (e.g., the Macquarie Arc), and interactions with Gondwana-derived clastics. Chiarenzelli utilized detrital zircons in an upper amphibolite facies terrain to document sediment provenance and basin evolution, and to provide initial temporal constraints on sedimentation. Das et al. have presented detrital records of sediment provenance and its shift in the Mesoproterozoic Singhora Group, central India. Sengupta et al. inferred sedimentary provenance, timing of sedimentation, and metamorphism from a suite of metapelites from the Chotanagpur Granite Gneiss Complex, eastern India, and discussed their implications for Proterozoic tectonics in the east-central part of the Indian shield. Mukhopadhaya et al. have undertaken SEM –CL fabric analysis of quartz framework population from the Mesoarchean Keonjhar Quartzite from Singhbhum Craton, eastern India. These authors have discussed implications of provenance analysis for the upper continental crustal evolution. Costa and Hofmann have undertaken provenance analysis of detrital pyrite in the Mesoarchaean Witwatersrand Basin of South Africa, the world's largest gold deposit. According to these authors, detrital pyrite is mainly derived from sedimentary sources and syn-sedimentary precipitates. Young has discussed the ice ages in earth history, "puzzling" paleolatitudes, and regional provenance of the ice sheets. According to Young, "the evolution of metazoans, climaxing with the 'Cambrian explosion,' may have been accelerated by rapid and radical environmental changes associated with glaciations." The world's oldest sedimentary structures are preserved in dolomitic carbonates, banded iron formations, volcaniclastic sedimentary rocks, and very rare sandstones and conglomerates in the 3.7–3.8 billion years old Isua supracrustal belt in North Atlantic craton (Greenland). The holistic appraisal of the Isua supracrustals by Nutman et al. indicates they formed *over a 100-million-year period in supra-subduction zone settings*.

I strongly believe that a state-of-the art exposition of sediment provenance analyses will help to identify key issues and gaps in the existing knowledge base and initiate new research to understand source rock characteristics, paleoweathering, paleoclimate, tectonics, and ultimately, the evolution of continental crust.

Acknowledgment

I am grateful to all contributors, reviewers, and colleagues at Elsevier, especially Tasha Frank and Marisa La Fleur, who supported me in various ways. I gratefully acknowledge infrastructural support provided by the Faculty of Engineering and Science, Curtin University, Sarawak, Malaysia. Professors Kenneth Eriksson, Patrick G. Eriksson, and Christopher Fedo critically commented on the original book proposal and helped me to organize the book.

References

Bhattacharya, J.P., Copeland, P., Lawton, T.F., Holbrook, J., 2016. Estimation of source area, river paleo-discharge, paleoslope, and sediment budgets of linked deep-time depositional systems and implications for hydrocarbon potential. Earth Science Reviews 153, 77–110.

Eduardo Garzanti, E., Vermeesch, P., Padoan, M., Resentini, A., Vezzoli, G., Andò, S., 2014. Provenance of passive-margin sand (Southern Africa). Journal of Geology 122, 17–42.

Eriksson, P.G., Altermann, W., Nelson, D.R., Mueller, W., Catuneanu, O., 2004. The Precambrian Earth, Tempos and Events. Elsevier Science, 966 p.

Haughton, P.D., Todd, S.P., Morton, A.C., 1991. Sedimentary provenance studies. In: Morton, A.C., Todd, S.P., Haughton, P.D.W. (Eds.), Developments in Sedimentary Provenance Studies, 57. Geological Society Special Publication No, pp. 1–11.

Wals, J.P., Wiberg, P.L., Aalto, R., Nittrouer, C.A., Kuehl, S.A., 2016. Source-to-sink research: economy of the Earth's surface and its strata. Earth Science Reviews 153, 1–6.

Weltje, G.J., Von Eynatten, H., 2004. Quantitative provenance analysis of sediments: review and outlook. Sedimentary Geology 171, 1–11.

CHAPTER

2

Evolution of Siliciclastic Provenance Inquiries: A Critical Appraisal

A. Basu

Indiana University, Bloomington, IN, United States

OUTLINE

1. INTRODUCTION

Curiosity about origin is a fundamental human urge. Investigating the provenance of siliciclastic debris and rocks is a subset of that curiosity. Henry Clifton Sorby sagaciously determined, more than 150 years ago, on the basis of optical petrography, that the quartz arenitic

Sediment Provenance
http://dx.doi.org/10.1016/B978-0-12-803386-9.00002-2

rock of the Millstone Grit in Yorkshire was derived from granitic grus: "The rock had been originally formed from a mixture of quartz sand and felspar sand, but, after deposition, the felspar having been decomposed into a clay-like material, has been forced by the pressure of the super-incumbent rocks into the spaces between the grains of quartz sand" (Sorby, 1859, p. 672). It still stands that siliciclastic rocks, formed by diagenetic preservation of the detritus from the lands and mountains that had been destroyed and only the ruins of which might have survived, are the only ancient repositories available for provenance analysis. The optical microscope was established by Sorby as the principal tool for provenance determination. It still is, although many other analytical techniques and tools have vastly contributed to a far better understanding of provenance analysis in the milieu of the Earth system. At present, it is common to use trace and rare earth element distributions, stable isotope systematics, robust U-Pb ages, magnetic resonance, Raman spectra, as well as optical and backscattered electron images of single minerals and whole rocks to infer provenance. Conceptually, investigations to solve local and somewhat regional problems (Groves, 1931; Mackie, 1897; Johnson, 1872) have evolved to addressing problems of global plate tectonics through time (Myrow et al., 2015; Burrett et al., 2014; Uddin et al., 2007; Argnani et al., 2004; Wombacher and Muenker, 2000; Kröner and Sengor, 1990) and to track crustal growth (Avigad et al., 2012; Bodet and Schärer, 2000). Yet, inferring what have been lost, i.e., temporal assemblages of parent rocks, from a body of left over, drifted, and modified detritus, remains inexact (e.g., Fitches et al., 1990). Pettijohn et al. (1972, p. 298) wrote: "The question of provenance is one of the most difficult problems the sedimentary petrographer is called on to solve." The Holy Grail of that omnipresent unique signature of provenance in siliciclastic material is still eluding sedimentary geologists (e.g., Garzanti, 2015; Artemeiva and Shulgin, 2015).

2. PURPOSE AND SCOPE

The purpose of this chapter is to present a critical appraisal of the conceptual evolution and the enhanced scope of inquiries into the provenance of siliciclastic sediments. The topic is popular. Tens of thousands of peer-reviewed papers have been published on the topic; in 2016 alone, the number has exceeded 1000 if not 2000! The scope of this paper is restricted to the inquiries that have forged fundamentally new insights into Earth processes and Earth history. A few predictions about lines of research are made, which are likely to continue for another 20 years (see Suttner, 1989 for comparison). Although methodology is not the primary focus of the paper, research in siliciclastic provenance has advanced in tandem with advances in new tools and new data processing capabilities. Hence, methodological advances are weaved into the discourse. The author does not apologize for not citing many remarkable works because this is not a comprehensive historical review but a short critique.

Six groundbreaking advances in provenance studies are recognized in this chapter (Fig. 2.1). Sorby (1859; also see quote above) related specific rocks to a sandstone body on the basis of petrography recognizing that detrital feldspars would lose their identity through diagenesis. Thus was born modern provenance studies. Mackie (1897) calculated the percent contribution of different source rocks to the proportion of minerals in sand and sandstones. That was the primary kernel of what would be known as quantitative provenance analysis (Weltje and von Eynatten, 2004; Basu and Hake, 1984). The importance of climate and rates

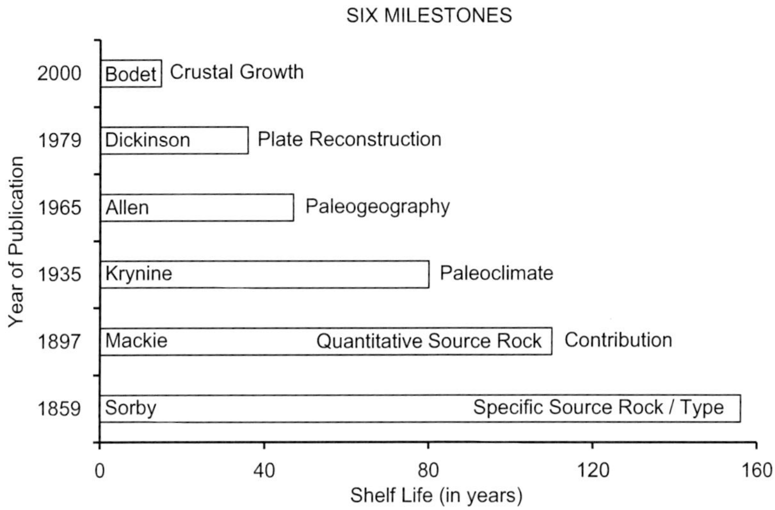

FIGURE 2.1 Six milestones in siliciclastic provenance research. The graph shows the year of significant publication and the number of years of their shelf-lives.

of erosion controlling the relative destruction of feldspars at the source (Krynine, 1935) added a new dimension to provenance investigations (Nesbitt and Young, 1982; Suttner et al., 1981; Ruxton, 1970). Allen (1965) deduced how different paleodrainage systems would give rise to coeval sedimentary provinces (Fig. 2.2) with different mineral compositions, thus adding petrographic constraints to reconstructions of paleogeography and sedimentary provinces (Suttner, 1974; Dickinson, 1970). In a giant leap, Dickinson and Suczek (1979) established a positive linkage between assemblages of rocks in various plate tectonic settings and modal compositions of sandstones derived from those plate tectonic associations (Dickinson, 1980, 1985; Dickinson et al., 1983). The Dickinsonian era of global tectonic provenance studies had begun and has swayed its scepter since then (Bhattacharyya and Das, 2015; Nagel et al., 2014; Uddin et al., 2007; Cawood and Nemchin, 2000; Ingersoll, 1990; Valloni and Zuffa, 1984; Bhatia, 1983). Whereas provenance studies have principally and overwhelmingly investigated the distribution of the earth's surficial rocks and climate, they are now reaching deep into the subsurface—exploring and tracking crustal growth (Bodet and Schärer, 2000).

3. MATERIALS AND RELEVANT PROPERTIES

The siliciclastic materials studied for provenance analyses belong to two groups: (1) samples of the whole rock or that of a size fraction, and (2) single grains of detrital minerals. For the former, petrographic modal analysis, chemical analysis for major and trace elements, and isotopic analysis for the determination of ε_{Nd} are the principal methods employed in provenance studies. Determination of the relative proportions of heavy minerals, and the presence or absence of diagnostic minerals, has been common, but is not as extensively used anymore.

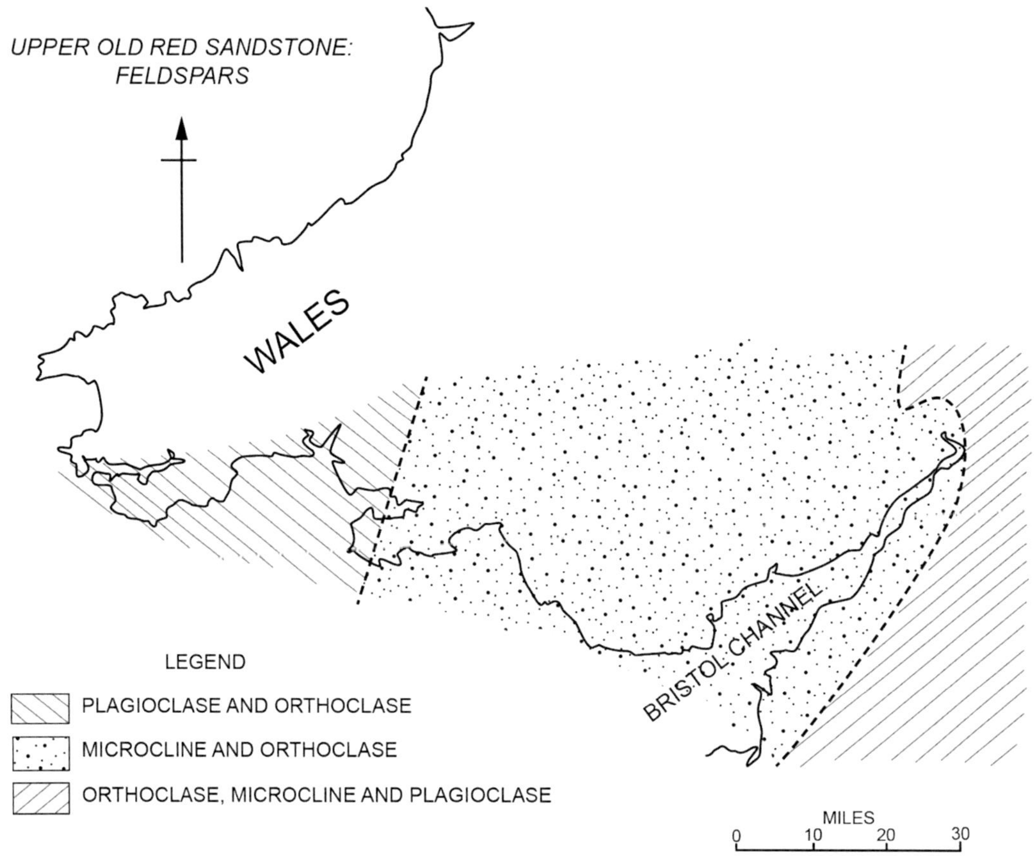

FIGURE 2.2 Paleogeographic reconstruction of the depositional basin of the Old Red Sandstone in southern Wales (UK) on the basis of field geology (mapping primary sedimentary structures and inferring paleocurrent direction) and the relative distribution of microcline, orthoclase, and plagioclase. Modified after Allen (1965).

For heavy minerals, (1) physical properties such as color, optical, and X-ray crystallography, Raman spectroscopy, and cathodoluminescence; (2) concentrations of major, minor, and trace elements; and especially (3) systematics of both stable and radioisotopes including absolute ages, are more in use.

4. INVESTIGATIVE TECHNIQUES AND INSIGHTFUL RESULTS

4.1 Optical Microscopy

Optical microscopy has been and continues to be the mainstay of provenance investigations for identification of mineral grains as small as ~20 μm in siliciclastic rocks. Objective and reproducible modal analyses of sandstones, however, were hampered for over a

100 years because "rock fragments" defied the traditional description of "two or more minerals in a grain of sand." Would a grain of rutilated quartz or a grain of perthite be counted as a rock fragment? Results of modal analyses are commonly plotted in triangular diagrams ostensibly for uniform communication with the three poles marked as Q, F, and L or R or RF. Three formal, fairly rigorous, but different definitions (≈counting methods) have been erected (Suttner et al., 1981; Folk, 1974; Dickinson, 1970; see Appendix I). Modal analyses by these three methods of the same thin section of a sandstone plot differently (Fig. 5 in Zuffa, 1985). The method by Dickinson, more popularly called the Gazzi-Dickinson (G-D) method, has proved to be the most useful and most widely used. Modal data, collected by the G-D method and plotted in the Dickinson diagram (Fig. 2.3), efficiently discriminate derivation of sand-sized siliciclastic detritus from different tectonic provenance (Dickinson, 1985; Dickinson et al., 1983; Dickinson and Suczek, 1979). All three methods, quite wisely, retained the identification of the original labile minerals such as feldspars as "feldspars" even if they were altered fully to clay minerals as long as the detrital grains retained their outlines and other characteristic features such as ghost twinning. Because experienced subjective judgment is necessary for such identification, automated analytical image analysis to determine the modal composition of sandstones is still not possible. But see Bangs-Rooney and Basu (1994) for a possible alternative.

4.2 Chemical Compositions of Bulk Rocks

Bhatia (1983) and Bhatia and Crook (1986) discovered that different sandstone suites from different tectonic settings in Australia, plot differently in CaO-Na_2O-K_2O, La-Th-Sc, Th-Sc-Zr, Ti/Zr versus La/Sc, and La/Y versus Sc/Cr spaces. They conducted statistical analysis of

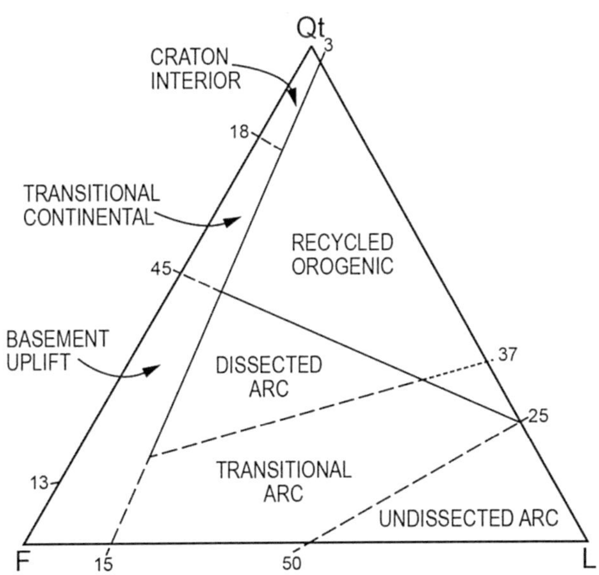

FIGURE 2.3 The basic QFL diagram to plot modal composition of sandstones, counted following Gazzi (1966) and Dickinson (1970). Many have assigned tectonic provenance of their modal data accordingly. Adapted from Dickinson (1985).

their data as did Roser and Korsch (1988) on additional data to validate the discriminatory power of the geochemical approach. Other follow-up studies bear them out. Because weathering and diagenesis convert rock-forming minerals into clay, chemical compositions of whole-rocks represent their mineral compositions at the time of their sampling and analysis. They do not require the subjective judgment of an operator to decide what should be counted as a precursor detrital grain (e.g., feldspar, mica, rock fragment). Elements that are relatively immobile under low-temperature aquatic alterations, and their elemental ratios, likely retain their original relative abundances in sedimentary rocks (Ali et al., 2014). One such plot of ppm Th-Sc-Zr/10 (Fig. 2.4) is very widely used as a template for provenance discrimination (Bhatia, 1983). Techniques for analyzing rock-material (e.g., XRF, INAA, ICPMS, etc.) have improved considerably in the last 30 years and many more elements can now be analyzed at ever-smaller concentrations and with ever-higher precision. The enlarged more precise chemical database has led to quite successful use of multidimensional discriminant function analysis to infer tectonic provenance of siliciclastic sedimentary rocks (Armstrong-Altrin, 2014).

These two avenues for tracking provenance, utilizing whole-rock samples, have been and are the most traveled, and likely to stay so, albeit with some adjustments (see Critique below).

4.3 Populations of Single Derital Minerals

Provenance-sensitive properties of populations of single grains of the same mineral have been determined by many different methods (e.g., optical microscopy, XRD, SEM with BSE and CL detectors, EPMA, LA-MC-ICPMS, SHRIMP, nanoSIMS) more to identify source rock types and petrologic provinces than to identify plate tectonic provenance. Quartz is an extremely durable and the most abundant detrital mineral in clastic sedimentary rocks. Its physical properties such as undulosity of optical extinction (Basu et al., 1975) and CL color

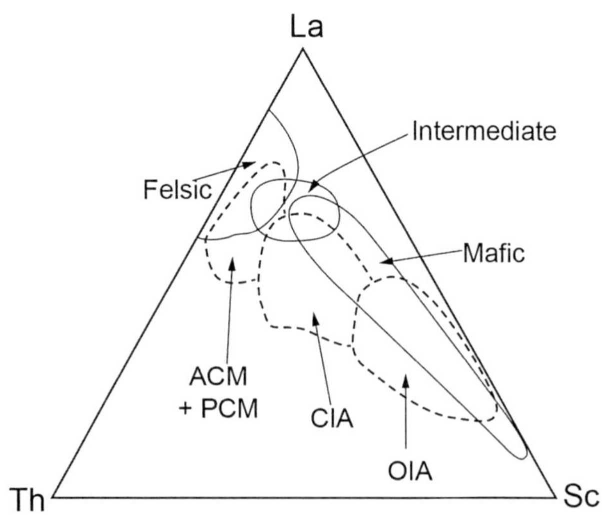

FIGURE 2.4 The standard La-Th-Sc plot to discriminate tectonic provenance of sandstones and shales (Bhatia and Crook, 1986). *PCM*, passive continental margin; *ACM*, active continental margin; *CIA*, continental island arc; *OIA*, oceanic island arc. Also potted are the fields of modern sediments from felsic, intermediate, and mafic source rocks in Colorado (USA) showing the inadequacy of the standard plot (Cullers, 2002). Adapted from Sinha et al. (2007).

(Augustsson and Reker, 2012), and chemical properties such as trace element concentrations (Götze, 2009; Dennen, 1967), have been used widely in provenance studies. Discounting diamond, zircon is the most durable detrital heavy mineral in sedimentary rocks. In situ anayses of individual detrital zircon grains by SHRIMP or LA-MC-ICPMS to determine their trace element characteristics, and isotopic distributions of U-Pb and Lu-Hf in them, have proven to be the most useful and most productive in investigating siliciclastic provenance in recent years (e.g., Fornelli et al., 2015; Fosdick et al., 2011; Grimes et al., 2007; Fedo et al., 2003; Compston and Pidgeon, 1986). Many detrital zircons, because of their durability in rock-forming systems, commonly have successive overgrowths on an igneous or metamorphic core. Absolute ages of the core and the overgrowths record the genetic history of their source rocks (e.g., Wintsch et al., 2007). Additionally, a large collection of detrital zircons yields a large number of absolute ages of parent rocks. The data are best viewed in plots of age versus frequency (Fig. 2.5A–B; Bickford et al., 2013, 2009). Inferring provenance of siliciclastic sediments from spectra of detrital zircon geochronology requires knowledge of the geology, including magmatic and metamorphic events, in the potential source area. For example, Fig. 2.5 shows two detrital zircon spectra from two different formations in two different Proterozoic basins with a dominant ~2.5 Ga peak but one with additional minor age peaks, which confirm two separate source domains. Dickinson and Gehrels (2010) used the ages of 5655 detrital zircons in Mesozoic sandstones in Colorado (USA) to infer the paleogeography and paleotectonics of North America. In a comparative study of the lithotectonic zones of the Himalayas and the Proterozoic–early Cambrian successions in the Indian peninsula, McKenzie et al. (2011) discovered "that rocks of similar depositional age bear strikingly similar detrital zircon age distributions." If so, detrital zircon age spectra thus becomes a robust tool for identifying iso-provenance sedimentary provinces. Although not as durable, but because of its lower blocking temperature, U-Pb ages of detrital monazites track the metamorphic history and the contribution from metamorphic rocks in association with granitic bodies (e.g., Hietpas et al., 2010). Crystallization ages of monazites can be obtained by the cheaper and easier CHIME method (Th-U-Pb) with dedicated EPMA, and are useful in tracking provenance (Pe-Piper et al., 2014; Suzuki and Kato, 2008). Dating rutile (U-Pb) is a new development (Bracciali et al., 2013). Following the trail of detrital zircons (>480 Ma) and rutile (~10 Ma), Braccialli et al. (2015) have discovered the "timing of river capture of the Yarlung Tsangpo by the Brahmaputra." These few examples show how the scope of sedimentary provenance studies has broadened in the last few years. Despite such success with detrital zircon geochronology, it is necessary to note that none of the most durable minerals (e.g., diamond, quartz, zircon, tourmaline, rutile, etc.) occur in all parent rocks of importance. Therefore, reliance on the properties of only one of these minerals, zircon geochronology for example, may be severely misleading in identifying source regions dominated by mafic volcanic rocks.

Major, minor, and trace element concentrations in many other detrital minerals, e.g., feldspar (Trevana and Nash, 1979), garnet (Morton, 1985), tourmaline (van Hinsberg et al., 2011), magnetite-ilmenite-hematite (Dill and Klosa, 2010; Grigsby, 1990), pyroxene (Cawood, 1991), and other heavy minerals (Mange and Wright, 2007), determined mostly with EPMA, have been widely used to solve mostly local and regional problems such as paleogeography, paleodrainage patterns, and stratigraphic correlations (Morton et al., 2013).

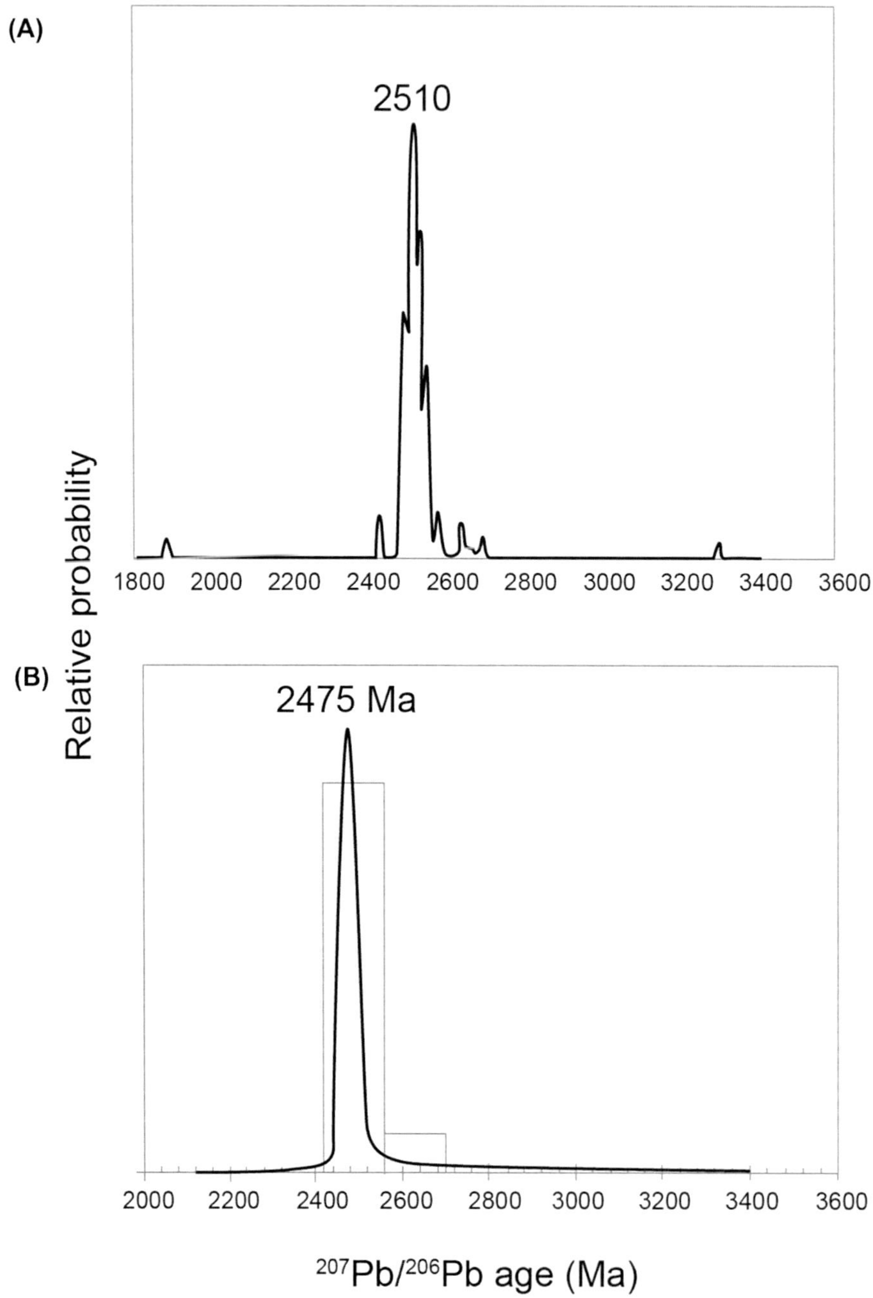

FIGURE 2.5 Detrital zircon age spectra of (A) Oak Shale (late Neoproterzoic? or late Mesoproterozoic?) in the Cuddapah basin, India, and (B) Kansapathar Sandstone (bracketed between 1000 and 1400 Ma) in the Chhattisgarh basin, India showing that the main source for both sedimentary units—some 600 km apart—are the ~2.5 Ga granitic rocks of two different cratons. Field geology precludes any correlation or a common provenance. After Bickford et al. (2009, 2013).

5. THE CRITIQUE

5.1 Bulk Mineralogical Compositions

Empirical studies led Dickinson and Suczek (1979), Dickinson et al. (1983), and Bhatia and Crook (1986) to identify tectonic provenances in North America and Australia in well-defined spaces in QFL and La-Th-Sc and additional/subsidiary plots. Because their sampling was geographically and temporally limited, it would be doubtful if their results could be taken as general templates. A few counter-examples to their perceived universal applicability are discussed below with some explanatory notes. One might note here in parenthesis, that statistical tests of the very datasets used to erect the QFL templates can achieve "success" up to 85% and no more (Molinaroli et al., 1991).

Climate is a significant factor in controlling the composition of sands at their origin. The large orographic barrier of the Himalayas has a much wetter and warmer climate to its south than to its north. Even a small orographic barrier in Jamaica has the same contrast (Gupta, 1975). Compositions of sands generated on the two sides of such orographic barriers are obviously different, although they have been sourced from the same mountain range ($\approx$orogen). Quartz enrichment at the source because of climatic effects has been well documented in modern sands and ancient sandstones (e.g., Garzanti et al., 2015; Mack, 1984; Suttner et al., 1981). Long-distance transport of sand with multiple storages in floodplains, and reworking on the beach, may produce "quartz sand" irrespective of its ultimate provenance. In contrast, beach sands in Papua after a very short transport down a steep slope, even in the hot humid climate, retain the quartz-poor character of their source of a volcanic island arc (Ruxton, 1970). Rivers, long or short, may also collect detritus en route, including recycled grains from older tectonic regimes, or cross other tectonic regimes, which compromise their QFL signature (e.g., Mack, 1984; Dickinson and Suzcek, 1979). Actually, compositions of some modern sands are shown to be affected by different degrees of weathering, systems of transport, and environments of deposition sufficiently enough to defy QFL-type expectations (e.g., Garzanti, 2015; Garzanti et al., 2015; and the extensive references therein). Diagenetic processes destroy labile grains in sandstones to different degrees and in the extreme may be flushed away by groundwater flow, leaving secondary pores and producing diagenetic quartz arenites that, of course, do not retain a QFL memory of their tectonic provenance (McBride, 1987). Diagenesis also produces pseudomatrix out of labile grains, especially feldspar and argillaceous grains (Dickinson, 1970; Sorby, 1859). If not converted fully to pseudomatrix, precursors (e.g., feldspars, volcanic lithic fragments, schist, shale) of some of the argillaceous grains may be identified and counted as such. But the preservation is variable. Hence, Heller et al. (1985) recommended that a sandstone with >20% pseudomatrix should not be included in the QFL-type provenance analysis.

In Dickinson's compilation of the petrography of Phanerozoic North American sandstones, carbonatic sand grains are insignificant and neglected. They are, however, quite profuse in sandstones derived from Mediterranean orogens (Zuffa, 1980). Whereas disregarding such sandstones in QFL-type provenance analysis (Dickinson, 1985, p. 336) would not necessarily invalidate tectonic inferences, it would leave out the provenance information contained in the carbonatic grains, especially those with fossils. They could also distort the modal data not envisaged in the QFL model. Additionally, QFL-type modal data could be distorted if a few sandstones had mixed heritage with recycled grains, and had suffered differential weathering under different climatic conditions that would produce erratic quartz concentrations (Mack, 1984).

Basaltic fragments and calcic plagioclase come not only from rocks in magmatic arcs but also from large intraplate igneous provinces (see map in Fig. 1 of Xia, 2014) that occur in "continental block" tectonic provenance. A somewhat unnoticed paper shows how the QFL compositions of sands derived principally from the largest flood basalt of the present time—the Deccan Traps in India—plot primarily in the magmatic arc provenance field and also in other fields (in response to quartz enrichment because of weathering under hot humid tropical climate) in the QFL diagram (see Figs. 2 and 3 of Garzanti, 2015, and Fig. 3 of Saha et al., 2010). The interpretative error is *potentially* enormous when Proterozoic and Archean (meta-) sandstones plotting in the magmatic arc fields are used as indicators of convergent boundaries of the past. Not recognizing "anorogenic magmatic" fields as substantial sources of volcanic fragments in siliciclastic sedimentary rocks is a deficiency of the Dickinsonian QFL-type provenance analysis (Garzanti, 2015). Sedimentary rocks and their metamorphic equivalents are abundant in orogens, especially in Phanerozoic orogens. Fragments of such rocks are prone to be argillaceous or rendered argillaceous through weathering and diagenesis. Thus, counted with the GD method, such grains would plot at the L-pole (Fig. 2.3) and indicate their recycled orogen provenance. However, uplifted continental blocks in many parts of the world cradle many flat-lying undeformed and unmetamorphosed sedimentary rocks such as in many of the Proterozoic and the Late Paleozoic—Mesozoic basins in the erstwhile Gondwana-Laurentia continents. Sedimentary lithic fragments derived from these basins, plotting at the L-pole, would strongly distort interpretations of tectonic provenance.

Many, many monomineralic quartz grains in siliciclastic sediments, this writer contends, are recycled fragments of sedimentary rocks. Detrital quartz grains with overgrowths are more commonly seen in modern sediments than in ancient sandstones where, in rare cases, abraded overgrowths are preserved (Basu et al., 2013; Critelli et al., 2003; Garzanti et al., 2003). Such rare quartz grains are recycled sedimentary rock fragments; but most others remain unidentified as such. QFL-type analyses miss the relevant provenance information. As of now, however, we have no other petrographic means to distinguish first-cycle quartz from recycled quartz.

5.2 Bulk Chemical Compositions

Chemical compositions of siliciclastic sedimentary rocks have the advantage of representing the bulk sediment and not only the sand-sized fraction as in the case of petrographic analyses although they lack the mineralogical information, i.e., any direct knowledge of the hosts of the chemical components. For example, quartz or calcite cemented quartz arenites will show anomalous enrichment of SiO_2 or CaO and associated trace elements over what was deposited originally. Likewise, a diagenetic quartz arenite with secondary pores after feldspar will show anomalously depleted Al_2O_3, Na_2O, K_2O, and associated trace elements. Barring such extremes, chemical compositions of the muddy parts of sandstones add to the information about the diagenetic products of labile detrital grains, which are now preserved as "matrix" *sensu latu*. If we make an assumption, as very eloquently and boldly stated by Ali et al. (2014), that weathering and diagenetic processes behave like a closed system with respect to a few critical and less mobile elements, then especially their ratios (e.g., La/Sc, Th/Sc, Cr/Th, Th/Co, La/Co, Eu/Eu*, Ba/Co, Nb/La, etc.) in binary or ternary plots would discriminate their tectonic provenance. In fact, all empirical chemical models for discriminating tectonic provenance (e.g., Roser and Korsch, 1986; Bhatia and Crook, 1986) are dependent on this assumption.

The general reservations about the QFL approach mentioned above also apply to the geochemical approach. The empirical data from the sample suites from Australia and New Zealand are not universally applicable. For example, chemical compositions of sands derived primarily from the Deccan basalts in the Indian peninsula plot all along the full stretch from the oceanic arc to the passive margin field in all commonly used geochemical tectonic provenance diagrams (Figs. 7 to 10 in Saha et al., 2010). In a series of papers, Cullers demonstrated that the discrimination between Oceanic Island Arc, Continental Island Arc, Active Continental Margins, and Passive Continental Margins is actually a discrimination between the relative dominance of ultramafic, mafic, intermediate, and felsic suites of rocks in source areas (e.g., Cullers, 2002 and references therein). Fig. 2.3 shows the common La-Th-Sc diagram of Bhatia and Crook (1986) in which Sinha et al. (2007) have plotted the rock-type fields of Cullers (1994). Chemical processes during weathering and diagenesis affect the ultimate chemical compositions of sedimentary rocks. Some elements or their ratios may be far less affected than others and retain their original parent rock characteristics (cf. Ali et al., 2014). Some others, although used in provenance determination, may be affected more. For example, redox conditions during pedogenesis and diagenesis affect the oxidation states and solubility of Fe, Cr, Eu, Ce, U, etc. (e.g., Maulana et al., 2014; Mukhopadhyay et al., 2014; Oze et al., 2004; Shields and Stille, 2001; Pan and Stauffer, 2000; Panahi et al., 2000; Sverjensky, 1984). This indicates that, for example, reliance on Eu and Ce anomalies as provenance indicators may have to be tempered.

It is clear that, by and large, chemical signatures of the source rock-types are preserved in their detritus. But, as in the case of mineralogical compositions, chemical compositions of siliciclastic detritus do not uniquely identify tectonic provenance (Basu et al., 2016). Indeed, PePiper et al. (2016) concludes "Detrital geochemistry alone shows too much variability to interpret provenance."

5.3 Properties of Single Minerals

Fresh, unaltered detrital minerals, individually or in an assemblage, preserve their parental identities. For example, simultaneous presence of high-Cr spinel, uvarovite, and Ni-rich forsterite in a sandstone would indicate derivation from ultramafic bodies such as kimberlite clan rocks. The rarity of such an association of minerals in a suite of heavy minerals makes the example rather unrealistic. In reality, all detrital minerals, other than diamond, quartz, zircon, and to some extent tourmaline, rutile, and garnet, are quite prone to differential preservation in the sedimentary milieu. Thus, although some of their physical, chemical, and isotopic compositions are diagnostic of their provenance, their absence does not necessarily exclude undetected provenances.

Even detrital zircon geochronology, despite the successes described above, has more than one nemesis. Small zircon grains (<30 μm) are not readily amenable to dating because the commonly used analyzing beams, laser or ion, are not much smaller. Hence a population of small detrital zircons may go unrepresented in the results. Because zircon is so durable, it is recycled many times with some mechanical attrition, and older detrital zircons tend to be smaller than younger zircons. Lawrence et al. (2011) have shown that different size fractions of detrital zircons may have different ages and inferences about provenance from just one size fraction may not be correct. Vermeesch (2004) calculates that, to be statistically adequate, at least 117 grains of detrital zircons should be dated. This number, of course, will go up if the diversity of ages goes up in a sample (see also Andersen, 2005). Similarity between two age spectra, given a high probability of a single source, may aid in stratigraphic correlation (McKenzie et al., 2011); but, the "similarity" must be tested statistically. It is easier to infer different provenance from even minor dissimilarity between age spectra (Fig. 2.5A–B). Zircons come from felsic rocks. For provenance studies, they miss mafic and ultramafic sources, for which baddeleyite must be sought. Therefore, zircons alone cannot comprehensively demarcate provenance. Similar reservations apply to populations of other single minerals.

6. DISCUSSION

For about 170 years, field geology and optical microscopic petrography have been and continue to provide the principal database for provenance interpretation of siliciclastic sedimentary rocks. In the last 50 years, chemical and isotopic analyses have supplemented such inquiries (e.g., Blatt, 1967; Middleton, 1960). Both methods and their subsidiaries have investigated bulk compositions and those of individual minerals. The scope has expanded from finding local or regional contexts of sandstone genesis to the plate tectonic regime(s) of provenance. It has become abundantly clear that no single method, or even a combination of a few methods, can always arrive at a unique solution. For example, Artemeiva and Shulgin (2015) showed that geophysical characteristics of the Ladoga Rift in the Baltics—the rift model widely accepted principally on the basis of chemical compositions of volcanic rocks—conform to craton-margin deformation and not a rift. The current trend is to move away from pigeonhole characterization of tectonic provenance and to weigh in the geological context *sensu latu*. Garzanti (2015) dispenses with the original QFL approach on the basis of extensive work on modern sediments. For petrographic modal analyses, Zuffa (personal communication) recommends counting about 50 grain types and 500 points strictly following Chayes (1956); but he still relies, very wisely, on the petrographic characteristics of each grain. The trend is also evident in the geochemical realm where the dominance and mixing of chemical characteristics of source rocks provide the first-order inference (e.g., Cullers, 2002, 2000). A consensus is emerging that source rock types inferred from mineralogical and chemical compositions of siliciclastic rocks alone do not uniquely identify tectonic provenances (cf. Nie et al., 2012).

One current trend is to use only the quantitative data collected with existing methodologies and applying robust multivariate statistical procedures to extract provenance information (e.g., Armstrong-Altrin, 2014; Weltje, 2012). The results look promising so far, but they are constrained by the flaws in the original premise and limited sampling, in successfully erecting universally applicable boundaries of templates for tectonic provenance determination.

7. THE FUTURE

For centuries, both curiosity and societal needs have inspired basic and applied scientific research. Search for the original source rocks or even intermediate "stop-overs" of economic placer deposits, such as of diamond and gold, are well-known time-honored examples (e.g., Oppenheim, 1943; Atkin, 1904). There is now a concentrated effort in the fossil fuel industry to predict the petrophysical properties of subsurface siliciclastic rocks on the basis of their inferred provenance and the estimated extent of their diagenesis (e.g., Heinz and Kairo, 2007). Such studies and predictive models will grow as needs for fossil fuel increase. Contemporary climatic change is a reality. Local and global paleoclimates of the last hundreds to thousands of years, as reflected in modern alluvial to deep-sea sediments (e.g., Asahara et al., 2012; Pal et al., 2012; Lugli et al., 2007), are clues to predicting the immediate future. Because the results require corrections and normalization for the source rock input, provenance studies of modern sediments will expand to decouple tectonic and climatic signatures.

The current trends in measurements and defining original characteristics of detrital minerals, which *survive* in the sedimentary milieu, are likely to gain prominence in the next 20 years or so (cf. Suttner, 1989). Determination of absolute ages of crystallization of individual mineral grains and the overgrowths on them, for example, zircon, monazite, rutile, feldspar, and others, are likely to increase manifold. If some of the mineral grains are recycled (e.g., zircon, rutile), then their histories, especially the records of postdepositional heating events, would help in "purifying" the process of identifying relevant provenance. The distributions of trace elements and stable isotopes (e.g., O, S, Si, Ti, Cr, Fe, Ni) locked up in minerals (e.g., zircon, quartz, rutile, pyroxene, etc.) are commonly indicative of the environments of their crystallization. In situ analyses for such clues (e.g., Hofmann et al., 2009; Götze et al., 2004) are likely to become common in the next decade or two.

Thus we follow Mackie (1897) in our optimistic yet cautious reasoning, and say: "The dust of the old lands has been built into the new. We have taken these tiny fragments—witnesses of a venerable past—and asked them to tell us something of the ancient world which they beheld," and confess, with humility, that provenance remains the most difficult problem for a sedimentary geologist to solve (Pettijohn et al., 1972).

8. CONCLUSIONS

Six giant conceptual leaps in the last 170 years constitute the foundations of contemporary provenance studies of siliciclastic sediments and sedimentary rocks. They have been evaluated, constrained, modified, and contested over the years. These new concepts have survived the tests of time and are likely to "go on forever" (Tennyson is gratefully acknowledged). However, there are caveats.

The revolutionary mineralogical (QFL) approach by Dickinson (1985) followed up by the chemical approach (elemental ratios) erected by Bhatia and Crook (1986), to determine the tectonic provenance of siliciclastic rocks, and thus unravel the geological histories of depositional basins, orogens, and plate movement, do not necessarily lead to unique solutions. Neglecting carbonatic detritus, ignoring the extent of recycled origin of detrital quartz, ignoring flood basalts as parts of uplifted continental/cratonic blocks, ignoring the diversity of uni-sourced detrital mineral assemblages under diverse climatic conditions, and not considering the effects of variable amounts of pore spaces and pore-filling cements, are some of the factors that have affected the empirical rubrics for inferring tectonic provenances.

Multivariate data analyses appear to discriminate a few tectonic provenances quite well. But it is not clear if incorporating new data from, for example, continental flood basalts, would still provide unique solutions.

Physical, chemical, and isotopic properties of single minerals are emerging as stronger discriminating parameters in provenance studies.

Provenance research has gone back to its roots of identifying rock types in their source areas instead of uniquely identifying tectonic provenance.

Acknowledgments

This paper is dedicated to the memory of William R. Dickinson, who revolutionized sedimentary provenance research.

Indiana University, NASA, and NSF have supported my research over the years. Dr. Rajat Mazumder kindly asked me to write this chapter. Reviews and feedback from Professor Daniela Fontana (Universitá Modena, Italy), Dr. Kasturi Bhattacharyya (IITKGP, India), Dr. Sarbani Patranabis-Deb (ISI, India), and especially Dr. Suzanne Kairo (ExxonMobil, USA) helped in correcting errors and omissions. I am grateful to all.

References

Ali, S., Stattegger, K., Garbe-Schönberg, D., Frank, M., Kraft, S., Kuhnt, W., 2014. The provenance of Cretaceous to Quaternary sediments in the Tarfaya basin, SW Morocco: evidence from trace element geochemistry and radiogenic Nd-Sr isotopes. Journal of African Earth Sciences 90, 64–76.

Allen, J.R.L., 1965. Upper Old Red Sandstone (Farlovian) paleogeography in south Wales and the Welsh borderland. Journal of Sedimentary Petrology 35, 167–195.

Andersen, T., 2005. Detrital zircons as tracers of sedimentary provenance: limiting conditions from statistics and numerical simulation. Chemical Geology 216, 249–270.

Argnani, A., Fontana, D., Stefani, C., Zuffa, G.G., 2004. Late Cretaceous carbonate turbidites of the Northern Apennines: shaking Adria at the onset of Alpine collision. Journal of Geology 112, 251–259.

Armstrong-Altrin, J.S., 2014. Evaluation of two multidimensional discrimination diagrams from beach and deep-sea sediments from the Gulf of Mexico and their application to Precambrian clastic sedimentary rocks. International Geology Review 57, 1446–1461.

Artemieva, I.M., Shulgin, A., 2015. Is the Proterozoic Ladoga Rift (SE Baltic Shield) a rift? Precambrian Research 259, 34–42.

Asahara, Y., Takeuchi, F., Nagashima, K., Harada, N., Yamamoto, K., Oguri, K., Tadai, O., 2012. Provenance of terrigenous detritus of the surface sediments in the Bering and Chukchi Seas as derived from Sr and Nd isotopes: implications for recent climate change in the Arctic regions. Deep Sea Research Part II: Topical Studies in Oceanography 61–64, 155–171.

Atkin, A.J.R., 1904. The genesis of the gold-deposits of Barkerville (British Columbia) and the vicinity. Quarterly Journal of the Geological Society (London) 60, 389–393.

Augustsson, C., Reker, A., 2012. Cathodolumenescence spectra of quartz as provenance indicators revisited. Journal of Sedimentary Research 82, 559–570.

Avigad, D., Gerdes, A., Morag, N., Bechstädt, T., 2012. Coupled UPb-Hf of detrital zircons of Cambrian sandstones from Morocco and Sardinia: implications for provenance and Precambrian crustal evolution of North Africa. Gondwana Research 21, 690–703.

Bangs-Rooney, C., Basu, A., 1994. Provenance analysis of muddy sandstones. Journal of Sedimentary Research A64, 2–7.

Basu, A., Hake, H., 1984. An experiment in quantitative provenance interpretation. Geological Society of America, Abstracts with Programs 16, 439.

Basu, A., Young, S.W., Suttner, L.J., James, W.C., Mack, G.H., 1975. Re-evaluation of the use of undulatory extinction and polycrystallinity in detrital quartz for provenance interpretation. Journal of Sedimentary Petrology 45, 873–882.

Basu, A., Schieber, J., Patranabis-Deb, S., Dhang, P.C., 2013. Recycled detrital quartz grains are sedimentary rock fragments indicating unconformities: examples from the Chhattisgarh Supergroup, Bastar craton, India. Journal of Sedimentary Petrology 83, 368–376.

Basu, A., Bickford, M.E., Deasy, R., 2016. Inferring tectonic provenance of siliciclastic rocks from their chemical compositions: a dissent. Sedimentary Geology 336, 26–35. http://dx.doi.org/10.1016/j.sedgeo.2015.11.013.

Bhatia, M.R., Crook, K.A.W., 1986. Trace element characteristics of graywackes and tectonic setting discrimination of sedimentary basins. Contributions to Mineralogy and Petrology 92, 181–193.

Bhatia, M.R., 1983. Plate tectonics and geochemical composition of sandstones. Journal of Geology 91, 611–627.

Bhattacharyya, K., Das, S., 2015. Sandstone petrology and geochemistry of the Kolhan Basin, eastern India: implications for basin tectonics. Journal of Geology & Geosciences 4. http://dx.doi.org/10.4172/2329-6755.1000196.

Bickford, M.E., Basu, A., Patranabis-Deb, S., Dhang, P., 2009. Depositional history of the Mesoproterozoic Chhattisgarh basin, central India: constraints from new SHRIMP zircon ages. Geological Society of America, Abstracts 41, 541.

Bickford, M.E., Saha, D., Schieber, J., Kamenov, G., Russell, A., Basu, A., 2013. New U-Pb ages of zircons in the Owk Shale (Kurnool Group) with reflections on Proterozoic porcellanites in India. Journal of the Geological Society of India 82, 207–216.

Blatt, H., 1967. Provenance determinations and recycling of sediments. Journal of Sedimentary Petrology 37, 1031–1044.

Bodet, F., Schärer, U., 2000. Evolution of the SE-Asian continent from U-Pb and Hf isotopes in single grains of zircon and baddeleyite from large rivers. Geochimica et Cosmochimica Acta 64, 2067–2091.

Bracciali, L., Parrish, R.R., Horstwood, M.S.A., Condon, D., Najman, Y., 2013. U-Pb LA-(MC)-ICP-MS dating of rutile: new reference materials and applications to sedimentary provenance. Chemical Geology 347, 82–101.

Bracciali, L., Najman, Y., Parrish, R.R., Akhter, S.H., Millar, I., 2015. The Brahmaputra tale of tectonics and erosion: early Miocene river capture in the Eastern Himalaya. Earth and Planetary Science Letters 415, 25–37.

Burrett, C., Zaw, K., Meffre, S., Lai, C.K., Khositanont, S., Chaodumrong, P., Udchachon, M., Ekins, S., Halpin, J., 2014. The configuration of Greater Gondwana: evidence from la ICPMS, U-Pb geochronology of detrital zircons from the Palaeozoic and Mesozoic of Southeast Asia and China. Gondwana Research 26, 31–51.

Cawood, P.A., Nemchin, A.A., 2000. Provenance record of a rift basin: U/Pb ages of detrital zircons from the Perth Basin, Western Australia. Sedimentary Geology 134, 209–234.

Cawood, P.A., 1991. Characterisation of intra-oceanic magmatic arc source terranes by provenance studies of derived sediments. New Zealand Journal of Geology and Geophysics 34, 347–358.

Chayes, F., 1956. Petrographic Modal Analysis, 113 p. Wiley, NY.

Compston, W., Pidgeon, R.T., 1986. Jack Hills, evidence of more very old detrital zircons in Western Australia. Nature 321, 766–769.

Critelli, S., Arribas, J., Le Pera, E., Tortosa, A., Marsaglia, K.M., Latter, K.K., 2003. The recycled orogenic sand provenance from an uplifted thrust belt, Betic Cordillera, Southern Spain. Journal of Sedimentary Research 73, 72–81.

Cullers, R.L., 1994. The controls on the major and trace element variation of shales, siltstones, and sandstones of Pennsylvanian-Permian age from uplifted continental blocks in Colorado to platform sediment in Kansas, USA. Geochimica et Cosmochimica Acta 58, 4955–4972.

Cullers, R.L., 2000. The geochemistry of shales, siltstones and sandstones of Pennsylvanian-Permian age, Colorado, USA: implications for provenance and metamorphic studies. Lithos 51, 181–203.

Cullers, R.L., 2002. Implications of elemental concentrations for provenance, redox conditions, and metamorphic studies of shales and limestones near Pueblo, CO, USA. Chemical Geology 191, 305–327.

Dennen, W.H., 1967. Trace elements in quartz as indicators of provenance. Geological Society of America Bulletin 78, 125–130.

Dickinson, W.R., Gehrels, G.E., 2010. Insights into North American Paleogeography and Paleotectonics from U-Pb ages of detrital zircons in Mesozoic strata of the Colorado Plateau, USA. International Journal of Earth Sciences 99, 1247–1265.

Dickinson, W.R., Suczek, C.A., 1979. Plate tectonics and sandstone compositions. AAPG Bulletin 63, 2164–2182.

Dickinson, W.R., Beard, L.S., Brakenridge, G.R., Erjavec, J.L., Ferguson, R.C., Inman, K.F., Knepp, R.A., Lindberg, A.F., Ryberg, P.T., 1983. Provenance of North American Phanerozoic sandstones in relation to tectonic setting. Bulletin of the Geological Society of America 94, 222–235.

Dickinson, W.R., 1970. Interpreting detrital modes of graywackes and arkose. Journal of Sedimentary Petrology 40, 695–707.

Dickinson, W.R., 1980. Plate tectonics and key petrologic associations. In: Strangway, D.W. (Ed.), The Continental Crust and Its Mineral Deposits. Geological Society of Canada Special Paper 20, pp. 341–360.

Dickinson, W.R., 1985. Interpreting provenance relations from detrital modes of sandstones. In: Zuffa, G.G. (Ed.), Provenance of Arenites. NATO ASI C-148, pp. 333–361.

Dill, H., Klosa, D., 2010. Heavy mineral-based provenance analysis of Mesozoic continental-marine sediments at the western edge of the Bohemian Massif, SE Germany: with special reference to Fe-Ti minerals and the crystal morphology of heavy minerals. International Journal of Earth Sciences 1–17. Online First.

Fedo, C.M., Sircombe, K.N., Rainbird, R.H., 2003. Detrital zircon analysis of the sedimentary record. Reviews in Mineralogy and Geochemistry 53, 277–303.

Fitches, W.R., Muir, R.J., Maltman, A.J., Bentley, M.R., 1990. Is the Colonsay-west Islay block of SW Scotland an allochthonous terrane? evidence from Dalradian tillite clasts. Journal of Geological Society (London) 147, 417–420.

Folk, R.L., 1974. Petrology of Sedimentary Rocks. Hemphill's, p. 184.

Fornelli, A., Micheletti, F., Langone, A., Perrone, V., 2015. First U-Pb detrital zircon ages from Numidian sandstones in southern Apennines (Italy): Evidences of African provenance. Sedimentary Geology 320, 19–29.

Fosdick, J.C., Romans, B.W., Fildani, A., Bernhardt, A., Calderón, M., Graham, S.A., 2011. Kinematic evolution of the Patagonian retroarc fold-and-thrust belt and Magallanes foreland basin, Chile and Argentina, 51°30'S. Geological Society of America Bulletin 123, 1679–1698.

Garzanti, E., Andò, S., Vezzoli, G., Dell'era, D., 2003. From rifted margins to foreland basins: investigating provenance and sediment dispersal across desert Arabia (Oman, U.A.E.). Journal of Sedimentary Research 73, 572–588.

Garzanti, E., Andò, S., Padoan, M., Vezzoli, G., El Kammar, A., 2015. The modern Nile sediment system: processes and products. Quaternary Science Reviews 130, 9–56.

Garzanti, E., 2015. From static to dynamic provenance analysis: sedimentary petrology upgraded. Sedimentary Geology 130, 9–56. http://dx.doi.org/10.1016/j.sedgeo.2015.07.010.

Gazzi, P., 1966. Le arenarie del flysch sopra cretaceo dell'Appennino modense: correlazioni con il flysch di Minghidoro. Mineralogica et Petrografica Acta 12, 69–97.

Götze, J., Plötze, M., Graupner, T., Hallbauer, D.K., Bray, C.J., 2004. Trace element incorporation into quartz: a combined study by ICP-MS, electron spin resonance, cathodoluminescence, capillary ion analysis, and gas chromatography. Geochimica et Cosmochimica Acta 68, 3741–3759.

Götze, J., 2009. Chemistry, textures and physical properties of quartz - geological interpretation and technical application. Mineralogical Magazine 73, 645–671.

Grigsby, J.D., 1990. Detrital magnetite as a provenance indicator. Journal of Sedimentary Petrology 60, 940–951.

Grimes, C.B., John, B.E., Kelemen, P.B., Mazdab, F.K., Wooden, J.L., Cheadle, M.J., Hanghøj, K., Schwartz, J.J., 2007. Trace element chemistry of zircons from oceanic crust: a method for distinguishing detrital zircon provenance. Geology 35, 643–646.

Groves, A.W., 1931. The unroofing of the Dartmoor Granite and the distribution of its detritus in the sediments of southern England. Quarterly Journal of the Geological Society 87, 62–96.

Gupta, A., 1975. Stream characteristics in eastern Jamaica, an environment of seasonal flow and large floods. American Journal of Science 275, 825–847.

Heins, W.A., Kairo, S., 2007. Predicting sand character with integrated genetic analysis. In: Arribas, J., Critelli, S., Johnsson, M.J. (Eds.), Sedimentary Provenance and Petrogenesis: Perspectives from Petrography and Geochemistry. Geological Society of America, Special Paper, 420, 345–379.

Heller, P., Peterman, Z.E., O'Neil, J.R., Shafiqullah, M., 1985. Isotopic provenance of sandstones from the Eocene tyee formation, Oregon coast range. Bulletin of the Geological Society of America 96, 770–780.

Hietpas, J., Samson, S., Moecher, D., Schmitt, A.K., 2010. Recovering tectonic events from the sedimentary record: detrital monazite plays in high fidelity. Geology 38, 167–170.

Hofmann, A.E., Valley, J.W., Watson, E.B., Cavosie, A.J., Eiler, J.M., 2009. Sub-micron scale distributions of trace elements in zircon. Contributions to Mineralogy and Petrology 158, 317–335.

Ingersoll, R.V., 1990. Actualistic sandstone petrofacies: discriminating modern and ancient source rocks. Geology 18, 733–736.

Johnson, M.H., 1872. Sources of sandstone. Nature 6, 26.

Kröner, A., Sengor, A.M.C., 1990. Archean and Proterozoic ancestry in late Precambrian to early Paleozoic crustal elements of southern Turkey as revealed by single-zircon dating. Geology 18, 1186–1190.

Krynine, P.D., 1935. Arkose deposits in the humid tropics. A study of sedimentation in southern Mexico. American Journal of Science 29 (Series 5), 353–363.

Lawrence, R.L., Cox, R., Mapes, R.W., Coleman, D.S., 2011. Hydrodynamic fractionation of zircon age populations. Geological Society of America Bulletin 123, 295–305.

Lugli, S., Dori, S.M., Fontana, D., 2007. Alluvial sand composition as a tool to unravel late Quaternary sedimentation of the Modena Plain, northern Italy. In: Arribas, J., Critelli, S., Johnsson, M.J. (Eds.), Sedimentary Provenance and Petrogenesis; Perspectives from Petrography and Geochemistry. Geological Society of America, Special Paper, 420, 57–72.

Mack, G.H., 1984. Exceptions to the relationship between plate tectonics and sandstone composition. Journal of Sedimentary Petrology 54, 212–220.

Mackie, W., 1897. The sands and sandstones of eastern Moray. Edinburgh Geological Society Transactions 7, 148–172.

Mange, M.A., Wright, D.T., 2007. Heavy Minerals in Use. In: Developments in Sedimentology, 58. Elsevier, 1283 pp.

Maulana, A., Yonezu, K., Watanabe, K., 2014. Geochemistry of rare earth elements (REE) in the weathered crusts from the granitic rocks in Sulawesi Island, Indonesia. Journal of Earth Science 25, 460–472.

McBride, E.F., 1987. Diagenesis of the Maxon sandstone (Early Cretaceous), Marathon Region, Texas: a diagenetic quartzarenite. Journal of Sedimentary Petrology 57, 98–107.

McKenzie, N.R., Hughes, N.C., Myrow, P.M., Xiao, S., Sharma, M., 2011. Correlation of Precambrian–Cambrian sedimentary successions across northern India and the utility of isotopic signatures of Himalayan lithotectonic zones. Earth and Planetary Science Letters 312, 471–483.

Middleton, G.V., 1960. Chemical compositions of sandstones. Geological Society of America Bulletin 71, 1011–1026.

Molinaroli, E., Blom, M., Basu, A., 1991. Methods of provenance determination tested with discriminant function analysis. Journal of Sedimentary Petrology 61, 900–908.

Morton, A., Hounslow, M.W., Frei, D., 2013. Heavy-mineral, mineral-chemical and zircon-age constraints on the provenance of Triassic sandstones from the Devon coast, southern Britain. Geologos 19, 67–85.

Morton, A.C., 1985. A new approach to provenance studies: electron microprobe analysis of detrital garnets from Middle Jurassic sandstones of the North Sea. Sedimentology 32, 553–566.

Mukhopadhyay, J., Crowley, Q.G., Ghosh, S., Ghosh, G., Chakrabarti, K., Misra, B., Heron, K., Bose, S., 2014. Oxygenation of the Archean atmosphere: New paleosol constraints from eastern India. Geology 42, 923–926.

Myrow, P.M., Hughes, N.C., Derry, L.A., McKenzie, N.R., Jiang, G., Webb, A.A.G., Banerjee, D.M., Paulsen, T.S., Singh, B.P., 2015. Neogene marine isotopic evolution and the erosion of Lesser Himalayan strata: implications for Cenozoic tectonic history. Earth and Planetary Science Letters 417, 142–150.

Nagel, S., Castelltort, S., Garzanti, E., Lin, A.T., Willett, S.D., Mouthereau, F., Limonta, M., Adatte, T., 2014. Provenance evolution during arc–continent collision. Sedimentary petrography of Miocene to Pleistocene Sediments in the western foreland basin of Taiwan. Journal of Sedimentary Research 84, 513–528.

Nesbitt, H.W., Young, G.M., 1982. Early Proterozoic climates and plate motions inferred from major element chemistry of lutites. Nature 299, 715–717.

Nie, J., Horton, B.K., Saylor, J.E., Mora, A.S., Mange, M., Garzione, C.N., Basu, A., Moreno, C.J., Caballero, V., Parra, M., 2012. Integrated provenance analysis of a convergent retroarc foreland system: U-Pb ages, heavy minerals, Nd isotopes, and sandstone compositions of the Middle Magdalena Valley basin, northern Andes, Colombia. Earth-Science Reviews 110, 111–126.

Oppenheim, V., 1943. Diamonds in the northeastern Bolivian Andes. Economic Geology 38, 658–661.

Oze, C., Fendorf, S., Bird, D.K., Coleman, R.G., 2004. Chromium geochemistry in serpentinized ultramafic rocks and serpentine soils from the Franciscan complex of California. American Journal of Science 304, 67–101.

Pal, D.K., Bhattacharyya, T., Sinha, R., Srivastava, P., Dasgupta, A.S., Chandran, P., Ray, S.K., Nimje, A., 2012. Clay minerals record from Late Quaternary drill cores of the Ganga Plains and their implications for provenance and climate change in the Himalayan foreland: Special Issue: Quaternary fluvial systems of Tropics. Palaeogeography, Palaeoclimatology, Palaeoecology 356–357, 27–37.

Pan, Y., Stauffer, M.R., 2000. Cerium anomaly and Th/U fractionation in the 1.85 Ga Flin Flon Paleosol: clues from REE- and U-rich accessory minerals and implications for paleoatmospheric reconstruction. American Mineralogist 85, 898–911.

Panahi, A., Young, G.M., Rainbird, R.H., 2000. Behavior of major and trace elements (including REE) during Paleoproterozoic pedogenesis and diagenetic alteration of an Archean granite near Ville Marie, Quebec, Canada. Geochimica et Cosmochimica Acta 64, 2199–2220.

Pe-Piper, G., Piper, D.J.W., Triantafyllidis, S., 2014. Detrital monazite geochronology, Upper Jurassic-Lower Cretaceous of the Scotian Basin: significance for tracking first-cycle sources. Geological Society, Special Publications 386, 293–311.

Pe-Piper, G., Piper, D.J.W., Wang, Y., Zhang, Y., Trottier, C., Ge, C., Yin, Y., 2016. Quaternary evolution of the rivers of northeast Hainan Island, China: tracking the history of avulsion from mineralogy and geochemistry of river and delta sands. Sedimentary Geology 333, 84–99.

Pettijohn, F.J., Potter, P.E., Siever, R., 1972. Sand and Sandstone. Springer-Verlag, p. 618.

Roser, B.P., Korsch, R.J., 1986. Determination of tectonic setting of sandstone-mudstone suites using SiO_2 content and K_2O/Na_2O ratios. Journal of Geology 94, 635–650.

Roser, B.P., Korsch, R.J., 1988. Provenance signatures of sandstone-mudstone suites determined using discriminant function analysis of major-element data. Chemical Geology 67, 119–139.

Ruxton, B.P., 1970. Labile quartz-poor sediments from young mountain ranges in northeast Papua. Journal of Sedimentary Petrology 40, 1262–1270.

Saha, S., Banerjee, S., Burley, S.D., Ghosh, A., Saraswati, P.K., 2010. The influence of flood basaltic source terrains on the efficiency of tectonic setting discrimination diagrams: an example from the Gulf of Khambhat, western India. Sedimentary Geology 228, 1–13.

Shields, G.A., Stille, P., 2001. Diagenetic constraints on the use of cerium anomalies as palaeoseawater redox proxies: an isotopic and REE study of Cambrian phosphorites. Chemical Geology 175, 29–48.

Sinha, S., Islam, R., Ghosh, S.K., Kumar, R., Sangode, S.J., 2007. Geochemistry of Neogene Siwalik mudstones along Punjab re-entrant, India: implications for source area weathering, provenance and tectonic setting. Current Science 92, 1103–1113.

Sorby, H.C., 1859. On the structure and origin of the Millstone-Grit in south Yorkshire. Proceedings of the Geological and Polytechnic Society of the West Riding of Yorkshire 3, 669–675.

Suttner, L.J., Basu, A., Mack, G.H., 1981. Climate and the origin of quartz arenites. Journal of Sedimentary Research 51, 1235–1246.

Suttner, L.J., 1974. Sedimentary Petrographic Provinces: An Evaluation. SEPM Special Pub. SP21, pp. 75–84.

Suttner, L.J., 1989. Recent advances in study of the detrital mineralogy of sand and sandstone; implications for teaching. Journal of Geoscience Education 37, 235–240.

Suzuki, K., Kato, T., 2008. CHIME dating of monazite, xenotime, zircon and polycrase: protocol, pitfalls and chemical criterion of possibly discordant age data. Gondwana Research 14, 569–586.

Sverjensky, D.A., 1984. Europium redox equilibria in aqueous solution. Earth and Planetary Science Letters 67, 70–78.

Trevena, A.S., Nash, W.P., 1979. Chemistry and provenance of detrital plagioclase. Geology 7, 475–478.

Uddin, A., Kumar, P., Sarma, J.N., 2007. Early orogenic history of the Eastern Himalayas; compositional studies of Paleogene sandstones from Assam, Northeast India. International Geology Review 49, 798–810.

Valloni, R., Zuffa, G.G., 1984. Provenance changes for arenaceous formations of the Northern Apennines, Italy. Bulletin of the Geological Society of America 95, 1035–1039.

van Hinsberg, V.J., Henry, D.J., Dutrow, B.L., 2011. Tourmaline as a petrologic forensic mineral: a unique recorder of its geologic past. Elements 7, 327–332.

Vermeesch, P., 2004. How many grains are needed for a provenance study? Earth and Planetary Science Letters 224, 441–451.

Weltje, G.J., von Eynatten, H., 2004. Quantitative provenance analysis of sediments: review and outlook. Sedimentary Geology 171, 1–11.

Weltje, G.J., 2012. Quantitative models of sediment generation and provenance: state of the art and future developments: actualistic models of sediment generation. Sedimentary Geology 280, 4–20.

Wintsch, R.P., Aleinikoff, J.N., Walsh, G.J., Bothner, W.A., Hussey, A.M., Fanning, C.M., 2007. SHRIMP U-Pb evidence for a Late Silurian age of metasedimentary rocks in the Merrimack and Putnam-Nashoba terranes, eastern New England. American Journal of Science 307, 119–167.

Wombacher, F., Muenker, C., 2000. Pb, Nd, and Sr isotopes and REE systematics of Cambrian sediments from New Zealand: implications for the reconstruction of the early Paleozoic Gondwana margin along Australia and Antarctica. Journal of Geology 108, 663–686.

Xia, L.-Q., 2014. The geochemical criteria to distinguish continental basalts from arc related ones. Earth-Science Reviews 139, 195–212.

Zuffa, G.G., 1985. Optical analysis of arenites: influence of methodology on compositional trends. In: Zuffa, G.G. (Ed.), Provenance of Arenites. NATO ASI C-148, pp. 165–189.

Zuffa, G.G., 1980. Hybrid arenites: their composition and classification. Journal of Sedimentary Petrology 50, 21–29.

APPENDIX I. DIVERSE CRITERIA FOR MODAL ANALYSIS OF SANDSTONES

Dickinson, 1970, Table 1, p. 698

Q is sum of:	(1) Monocrystalline quartz grains; (2) polycrystalline quartz or chalcedony fragments; (3) cryptocrystalline opaline fragments;* (4) quartz within microphanteritic lithic fragments;* (5) microphenocrystic quartz within aphanite lithic fragments.
F is sum of:	(1) Monocrystalline feldspar grains;* (2) feldspar within microphaneritic lithic fragments;* (3) microphenocrystic within aphanite lithic fragments.
L is aphanatic rock fragments *less*:	(1) Quartzose, chalcedonic, and opaline aphanite fragments;* (2) microphenocrysts.

**Optionally, count only sand-sized crystals.*

Folk, 1974, p. 127

Q-pole:	All types of quartz including metaquartzite (but *not* chert).
F-pole:	All single feldspar (K or NaCa), plus granite and gneiss fragments (plutonic and coarse grained, deep-crustal rocks).
RF-pole:	All other fine-grained rock fragments (supracrustal): chert, slate, schist, volcanics, limestone, sandstone, shale, etc.

Traditional Method. Formalized by Suttner et al. (1981), footnote p. 1236

A *rock-fragment* is a grain with two or more phases or crystal where (1) no single crystal is $>$ 90 percent of the total volume of the particle, or (2) at least two phases or crystals are both $>$0.063 mm in size (usually applicable in coarser size fractions). Polycrystalline quartz is a special rock fragment.

CHAPTER

3

Tracing the Source of the Bio/Siliciclastic Beach Sands at Rosa Marina (Apulian Coast, SE Italy)

A.J. (Tom) Van Loon[1], *M. Moretti*[2], *M. Tropeano*[2], *P. Acquafredda*[2], *R. Baldacconi*[3], *V. Festa*[2], *S. Lisco*[2], *G. Mastronuzzi*[2], *V. Moretti*[4], *R. Scotti*[3]

[1]Geocom Consultants, Benitachell, Spain; [2]Università degli Studi di Bari, Bari, Italy; [3]Freelancer, Taranto, Italy; [4]Regione Puglia – Servizio Ecologia – Ufficio Programmazione, Politiche Energetiche, Bari, Italy

OUTLINE

Sediment Provenance
http://dx.doi.org/10.1016/B978-0-12-803386-9.00003-4

1. INTRODUCTION

Because wave erosion of coasts, whether or not related to sea-level rise, poses a significant hazard for many areas, beaches nowadays attract much interest from earth scientists (Schwartz, 2005), and particularly from sedimentologists (Greenwood and Davies, 1984) and geomorphologists (Bird, 2008). It appears that the evolution of the beaches is quite complex, but understanding of the processes involved may have an important economic and social impact; numerous scientific articles and books therefore focus on a variety of geological features concerning beaches such as shoreline dynamics (Ingle, 1966; Fredsøe and Deigaard, 1994; Anthony, 2009), coastal erosion (Charlier and De Meyer, 1989; Uda, 2010; Van Rijn, 2011; Manca et al., 2013), management and monitoring of coastal areas (NRC, 1989; Kay and Alder, 2002), and beach nourishment (Finkl, 1981; NRC, 1989; Nordstrom, 2005).

Here we present a study of a beach using a multidisciplinary approach in order to describe the relationship between lithoclastic and bioclastic sediment in the Rosa Marina coastal area, along the Adriatic sector of the Apulian region. We present for the purpose a methodology for beach-sediment analysis, aimed at both textural/petrographical characterization of the sands and at the definition of the bioclast content as related to benthic populations and their relationships with the sandy or rocky substratum.

1.1 Beaches as Multiprocess Environments

Coastal sediments may result from the redistribution (due to waves, tides and currents) of the material supplied by rivers and/or eroded from rocks in the coastal area (both types form terrigenous clastic material) and/or from the production of bioclastic particles in the sea. Sediment generated by these biotic processes are most commonly carbonates; beach sediments therefore tend to contain a variable percentage of carbonates, derived from bioclasts (i.e., shells or fragments of organisms living at depth), in combination with siliciclastic material that has been supplied by whatever physical process.

As a consequence, not only a wide variety of physical processes, but also a variety of biological processes could influence beach development. It is therefore remarkable that only a few studies describe the interactions between the biological processes and the sedimentary dynamics in coastal areas (NRC, 1994a,b). The present contribution focuses on the interaction between the organisms living in the near-shore sea, and on the physical processes connected to the local sedimentation. Both aspects are closely related, particularly as far as sediment production is concerned.

1.2 Characteristics of Beach Sands

The interactions between physical and biological processes affecting beaches are so complex that studies of a wide variety of processes are required. The information thus gathered may provide key information for monitoring, protection, and restoration of coastal areas. For example, beach nourishment by supply of sand that replaces eroded sediment (Chiocci and La Monica, 1999; Van der Salm and Unal, 2003; Nicoletti et al., 2006; APAT-ICRAM, 2007; Anfuso et al., 2011) needs, first of all, a physical characterization of the site of interest (bathymetry, wave conditions, coastal shipping, historical analysis, etc.) aimed at restoring the preexisting landscape before erosion took place (slope, extension, articulation, and aesthetics of the rebuilt beach); secondly, it needs a chemical characterization of the sediments to be supplied (with particular attention to their organic and inorganic pollution); and finally, it needs a set of detailed biological and ecological analyses (Colosio et al., 2011) so that insight is gained into the main benthic populations and the presence of seagrasses, as well as insight into the impact of beach nourishment on the biotope.

It is important to note with regard to the situation of the area where replenishment of eroded sand (beach nourishment) is required, that aspects of biological safeguarding are interconnected with physical beach protection, because part of the grains composing the sediment results from the presence in the coastal area of organisms with calcareous shells, possibly contributing a volumetrically significant component to the physical balance of the system.

The sand that is used for beach nourishment must be "compatible" with the original coastal sands, if it is not to be eroded soon again; the definition of compatibility requires a quantitative evaluation of the textural, petrographical, and mineralogical parameters of the main (both lithoclastic and bioclastic) sediment constituents such as color, grain size (to be determined with sieve-size intervals of 1/2 φ), main mineralogical characteristics, etc. It is therefore of great importance to trace the source area(s) of the beach sands.

2. SETTING OF THE STUDY AREA

The area under study, Rosa Marina beach (N40°50′, E17°50′), is located north of Brindisi, along the southeastern coast of Italy (Fig. 3.1). It is now under pressure from a lot of tourism, but it has a great natural and environmental value, which is well recognized by the authorities, because it is included in the Regional Natural Park of the Coastal Dunes between Torre Canne and Torre San Leonardo (established by Regional Law 31, dated 26-10-2006).

2.1 Geographical Aspects

This coastal area includes small catchment areas through which ephemeral streams run (the Pilone Vallone and the Rosa Marina Lama; Fig. 3.1) that are capable of carrying only moderate amounts of sediments to the sea, usually during intense weather events. This supply of terrigenous particles builds a thin coastal wedge in a microtidal, wave-dominated area. The wave conditions, deduced from the 1968–2008 data of a tide gauge and wave measurements at Monopoli (N40°58′30.0″, E17°22′36.1″, less than 25 km from the beach under study), indicate that the prevailing direction of sea storms is from the northwest. The sediment transport by traction along the coast should, therefore, mainly occur in a direction from northwest to southeast.

The coastline of this sector of the Apulian coast is characterized by low-elevation, active sandy coastal dunes and well-developed backshore marsh areas (south of Torre Canne). From Pilone to the southern part of Monticelli (an area including Rosa Marina), the coast

FIGURE 3.1 Location of the study area. Rosa Marina is situated along the Adriatic coast between Bari and Brindisi. It is the site of a large tourist complex. Note the presence of two narrow stream incisions ("Lama" and "Vallone" are two Italian terms for "stream").

becomes more irregular with rocky areas expressed as local headlands. Inactive and/or erosion-affected rows of dunes are preserved in some sandy pocket beaches (narrow beaches between two headlands).

The coastal area under study is characterized by intense urbanization (Puglia Region, 2012), but direct interventions on the coast have been restricted until now to a transverse barrier at the mouth of the Rosa Marina Lama stream and a short breakwater pier in the same area. Yet the coastal area here is affected by erosion; phenomena of coastal erosion have been reported (Annese et al., 2003) for the Torre Canne area (about 7 km north of Rosa Marina), but the shoreline seems stable (Puglia Region, 2006) in the Pilone area (Fig. 3.1).

2.2 Stratigraphy and Sedimentology

We investigated the geological setting and the sedimentological characteristics of the study area briefly, with a focus on the recognition of the main outcropping sedimentary units, the location of their stratigraphic boundaries with reference to the average sea level, the lateral variations in their facies and/or thickness, and their state regarding erosion. The geological setting (Fig. 3.2) has been described in detail earlier (Ciaranfi et al., 1988; Mastronuzzi et al., 2001; Tropeano and Spalluto, 2006).

The Calcarenite di Gravina Formation (Pliocene—Early Pleistocene) is the oldest sedimentary unit cropping out in this area; it represents the substratum of a series of transitional and marine terrace-forming units deposited during the Middle Pleistocene to Holocene (Ciaranfi et al., 1988). The formation is extensively exposed throughout the coastal area under study, especially in

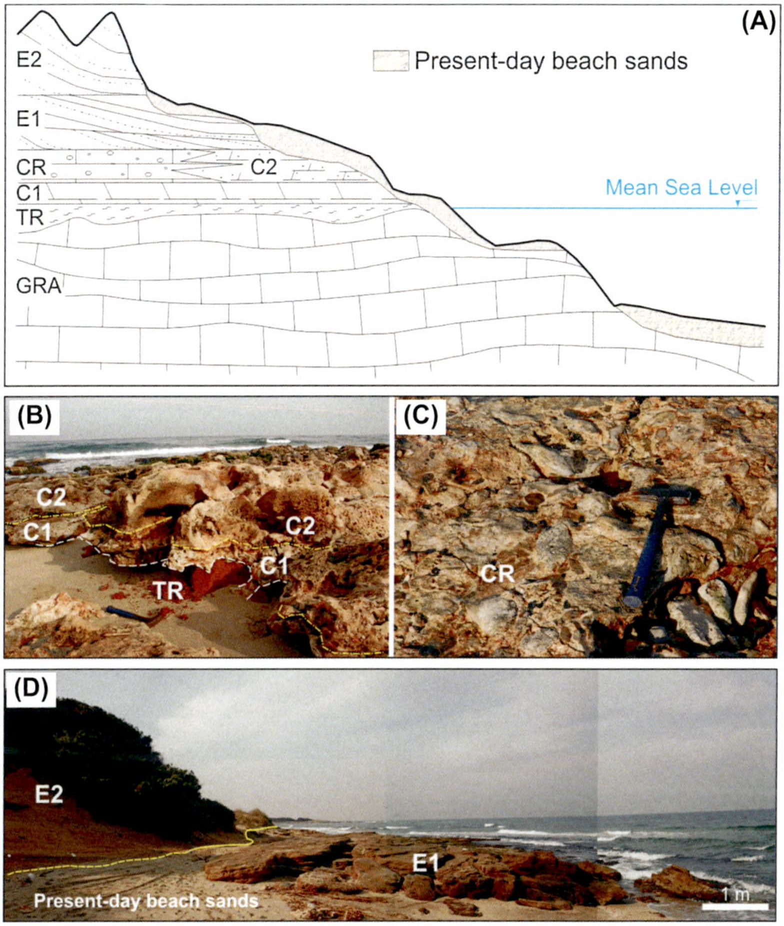

FIGURE 3.2 Geology of the Rosa Marina area. (A) Schematic stratigraphy of the Rosa Marina area. See the text for the complete names and ages of the sedimentary units. (B) Stratigraphical contact (in white) between the red soil unit (TR, at the bottom) and the overlying lagoonal calcarenite (C1). The *yellow line* indicates the contact between the lagoonal unit (C1) and the overlying transgressive calcarenite (C2). (C) Macroscopic features of the calcirudite unit (CR), which passes laterally to the C2 calcarenites. (D) Contact between the middle Holocene aeolian unit (E1) and the overlying younger dune (E2). Note that, in the foreground, unit E1 undergoes considerable erosion, with localized falls and block rotation; in the background, unit E2 is directly exposed to wave erosion.

submerged sectors, where it is eroded by storm waves (GRA in Fig. 3.2A). In this area, the Calcarenite di Gravina Formation is made up of thick beds of medium- and coarse-grained calcarenites, with intense bioturbation (Moretti et al., 2011). Bioclasts (mostly red algae) are the most common constituent.

A thin red soil unit has been recognized on top of the Calcarenite di Gravina Formation. It overlies a continental erosional surface, approximately at sea level (TR in Fig. 3.2A and B) and it passes upwards into marly limestones (C1). Unit C1 is made up of parallel-laminated, fine-grained limestones with abundant ostracods and with rare clay chips of red soils. On top of the TR unit, a coarse-grained calcarenite unit crops out, a few centimeters above sea level (C2 in Fig. 3.2A and B). It is made up of parallel-laminated calcarenites to calcirudites with a high bioclast content (mainly fragments of bivalves and gastropods). Laterally, close to the Rosa Marina Lama stream, unit C2 passes into a calcirudite (unit CR) which contains some gastropod remains and abundant pebbles of micritic limestone and calcarenite (Fig. 3.2A and C).

Overlying older units, an arenaceous unit (E1) crops out between 10 and 30 cm above sea level (Fig. 3.2A and C). It is made up of alternations of well-sorted sandstone with high-angle cross-laminated sandstones; the sandstones have a mixed composition, with quartz and carbonates in almost similar quantities. The topmost outcropping unit is sandy; its base is situated 40 cm to 2 m above sea level. It is a subrecent aeolian unit (E2), representing a coastal dune that is no longer active nowadays, but instead exposed to strong erosion (Fig. 3.2A and C).

The informal units, TR, C1, C2, and E1, are considered to have been deposited on top of the Calcarenite di Gravina Formation, during a transgressive/regressive cycle, recorded by units of coastal lagoon/backshore sediments (TR and C1), passing upwards into shoreface transgressive deposits (C2, laterally fed by the terrigenous carbonates supplied by the Rosa Marina Lama stream, CR). The aeolian unit (E1) represents the regressive part of the succession: it is considered to have been deposited in this area, as in many other coastal areas of the Apulian Foreland, during the middle Holocene (about 6000 years ago: Mastronuzzi et al., 2001; Mastronuzzi and Sanso, 2002). According to Mastronuzzi et al. (2001), the more recent coastal dune E2 was deposited presumably during the late Holocene.

3. METHODS

In order to trace the source area(s) of the present-day beach sands of Rosa Marina, the characteristics of the beach sand and of the sandy or rocky sea bottom were investigated, as well as the marine life forms—up to a water depth of 6 m—that contribute bioclasts to the beach sand.

3.1 Sampling of the Beach Sand

The sands of the present-day beach were sampled in both emerged and submerged areas (Fig. 3.3): (1) in the shoreface, along a transect perpendicular to the coast, from the shoreline to a depth of 6 m (the local storm-wave base); and (2) in the backshore and foreshore, where sand samples were taken every 5 m from the shoreline, taking care to sample both the ordinary and winter berms (in lateral and less frequented areas) until the base of the E2 aeolian sands.

The samples were collected by driving a cylinder sampler, which was tightly closed in order to avoid loss of finer sediments. In the laboratory, the samples were washed with distilled water, dried, and weighed. They were processed with hydrogen peroxide and subsequently passed through a 0.063-mm sieve in order to determine the percentage of organic matter and fine sediment (both present in negligible percentages). The results were processed with the

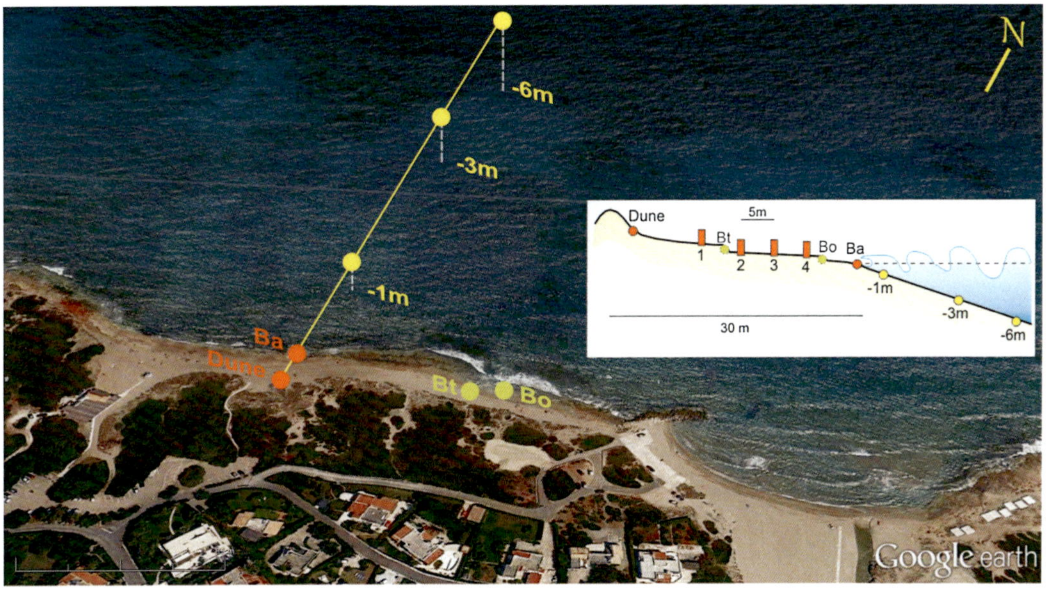

FIGURE 3.3 Locations of the sampling stations. In the backshore (Ba), samples 1 to 4 are located at 20, 15, 10, and 5 m from the swash zone, respectively. Sample "dune" was collected at the base of the E1 dune unit, whereas samples Bo and Bt (ordinary and winter berm) were collected laterally (to the southeast of the transect) in an area of the beach where such morphosedimentary steps were still recognizable.

specific Gradistat v8 software (Blott and Pye, 2001), which yielded distribution histograms, cumulative curves, and the automatic evaluation of the following textural parameters: median grain size (D_{50}), sorting (sg), skewness (Sk), and kurtosis (Kg).

Of course, more reliable data might have been obtained if samples had been collected repeatedly throughout a year, and preferably during several years. Reference data with which our data can be compared are available, however, for adjacent areas (Pilone beach resort; Fig. 3.1); data are taken from Puglia Region (2006). The Pilone beach is located in a stable coastal sector showing subsymmetric distributions (Sk about 0) and mesokurtic to leptokurtic curves (Table 3.1).

3.2 Petrographical Methods

Samples were collected from the various beach subenvironments and analyzed petrographically with a binocular optical microscope and in thin sections. All sedimentary units cropping out in the study area (also submarine, to a depth of 6 m) have also been sampled and studied petrographically for comparison in order to obtain more information on the role of erosion of the local substratum as a source for the sands building the present-day beach.

The various size fractions of the beach sands were analyzed in order to reveal the percentages of the bioclastic particles and any variations in the composition and/or concentration of specific minerals in the different granulometric classes. First the bioclast content was

TABLE 3.1 Grain-Size Parameters Determined for the Adjacent Pilone Beach

Sample	Sampling Depth	D_{50} (φ)	Sorting (σφ)	Description	Skewness (Skφ)	Shape	Kurtosis (kg)	Shape
2006 3_15_7	+1.39	1.56	0.39	Medium sand, well sorted	0	Symmetrical	0.93	Mesokurtic
2006 3_15_8	+0.58	1.85	0.48	Medium sand, well sorted	0.03	Symmetrical	1.04	Mesokurtic
2006 3_15_15	−2.90	2.08	0.43	Fine sand, well sorted	−0.07	Symmetrical	1.33	Leptokurtic
2006 3_15_16	−4.58	2.27	0.32	Fine sand, very well sorted	0.09	Symmetrical	1.54	Strongly leptokurtic

determined quantitatively (the quality is discussed in Section 5.2). All bioclasts (shells or fragments of shells) have been separated and placed in special Petri dishes using a binocular optical microscope; they were weighted and their total fraction was calculated (Table 3.2) as the weight percentage of the total sample.

A simple conversion was used to calculate the bioclast percentage by volume. Siliciclastic particles and nonorganic carbonate particles have more or less the same density: the quartz grains have a density of 2.66 g cm^{-3} whereas that of the carbonates is about 2.70 g cm^{-3}. Bioclastic carbonates can, however, have different densities (Schlager, 2005) according to their composition (2.94 g cm^{-3} for aragonite; 2.72 g cm^{-3} for calcite; 2.89 g cm^{-3} for dolomite), and the structure of the shells and other fragments of marine organisms usually

TABLE 3.2 Bioclast/Lithoclast Percentages (in Weight and Volume) in the Beach Sand

Sample	Bioclasts (% Weight)	Lithoclasts (% Weight)	Bioclasts (% V_{max})	Lithoclasts (% V_{min})
Dune (E2)	0.74	99.26	1	99
1 (20 m from the shoreline)	1.98	98.02	2.67	97.33
2 (15 m from the shoreline)	4.71	95.29	6.35	93.65
3 (10 m from the shoreline)	3.05	96.95	4.11	95.89
4 (5 m from the shoreline)	3.04	96.96	4.1	95.9
Ordinary berm (Bo)	16.84	83.16	22.73	77.27
Swash zone (Ba)	54.58	45.42	73.68	26.32
−1 m	1.06	98.94	1.43	98.57
−3 m	2.42	97.58	3.27	96.73
−6 m	11.36	88.64	15.34	84.66

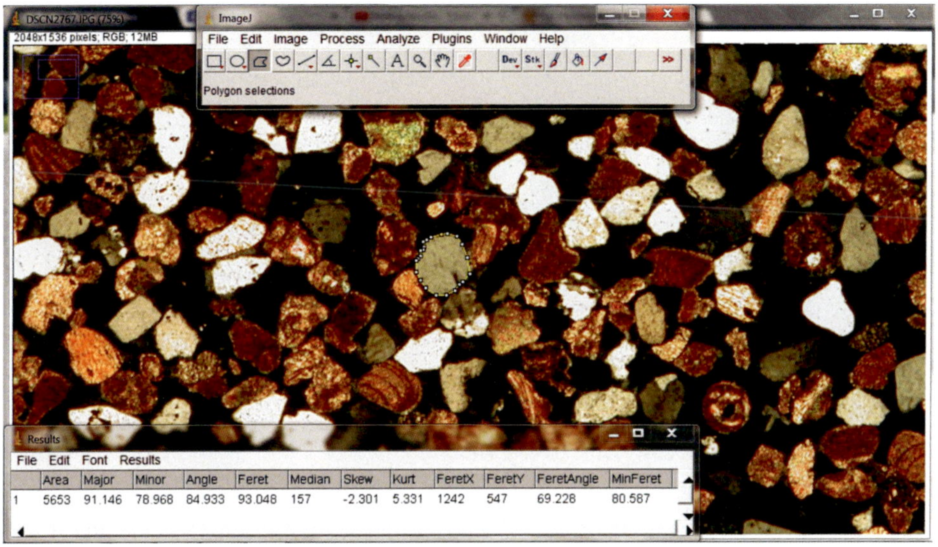

FIGURE 3.4 Image analysis (Image J) software screen. It contains some intuitive but powerful tools (top), allowing the selection or drawing of individual sand grains: the software directly shows the surface area, the diameter, and the shape parameters of the selected grain.

have a varying but relatively high porosity (intraparticle porosity of Choquette and Pray, 1970), resulting in a total bulk volume that varies from a maximum value of 2.7 g cm^{-3} (porosity = 0) to a minimum of 2.0 g cm^{-3} (Jackson and Richardson, 2007). The maximum volume percentage of the bioclasts (the maximum difference from the percentages calculated considering their weight; Table 3.2) was obtained by taking the bulk density of the more porous skeletal material (the "unfilled shells" of Choquette and Pray, 1970) into account.

More quantitative data on the composition of these sands were obtained by analyzing five thin sections (after cementing the grains with epoxy resin); the thin sections cover both emerged and submerged subenvironments. In addition, high-resolution microscopic photos were taken using low magnifications ($\times 1$ and $\times 2$). The raster images that were thus obtained were imported into image-analysis freeware (ImageJ, version 1.49; Fig. 3.4), with the help of which it is possible to recognize, select, and draw (Fig. 3.4) individual grains and to assign them to carbonates, quartz, or "other" minerals, as these form the three main compositional classes of the sands.

4. CHARACTERISTICS OF THE BEACH SAND AND OLDER UNITS

4.1 Composition of the Beach Sand

The beach sand is built of clasts (Fig. 3.5) that consist mainly of carbonates (either lithoclasts or bioclasts), quartz, and other minerals that are present in negligible percentages (such as

FIGURE 3.5 Petrographical features of the Rosa Marina sands observed with an optical binocular microscope. (A) Grains with a diameter of >500 μm: limestone lithoclasts and several rounded quartz grains. (B) Piece of polished metal from some jewelry (earrings, necklace?). (C) Large fragment of transparent glass (from a bottle).

pyroxene and feldspar); there are also rare fragments of siliciclastic rocks and of anthropogenic material (Fig. 3.5B and C).

The terrigenous portion of the sands is dominated by carbonate lithoclasts. They are monomineralic fragments of older rocks (micritic limestones and rarely calcarenites). The second terrigenous component in the beach sands is quartz. It is almost exclusively crystalline quartz and only rarely chert or chalcedony. The quartz grains always are well rounded (Fig. 3.6B). Furthermore, some rare feldspars are present: either potassium feldspar or plagioclase. The potassium feldspar grains (mainly microcline) contain inclusions (zircon and plagioclase; Fig. 3.6C) and show alteration rims in the form of clay minerals (Fig. 3.6D). The dark-colored minerals consist mainly of pyroxene (Fig. 3.6E), which shows up light green under crossed nichols and yellowish under parallel nichols; they often are well rounded and they may contain inclusions of volcanic glass (Fig. 3.6E).

The bioclast content varies in the different grain-size fractions, but the shells and fragments of shells are most characteristic of the coarse fractions (>500 μm). As a rule, the bioclast content of the beach sand at Rosa Marina is low: in the backshore and foreshore it is low, but it increases linearly from the dune to the shoreline, where it becomes predominant (Table 3.2); in the shoreface it increases with water depth.

Using photos of 10 scanned thin sections, we obtained the values shown in Table 3.3. The mean values thus obtained with this procedure can be considered as representative (note that values do not differ significantly) and indicate relative frequencies of carbonates as 62% (lithoclasts and/or bioclasts), of quartz as 34%, and of other minerals as 4%.

4.2 Grain-Size Characteristics of the Beach Sand

Using the analytical procedures described in Section 3, we found that, from a granulometric point of view, the sampled sediments are medium- to coarse-grained sands (Table 3.4 and Fig. 3.7).

Relatively high D_{50} values are typical for the berms; the values decrease in the shoreface environments with increasing water depth (Fig. 3.8). No bars or sediment accumulation areas have been detected in the submerged sectors (Fig. 3.8). The sands are mostly well-sorted; sorting decreases with water depth. The skewness is equal to zero on the shoreline (Fig. 3.3) and

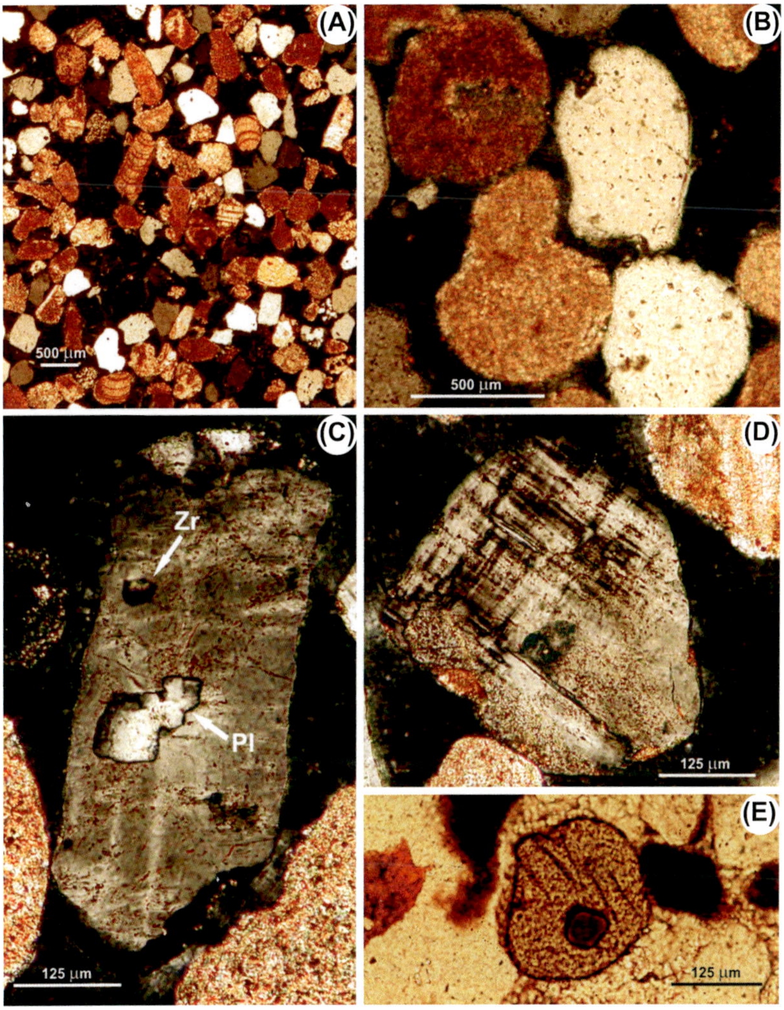

FIGURE 3.6 .Petrographical features of the Rosa Marina sands in thin section. (A) Sands with a mixed composition (mainly carbonates and quartz). Large fragments of red algae are clearly visible in the carbonate fraction (crossed nichols). (B) Grains of calcite and well-rounded quartz; the quartz grains with magmatic loops are derived from volcanic rocks (crossed nichols). (C) Microcline (K-feldspar) grain with inclusions of zircon (Zr) and plagioclase (Pl) (crossed nichols). (D) Microcline twinning albite and pericline (crossed nichols). (E) Yellow-greenish pyroxene with an inclusion of volcanic glass (parallel nichols).

negative on the backshore profile, resulting in a tail of coarse material (mean left of the median); at the same points, the kurtosis (Kg, i.e., the ratio between the width of the central part of the diagram and that of the tail) shows high values (leptokurtic type), but sample Bt (curve of the platykurtic type) forms an exception.

Also essentially qualitative information on the evolution of a beach with particular reference to the susceptibility to erosion can be obtained from the grain-size parameters

TABLE 3.3 Composition (in Percentages) of the Rosa Marina Sands

	Carbonate (%)	Quartz (%)	Other (%)
1	59.33	36.42	4.25
2	62.06	32.73	5.21
3	63.24	33.4	3.36
4	63.43	35.35	1.22
5	62.91	32.21	4.88
6	58.05	37.34	4.61
7	62.43	33.26	4.31
8	61.78	34.86	3.36
9	60.73	35.29	3.98
10	64.67	33.91	1.42
Mean Value	**61.86%**	**34.48%**	**3.66%**

The "other" class contains feldspar, pyroxene, and other minerals present in negligible percentages.

(Dal Cin, 1969). With the same mean diameter, prograding beaches tend to show curves of a platykurtic type (low kurtosis values) and to be less sorted than beaches undergoing erosion; prograding beaches also show a positive skewness (or less well marked negative asymmetries) and have tails toward fine-grained sediments, if compared with those being eroded. In other words: the finer-grained fractions are easily removed by waves in erosion-affected beaches, and this process relatively enriches the coarser fractions of the sand, increasing its sorting, because of depletion of the finer particles.

The Rosa Marina beach samples commonly show values of the various textural parameters which indicate an erosional-regressive evolutionary trend: (1) sorting is clear in the backshore-foreshore-shoreface sectors; (2) Sk is mainly negative (it is a clear record of severe storm-wave hydrodynamics); and (3) the distribution curves are often leptokurtic or even very leptokurtic, with tails shifted mostly toward the coarse-grained material.

4.3 Older Sedimentary Units

Different erosional features characterize the older sedimentary units that crop out in the study area. We sampled all these units in order to determine their contribution to the sands of the present-day beach. In particular, the relatively thick units along the investigated shoreline that are affected by erosion (Fig. 3.2) are formed by the Calcarenite di Gravina Formation (GRA) and by the recent aeolian dunes (E1 = middle Holocene and E2 = late Holocene). The Calcarenite di Gravina Formation consists locally of a massive biocalcarenite with a packstone texture (Fig. 3.9A); it consists almost entirely of red algae (with rare lithoclasts, fragments of bivalves, and benthic foraminifers). The middle Holocene aeolian unit

TABLE 3.4 Grain-Size Parameters Determined for the Samples From the Beach Sands

Sample	D_{50} (μm)	Sorting (σφ)	Description	Skewness (Skφ)	Shape	Kurtosis (kg)	Shape
Dune (E2)	399.695	0.47	Medium sand, well sorted	0.568	Tail to the fine fraction	1.676	Strongly leptokurtic
1	417.618	0.365	Medium sand, well sorted	0.051	Symmetrical	2.482	Strongly leptokurtic
Winter berm (Bt)	782.277	0.498	Coarse sand, well sorted	0.562	Tail to the fine fraction	0.631	Strongly leptokurtic
2	422.702	0.416	Medium sand, well sorted	−0.006	Symmetrical	2.636	Strongly leptokurtic
3	433.633	0.577	Medium sand, moderately well-sorted	−0.299	Tail to the coarse fraction	2.446	Strongly leptokurtic
4	429.620	0.302	Medium sand, very well sorted	−0.292	Tail to the coarse fraction	1.77	Strongly leptokurtic
Ordinary berm (Bo)	800.881	0.467	Coarse sand, well sorted	0.579	Tail to the fine fraction	1.737	Strongly leptokurtic
Backshore (Ba)	838.784	0.155	Coarse sand, very well sorted	0	Symmetrical	0.738	Platykurtic
−1 m	392.986	0.576	Medium sand moderately to well-sorted	0.319	Tail to the very fine fraction	0.836	Platykurtic
−3 m	397.568	0.825	Medium sand, moderately to well-sorted	−0.04	Symmetrical	0.847	Platykurtic
−6 m	249.487	0.662	Medium sand moderately to well-sorted	−0.558	Tail to the very coarse fraction	0.850	Platykurtic

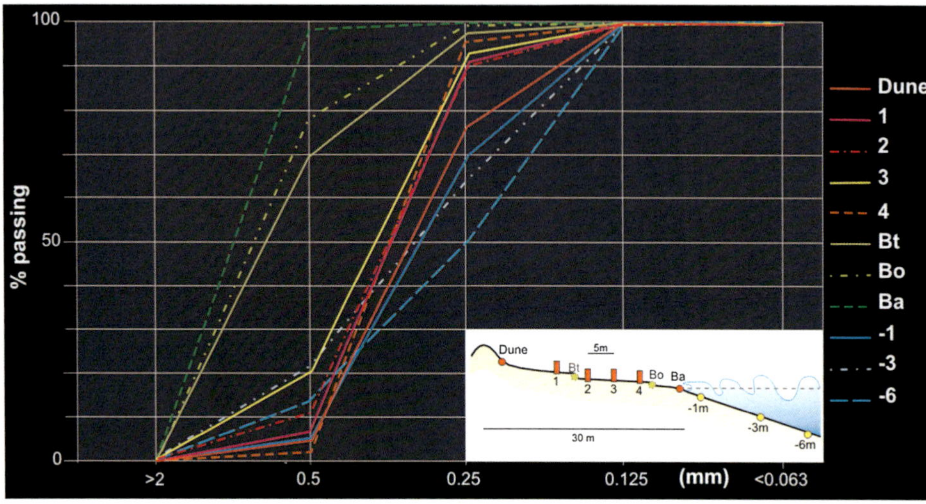

FIGURE 3.7 Cumulative grain-size curves of the samples from the Rosa Marina beach.

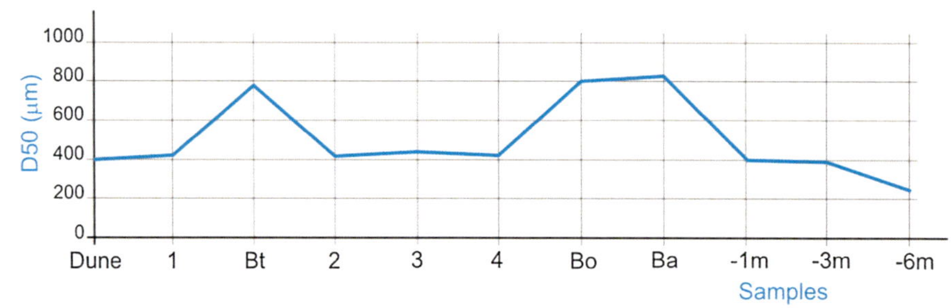

FIGURE 3.8 Variation of D_{50} from the backshore to the lower shoreface/offshore transition. Note the increase in grain size in the backshore, for both the ordinary and the winter berm.

(E1, Fig. 3.9B) has the same petrographic characteristics as the present-day sands except for the presence of a calcite cement. Carbonate clasts (which form over 60% of the rock), quartz, and some pyroxene (with petrographic characteristics very similar to those of the present-day sands) are present.

4.4 Classification of the Beach Sand

The present-day sands clearly show a compositional affinity with the recent aeolian-dunes (E1 and E2) and contain also numerous clasts of fossil red algae that probably come from the Calcarenite di Gravina Formation. In other words, the lithoclast content of the Rosa Marina sands records the erosion of the thickest (and well exposed to wave action) sedimentary units in this coastal sector, showing that explanation of the composition of the present-day beach sand does not need supply by longshore transport from other areas.

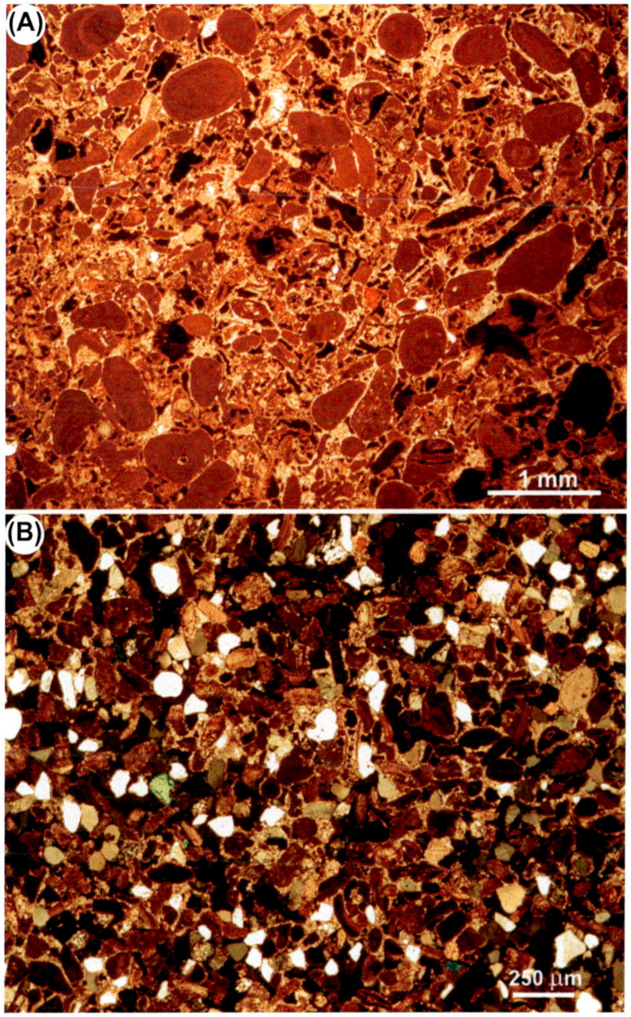

FIGURE 3.9 Petrographical details in thin section. (A) The Calcarenite di Gravina Formation is made up by large clasts of red algae (crossed nichols). (B) The aeolian unit E1 shows the same petrographical features as the present-day beach sands: carbonates, quartz, and some feldspar and pyroxene (bright interference colors) (crossed nichols).

The classification of sands containing siliciclastic and carbonate grains (either lithoclasts or bioclasts) is related to the definition of "hybrid sands" (Zuffa, 1980, 1985), corresponding to the "mixed sand" of Mount (1985) and the "miscellaneous sand" of Pettijohn (1975). Classification of the Rosa Marina mixed sands requires the recognition of the following components: (1) carbonate lithoclasts (CE = carbonate extraclasts, i.e., terrigenous carbonate particles eroded from older limestones); (2) bioclasts (B, i.e., only carbonate particles derived

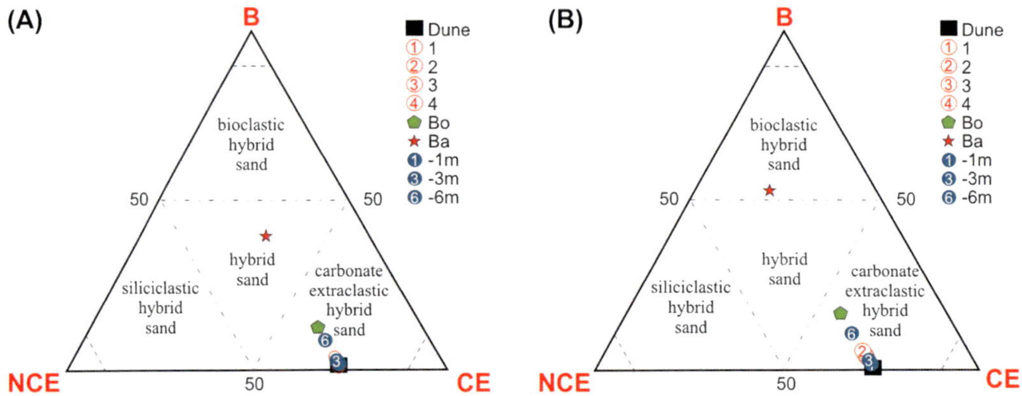

FIGURE 3.10 Composition-based classification of the Rosa Marina sands (*B*, bioclasts; *CE*, carbonate extraclasts; *NCE*, noncarbonate extraclasts). Note that the classification of the analyzed samples does not depend significantly on measurement of weight or volume. (A) Percentages by weight, as determined in the laboratory. (B) Percentages by volume, as obtained using the minimum bulk density for bioclasts.

from present-day organisms), and (3) siliciclastic particles (NCE = noncarbonate extraclasts, i.e., terrigenous particles eroded from noncarbonate older rocks/sediments).

Fig. 3.10 shows the classification of the sands (after Folk, 1959; Zuffa, 1980, 1985; Mount, 1985; Flügel, 2004) as a function of the percentages of these components by weight (Fig. 3.10A) and by volume (Fig. 3.10B) of their bioclast content; the differences are slight. The samples can commonly be classified as carbonate extraclastic hybrid sand. Two samples, viz. Bo (ordinary berm) and the sample collected at a water depth of 6 m, are close to the field of hybrid sand. The sample from the shoreline (Ba) has a bioclast content that classifies it as a hybrid sand if we consider the weight percentage, and as a bioclastic hybrid sand on the basis of the volume percentage.

5. MARINE LIFE FORMS CONTRIBUTING TO THE BEACH SAND

Bioclasts component of the Rosa Marina beach sands and consequently it is important to trace where these bioclasts come from. For the purpose, we investigated the main biocenoses in both sandy areas and in the hard-rock substratum. Single bioclasts of organisms that contribute with their remains to the sands have been collected and identified (commonly at genus level). The distribution of the living organisms along the coastal profile yields evidence regarding the origin of the bioclast content in terms of bathymetry.

5.1 The Main Biocenoses

A biological survey was carried out through diving at the investigated depth (up to −6 m) and through sampling of organisms from both the sandy and the hard substratum. The analyzed area is situated between the supralittoral and the upper infralittoral zones (sensu Peres and Picard, 1964); it can be divided into (1) the backshore, corresponding to the supralittoral zone; (2) the foreshore, corresponding to the mesolittoral zone; and (3) the shoreface until the wave base (−6 m), corresponding to the upper infralittoral zone. In all parts of the study area, soft sediment and hard substratum alternate. In the supralittoral zone, the biocenosis of the supralittoral sands is characterized by the presence of bioclastic material deposited during severe storms (algae, seagrass, remains of terrestrial plants, remains of marine and terrestrial invertebrates). This biocenosis alternates with that of supralittoral rocks colonized by few organisms such as gastropods.

In the mesolittoral zone, the upper intertidal rock contains a biocenosis that is characterized by deposits of cyanobacteria, crustaceans, barnacles, and gastropods. In the lower intertidal zone, vermetids occur, indicating a priority habitat for the marine conservation in the SPA/BIO Protocol (Specially Protected Areas and Biological Diversity in the Mediterranean) (Barcelona Convention, 1997; Relini and Giaccone, 2009). The facies is characterized by bioconstructions of a sessile gastropod that builds complexes which induce a large increase in animal biodiversity (especially annelids, mollusks, crustaceans, echinoderms, and small benthic fish) and vegetation (calcareous seaweed thallus, algal mats, and leafy algae). This highly diversified habitat is particularly sensitive to oil pollution and surfactants as well as to mechanical destruction related to the harvesting of some bivalves.

In the infralittoral zone, the soft-sediment sea floor is locally characterized by a biocenosis of infralittoral algae. At intermediate depths (−3 m), the bedrock is only sparsely inhabited, probably as a consequence of wave-induced erosion. At a depth of −6 m, the biocenosis is more diversified and characterized by encrusting and leafy algae, sponges, polychaete serpulids, mollusks, vermetids, and decapod crustaceans (especially hermit crabs).

The bioclast content (Table 3.3) of the beach sands has been analyzed in the remains of the organisms, which were identified and classified, as far as possible, from a taxonomical point of view. The bioclasts consist of fragments of (1) rhizopods, (2) mollusk shells, (3) thorns or fragments of echinoderm exoskeletons, (4) bryozoans, and (5) fragments coming from less frequent organisms (this fifth category, called "other," includes the remains of algae, spicules of sponges, fragments of serpulid pipes, and fragments of barnacles and other crustaceans).

The largest contribution comes from mollusks, in particular from gastropods and bivalves, although the relative percentages are highly variable in the analyzed subenvironments (Fig. 3.11). The number of identified mollusk taxa is 55; the other bioclasts come from 39 taxa of gastropods and 16 taxa of bivalves. The number of taxa (Fig. 3.12) tends to increase gradually from the dunes (only one taxon) to the shoreline (52 taxa), to decrease again with water depth; it finally increases again at a depth of −6 m (18 taxa).

The recognized mollusk taxa are not typical of a single area, but rather come from slightly different ecological and bathymetric subenvironments (Fig. 3.13). A conspicuous part of the mollusks (36.4% of the total) is typical of rocky bottoms. Another large fraction (23.6% of the total) is characteristic of sandy bottoms. Another fraction (10.9% of the total) consists of mollusks that can live on both sandy and rocky bottoms.

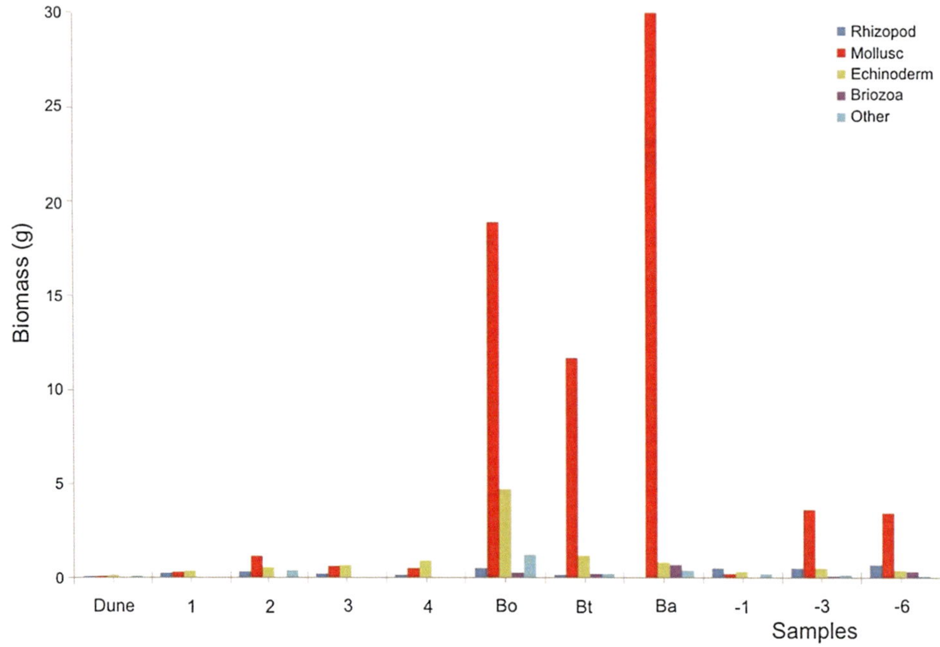

FIGURE 3.11 Mean values of biomass (in grams) in sediments from different sectors of the beach.

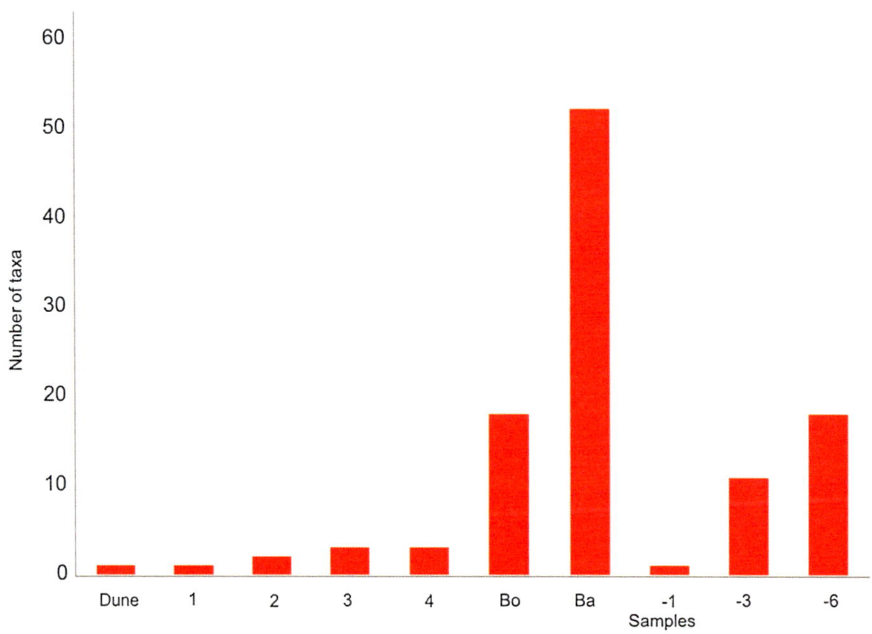

FIGURE 3.12 Number of taxa in the various beach subenvironments.

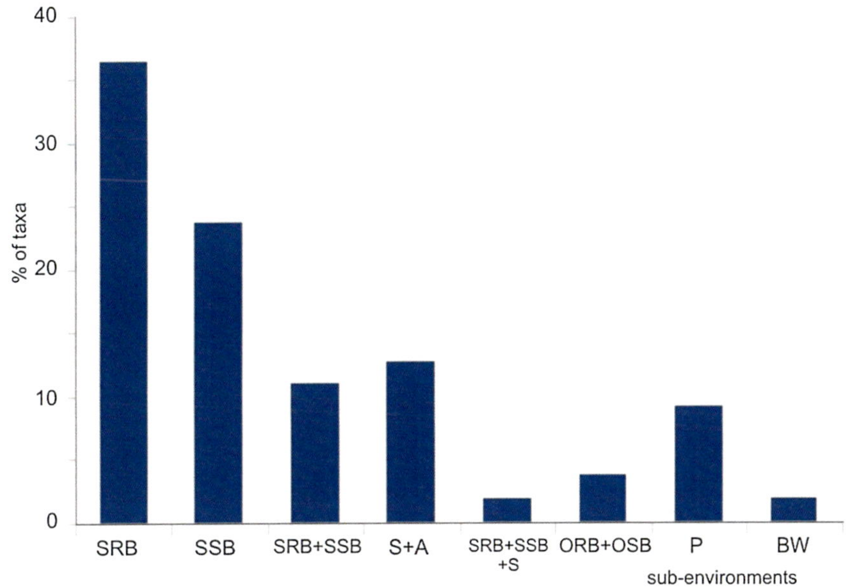

FIGURE 3.13 Subenvironments that form source areas for the mollusk fragments in the Rosa Marina beach sands. *SRB*, shallow rocky bottom; *SSB*, shallow sandy bottom; *S*, seagrass; *A*, algae; *ORB*, offshore rocky bottom; *OSB*, offshore soft bottom; *P*, parasites; *BW*, brackish water.

There are also seven taxa of gastropods (12.7% of the total) that prefer to live in the grasslands of seagrass or algae. One taxon, the genus *Bittium*, is practically ubiquitous and can live on shallow rocky bottoms covered with vegetation, in the grasslands of marine plants, and also on soft substratum. In particular, shell fragments of *Bittium* have been found in all analyzed samples. Only two taxa (3.6% of the total) are typical of slightly deeper environments. Five taxa (9.1% of the total) are parasites of gastropods or of other organisms. Finally, one taxon, the genus *Hydrobia*, is typical of brackish environments.

6. CONCLUSIONS

Beach sands from Rosa Marina, located along the Adriatic coast of the Apulian Region (north of Brindisi), have been investigated with the objective to characterize the lithoclast and bioclast components, so that the source of the beach sands could be traced. The beach was analyzed for the purpose from both a physical point of view (geomorphology and sedimentology) and a biological point of view. This approach was considered the most appropriate because beaches are the result of the interaction at various scales and at different times of several physical processes (erosion, transport, and sedimentation) and biological agents.

The Rosa Marina beach is now distinctly subjected to erosion; evidence consists of the local erosion of the coastal dunes, and of the lack of detectable depositional bars in the shoreface.

Moreover, the calculation of all statistical parameters regarding the grain sizes of the sediments at various places along and perpendicular to the coast indicate erosion, although more stable conditions seem to exist in adjacent coastal sectors.

A petrographical analysis indicates that the beach sands are made up of calcium carbonates (about 62%), quartz (about 34%), and other minerals (K-feldspar, plagioclase, pyroxene, etc., constitute about 4%). Comparing this petrographical composition with that of the rocks cropping out in the coastal area, we come to the conclusion that the lithoclast component of the present-day beach sands is controlled by the storm-wave erosion of the rocky substratum, without a necessary contribution from longshore sediment transport. Furthermore, the sedimentary units of the recent coastal dunes have a composition that is highly similar to the present-day beach sands; this conclusion seems only logical considering the unchanged or similar palaeogeographical situation during the late Pleistocene–Holocene. Nevertheless, these data show the importance of the cannibalization processes for the recent formation and evolution of coastal deposits in the Apulian Foreland. Previous research (Tropeano et al., 2002; Gallicchio et al., 2014; Gioia et al., 2014) came to comparable conclusions.

The bioclast fraction varies from a minimum of 1% to a maximum of over 50% by weight (this equals some 70% by volume). The distinction between bioclastic and lithoclastic particles indicates that the beach sands should be classified as carbonate extraclastic hybrid sands, except for one sample that can be classified as a hybrid sand or a bioclastic hybrid sand, depending on whether the weight or the volume is considered for the various components.

Regarding the coastal erosion, an important conclusion can be drawn from the bioclast content: also in beaches with an insignificant bioclastic content (in one sample it was close to only 1%), a considerable coastal retreat can occur in only a few years by the loss or by a drastic decrease of the bioclast input. This conclusion made us decide to study in more detail the bioclastic components of the sands by identification of the taxa that supply sedimentary material that build up the beach sands.

Our biological survey, carried out in the various near-coast, shallow-marine subenvironments, made clear which are the source areas of the various kinds of bioclasts that are present in the sands. Most of the bioclasts are shells and fragments of bivalves and gastropods. Forty taxa (72.7% of the total) come from sandy and rocky submerged shallow environments, in particular from the biocenosis of infralittoral algae and from the facies with vermetids. A smaller number of taxa (16.4% of the total) come from deeper zones, especially from coralligenous complexes and from seagrasses.

A group of mollusk shells belongs to parasites that live in various subenvironments; only one taxon is typical of brackish environments. It seems evident that there are multiple sources of bioclasts and that these sources are characterized by different ecological parameters (salinity, depth, light, type of substratum, presence of vegetation, presence of bioconstructors). Nevertheless, our study shows that the most important source areas are located within the shoreface and offshore transition environments (only 16.4% comes from deeper environments). The natural bioclast supply on the beach depends on the health state of these environments.

This study indicates in what way effective and less expensive procedures can be developed to prevent or minimize coastal erosion. In this context we think it of presumably great social importance to emphasize the enormous potential of geological/biological multidisciplinary approaches in the study of beaches: most important are (1) the characterization of the physical/biological/ecological environment, (2) the analysis and monitoring of the phenomena of parameters involved in coastal retreat, (3) the investigation and characterization of materials that can be useful for beach nourishment, and (4) the analysis of ecological impacts in the widest sense at every stage of beach nourishment. It is interesting to realize that these potentially socially important conclusions can be drawn on the basis of a study that aimed primarily at a sedimentological investigation concerning the possible transport of sedimentary particles from source to sink.

References

Anfuso, G., Pranzini, E., Vitale, G., 2011. An integrated approach to coastal erosion problems in Northern Tuscany (Italy). Littoral Morphological Evolution and Cell Distribution Geomorphology 129, 204–214.

Annese, R., De Marco, R., Gianfreda, F., Mastronuzzi, G., Sanso, P., 2003. Caratteri morfo-sedimentologici della Baia di Torre Canne (Brindisi, Puglia) (Morpho-sedimentary characteristics of the Torre Canne Bay (Brindisi, Puglia)). Studi Costieri 7, 3–19.

Anthony, E.J., 2009. Shore processes and their paleoenvironmental applications. Developments in Marine Geology 4, 519.

APAT-ICRAM, 2007. Manuale di movimentazione dei sedimenti marini (Handbook of marine sediment transport), 77 pp. http://www.isprambiente.gov.it/contentfiles/00006700/6770-manuale-apaticram-2007.pdf/view.

Barcelona Convention, 1997. Convention for the Protection of the Marine Environment and the Coastal Region of the Mediterranean, 22 pp. http://195.97.36.231/dbases/webdocs/BCP/bc95_Eng_p.pdf.

Bird, E., 2008. Coastal Geomorphology – An Introduction, second ed. J. Wiley and Sons. 411 pp.

Blott, S.J., Pye, K., 2001. GRADISTAT: a grain size distribution and statistics package for the analysis of unconsolidated sediments. Earth Surface Processes and Landforms 26, 1237–1248.

Charlier, R.H., De Meyer, C.P., 1989. Coastal erosion. In: Lecture Notes in Earth Sciences, 70. Springer, 343 pp.

Chiocci, F.L., La Monica, G.B., 1999. Individuazione e caratterizzazione dei depositi sabbiosi presenti sulla piattaforma continentale della Regione Lazio e valutazione di un loro utilizzo ai fini del ripascimento dei litorali in erosione – Rapporto finale della I fase (Identification and characterisation of sand deposits on the continental shelf of the Lazio region and assessment of their use in the nourishment of retreating coastlines – First phase final report). Università degli Studi di Roma 'La Sapienza', Dipartimento di Scienza della Terra – Regione Lazio, Assessorato Opere e Reti di Servizio e Mobilità, Rome, 110 pp.

Choquette, P.W., Pray, L.C., 1970. Geologic nomenclature and classification of porosity in sedimentary carbonates. American Association of Petroleum Geologists Bulletin 54, 207–250.

Ciaranfi, N., Pieri, P., Ricchetti, G., 1988. Note alla carta geologica delle Murge e del Salento (Puglia centromeridionale) (Notes to the geological map of Murgia and Salento (south-central Puglia)). Memorie della Società Geologica Italiana 41, 449–460.

Colosio, F., Abbiati, M., Arnoldi, L., 2011. Effects of beach nourishment on sediments and benthic assemblages. Marine Pollution Bulletin 54, 1197–1206.

Dal Cin, R., 1969. Distinzione tra spiagge in erosione ed in avanzamento mediante metodo granulometrico (Distinction between retreating and advancing beaches through particle size method). Rivista Italiana di Geotecnica 4, 227–233.

Finkl, W., 1981. Beach nourishment, a practical method of erosion control. Geo-Marine Letters 1, 155–161.

Flügel, E., 2004. Microfacies of Carbonate Rocks. Springer, 976 pp.

Folk, R.L., 1959. Practical petrographic classification of limestones. American Association of Petroleum Geologists Bulletin 43, 1–38.

Fredsøe, J., Deisgaard, R. (Eds.), 1994. Mechanics of Coastal Sediment Transport. Advanced Series on Ocean Engineering, vol. 3, 369 pp.

Gallicchio, S., Moretti, M., Spalluto, L., Angelini, S., 2014. Geology of the middle and upper Pleistocene marine and continental terraces of the northern Tavoliere di Puglia plain (Apulia, Southern Italy). Journal of Maps 10, 569–575.

Gioia, D., Gallicchio, S., Moretti, M., Schiattarella, M., 2014. Landscape response to tectonic and climatic forcing in the foredeep of the southern Apennines, Italy: insights from Quaternary stratigraphy, quantitative geomorphic analysis, and denudation rate proxies. Earth Surface Processes and Landforms 39, 814–835.

Greenwood, B., Davis Jr., A.R. (Eds.), 1984. Hydrodynamics and Sedimentation in Wave-dominated Coastal Environments. Developments in Sedimentology, 39. Elsevier, Amsterdam, 473 pp.

Ingle Jr., J.C. (Ed.), 1966. The Movement of Beach Sand. Developments in Sedimentology, 5. Elsevier, Amsterdam, 221 pp.

Jackson, D.R., Richardson, M.D., 2007. High-Frequency Seafloor Acoustics. Springer, 616 pp.

Kay, R., Alder, J., 2002. Coastal Planning and Management. Taylor & Francis, 387 pp.

Manca, E., De Pascucci, V.L., Cossua, A., 2013. Shoreline evolution related to coastal development of a managed beach in Alghero, Sardinia, Italy. Ocean & Coastal Management 85 A, 65–76.

Mastronuzzi, G., Sanso, P., 2002. Holocene coastal dune development and environmental changes in Apulia (Southern Italy). Sedimentary Geology 150, 139–152.

Mastronuzzi, G., Palmentola, G., Sanso, P., 2001. Evoluzione morfologica della fascia costiera di Torre Canne (Puglia adriatica) (Morphological evolution of the Torre Canne coastline (Adriatic Apulia)). Studi Costieri 4, 19–31.

Moretti, M., Owen, G., Tropeano, M., 2011. Soft-sediment deformation induced by sinkhole activity in shallow marine environments: A fossil example in the Apulian Foreland (Southern Italy). Sedimentary Geology 235, 331–342.

Mount, J., 1985. Mixed siliciclastic and carbonate sediments: aproposed first-order textural and compositional classification. Sedimentology 32, 435–442.

Nicoletti, L., Paganelli, D., Gabellini, M., 2006. Aspetti ambientali del dragaggio di sabbie relitte a fini di ripascimento: proposta di un protocollo di monitoraggio (Environmental aspects of relict sand dredging for beach nourishment: proposal for a monitoring protocol). Quaderno ICRAM 5, 150 pp.

Nordstrom, K.F., 2005. Beach nourishment and coastal habitats: research needs to improve compatibility. Restoration Ecology 13, 215–222.

NRC (National Research Council), 1989. Measuring and Understanding Coastal Processes. National Academies Press. http://www.nap.edu/catalog/1445.html.

NRC (National Research Council), 1994a. Environmental Science in the Coastal Zone – Issues for Further Research. http://www.nap.edu/catalog/2249.html.

NRC (National Research Council), 1994b. Priorities for Coastal Ecosystem Science. http://www.nap.edu/catalog/4932.html.

Peres, J.M., Picard, J., 1964. Nouveau manuel de bionomie benthique de la Mer Méditerranée (A new manual for the bentic bionomics in the Mediterranean Sea). Recueil des Travaux de la Station Marine d'Endoume 47, 1–37.

Pettijohn, F.J., 1975. Sedimentary Rocks, third ed. Harper and Row. 628 pp.

Puglia Region, 2006. Miglioramento delle conoscenze di base, adeguamento e ampliamento del sistema di monitoraggio del suolo, dei corpi idrici superficiali, sotterranei e costieri (Improvement of the basic knowledge, adaptation and expansion of the monitoring system of soil, surface, underground and coastal water bodies). POR 2000–2006. Misura 1.3. Area d'azione 2 e 4, Pilone, 3 pp.

Puglia Region, 2012. Attività finalizzate alla redazione del Piano Regionale delle Coste (P.R.C.) della Regione Puglia (Activities for the preparation of the Regional Plan of Coste (PRC) of the Puglia Region]. POR 2012, All. n. 7.1.2, L'erosione costiera in Europa, in Italia e in Puglia (Coastal erosion in Europe, Italy and Puglia)). Bollettino Ufficiale Regione Puglia 1, 6123–6142. http://www.regione.puglia.it/index.php?page=documenti&opz=getdoc&id=229.

Relini, G., Giaccone, G., 2009. Gli habitat prioritari del protocollo SPA/BIO (Convenzione di Barcellona) presenti in Italia, Schede descrittive per l'identificazione (The priority habitats of the Protocol SPA/BIO (Barcelona Convention) in Italy, Illustrative sheets for identification). Biologia Marina Mediterranea 16, 1–372.

Schlager, W., 2005. Carbonate Sedimentology and Sequence Stratigraphy. Concepts in Sedimentology and Paleontology, 8. SEPM, 200 pp.

Schwartz, M.L., 2005. Encyclopedia of Coastal Science. Springer, 1211 pp.

Tropeano, M., Spalluto, L., 2006. Present-day temperate-type carbonate sedimentation on Apulia shelves (Southern Italy). GeoActa 5, 129–142.

Tropeano, M., Sabato, L., Pieri, P., 2002. Filling and cannibalization of a foredeep: the Bradanic trough (Southern Italy). In: Jones, S.J., Frostick, L.E. (Eds.), Sediment Flux to Basins: Causes, Controls and Consequences, 191. Geological Society of London, Special Publications, pp. 55–79.

Uda, T., 2010. Japan's beach erosion – reality and future measures. In: Advanced Series on Ocean Engineering, 31, 418 pp.

Van der Salm, J., Unal, O., 2003. Towards a common Mediterranean framework for beach nourishment projects. Journal of Coastal Conservation 9, 35–42.

Van Rijn, L.C., 2011. Coastal erosion and control. Ocean and Coastal Management 54, 867–887.

Zuffa, G.G., 1980. Hybrid arenites: their composition and classification. Journal of Sedimentary Petrology 50, 21–29.

Zuffa, G., 1985. Optical analysis of arenites: influence of methodology on compositional results. In: Zuffa, G.G. (Ed.), Provenance of Arenites. Reidel, Dordrecht, pp. 165–189.

CHAPTER

4

Changes in the Heavy-Mineral Spectra on Their Way From Various Sources to Joint Sinks: A Case Study of Pleistocene Sandurs and an Ice-Marginal Valley in Northwest Poland

A.J. (Tom) Van Loon[1], *M. Pisarska-Jamroży*[2]

[1]Geocom Consultants, Benitachell, Spain; [2]Geological Institute, Adam Mickiewicz University, Poznań, Poland

OUTLINE

Sediment Provenance
http://dx.doi.org/10.1016/B978-0-12-803386-9.00004-6

1. INTRODUCTION

Heavy-mineral analysis is a useful tool for the recognition of source areas (e.g., Dick, 1887; Mange and Wright, 2007; Woronko et al., 2013), for the reconstruction of the transport history in a fluvial system (Weckwerth and Chabowski, 2013), and for obtaining an insight into the paleoclimate (Derkachev and Nikolaeva, 2013; Wachecka-Kotkowska and Ludwikowska-Kedzia, 2013). These analyses can thus help to unravel the often complicated development of glaciation-related sedimentary successions. Such successions were formed due to the ice caps that covered large parts of the northern hemisphere during the Pleistocene glaciations. These ice caps drained huge quantities of sediment-laden meltwater. The Scandinavian ice sheet thus largely affected the mineralogical composition of Pleistocene sediments in northern Europe. The meltwater built extensive sandurs (outwash plains) and ice-marginal valleys (pradolinas) that now still cover areas of considerable size.

According to Racinowski (2010), Quaternary sediments in Poland—regardless of their geographical position, age, and type—are dominated by amphibole and garnet, supplemented by epidote, biotite, and pyroxene. Woronko et al. (2013) also found that glacial and periglacial sediments in eastern Poland are dominated by amphibole and garnet. Similar spectra were obtained by Pisarska-Jamroży et al. (2015a) from sandur and ice-marginal-valley sediments in northwest Poland, where the sediments are dominated by amphibole, limonite, an opaque rest group (magnetite, other iron oxides and hydroxides), garnet, epidote, and biotite. This simplified picture actually is more complex, but hardly any data are available regarding the precise reasons for differences in the heavy-mineral spectra.

The main objective of the present contribution is to show, on the basis of recently collected heavy-mineral samples, the relationships influencing the spectra of the heavy minerals in glacigenic sediments at different sites. As shown earlier by Pisarska-Jamroży et al. (2015a,b), the mere spectra of heavy minerals in Pleistocene sediments are insufficient by themselves to unravel the pathway from their source to their sink. This is because the heavy-mineral spectrum in the sediments at a specific site is the result of numerous factors, including travel distance, flow regime of the streams that transported the particles, erosion of the substratum, and postdepositional processes. In order to deepen the insight into all these factors, we investigated the heavy-mineral compositions of sediments from some sandurs and an ice-marginal valley that were deposited during the same time. The sediments in the ice-marginal valley are commonly considered to be supplied almost exclusively from the sandurs to the north, including the two sandurs dealt with in the present study. We thus could reconstruct the effect of the transport distance, the reworking, and the flow regime in glaciofluvial and fluvial environments, and establish the effect on grain sorting—and thus on the heavy-mineral composition. Moreover, the variation in grain size of the sediments in the sandurs and the ice-marginal valley under study (ranging from silt to sand to gravel, including diamictons) allowed to distinguish between heavy-mineral species that are more susceptible to postdepositional processes than other species, as reflected by the alteration of the various heavy-mineral species. Particularly because sediments, including heavy minerals derived from the Fennoscandian Shield, were deposited in front of the ice first on sandurs, which were built up by north–south (N–S) running meltwater streams that also eroded again the sandur material to deposit it later in the ice-marginal valley, the study offers a rare opportunity to study the effect on heavy-mineral spectra of successive depositional and erosional phases, and of transport under various flow conditions. These flow conditions have been dealt with in detail by Pisarska-Jamroży (2015) and Pisarska-Jamroży et al. (2015a,b).

2. GEOGRAPHICAL SETTING

During the Pleistocene, the Scandinavian ice sheet drained huge amounts of sediment-laden meltwater (see Brodzikowski and Van Loon, 1992, for the various depositional processes and the resulting lithofacies). The sedimentary particles were deposited in front of the ice mainly on sandurs but partially also in the ice-marginal valleys of the central European lowlands (Fig. 4.1) that ran parallel to the ice-sheet margin and perpendicular to the prograding sandurs. The streams in the ice-marginal valleys flowed to the west because the area sloped to the north-northwest, and because the sandurs and the ice mass prevented a current direction to the north. The ice-marginal valleys were fed also by extraglacial rivers flowing northward from the more elevated areas in the south, as already found by, among others, Woldstedt (1950), Galon (1961), and Kozarski (1962).

The main ice-marginal valleys of Poland and Germany are the Wrocław-Magdeburg-Bremen, the Głogów-Baruth-Hamburg, the Vilnius-Warsaw-Poznań-Berlin and the Toruń-Eberswalde ice-marginal valleys (Fig. 4.2). The Toruń-Eberswalde ice-marginal valley, also referred to as the Noteć-Warta ice-marginal valley, is the longest one. It can be divided geographically into several basins and valleys, viz. the Toruń Basin, the Middle Noteć valley, the Gorzów Basin, and the Eberswalde valley. The part under study here is 300 km long and comprises the middle and western parts of the Toruń-Eberswalde ice-marginal valley, that is, the Middle Noteć valley and the Gorzów Basin. In the following, the abbreviation IMV will be used for this specific part of this specific ice-marginal valley.

The sites from where the heavy-mineral composition was analyzed are three gravel pits on the Drawa and Gwda sandurs and five gravel pits in the Toruń-Eberswalde IMV (Fig. 4.3). At two sites in the IMV samples were taken from both terrace sediments and the Pleistocene

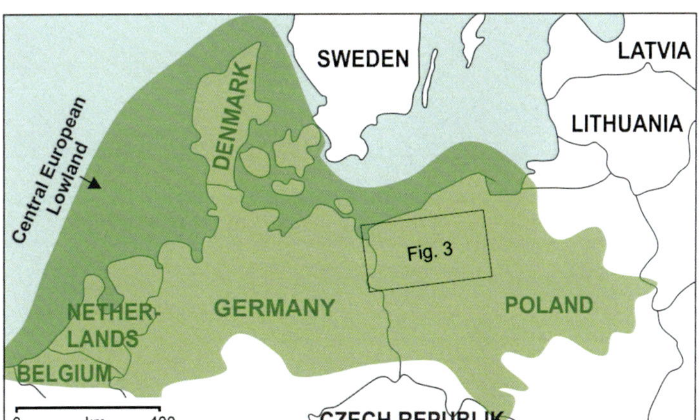

FIGURE 4.1 Position of the study area (*rectangle* of Fig. 3) within the Central European Lowland.

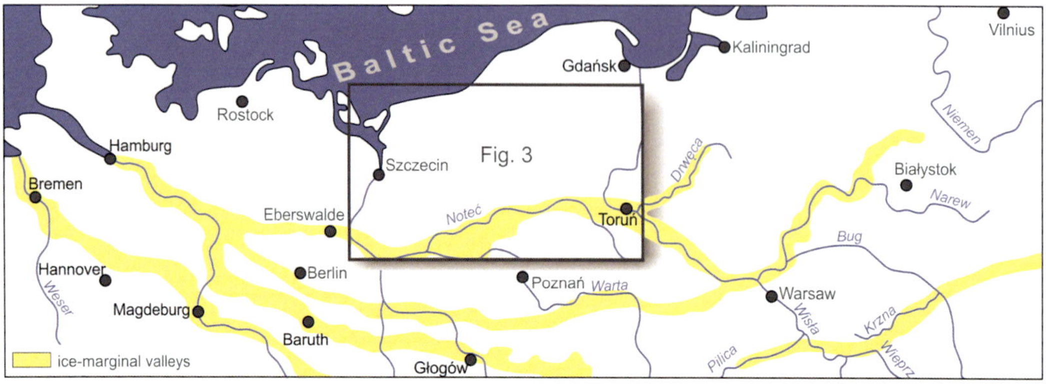

FIGURE 4.2 Distribution of ice-marginal valleys in Europe, showing the study area (*rectangle* of Fig. 3) as the middle part of a much larger system. Modified after Pisarska-Jamroży (2015).

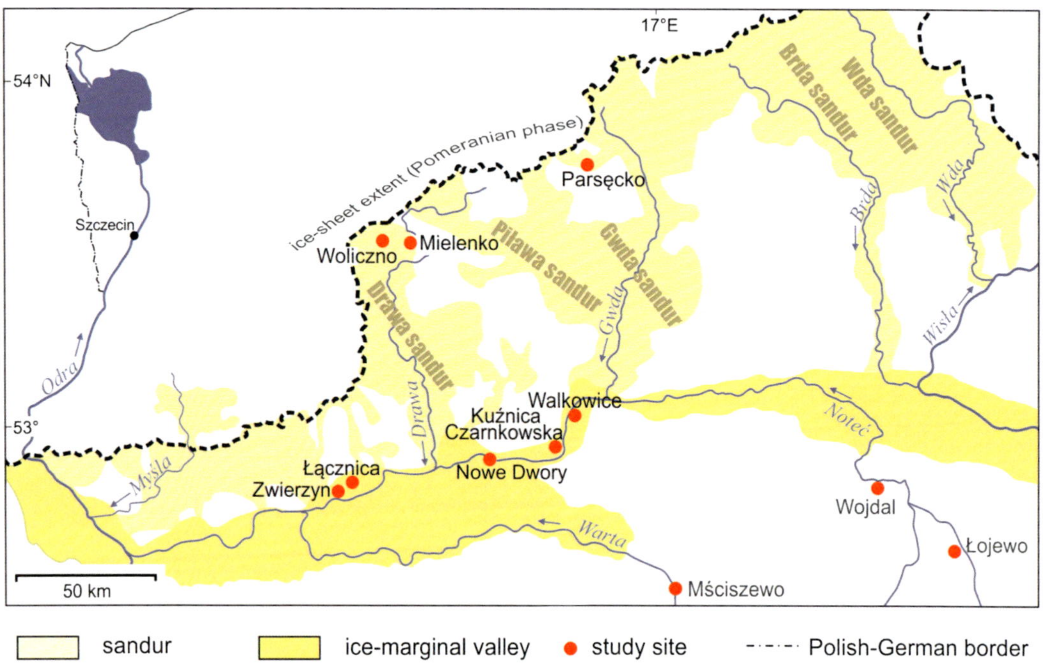

FIGURE 4.3 Regional setting of the Drawa and the Gwda sandurs and the Toruń-Eberswalde ice-marginal valley, with sampling sites.

substratum. Both sandurs and the IMV were fed by meltwater streams; the meltwater streams on the sandurs in turn fed the Toruń-Eberswalde IMV. As mentioned earlier, the IMV was fed also by extraglacial rivers running from the south, which slightly changed the proportion of some heavy minerals (Pisarska-Jamroży, 2015; Pisarska-Jamroży et al., 2015a). For this reason, the heavy minerals from three sites on terraces of the Pomeranian phase along these southern rivers were also analyzed. Obviously, the sediments at the study sites in the IMV were partly also derived from the catchment area of the IMV farther to the east. In addition, the proglacial and extraglacial areas in front of the ice formed a source of fine particles that were carried along by winds and that partly were deposited on the sandurs and in the IMV.

Both the two sandurs and the terrace under study in the Toruń-Eberswalde IMV date from the Pomeranian phase of the Weichselian glaciation, when the Scandinavian Ice Sheet almost reached the area (16–17 ka; Marks, 2012). The Drawa and Gwda sandurs (Fig. 4.3) are large examples (80 and 110 km long, respectively), and the Toruń-Eberswalde IMV is the largest (>500 km long, 2–20 km wide) IMV of the European lowlands; it runs from eastern Poland to Germany.

3. METHODS

The heavy-mineral analysis was carried out for grains of the 0.125–0.25 mm fraction from 90 samples (Fig. 4.3), collected from three sites on the sandurs, five sites from a single terrace of the IMV, and three from terraces along the southern rivers, as mentioned before, all dating from the Pomeranian phase. The reason for taking (on average) some 10 samples from each site was that we wanted to investigate samples from material deposited under different flow regimes, so as to find out about the influence of the flow conditions on the heavy-mineral spectra.

Separation from the light minerals was done at a density of 2.8 g cm^{-3} following Mange and Maurer (1992). The percentage of each heavy mineral was determined by counting on average 700 transparent and opaque grains per slide; this relatively large number was chosen to ascertain that at least 300 transparent grains were included in the counting.

The various heavy-mineral species were identified with a petrographic microscope. To confirm the identification of selected heavy minerals, polished thin sections were prepared and analyzed by scanning electron microscopy and energy dispersive spectroscopy. The following transparent minerals were thus recognized: andalusite (An), rutile (R), zircon (Z), kyanite (K), staurolite (S), tourmaline (T), clinozoisite (Cl), epidote (E), garnet (G), sillimanite (Si), amphibole (A), orthopyroxene (O), clinopyroxene (C), glauconite (Gl), muscovite (M), biotite (B) and chlorite (Ch). Among the opaque minerals, limonite (L) and pyrite (P) were distinguished; the other opaque minerals—other iron (hydr)oxides and magnetite—were taken jointly as the opaque rest group (RO). The term "all opaques" is used in the following, unless indicated otherwise, for all opaque heavy minerals together (limonite, pyrite, and the opaque rest group).

4. HEAVY-MINERAL SPECTRA

Six groups of heavy minerals (amphibole, garnet, limonite, the opaque rest group, epidote and biotite) dominate the sediments of the three sandur sites, the five sites of the Toruń-Eberswalde IMV, and the three sites along the southern rivers (Table 4.1). The most common

TABLE 4.1 Average Heavy-Mineral Percentages of All Samples From the Sandurs, All Samples From the IMV Terraces, the Pleistocene Substratum of the Ice-Marginal Valley (IMV), and the Terraces of the Pleistocene South–North Running Rivers South of the IMV

	Heavy-Mineral Composition (%)			
Heavy Minerals	**Sandurs**	**Ice-Marginal Valley**	**Pleistocene Substratum**	**Pre-Warta and Pre-Noteć Rivers**
Transparent	**58.2**	**65.5**	**58.4**	**64.5**
Andalusite	0	0	0.2	0.1
Rutile	1.0	0.8	1.2	1.1
Zircon	1.4	1.1	2.2	2.9
Kyanite	0.7	0.5	0.9	0.5
Staurolite	0.7	1.1	1.7	1.1
Tourmaline	1.1	1.1	2.5	1.0
Clinozoisite	1.9	2.0	2.1	1.9
Epidote	7.4	5.8	5.3	9.4
Garnet	12.4	13.5	14.0	17.1
Sillimanite	0.6	0.5	0.9	0.5
Amphibole	20.9	22.6	16.0	21.3
Orthopyroxene	0.2	0.4	0.1	0.5
Clinopyroxene	1.2	1.9	1.7	1.2
Glauconite	0.3	1.3	1.6	1.4
Muscovite	0.3	0.3	1.2	1.5
Biotite	1.6	4.1	6.3	2.6
Chlorite	0.2	0.3	0.3	0.9
Opaque	**41.8**	**34.5**	**41.6**	**35.5**
Limonite	18.0	23.8	26.9	9.7
Pyrite	0	0	0.1	0.4
Opaque rest group[a]	23.8	11.7	14.7	25.5

[a]*Magnetite and other iron oxides.*

(in order of frequency) minerals from the sandurs are L > A > G > RO > E > B; from the middle part of the IMV they are RO > A > L > G > E, and from the southern rivers they are RO > A > G > L > E.

The sediments from the sandurs (Fig. 4.4) and from the middle part of the IMV (Fig. 4.5) show comparable overall heavy-mineral spectra, which suggests a similar source of the sediments. There are, however, slight differences. Part of these may be explained by commonly

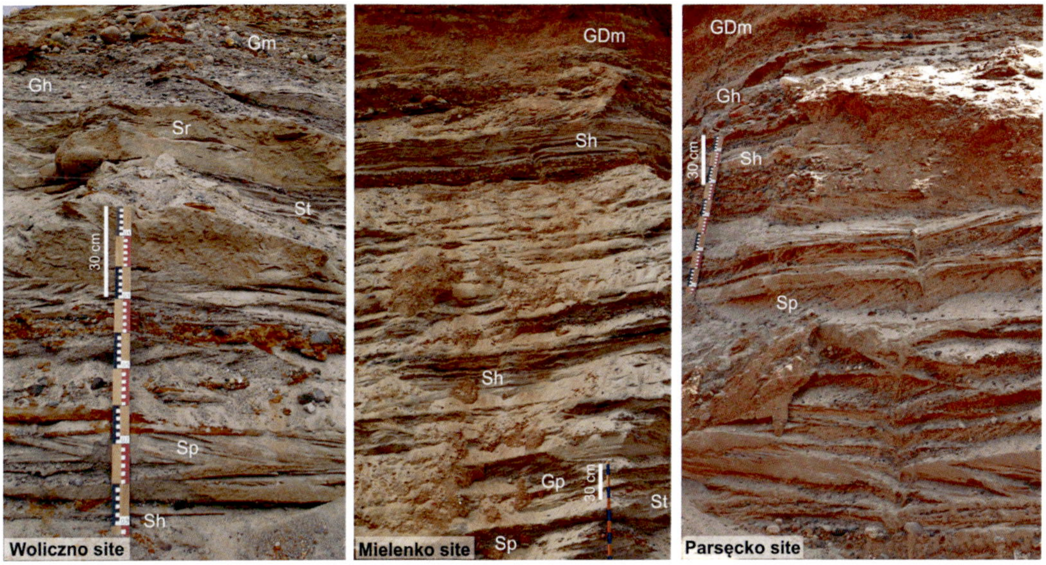

FIGURE 4.4 Sediments from the sandurs. Note the differences in lithology at each site, as well as the various sedimentary structures; these affect the heavy-mineral spectra. Lithofacies codes: *GDm*, massive diamictic gravel; *Gm*, massive gravel; *Gh*, horizontally stratified gravel; *Gp*, planar cross-stratified gravel; *Sh*, horizontally-stratified sand; *St*, trough cross-stratified sand; *Sp*, planar cross-stratified sand; *Sr*, ripple cross-laminated sand. For more details, see Pisarska-Jamroży (2015) and Pisarska-Jamroży et al. (2015a,b).

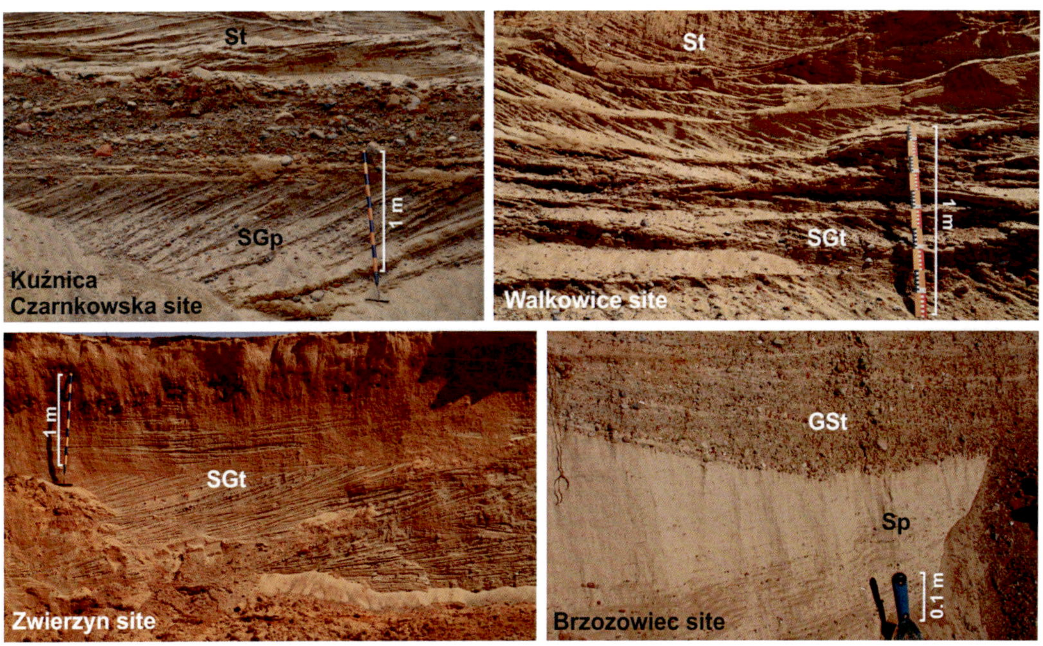

FIGURE 4.5 Sediments from the middle part of the Toruń-Eberswalde ice-marginal valley. Note the differences in lithology at each site, as well as the various sedimentary structures; these affect the heavy-mineral spectra. Lithofacies codes: *GSt*, trough cross-stratified sandy gravel; *SGt*, trough cross-stratified gravelly sand; *SGp*, planar cross-stratified gravelly sand; *St*, trough cross-stratified sand; *Sp*, planar cross-stratified sand. For more details, see Pisarska-Jamroży (2015) and Pisarska-Jamroży et al. (2015a,b).

considered factors such as differences in (1) their source sediments (melted sediment-laden ice cap, eroded Pleistocene and older substratum, catchment area of extraglacial rivers, upstream sediments of the IMV), and (2) postdepositional (diagenetic) processes affecting some heavy minerals. These factors, however, cannot explain several of the differences in the heavy-mineral spectra, particularly for samples taken from the same site.

4.1 Northern Sources of the Heavy Minerals

The heavy-mineral spectra of the sandurs and the Toruń-Eberswalde IMV suggest that the sediments came mainly from the north (Fennoscandian Shield); they were transported roughly southward by the ice cap. Racinowski (2010) states that the less-resistant minerals in Polish glacial sediments originate from eroded non-weathered crystalline rocks and that the resistant species were derived from the pre-Quaternary bedrock. Rappol and Stoltenberg (1985) and Vareikienė et al. (2007) suggest, however, that the initial sources of the heavy minerals in the Quaternary deposits of The Netherlands, northern Germany, and Lithuania are derived from East Central Baltic sedimentary rocks belonging to the Fennoscandian Shield, which consist of Archaean-Proterozoic crystalline rocks and younger, recycled sedimentary rocks that range in age from Cambrian to Paleogene.

4.2 The IMV Substratum as a Source

The erosion by the Scandinavian ice cap and the sub-, supra-, and proglacial meltwater flow conditions influenced the heavy-mineral composition of the sediments. Small differences in the proportions of some mineral species, particularly limonite and glauconite, between the sandurs and the Toruń-Eberswalde IMV might be ascribed to the presence of previously eroded Miocene and Pleistocene material from the IMV substratum.

Limonite constitutes a substantial portion of the heavy minerals in the pre-Quaternary deposits within the Toruń-Eberswalde IMV; it probably comes from eroded terrestrial Miocene deposits and from weathered Pleistocene sediments (Weckwerth and Pisarska-Jamroży, 2015; see also Pisarska-Jamroży and Zieliński, 2011) eroded from the valley walls. Part, however, should be ascribed to intensive and long-lasting weathering.

Glauconite occurs in larger proportions in the sediments from the middle part of the Toruń-Eberswalde IMV than in those from the sandurs. This mineral must have been derived from fluvial downward erosion by the streams in the IMV; glauconite is present in the IMV substratum in Miocene deposits, which are also exposed in the IMV nowadays (Bartczak, 2006).

4.3 Source Areas in the South

The IMVs in Europe were fed not only by meltwater streams from the north but also by extraglacial rivers with a catchment area in the south. Their sediment load, in principle, will have affected the heavy-mineral spectra in the east—west (E—W) running IMVs. Some minerals in the Toruń-Eberswalde IMV, indeed, may have been supplied by these extraglacial rivers such as the Noteć and Warta rivers (indicated in the following as the pre-Noteć and pre-Warta, respectively, although this term has been reserved by some authors—Dyjor

(1987)—for pre-Weichselian streams); this concerns in particular epidote and amphibole (see Pisarska-Jamroży et al., 2015a). The epidote proportion is higher in a terrace along the S–N running pre-Noteć River than in the Toruń-Eberswalde IMV, and it is also slightly higher in the terrace sediments of the S–N running pre-Warta River (Table 4.1).

Epidote and amphibole, which are common minerals in the Polish glacigenic sediments, may have been supplied mainly from the catchment areas of the pre-Noteć and pre-Warta rivers in the south (see Racinowski, 2010).

4.4 Postdepositional Processes Affecting the Heavy-Mineral Spectra

Diagenetic processes in the form of alteration transform some mineral species into other species. The increasing content of chlorite in the Toruń-Eberswalde IMV from east to west, together with a decreasing content of biotite in the same direction, was probably caused by alteration of biotite into chlorite.

A distinct (and common) result of alteration is the increase of limonite with time as a result of oxidation processes in a water-rich sediment. Our data show, however, that another factor strongly influences the limonite content. As the content of limonite in gravel was found to be significantly higher in gravel than in sand, even if samples were collected closely together at the same site, it must be deduced that the higher porosity of gravel plays a role, which implies that mineral alteration into limonite is more effective in gravel than in sand. We ascribe this to a better water percolation in the gravel than in the sand because of a higher permeability (Pisarska-Jamroży et al., 2015b). This implies that the current energy of the depositional stream influences the proportion of limonite, and thus of the overall heavy-mineral spectrum.

4.5 Sorting as an Important Factor Controlling the Heavy-Mineral Spectra

The relative proportions of the various heavy minerals can vary with the transport distance (see Van Andel, 1950; Lowright et al., 1972). Transport in a braided system on a sandur and in an IMV is nowhere regular, so erosional and depositional phases change frequently (e.g., Boothroyd and Ashley, 1975; Church and Gilbert, 1975; Pisarska-Jamroży and Zieliński, 2011, 2014). Short-lived and fast-changing currents, such as the meltwater streams, will cause intensive mechanical abrasion of some fragile heavy minerals (e.g., platy minerals), which thus will commonly become smaller than the analyzed fraction. The main responsible sorting factors thus are the type of transport (suspension, saltation, traction) and the duration of transport (Pisarska-Jamroży et al., 2015b).

The stream in the Toruń-Eberswalde IMV ran from east to west, and the percentages of amphibole and biotite change over the distance of 90 km of the IMV. Fluvial sorting processes influenced the relative proportions of some mineral species in the Toruń-Eberswalde IMV; for example, biotite decreases downstream because of intensive mechanical abrasion during transport. Some of the platy minerals were abraded and consequently fragmented during transport and thus possibly increased the percentage in the fractions that are finer than the analyzed fractions. The amphibole content, in contrast, slightly increases westward because this mineral is much less susceptible to mechanical abrasion than biotite and many other heavies.

5. DISCUSSION

The changes in the heavy-mineral spectra during the sediment transport from source to sink cannot be ascribed exclusively to the transport distance, although some heavy-mineral species are more resistant to attrition than other species. The processes behind the development can be understood only if the numerous processes involved are reconstructed, and if it can be determined what was the relative importance of these processes. Only such an analysis can result in a reliable reconstruction of the source areas, and thus of the changes in the heavy-mineral spectra on their way from source to sink.

5.1 Factors Influencing Heavy-Mineral Spectra

Tracing sediments from source to sink (or better: from sink to source) on the basis of heavy minerals meets numerous problems (Mange and Wright, 2007), as the heavy-mineral spectrum in the sink does not necessarily reflect the heavy-mineral spectrum in the source area. A well-known aspect in this context is the influence of diagenetic changes that not only may lead to (either or not partial) disappearance of mineral species (Milliken, 2007; Turner and Morton, 2007; Van Loon and Mange, 2007; Velbel, 2007), but also to the appearance of new species due to authigenesis (Bateman and Catt, 2007).

There are, however, several more factors that influence the heavy-mineral spectrum during their transport from the parent material to the final depositional site. Hydrological sorting is probably the most important (Cascalho and Fradique, 2007; Frihy, 2007; Komar, 2007; Pisarska-Jamroży et al., 2015a), mainly because of (1) changes in current energy that result in deposition of the largest and/or heaviest grains first, (2) gradual abrasion of grains during transport, and (3) destruction of non-resistant grains. Hydrological sorting, however, is not the only factor that is related to the flow regime. Earlier analyses (Pisarska-Jamroży et al., 2015a,b) of the heavy minerals under study here indicated that the flow regime also plays an important role: sediments from a gravelly layer show a different heavy-mineral spectrum than the sediments from a sandy layer just above or below the gravelly layer, because specific minerals become trapped in the spaces between gravel clasts more easily than between sand grains. Moreover, some species become trapped in gravel more easily than other mineral species. Only taking all these findings into account, the significance of a heavy-mineral spectrum for a reconstruction of the source area(s) can be assessed.

5.2 Sources of the Sediments Under Study

On the basis of the processes involved in the formation of sandurs, it is to be expected that the sediments of the two sandurs under study—and thus their heavy-mineral content—are derived from source rocks cropping out in the north. This appears to be true: most heavy minerals can be traced back to rocks of the Fennoscandian Shield. There are, however, also minerals that were picked up by the advancing ice masses by erosion of the mainly soft-sediment subsoil. This is consistent with what was known already from other investigations, including analysis of erratics (see e.g., Górska-Zabielska, 2008).

The situation appears different, however, for the middle part of the Toruń-Eberswalde IMV: it was taken for granted thus far that the sediments in this ice-marginal valley were supplied for a significant part by extraglacial rivers running northward from source areas in the south (cf. Kozarski, 1965; Galon, 1968). It was found now, admittedly, that the heavy minerals from the pre-Warta and pre-Noteć river terraces are largely comparable to those of the Drawa and Gwda sandurs and the Toruń-Eberswalde IMV (they are all characterized by the same limited number of mineral species), but it appeared also that there are exceptions. For instance, the order of the relative frequency of the heavy-mineral species is variable; significant differences occur particularly with respect to epidote, limonite, and glauconite.

The proportion of glauconite in the IMV is higher than in the sandurs, which suggests an origin from an eroded subsoil. It is also interesting that the percentages of garnet, glauconite, and limonite tend to fluctuate more in the samples from the IMV than in those from the sandurs. This might also be ascribed to inherited material: sandurs from successive glaciations may occur stacked upon each other, and erosion during the Pomeranian phase by meltwater floods may have set free older sandur sediments that were enriched in mineral grains from still older sandurs (the samples for our study were taken from the upper 10–30 m, whereas the thickness of the sandur complexes is 40 m).

In addition, the braided river in the IMV eroded older bedrock (e.g., Miocene marine sediments with diagenetically formed glauconite, as well as Miocene terrestrial sediments with reworked glauconite) so that minerals from these sediments became incorporated in the IMV terrace sediments. The high percentage of limonite in the IMV in comparison to the sandurs was caused by the relatively frequently occurring weak currents and flow stagnation in abandoned channels and/or overbank basins of the braided-river system of the IMV, so that the heavy minerals were exposed much longer to weathering in the IMV than on the very-fast aggrading sandurs (Pisarska-Jamroży, 2015).

An interesting finding was also the proportion of epidote in the heavy-mineral spectra. The proportion was comparable in the sandur and IMV sediments, but it was significantly higher in the terraces of the S–N running extraglacial rivers. As these rivers fed the IMV but did not increase the epidote proportion in the IMV, it must be deduced that, in contrast to what was commonly thought before, the southern rivers supplied such limited amounts of sediment to the IMV that their contribution to the heavy-mineral spectra of the IMV sediments was negligible.

Finally, fine-grained material must unavoidably have been deposited by winds in front of the ice cap, also on the sandurs, in the IMV, and in the pro- and extraglacial areas to the south. The wind-blown material, however, must have been too fine as a rule to be included in the grain sizes dealt with in the heavy-mineral analyses; moreover, both the sandur and the IMV samples contain a small amount of wind-transported quartz grains (Woronko et al., 2015). The fairly limited presence of wind-affected grains must be ascribed to the high sedimentation rate, which led to burial before once deposited grains could undergo more phases of aeolian sand transport (see Pisarska-Jamroży et al., 2015a). The reader is referred to Weckwerth (2013) for more details about the influence on the heavy-mineral spectra of fluvial lateral erosion of frozen and weathered Pleistocene fluvial and glacial sediments that build the walls of the IMV (the discharge from the Toruń Basin was significant, with an average of 10,000 $m^3 s^{-1}$ and a maximum of 18,000 $m^3 s^{-1}$) during the Pomeranian phase of the Weichselian glaciation.

6. CONCLUSIONS

Heavy-mineral analysis can be a useful tool for reconstructing the pathways followed by sedimentary particles from source to sink. This reconstruction should be based not only on comparison of the heavy-mineral spectrum at the depositional site with that in the source area(s), but also on a variety of parameters that affect both the various grain sizes and the various mineral species in different degrees, sometimes in a way that hardly—if at all—has been recognized thus far.

The case study presented here makes this clear by analyses of the heavy-mineral spectra of Pleistocene sediments that are derived from eroded rocks in Scandinavia and the East Central Baltic, as well as from erosion of older Pleistocene glacigenic sediments, interglacial sediments, and Miocene and Pliocene sediments. The impact on the heavy-minerals spectra by extraglacial rivers coming from the south was only slight, indicating a much lower sediment supply to the IMV by southern rivers than hitherto commonly assumed. Sorting and postdepositional processes affected the spectra much more. It thus is shown that heavy-mineral analysis is useful for tracing sediments from sink to source (and thus from source to sink) if the numerous pitfalls are taken into account.

References

Bateman, R.M., Catt, J.A., 2007. Provenance and palaeoenvironmental interpretation of superficial deposits, with particular reference to post-depositional modification of heavy-mineral assemblages. In: Mange, M.A., Wright, D.T. (Eds.), Heavy Minerals in Use, Developments in Sedimentology, vol. 58. Elsevier, Amsterdam, pp. 151–188.

Bartczak, E., 2006. Objaśnienia do Szczegółowej Mapy Geologicznej w skali 1:50 000. Arkusz Piła (Explanations to the Detailed Geological Map of Poland Scale 1:50 000. Sheet Piła). PGI Press, Warsaw.

Boothroyd, J.C., Ashley, G.M., 1975. Processes, bar morphology and sedimentary structures on braided outwash fans, northeastern Gulf of Alaska. In: Jopling, A.V., McDonald, B.C. (Eds.), Glaciofluvial and Glaciolacustrine Sedimentation, vol. 23. Society of Economic Paleontologists and Mineralogists Special Publication, pp. 193–222.

Brodzikowski, K., Van Loon, A.J., 1992. Glacigenic Sediments. In: Developments in Sedimentology, vol. 49. Elsevier, Amsterdam, 674 pp.

Cascalho, J., Fradique, C., 2007. The sources and hydraulic sorting of heavy minerals on the northern Portuguese continental margin. In: Mange, M.A., Wright, D.T. (Eds.), Heavy Minerals in Use, Developments in Sedimentology, vol. 58. Elsevier, Amsterdam, pp. 75–110.

Church, M., Gilbert, R., 1975. Proglacial fluvial and lacustrine sediments. In: Jopling, A.V., McDonald, B.C. (Eds.), Glaciofluvial and Glaciolacustrine Sedimentation, vol. 23. Society of Economic Paleontologists and Mineralogists Special Publication, pp. 22–100.

Derkachev, A.N., Nikolaeva, N.A., 2013. Possibilities and restrictions of heavy-mineral analysis for the reconstruction of sedimentary environments and source areas. Geologos 19, 147–158.

Dick, A.B., 1887. On zircons and other minerals contained in sand. Nature 36, 91–92.

Dyjor, S., 1987. Systemy kopalnych dolin Polski zachodniej i fazy ich rozwoju w młodszym neogenie i eoplejstocenie (Buried valley systems and phases of their development during the younger Neogene and Eopleistocene in western Poland). In: Jahn, A., Dyjor, S. (Eds.), Problemy młodszego neogenu i eoplejstocenu w Polsce (The Younger Neogene and Eopleistocene in Poland). Ossolińskich Press, Wrocław, pp. 85–101.

Frihy, O.E., 2007. The Nile delta: processes of heavy mineral sorting and depositional patterns. In: Mange, M.A., Wright, D.T. (Eds.), Heavy Minerals in Use, Developments in Sedimentology, vol. 58. Elsevier, Amsterdam, pp. 49–74.

Galon, R., 1961. Morphology of the Noteć-Warta (or Toruń-Eberswalde) ice marginal streamway. Prace Geograficzne 29, 7–115 (In Polish, with English summary).

Galon, R., 1968. New facts and problems pertaining to the origin of the Notec-Warta pradolina and the valleys linked with it. Przeglad Geograficzny 40, 307–315.

Górska-Zabielska, M., 2008. Fennoskandzkie obszary alimentacyjne osadów akumulacji glacjalnej i glacjofluwialnej lobu Odry (Fennoscandian Source Areas of Glacial and Glaciofluvial Deposits of the Odra Lobe (North-Western Poland and North-Eastern Germany)). Adam Mickiewicz University Press, Poznań, 78, 330 pp. (In Polish with English summary).

Komar, P.D., 2007. The entrainment, transport and sorting of heavy minerals by waves and currents. In: Mange, M.A., Wright, D.T. (Eds.), Heavy Minerals in Use, Developments in Sedimentology, vol. 58. Elsevier, Amsterdam, pp. 3–48.

Kozarski, S., 1962. Recesja ostatniego ladolodu z północnej cześci Wysoczyzny Gnieźnieńskiej a kształtowanie sie Pradoliny Noteci-Warty (The problem of the recession of the last ice sheet from the northern part of the Gniezno Plateau and development of the Noteć-Warta ice-marginal valley). Prace Komisji Geografizno-Geologicznej 2, 3Poznań, 145 pp.

Kozarski, S., 1965. Zagadnienie drogi odpływu wód pradolinnych z zachodniej cześci Pradoliny Noteci-Warty (The problem of the outflow from the western part of Noteć-Warta Pradolina). Prace Komisji Geografizno-Geologicznej 5, 1–87.

Lowright, R., Williams, E.G., Dachille, F., 1972. An analysis of factors controlling deviations in hydraulic equivalence in some modern sands. Journal of Sedimentary Petrology 42, 635–645.

Mange, M.A., Maurer, H.F.W., 1992. Heavy Minerals in Colour. Chapman and Hall, London, 147 pp.

Mange, M.A., Wright, D.T. (Eds.), 2007. Heavy Minerals in Use. Developments in Sedimentology, vol. 58. Elsevier, Amsterdam, 1283 pp.

Marks, L., 2012. Timing of the Late Vistulian (Weichselian) glacial phases in Poland. Quaternary Science Reviews 44, 81–88.

Milliken, K.L., 2007. Provenance and diagenesis of heavy minerals, Cenozoic units of the northwestern Gulf of Mexico sedimentary basin. In: Mange, M.A., Wright, D.T. (Eds.), Heavy Minerals in Use, Developments in Sedimentology, vol. 58. Elsevier, Amsterdam, pp. 247–261.

Pisarska-Jamroży, M., 2015. Factors controlling sedimentation in the Toruń-Eberswalde ice-marginal valley during the Pomeranian phase of the Weichselian glaciation: an overview. Geologos 21, 1–29.

Pisarska-Jamroży, M., Zieliński, T., 2011. Genesis of till/sand breccia (Pleistocene, Noteć valley near Atanazyn, central Poland). Sedimentary Geology 236, 109–116.

Pisarska-Jamroży, M., Zieliński, T., 2014. Pleistocene sandur rhythms, cycles and megacycles: interpretation of depositional scenarios and palaeoenvironmental conditions. Boreas 43, 330–348.

Pisarska-Jamroży, M., Van Loon, A.J., Woronko, B., Sternal, B., 2015a. Heavy-mineral analysis as a tool to trace the source areas of sediments in an ice-marginal valley, with an example from the Pleistocene in northwest Poland. Netherlands Journal of Geosciences 94, 185–200.

Pisarska-Jamroży, M., Van Loon, A.J., Woronko, B., 2015b. Sorting of heavy minerals in sediments deposited at a high accumulation rate, with examples from sandurs and an ice-marginal valley in NW Poland. GFF 137, 126–140.

Racinowski, R., 2010. Główne przeźroczyste minerały cieżkie w osadach czwartorzedowych Polski (Main transparent heavy minerals in Quaternary deposits). Biuletyn PIG 438, 99–106.

Rappol, M., Stoltenberg, H.M.P., 1985. Compositional variability of Saalian till in the Netherlands and its origin. Boreas 14, 33–50.

Turner, G., Morton, A.C., 2007. The effects of burial diagenesis on detrital heavy mineral grain surface textures. In: Mange, M.A., Wright, D.T. (Eds.), Heavy Minerals in Use, Developments in Sedimentology, vol. 58. Elsevier, Amsterdam, pp. 393–412.

Van Andel, T.J., 1950. Provenance, Transport and Deposition of Rhine Sediments (Ph.D. thesis). Groningen University, Groningen, 129 pp.

Van Loon, A.J., Mange, M.A., 2007. 'In situ' dissolution of heavy minerals through extreme weathering, and the application of the surviving assemblages and their dissolution characteristics to correlation of Dutch and German silver sands. In: Mange, M.A., Wright, D.T. (Eds.), Heavy Minerals in Use, Developments in Sedimentology, vol. 58. Elsevier, Amsterdam, pp. 189–213.

Vareikienė, O., Marmo, J., Chernet, T., Laukkanen, J., 2007. Results of heavy mineral pre-concentration by the Knelson for the geochemical study of soil: a case study in Lithuania. Geologija 60, 1–9.

Velbel, M.A., 2007. Surface textures and dissolution processes of heavy minerals in the sedimentary cycle: examples from pyroxenes and amphiboles. In: Mange, M.A., Wright, D.T. (Eds.), Heavy Minerals in Use, Developments in Sedimentology, vol. 58. Elsevier, Amsterdam, pp. 113–150.

Wachecka-Kotkowska, L., Ludwikowska-Kedzia, M., 2013. Heavy-mineral assemblages from fluvial Pleniglacial deposits of the Piotrków Plateau and the Holy Cross Mountains – a comparative study. Geologos 19, 131–146.

Weckwerth, P., 2013. Ewolucja fluwialnych systemów depozycyjnych i jej uwarunkowania paleośrodowiskowe w Kotlinie Toruńskiej podczas zlodowacenia wisły (The Evolution of Fluvial Depositional Systems and Their Paleoenvironmental Controls in the Toruń Basin During the Weichselian Glaciation). Mikołaj Kopernik University Press, Toruń, 205 pp. (In Polish with English summary).

Weckwerth, P., Chabowski, M., 2013. Heavy minerals as a tool to reconstruct river activity during the Weichselian glaciation (Toruń Basin, Poland). Geologos 19, 25–46.

Weckwerth, P., Pisarska-Jamroży, M., 2015. Periglacial and fluvial factors controlling the sedimentation of Pleistocene breccia, NW Poland. Geografiska Annaler 97A, 415–430.

Woldstedt, P., 1950. Norddeutschland und angrenzende Gebiete im Eiszeitalter. Koehler Verlag, Stuttgart, 464 pp.

Woronko, B., Rychel, J., Karasiewicz, M.K., Ber, A., Krzywicki, T., Marks, L., Pochocka-Szwarc, K., 2013. Heavy and light minerals as a tool for reconstructing depositional environments: an example from the Jałówka site (northern Podlasie region, NE Poland). Geologos 19, 47–66.

Woronko, B., Pisarska-Jamroży, M., Van Loon, A.J., 2015. Reconstruction of sediment provenance and transport processes from the surface textures of quartz grains from Late Pleistocene sandurs and an ice-marginal valley in NW Poland. Geologos 21, 105–115.

CHAPTER

5

Reconstructions of Paleohydraulic Conditions From Primary Sedimentary Structures: Problems and Implications for Sediment Provenance

P. Dasgupta

Durgapur Government College, Durgapur, India

OUTLINE

1. INTRODUCTION

Sedimentary structures have long been recognized as the fundamental tool for understanding the depositional processes. The prodromus of Nicolaus Steno's dissertation (*De solido intra solidum naturaliter contento dissertationis prodromus*) published in 1669 depicts

http://dx.doi.org/10.1016/B978-0-12-803386-9.00005-8

that the formal journey of earth science started with the identification of sedimentary rocks as the product of earth's surface processes. It won't be unwise to speculate that the presence of certain sedimentary structures, which were identified with the active surface processes, guided in deciphering the process–product relationship, and the host rocks, in general, were considered as of sedimentary origin. The basis of the classification of rocks of the earth's crust as proposed by Jahann Gottlob Lehmann in 1756 indicates that this idea persisted until the middle of the 18th century, and reached its maxima when Abraham Gottlob Werner put forward his dogmatic idea of *Neptunism*. The approach was more logical when James Hutton proposed the *Law of Uniformitarianism* based on the close resemblance between the features (presumably structures) within the recent sediments and their consolidated counterparts. Gradually the concept of stratigraphy evolved out of these radical ideas.

In the year 1879, through his presidential address delivered before the Geological Society of London, Henry Clifton Sorby (Sorby, 1879) introduced the importance of the study of sedimentary processes as a separate discipline. Although sedimentology was formally introduced as a specialized branch of geology through this landmark event, its foundation was laid a few decades back (Potter and Pettijohn, 1977). Hall (1843a,b,c) for the first time described with proper interpretation the primary directional structures like flute casts, groove casts, current crescents, sand shadows, oriented fossils, cross-bedding, and ripple marks. Sorby (1851, 1856, 1857, 1858, 1859a,b) not only enriched the knowledge bank with the description and proper interpretation of a few more structures like parting lineation and flame structure, he moved forward with the introduction of measurement of paleocurrent directions and establishment of its importance in paleogeographic reconstruction. His opinion "… and that such observations would enable us to learn the quarter from whence their materials were drifted" (Sorby, 1859a, p. 145) also opened a new window, the concept of provenance study. These, in conjunction with the concept of flow regime introduced by Sorby (1851), were indeed the prelude to the present day basin analysis. Again it was Sorby (1859a, 1908), who introduced the quantitative and experimental approach to sedimentological interpretation. It is a fact that the contribution of those pioneers cannot even be listed within a few printed pages, and any evaluation thereof would just be a pygmy's effort to garland the giants.

For a long time, the emphasis was given on the reconstruction of paleocurrent pattern from the directional structures, mainly to determine the provenance and broad sediment dispersal pattern for reconstruction of the basin evolution. Potter and Pettijohn (1977) elaborately discussed the stages of development of this concept through the decades. It took about a century to realize the importance of contributions to be made from the field of fluid mechanics for a better appreciation of the pattern of disposition of the constituent particles of a sedimentary body in the light of paleohydraulic condition. This, in turn, would help in understanding the flow dynamics responsible for changes in composition of the sediments from source to sink in a more objective way. Precise estimation of the paleohydraulic condition can only guide us to assess the approximate distance traveled by the sediments before deposition. It gives an idea about the primary derivation from the source rock, possibility of reworking and mixing, sediment budget of the transporting medium, and its seasonal fluctuation. Since the middle of 20th century, different experimental and theoretical methods had been adopted for explaining the nature of interplay between the solid particles and the fluid medium in producing specific types of sedimentary structure. The present discussion aims at critical evaluation of the applicability of these contributions in reconstruction of the paleohydraulic condition for a better understanding of the depositional processes.

2. ENTRAINMENT AND TRANSPORTATION

The mechanism of incorporation of a particle into the flowing system is of fundamental importance in understanding the dynamics of the depositional system. Lifting, sliding, and rolling are the common mechanisms of entrainment depending on the size, shape, and density of the grain. Thin, flaky grains of low-density material are often entrained through lifting. According to Bernoulli's theorem, the interstitial fluid below such grains is static in nature and has the pressure higher than the fluid flowing above the grain. When the pressure gradient thus generated across the grain exceeds the weight of the grain, the grain is lifted and carried in suspension by the flowing fluid.

Sliding or rolling is the common entrainment mechanism for bed load material. Probe into these mechanisms elucidates the particle–fluid interaction and the ultimate controlling factor for deposition. When a static particle comes on the way of the flowing fluid, it tends to obstruct the flow. The particle, by virtue of its weight, exerts some force on the flow lines colliding against it, and the fluid column tends to suffer shear deformation. As a natural response some stress develops within the flowing fluid. It is the shear stress that defines whether the fluid column would suffer deformation or not. If the shear stress fails to withstand this deforming force, the flow lines are deflected. Otherwise, the particle is entrained, keeping the flow lines undeformed. The initiation of movement of the oblate (pancake-shaped) or tabular grains normally takes place through sliding because of the fact that to make such particle rolling the center of gravity has to be raised. The equant grains, however, start rolling, as do the prolate (cigar-shaped) ones, with their long axes oriented perpendicular to the flow at onset of entrainment. Dasgupta and Manna (2011) elaborately discussed the behavior of rolling particles in a flowing system (Fig. 5.1), and demonstrated that the critical shear stress (τ_{cr}) for entrainment of a grain (of diameter D_1) through rolling is given by:

$$\tau_{cr} = \frac{\omega \cos \beta(\tan \alpha - \tan \beta)}{\pi\left(\frac{D_1}{2}\right)^2} \text{ Nm}^{-2} \qquad (5.1)$$

In this equation, ω is the immersed weight of the grain (of diameter D_1) resting on an inclined surface that makes an angle β with the horizon, and α is the angle, the "line of easiest movement" (Middleton and Southard, 1978), that the common tangent between the grains of diameter D_1 and D_2 (Fig. 5.1A) makes with the slope of the substratum.

When $\alpha = \beta$, according to Eq. (5.1), the critical value of τ_{cr} becomes zero. Under such condition the common tangent becomes horizontal and the motion of the grain is impending (Fig. 5.1C). Now if the adjacent grains are of equal diameter (Fig. 5.1D), the value of α becomes 90 degree (the common tangent becomes perpendicular to the inclined substratum) and the total downslope force on the line of easiest movement, $\{\tau\pi(D_1/2)^2 + \omega \sin \beta\} \cos \alpha$, becomes zero. As a result, under the influence of slope parallel forces, the rear grain can only push forward the adjacent grain in the downslope direction. If the diameter of the rear grain is less than that of the grain in the downslope direction (Fig. 5.1E), the value of α exceeds 90 degree, and the total downslope force on the line of easiest movement, $\{\tau\pi(D_2/2)^2 + \omega \sin \beta\}\cos \alpha$, becomes negative, i.e., it acts downward along the common tangent. So the particle, under any circumstances, cannot climb up the adjacent larger grain in the downslope direction. So a larger grain can roll over a smaller one on the downflow

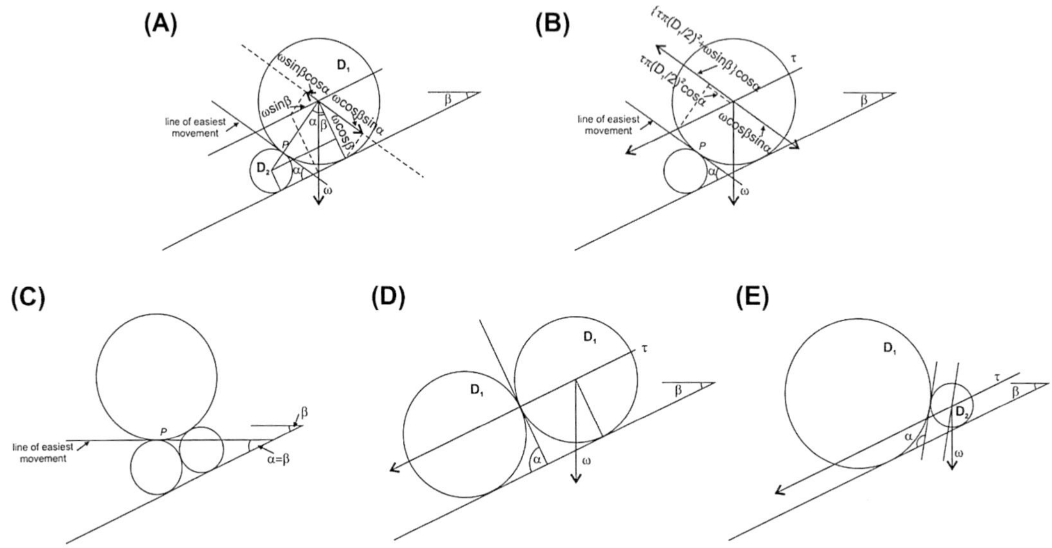

FIGURE 5.1 Definition diagrams showing (A) the components of gravitational pull, (B) combination of shear force and gravitational pull acting on a grain of diameter D_1 in contact with a smaller grain (of diameter D_2) lying in the downslope direction; (C to E) the possible variations in the orientation of the line of easiest movement: (A) horizontal and the motion of the grain is impending. (C) and (D) depict situations unfavorable for climbing of grains during a shear flow. (After Dasgupta and Manna, 2011.)

direction only when the shear stress of the flow attains this critical value. This attests the conclusions drawn by Kramer (1934) and Chang (1939) that a higher proportion of the coarser constituents are entrained at the threshold velocity. Now putting $\omega = \frac{4}{3}\pi\left(\frac{D_1}{2}\right)^3(\sigma - \rho)g$ (σ being the density of the grain, ρ is the density of the fluid medium, g is the acceleration due to gravity) in Eq. (5.1):

$$\tau_{cr} = \frac{2}{3}D_1(\sigma - \rho)g\cos\beta(\tan\alpha - \tan\beta)\ \mathrm{Nm}^{-2} \tag{5.2}$$

If the bed shear stress falls below this critical value, the particles being transported in the bed load population tend to get deposited. Now the value of bed shear stress τ is given by γhS, where γ is the specific weight of water ($=g\rho$), h is the hydraulic radius ($\approx$depth of flow), and S is the slope ($\approx\sin\beta$). Conventionally, in the literature on hydraulics, S is expressed as $\tan\beta$ (Grant, 1997), but since it defines the slope parallel component of the gravitational pull, it must be $\sin\beta$. In different hydraulic analysis, this error has not been identified because of the low value of β, for which the sine and tangent values are very close to each other. Putting $\tau = g\rho h\sin\beta$ in Eq. (5.2):

$$h_{cr} = \frac{2}{3\rho}D_1(\sigma - \rho)(\cot\beta\tan\alpha - 1)\ \mathrm{m} \tag{5.3}$$

where h_{cr} is the critical depth of deposition. This relation depicts that if the slope remains constant, the size of the depositing grain is directly proportional to the depth of flow. It further explains that since the hydraulic radius gradually decreases during continuous sedimentation from a unidirectional flow (e.g., riverine flow), the resultant deposit shows fining upward character.

After incorporation into the flowing system, the movement of a particle is governed by the resultant of two vectors: (1) the shear velocity (U^*) of the flow that drives the particle along the direction of flow and (2) settling or terminal fall velocity of the particle (ϖ) acting vertically downward in response to the gravitational pull. Now, in a flowing system if $U^* > \varpi$ for a particular grain lifted above the flow base, the resultant of these two vectors makes very low angle with the slope and the particle remains in suspension. Under the reverse situation, when $\varpi > U^*$, the resultant becomes subvertical and the particle tends to confine to the flow base and moves in bed load. For the intermediate situation, i.e., when $U^* = \varpi$, the lifted particle collides against the substratum obliquely and bounces back, resulting in saltation mode of transportation, which is considered as part of the bed load. Lane and Kalinske (1939), based on field and laboratory studies, inferred that the grains giving values of ϖ/U^* of 1 or more can never be found in suspension.

3. BED FORM STABILITY

The experimental studies by Simons et al. (1965) and Guy et al. (1966) made significant contributions in understanding the relation between hydraulic conditions, grain properties, and nature of the resultant bed forms. Simons et al. (1965, Fig. 21). arrived at a simple relation between stream power, median fall velocity of bed material, and form roughness. They evaluated this "stream power" as the product of the bed shear stress (τ) and the depth-averaged flow velocity $(\overline{U})$. According to them, this could be used to anticipate the form of bed roughness given the depth, slope, velocity, and fall diameter of bed material. Allen (1968), based on the experimental results of Guy et al. (1966), modified the bed form stability diagram proposed by Simons et al. (1965) with a view to identify the limiting factors for the development and preservation of different bed forms. Bagnold (1966), in an attempt to justify "stream power" as an expression for the flow energy, pointed out that the available power to the unit length of a flowing stream is the time rate of liberation in kinetic form of the liquid's potential energy as it descends the gravity slope. Since stream power is the expression of energy, Allen (1968, Fig. 6.9) also delineated the stability fields of different bed forms based on the median fall diameter of the constituent particles of respective bed forms and the stream power. Van Den Berg and Van Gelder (1993) were of the opinion that the bed form stability diagram is useful in predicting the bed state for given steady flow and sediment conditions, and since $\tau = \rho g h S = \rho g \overline{U}^2 C^{-2}$ (where ρ = fluid density, g = acceleration due to gravity, h = hydraulic radius, S = channel slope, and C is the Chezy coefficient), and no information regarding the energy gradient or alluvial roughness can be obtained from the bed form geometry, these stability relations are of limited value in paleohydraulic analysis. With a view to overcome this problem they compiled all available experimental data including their own (Van Den Berg and Van Gelder, 1993) to propose a new bed form stability diagram, which was claimed to be applicable without knowledge of bed form roughness or energy slope. In this bed form stability diagram the stability fields were defined with reference to *modified mobility parameter*

(θ') and *nondimensional particle diameter* (D^*). Van Rijn (1984) defined the modified mobility parameter (θ') as:

$$\theta' = \frac{\rho \overline{U}^2}{(\sigma - \rho) C'^2 D_{50}}$$

where σ = grain density, D_{50} = 50th percentile of the grain size population, and C' = the Chezy coefficient to grain roughness, which is given by:

$$C' = 18 \log \frac{4h}{D_{90}}$$

The nondimensional particle diameter (D^*) was proposed by Bonnefille (1963) as:

$$D^* = D_{50} \left[\frac{(\sigma - \rho) g}{\rho \nu^2} \right]^{\frac{1}{3}}$$

where ν = kinematic viscosity.

These relations may apparently justify the theoretical validity of the claim by Van Den Berg and Van Gelder (1993). The wide spread of the stability fields for each phase (Fig. 5 of Van Den Berg and Van Gelder, 1993), even for the same D^*, however, rules out the applicability of this stability diagram in precise determination of paleohydraulic parameters from any specific bed form. Van Den Berg and Van Gelder (1993) did not specifically address this limitation of their proposed diagram. Allen's diagram, on the other hand, appears to be more convenient in understanding the genetic relations between different bed forms, the nature of phase transition with flow regime, and average grain size of the deposit. Specific hydrodynamic parameters cannot be determined from any proposed bed form stability diagram.

3.1 Froude Number and the Bed Form Geometry

Although different authors defined the stability fields of bed forms with reference to different parameters, scanning of the available hydrodynamic parameters reveals a direct relation between bed form geometry and the Froude number (F) of the flow, which is given by $F = \frac{\overline{U}}{\sqrt{gh}}$, where $\overline{U}$ is the mean (depth-average) flow velocity, h is the hydraulic radius, and g is the acceleration due to gravity. Experimental data depict how the bed forms may broadly be classified with reference to critical flow ($F = 1$) condition (Fig. 5.2), and the stability limits of different phases with reference to the F of the flow (Fig. 5.3). Flow separation is perhaps the most important natural phenomenon, which has a direct control on the formation and geometry of different bed forms. This, in turn, depends solely on the F of the flow. Allen (1968, 1982) made an elaborate discussion on the mechanism of flow separation. Proper understanding of the different stages from flow separation to bed form migration is the key to the paleohydraulic analysis.

Fig. 5.4 depicts the different stages of flow separation on a flat frictional surface. The flowing fluid is considered as a stack of infinitesimally thin films. The lower part of the fluid column experiences maximum frictional resistance and the lowermost film adheres to the substratum due to adhesive force. The immediate overlying one moves under the influence of the resultant of two

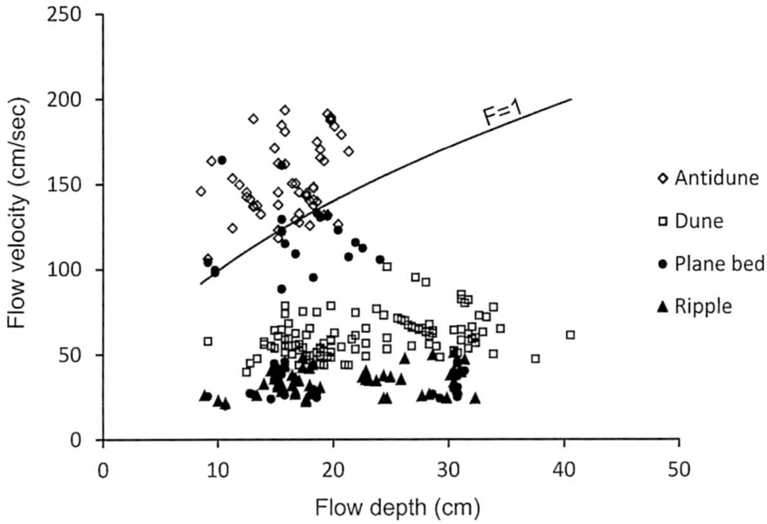

FIGURE 5.2 Experimental data (recalculated from Guy et al., 1966) plots depicting bed forms in relation to flow velocity and flow depth. Broad classification with reference to critical flow condition ($F = 1$) is apparent.

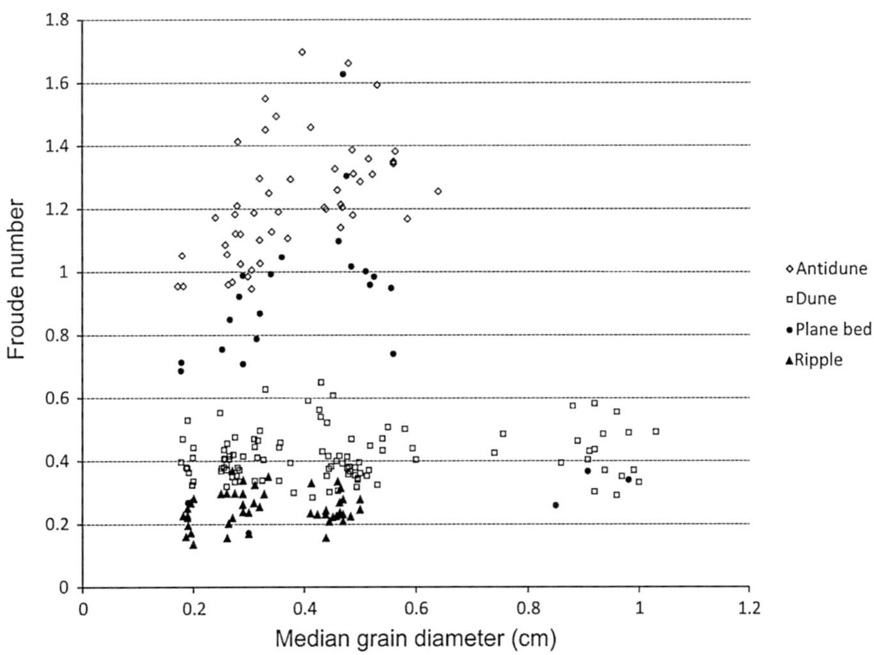

FIGURE 5.3 Froude number versus median grain size plots defining stability limits of different bed forms (data recalculated from Guy et al., 1966.)

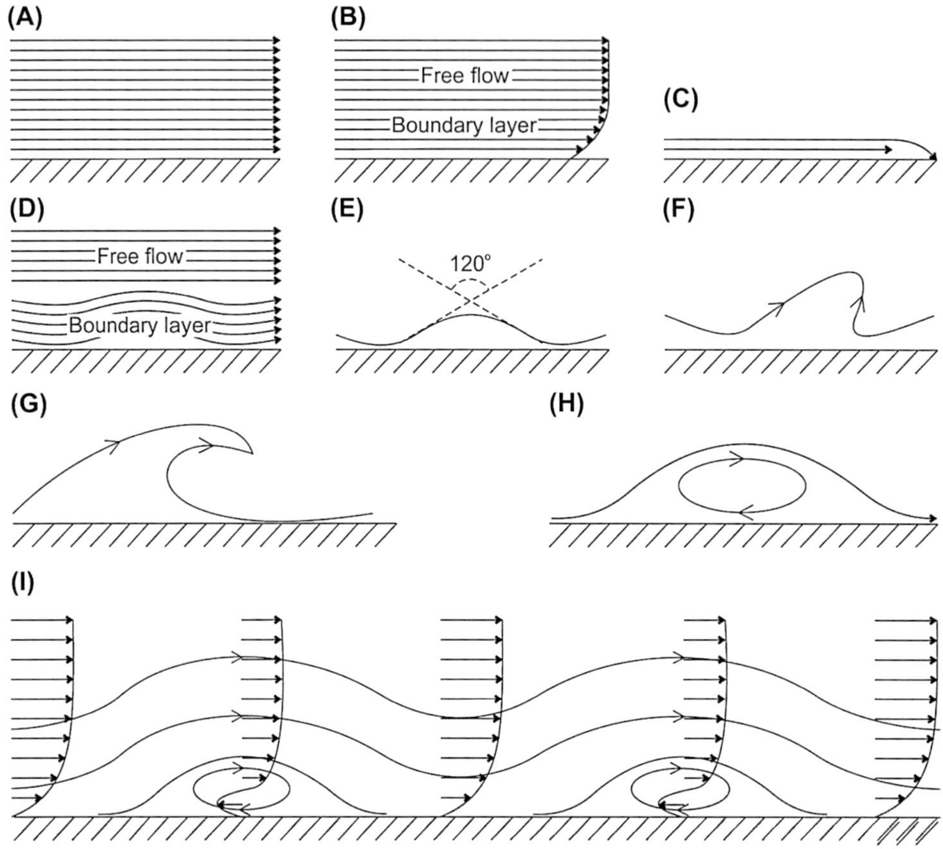

FIGURE 5.4 Different stages of flow separation on a flat frictional surface. (A) Stack of flow lines at the onset of flow. (B) In consequence of differential retardation the velocity profile of the boundary layer becomes parabolic, and on further progression (C) the upper flow line plunges down. (D) Development of a wave form within boundary layer. (E) Waveform grows to reach the critical apical angle of 120 degree, on further growth instability sets in (F) and the waveform breaks (G) leading to the formation of the separation bulb (H). Continuation of this process leads to formation of separation bulbs at regular interval (I).

forces: the retarding cohesive force exerted by the lower quasi-static film and the accelerating downslope gravitational pull. This process continues upward through successive films, and the retarding force gradually dies out upward. The lower part of the flowing fluid directly affected by the retarding force is called boundary layer, the thickness of which depends on the magnitude of the frictional resistance exerted by the substratum. The upper part of the flow is called the free flow. In consequence of the differential retardation, the boundary layer is characterized by a parabolic velocity profile (Fig. 5.4B), and the upper flow lines in the boundary layer moves forward and plunges down before the lower ones reach the point (Fig. 5.4C). The lower flow lines on reaching this higher pressure point (because it has already been occupied by the fluid from the upper flow layers, which has become slower due to frictional resistance from the substratum) move upward following the pressure gradient and surpass the high pressure point resulting development of a waveform (Fig. 5.4D). Consequently, the flow passage within the boundary layer no longer remains uniform in thickness. The gap between the top of the waveform and base of the free flow is constricted (Fig. 5.4D), and following the law of continuity the flow becomes faster during passage through this narrower path. As the flow becomes faster above the waveform, the waveform below tends to grow vertically in response to this local pressure gradient (Bernoulli's law). On further increase in the amplitude, the apical angle of the waveform reaches the critical value of 120 degree (Milne-Thompson, 1976) and the Kelvin–Helmholtz instability is impending (Fig. 5.4E). On further growth of the waveform as the apical angle tends to decrease, the instability sets in and the waveform breaks, leading to the formation of the separation bulb, a closed region of arrested fluid (Fig. 5.4H). Since the original character of the flow is restored just beyond the point of attachment toward the downstream side of the separation bulb, the next separation bulb is formed after a certain distance in the same process, and finally at periodic interval (Fig. 5.4I). The movement of noncohesive sediments transported in bed load is obstructed on reaching the separation bulb and heaved toward its upstream side. The shape of the heap is modified by the main flow and the separation bulb through time, leading to the formation of train of ripple or dune.

The most important prerequisite for flow separation is the vertical growth of the waveform developed within the boundary layer, and it is possible only in case of subcritical flow. That is why the ripples and dunes are formed under subcritical condition only (Figs. 5.2 and 5.3). Allen (1968) demonstrated that the ripple geometry shows a definite trend of variation from continuous straight-crested to discontinuous lunate or linguoid through sinuous and catenary with increase in flow velocity or decrease in the flow depth. Harms (1969) also advocated a similar variation in ripple geometry with energy condition. Hence, this may be concluded that the specific ripple geometry is a function of F. Baas (1994), however, is of the opinion that straight and sinuous ripples are nonequilibrium bed forms at all flow velocities and gradually become discontinuous and arcuate for attaining equilibrium. Allen (1969) experimentally demonstrated that the ripples grow more disordered as the flow is made shallower, but did not address the basic reason behind this relation. In the experimental studies, of which the data were used by Allen (1969), the average depth was estimated, and the relief of the depositional surface was not taken into consideration. In deeper flow, the local variation in depth might have been negligible in comparison with the average flow depth. However, this variation possibly plays a vital role in local variation of F and other hydrodynamic parameters on shallowing of the flow and that finally caused discontinuity in crestline and development of the arcuate ripple forms.

In case of turbidity current, the flow dynamics is explained with reference to densimetric Froude number (F'), which is given by $F' = \frac{\overline{u}}{\sqrt{g'h}}$, where $g' = g\Delta\rho/\rho$ (ρ is the density of the flow and $\Delta\rho$ is the density contrast between the flow and the ambient fluid) (Komar, 1971; Hand, 1974; Waltham, 2004; Postma and Cartigny, 2014). Postma and Cartigny (2014) furnished an account of the facies characteristics of the products of subcritical and supercritical turbidity currents. Experimental study by Cartigny et al. (2011) revealed that the internal structures of waveforms generated by turbidity currents readily distinguish between sub- and supercritical bed forms. The higher resolution internal structures further discriminate between the antidunes and the cyclic steps. It was also demonstrated that sediment waves form only over a certain range of aspect ratios. Cartigny et al. (2011) plotted various aspect ratios in numerically obtained stability diagrams and argued that the basic numerical modeling can be used to calculate the Froude number and discharge from the internal structure of waveforms and geometry of cyclic steps.

4. ESTIMATION OF PALEOHYDRAULIC PARAMETERS FROM DIFFERENT STRUCTURES

4.1 Dunes and Ripples

Knoroz (1959) (in Jopling, 1966c) proposed the following empirical relationship between the mean height and length of dunes, and velocity, depth of flow, and the particle size:

$$H = 3.5h\frac{\left(U - \frac{U_c}{U}\right)^{2/3}}{\frac{\log_{10}h}{D} + 6}$$

$$L = \frac{2.8hU}{U - U_c}$$

where H is the dune height, L is the dune length, and U_c is the threshold velocity for entrainment of the particle of diameter D. Jopling (1966c) further mentioned that "these equations also apply moderately well for the prediction of the height and length of bed forms in sands of diameter less than 0.5 mm." However, the results obtained from these equations for median grain size below 0.5 mm strongly mismatched with the measured values of dune height and length published by Guy et al. (1966). This disproves the generality of these equations.

The ripple patterns obtained in a flume experiment by Allen (1969) depicted a well-defined streamwise spacing (wavelength) between the ripples, and longitudinal features (ridges and spurs) with a characteristic wavelength measurable transversely to flow. These observations together with theoretical considerations led Allen (1969) to propose an empirical formula to describe the plan form of the ripples in relation to flow condition:

$$\frac{\lambda_x}{\lambda_z} = 6.4\left(F\frac{H}{h}\right)^{0.27}$$

where λ_x is the mean ripple wavelength measured parallel to the flow direction, λ_z is the mean wavelength of the longitudinal features measured transversely to the flow direction, F is the Froude number, H is the mean ripple height, and h is the mean flow depth (hydraulic radius). Precision of the values of λ_z (Allen, 1969, Table 8 and Fig. 36), claimed to have been acquired from photographs published in Guy et al. (1966), used in establishing the validity of the proposed equation is questionable. It is because of the fact that out of the total 93 photographs of flume experiments published by Guy et al. (1966) only one (Fig. 51 of Guy et al., 1966) is orthographic plan view of the ripples with scale and suitable for such measurement; four others (Figs. 8, 10, 14, and 42 of Guy et al., 1966) are apparently suitable but without any scale; and the remaining 88 are either sections or perspective views (these also without scale). Then how could the 18 (Table 8 of Allen, 1969) values of λ_z be determined and be related with the corresponding λ_x? Banks and Collinson (1975) based on results obtained from similar flume experiments opined that a measure of ripple shape (the ratio of wavelengths of transverse to streamwise features) has a more complex relationship with the flow property than was previously proposed by Allen (1969). Admitting this discrepancy, Allen (1977) pointed out that the values of flow depth assigned to the proposed equation were larger than they should be, by an amount depending on the flow width relative to depth, and proposed a pair of equations after necessary modifications:

$$\frac{\lambda_x}{\lambda_z} = 5.85\left(F\frac{H}{h}\right)^{0.412}\left(1+\frac{h}{w}\right)^{1.71}$$

$$F\frac{H}{h} = 0.026\left(\frac{\lambda_x}{\lambda_z}\right)^{1.73}\left(1+\frac{h}{w}\right)^{-5.20}$$

where w is the channel width.

Allen (1977) argued that these equations can be used in paleohydraulic reconstructions, at least for rocks of fluviatile origin. In the proposed equations, three parameters related to ripple geometry, i.e., λ_x, λ_y, and H, can be obtained from the rock exposure, and three hydrodynamic parameters (U, h, and w) are to be estimated or determined for paleohydraulic reconstruction. Allen (1977) suggested three methods for the estimation of depth and width of the flow. In this context, the statements made by Allen (1977, pp. 60–61) like "It is commonly assumed with regard to coarse grade members of fining-upward cyclothems that the thickness of the coarse-grade member is equal to bankfull channel depth" or "the vertical distance between the rippled surface and the top of the sandstone member may be regarded as an estimate of flow depth" appear to be too speculative and unrealistic. If the thickness of the coarse-grade member is equal to bankfull channel depth, then how could the finer fractions have been deposited subsequently? In reality, the deposition of the finer fraction in any cyclothem takes place on deceleration of the flow and that also from shallower flow due to fall in discharge. Even the units like clay with desiccation cracks indicating subaerial exposure were deposited in subaqueous condition and subsequently exposed. The second proposal, of "calculation of the limiting velocities that could have prevailed during the construction of the ripples" from the bed form stability diagram, and the third proposal, of assumption of plausible value for the Darcy–Weisbach f for estimation of mean velocity, appear unrealistic. This type of approximation cannot give any realistic picture. However, an illustrative explanation could have been furnished in support of these proposed methods.

Baas (1994) carried out a series of steady-flow flume experiments to work out the equilibrium geometry of small-scale, unidirectional bed forms, and demonstrated that the time required for bed forms to develop toward equilibrium dimensions shows an inverse power relation with flow velocity for both bed form height and bed form wavelength. The best fit power functions obtained by Baas (1994) through nonlinear regression analysis of the experimental results are as follows:

$$U_{10} = 0.233 + 0.225{T_{eH}}^{-0.442}$$

$$U_{10} = 0.233 + 0.297{T_{eL}}^{-0.47}$$

where U_{10} is the 10°C equivalent flow velocity, 0.233 m s^{-1} is the threshold velocity of sediment motion (calculated from Miller et al., 1977 for median grain size of 0.095 mm), T_{eH} is the time required by the bed form to reach equilibrium height, and T_{eL} is the time required by the bed form to reach equilibrium length. The following generalized relations can be deduced from the above equations:

$$T_{eH} = \left(\frac{0.225}{U - U_c}\right)^{\frac{1}{0.442}}$$

$$T_{eL} = \left(\frac{0.297}{U - U_c}\right)^{\frac{1}{0.47}}$$

where U is the average flow velocity and U_c is the threshold velocity for entrainment of the median particle diameter.

With a view to apply these equations for estimation of the growth rate of the bed forms produced by Guy et al. (1966), the threshold velocity for entrainment of each size fraction was determined using the relation proposed by Miller et al. (1977) and the growth rate was determined by:

$$G_H = \frac{1}{H}\left(\frac{0.225}{U_{100} - 122.6{D_{50}}^{0.29}}\right)^{\frac{1}{0.442}}$$

$$G_L = \frac{1}{L}\left(\frac{0.297}{U_{100} - 122.6{D_{50}}^{0.29}}\right)^{\frac{1}{0.47}}$$

where G_H is the growth rate of the equilibrium height of the bed form, G_L is the growth rate of the equilibrium length of the bed form, U_{100} is the flow velocity measured 100 cm above the sediment bed, and the expression $122.6{D_{50}}^{0.29}$ represents the threshold velocity for entrainment of the median particle diameter D_{50} (finer than 2 mm) (Miller et al., 1977). Since Miller et al. (1977) measured the threshold velocity for entrainment at 100 cm above the sediment bed, the average flow velocity at the same level (U_{100}) was determined for each run through extrapolation of the velocity profile provided by Guy et al. (1966, Tables 12–21). The regression analysis of 116 sets of values of G_H and G_L thus obtained from the data of Guy et al. (1966) gives the best fit $(R^2 = 0.9959)$ function (Fig. 5.5) as:

$$G_H = 8.0553G_L + 8 \times 10^{-8} \tag{5.4}$$

Now, this relation can be used for determination of the flow velocity from the values of height and length of the bed form and the median grain diameter obtained from the rock record. Proper care has to be taken for acquisition of the median grain diameter of the lithified sediment. It is a potential source of error, as the grain size distribution pattern is badly affected during diagenesis and additional error is often imposed during disintegration of the rock samples for grain size analysis. Microscopic method for determination of grain size from thin section is, however, not recommended for the purpose, because the probability of the thin section to pass through the longest diameter of all the grains is definitely low and thus the results cannot portray the actual grain size distribution pattern.

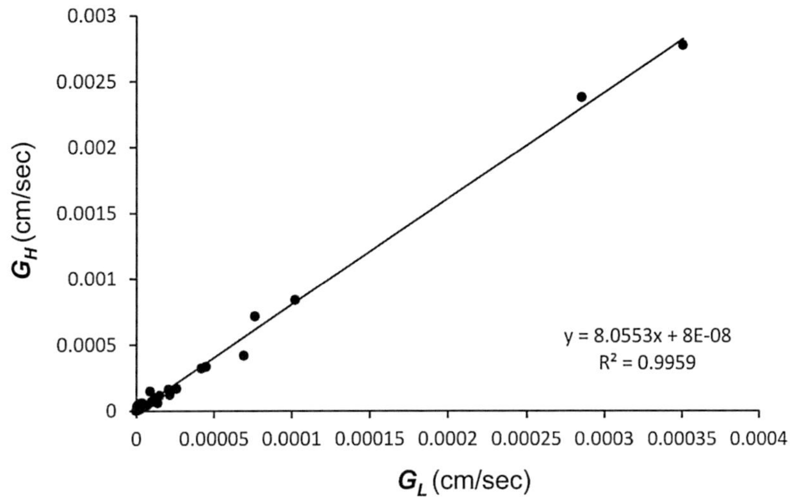

FIGURE 5.5 Relation between the growth rates of ripple height (G_H) and ripple length (G_L) calculated from the experimental data of Guy et al. (1966).

4.2 Cross-Stratification

Cross-stratification has long been recognized as the principal directional structure for the reconstruction of paleocurrent pattern, the essential component for provenance study (Pettijohn and Potter, 1964; Pettijohn, 1975; Potter and Pettijohn, 1977). Different propositions on descriptive and genetic classifications of cross-stratifications (McKee and Weir, 1953; Allen, 1963; Elliott, 1964) in conjunction with some experimental studies (McKee, 1957; Jopling, 1963, 1966a,b) enriched the ideas about the depositional mechanisms involved in the origin of such stratifications. These are typically formed by the primary current and correlates strongly with the flow direction (Harms and Fahnestock, 1965; Wermund, 1965; Meckel, 1967; Williams, 1968; Barrett, 1970; McGowen and Garner, 1970; Dott, 1973; Michelson and Dott, 1973; High and Picard, 1974). Emphases have also been given on precise acquisition of paleocurrent data from cross-stratifications of different geometry (Slingerland and Williams, 1979; DeCelles et al., 1983; Dasgupta, 1995, 2002). The ideas pertaining to the hydraulic controls on the shape of the cross-stratifications, evolved through the experimental studies by Jopling (1965a), added a new dimension in the interpretation of cross-stratifications. Jopling (1965a) demonstrated that with the increase in stream velocity and depth ratio (ratio of stream depth to basin depth), the frontal profile of foreset evolves from angular (planar tabular) to low-angle concave through incipiently tangential, tangential, and strongly tangential stages. The same trend was observed with decrease in the relative fall velocity (ratio of the median fall velocity of the sediment particles to the average velocity of streamflow) of the transported particles. Jopling (1965a) further pointed out that the influence of wave action and small oscillations of base level may produce either convex or sigmoidal profiles. Jopling (1966c) correlated certain foreset characters like decrease in angle of dip of foresets or increase in laminae frequency with increase in flow velocity. According to Jopling

(1966c), with increase in flow velocity the foreset laminae tend to become less distinct. However, none of these parameters were duly quantified, and such significant observations only contributed in refinement of comparative interpretation.

The nature of development of successive foresets during ripple or dune migration (Reineck, 1961; Brush, 1965; Jopling, 1965b, 1966c, 1967; Basumallic, 1966; Jopling and Walker, 1968) leads to a definite grain size distribution pattern in cross-stratified deposits. The periodic interruption of suspension fallout by avalanching of bed load material along the slip face causes textural (and may be compositional also) contrast between successive foresets. This textural contrast actually makes it possible to distinguish between the bed load and suspension load populations composing alternate foreset strata. As indicated by Jopling (1966c), the relation between the shear velocity (U^*), particle settling velocity (ϖ), and mode of transportation (Lane and Kalinske, 1939) can be utilized for paleohydraulic reconstruction. Fig. 5.6 illustrates the procedure of acquisition of basic data from a succession of cross-stratified sandstone. Samples are collected for grain size analysis from a co-set of three planar cross-bed sets (A, B, and C) (Fig. 5.6A). After plotting the grain size data on an arithmetic probability paper, the intersection point between the bed load and suspension load is obtained. The corresponding grain size may be termed as critical size for which $\frac{\varpi}{U^*} = 1$ (Lane and Kalinske, 1939). The terminal fall velocity (ϖ) for the grain of critical size may be obtained from the equation proposed by Gibbs et al. (1971). It is assumed that the flow top was at f–f' during deposition of this co-set. Following the depth-ratio concept of Jopling (1965a), the flow base during deposition of cross-bed sets A, B, and C are marked by a–a', b–b', and c–c', respectively (Fig. 5.6A). The shear velocity $U^* = \sqrt{gSh}$, where g is the acceleration due to gravity, h is the hydraulic radius, and $S = \sin\beta$, where β is the angle of slope. Now, for critical size, since $\frac{\varpi}{U^*} = 1$, for cross-bed set A we obtain (Fig. 5.6A):

$$\varpi_A = \sqrt{gSh_0}$$

or

$$S = \frac{\varpi_A^2}{gh_0} \tag{5.5}$$

where ϖ_A is the calculated terminal fall velocity of the critical grain size of set A.

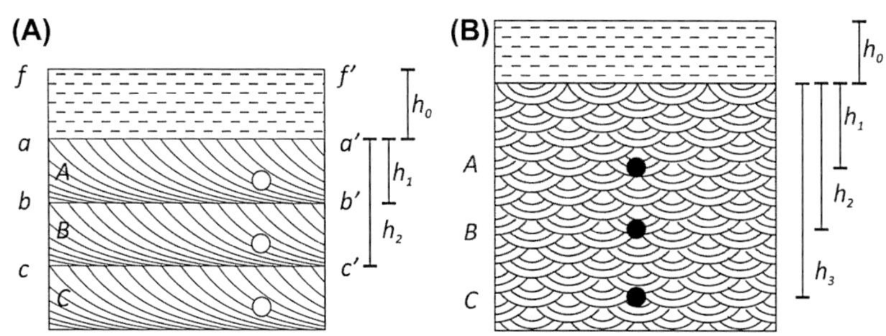

FIGURE 5.6 Definition diagrams illustrating the method for collection of samples from (A) co-set of planar cross-stratified sandstone (*white circles* represent the sample positions) and (B) trough cross-stratified sandstone (*black circles* represent the sample positions) for estimation of the shear velocity from critical grain size.

Similarly for sets B and C we get:

$$S = \frac{\varpi_B^2}{g(h_0 + h_1)} \tag{5.6}$$

$$S = \frac{\varpi_B^2}{g(h_0 + h_2)} \tag{5.7}$$

From Eqs. (5.5) and (5.6) the value of h_0 can be obtained by eliminating S:

$$h_0 = h_1 \bigg/ \left(\frac{\varpi_B^2}{\varpi_A^2} - 1\right) \tag{5.8}$$

Now, for verification of the precision of the obtained value of h_0, the same may be determined either from Eqs. (5.5) and (5.7) or (5.6) and (5.7). After getting the value of h_0, the value of S may also be obtained from Eq. (5.5). It is to be kept in mind that samples from close vertical interval may give very close value of the critical grain size, so it is advisable to collect samples at a larger interval from available thickest co-set. The same methodology can be applied within thick unit of trough cross-stratified unit or even within climbing ripple laminated sandstone, excepting the supercritical variety (Allen, 1982), which is formed through dominant suspension fallout. Since the trough cross-stratified sandstone is formed through subaqueous dune migration, in this case the flow depth may be recorded in a different way, as shown in Fig. 5.6B. For the sample collected from level A, the relation would be:

$$S = \frac{\varpi_A^2}{g(h_0 + h_1)} \tag{5.9}$$

Similarly, for samples from level B and C, the relation would be:

$$S = \frac{\varpi_B^2}{g(h_0 + h_2)} \tag{5.10}$$

and

$$S = \frac{\varpi_C^2}{g(h_0 + h_3)} \tag{5.11}$$

From Eqs. (5.9) and (5.10), we get:

$$h_0 = \frac{\varpi_B^2 h_1 - \varpi_A^2 h_2}{\varpi_A^2 - \varpi_B^2} \tag{5.12}$$

Putting this value of h_0 in Eq. (5.11), we may get the value of S.

4.3 Antidune

Antidune is considered to be a bed form developed under supercritical flow condition (Gilbert, 1914; Allen, 1968, 1982). The experimental results, however, show some deviations from this general trend. The antidunes generated by Guy et al. (1966) were at F ranging between 0.83 and 1.7, while those of Alexander et al. (2001) were formed at F between 1.5 and 1.71. Critical analyses of the mechanics of antidunes by Kennedy (1963), Allen (1976), and Grant (1997) led to infer the following stages of formation and destruction of antidune with change in F of the flow: (1) The initially plane bed is deformed into a series of antidunes by accelerating near-critical flow and the antidunes tend to steepen on further increase in F. (2) As the flow becomes supercritical in the troughs, instability sets in and the antidunes break upstream as hydraulic jumps. (3) The downward flux of momentum from the breaking hydraulic jumps erodes the antidunes and the plane beds are restored with abrupt reduction in velocity. (4) Lowered velocities coupled with increased depths, as water previously stored in the stationary waves is released, causes the flow to become subcritical again and the cycle to repeat itself. According to Grant (1997), "the high-amplitude bed configuration caused by increasing flow velocity induces flow instability at flow slightly above critical, leading to very rapid energy dissipation and erosion of bed forms. This feedback results in unsteady non-uniform flow around $F = 1$ and a cyclic creation-destruction sequence of bed forms." This explains well the very low preservation potential of antidune that makes the bed form unsuitable for paleohydraulic analysis. The equations developed by Kennedy (1963) and Reynolds (1965) defining the relationship between the dimensions of antidune and the flow parameters did not produce matching results between the antidune dimensions and corresponding flow parameters produced by Guy et al. (1966). Shaw and Kellerhals (1977) also experienced limited applications of these equations in paleohydraulic computations from antidune dimensions. Alexander et al. (2001), however, carried out flume studies with a view to develop a relationship between the three-dimensional geometry of the internal structures and the formative bed forms under supercritical flow condition. It was concluded that "the length and maximum thickness of the lenticular laminasets are approximately half of the length and height of formative antidunes, providing a potentially useful tool for palaeohydraulic reconstructions." However, specific mathematical relationships are yet to be established.

5. RANDOMNESS OF EXPERIMENTAL RESULTS

The preceding discussion reveals that the attempts made so far to formulate the relations between hydrodynamic parameters and bed form geometry derived from experimental results for computation of paleohydraulic parameters are mainly based on best-fit regression relations, but without any mention of the corresponding R^2 value. So the range of possible variations cannot be assessed. Application of these equations on different sets of similar experimental data often gives unrealistic results, and generality of the proposed formulation does not stand valid. The basic problem with most of these equations is the lack of any apparent theoretical justification, and thus the specific role of individual variable cannot be worked out. A critical analysis of the largest experimental data provided by Guy et al. (1966) reveals that most of the results are absolutely random. For example, on examining a set of results of 48 runs (providing the complete information) related to ripple, it was observed that in the results of 18 runs all the variables are independent. The remaining 30 runs were classified according to the ripples of identical dimensions produced into 10 sets, out of which five sets are represented by two runs, two sets by three runs, one set by four runs, and two sets by five runs. The ripples of same dimension were found to show varied relations with other parameters (Table 5.1). As a result, no definite inference could be drawn on the basic factor(s) controlling the dimension of the ripples. Similar randomness was observed in the results of other runs.

TABLE 5.1 Details of Experimental Data Related to Generation of Ripple (Recalculated From the Data Furnished by Guy et al., 1966)

	Experimental Setting			Flow Parameters				Median Grain Size	Ripple Dimensions	
Run	Slope	Depth (cm)	Width (cm)	U (cm s^{-1})	U^* (cm s^{-1})	τ (dyne cm^{-2})	F	D_{50} (cm)	L (cm)	H (cm)
25	0.00018	28.3464	**243.84**	26.5176	2.22504	4.788	0.16	**0.019**	**12.192**	**0.9144**
23	0.00062	13.4112	**243.84**	25.908	2.83464	8.1396	0.23	**0.019**	**12.192**	**0.9144**
10	0.00041	17.9832	**243.84**	26.8224	2.68224	7.182	0.2	0.028	**15.24**	**0.9144**
27	0.00057	16.764	**243.84**	28.3464	3.048	9.576	**0.22**	0.019	**15.24**	**0.9144**
3	0.00026	18.8976	**60.96**	30.48	2.19456	4.788	**0.22**	0.054	**15.24**	**0.9144**
6	0.00047	15.5448	**60.96**	32.3088	2.68224	7.182	0.26	0.033	**15.24**	**0.9144**
2	0.00015	32.3088	**243.84**	24.0792	2.19456	4.74012	0.14	**0.019**	**18.288**	**0.9144**
30	0.00028	30.48	**243.84**	33.8328	2.8956	8.1396	0.2	**0.019**	**18.288**	**0.9144**
31	0.00043	31.0896	**243.84**	39.624	3.62712	**12.9276**	0.23	**0.019**	**18.288**	**1.2192**
28	0.00079	16.4592	**243.84**	31.6992	3.56616	**12.9276**	0.25	**0.019**	**18.288**	**1.2192**
29	0.00084	17.0688	**243.84**	34.4424	3.74904	13.8852	0.27	**0.019**	**18.288**	**1.2192**
18	0.00031	17.6784	**243.84**	23.7744	2.31648	5.2668	0.18	0.045	**21.336**	**1.2192**
5	0.00045	30.48	**243.84**	40.8432	3.6576	13.4064	0.24	0.028	**21.336**	**1.2192**
3	0.00092	16.764	**243.84**	35.9664	3.90144	15.3216	0.28	0.019	**21.336**	**1.2192**
16	0.00021	24.6888	**243.84**	24.0792	2.04216	4.11768	0.15	**0.045**	**21.336**	**1.8288**
9	0.0004	16.764	**243.84**	26.8224	2.56032	6.7032	0.21	**0.045**	**21.336**	**1.8288**
8	0.0006	15.5448	**243.84**	28.3464	3.01752	9.0972	0.23	**0.045**	**21.336**	**1.8288**
11	0.00049	10.668	**243.84**	21.336	2.25552	5.2668	0.21	0.045	**24.384**	**1.8288**
4	0.00069	26.2128	**243.84**	47.5488	4.20624	17.7156	0.3	0.028	**24.384**	**1.8288**
5	0.00058	31.3944	**243.84**	46.9392	4.23672	17.7156	0.27	0.019	**27.432**	**1.2192**
12	0.00108	17.3736	**243.84**	48.1584	4.29768	18.1944	0.37	0.028	**27.432**	**1.2192**
5	0.00088	14.9352	60.96	35.6616	3.59664	12.9276	0.29	0.033	**27.432**	**1.2192**
87	0.00046	22.86	**243.84**	35.3568	3.2004	10.5336	0.24	0.047	**27.432**	**1.8288**
12	0.00106	8.8392	**243.84**	25.908	3.01752	9.0972	0.28	0.045	**27.432**	**1.8288**
88	0.00049	22.5552	**243.84**	36.576	3.29184	11.0124	0.25	0.047	**30.48**	**2.1336**
5	0.00047	22.86	**243.84**	40.2336	3.26136	10.5336	0.27	0.045	**30.48**	**2.1336**
2	0.00036	24.9936	**243.84**	36.576	2.95656	8.6184	**0.23**	0.045	**36.576**	**2.1336**

(Continued)

TABLE 5.1 Details of Experimental Data Related to Generation of Ripple (Recalculated From the Data Furnished by Guy et al., 1966)—cont'd

	Experimental Setting			Flow Parameters				Median Grain Size	Ripple Dimensions	
Run	Slope	Depth (cm)	Width (cm)	U (cm s^{-1})	U^* (cm s^{-1})	τ (dyne cm^{-2})	F	D_{50} (cm)	L (cm)	H (cm)
85	0.00047	23.7744	**243.84**	34.4424	3.32232	11.0124	**0.23**	0.047	**36.576**	**2.1336**
b90	0.00053	**18.288**	**243.84**	44.196	3.07848	9.576	**0.33**	**0.047**	**48.768**	**1.8288**
b89	0.00065	**18.288**	**243.84**	44.8056	3.44424	11.4912	**0.33**	**0.047**	**48.768**	**1.8288**

For each set of identical ripples, the similarity in other parameters are shown in bold figures. It is noteworthy that no definite relationship is apparent between any two parameters.

These disparities can be attributed to hysteresis (Simons and Richardson, 1962; Allen, 1973). Since the natural flows are inherently unsteady, the experimental results based on steady hydrodynamic conditions may not provide adequate information for hydraulic interpretation of a bed form and the internal organization (Allen, 1973). Simons and Richardson (1962) experimentally demonstrated the qualitative variation in the resultant bed forms with systematic variation in flow depth between rising and falling stages. It was further demonstrated that excepting for upper-stage plane beds and antidunes (Fig. 9, Simons and Richardson, 1962), the bed forms always lagged in development of the change of flow, and the depth to discharge relations produced hysteresis loops (Fig. 4–8, Simons and Richardson, 1962). According to Simons and Richardson (1962), outflow from (during rising stage) or inflow to (during falling stage) the alluvial channels through the bank and bed material causes seepage forces, which influence the bed form stability and the median grain size. Depending on the seepage force in effect, bed materials of different median grain size may be molded into a particular bed form under the same flow condition. Simons and Richardson (1962) further explained how the viscosity of the flowing fluid has a direct bearing on fall velocity and consequently mobility of the bed material. Accordingly, difference in temperature or concentration (and composition) of the suspended material may cause development of different bed forms within the same bed material by the flows with same slope and discharge.

6. DISCUSSION AND CONCLUSIONS

Proper understanding of the paleohydraulic condition is important for determination of the sediment dispersal pattern and the provenance, the essential components of basin analysis. It is the prevailing hydrodynamic conditions that determine the transportation of sediments from source to sink. Estimation of paleohydraulic parameters thus plays an important role in basin analysis. Depositional sedimentary structures, the product of fluid–solid interplay in this course of their journey, are considered to be the basic source for determination of the paleohydraulic condition. Laboratory studies have elucidated the physical explanations of different bed forms and associated structures. The stability limits of different bed forms were defined based on the empirical relations between hydrodynamic parameters and bed form geometry derived from experimental results, which definitely improved the resolution of qualitative interpretation of the depositional structures. The nature of phase transitions inferred therefrom also enriched the comparative interpretation of the prevailing paleohydraulic conditions. Most of the equations defining relationship between the hydrodynamic and bed form parameters derived from the laboratory studies were found incompatible with the data derived from different sets of experiments. These incompatibilities may be attributed to hysteresis, the role of seepage force, and the factors like variation in viscosity of the flowing fluid due to temperature fluctuations and presence or absence of suspension cloud. Critical review of the experimental studies, however, identifies some other serious limitations, which raise questions regarding the acceptability of the results.

1. In scientific research, the experimental result is only acceptable when the same result is obtained by repeated runs under identical conditions. Unfortunately, the available published data were produced mostly from single runs.
2. In case of multivariate systems, the experiments are supposed to be carried out with systematic change in each of the preassigned variables (e.g., flume width, slope, flow depth, flow velocity, grain size, run time, etc.), keeping the rest constant. Only then can the contribution of each variable be worked out. The available results were not produced in such a systematic manner.
3. In most of the results (excepting those of Baas, 1994), there is no mention of the run time. So it is not clear whether the bed forms attained equilibrium with the prevailing hydrodynamic condition or not, and what was the nature of bed form modification with time.
4. There are no data available about the surface relief of the granular bed at the initiation of the run. This is a very important parameter, particularly in case of shallow flow. This would cause local variation in the value of the Froude number and the bed shear stress, which play important roles in the generation and stability of a particular bed form.

So it may be concluded that the available experimental results can only be used in qualitative assessment of the paleohydrodynamic conditions. Any attempt for quantitative reconstruction of the paleohydraulic condition based on these data may lead to an erroneous conclusion. So at present, qualitative determination of the provenance with reference to the regional slope inferred from the directional structure can only help in basin analysis. Precise estimation of the sediment dispersal pattern and gradual variation in sediment composition from source to sink requires establishment of a more specific process–product relationship between the hydraulic and bed form parameters.

References

Alexander, J., Bridge, J.S., Cheel, R.J., Leclair, S.F., 2001. Bed forms and associated sedimentary structures formed under supercritical water flows over aggrading sand beds. Sedimentology 48, 133–152.

Allen, J.R.L., 1963. The classification of cross-stratified units with notes on their origin. Sedimentology 2, 93–114.

Allen, J.R.L., 1968. Current Ripples. North-Holland, Amsterdam.

Allen, J.R.L., 1969. On the geometry of current ripples in relation to stability of fluid flow. Geografiska Annaler 51, 61–96.
Allen, J.R.L., 1973. Phase differences between bed configuration and flow in natural environments, and their geological relevance. Sedimentology 20, 323–329.
Allen, J.R.L., 1976. Bed forms and unsteady processes: some concepts of classification and response illustrated by common one-way types. Earth Surface Processes 1, 361–374.
Allen, J.R.L., 1977. The plan shape of current ripples in relation to flow conditions. Sedimentology 24, 53–62.
Allen, J.R.L., 1982. Sedimentary Structures: Their Character and Physical Basis, vols. I & II. Elsevier, Amsterdam.
Baas, J.H., 1994. A flume study on the development and equilibrium morphology of current ripples in very fine sand. Sedimentology 41, 185–209.
Bagnold, R.A., 1966. An approach to the sediment transport problem from general physics. Professional Paper U.S. Geological Survey 422-I, 1–37.
Banks, N.L., Collinson, J.D., 1975. The size and shape of small-scale current ripples: an experimental study using medium sand. Sedimentology 22, 583–599.
Barrett, P.J., 1970. Paleocurrent analysis of the mainly fluviatile Permian and Triassic Beacon rocks, Beardmore Glacial area, Antarctica. Journal of Sedimentary Petrology 40, 395–411.
Basumallic, S., 1966. Size differentiation in a cross-stratified unit. Sedirnentology 6, 35–68.
Bonnefille, R., 1963. Essaia de synthese des lois du début d'entrainement des sédimentssous l'actiond'un courant en régime continu. Bullettin du Centre de Recherches et d'essais de Chatou 5, 17–22.
Brush, L.M., 1965. Sediment sorting in alluvial channels. In: Middleton, G.V. (Ed.), Primary Sedimentary Structures and Their Hydrodynamic Interpretation, Soc. Econ. Paleontol. Mineral. Special Publication, 12, pp. 25–33.
Cartigny, M.J.B., Postma, G., Van Den Berg, J.H., Mastbergen, D.R., 2011. A comparative study of sediment waves and cyclic steps based on geometries, internal structures and numerical modelling. Marine Geology 280, 40–56.
Chang, Y.L., 1939. Laboratory investigation of flume traction and transportation. American Society of Civil Engineers Transactions 104, 1246–1313.
Dasgupta, P., 1995. A new model of inclinometer and its application in determination of attitude of foreset planes. Journal of Sedimentary Research A65, 562–563.
Dasgupta, P., 2002. Determination of paleocurrent direction from oblique sections of trough cross-stratification – a precise approach. Journal of Sedimentary Research 72, 217–219.
Dasgupta, P., Manna, P., 2011. Geometrical mechanism of inverse grading in grain-flow deposits: an experimental revelation. Earth-Science Reviews 104, 186–198. http://dx.doi.org/10.1016/j.earscirev.2010.10.002.
DeCelles, P.G., Langford, R.P., Schwartz, R.K., 1983. Two new methods of paleocurrent determination from trough cross-stratification. Journal of Sedimentary Petrology 53, 629–642.
Dott Jr., R.H., 1973. Paleocurrent analysis of trough cross-stratification. Journal of Sedimentary Petrology 43, 779–783.
Elliott, R.E., 1964. A classification of cross-stratification structures. Nature 202, 451–452.
Gibbs, R.J., Matthews, M.D., Link, D.A., 1971. The relationship between sphere size and settling velocity. Journal of Sedimentary Petrology 41, 7–18.
Gilbert, G.K., 1914. The transportation of debris by running water. Professional Paper U.S. Geological Survey 86.
Grant, G.E., 1997. Critical flow constrains flow hydraulics in mobile-bed streams: a new hypothesis. Water Resources Research 33, 349–358.
Guy, H.P., Simons, D.B., Richardson, E.V., 1966. Summary of alluvial channel data from flume experiments, 1956–61. Professional Paper US Geological Survey 462-I, 1–96.
Hall, J., 1843a. On wave lines and casts of mud furrows (abstract). American Journal of Science 45, 148–149.
Hall, J., 1843b. Remarks upon casts of mud furrows, wave lines, and other markings upon rocks of the New York system. Report of the Association of American Geologists and Naturalists 422–423.
Hall, J., 1843c. Geology of New York, Part IV, Survey of the 4th District. Carroll and Cook, Albany.
Hand, B.M., 1974. Supercritical flow in density currents. Journal of Sedimentary Research 44, 637–648.
Harms, J.C., 1969. Hydraulic significance of some sand ripples. Bulletin of the Geological Society of America 80, 363–396.
Harms, J.C., Fahnestock, R.K., 1965. Stratification, bed forms and flow phenomena (with an example from Rio Grande). In: Middleton, G.V. (Ed.), Primary Sedimentary Structures and Their Hydrodynamic Interpretation – A Symposium: Soc. Econ. Paleontologists and Mineralogists. Spec. Publ., vol. 12, pp. 84–115.

High Jr., L.R., Picard, M.D., 1974. Reliability of cross-stratification types as paleocurrent indicators in fluvial rocks. Journal of Sedimentary Petrology 44, 158–168.
Jopling, A.V., 1963. Hydraulic studies on the origin of bedding. Sedimentology 2, 115–121.
Jopling, A.V., 1965a. Hydraulic factors controlling the shape of laminae in laboratory deltas. Journal Sedimentary Petrology 35, 777–791.
Jopling, A.V., 1965b. Laboratory study of the distribution of grain sizes in cross-bedded deposits. In: Middleton, G.V. (Ed.), Primary Sedimentary Structures and Their Hydrodynamic Interpretation–A Symposium: Soc. Econ. Paleontologists and Mineralogists. Spec. Publ., vol. 12, pp. 53–65.
Jopling, A.V., 1966a. Origin of cross-laminas in a laboratory experiment. Journal of Geophysical Research 71, 1123–1133.
Jopling, A.V., 1966b. Some applications of theory and experiment to the study of bedding genesis. Sedimentology 7, 71–102.
Jopling, A.V., 1966c. Some principles and techniques used in reconstructing the hydraulic parameters of a paleo-flow regime. Journal of Sedimentary Petrology 36, 5–49.
Jopling, A.V., 1967. Origin oflaminae deposited by the movement of ripples along a streambed: a laboratory study. Journal of Geology 75, 287–305.
Jopling, A.V., Walker, R.G., 1968. Morphology and origin of ripple-drift cross lamination, with examples from the Pleistocene of Massachusetts. Journal of Sedimentary Petrology 38, 971–984.
Kennedy, J.F., 1963. The mechanics of dunes and antidunes in erodible-bed channels. Journal of Fluid Mechanics 16, 521–544.
Komar, P.D., 1971. Hydraulic jumps in turbidity currents. Geological Society of America Bulletin 82, 1477–1487.
Kramer, H., 1934. Sand mixtures and sand movement in fluvial models. American Society of Civil Engineers 60, 443–454.
Lane, E.W., Kalinske, A.A., 1939. The relation of suspended to bed material in rivers. American Geophysical Union Transactions 20, 637–641.
McGowen, J.H., Garner, L.E., 1970. Physiographic features and stratification types of coarse-grained point bars: modern and ancient examples. Sedimentology 14, 77–111.
McKee, E.D., 1957. Flume experiments on the production of stratification and cross-stratification. Journal of Sedimentary Petrology 27, 129–134.
McKee, E.D., Weir, G.W., 1953. Terminology for stratification and cross-stratification. Geological Society of America Bulletin 64, 381–390.
Meckel, L.D., 1967. Tabular and trough cross-bedding comparison of dip azimuth variability. Journal of Sedimentary Petrology 37, 80–86.
Michelson, P.C., Dott Jr., R.H., 1973. Orientation analysis of trough cross-stratificationin Upper Cambrian sandstones of western Wisconsin. Journal of Sedimentary Petrology 43, 784–794.
Middleton, G.V., Southard, J.B., 1978. Mechanics of sediment movement. SEPM Pacific Section Short Course 3.
Miller, M.C., McCave, I.N., Komar, P.D., 1977. Threshold of sediment motion under unidirectional currents. Sedimentology 254, 507–527.
Milne-Thompson, L.M., 1976. Theoretical Hydrodynamics. The MacMillan Press Limited, London.
Pettijohn, F.J., 1975. Sedimentary Rocks, third ed. Harper & Row, Publishers, New York.
Pettijohn, F.J., Potter, P.E., 1964. Atlas and Glossary of Primary Sedimentary Structures. Springer, Berlin.
Postma, G., Cartigny, M.J.B., 2014. Supercritical and subcritical turbidity currents and their deposits – a synthesis. Geology 42, 987–990.
Potter, P.E., Pettijohn, F.J., 1977. Paleocurrents and Basin Analysis. Springer-Verlag, Berlin.
Reineck, H.-E., 1961. Sedimentbewegungen an Kleinrippeln im Watt. Sencken bergiana Lethaea 42, 51–61.
Reynolds, A.J., 1965. Waves on the erodible bed of an open channel. Journal of Fluid Mechanics 22, 114–133.
Shaw, J., Kellerhals, R., 1977. Paleohydraulic interpretation of antidune bed forms with applications to antidunes in gravel. Journal of Sedimentary Petrology 47, 257–266.
Simons, D.B., Richardson, E.V., 1962. The effect of bed roughness on depth-discharge relations in alluvial channels. Water-Supply Papers 1498-E, USGS 26.
Simons, D.B., Richardson, E.V., Nordin Jr., C.F., 1965. Sedimentary structures generated by flow in alluvial channels. In: Middleton, G.V. (Ed.), Primary Sedimentary Structures and Their Hydrodynamic Interpretation, SEPM Special Publication, 12, pp. 34–52.

Slingerland, R.L., Williams, E.G., 1979. Paleocurrent analysis in light of trough cross-stratification geometry. Journal of Geology 87, 724–732.

Sorby, H.C., 1851. On the oscillation of the currents drifting the sandstone beds of the southeast of Northumberland and on their general direction in the coal field in the Neighborhood of Edinburgh. Proceedings West Yorkshire Geological Society 3, 232–240.

Sorby, H.C., 1856. On the physical geography of the Old Red Sandstone sea of the Central District of Scotland. Edinburgh New Philosophical Journal 3, 112–122.

Sorby, H.C., 1857. On the physical geography of the Tertiary estuary of the Isle of Wight. Edinburgh New Philosophical Journal 5, 275–298.

Sorby, H.C., 1858. On the ancient physical geography of the southeast of England. Edinburgh New Philosophical Journal 7, 226–237.

Sorby, H.C., 1859a. On the structures produced by the current present during the deposition of stratified rocks. The Geologist 2, 137–147.

Sorby, H.C., 1859b. On the contorted stratification of the drift of the coast of Yorkshire. In: Proceedings of the Geological and Polytechnic Society West Riding of Yorkshire 1849–1859, pp. 220–224.

Sorby, H.C., 1879. Presidential address. Quarterly Journal of Geological Society of London 35, 56–77.

Sorby, H.C., 1908. On the application of quantitative methods to the study of the structure and history of rocks. Quarterly Journal of Geological Society of London 64, 171–233.

Van Den Berg, J.H., Van Gelder, A., 1993. A new bed form stability diagram, with emphasis on the transition of ripples to plane bed in flows over fine sand and silt. International Association of Sedimentologists Special Publication 17, 11–21.

Van Rijn, L.C., 1984. Sediment transport, part 3: bed forms and alluvial roughness. Journal of Hydraulic Engineers 110, 1733–1754.

Waltham, D., 2004. Flow transformation in particulate gravity currents. Journal of Sedimentary Research 74, 129–134.

Wermund, E.G., 1965. Cross-bedding in the Meridian sand. Sedimentology 5, 69–79.

Williams, G.E., 1968. Formation of large scale trough cross-stratification in a fluvial environment. Journal of Sedimentary Petrology 38, 136–140.

CHAPTER

6

Physico-Chemical Characteristics of the Barremian-Aptian Siliciclastic Rocks in the Pondicherry Embryonic Rift Sub-basin, India

N. Chakraborty[1], S. Sarkar[1], A. Mandal[1], W. Mejiama[2], H.A. Tawfik[3], R. Nagendra[4], P.K. Bose[1], P.G. Eriksson[5]

[1]Jadavpur University, Kolkata, India; [2]Osaka City University, Osaka, Japan; [3]Tanta University, Tanta, Egypt; [4]Anna University, Chennai, India; [5]University of Pretoria, Pretoria, South Africa

OUTLINE

Sediment Provenance
http://dx.doi.org/10.1016/B978-0-12-803386-9.00006-X

1. INTRODUCTION

Presedimentation history of detrital rocks can be reconstructed with the analytical results of geochemistry and detrital mineralogy of sedimentary deposits. While detrital mineralogy reveals the type of source rock which attribute to the tectonic setting (Krynine, 1948; Folk et al., 1970; Basu, 1985; Basu et al., 2013), geochemistry interprets the extent of source area weathering, rainfall, and paleotemperature (Suttner and Dutta, 1986; Corcoran and Mueller, 2002; Roy and Roser, 2013). The constitutional elements within sediment contribute to infer the presedimentation history. Trace elements, being more immobile, are preferred major elements, and shales are ideal than sandstone for being more homogenized and being the favored host of trace elements (Blatt, 1985; Graver and Scott, 1995; Hastie et al., 2007). This is true when crustal evolution or tectonic setting is the moot question. Major elements, on the other hand, are preferred indicators for source rock type and sandstones are useful to pervasive the transportational effect, if there is any (Corcoran et al., 1998).

Its potential notwithstanding, sediment geochemistry may not provide a clear presedimentation record. Sediment transport and diagenesis alter the depositional record. Trace elements generally concentrate within clay and hence in the distal part of the transport systems. On the contrary, if embedded within detrital heavy minerals, they are likely to concentrate in the proximal part (Morton and Hallsworth, 2007). Further complication arises as heavy minerals can undergo postdepositional dissolution and can also be generated diagenetically (Pettijohn et al., 1987). The major element budget may also alter significantly because of precipitation of diagenetic minerals. A sound background of paleogeography, petrography, and mineralogy, therefore, is a prerequisite for proper utilization of elementary chemistry of sediments.

This chapter focuses principally on bulk chemistry of the initial siliciclastic infill of the Pondicherry Sub-basin within the Cauvery Basin, India (Fig. 6.1), and deciphers the extrabasinal history of the sedimentary rocks. It also investigates the potential alteration of source-inherited sediment geochemistry by intrabasinal processes of transportation, deposition, and diagenesis. In order to do so, a high-resolution process- and paleogeography-related sedimentary facies analysis is presented (the first such analysis on these deposits). Petrographic and petrogenetic analysis further aided thesource-related interpretations of sediment geochemistry. The work was done in three

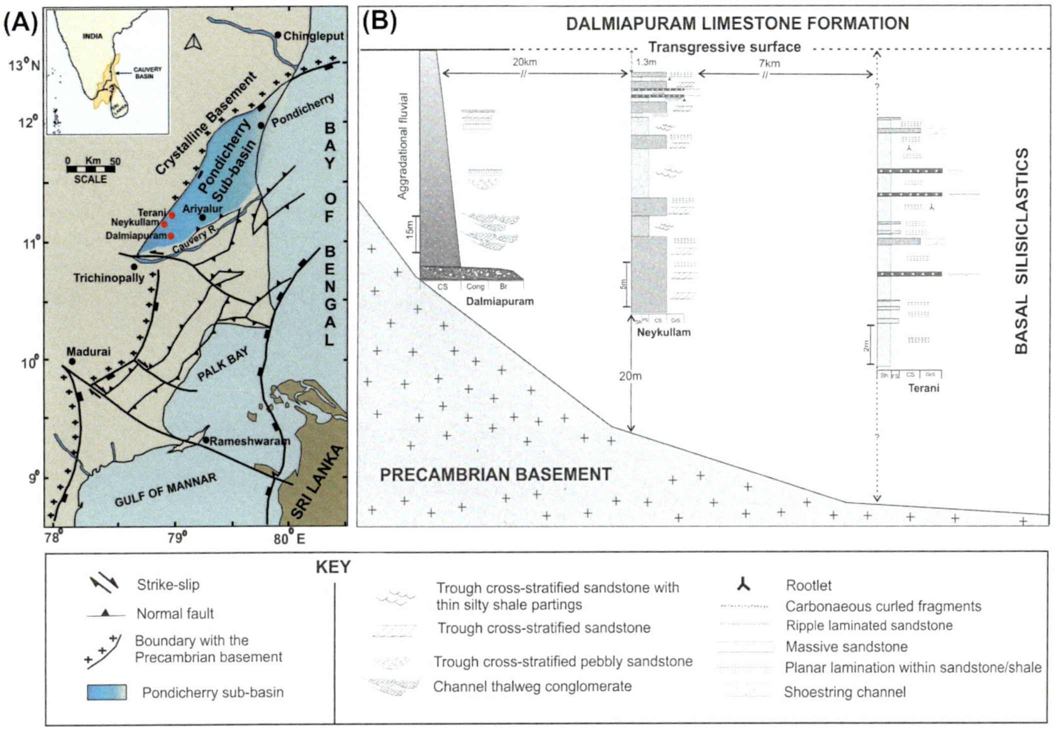

FIGURE 6.1 Location and structural definition of the Cauvery Basin and within it the Pondicherry subbasin extending into the Bay of Bengal (modified after Sastri et al., 1973) (A). Lithologs with sedimentary structures in the three studied locations (B).

isolated open-cut quarries at Dalmiapuram, Neykulam, and Terani (Fig. 6.1). Distinct variations in facies composition and paleocurrent direction reflect paleogeographic variation from the basin-margin to its interior clearly, allowing appreciation of the transportation effect on the chemistry. Interpretation of the sediment source, sediment dynamics, paleogeography, related variations in paleocurrents, major diagenetic modifications, and of the mean annual average of provenance rainfall and temperature are the primary aims of this chapter. The spatial distribution of facies and paleocurrent variability elicit a ideal example of initial filling of an embryonic rift basin engendered by the Mesozoic break-up of the Gondwaland.

2. GEOLOGICAL BACKGROUND

The siliciclastic formation under consideration comprises the initial filling of the Cauvery Rift Basin, one of the many intracratonic basins that formed as the eastern Gondwana broke up into India, Antarctica, and Australia during the Late Jurassic-Early Cretaceous (Sastri et al., 1981; Powell et al., 1988; Narasimha Chari et al., 1995; Li and Powell, 2001; Santosh et al., 2009; Nagendra et al., 2011; Chatterjee et al., 2013). Normal faults trending parallel to the Precambrian Eastern Ghat trend (NNE–SSW) gave rise to a horst-graben system of linear geometry, within which the Cauvery Basin lies (Fig. 6.1). WNW–ESE and WSW–ENE trending conjugate normal faults further subdivided the Cauvery Basin into multiple sub-basins (Murthy et al., 2008). In the Pondicherry Sub-basin, the unnamed formation examined here is considered by many as an Upper Gondwana-equivalent (Barremian to Aptian) unit (e.g., Tewari et al., 1996; Watkinson et al., 2007; Sundaram and Rao, 1986). We term this unit, informally, as the Basal Siliciclastic formation (BS formation); stratigraphically it lies between the Precambrian basement and the Dalmiapuram Limestone of Albian age (Fig. 6.1; Tewari et al., 1996; Watkinson et al., 2007; Sarkar et al., 2014). The basement in the immediate neighborhood is principally granitic, enclosing small patches of amphibolites. The BS formation is probably of Barremian to Aptian age as it incorporates plant fossils, such as *Ptillophyllum acutifolium*, *P. Cutchense*, *Taeniopteris spatulata*, *Taeniopteris* sp., *T. Lata*, *Dioctyozamites* sp., *Sphenopteris* sp., *Cladophlebis indica*, *Elatocladus plana*, *E. Conferta*, *Ginkgoites cf. rajmahalensis* (Ramasay and Bannerji, 1991; Venkatachalapathy and Ragothaman, 1995; BouDagher-Fadel et al., 1997), palynological fossils like *Microcachyidites*, *Cooksonites*, and *Aequitriradites* (Singh and Venkatachala, 1988), and a dinocyst assemblage (Garg et al., 1988). In absence of high resolution sedimentological analysis, the siliciclastic formation has been limply interpreted as of fluvial origin (Blandford, 1862; Pascoe, 1959; Banerji, 1983). Lying between a basal nonconformity on the basement and a transgressive surface above, the BS formation has been received less attention and lacks a defining name, reflecting its sparse exposures, none exceeding 70 m in thickness (Fig. 6.1), from under extensively mined younger rocks and thick alluvium.

3. METHODOLOGY

The present investigation involves a detailed major and trace element analysis of studied sedimentary rocks supplemented by petrographic observations. Care was taken during sampling to precisely identify the sedimentary facies and to collect fresh samples from unweathered parts of the beds. Collected samples were kept in airtight ziplock plastic sample bags to avoid contamination.

Araldite impregnation with help of a Buehler Vacuum Impregnation Unit at the Sedimentology Laboratory, Department of Geological Sciences, Jadavpur University, was utilized for thin sectioning of many of the friable samples. A Leica DMLP polarizing microscope attached to a Leica DFC320 digital camera was also used for the petrographic study. The sandstone modal composition was quantified on counting 500 points in each of 37 thin sections, using the Leica Point Counting System with CVS Petrog software version 2.62 (Table 6.1).

Major and trace element compositions of 25 sandstone and shale samples were determined by a RIGAKU RIX 2100 X-ray Fluorescence Spectrometer, equipped with Rh/W dual-anode X-ray tube. The analyses were performed on the whole rock specimens under 50 kV and 50 mA accelerating voltage and tube current, respectively. Fused glass beads were prepared by mixing 1.8 gm of powdered sample (dried to 110°C for 4 h), 3.6 gm of spectroflux ($Li_2B_4O_7$ 20%, $LiBO_2$ 80%, dried at 450°C for 4 h), 0.54 gm of oxidant $LiNO_3$ (dried at 110°C for 4 h), and traces of LiI. The mixture is fused at 800°C for 120 s and

TABLE 6.1 Results of Modal Analysis for Mineralogical Composition of the Studied Samples

Samples	Monocrystalline Quartz (Qm)	Polycrystalline Quartz (Qp)	Polycrystalline Quartz Stretched (Qps)	Polycrystalline Quartz Unstretched Qpus	Quartz Total (Qt)	Feldspar (F)	Lithic Fragment Granite (Lg)	Lithic Fragment Amphibole (La)	Lithic Fragment Total (Lt)	Total
				Dalmiapuram						
G/D/14-1	44.650	2.850	1.600	1.250	47.5	36.3	14.009	2.492	16.5	100.3
G/D/14-2	40.125	13.375	3.500	9.875	53.5	29.5	14.535	2.565	17.1	100.1
G/D/14-3	40.793	11.708	3.100	8.608	52.5	31.8	13.471	2.229	15.7	100
G/D/14-4	41.040	4.560	1.300	3.260	45.6	39.5	13.395	1.505	14.9	100
G/D/14-5	42.499	18.301	3.400	14.901	60.8	25.4	12.213	1.587	13.8	100
G/D/14-6	33.480	11.520	1.900	9.620	45	44	10.283	1.117	11.4	100.4
G/D/14-7	38.836	11.864	3.200	8.664	50.7	31.4	15.116	2.584	17.7	99.8
G/D/14-8	42.986	7.114	2.500	4.614	50.1	38.3	10.430	1.670	12.1	100.5
G/D/14-9	42.576	5.424	2.100	3.324	48	40.2	10.144	1.556	11.7	99.9
G/D/14-10	39.921	15.679	6.100	9.579	55.6	29.2	14.272	1.928	16.2	101
				Neykulam						
G/N/14-11	62.364	9.236	3.100	6.136	71.6	25.7	2.945	0.155	3.1	100.4
G/N/14-12	67.361	8.839	2.200	6.639	76.2	20.1	3.189	0.211	3.4	99.7
G/N/14-13	65.015	7.385	3.400	3.985	72.4	23.6	2.373	0.128	2.5	98.5
G/N/14-14	71.060	4.940	1.100	3.840	76	22	1.908	0.092	2	100
G/N/14-15	69.352	4.348	1.300	3.048	73.7	23.4	2.951	0.149	3.1	100.2
G/N/14-16	70.237	4.963	2.000	2.963	75.2	22.6	2.408	0.093	2.5	100.3
G/N/14-17	73.427	7.173	4.100	3.073	80.6	17.7	1.433	0.068	1.5	99.8
G/N/14-18	77.043	6.157	1.500	4.657	83.2	15.5	1.222	0.078	1.3	100
G/N/14-19	69.462	7.038	2.600	4.438	76.5	17.2	6.657	0.143	6.8	100.5
G/N/14-20	70.915	7.185	2.500	4.685	78.1	20.5	1.300	0.000	1.3	99.9
G/N/14-21	63.461	6.739	1.800	4.939	70.2	22.8	6.330	0.070	6.4	99.4
G/N/14-22	70.122	7.878	2.300	5.578	78	16.9	4.997	0.203	5.2	100.1

1200°C for 200 s (Furuyama et al., 2001; Tawfik et al., 2011). The analyses were carried out in the Department of Geosciences, Osaka City University, Japan. The major oxide data have been presented in normalized values.

4. FACIES ANALYSIS

Breccias-conglomerates and pebbly sandstones constitute the major part and coarse sandstone the balance of the BS formation in the Dalmiapuram section (Fig. 6.1). Muddy material is restricted to clay cutans around detrital grains, thin stringers, and minor interstitial matrix. At the Neykulum section, sandstone and mudrock beds alternate in equal proportions; in contrast, mudrock dominates at Terani, with fine-grained sheet and channel-form sandstone, and a scanty lenticular granular sandstone interbeds. At the Dalmiapuram section the wide lithological spectrum is divisible into four different facies associations (I–IV). The wide difference in the bedload-suspended load ratios in the finer-grained sediment fraction between the Neykulam and Terani sections led to identification of two more associations (respectively, V and VI). The Dalmiapuram section rests directly on the Precambrian basement. At Neykulam there is an exposure gap of ~20 m between the Precambrian basement and the BS formation siliciclastic rocks, and no Precambrian exposure could be traced within a 2 km radius from the Terani section. The three sections of distinctive sedimentlogical characters are thus considered to be at variable stratigraphic heights above the basement. The Dalmiapuram section is inferred as the proximal, Neykulam the intermediate, and Terani the distal with respect to the basin-margin (Fig. 6.1).

The sedimentary facies classification applied based on lithology, sedimentary structures, sediment body geometry, lateral extent, thickness, and association. The identified facies primarily relate to sedimentation dynamics and their associations to inferred paleogeography. In each of the six associations component facies are described in the following sections in order of their relative abundance.

4.1 Facies Association I, Scree Cone

This is the coarse-grained association, dominated by breccias (facies Ia) interspersed with thin sandstones (facies Ib) and conglomerates (facies Ic). Repeated alternations of facies Ia and Ib give rise to wedge-shaped bodies, of maximum thickness around 8 m. The breccia beds attain thicknesses up to 75 cm and are also wedge-shaped and stacked one above another unless locally interleaved by facies Ib with its sheet-like geometry. The top of the breccia beds are irregular as the clast edges protrude above bed surfaces. The bases of the breccia at places may be even more irregular and jagged if underlain by the sandstone beds (Fig. 6.2A); some subvertical clasts have their basal portions extending down into the underlying sandstone laminae (Fig. 6.2B).

The breccia clasts are sharply angular. In size they are generally of pebble grade, but may reach up to the boulder grade, maximum length measuring up to 45 cm. These clasts are mostly traceable into the local granitic/amphibolite basement. They are generally haphazardly oriented, but in mutual contact with each other, though their interstitial spaces are filled by smaller granule-rich sand of the same composition. In one instance, the relationship

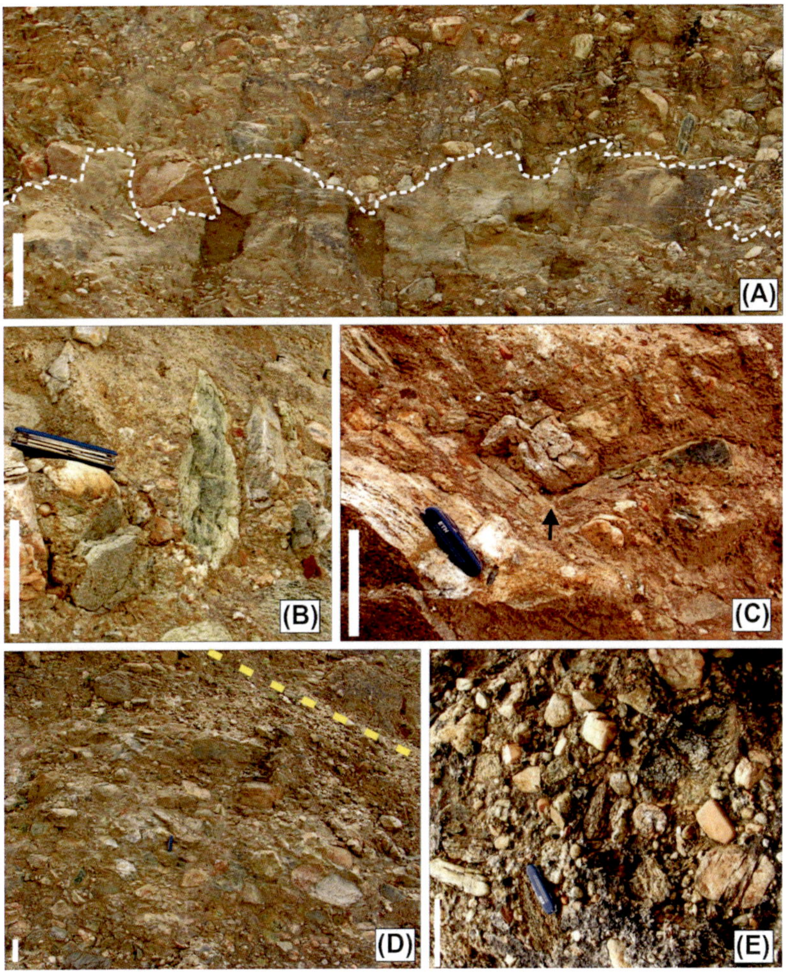

FIGURE 6.2 Facies association I: Jagged boundary between massive hillwash sandstone, facies Ib and scree breccia, facies Ia overlying it. Note that breccia clasts have readily penetrated the bed contact (A). Some of the breccia clasts are vertically oriented (B). A broken (*arrow*) clast wraps the bottom of another clast resting on it: breakage of the clast had presumably taken place because of impact of another clast with significant energy (C). Inclined contact (*dashed*) between two vertically juxtaposed scree bodies (facies Ia); tabular clasts lie parallel to it and stack one above another immediately under the contact though otherwise being chaotically oriented (D). Bimodal clast-size distribution in a clast-supported conglomerate (facies Ic): roundness is high in the smaller clast population, but low in the larger (E). (Bars equal to 10 cm.)

between two large clasts is such that one seems to have landed on the other, breaking it and pushing it downward (Fig. 6.2C). Tabular clasts, at places, are found reclining on inclined surfaces of the breccia beds (Fig. 6.2D).

The sandstone facies Ib is poorly sorted, massive, with thin beds not exceeding 15 cm, and laterally impersistent; outcrop width measures up to 1.5 m (Fig. 6.2A). The top parts of the beds may locally bear planar laminae, but at places are ruptured and eroded out around breccia clasts that intruded into them from above. In comparison to sandstone bed-tops, their bases are broadly smooth tending to retain the configuration of the underlying breccia bed surfaces.

Down the inferred paleoslope (viz., basinward) the facies Ia–Ib combination gives way to the nonrecurrent conglomerate facies Ic, generally clast-supported (Fig. 6.2E). These beds are more or less tabular in geometry and around 80 cm thick. As in facies Ia, clasts constituting this facies are also derived from the granite-amphibolite basement exposed in close proximity. Strikingly, the conglomerate clasts are distinctly bimodal both in size and shape without any apparent correlation with composition. The comparatively larger clasts, smaller though with respect to the clasts of facies Ia, are angular and tabular in shape, while the smaller clasts are well rounded, often highly spherical (Fig. 6.2E). No compositional preference is apparent between the two size populations. Although the beds are internally massive, the long axes of the larger tabular clasts within them are dominantly aligned parallel to the bed.

4.1.1 *Interpretation*

Considering its occurrence directly on the basement, clast derivation from local basement, rapidly wedging bed geometry and internal characteristics, facies Ia is identified as a scree or rock fall deposit formed at steep basin-margin localities (cf., Selley, 1965). Clast angularity reflects little bed load transport. That the clasts might often have been dropped from above and have fallen through the air is apparent from the observation of broken clasts under some of them. Wrapping of underlying sandstone laminae around the bottom of many clasts and jagged lower boundaries of breccia beds are also consistent with this contention (Bose et al., 2008). Tabular clasts reclining on breccia bed-surfaces possibly reflect some degree of sliding of those clasts along inclined scree slopes.

The thin and laterally impersistent sandstone beds of facies Ib sparsely interspersed with the scree deposits make drastic alteration of sedimentation dynamics apparent, without any change in the inferred basin-margin paleogeography. These generally massive beds indicate rapid deposition and their tendency to retain primary configuration of underlying surfaces testifies to the settling of sand grains dominantly from suspension. Sheet-like geometry of the beds and local occurrence of planar laminae at their tops point to sheet flow on rapid reduction of flow depth, logically due to water percolation through porous underlying breccias. The sandstone beds were possibly deposited as hill-wash during rain storms (Bose et al., 2008). Violent rock fall incidents understandably rendered their preservation difficult.

The conglomerate facies Ic found only at the interpreted downslope fringe of this association presents a clear case of textural inversion (Pettijohn, 1975). Without any correlation with clast composition this textural inversion is likely to be a manifestation of sediment supply from dual sources. The angular larger clasts presumably had a proximal source, possibly the scree cone upslope. On the contrary, the better rounded smaller clasts possibly came from a relatively distal source, possibly being transported along the base of the scree cones. Bed-load movement is manifested in bed-parallel elongation of the clasts in this clast-supported conglomerate. Since the facies appears to skirt the postulated scree cone, it is identified as an apron deposit. Lateral shift of the flow in response to scree fan progradation can explain the tabular geometry of the conglomerate bodies.

4.2 Facies Association II, Alluvial Fan Apex

Lithology of this association varies from sandy conglomerate to pebbly sandstone, of lenticular geometry, and in which the clasts are distinctly less angular than those within the Association I breccias. Grain-size distribution in these rocks is bimodal, one mode for the clast population and the other for the sandy matrix. The clast population is moderately sorted, but the matrix is poorly sorted, having significant mud content.

The conglomerates are dominantly matrix-supported. Most have lenticular geometry with bases of conglomerate beds being flat and their tops convex-upward. Their thickness maxima range up to ~1.25 m (Fig. 6.3A). With respect to clast composition they do not differ

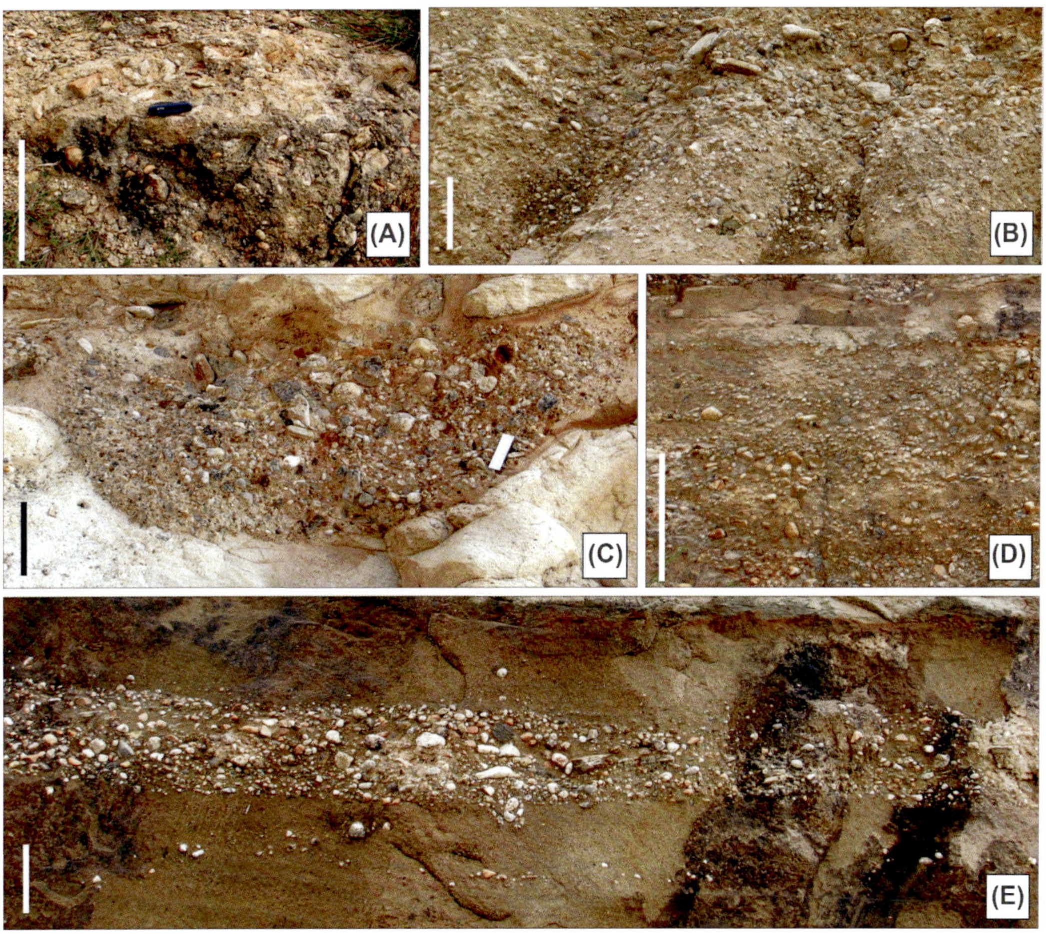

FIGURE 6.3 Facies association II: Matrix-supported debris flow conglomerate (facies IIa) (A). Reverse graded conglomerate (facies IIb) from unconfined flow (B). Channel-confined reverse graded conglomerate (facies IIb) (C). Crudely cross-stratified conglomerate giving way upward gradationally into sandstone (sieve deposit, facis IIc) (D). Sheet conglomerate (facies IId) encased below and above by massive sandstone (E). (Bars equal to 20 cm.)

significantly from the breccias, although their maximum clast size is reduced to 8 cm. Internally, they are massive, clasts floating haphazardly within fine-grained matrix; some clasts protrude above the bed surface (facies IIa).

Some of the lenticular conglomerates are matrix- to clast-supported, but reverse graded in coarse-tailed fashion (Fig. 6.3B; Middleton and Hampton, 1976) and have maximum thickness of ~1 m (facies IIb). These conglomerates also generally have convex upward bed geometry, although a few have scoured bases. In one instance such a conglomerate was found within the thalweg of a small channel form, but in that case the conglomerate body had a flat top and a concave upward base (Fig. 6.3C). The maximum clast size of this facies is comparable to that of the preceding facies, but in clast composition a significant decrease in amphibolite content is recorded (see later).

Some other convex-upward conglomerate bodies are clast-supported in the basal part of these beds, but gradually turn matrix-supported upward and may eventually even become sandy (Fig. 6.3D; facies IIc). Both the bed-bounding surfaces are sharp. The pebbly lower part of these bodies is generally massive, but may also bear crudely defined cross-strata. Maximum cross-set thickness and maximum clast-size recorded in this facies are around 8 and 5.5 cm, respectively. Still other conglomerates, generally, though not always, encased by pebbly sandstone (the latter belonging to facies association III), are tabular in geometry, but locally have small scours at their bases (Fig. 6.3E; facies IId). Their clasts are smaller than those constituting the preceding varieties of conglomerates, the maximum clast length measuring up to 8 cm. No grading was observed, but tabular clasts are dominantly bed-parallel and locally imbricated.

4.2.1 Interpretation

This facies association having inferred basin-margin scree deposits in lateral contiguity appears to represent alluvial fan deposits. Facies IIa is interpreted as a debris flow that underwent sudden freezing of matrix with expulsion of fluid (cf., Blair and McPherson, 1994). Where the flow viscosity did not allow the clasts to sink, they protruded above the flow surface (cf., Lowe, 1982; Blair and McPherson, 1994; Blair, 1999). Reverse grading in facies IIb manifests collision between clasts and freezing of the flow through clast–clast interlocking (cf., Lowe, 1976; Middleton and Hampton, 1976; Schultz, 1984; Mack and Rasmussen, 1984; Nemec and Postma, 1993; Mulder and Alexander, 2001; Davis et al., 2002; Gani, 2004). Facies IIc units appear to be sieve deposits manifesting freezing bedforms, possibly because water readily percolated away through the porous sediment. As the flow waned sand infiltrated on top of the bedforms (cf., Todd, 1989). In contrast, facies IId appears to have been deposited from a more fluidal sheet-flow, the basal scours manifesting initial turbulence. Dominant bed-parallel orientation of clasts (Facies IIc), indicates rapid suppression of turbulence because of enhancement of sediment load as a possible consequence of dewatering; the flow eventually turned strongly sheared. Because of the presumed fluidal nature of the flow the interpreted site of deposition of facies IIc is likely to have been immediately below the level where the water percolated at the fan apex reemerged on the fan surface, or in other words, where the groundwater table intersected the fan surface (cf., Enos, 1977; Hein, 1982; Bose and Sarkar, 1991; Bose et al., 2008). Pebbly sandstones generally enclosing them and interpreted under facies association III further corroborate this contention (see next).

4.3 Facies Association III, Alluvial Fan Base

This association is pervasively characterized by mutually cross-cutting small channel bodies not exceeding 3 m in exposure width and 1.5 m in thickness; they are filled by minor conglomerate at their bases, generally clast-supported and by dominant sandstone above, which can be massive, planar laminated, cross-stratified, and locally rippled. The conglomerate (facies IIIa) occupies the deepest part of the channel forms and has concave upward bases and flat tops. Multiples of such beds, not exceeding 60 cm in total thickness, may amalgamate. Pebble size, however, decreases upward through such a stack, the maximum clast-size attaining 10 cm (Fig. 6.4A). Pebbly sandstone, locally crudely trough cross-

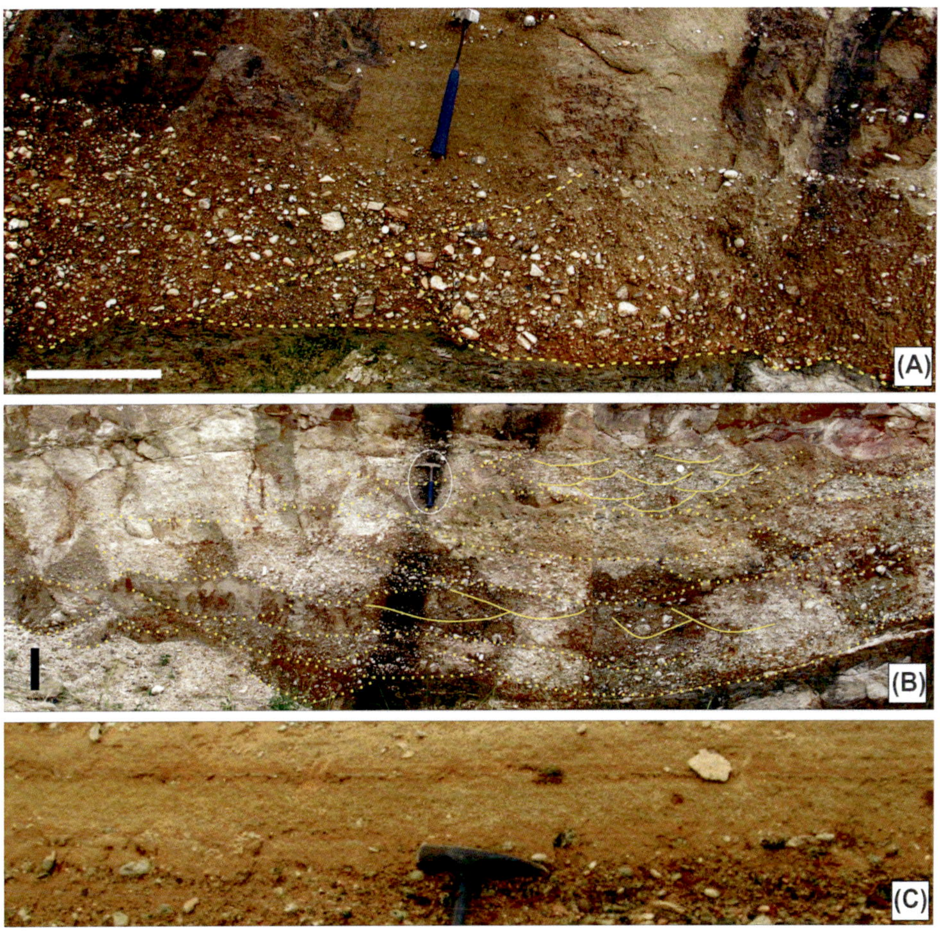

FIGURE 6.4 Facies association III: Conglomerate concentrated within nested channels (demarcated) passes upward into planar laminated sandstone, locally pebbly (facies IIIa) (A). Mutually cross-cutting small channel deposits in pebbly sandstone, outlined by dotted boundaries, internally exhibiting trough co-sets (see on right of the hammer; facies IIIb). Note that pebbles concentrate along set and co-set boundaries (B). Faintly planar-laminated sandstone (facies IIIc) (C). (Bars equal to 30 cm.)

stratified (facies IIIb, Fig. 6.4B) is the other common constituent of this association. The cross-sets do not exceed 15 cm in thickness while their co-sets range up to 50 cm. The pebbles generally concentrate along foreset bases and cross-set boundaries. The cross-stratified facies IIIb is generally overlain by the dominant planar laminated sandstone (Fig. 6.4C; facies IIIc). Being channel-confined, exposure width of the IIIc facies bodies does not exceed 3 m and their thicknesses range up to 22 cm. In a few other instances, however, facies IIIb gives way upward to small ripples, not exceeding 3 cm in height, and draped by a thin film of mud.

4.3.1 *Interpretation*

Facies association III appears to have developed with an abundance of water, amid rapidly shifting channels. The conglomerate beds were deposited from strong turbulent flows preferably along the channel thalwegs (Bose et al., 2008). The upward passage of the conglomerates to cross-stratified pebbly sandstone clearly manifests progressive decline in flow shear; nonetheless, the upward passage of such cross-strata into ripple forms draped by mud, though rare, records instances of flooding (Long, 2011; Miall, 1996). Considering its lateral contiguity with facies association II, apparently intertonguing with facies IId, this facies association perhaps belongs to the base of the alluvial fans, well below the level of intersection between the groundwater table and the fan surfaces.

4.4 Facies Association IV, Axial Channel Association

This facies association, unlike the preceding association, seldom carries pebbles. The dominant sandstone lithology is characterized by a mosaic of channel-fill bodies, considerably larger in dimensions than their counterparts in the facies association III. The measurable and thus possibly the comparatively smaller channels are ~7 m in outcrop width. On the other hand, the maximum thickness of the individual channel-fills measures up to 2.5 m (Fig. 6.5A). The channel-fills are often multistoreyed, each storey generally having massive sandstone at the base (Fig. 6.5B; facies IVa), followed by planar laminated (Fig. 6.5C; facies IVb) and large-scale cross-stratified sandstone (Fig. 6.5D; facies IVc) above, and locally ripple laminated sandstone (facies IVd) on top. The cross-set (>20 cm) and co-set (~1.2 m) thicknesses in this association are also considerably larger than their counterparts in the facies association III. The channel-fill mosaic is locally disrupted by the presence of isolated wedge-shaped or broadly tabular and rust-colored granular sandstone bodies, internally devoid of structure (facies IVe).

4.4.1 *Interpretation*

This association, in lateral contiguity with the previous three facies associations, is interpreted to have accumulated at the base of the basin-margin slope amid a network of channels. Rapid avulsions characterized the channels that were considerably larger in dimensions than their counterparts in association III. The massive facies IVa attests to rapid deposition presumably because of hydraulic jump induced by sharp decrease in gradient at the fan base. The planar laminated facies (IVb), when coarse-grained and resting on channel base IVa deposits, is thought to have formed either as linguoid bars or cross-channel bars (cf., Allen, 1968; Collinson, 1978; Blodgett and Stanley, 1980; Miall, 1996). Overlying facies IVc (large scale cross-stratified sandstone) and being comparatively finer grained, facies IVd probably

FIGURE 6.5 Large channel-fill facies association IV: General view (A), massive sandstone (B), planar laminated sandstone (C), and trough cross-stratified sandstone (D). (Bars equal to 60 cm; Marker equal to 14 cm.)

manifests bar-top reworking during low water stage. Declining flow strength is reflected in the vertical sequences of sedimentary structures within the channel-fill sections. Migratory bedforms possibly progressively blocked the upstream entry of the flow into the channels (cf., Bridge, 2006). Significant diversion in orientation of cross-strata between this and the preceding fan associations suggests that the river channel system was axial, running along the base of the fan complex. The wide span of paleocurrent data in this facies suggests a braided nature of the river system (Fig. 6.6). The granule-rich facies IVe stands out as aberrant within the association and possibly owes its origin to intermittent massflows. Seismicity, overloading, and undercutting are the most likely mechanisms to trigger such flows (Sarkar et al., 2014; Seth et al., 1990).

4.5 Facies Association V, Intermediate Flood-Plain Association

The facies association V, comprising the Neykulam section, is made up of repeated alternations between mudrock and sandstone beds or bed-sets with rare intercalations of significantly calcareous beds. Detailed observations helped identify seven different facies within the association.

Among the sandstones most are trough cross-stratified, poorly sorted, and coarse-grained, though pebble-free (Fig. 6.7A; facies Va). They exhibit broadly lenticular geometry with

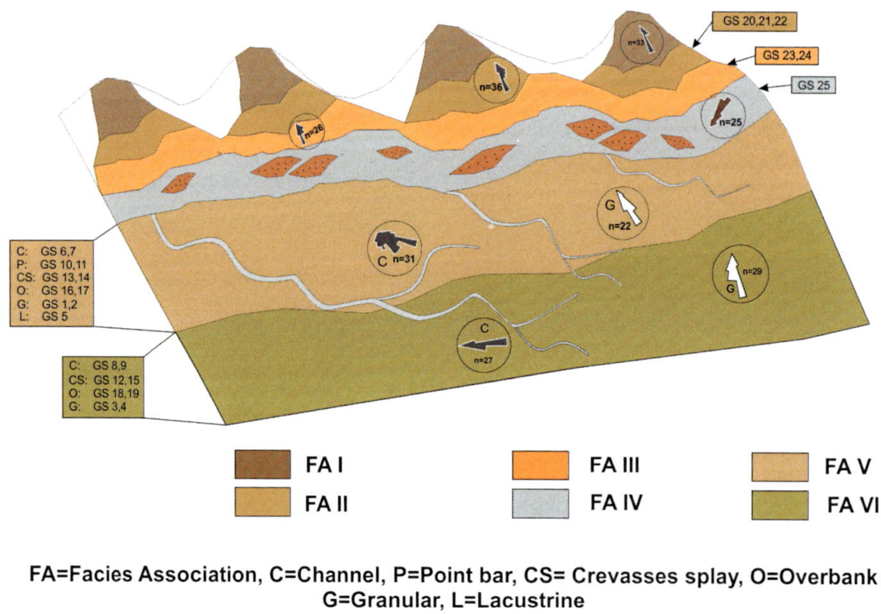

FIGURE 6.6 Schematic depiction (not to scale) of paleogeographic distribution of facies associations comprising the studied formation. General paleocurrent directions in each of them are displayed in dark current roses; white roses are for the granular sandstones and the yellow one is for the inclination direction of scree surfaces. Paleocurrent directions are measured, in general, from cross-strata and only in the case of the granular sandstones have gutter orientations been used. Facies associationwise distribution of samples used for geochemical analyses is given in boxes on the sides.

FIGURE 6.7 Facies association V: Leftward pinching of a channel-fill sandstone (facies Va) (A). Coarse-grained planar-laminated sandstone of facies Vb (B). Co-sets of trough cross-strata bounded below and above by gently inclined silty shale accretionary laminae (facies Vc) (C). Planar-laminated sheet sandstone (center, facies Vd) encased by mudrock (facies Ve) below and granular sandstone (facies Vg) above (D). Carbonate-rich mudstone (facies Vf) bearing rows of light-colored mud clasts defining sagging interlaminae in the lower part as well as darker carbonaceous interlaminae in the upper part (*arrows*) (E). (Bars equal to 20 cm, coin diameter equal to 2 cm.)

concave-up bases and maximum thickness around 1.85 m. Some coarse-grained sandstones have planar erosional bases and internal planar laminae, and may be slightly convex upward (Fig. 6.7B; facies Vb). They have tabular body geometry and measure up to ~1.4 m in thickness. Some trough cross-stratified sandstone beds having flat bases and convex-up tops also bear thin silty mudstone partings inclined at a high angle to the orientation of the associated trough cross-strata (Fig. 6.7C; facies Vc). The cross-set thickness is, on average, 16 cm although it discernibly decreases upward within individual co-sets. Some sandstone bodies, distinctly finer grained, thinner (<3 cm), and sheet-like in geometry, are generally found interbedded with comparatively thicker (>23 cm) mudrocks (Fig. 6.7D; facies Vd). They are typically planar laminated, but may have small ripples on top of them. The bounding mudrocks also bear planar laminae manifested in thin silt stringers. Rootlet marks occur locally on bed-tops and burrows, often compressed, within the beds. Mudrock beds are comparatively thicker (>40 cm; Fig. 6.7D; facies Ve) and darker in color. Under the microscope they reveal crinkled carbonaceous laminae, stray rafted oversized sand grains, and disseminated pyrite. Dark gray carbonate mudstone (facies Vf), broadly lenticular and substantially thick (~70 cm), occurs rarely within encasing mudrock. Dark carbonaceous, often curled fragments concentrate preferably in the lower part of these argillaceous bodies (Fig. 6.7E). The granule-rich facies in the Neykulam section is nonrecurrent, tabular in shape; it maintains a uniform thickness of over 25 cm (facies Vg; Fig. 6.7D). This seventh facies has a sharp base and less distinct top contact, and is internally massive.

4.5.1 *Interpretation*

The poorly sorted trough cross-stratified sandstone facies, Va, is interpreted as a river channel deposit (cf., Miall, 1996; Long, 2011). However, the other cross-stratified sandstone facies (Vb), comparatively finer-grained with muddy siltstone partings inclined in a direction at a high angle to the troughs, is inferred as reflecting a point bar (cf., Chung et al., 2005). The inclined mud laminae were presumably deposited during successive floods. The coarse-grained planar laminated sandstone facies, Vc, resting on major erosion surfaces, possibly formed as linguoid or cross-channel bars on the channel-floor (cf., Allen, 1968; Collinson, 1978; Blodgett and Stanley, 1980). The fine-grained sandstone facies (Vd) interbedded with mudrock is most probably of overbank crevasse-splay origin. The sand-laden water spilled over from the fixed channels (e.g., Bristow et al., 1999; Farrell, 2001) during floods had possibly given rise to sheet flows. Later reworking might have generated the ripples on the bed-tops. Rootlets on top and burrows within the beds corroborate the suggested overbank origin of the facies. The inferred scenario manifests slow and discontinuous mud deposition. Facies Ve, in comparison, was presumably deposited farther away from the river channels. Abundant occurrence of crinkled carbonaceous laminae of possible microbial mat origin (following Schieber et al., 2007) supports an even slower rate of sedimentation. Curled mat fragments at certain levels suggest episodic deposition. Pyrite presumably precipitated during early diagenesis because of mat decomposition (cf., Schieber et al., 2007). The micritic carbonate facies (Vf) in close association with the inferred overbank facies possibly formed in isolated lakes or ponds, where prolonged evaporation helped achieve Ca^{+2} and HCO_3^{-2} ionic concentration high enough to precipitate calcite. The granular facies, Vg, is an obvious aberration in the general fine-grained motif of this facies association. Its inordinately coarser grain-size, internal massive nature, and the observed facies bases being sharper than their tops can be interpreted as rapid deposition from waning sediment gravity flow. Absence of overbank deposits in the previously described Dalmiapuram section and their substantial presence in the Neykulam section indicates a relatively higher rate of accommodation space generation at the latter site (cf., Bridge, 2006). An inferred low-sinuosity river system at Dalmiapuram presumably branched out into high sinuosity channels at Neykulam (Fig. 6.6). Fixed channels were favored with the greater abundance of mud.

4.6 Facies Association VI, Distal Flood-Plain Association

The facies association making up the section at Terani is mudstone-dominated and consists of four distinctive facies. The dominant facies is a light gray-colored mudstone (facies VIa; Fig. 6.8A) thicker than 50 cm, incorporating siltstone stringers (<3 mm thick), and less frequently thin, sheet-like, and fine-grained sandstone interlaminae (<5 mm thick). Polygonal cracks filled by silt or sand are encountered rarely because of rare availability of bedding plane sections in the mines. Corresponding V-shaped cracks filled by coarser clastics, nonetheless, are not uncommon in vertical sections. At certain levels burrows and rootlet

FIGURE 6.8 Facies association VI: Internally planar laminated mudrock (facies VIa) (A), locally bearing iron-stained rootlet marks (B). Sheet sandstones (facies VIb) encased by mudrock (C). Channel sandstone (facies VIc) encased by mudrock (D). Granular sandstone (facies VId) having base sharper than its top; mud-clasts concentrate along its base (E). A series of gutters are present at the sole of this granular sandstone bed (F). (Bars equal to 15 cm.)

structures are present (Fig. 6.8B). Next in order of abundance is a mudrock-sandstone interbedded facies (VIb). The shales are thinner than 20 cm, averaging 12 cm, while the fine-grained sandstone interbeds are comparatively thinner, not exceeding 2.5 cm. The sandstone interbeds are sheet-like in geometry and internally planar laminated (Fig. 6.8C); rarely, they bear minute asymmetric ripples on their tops. Both the tops and bottoms of these sandstone beds are generally sharp, although bases locally appear a bit irregular because of filling cracks in underlying mudrock beds. The mudrock-dominated character of the Terani section is disrupted by intermittent occurrence of medium- to fine-grained, poorly sorted sandstones of lenticular geometry (facies VIc; Fig. 6.8D). The lenses are, on average, ~25 cm thick and are internally trough cross-stratified. These sandstone beds, however, may amalgamate laterally as well as vertically. Vertical amalgamation may produce thicknesses up to 45 cm to the sandstone bodies and their contacts with the encasing mudrock then occur at a high angle. The most rare though conspicuous facies is of reddish granule-rich beds, sheet-like in geometry (facies VId; Fig. 6.8E). These beds having thickness <12 cm are thinner than their counterparts in other associations (i.e., facies IVe in Dalmiapuram and Vg in Neykulam sections). They are internally massive in most cases, but locally bear crude planar laminae or low angle cross-laminae. Both their upper and lower surfaces are sharp, but the latter are relatively sharper in most cases. Their lower surfaces, unlike the upper, are irregular because of frequent presence of gutters (Fig. 6.8F). It is significant that orientation of these gutters is at high angles to the trough cross-strata in facies VIc of the same association (Fig. 6.6).

4.6.1 *Interpretation*

Amid emergence features in the association, facies VIa is interpreted to have settled from suspension, and was likely to have been an overbank deposit. The sheetlike sandstones of facies VIb with sharp upper and lower contacts, enclosed by VIa, are inferred to have been crevasse splay deposits. On the other hand, the lenticular poorly sorted and coarser-grained sandstone facies (VIc) suggests deposition within shoestring channels (cf., Bridge, 2006). Vertical stacking of such channel sandstones within the background of inferred over-bank fines suggests a tendency of these channels to have anastomosed (e.g., Makaske, 2001). The granular facies VId is an aberrant member within the interpreted paleoenvironment, manifesting episodic enhancement of depositional energy. Its deposition evidently took place from a waning current. The flow driving the granules initially had been intensely turbulent, as manifested in gutters present at the bed-soles. The generally massive character of this facies, nonetheless, indicates high sediment load in the flows. Orientation of the gutters, making a distinct angle to the channel-filling trough cross-strata directions at all three localities studied here suggests derivation of these granular materials from basin-flanks presumably during seismic events (Fig. 6.8). The thin nature of the facies VId beds in comparison to their look-alike facies IVe and Vf in the other two study locations further corroborates the contention that the present site was farthest from the basin-margin (Fig. 6.6).

5. SANDSTONE PETROGRAPHY

The rocks are arkosic/subarkosic arenites containing 46–90% quartz, 8–44% feldspar, 0–12% lithic fragments of granite and amphibolites, plus polycrystalline quartz of both sheared and nonsheared varieties and <15% matrix with 5–12% mica, especially biotite (Table 6.1). The feldspar population is dominantly potassic, mostly microcline, although generally they are heavily decomposed, partially or completely (Fig. 6.9A). The biotite grains are often strongly leached and account for common profusion of ferruginous cement (Fig. 6.9B). Clay cutans around framework grains are common (Fig. 6.9C). Sorting of the framework grain population is generally poor and worst within the granule-rich facies (Fig. 6.9C). Multigranular rock fragments are dominantly granitic in composition, making up, on average, 40% of fragments; only 5% of rock fragments are amphibolites, the rest being polycrystalline quartz, 39% unstretched and 16% stretched (Fig. 6.9D). Percentage of labile multigranular components decreases drastically from the Dalmiapuram section through Neykulam to Terani section; at the last site, the multigranular framework population includes only 5% granite fragments, while the rest are polycrystalline quartz (Fig. 6.9D).

The poor sorting of the studied sediment samples is consistent with the interpreted fluvial origin of the rocks. Their mineralogical immaturity evinces restricted weathering, but heavy in situ decomposition of feldspar grains, leaching of biotite grains, while the presence of clay cutans around many detrital grains, especially highly spherical rock fragments, does not reflect any dearth of water.

A QFL plot (Fig. 6.9E; Dickinson and Suczec, 1979) indicates a continental provenance as the source for the sediments. The plot also documents increasing mineralogical maturity for samples from Dalmiapuram through Neykulam to Terani sections in consistency with the

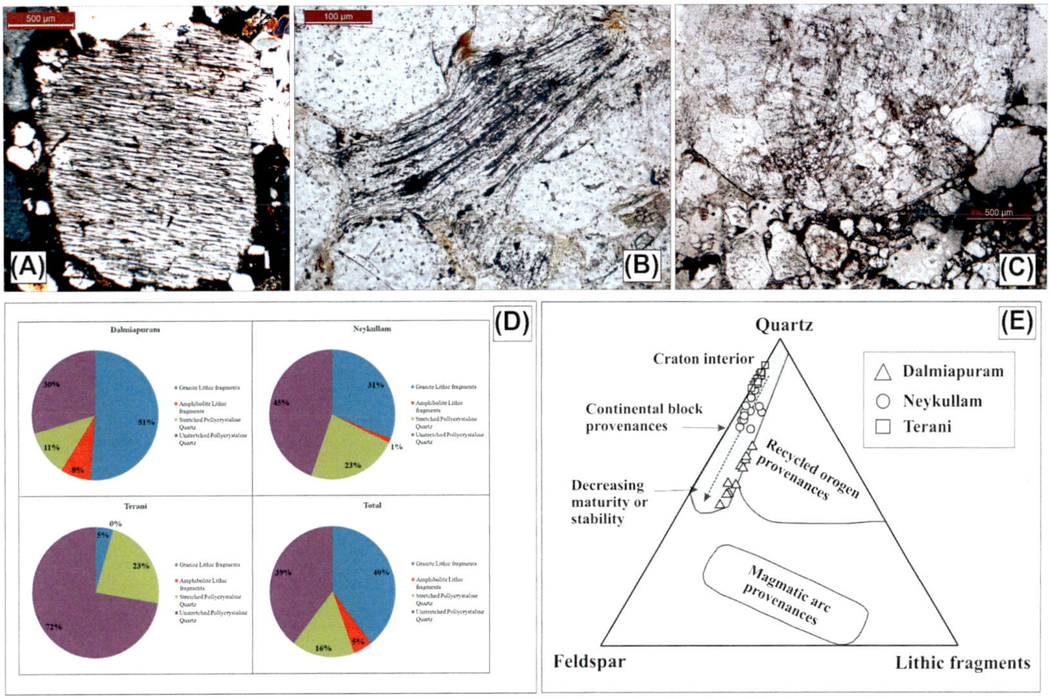

FIGURE 6.9 In situ cleavage-parallel decomposition of feldspar (A), leaching of biotite (B), clay cutan around a large rock fragment within a poorly sorted sandstone (C). Areawise and general compositional variations in terms of rock fragments (D); QFL plot shows preference for continental block provenance and increasing mineralogical maturity from Dalmiapuram through Neykulam to Terani study locations.

postulated paleogeographic transition from the basin-margin to the basin-interior (Fig. 6.6). Intrabasinal transport accounts well for the increasing mineralogical maturity in this transition. The Terani samples thus approach closest to the presumed cratonic interior mineralogy.

6. GEOCHEMISTRY

6.1 Results of Major and Trace Element Analysis

SiO_2 (av. 68%) is the dominant major element oxide in the majority of samples (Table 6.2), and in the rest its deficiency is generally compensated by iron (11–40%). Al_2O_3 content is also generally high, up to ~31%. MgO and MnO contents are low, but the former invariably exceeds the latter. P_2O_5 content is invariably low, and Ba and W contents are generally high. Ba shows positive correlation with K_2O (Fig. 6.10). An inordinate increase in Ba corresponds to complete absence of detectable W as well as La, Ce, and Nd. Sr content is high, invariably exceeding that of Rb; the Rb/Sr ratio ranges between 0.03 and 0.49. V, Cr, Ni, Pb, Zn, and Zr are also present in substantial quantities (Tables 6.2 and 6.3).

TABLE 6.2 Contents (wt.%) of Major Oxides Within the Studied Samples

Samples	SiO_2	TiO_2	Al_2O_3	T-Fe_2O_3	MnO	MgO	CaO	Na_2O	K_2O	P_2O_5	Total
GS 1	69.09	0.44	5.72	1.92	0.04	0.78	8.95	0.23	1.63	0.01	88.80
GS 2	70.62	0.49	12.29	4.83	0.02	0.54	0.91	0.27	2.07	0.02	92.05
GS 3	88.68	0.06	3.36	1.28	0.00	0.05	0.10	0.05	1.35	0.01	94.94
GS 4	87.17	0.08	3.94	2.96	0.01	0.08	0.22	0.05	1.33	0.03	95.87
GS 5	70.39	0.53	14.11	1.72	0.01	0.77	0.80	0.38	2.30	0.01	91.02
GS 6	59.76	0.64	14.86	9.86	0.02	1.14	0.42	0.31	2.38	0.05	89.44
GS 7	47.11	0.66	13.50	23.34	0.03	0.58	0.38	0.21	1.65	0.14	87.60
GS 8	46.22	0.55	12.04	20.54	0.04	1.79	0.74	0.68	1.41	0.24	84.24
GS 9	30.55	0.70	14.85	32.90	0.00	0.29	0.50	0.00	0.79	0.72	81.32
GS 10	77.15	0.95	7.74	5.39	0.06	0.31	0.36	0.61	2.60	0.04	95.22
GS 11	51.81	1.29	27.64	1.63	0.00	0.62	0.59	0.00	1.27	0.04	84.90
GS 12	40.89	0.95	20.83	17.04	0.00	0.47	2.48	0.00	1.03	0.33	84.01
GS 13	56.77	1.24	22.79	1.67	0.01	1.23	0.79	0.01	0.94	0.02	85.47
GS 14	51.97	1.08	20.41	1.79	0.01	1.47	5.44	0.00	0.77	0.02	82.97
GS 15	52.38	1.30	27.87	1.54	0.00	0.67	0.63	0.00	1.29	0.05	85.75
GS 16	56.85	0.98	18.05	1.85	0.01	1.29	4.44	0.05	1.19	0.02	84.72
GS 17	41.94	0.62	12.54	1.00	0.01	1.47	18.47	0.04	1.09	0.02	77.19
GS 18	76.94	0.40	11.14	1.47	0.01	0.44	0.41	0.72	2.90	0.01	94.44
GS 19	53.48	1.31	27.14	1.61	0.00	0.54	0.57	0.01	1.38	0.04	86.08
GS 20	70.65	0.32	13.26	2.05	0.02	0.48	1.03	2.39	5.06	0.01	95.27
GS 21	72.09	0.31	13.57	1.98	0.01	0.46	1.06	2.47	5.13	0.02	97.10
GS 22	67.55	0.32	12.38	2.01	0.01	0.46	1.02	2.25	4.87	0.01	90.89
GS 23	71.34	0.25	13.57	1.51	0.01	0.42	1.05	2.18	5.95	0.03	96.32
GS 24	70.19	0.30	13.93	1.73	0.01	0.44	1.21	2.46	5.56	0.04	95.88
GS 25	71.13	0.39	14.08	2.02	0.01	0.55	1.34	2.41	5.16	0.05	97.15
Average	62.11	0.65	14.86	5.83	0.02	0.69	2.16	0.71	2.44	0.08	89.55

Paleogeographic context of the samples is given in Fig. 6.15.

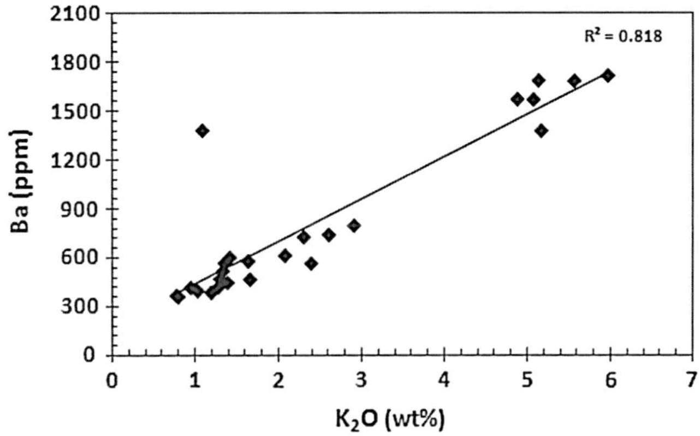

FIGURE 6.10 Plots of K_2O versus Ba. Note the positive correlation.

TABLE 6.3 Trace Element Contents (ppm) and Some Ratios Used in This Chapter

Samples	V	Cr	Co	Ni	Cu	Zn	Rb	Sr	Y	Zr	Nb
GS 1	41.94	43.97	16.91	266.89	76.42	48.18	31.50	411.98	10.26	243.32	5.55
GS 2	81.69	73.30	73.45	41.66	17.09	49.10	39.06	165.66	7.82	187.34	6.75
GS 3	17.17	7.99	4.46	19.42	13.29	19.99	20.38	92.13	8.52	54.14	1.36
GS 4	31.19	13.74	68.52	57.16	39.88	35.99	22.80	106.80	9.14	47.15	2.49
GS 5	80.80	64.44	21.15	229.64	28.71	23.58	43.24	215.05	9.01	176.70	8.56
GS 6	148.40	111.53	26.05	62.84	40.67	124.42	59.38	119.85	13.79	164.05	9.11
GS 7	327.25	119.54	74.19	1208.77	76.45	95.70	29.00	109.24	27.21	298.99	9.46
GS 8	65.18	90.16	93.33	339.59	55.35	1263.12	82.43	131.62	79.57	452.31	7.00
GS 9	568.73	171.29	23.96	181.40	130.47	320.39	15.17	94.57	24.87	129.55	12.59
GS 10	82.77	42.92	10.63	25.51	13.01	37.74	49.52	165.32	22.39	979.79	11.70
GS 11	139.07	170.65	11.92	65.28	78.13	92.29	22.92	151.94	35.72	168.01	19.30
GS 12	363.72	177.73	17.31	164.09	98.24	210.77	18.06	108.80	29.57	159.99	15.04
GS 13	143.52	149.32	18.61	54.56	56.70	39.26	19.35	297.80	16.52	268.56	17.98
GS 14	135.34	127.11	8.89	64.26	57.75	37.99	18.62	557.83	14.91	198.78	15.57
GS 15	141.09	174.41	12.47	84.53	89.33	95.39	23.57	175.24	55.65	160.59	17.47
GS 16	131.63	127.52	44.23	125.07	64.26	44.13	24.39	243.05	14.45	143.16	14.12
GS 17	78.32	58.93	7.47	29.17	23.27	18.43	19.96	822.25	11.69	253.18	9.79
GS 18	39.39	39.32	87.70	32.84	14.00	16.24	53.30	181.62	8.44	255.39	5.99
GS 19	144.21	163.91	14.97	91.93	65.08	149.70	24.31	149.30	20.95	192.52	18.78
GS 20	35.60	23.93	2.47	11.45	5.60	21.36	87.49	239.39	10.21	455.78	9.48
GS 21	34.22	24.98	4.62	12.45	4.71	21.54	87.29	252.01	10.21	409.23	9.33

(Continued)

TABLE 6.3 Trace Element Contents (ppm) and Some Ratios Used in This Chapter—cont'd

Samples	V	Cr	Co	Ni	Cu	Zn	Rb	Sr	Y	Zr	Nb
GS 22	36.12	23.73	3.76	12.87	5.19	21.64	80.91	230.45	10.59	396.88	9.54
GS 23	34.61	29.85	2.32	7.54	8.68	34.28	96.76	214.61	10.32	239.93	7.76
GS 24	38.55	33.89	2.63	8.60	10.10	30.25	88.13	215.32	13.16	332.66	9.45
GS 25	45.86	42.39	3.06	13.99	8.56	34.59	86.43	215.75	14.91	426.39	12.34
Average	119.45	84.26	26.20	128.46	43.24	115.44	45.76	226.70	19.60	271.78	10.66

Samples	Ba	La	Ce	Nd	W	Pb	Th	TiO_2/Al_2O_3	TiO_2/Nb	Zr/TiO_2	Rb/Sr
GS 1	578.69	8.82	23.70	7.99	827.07	0.00	0.00	0.08	0.08	556.08	0.08
GS 2	612.75	12.27	27.78	8.53	881.81	0.00	2.30	0.04	0.07	381.71	0.24
GS 3	566.08	4.61	12.72	3.30	0.10	0.00	0.00	0.02	0.04	980	0.22
GS 4	517.67	12.25	19.68	6.82	2924.51	0.00	0.25	0.02	0.03	575.39	0.21
GS 5	727.13	20.11	54.44	12.05	1262.88	0.00	4.33	0.04	0.06	330.75	0.20
GS 6	564.93	19.84	47.86	12.29	0.81	0.00	4.18	0.04	0.07	255.25	0.50
GS 7	465.59	17.98	48.40	9.98	115.94	0.00	4.55	0.05	0.07	453.36	0.27
GS 8	600.39	45.35	81.77	35.36	0.00	0.00	13.71	0.05	0.08	825.75	0.63
GS 9	359.28	26.51	32.16	18.02	0.00	0.00	11.53	0.05	0.06	184.62	0.16
GS 10	741.13	38.54	77.78	25.34	1.25	0.00	16.48	0.12	0.08	1027.99	0.30
GS 11	418.08	52.21	68.05	27.40	0.81	0.39	15.37	0.05	0.07	130.43	0.15
GS 12	396.05	49.01	86.32	49.36	0.00	0.00	15.02	0.05	0.06	168.45	0.17
GS 13	415.93	33.91	84.73	25.28	23.06	2.12	13.41	0.05	0.07	216.2	0.06
GS 14	366.40	31.87	55.83	24.13	0.97	1.80	9.29	0.05	0.07	184.4	0.03
GS 15	470.54	53.80	73.21	32.33	0.30	2.64	17.96	0.05	0.07	123.85	0.13
GS 16	385.16	39.93	88.63	31.66	164.03	0.75	9.05	0.05	0.07	146.02	0.10
GS 17	1381.90	16.36	32.76	11.74	0.00	0.00	7.03	0.05	0.06	409.7	0.02
GS 18	797.33	13.44	26.18	10.50	1425.13	0.00	2.62	0.04	0.07	638.94	0.29
GS 19	446.57	58.99	76.91	34.87	0.00	0.00	16.06	0.05	0.07	147.34	0.16
GS 20	1570.51	0.00	0.00	0.00	0.00	26.14	5.51	0.02	0.03	1410	0.37
GS 21	0.00	0.00	0.00	0.00	1.12	24.09	6.97	0.02	0.03	1307.8	0.35
GS 22	1573.71	0.00	0.00	0.00	0.00	21.06	6.34	0.03	0.03	1226.6	0.35
GS 23	1717.87	0.00	0.00	0.00	0.00	26.97	5.92	0.02	0.03	972.6	0.45
GS 24	0.00	0.00	0.00	0.00	0.00	23.10	7.00	0.02	0.03	1092.3	0.41
GS 25	0.00	0.00	0.00	0.00	2.60	21.89	5.71	0.03	0.03	1091.1	0.40
Average	626.95	0.00	0.00	0.00	305.30	6.04	8.02	0.04	0.06	593.47	0.25

Paleogeographic context of the samples is given in Fig. 6.15.

6.2 Discussion: Implications of Geochemical and Physical Characteristics for Sediments

6.2.1 Provenance and Tectonic Setting

Silica being immobile, its abundance, averaging 68%, indicates both mineralogical as well as chemical maturity of the studied sedimentary rocks if we go by Crook's (1974) classification. This maturity could have been inherited from the source rock or attained through its weathering. Pronounced negative correlation with Al_2O_3 identifies silica as principally detrital and this fact is in good agreement with the dominant granitic composition of the basement, and also supported by the QFL plot depicting a cratonic source (Fig. 6.9D). The K_2O/Na_2O versus SiO_2 plot indicates little albitization of feldspar and thereby suggests lack of tectonic activity on the craton (Fig. 6.11A; cf., Roser and Korsch, 1986). On the other hand, significant variability in the Ti/Nb ratio is not in support of this contention (Table 6.4; Pearce et al., 1984; Hofmann, 1988). These elements are mutually not interchangeable, both being immobile and of high field strength (Cramer and Nesbitt, 1983; Bonjour and Dabard, 1991). Moreover, the Ti/Nb ratio does not generally fractionate on weathering or diagenesis (Bonjour and Dabard, 1991). Nb having large ionic charge and comparatively higher field strength concentrates in granitic melts, and Ti having comparatively lower ionic charge

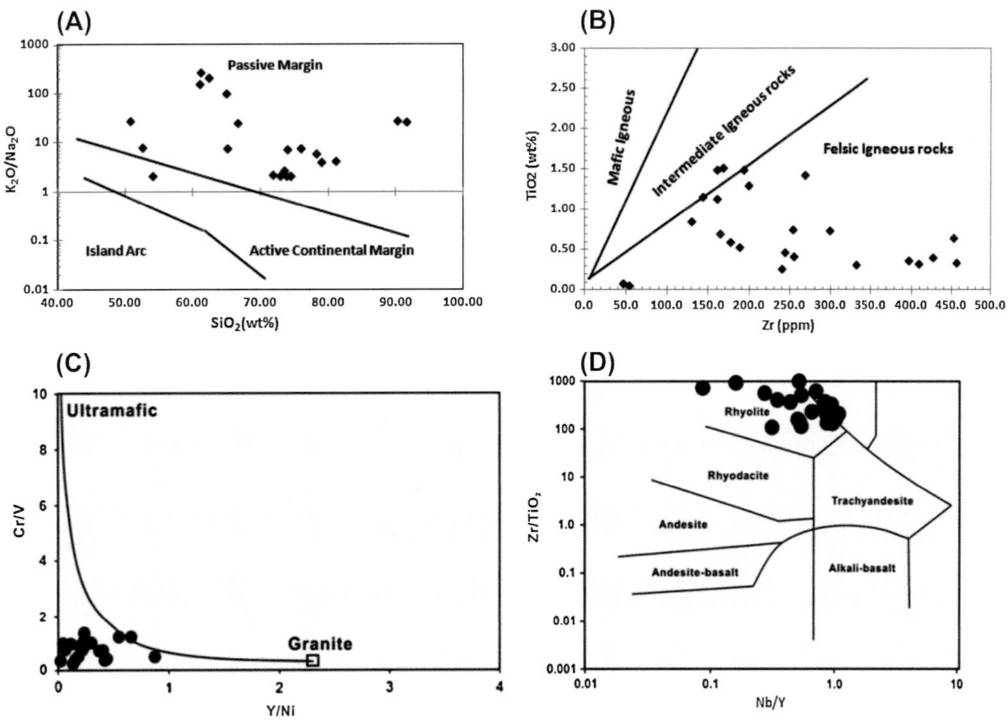

FIGURE 6.11 The SiO_2 versus K_2O/Na_2O, Zr versus TiO_2, Y/Ni versus Cr/V, and Nb/Y versus Zr/TiO_2 variation diagrams. All suggest derivation of sediments from felsic source in a stable platform setting.

TABLE 6.4 Weathering Indices Derived From the Raw Data in Table 6.2

Samples	Chemical Index of Weathering	Chemical Index of Alteration	Plagioclase Index of Alteration
GS 1	38.4	34.61	30.84
GS 2	91.27	79.11	89.68
GS 3	95.67	69.07	92.96
GS 4	93.58	71.13	90.62
GS 5	92.26	80.21	90.89
GS 6	95.35	82.69	94.5
GS 7	95.87	85.79	95.32
GS 8	89.46	80.96	88.22
GS 9	96.71	91.95	96.53
GS 10	88.88	68.46	84.16
GS 11	97.89	93.67	97.79
GS 12	89.37	85.6	88.88
GS 13	96.59	92.87	96.45
GS 14	78.94	76.65	78.29
GS 15	97.77	93.53	97.67
GS 16	80.1	76.09	78.99
GS 17	40.39	39.03	38.24
GS 18	90.74	73.39	87.87
GS 19	97.9	93.25	97.79
GS 20	74.77	60.98	70.56
GS 21	79.36	61.05	70.52
GS 22	79.1	60.33	69.67
GS 23	80.75	59.62	70.19
GS 24	79.13	60.15	69.51
GS 25	78.95	61.24	70.39
Average	84.768	73.2572	81.4612

and radius is favored by mafic sources. The Ti/Nb ratio variability thus evinces frequent source shifting, a postulate that is consistent with active intracratonic rifting (Young, 1983; Young and Nesbitt, 1985). The TiO_2/Zr ratio, though mostly low, also varies widely, some values ranging from 59 to 81 while <55 is typical for felsic rocks (Hayashi et al., 1997).

A variable degree of admixing of contributions from amphibolite bodies with a granite-derived main sediment supply seems apparent (Fig. 6.11B; Saha et al., 2010). Strikingly there is not much difference between the sandstones and shales in this respect. An inferred dominant contribution from the basal granite, nonetheless, is amply supported by plots within the Cr/V-Y/Ni (Fig. 6.11C) and Zr/TiO_2-Nb/Y (Fig. 6.11D) diagrams proposed by Floyd et al. (1990) and Hayashi et al. (1997), respectively. Comparative enrichment of the studied sedimentary rock samples with respect to W further strengthens this view. While the high content of K_2O corroborates well with the dominant felsic nature of the source rock, the positive correlation between K_2O and Ba suggests derivation of the latter from feldspars; Ba readily replaces K because of mutual compatibility (McLennan et al., 1983; Rahman and Suzuki, 2007, Fig. 6.10). The two discrete clusters in the figure are reflective of paleogeographic variations; contents of both the elements are considerably higher at the basin-margin than within the basin. Furthermore, the observation that the Al_2O_3/TiO_2 ratio is higher than 21 in a majority of cases (av. 28.5) further supports derivation of the studied sediments from a dominantly felsic rock (Table 6.4; Hayashi et al., 1997). However, the high degree of variability in the Al_2O_3/TiO_2 ratio, from 8 to 61, hints at contribution from sources of different characters too.

6.2.2 Comparison With Upper Cratonic Crust Standards

As a further support of our postulate that the studied Cretaceous sedimentary rocks were derived from upper cratonic crust (UCC), their trace element contents are compared with the standard UCC values as revised by McLennan (2001). Trace elements, unlike many major elements, being immobile as well as irreplaceable serve as better proxies for sediment source estimations (Hastie et al., 2007). More efficient partitioning during magma fractionation renders trace elements better guides for source identification (Bhatia and Crook, 1986; Bonjour and Dabard, 1991; Taylor and McLennan, 1995). Plots of La content, the most incompatible element, against the contents of the other trace elements shows an overall increasing trend in correspondence to increasing order of ionic potential for both the channel sandstones and the inferred overbank deposits (Fig. 6.12; cf., McLennan, 2001; Paikaray et al., 2008). Such trends are commensurate with the fact that high field strength elements preferably concentrate in the upper cratonic crust (Condie, 2005). The overbank sample population has its exponential trend line parallel to that of the UCC, but this is not the case for the channel sandstones. Greater homogenization achieved in overbank deposits presumably made them more comparable to the UCC. Comparative immaturity of these first-generation sediments can explain the lateral shifts of both of their trend lines from that of the UCC (Fig. 6.12).

6.2.3 Weathering Intensity

With the SiO_2-enrichment of the studied sedimentary rocks generally being comparable to that of the UCC, a moderate degree of weathering is inferred (cf., McLennan, 2001). Application of weathering indices like Chemical Index of Alteration (CIA), Chemical Index of Weathering, and Plagioclase Index of Alteration, formulated on the basis of major oxide contents (Nesbitt and Young, 1982; Harnois, 1988; Fedo et al., 1995), also generally indicate a moderate degree of weathering (Table 6.4; Young and Nesbitt, 1998). Two samples, viz., GS 1 and GS 17, nonetheless, have significantly low indices and they differ from the others by

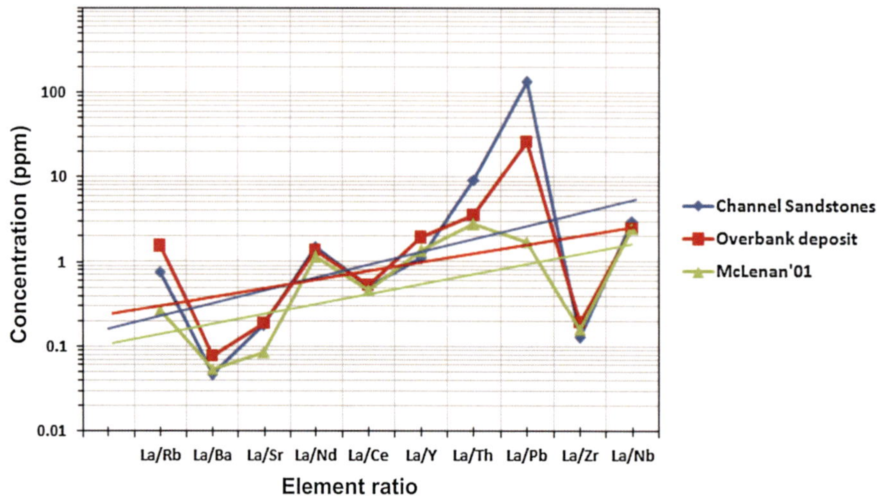

FIGURE 6.12 Variation in La versus other elements in channel sandstones and overbank deposits with reference to upper cratonic crust as revised by McLenan (2001). Trend lines in all three cases show higher concentration in high field strength elements.

incorporating an unusually high amount of $CaCO_3$ cement. Alteration of the weathering record by diagenesis is therefore apparent; the higher the CaO content, the lower the weathering indices (Fig. 6.13A). Sr because of its smaller ionic radius leaches out more readily than Rb on weathering and Rb/Sr can thus be taken as another proxy for weathering at sediment source (Chen et al., 2001; Jin et al., 2001). However, the typically low Rb/Sr ratios in the studied samples are in distinct disagreement with the aforementioned moderate weathering indicated from the indices (Table 6.4). A CIA versus Rb/Sr ratio plot, with exclusion of the two CaO-enriched samples, shows a weak negative correlation (Fig. 6.13B). Xu et al. (2010) reported such a negative correlation at CIA values greater than 75 and attributed this relation to stronger activity of Rb with respect to Sr. In the case under study, however, the negative correlation exists at CIA value as low as 60. Evidently the relation between the CIA and Rb/Sr ratio is not straightforward and the changes in distribution of Rb and Sr can also be ascribed to various other factors like modes of occurrence, medium conditions, and the formation of secondary minerals like carbonates (Xu et al., 2010). As for the last factor, the CaO versus Rb/Sr plot (excluding the samples with unusually high content of CaO) elicits a tendency of fall in the ratio with increase in CaO content (Fig. 6.13C). Apparently, with respect to UCC (McLennan, 2001), the Rb/Sr ratio increased with depletion in Sr content within the weathering residue at source, but decreased as chemogenic and/or biogenic precipitates trapped Sr at the depositional site (cf., Xu et al., 2010).

6.2.4 Rainfall and Temperature

Correlation of elemental compositions and ratios in soils provides a robust means of estimating rainfall and moderately well enables inference of temperature (Marbut, 1935); these techniques have been successfully adopted to determine paleoprecipitation rate and rainfall

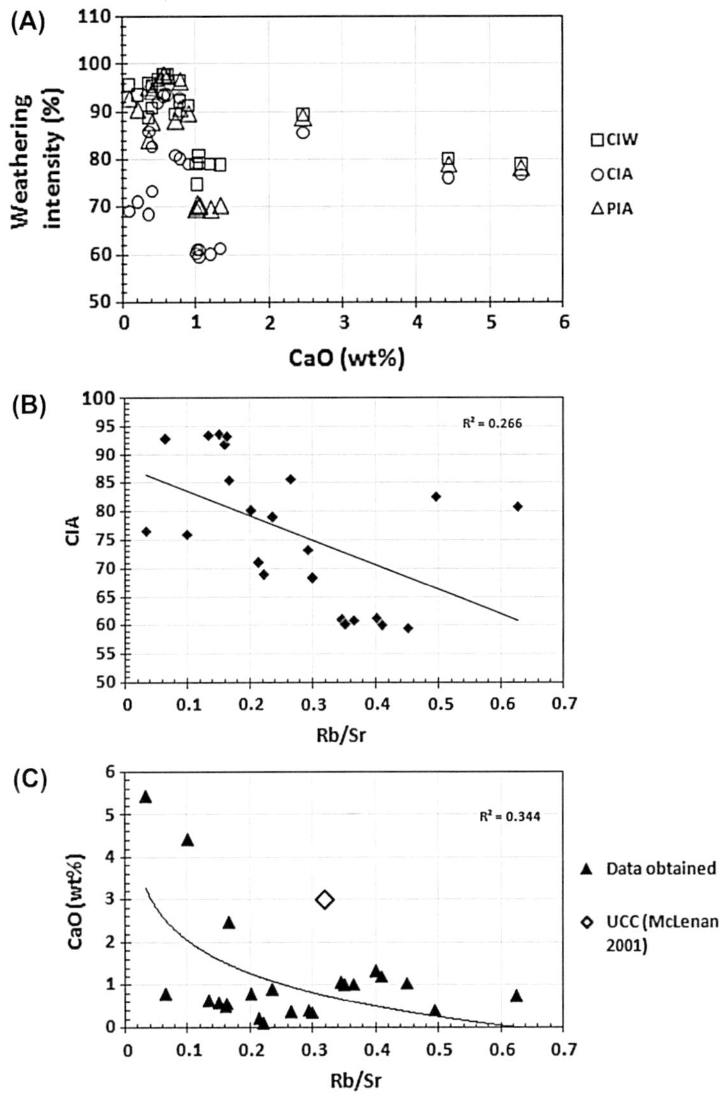

FIGURE 6.13 Plots of weathering index values with respect to CaO contents. Note deviation from the cluster is explicitly influenced by unusually high CaO content (A). CIA versus Rb/Sr plot (excluding two samples having highest CaO concentration); note negative correlation (B). Plot of CIA versus Rb/Sr ratios also shows negative correlation (C).

(Sheldon et al., 2002). Different formulae based on the same premise that alkali and alkaline earth elements (Ca, Mg, Na, K) are discriminated against Al have been put to use:

$$\text{P(Mean annual precipitation)} = 14.265(\text{CIA} - \text{K}) - 37.632 \quad \text{Maynard (1992)} \qquad (6.1)$$

$$\text{P(Mean annual precipitation)} = -259.34 \ln(B) + 759.05 \quad \text{Maynard (1992)} \qquad (6.2)$$

where B = molar ratio of bases (CaO, MgO, NaO, K_2O) to Al_2O_3

$$= -130.93\ \ln(C) + 467.4 \quad \text{Retallack (2001)} \tag{6.3}$$

where C = molar ratio of CaO to Al_2O_3

The studied samples, excluding GS 1 and GS 17 containing substantial amounts of diagenetic carbonates, yield values of mean annual precipitation of 1023, 1060, and 844 mm/year, using the earlier three formulae, respectively (Table 6.5). The results are comparable

TABLE 6.5 Mean Annual Precipitation (MAP) and Mean Annual Temperature (MAT) Estimated From Sample Compositions

Samples	MAP 1 (mm/year)	MAP 2 (mm/year)	MAP 3 (mm/year)	MAT (°C)
GS 2	1066.41	1064.85	808.50	13.78
GS 3	931.68	958.68	927.61	9.56
GS 4	961.28	980.66	843.77	10.83
GS 5	1079.48	1070.17	842.93	13.78
GS 6	1113.77	1083.53	935.76	13.94
GS 7	1166.72	1165.30	936.55	14.75
GS 8	1100.62	1007.48	832.58	14.08
GS 9	1264.67	1338.35	910.11	16.31
GS 10	908.20	938.21	868.50	9.63
GS 11	1283.48	1383.88	970.77	16.44
GS 12	1171.43	1189.06	746.23	16.39
GS 13	1276.08	1287.16	906.84	16.52
GS 14	1046.74	1012.37	640.50	16.60
GS 15	1281.20	1374.04	962.74	16.44
GS 16	1033.70	1006.17	651.18	16.03
GS 18	975.09	995.56	899.07	11.27
GS 19	1276.21	1377.44	972.90	16.35
GS20	772.40	860.64	802.39	6.89
GS 21	772.68	862.19	801.17	6.94
GS 22	765.45	853.51	794.11	6.65
GS 23	742.50	848.61	802.29	6.20
GS 24	754.72	853.76	787.29	6.64
GS 25	774.96	862.27	775.20	7.35
Average	1022.59	1059.73	844.30	12.32

Three MAP values for three different formulae used within the text.

to the estimation made by Chatterjee et al. (2013) for the same geographic setting in a similar timeframe. A humid paleoclimate is implied and the range of precipitation calculated satisfies the prerequisite rate of 200–1600 mm/year for safe application of these formulae (Sheldon et al., 2002).

Mean annual temperature can be estimated from the formula:

$$T(\text{Mean annual temperature}) = -18.51(S) + 17.2989 \quad \text{Sheldon et al.}(2002) \qquad (6.4)$$

where S is the molecular ratio of Na_2O and K_2O to Al_2O_3

The annual average of paleotemperature yielded by the studied samples is around 12.5°C. Again its range (Table 6.5) suits safe application of the formula. Nevertheless, an important caveat to bear in mind is that none of our samples represents a paleosol. However, it is noted that alkali and alkaline earth elements are present within the samples in excess to contents of typical soils. So our estimates can be seen as rather conservative and possibly represent the minimal value in each case. Contextually, the annual average of the ocean water temperature had possibly been around 12°C in the same latitude of ~60°S where the study area was located during the Barremian-Aptian time (Barron, 1983, 1987; Ronov et al., 1989; Hay et al., 1999). The annual average land temperature had presumably been somewhat higher. Overall, a postulate can thus be made that an apparently warm temperate climate prevailed in the study area during the time of deposition of the studied BS formation (Fig. 6.14).

6.2.5 Paleogeographic Overprint

Both Si and Al are lithophilic and immobile, but the former oxide is generally inherited and the latter is acquired through weathering. However, the observed decrease in SiO_2/Al_2O_3 ratio from the interpreted fan apex to fan base deposit samples and within the channels from samples taken from proximal through intermediate to distal sites was most probably related to preferred concentration of granular materials nearer the source and that of the

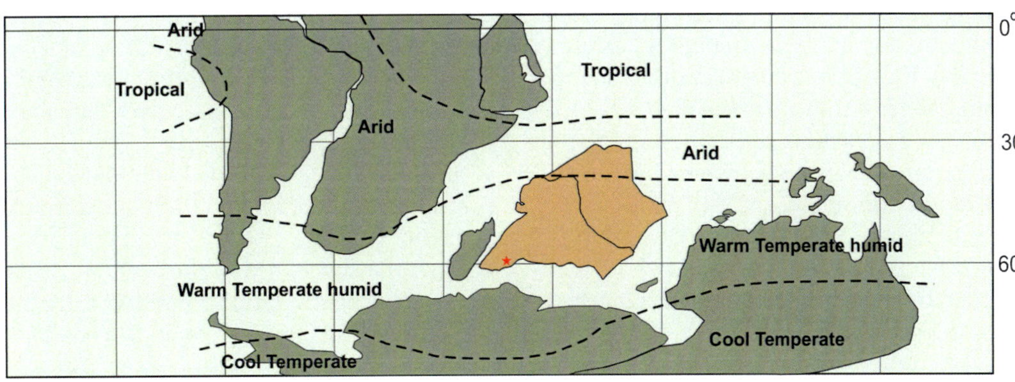

FIGURE 6.14 Position of the study locality (*dot*) shown against the background of global paleoclimatic belts. *Modified after Hay et al. (1999), Kent and Muttoni (2013), and Scotese (2001).*

clay fraction in the distal association (Fig. 6.15A). Despite this fact, increase in the SiO_2/K_2O ratio within samples taken from channel sandstones, in the proximal to distal transition (Fig. 6.15B), suggests rapid loss of cleavable and chemically weak feldspar during transport with respect to quartz (cf., Corcoran et al., 1999). The estimated loss of feldspar grains on bed-load transport under warm and humid climate is around 0.33% per kilometer (McBride et al., 1996). Negative correlation between K_2O and Al_2O_3 also supports transportational fractionation and/or large-scale diagenetic clay generation through postdepositional alteration of feldspar.

Progressive decline in the proportion of SiO_2 with respect to that of Al_2O_3 (Fig. 6.15A) and TiO_2 (Fig. 6.15C) through transport along the same channel system is to be expected. Both Ti and Al being immobile during weathering and diagenesis (Law et al., 1991; Nesbitt and Wilson, 1992), their ratio is fairly constant in sediments and thus used by many to identify sediment provenance (Wintsch and Kvale, 1994).

However, contrary to such an expectation in the studied formation, the observed TiO_2/Al_2O_3 ratio is not constant (Table 6.4) and increases down the channel system and in the channel to flood-plain transition (Fig. 6.15D). In the soil profile such an increase has been explained by preferred transportation of Al_2O_3 because of its greater mobility (Plank and Langmuir, 1998). However, removal of Al_2O_3 from these nonsoil materials is a difficult proposition and further extensive diagenetic alteration of the feldspar may have added Al_2O_3 to what had been inherited from the sediment source. A more likely explanation is diagentic production of rutile (Pettijohn et al., 1987). Akul'shina (1976) indicated replacement of Al by Ti in the clay mineral lattice. The wide variability in the Al_2O_3/TiO_2 ratio in sandstones observed in this study, assuaged to a constricted range around 21 within the overbank deposits, testifies to increasing sediment homogenization down the depositional energy spectrum (Table 6.4). Maynard (1992) and Young and Nesbitt (1998) noted such transportational fractionation only in association with a high degree of weathering. In the Cretaceous formation studied here the differential mobility between Al_2O_3 and TiO_2, nonetheless, is readily recognizable despite the inferred moderate scale of weathering.

Despite its source-sensitivity the measured Ti/Nb ratio steadily increases within the channel sandstones in proximal to distal transition (Fig. 6.15E) and provides evidence for transportational fractionation of these elements, as hinted at by McLennan (2001). Incorporation of these elements in the heavy mineral suite is an imperative. Because of its much lighter atomic weight Ti is more readily transportable than Nb and further, some Ti-bearing minerals, such rutile and sphene, can be diagenetically produced within the clay fraction of the deposit.

Equidistant plots of weathering indices collected from the fan apex down to the most distal sector as determined from our facies association studies, including its overbank deposits (excluding the samples unusually rich in diagenetic $CaCO_3$), record steady increase (Fig. 6.15F). The moot factor had been the Al_2O_3 content that increased in proportion during transport away from the source. Diagenesis might have also abetted in this apparent disruption of the weathering record. Pertinently, a similar increase in weathering index has been noted down the transport direction from margin to the center of a modern lake surrounded by hills (Xu et al., 2010). We thus underline that consideration of intrabasinal processes is obligatory while utilizing geochemical characteristics of sedimentary rocks for reconstructing their extrabasinal history. The mutual compliment between interpretations of facies and geochemical characteristics in the studied formation is critical.

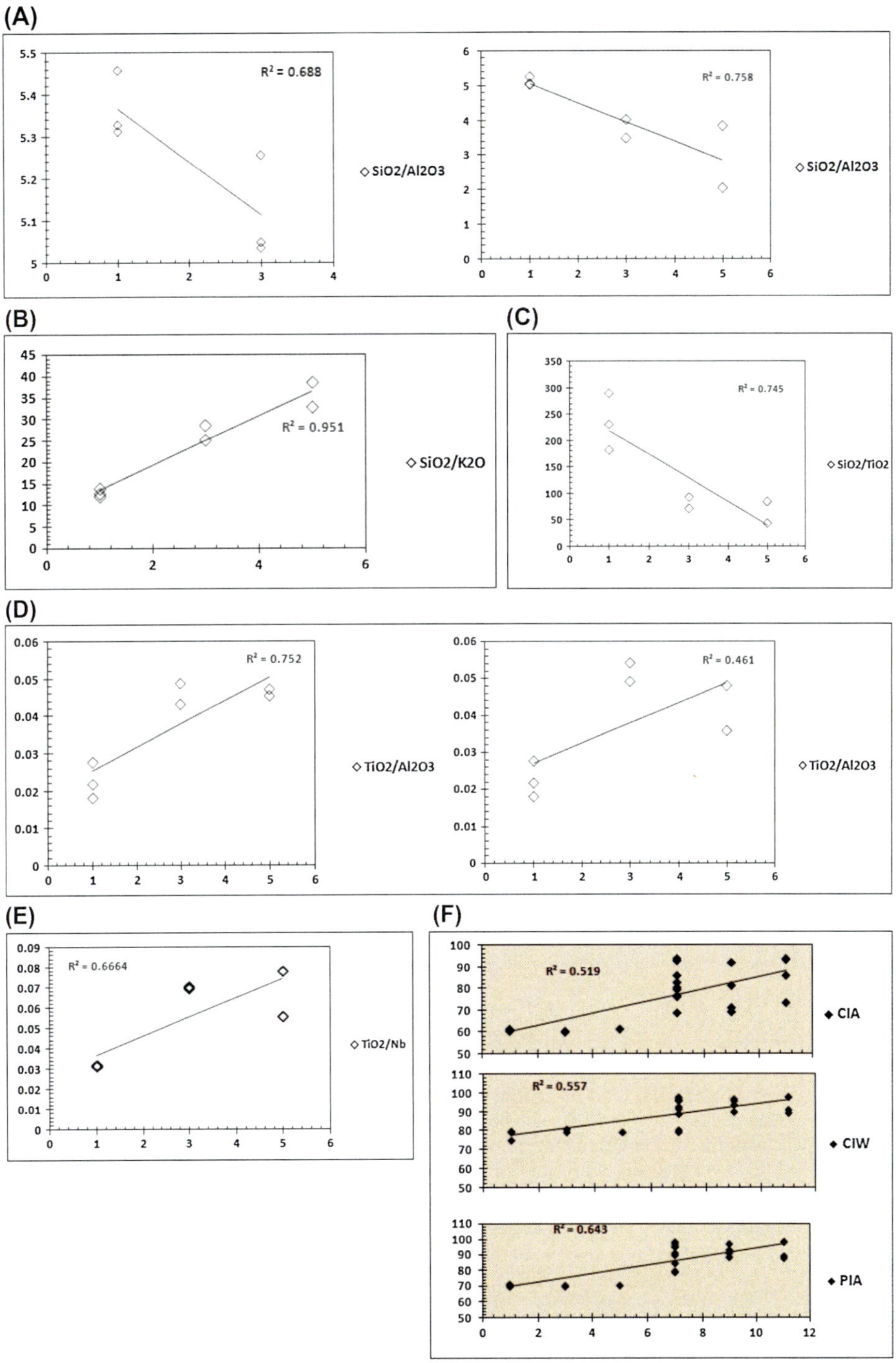

FIGURE 6.15 Compositional change with reference to paleogeography: SiO_2/Al_2O_3 ratio variation from apex to base and along channel system (A); SiO_2/K_2O ratio variation along the channel system (B); SiO_2/TiO_2 ratio variation along channel system (C); TiO_2/Al_2O_3 ratio variation along the channel system and channel-to-flood plain transition (D); TiO_2/Nb ratio variation along the channel system (E); and variation in CIA, CIW, and PIA from fan apex to distal overbank through fan base, axial channel, channel in intermediate position, and distal channel (F).

7. CONCLUSIONS

Bulk geochemistry, despite having overprints of intrabasinal processes, reveals details of the source of the Barremian-Aptian siliciclastic formation (here termed informally the Basal Siliciclastic formation), which occurs stratigraphically between a basal unconformity and an overlying transgressive surface, on the floor of the Pondicherry sub-basin of the Cauvery Basin, India. Both felsic and mafic components of the immediate basement evidently contributed to the sediment budget, which is corroborated by detrital mineralogy. The combined scenario of sediment distribution and paleocurrent variation, derived from a detailed study of facies and facies associations, reveals formation of an axial river skirting the basin-margin scree/alluvial fans and oblique encroachment of its minor branches into the distal floodplain along the margin of an intracratonic rift basin.

At the inferred basin interior, while the detrital mineralogy gains maturity progressively, the geochemistry still reflects an admixture of contributions made by compositionally different sources. The implied source shifting is consistent with postulated ongoing rifting of the intracratonic basin at the floor of which this formation accumulated. Despite the estimated paleoclimate having been reasonably humid with estimated rate of precipitation around 1000 mm/year, the weathering intensity remained moderate because of domination of erosion at the steep basin margin. Average annual paleotemperature, however, had possibly been moderate due to the high Barremian-Aptian paleolatitude of the depositional site.

Transportation, on the other hand, did affect lateral distribution of elements, whether major or trace, and their ratios. Preferred concentration of detrital framework elements, including heavy minerals, closer to the basin-margin, and clay minerals in the distal part is one aspect that distinctly affected even the weathering indices. This complexity was compounded further by carbonate cementation, as well as diagenetic addition of Al_2O_3 and authigenic heavy mineral formation. Bulk geochemistry in concert with detrital mineralogy provides the best key to reconstruct extrabasinal history of sediments, but their source implications, as the present study highlights, should always be investigated against the background of physical characteristics of the sediments, usually determined through a detailed facies/facies association study.

Acknowledgments

SS acknowledges the Center of Advanced Study (CAS Phase V) and University with Potential for Excellence (UPE II) programs of Jadavpur University. Geochemical work was carried out mostly when SS visited Osaka University under the INSA-JSPS exchange program. NC and AM acknowledge UGC and CSIR, respectively, for their financial support. The authors are grateful to the two reviewers on behalf of the journal for their critical and inspiring comments. They are indebted to their respective departments for infrastructural facilities. We thankfully acknowledge the kind hospitality of the Dalmia Cement (Bharat) Mines authority during field work.

References

Akul'shina, E.P., 1976. Methods for determining weathering conditions, sedimentation and post-sedimentary transformations according to clay minerals. In: Akul'shina, E.P. (Ed.), Glinistye Mineral Kak Pokazateli Litogeneza (Clay Minerals as Indicators of Rock Forming Conditions). Naukka, Novosibirsk, pp. 9–37 (in Russian).

Allen, J.R.L., 1968. Current Ripples. North-Holland Publishing Co., Amsterdam, p. 433.
Banerji, R.K., 1983. Evolution of the Cauvery Basin during Cretaceous. Cretaceous of India. Indian Association of Palynostratigraphy, Luclnow.
Barron, E.J., 1983. A warm, equable Cretaceous: the nature of the problem. Earth-Science Reviews 19, 305–338.
Barron, E.J., 1987. Cretaceous plate tectonic reconstructions. Palaeogeography, Palaeoclimatology, Palaeoecology 59, 3–29.
Basu, A., 1985. Influence of climate and relief on compositions of sands released at source areas. In: Zuffa, G.G. (Ed.), Provanance of Arenites, NATO ASI Series, vol. 148. Springer, pp. 1–18.
Basu, A., Schieber, J., Patranabis-deb, S., Dhang, P.C., 2013. Recycled detrital quartz grains are sedimentary rock fragments indicating unconformities: examples from the Chhattisgarh Supergroup, Bastar Craton, India. Journal of Sediment Research 83, 368–376.
Bhatia, M.R., Crook, K.A.W., 1986. Trace element characteristics of graywackes and tectonic setting discrimination of sedimentary basins. Contributions to Mineralogy and Petrology 92, 181–193.
Blair, T.C., 1999. Cause of dominance by sheet flood vs. debris-flow processes on two adjoining fans, Death Valley, California. Sedimentology 46, 1015–1028.
Blair, T.C., McPherson, J.G., 1994. Alluvial fans and their natural distinc-tion from rivers based on morphology, hydraulic processes, sedimentary processes, and facies assemblages. Journal of Sedimentary Research 64, 450–489.
Blanford, H.F., 1862. Cretaceous and other rocks of South Arcot and Trichinopollly districts, Madras. Memoirs of the Geological Survey of India 4 (1), 217.
Blatt, H., 1985. Provenance studies and mud rocks. Journal of Sedimentary Petrology 55, 69–75.
Blodgett, R.H., Stanley, K.O., 1980. Stratification, bedforms, and discharge relations of the platte braided river system, Nebraska. Journal of Sedimentary Research 50, 139–148.
Bonjour, J.L., Dabard, M.P., 1991. Ti/Nb ratios of clastic terriginous sediments used as an indicator of provenance. Chemical Geology 91, 257–267.
Bose, P.K., Sarkar, S., 1991. Basinal autoclastic massflow regime in precambrian chanda limestone formation, Adilabad, India. Sedimentary Geology 73, 299–315.
Bose, P.K., Sarkar, S., Mukhopadhyay, S., Saha, B., Eriksson, P., 2008. Precambrian basin-margin fan deposits: mesoproterozoic bagalkot group, India. Precambrian Research 162, 264–283.
BouDagher-Fadel, M.K., Banner, F.T., Whittaker, J.E., 1997. Chapman & Hall. In: The Early Evolutionary History of Plantonic Foraminifera. London, p. 269.
Bridge, J.S., 2006. Fluvial facies models. In: Posamentier, H., Walker, R.G. (Eds.), Facies Models Revisited. SEPM, Special Publication, vol. 84, pp. 85–170.
Bristow, C.S., Skelly, R.L., Ethridge, F.G., 1999. Crevasse splays from the rapidly aggrading, sand-bed, braided Niobrara River, Nebraska: effect of base-level rise. Sedimentology 46, 1029–1047.
Chatterjee, S., Goswami, A., Scotese, C.R., 2013. The longest voyage: tectonic, magmatic, and paleoclimatic evolution of the Indian plate during its northward flight from Gondwana to Asia. Gondwana Research 23, 238–267.
Chen, J., An, Z., Liu, L., Ji, J., Yang, J., Chen, Y., 2001. Variations in chemical compositions of the eolian dust in Chinese Loess Plateau over the past 2.5 Ma and chemical weathering in the Asian inland. Science in China Series D: Earth Sciences 44 (5), 403–413.
Chung, G.S., Lee, J.Y., Yang, D.Y., Kin, J.Y., 2005. Architectural elements of the fluvial deposits of meander bends in midstream of the Yeongsan River, Korea. Journal of the Korean Earth Science Society 26 (8), 809–820.
Collinson, J.D., 1978. Vertical sequence and sand body shape in alluvial sequences. Canadian Society of Petroleum Geologists Memoir 5, 577–586.
Condie, K.C., 2005. High field strength element ratios in Archean basalts: a window to evolving source of mantle plumes? Lithos 79, 491–504.
Corcoran, P.L., Mueller, W.U., 2002. The effects of weathering, sorting and source composition in Archaean high-relief basins: examples from the Slave Province, Northwest Territories, Canada. In: Alterman, W., Corcoran, P.L. (Eds.), Precambrian Sedimentary Environment: A Modern Approach to Ancient Depositional System, Special Publications of the International Association of Sedimentologists, vol. 33, pp. 183–211.

Corcoran, P.L., Mueller, W.U., Chown, E.H., 1998. Climatic and tectonic influences on fan deltas and wave to tide-controlled shoreface deposits: evidence from the Archean Keskarrah Formation, Slave Province, Canada. Sedimentary Geology 120, 125–152.

Corcoran, P.L., Mueller, W.U., Padgham, W., 1999. Influence of tectonism and climate on lithofacies distribution and sandstone and conglomerate composition in the Archean Beaulieu Rapids Formation, Northwest Territories, Canada. Precambrian Research 94, 175–204.

Cramer, J.J., Nesbitt, H.W., 1983. Mass-balance relations and trace-element mobility during continental weathering of various igneous rocks. Sciences Géologiques - Memoires 73, 63–73.

Crook, K.A.W., 1974. Lithogenesis and geotectonics: the significance of compositional variations in flysh arenites (graywackes). In: Dott, R.H., Shaver, R.H. (Eds.), Modern and Ancient Geosynclinal Sedimentation, SEPM, Special Publication, vol. 19, pp. 304–310.

Davis, A.M., Aitchison, J.C., Badengzhu, L.H., Zyabrev, S., 2002. Paleogene island arc collision-related conglomerates, Yarlung-Tsangpo suture zone, Tibet. Sedimentary Geology 150, 247–273.

Dickinson, W.R., Suczec, C.A., 1979. Plate tectonics and sandstone compositions. AAPG Bulletin 63, 2–31.

Enos, P., 1977. Flow regimes in debris flow. Sedimentology 24, 133–142.

Farrell, K.M., 2001. Geomorphology, facies architecture, and high-resolution, non-marine sequence stratigraphy in avulsion deposits, Cumberland Marshes, Saskatchewan. Sedimentary Geology 139, 93–150.

Fedo, C.M., Nesbitt, H.W., Young, G.M., 1995. Unraveling the effects of K-metasomatism in sedimentary rocks and paleosols, with implica-tions for paleoweathering conditions and provenance. Geology 23, 921–924.

Floyd, P.A., Franke, W., Shail, R., Dörr, W., 1990. Provenance and depositional environment of Rhenohercynian syn-orgenic greywacke from the Giessen nappe, Germany. Geologische Rundschau 79, 611–626.

Folk, R.L., Andrews, P.B., Lewis, D.W., 1970. Detrital sedimentary rock classification and nomenclature for use in New Zealand. New Zealand Journal of Geology and Geophysics 13, 937–968.

Furuyama, K., Hari, K.R., Santosh, M., 2001. Crystallisation history of primitive decan basalat from Pavagadh hill, Gujarat, Western India. Gondwana Research 4, 427–436.

Gani, M.R., 2004. From turbid to lucid: a straightforward approach to sediment gravity flows and their deposits. Sedimentary Record 2 (3), 4–8.

Garg, R., Ateequzzaman, K., Jain, K.P., 1988. Jurassic and lower cretaceous dinoflagellate cysts with some remarks on the concept of upper Gondwana. The Palaeobotunist 36, 254–267.

Graver, J.I., Scott, T.J., 1995. Trace elements in shale as indicators of crustal provenance and terrain accretion in south Canadian Cordillera. Geological Society of America Bulletin 107, 440–453.

Harnois, L., 1988. The CIW index: a new chemical index of weathering. Sedimentary Geology 55, 319–322.

Hastie, A.R., Kerr, A.C., Pearce, J.A., Mitchell, S.F., 2007. Classification of altered volcanic island arc rocks using immobile trace elements: development of Th-Co discrimination diagram. Journal of Petrology 48 (12), 2341–2357.

Hay, W.W., DeConto, R.M., Wilson, K.M., Voigt, S., Schulz, M., Wold-Rossby, A., Dullo, W.-C., Ronov, A.B., Balukhovsky, A.N., Söding, E., 1999. Alternative global Cretaceous paleogeography. Geological Society of America Special Papers 332, 1–46.

Hayashi, K., Fujisawa, H., Holland, H.D., Ohmoto, H., 1997. Geochemistry of 1.9 Ga sedimentary rocks from northeastern Labrador, Canada. Geochimica et Cosmochimica Acta 61, 4115–4137.

Hein, F.J., 1982. Depositional mechanism of deep-sea coarse clastic sediments, Cap Enruge Formation. Quebec. Canadian Journal of Earth Sciences 19, 267–287.

Hofmann, A.W., 1988. Chemical differentiation of the earth: the relationship between mantle, continental crust, and oceanic crust. Earth and Planetary Science Letters 90, 297–314.

Jin, Z., Wang, S., Shen, Ji, Zhang, E., Ji, J., Li, F., 2001. Weak chemical weathering during the little ice age recorded by lake sediments. Science in China (Series D) 44 (7), 652–658.

Kent, D.V., Muttoni, G., 2013. Modulation of late cretaceous and cenozoic climate by variable drawdown of atmospheric pCO_2 from weathering of basaltic provinces on continents drifting through the equatorial humid belt. Climate of the Past 9, 525–546.

Krynine, P.D., 1948. The megascopic classification of sedimentary rocks. The Journal of Geology 56, 130–165.

Law, K.R., Nesbitt, H.W., Longstaffe, F.J., 1991. Weathering of granitic tills and the genesis of a podzol. American Journal of Science 291, 940–976.

Li, Z.X., Powell, C.M., 2001. An outline of the palaeogeographic evolution of the Australian region since the beginning of the neoproterozoic. Earth-Science Reviews 53, 237–277.

Long, D.G.F., 2011. Architecture and depositional style of fluvial systems before land plants: a comparison of Precambrian, early Paleozoic, and modern river deposits. In: North, C. (Ed.), From River to Rock Record: The Preservation of Fluvial Sediments and Their Subsequent Interpretation, SEPM, Special Publication, vol. 97, pp. 37–61.

Lowe, D.R., 1976. Grain flow and grain flow deposits. Journal of Sedimentary Petrology 46, 188–190.

Lowe, D.R., 1982. Sediment gravity fows II. Depositional models with special reference to the deposits of high-density turbidity currents. Journal of Sedimentary Petrology 52, 279–297.

Mack, G.H., Rasmussen, K.A., 1984. Alluvial-fan sedimentation of the cutler formation (Permo-Pennsylvanian) near Gateway Colorado. Geological Society of America Bulletin 95, 109–116.

Makaske, B., 2001. Anastomosing rivers: a review of their classification, origin and sedimentary products. Earth-Science Reviews 53, 149–196.

Marbut, C.F., 1935. Atlas of American Agriculture. III. Soils of the United States. Government Printing Office, Washington, D.C.

Maynard, J.B., 1992. Chemistry of modern soils as a guide to interpreting Precambrian paleosols. The Journal of Geology 100, 279–289.

McBride, E.F., Abel-Wahab, A., McGilvery, T.A., 1996. Loss of sand-size feldspar and rock fragments along the South Texas Barrier Island, USA. Sedimentary Geology 107, 37–44.

McLennan, S.M., 2001. Relationships between the trace element composition of sedimentary rocks and upper continental crust. Geochemistry, Geophysics, Geosystems 2, 1–24.

McLennan, S.M., Taylor, S.R., Errikson, K.A., 1983. Geochemistry of Archean shales from the Pilbara supergroup, western Australia. Geochimica et Cosmochimica Acta 47, 1211–1222.

Miall, A.D., 1996. The Geology of Fluvial Deposits: Sedimentary Facies, Basin Analysis and Petroleum Geology. Springer, New York, p. 582.

Middleton, G.V., Hampton, M.A., 1976. Subaquous sediment transport and deposition by sediment gravity flows. In: Stanley, D.J., Swift, D.J.P. (Eds.), Marine Sediments, Transport and Environmental Management. John Wiley, New York, pp. 197–219.

Morton, A.C., Hallsworth, C.R., 2007. Stability of detrital heavy minerals during burial diagenesis. In: Mange, M.A., Wright, D.T. (Eds.), Heavy Minerals in Use Developments in Sedimentology, p. 58.

Mulder, T., Alexander, J., 2001. The physical character of subaqueous sedimentary density flows and their deposits. Sedimentology 48, 269–299.

Murthy, K.S., Chaudhuri, A., Ramana, L.V., Rao, M.V., Dobriyal, J.P., 2008. Hydrocarbon exploration of syn rift sediments in Nagapattinam Sub Basin, Cauvery Basin - a case study. In: 7th Biennial International Conference & Exposition on Petrolelum Geophysics, p. 443.

Nagendra, R., Kamalakkannan, B.V., Sen, G., Gilbert, H., Bakkiaraj, D., Nallapa Reddy, A., Jaiprakash, B.C., 2011. Sequence surfaces and paleobathymetric trends in Albian to Maastrichtian sediments of Ariyalur area, Cauvery Basin, India. Journal of Marine and Petroleum Geology 28, 895–905.

Narasimha Chari, M.V., Sahu, J.N., Banerjee, B., Zutshi, P.L., Chandra, K., 1995. Evolution of the Cauvery basin, India from subsidence modelling. Marine and Petroleum Geology 12, 667–675.

Nemec, W., Postma, G., 1993. Quaternary alluvial fans in southwestern crete: sedimentation processes and geomorphic evolution. In: Marzo, M., Puigde-fabregas, C. (Eds.), Alluvial Sedimentation, International Association of Sedimentologists Special Publication, 17, pp. 235–276.

Nesbitt, H.W., Wilson, R.E., 1992. Recent chemical weathering of basalts. American Journal of Science 292, 740–777.

Nesbitt, H.W., Young, G.M., 1982. Early Proterozoic climates and plate motions inferred from major element chemistry of lutites. Nature 299, 715–717.

Paikaray, S., Banerjee, S., Mukherji, S., 2008. Geochemistry of shales from the paleoproterozoic to neoproterozoic vindhyan supergroup: implications on provenance, tectonics and paleoweathering. Journal of Asian Earth Sciences 32, 34–48.

Pascoe, E.H., 1959. A Manual of Geology of India and Burma, third ed. Government of India Press, Calcutta, India. p. 3.

Pearce, J.A., Harris, N.B.W., Tindle, A.G., 1984. Trace element discrimination diagrams for the tectonic interpretation of granitic rocks. Journal of Petrology 25, 956–983.

Pettijohn, F.J., 1975. Sedimentary Rocks, third ed. Harper and Row, New York, pp. 1–628.

Pettijohn, F.J., Potter, P.E., Seiver, R., 1987. Sand and Sandstone, second ed. Springer-Verlag, New York, NY.
Plank, T., Langmuir, C.H., 1998. The chemical composition of subducting sediment and its consequences for the crust and mantle. Chemical Geology 145, 325–394.
Powell, C.M., Roots, S.R., Veevers, J.J., 1988. Pre-breakup of continental extension in east Gondwanaland and the early cretaceous opening of the eastern Indian Ocean. Tectonophysics 155, 261–283.
Rahman, M.J.J., Suzuki, S., 2007. Geochemistry of sandstone from the Miocene Surma Group, Bengal Basin, Bangladesh: implications for provenance, tectonic setting and weathering. Geochemical Journal 41, 415–428.
Ramasamy, S., Banerji, R.K., 1991. Geology, petrography and systematic stratigraphy of the pre-Ariyalur sequence in Trichinopoly district, Tamil Nadu, India. Journal of the Geological Society of India 37, 577–594.
Retallack, G.J., 2001. Cenozoic expansion of grasslands and cli-matic cooling. The Journal of Geology 109, 407–426.
Ronov, A.B., Khain, V.E., Balukhovsky, A.N., 1989. In: Barsukov, V.L., Laviorov, N.P. (Eds.), Atlas of Lithological-paleogeographical Maps of the World: Mesozoic and Cenozoic of Continents and Oceans. Editorial Publishing Group VNII Zarubezh-Geologia, Moscow, p. 79.
Roser, B.P., Korsch, R.J., 1986. Determination of tectonic settings of sandstone-mudstone suits using SiO_2 content and K_2O/Na_2O ratio. The Journal of Geology 94, 635–650.
Roy, D.K., Roser, B.P., 2013. Climatic control on the composition of Carboniferous–Permian Gondwana sediments, Khalaspir basin, Bangladesh. Gondwana Research 23, 1163–1171.
Saha, S., Banerjee, S., Burley, S.D., Ghosh, A., Saraswati, P.K., 2010. The influence of flood basaltic source terrains on the efficiency of tectonic setting discrimination diagrams: an example from the Gulf of Khambhat, western India. Sedimentary Geology 228, 1–13.
Santosh, M., Maruyama, S., Yamamoto, S., 2009. The making and breaking of supercontinents: some speculations based on superplumes, super downwelling and the role of tectosphere. Gondwana Research 15, 324–341.
Sarkar, S., Chakraborty, N., Mandal, A., Banerjee, S., Bose, P.K., 2014. Siliciclastic-carbonate mixing modes in the river-mouth bar palaeogeography of the Upper Cretaceous Garudamangalam Sandstone (Ariyalur, India). Journal of Palaeogeography 3 (3), 233–256.
Sastri, V.V., Sinha, R.N., Singh, G., Murti, K.V.S., 1973. Stratigraphy and tectonics of sedimentary basins on east coast of peninsular India. American Association of Petroleum Geologists Bulletin 57, 655–678.
Sastri, V.V., Venkatachala, B.S., Narayanan, V., 1981. The evolution of the east coast of India. Palaeogeography, Palaeoclimatology, Palaeoecology 36, 23–54.
Schieber, J., Bose, P.K., Eriksson, P.G., Banerjee, S., Sarkar, S., Altermann, W., Catuneanu, O., 2007. Atlas of microbial mat features preserved within the siliciclastic rock record. In: Atlases in Geoscience, 2. Elsevier, Amsterdam p. 311.
Schultz, A.W., 1984. Subaerial debris flow deposition in the upper Paleozoic Cutler Formation, Western Colorado. Journal of Sedimentary Petrology 54, 759–772.
Scotese, C.R., 2001. Atlas of earth history. In: Paleogeography, PALEOMAP Project, Arlington, Texas, 1, p. 6.
Selley, R.C., 1965. Diagnostic characters of fluviatile sediments of the Torridonian formation (Precambrian) of North-west Scotland. Journal of Sedimentary Research 35, 366–380.
Seth, A., Sarkar, S., Bose, P.K., 1990. Synsedimentary seismic activity in an immature passive margin basin, lower member of Katrol Formation, Upper Jurassic, Kutch, India. Sedimentary Geology 68, 279–291.
Sheldon, N.D., Retallack, G.J., Tanaka, S., 2002. Geochemical Climofunctions from North American soils and application to paleosols across the Eocene-Oligocene boundary in Oregon. The Journal of Geology 110, 687–696.
Singh, H.P., Venkatachala, B.S., 1988. Upper Jurassic-Lower Cretaceous spore-pollen assemblages in the peninsular India. The Palaeobotanist 36, 168–176.
Sundaram, R., Rao, P.S., 1986. Lithostratigraphy of Cretaceous and Paleocene rocks of Trichinopolly district, Tamil Nadu, south India. Records of the Geological Survey of India 116, 11–23.
Suttner, L.J., Dutta, P.K., 1986. Alluvial sandstone composition and paleoclimate, I. Framework mineralogy. Journal of Sedimentary Petrology 56, 329–345.
Tawfik, H.A., Ghandour, I.M., Maejima, W., 2011. Petrography and geochemistry of the Lower Paleozoic Araba Formation, northern eastern Desert, Egypt: implications for provanance, tectonic setting and weathering signature. Journal of Geosciences, Osaka City University 54 (1), 1–16.
Taylor, S.R., McLennan, S.M., 1995. The geochemical evolution of the continental crust. Reviews of Geophysics 33 (2), 241–265.

Tewari, A., Hart, M.B., Watkinson, M.P., 1996. A revised lithostratigraphical classification of the Cretaceous rocks of Trichinopoly district, Cauvery Basin, Southeast India. In: Pandey, J., Azmi, R.J., Bhandari, A., Dave, A. (Eds.), Contributions XVth Indian Colloquium on Micropaleontology and Stratigraphy, Dehra Dun, pp. 789–800.

Todd, S.P., 1989. Sream-driven, high-density gravelly traction carpets: pos-sible deposits in the Trabeg Conglomerate Formation, SW Ireland and some theoretical considerations of their origin. Sedimentology 36, 513–530.

Venkatachalapathy, R., Ragothaman, V., 1995. A foraminiferal zonal scheme for the mid-Cretaceous sediments of the Cauvery Basin, India. Cretaceous Research 16, 415–433.

Watkinson, M.P., Hart, M.B., Joshi, A., 2007. Creatceous tectonostratigraphy and the development of the Cauvery Basin, southern India. Geological Society London Special Publications 13, 181–191.

Wintsch, R.P., Kvale, C.M., 1994. Differential mobility of elements in burial diagenesis of siliciclastic rocks. Journal of Sedimentary Research 64A, 349–361.

Xu, H., Liu, B., Wu, F., 2010. Spatial and temporal variations of Rb/Sr ratios of the bulk surface sediments in Lake Qinghai. Geochemical Transactions 11 (3), 1–5.

Young, G.M., 1983. Tectono-sedimentary history of early proterozoic rocks of the northern great lakes region. In: Medaris Jr., L.G. (Ed.), Early Proterozoic Geoloogy of the Great Lakes Region: Geological Society of America Memoirs, vol. 160, pp. 15–32.

Young, G.M., Nesbitt, H.W., 1985. The Gowganda Formation in the southern part of the Huronian outcrop belt, Ontario, Canada: stratigraphy, depositional environments and regional tectonic significance. Precambrian Research 29, 265–301.

Young, G.M., Nesbitt, H.W., 1998. Processes controlling the distribution of Ti and Al in weathering profiles, siliciclastic sediments and sedimentary rocks. Journal of Sedimentary Research 68 (3), 448–455.

CHAPTER

7

Petrological and Geochemical Constraints on Provenance, Paleoweathering, and Tectonic Setting of Clastic Sediments From the Neogene Lambir and Sibuti Formations, Northwest Borneo

R. Nagarajan[1], *J.S. Armstrong-Altrin*[2], *F.L. Kessler*[3], *J. Jong*[4]

[1]Curtin University, Miri, Sarawak, Malaysia; [2]Universidad Nacional Autónoma de México, México D.F., México; [3]Goldbach Geoconsultants O & G, Glattbach, Aschaffenburg, Germany; [4]JX Nippon Oil and Gas Exploration (Deepwater Sabah) Limited, Kuala Lumpur, Malaysia

OUTLINE

Sediment Provenance
http://dx.doi.org/10.1016/B978-0-12-803386-9.00007-1

1. INTRODUCTION

Provenance studies involve the interpretation of the lithologic source of sediments and/or sedimentary rocks. Fine-grained sediments are often the final product of preexisting clastic sediments, since these particles are stable weathering products and can be recycled through several episodes of burial, uplift, and erosion, depending on the tectonic processes of the region (Potter et al., 2005). The provenance of sediments and/or sedimentary rocks can be reconstructed based on their geochemical and mineralogical compositions. The study of the provenance of siliciclastic sediments is a tool to investigate the evolution of ancient sedimentary basins. Major and trace element geochemistry of siliciclastic sediments provides information on the type of source rocks, paleoweathering conditions, hydraulic sorting, and extent of recycling in the tectonic development of sedimentary basins (Nesbitt and Young, 1982; Cullers, 1995; Armstrong-Altrin et al., 2004; Nagarajan et al., 2007a,b; Nagarajan et al., 2014; Armstrong-Altrin, 2015).

Trace elements such as Th, Zr, Hf, Nb, Sc, Y, Cr and rare earth elements (REEs) such as La, Ce, Nd, Gd, and Yb are suited for the discrimination of provenance and tectonic setting since these elements have relatively low mobility during sedimentary processes and have short residence time in seawater (Taylor and McLennan, 1985). It is commonly assumed that these elements are transferred quantitatively into detrital sediments during the sedimentary process and their concentration reflects the signature of the source rock composition (McLennan et al., 1980, 1993; Bhatia and Crook, 1986; Condie, 1993; Bakkiaraj et al., 2010; Armstrong-Altrin et al., 2004, 2012, 2013, 2014, 2016; Nagarajan et al., 2014, 2015). Early studies from Northwestern (NW) Borneo were focused on lithology, stratigraphy, and tectonic evolutions on a regional scale (Zin, 1996; Hutchison, 2005; Morley et al., 2008; Hall et al., 2008; Hall, 2013; references therein). Recent studies are focused on NW Borneo clastic systems (Viet, 2014; Togunwa et al., 2015). However, there is a gap in research on the provenance of clastic sediments in NW Borneo (Hall and Nichols, 2002; Van Hattum et al., 2013; Nagarajan et al., 2014, 2015). We present new petrographic and geochemical data on the Miocene clastic sediments of the Lambir and Sibuti Formations, collected from outcrops along the beaches between Miri and Bekenu in North Sarawak. The objective is to investigate the provenance, paleoweathering, and probable tectonic setting of the Miocene age Lambir and Sibuti Formations.

2. SEDIMENTOLOGICAL CONSIDERATIONS AND TECTONICS

2.1 Sediment Chronology and Depositional Environment

The North Sarawak region, located at the north and eastern parts of the Rajang-Baram watershed of the NW Borneo Basin (Banda, 1998, Fig. 7.1), consists of thick (600 m), shallow, and deep marine sediment sequences comprising sandy and shaly formations of Neogene age. The Setap and Sibuti Formations were deposited in an open marine environment, while the Belait, Lambir, Tukau, and Miri Formations were deposited in a shallow marine to intertidal and coastal environment (Banda, 1998; Kessler, 2009). The litho-stratigraphy of the study area is summarized in Fig. 7.1 (after Kessler and Jong, 2015a).

The Tukau Formation is a coal-bearing sequence, which underlines the paralic influence on the sequence. The outcrops of the Lambir Formation (LF) are located in the Lambir Hill area rising to 508 m. These sediments were deposited from Middle to Late Miocene times (Langhian to Messinian) (Fig. 7.1). The Middle-Late Miocene LF marks a shift from carbonate (Sibuti Formation; SF) to clastic sediments. The depositional transition is recognized at the southern margin of Lambir Hill (Hutchison, 2005; Kessler and Jong, 2015b). Fossil evidence, such as gastropods and crabs, suggest that the SF was deposited in a shallow marine environment during the Early Miocene (Aquitanian) to late Middle Miocene (Serravallian) (Hutchison, 2005; Simon et al., 2014). The SF contains shaly and calcareous layers with thin lenses of limestone (Hutchison, 2005). In a few areas, marly limestone deposits occur, such as in the Opak Quarry (Simon et al., 2014).

The LF is comprised of sandstone and sandy intercalations with shale and siltstones. The sandstones are fine- to medium-grained with lignite laminations (0.1 and 0.9 cm). The basal part of the LF is erosive and overlies dark to light gray mudstones of the SF. The sediments were deposited under shallow marine to coastal conditions as evidenced by hummocky cross-bedding and low-angle planar cross-bedding (Hutchison, 2005). According to the author, the basal part of the LF comprises well-sorted sandstones in a number of cycles with hummocky cross-bedding, but the upper part consists of low-angle, planar cross-bedding, with the occurrence of foraminifera and ophiomorpha reflecting a beach environment. Lambir sandstones have permeability of 1105–3018 mD and 25.3–28.7% porosity (Viet, 2014). The source rock properties of organic rich sediments from NW Borneo, including the LF, were identified as occasionally organic rich sediment, with Type III kerogen (Togunwa et al., 2015).

2.2 Tectonics

Jong et al. (2015) concluded that the tectonic activity within the Sundaland Plate occurred along defined lineaments, which represent zones of weakness in the plate and often have the character of fold and thrust belts. In the study area, deformation is related to the Baram Line fault system. During the Oligocene the development of an SW–NE trending synclinorium, or regional sag, known as the NW Borneo Foredeep was documented by Jong et al. (2015). There are indications that strike-slip movements occurred along the Baram Line, possibly in connection with rotation of parts of the greater Borneo area.

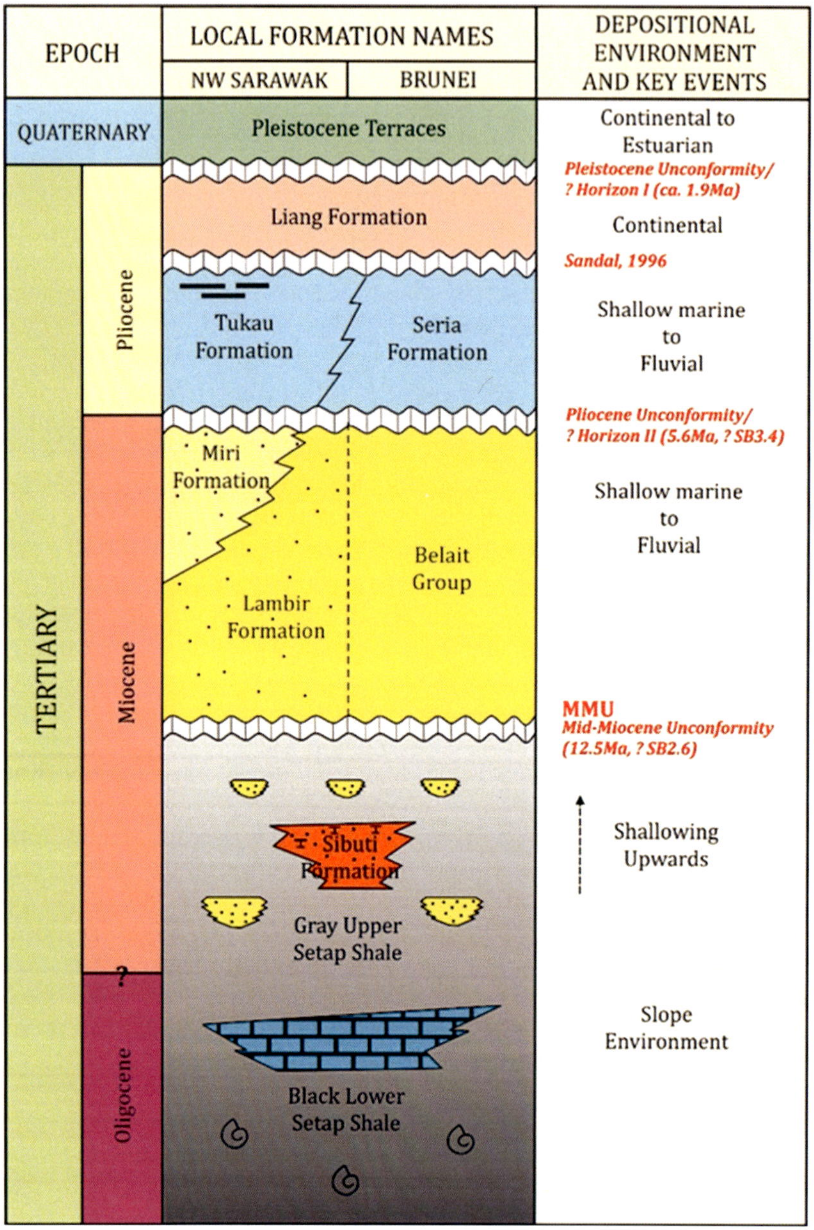

FIGURE 7.1 Simplified litho-stratigraphy scheme of the study area. The nomenclature of Miri Formation is generally used in the Greater Miri area and is age-equivalent to the upper section of Lambir Formation, Sandal (1996); however, placed the formation partially age-equivalent to the lower Tukau Formation. Likewise, the mid Early Miocene Sibuti Formation is more locally confined with the Subis Limestone Member in the lower part of the formation located along the central anticlinorium of the Sibuti Formation (Banda and Honza, 1997). Carbonates are also widespread in the Paleogene section, and are seen in a number of outcrops and wells (e.g., Batu Niah, Engkabang-1; Jong et al., 2016). Note the unconformity between Tukau/Seria and Liang Formations was not observed in this study but in Brunei had been documented by Sandal (1996). Reprinted from Kessler and Jong (2015a) with permission from Geological Society of Malaysia.

From around the Middle Miocene, tectonic activity within the Sundaland Plate switched from extensional to compression, which caused uplift, and was often associated with strike-slip tectonism (Jong et al., 2014, 2015; Kessler and Jong, 2015b). From the Middle Miocene, parts of the central Borneo hinterland were exhumed (Kessler and Jong, 2015b, leading to a pulse of clastic deposition onto the Sarawak and Sabah shelves. The total uplift of the Borneo hinterland, up to the present day, could be in the order of 6000 m (Kessler and Jong, 2015b).

The sampled Neogene sediments from the onshore part of the NW Borneo Basin experienced folding and uplift, leading to a Plio-Pleistocene unroofing of the sequence. Lambir Hill is an inversion (pop-up) structure between branches of the Baram Line. There is an angular unconformity reported, which is older than the Late Pleistocene (Kessler and Jong, 2015b). Post-Pleistocene, C^{14} dated sediments are found at *ca* 20 m above the current shoreline, and suggest a coastal uplift that has exceeded eustatic sea level rise (Kessler and Jong, 2014a,b).

3. METHODOLOGY

Fifty-five samples were collected from the outcrops of the Sibuti and LFs at different stratigraphic intervals in order to cover all litho-sections. Based on the lithology and locality in the stratigraphic column, 30 fresh or unweathered rock samples were selected for geochemical analysis. Twenty-seven samples from the LF and three samples from the SF were selected. The samples were washed with distilled water in order to remove salts because the samples were collected from outcrops along the beach. Samples were then oven-dried for 24 h at 60°C and ground to a size of 63 μm using an agate-mortar. Geochemical analysis was carried out at Activation Laboratories Ltd, Canada, using Code 4LITHO (11+) Major Element Fusion ICP (WRA)/Trace Element Fusion ICP/MS (WRA4B2) packages.

The samples were mixed with a flux of lithium metaborate/lithium tetraborate and fused in an induction furnace with platinum crucibles. The resulting molten bead was rapidly digested in a weak nitric acid solution in a glass disc. XRF analysis was then carried out for major oxides and the same technique was also employed for trace element and REE analyses. The application of fusion technique ensured that the entire sample was dissolved and the analysis was carried out by inductively coupled plasma (ICP) and ICP mass spectrometry (ICP-MS). Certified reference materials NIST 694, W-2a, BIR-1a (for major and trace elements) and NCS DC70014, LKSD 3, W-2a (for REE) were used to ensure the accuracy and precision of the geochemical analysis, which was better than ± 5%.

For the discussion of REE results, the Upper Continental Crust (UCC), Post-Archaean Australian Shale (PAAS), and chondrite normalization factors listed in Taylor and McLennan (1985) were used. Eu anomaly (Eu/Eu*) was calculated using the formula ($Eu/Eu^* = Eu_{CN}/(Sm_{CN}/Gd_{CN})_{1/2}$, where CN refers to the chondrite normalized values (McLennan, 1989).

Eight thin sections were prepared using a standard technique. Point counts were undertaken for six sandstone samples using both Gazzi-Dickinson (Gazzi, 1966; Dickinson, 1970) and standard methods. In each thin section, at least 300 framework grains were counted for quartz [Q = all quartz grains], total feldspar [F = potash feldspar (k) + plagioclase (P)], lithics (L) [volcanic (Lv), sedimentary (Ls), metamorphic (Lm), plutonic (Lp)], and heavy minerals (HM). X-ray diffraction analysis was performed on four samples in a Panalytical

X'Pert Pro diffractometer at Activation Laboratories Ltd, Canada, equipped with a Cu X-ray source and an X'celerator detector, operating at the following conditions: voltage: 40 kV; current: 40 mA; range: 5–70 deg 2θ; step size: 0.017 deg 2θ; time per step: 50.165 sec; divergence slit: fixed, angle 0.5 deg. Corundum was used as an internal standard, to determine the amount of X-ray amorphous material. The crystalline mineral phases were identified in X'Pert HighScore Plus using the PDF-4 Minerals 2014 ICDD database. The quantities of the crystalline minerals were determined using the Rietveld method, which is based on the calculation of the full diffraction pattern from crystal structure information.

4. RESULTS

4.1 Petrographic Description

The studied rocks are dominated by quartz with feldspar, lithic fragments, and accessory minerals such as muscovite, biotite, zircon, rutile, and opaque grains (magnetite and ilmentite). Cement types vary from argillaceous to calcareous in nature. Based on the petrography, the sandstones are classified as quartz arenites and sublitharenites (Fig. 7.2A; Table 7.1). Quartz arenites consist of angular to subangular monocrystalline quartz with minor lithic fragments and other accessory minerals. Quartz is the most dominant mineral (81–99%) followed by lithic fragments (3–14%), opaque minerals (up to 5%), feldspar (1–2%), muscovite (trace), and heavy minerals (zircon and rutile; trace to 1%). Significant iron leaching was identified along the pore edges. The sediments are fine-grained to medium-grained. Sublitharenites are composed of monocrystalline quartz grains

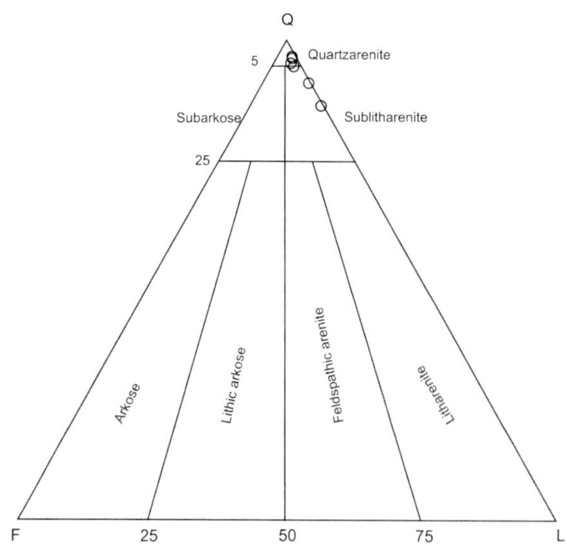

FIGURE 7.2A Q-F-L ternary plot: Petrographic classification of sandstones (Folk, 1974).

TABLE 7.1 Modal Analysis Data for the Sandstones of Lambir Formation

S.No	Quartz	Feldspar	Lithics	Biogenic	Mica	HM
A4	304	4	10	0	1	5
A19	310	6	10	0	0	11
A23	290	5	12	0	0	11
A24	263	2	40	0	0	0
A28	292	2	27	0	1	10
A7	257	2	7	29	0	11

HM, heavy minerals.

with calcareous cement with lithic fragments and accessory minerals. Two ferruginous sandstones are carbonate cemented and the grains are coated with Fe-oxides. Fossils are also common in these sandstones.

Calcareous sandstones consist of a high percentage of calcareous cement with a considerable amount of shell fragments. The grains are subangular with concavo-convex contacts between the quartz grains and *foraminifera* sp. This suggests that the quartz grains are detrital rather than a product of recrystallization. The foraminifers (*Nummulites* sp, *Lepidocyclina* sp, *Globigerinoid*, and uniserial benthic foraminifera), bivalve fragments, and rugose coral fragments indicate a shallow marine depositional environment. Based on the characteristics and features of the detrital grains, it is interpreted that the clastic sediments were deposited in a high-energy environment with little compaction.

Four samples were analyzed with X-ray diffraction (XRD) to identify the major mineral phases of each rock type. The calcareous sandstone consists of calcite (33.9%), ankerite (15.5%), quartz (14.2%), chlorite (7%), illite/muscovite (5%), and a trace of aragonite (1.2%). The LF sandstones consist of quartz (50.6–88.8%), illite/muscovite (1.5–16.9%), plagioclase (0.5–0.9%), and amorphous phases (8.8–27.5%). The Mudstone of the SF consists of quartz (32.7%), illite/muscovite (23.45), chlorite (7%), plagioclase (6.8%), and calcite (3.3%). Significant amorphous phases are recorded between 8.8% and 27.5%.

4.2 Chemical Composition

The major, trace, and rare earth element concentrations are presented in Table 7.2, arranged by rock type, based on the geochemical classification of Herron (1988). On a geochemical classification diagram from Herron (1988; Fig. 7.2B), data from 27 samples of the LF were plotted on a geochemical classification diagram from Herron (1988; Fig. 7.2B). Of these, 10 are plotted in the sublitharenite and litharenite fields, 4 are in the subarkose field, 10 are in shale and wacke fields, 1 is in the quartz arenite field, and 2 are in the Fe-sand field (Ferruginous sandstones). Three samples from the SF are plotted in the wacke field. Samples plotted in shale, sublitharenite, and subarkose fields are combined and named as wacke ($n = 3$ in SF; $n = 10$ in LF), litharenite ($n = 10$), and arkose ($n = 4$) types.

TABLE 7.2 Statistical Summary (Range, Mean and Standard Deviation) of Geochemical Data of the Siliciclastic Sediments of the Lambir and Sibuti Formations

Rock Name	Wackes (SF; n = 3)				Litharenites (LF; n = 10)				Arkoses (LF; n = 4)			
Elements	Min	Max	Avg	St. Dev	Min	Max	Avg	St. Dev	Min	Max	Avg	St. Dev
SiO_2	41.60	61.90	51.77	10.15	73.65	91.98	80.03	7.16	87.53	94.13	89.49	3.14
$SiO_{2(adj)}$	53.04	68.38	60.76	7.67	79.68	94.62	84.48	5.55	90.02	95.45	92.34	2.34
Al_2O_3	6.77	14.32	10.97	3.85	3.07	11.81	8.73	3.15	2.75	6.45	5.04	1.63
Fe_2O_3(T)	2.97	4.90	3.89	0.97	1.00	3.66	2.49	0.94	0.43	1.08	0.64	0.31
MnO	0.05	0.06	0.05	0.01	0.01	0.03	0.01	0.01	0.00	0.01	0.01	0.00
MgO	1.14	1.94	1.61	0.42	0.22	1.22	0.75	0.35	0.15	0.40	0.25	0.11
CaO	3.75	23.57	13.23	9.94	0.02	0.78	0.13	0.23	0.04	0.08	0.05	0.02
Na_2O	0.60	0.81	0.71	0.11	0.04	0.76	0.35	0.26	0.08	0.10	0.09	0.01
K_2O	1.25	2.26	1.84	0.53	0.55	2.09	1.51	0.52	0.60	1.10	0.88	0.25
TiO_2	0.37	0.69	0.54	0.16	0.30	0.73	0.57	0.16	0.25	0.62	0.44	0.16
P_2O_5	0.07	0.10	0.09	0.02	0.02	0.05	0.03	0.01	0.02	0.02	0.02	0.00
LOI	9.43	21.01	15.01	5.80	1.51	8.64	5.20	2.32	1.33	4.00	2.81	1.12
Total	99.44	99.95	99.70	0.26	98.67	100.80	99.82	0.69	98.68	100.50	99.70	0.76
K_2O/Al_2O_3	0.16	0.18	0.17	0.01	0.16	0.19	0.17	0.01	0.14	0.22	0.18	0.03
Al_2O_3/TiO_2	18.35	21.67	20.26	1.71	10.10	18.07	14.87	2.43	9.55	13.86	11.75	1.87
ICV	0.99	4.42	2.40	1.79	0.54	0.86	0.68	0.12	0.37	0.63	0.48	0.11
CIA	67.05	73.70	71.05	3.52	69.51	82.99	78.71	4.48	76.46	82.78	80.82	2.98
Sc	7.00	13.00	10.00	3.00	3.00	10.00	7.60	2.80	2.00	5.00	4.00	1.41
Be	1.00	2.00	1.50	0.71	1.00	1.00	1.00	0.00	—	—	—	—
V	50.00	101.00	77.67	25.77	27.00	97.00	70.50	25.24	22.00	50.00	38.25	13.62
Ba	145.00	243.00	197.00	49.27	86.00	256.00	175.60	52.03	92.00	145.00	119.50	27.38
Sr	363.00	576.00	494.33	114.86	18.00	82.00	39.60	17.75	16.00	28.00	24.25	5.68
Y	17.00	24.00	19.33	4.04	12.00	25.00	18.70	4.45	10.00	27.00	16.75	7.63
Zr	181.00	291.00	223.67	59.00	202.00	430.00	312.20	77.67	252.00	1019.00	515.00	348.90
Cr	40.00	60.00	53.33	11.55	30.00	60.00	47.00	13.37	40.00	80.00	60.00	28.28
Co	5.00	12.00	9.00	3.61	2.00	12.00	6.10	3.51	2.00	4.00	3.00	1.00
Ni	20.00	40.00	30.00	10.00	20.00	20.00	20.00	0.00	—	—	—	—
Cu	40.00	50.00	46.67	5.77	10.00	380.00	129.00	146.85	10.00	90.00	47.50	35.00
Zn	50.00	80.00	66.67	15.28	40.00	220.00	85.00	58.80	40.00	50.00	43.33	5.77

TABLE 7.2 Statistical Summary (Range, Mean and Standard Deviation) of Geochemical Data of the Siliciclastic Sediments of the Lambir and Sibuti Formations—cont'd

Rock Name	Wackes (SF; n = 3)				Litharenites (LF; n = 10)				Arkoses (LF; n = 4)			
Elements	**Min**	**Max**	**Avg**	**St. Dev**	**Min**	**Max**	**Avg**	**St. Dev**	**Min**	**Max**	**Avg**	**St. Dev**
Ga	8.00	19.00	13.67	5.51	4.00	16.00	11.10	4.09	4.00	11.00	7.50	2.89
Ge	1.00	1.00	1.00	0.00	2.00	2.00	2.00	0.00	2.00	2.00	2.00	0.00
As	8.00	8.00	8.00	—	6.00	23.00	13.33	5.92	7.00	12.00	9.50	3.54
Rb	49.00	103.00	80.33	28.02	24.00	91.00	64.50	23.95	21.00	44.00	35.25	10.72
Nb	8.00	12.00	10.33	2.08	5.00	10.00	7.80	1.55	5.00	26.00	12.25	9.50
Ag	2.20	3.10	2.67	0.45	1.90	5.00	3.85	1.13	3.10	11.10	5.80	3.77
Sn	4.00	4.00	4.00	0.00	3.00	18.00	8.50	5.13	2.00	7.00	4.00	2.16
Cs	3.20	8.30	6.10	2.62	1.30	6.30	4.04	1.82	1.00	2.60	2.00	0.71
Hf	4.00	6.40	5.03	1.23	4.40	9.30	7.03	1.73	5.70	22.50	11.18	7.88
Ta	0.50	0.80	0.67	0.15	0.50	0.90	0.75	0.16	0.30	0.90	0.60	0.26
W	2.00	2.00	2.00	0.00	1.00	2.00	1.75	0.46	1.00	1.00	1.00	—
Tl	0.40	0.70	0.57	0.15	0.10	0.50	0.38	0.13	0.30	0.70	0.43	0.23
Pb	13.00	24.00	18.67	5.51	11.00	23.00	17.90	4.41	9.00	28.00	16.00	8.29
Th	6.40	11.70	9.23	2.67	5.10	11.90	8.95	2.42	4.90	10.90	7.48	2.59
U	2.20	3.00	2.60	0.40	2.00	3.70	2.69	0.59	1.40	4.60	2.85	1.32
La	22.00	32.40	26.20	5.48	15.70	31.60	23.89	6.15	12.80	27.50	19.75	6.19
Ce	44.00	62.90	51.23	10.20	29.80	60.80	46.22	11.92	23.80	52.10	37.70	11.96
Pr	4.97	7.40	5.86	1.34	3.36	6.79	5.17	1.32	2.73	5.82	4.26	1.31
Nd	18.90	27.70	21.93	5.00	12.70	25.00	19.04	4.79	9.50	21.20	15.60	4.91
Sm	3.80	5.70	4.57	1.00	2.30	4.80	3.65	0.92	1.80	4.00	2.88	1.00
Eu	0.79	1.15	0.95	0.18	0.42	0.92	0.67	0.17	0.32	0.65	0.51	0.16
Gd	3.00	4.90	3.70	1.04	2.00	4.50	3.10	0.80	1.60	3.30	2.38	0.90
Tb	0.50	0.80	0.60	0.17	0.30	0.80	0.55	0.15	0.30	0.60	0.43	0.15
Dy	2.80	4.70	3.50	1.04	2.00	4.60	3.31	0.83	1.70	3.90	2.68	1.01
Ho	0.60	0.90	0.70	0.17	0.40	1.00	0.72	0.19	0.40	0.90	0.60	0.24
Er	1.70	2.70	2.07	0.55	1.20	2.80	2.13	0.54	1.20	2.70	1.78	0.72
Tm	0.25	0.39	0.30	0.08	0.21	0.44	0.34	0.08	0.19	0.47	0.30	0.13
Yb	1.80	2.50	2.03	0.40	1.40	2.90	2.33	0.55	1.30	3.40	2.13	0.95
Lu	0.28	0.39	0.32	0.06	0.24	0.46	0.37	0.08	0.21	0.61	0.36	0.18

(Continued)

TABLE 7.2 Statistical Summary (Range, Mean and Standard Deviation) of Geochemical Data of the Siliciclastic Sediments of the Lambir and Sibuti Formations—cont'd

Rock Name	Wackes (SF; n = 3)				Litharenites (LF; n = 10)				Arkoses (LF; n = 4)			
Elements	**Min**	**Max**	**Avg**	**St. Dev**	**Min**	**Max**	**Avg**	**St. Dev**	**Min**	**Max**	**Avg**	**St. Dev**
$\sum$REE	106.61	154.53	123.96	26.55	72.14	147.39	111.49	28.34	57.85	127.15	91.33	29.56
Cr/V	0.59	0.80	0.71	0.10	0.52	1.11	0.71	0.18	0.82	1.60	1.21	0.55
Eu/Eu*	0.66	0.76	0.71	0.05	0.56	0.71	0.61	0.04	0.55	0.72	0.61	0.07
$(La/Yb)_{CN}$	8.26	9.09	8.70	0.42	5.73	7.58	6.93	0.65	5.47	7.93	6.56	1.03
$(La/Sm)_{CN}$	3.30	4.01	3.63	0.36	3.86	4.31	4.12	0.13	3.91	4.82	4.38	0.38
$(Gd/Yb)_{CN}$	1.35	1.59	1.46	0.12	0.85	1.26	1.08	0.12	0.79	1.06	0.93	0.12

Rock Name	Wackes (LF; n = 10)					Fe-Sand (LF; n = 2)			
Elements	**Min**	**Max**	**Avg**	**St. Dev**	**Quartz arenite (n = 1)**	**Min**	**Max**	**Avg**	**St. Dev**
SiO_2	26.72	73.77	62.60	14.74	95.03	55.58	73.44	64.51	12.63
$SiO_{2(adj)}$	37.69	78.78	69.83	13.17	97.00	67.58	77.72	72.65	7.17
Al_2O_3	7.04	18.16	13.43	3.01	1.88	3.80	7.89	5.85	2.89
$Fe_2O_3(T)$	1.61	5.81	4.07	1.41	0.22	3.81	7.01	5.41	2.26
MnO	0.01	0.14	0.04	0.04	0.00	0.06	0.12	0.09	0.04
MgO	0.51	3.04	1.42	0.74	0.10	1.36	5.52	3.44	2.94
CaO	0.04	28.43	3.73	8.95	0.04	1.59	12.26	6.93	7.54
Na_2O	0.12	0.60	0.27	0.16	0.13	0.16	0.73	0.45	0.40
K_2O	1.19	3.07	2.23	0.50	0.27	0.72	1.32	1.02	0.42
TiO_2	0.27	0.80	0.69	0.15	0.30	0.23	0.50	0.36	0.19
P_2O_5	0.03	0.30	0.08	0.08	<0.01	0.05	0.61	0.33	0.40
LOI	5.80	28.53	11.30	6.64	1.34	5.51	17.27	11.39	8.32
Total	98.91	100.70	99.86	0.64	99.31	99.51	100.00	99.76	0.35
K_2O/Al_2O_3	0.15	0.18	0.17	0.01	0.14	0.17	0.19	0.18	0.02
Al_2O_3/TiO_2	16.73	25.87	19.96	3.30	6.25	15.88	16.52	16.20	0.46
ICV	0.41	5.24	1.13	1.45	0.56	1.59	5.97	3.78	3.10
CIA	75.36	85.25	80.68	3.16	77.19	66.02	74.43	70.22	5.94
Sc	6.00	17.00	12.10	3.03	2.00	4.00	10.00	7.00	4.24
Be	1.00	2.00	1.89	0.33	<1	1.00	1.00	1.00	—
V	61.00	160.00	107.60	26.03	18.00	36.00	78.00	57.00	29.70

TABLE 7.2 Statistical Summary (Range, Mean and Standard Deviation) of Geochemical Data of the Siliciclastic Sediments of the Lambir and Sibuti Formations—cont'd

Rock Name	Wackes (LF; n = 10)					Fe-Sand (LF; n = 2)			
Elements	**Min**	**Max**	**Avg**	**St. Dev**	**Quartz arenite (n = 1)**	**Min**	**Max**	**Avg**	**St. Dev**
Ba	112.00	275.00	221.20	44.58	46.00	86.00	157.00	121.50	50.20
Sr	49.00	786.00	175.80	250.92	14.00	145.00	151.00	148.00	4.24
Y	12.00	30.00	23.10	4.77	9.00	10.00	26.00	18.00	11.31
Zr	112.00	268.00	192.50	46.68	321.00	165.00	239.00	202.00	52.33
Cr	40.00	80.00	64.00	11.74	<20	30.00	40.00	35.00	7.07
Co	3.00	29.00	12.10	7.05	2.00	4.00	7.00	5.50	2.12
Ni	20.00	40.00	28.89	9.28	<20	—	—	—	—
Cu	10.00	100.00	51.00	30.35	100.00	30.00	50.00	40.00	14.14
Zn	40.00	130.00	82.22	27.28	60.00	50.00	50.00	50.00	0.00
Ga	8.00	22.00	16.80	4.05	3.00	5.00	10.00	7.50	3.54
Ge	1.00	3.00	1.67	0.71	2.00	1.00	2.00	1.50	0.71
As	7.00	35.00	15.90	9.45	<5	11.00	12.00	11.50	0.71
Rb	53.00	143.00	99.70	24.17	11.00	29.00	56.00	42.50	19.09
Nb	6.00	26.00	11.67	5.72	5.00	4.00	8.00	6.00	2.83
Ag	1.00	3.40	2.03	0.71	3.70	1.70	2.10	1.90	0.28
Sn	3.00	10.00	5.40	2.41	6.00	3.00	4.00	3.50	0.71
Cs	3.70	10.30	7.11	2.02	0.60	1.70	3.10	2.40	0.99
Hf	2.70	6.00	4.28	1.03	7.00	3.90	5.20	4.55	0.92
Ta	0.30	1.00	0.86	0.21	0.40	0.30	0.60	0.45	0.21
W	1.00	7.00	2.78	1.86	<1	—	—	—	—
Tl	0.20	0.90	0.45	0.20	0.20	—	—	—	—
Pb	14.00	43.00	20.50	8.64	9.00	13.00	21.00	17.00	5.66
Th	6.60	14.10	11.21	2.06	4.20	4.80	7.60	6.20	1.98
U	2.50	4.30	3.12	0.57	1.50	1.70	3.30	2.50	1.13
La	16.30	39.40	30.51	6.20	11.40	12.70	22.60	17.65	7.00
Ce	31.50	78.40	59.97	12.44	21.00	25.80	48.10	36.95	15.77
Pr	3.57	8.89	6.77	1.41	2.34	2.93	5.55	4.24	1.85
Nd	14.50	33.80	25.51	5.11	8.40	11.10	21.90	16.50	7.64
Sm	3.00	7.00	5.15	1.16	1.50	2.40	5.20	3.80	1.98

(Continued)

TABLE 7.2 Statistical Summary (Range, Mean and Standard Deviation) of Geochemical Data of the Siliciclastic Sediments of the Lambir and Sibuti Formations—cont'd

Rock Name	Wackes (LF; n = 10)					Fe-Sand (LF; n = 2)			
Elements	**Min**	**Max**	**Avg**	**St. Dev**	**Quartz arenite (n = 1)**	**Min**	**Max**	**Avg**	**St. Dev**
Eu	0.68	1.45	1.04	0.24	0.28	0.45	1.23	0.84	0.55
Gd	2.50	5.60	4.28	1.08	1.40	1.90	5.20	3.55	2.33
Tb	0.40	0.90	0.69	0.18	0.20	0.30	0.80	0.55	0.35
Dy	2.20	5.40	4.14	0.93	1.30	1.60	4.50	3.05	2.05
Ho	0.40	1.00	0.81	0.17	0.30	0.30	0.90	0.60	0.42
Er	1.10	3.00	2.39	0.52	1.00	1.00	2.40	1.70	0.99
Tm	0.17	0.46	0.38	0.08	0.17	0.14	0.33	0.24	0.13
Yb	1.20	3.00	2.51	0.52	1.20	1.00	2.20	1.60	0.85
Lu	0.21	0.47	0.39	0.07	0.19	0.18	0.35	0.27	0.12
$\sum$REE	77.73	187.02	144.53	29.23	50.68	61.80	121.26	91.53	42.04
Cr/V	0.47	0.67	0.60	0.07	—	0.51	0.83	0.67	0.23
Eu/Eu*	0.59	0.76	0.68	0.04	0.59	0.64	0.72	0.68	0.06
$(La/Yb)_{CN}$	7.22	10.24	8.30	1.00	6.42	6.94	8.58	7.76	1.16
$(La/Sm)_{CN}$	3.20	5.02	3.77	0.53	4.78	2.74	3.33	3.03	0.42
$(Gd/Yb)_{CN}$	1.01	1.75	1.39	0.23	0.95	1.54	1.92	1.73	0.27

SF, Sibuti Formation; *LF*, Lambir Formation; — Not determined due to below detection limit.

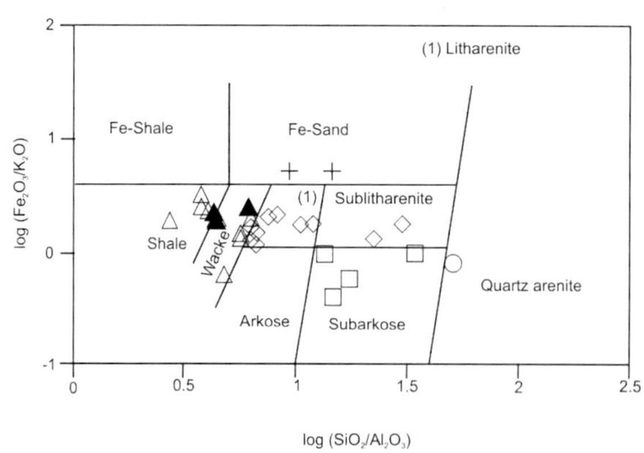

FIGURE 7.2B Geochemical classification of sandstones using log (SiO_2/Al_2O_3) – log (Fe_2O_3/K_2O) (Herron, 1988).

4.3 Elemental Variations

4.3.1 Major Oxides

SiO_2 content of the samples is higher than other major oxides, ranging between 26.72 and 95.03 wt% with adjusted values (LOI free basis) ranging from 37.69 to 97.00 wt%. Quartz arenite (LF) has the highest SiO_2 content (95.03 wt%; n = 1) followed by arkose (LF) (avg. 89.49 ± 3.14 wt%; n = 4), litharenite (LF) (80.03 ± 7.06 wt%; n = 10), and Fe-sand (LF) (64.51 ± 12.63 wt%; n = 2). The lowest silica content is in wackes from both LF (62.60 ± 14.74 wt%; n = 10) and SF (51.77 ± 10.15 wt%; n = 3). The SiO_2 has a negative correlation with major oxides such as SiO_2 versus Al_2O_3, Fe_2O_3, MnO, MgO, CaO, and K_2O (r = −0.46, −0.64, −0.80, −0.68, −0.81, and −0.46, respectively; n = 30).

Al_2O_3 is recorded as the second most common major oxide in all the samples. Al_2O_3 is high in wacke types of LF (13.43 ± 3.01 wt%) and SF (10.97 ± 3.85 wt%). The sandstones have less Al_2O_3 in litheranites (LF) (8.73 ± 3.15 wt%), arkose (5.04 ± 1.63 wt%), Fe-sand (5.85 ± 2.89 wt%), and quartz arenite (1.88 wt%). The higher content of Al_2O_3 in clastic sediments is associated with clay minerals, whereas SiO_2 is associated with quartz content. The correlation between Al_2O_3 and K_2O is statistically significant (r = 0.99; n = 30), indicating their association with clay minerals. High CaO content recorded in wacke (SF) (13.23 ± 9.94 wt%), wacke (LF) (3.73 ± 8.95 wt%), and Fe-sand (LF) (6.93 ± 7.54 wt%) was due to fossil content and calcite cement. These rocks are considered to be chemically immature.

Major oxides of sandstones were normalized against UCC (McLennan, 2001) (Fig. 7.3). The SiO_2 shows enrichment in all samples except for the wackes from SF and LF, and Fe-sands from LF. Quartz arenite, arkoses, and litharenites of LF have highly depleted MgO, Fe_2O_3, CaO, and Na_2O, and moderately depleted Al_2O_3 and K_2O compared to wackes of SF. Wackes

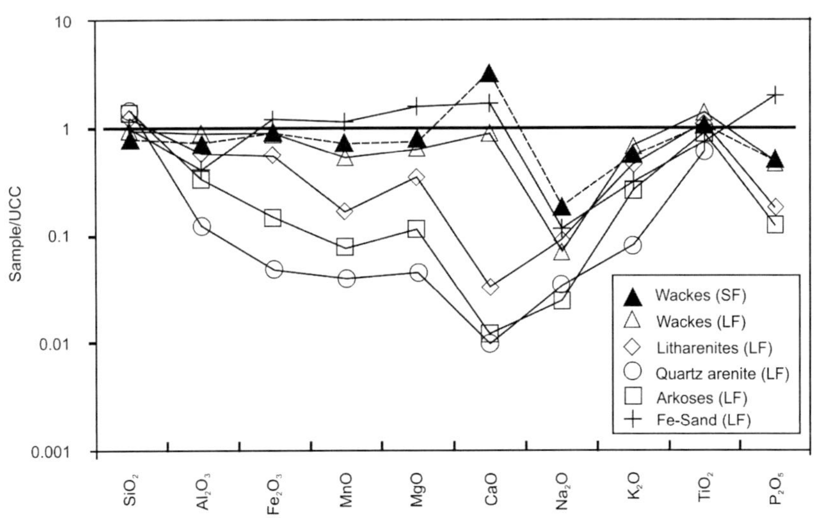

FIGURE 7.3 Multielement (major element) normalized diagram for the Lambir and Sibuti sandstones, normalized against average upper continental crust values (UCC; McLennan, 2001).

from both formations (LF and SF) are relatively similar to UCC, except in CaO and Na_2O contents. The wackes and Fe-sand have a high CaO content compared to sandstones due to the concentration of fossils and calcareous cement. On the other hand, the Fe-sand in LF has high P_2O_5 content compared to other sandstones. The high SiO_2 content in quartz arenite (LF) compared to wackes (SF) indicates higher maturity. All the samples of LF and SF show relatively similar values of TiO_2 and are comparable with UCC, except the quartz arenite (LF). However, TiO_2 enrichment in wackes (SF and LF) and litharenites (LF) may be related to the concentration of rutile in the heavy mineral phase or enrichment of phyllosilicates in residual phase.

4.3.2 *Trace Elements*

Rb is abundant in wackes of SF and LF (80 ± 28 ppm and 100 ± 24 ppm) and indicates the abundance of fine-grained clay size sediments. The Rb content recorded is lower than UCC in sandstones except for wackes, which is comparable with UCC. Sr has a strong positive correlation with CaO ($r = 0.91$) in wacke (SF and LF) and litharenites (SF) due to the high content of carbonate minerals. The possible reason for the low Sr content in arenite (14 ppm) is probably due to intensive weathering in the source area.

The Ba content is depleted in quartz arenite (LF) (46 ppm) and enriched in wackes (LF) (221 ± 45 ppm). The difference in Ba content between quartz arenite and wacke is probably due to the variation in clay minerals as wacke should have higher clay content than arenite. Transition trace elements (Sc, Cs, Rb) are compatible throughout magmatic fractionation and they show positive correlation with Al_2O_3 ($r = 0.98$, $r = 0.13$, $r = 0.99$, respectively; $n = 30$) indicating that these elements are controlled by phyllosilicates (Ali et al., 2014). The high field strength elements (Y, Zr, Nb, Th, Hf) have small ionic radii but with higher charge and hence are considered to be a good indicator of source rock characteristics and sorting effects during depositional processes (Armstrong-Altrin et al., 2013).

Arkose (LF) type is high in Zr (515 ± 349 ppm; $n = 4$) content in comparison to wackes, litharenites, Fe-sand, and quartz arenite. The second highest Zr content is noted in quartz arenite (LF) (321 ppm). The enrichment of Zr content in arkose and arenite types is due to the concentration of zircon and indicates that the region is dominated by a felsic source. Fig. 7.4 indicates the likely presence of a fair amount of zircon, which infers a sediment sorting effect during deposition. As expected, Zr and Hf show positive correlation ($r = 0.99$; $n = 30$), indicating the presence of heavy mineral zircon, especially in arkose and arenite types. This is further supported by petrography.

The Nb content varies from 26 to 5 ppm and its concentration is high in arkoses (12 ± 10 ppm) and low in quartz arenite (5 ppm). Thorium content in wackes from both LF (11 ± 2 ppm) and SF (9 ± 3 ppm) is enriched, whereas Th content is higher in clay and silt sediments compared to sand. Correlation between Al_2O_3 with Zr and Hf ($r = -0.38$, $r = -0.38$, respectively; $n = 30$) implies that these elements are not controlled by clay minerals (Armstrong-Altrin et al., 2013). Arsenic and Ag contents in the studied samples are enriched 10- to 100-fold, respectively, compared to UCC, which is related to a high content of the sulphur-bearing mineral pyrite, common in NW Borneo basins as macro and micro concretions.

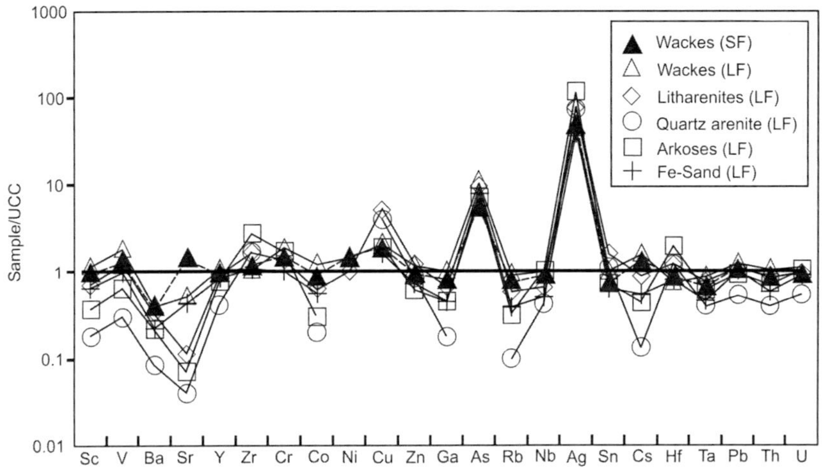

FIGURE 7.4 Multielement (trace element) normalized diagram for the Lambir and Sibuti sandstones, normalized against average upper continental crust values (UCC; McLennan, 2001).

4.3.3 *Rare Earth Elements*

REEs (La-Lu) are the least soluble trace elements and are relatively immobile during weathering, low-grade metamorphism, and hydrothermal alteration (Rollinson, 1993, p.137). The chondrite normalized REE pattern for the Lambir and the Sibuti sediments shows an enrichment of light REE (LREE) and flat heavy REE (HREE) with a negative Eu anomaly (Fig. 7.5). The enrichment of HREE in a few samples of the Lambir sediments compared to UCC and PAAS indicates a heavy mineral phase (e.g., zircon) controlling HREEs. ΣREE levels are higher in wackes (avg. 145 ppm in LF and 124 ppm in SF) than in litharenites (112 ppm in LF), arkoses (91 ppm in LF), Fe-sand (92 ppm in LF), and quartz arenite (51 ppm in LF). The LF shows a similar REE pattern compared with UCC and PAAS. However, LF differs in Eu anomaly value when compared with SF, probably caused by fractionation involving substantial plagioclase.

5. DISCUSSION

5.1 Statistical Analysis

SiO_2 shows a negative correlation with Fe_2O_3, MnO, MgO, Sr, Ge, Cs, and Eu, which indicates the association of SiO_2 with quartz (Osman, 1996). Al_2O_3 concentration in sediments is the detrital indicator and their relationship with other elements suggests their association either with clay or phyllosilicates. In the present study, Al_2O_3 exhibits significant correlation with K_2O, TiO_2, Sc, V, Ba, Y, Ga, Rb, Cs, Th, and REEs, which indicates their association with clay minerals (particularly illite), phyllosilicates, and accessory phases (rutile and monazite) (Condie et al., 2001). CaO shows a significant relationship with Sr, LOI, and Ge indicating that these elements are related to carbonate minerals.

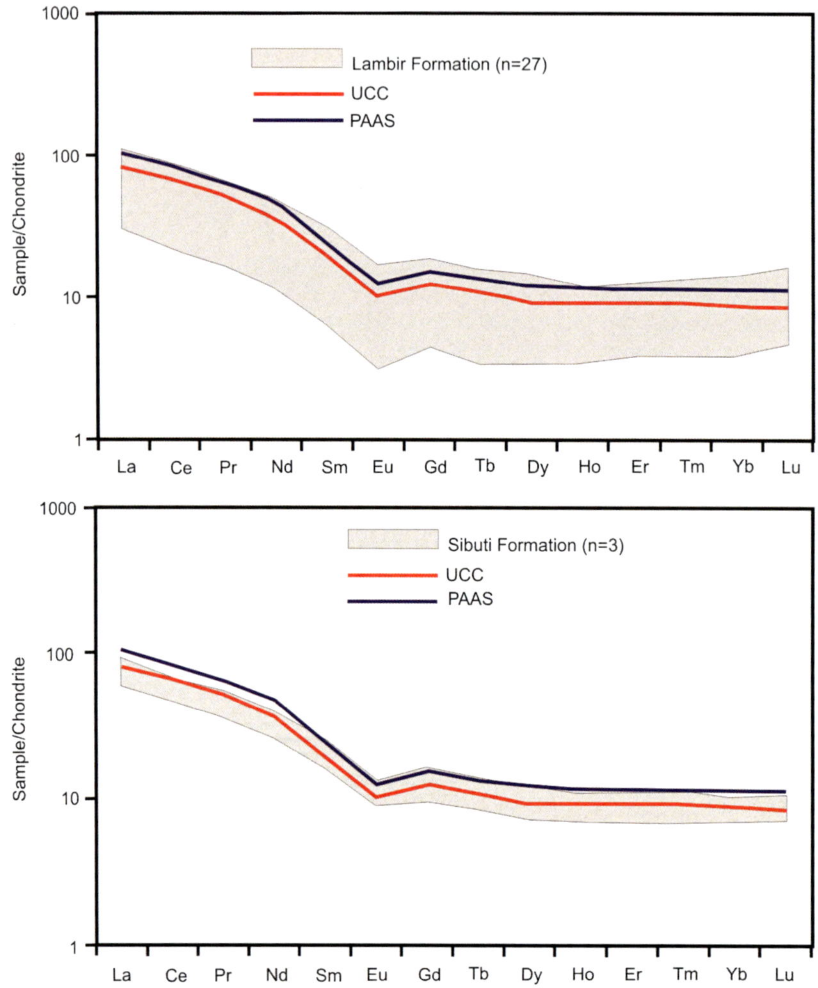

FIGURE 7.5 Chondrite normalized REE pattern for the siliciclastic sediments of Lambir Formation and Sibuti Formation (chondrite normalized values are from Taylor and McLennan, 1985). Average upper continental crust values (UCC) and Post-Archaean Australian Shale (PAAS) (McLennan, 2001) are also included for comparison.

The Varimax Rotation method (Gotelli and Ellison, 2004) was used to maximize the sum of variance of factor coefficients. All the variables for Lambir and Sibuti clastic sediments are distributed in four factors. Factor 1 accounted for 49.4% of the total variance characterized by high positive loadings of Al_2O_3, Fe_2O_3, K_2O, TiO_2, Sc, V, Ba, Y, Ga, Rb, Cs, Th, and REEs; and moderate negative loadings of P_2O_5, Be, and W. This factor can be considered as clay and phyllosilicate, which primarily control the trace and rare earth element concentrations in the siliciclastic sediments with minor input from heavy minerals.

Factor 2 accounts for 16.2% of the total variance and is supported by moderate to high positive loadings of MnO, MgO, CaO, LOI, Sr, and Ge, and a negative loading of SiO_2 and Sn.

This factor is controlled by carbonate minerals, which can be defined by a strong positive relationship of these elements with CaO and Sr. This prediction has been confirmed petrographically. Calcareous cements were observed in mudstones and some samples contain microfossils and fossil fragments.

Factor 3 accounted for 7.40% of the total variation and is controlled by Zr and Ag. Zr and Ag are negatively loaded and show significant correlation ($r = 0.97$; $n = 30$) indicating that Ag was derived from the same source as Zr.

Factor 4 accounts for 4.54% of the total variance and is characterized by a strong positive loading of Cr and moderate negative loadings of Cu, Zn, and Sn. Cr is controlled by heavy mineral chromite, which is common in the recycled sediments of NW Borneo (Nagarajan et al., 2014), but the concentration may be much less in LF since these sediments are felsic in nature. Chalcophile elements Cu and Zn are associated with Sn, which may be derived from pyrite and cassiterite. The latter is common in NW Borneo sediments derived from Sn-bearing granites (e.g., Van Hattum et al., 2013). Sn shows bimodal distribution in Factors 2 and 4, and indicates their association with pyrite and cassiterite.

5.2 Paleoweathering

Stronger chemical weathering is associated with warm and humid climates, whereas arid climates are associated with relatively weak chemical weathering (Nesbitt and Young, 1982). The intensity of chemical weathering of sedimentary rocks can be determined by plotting on an $Al_2O_3 - CaO^* + Na_2O - K_2O$ (A-CN-K) ternary diagram and Chemical Index of Alteration (CIA: Nesbitt and Young, 1984). The CIA value increases from 50 in unweathered igneous rocks, to 100 during weathering to residual clays. On average, shale CIA values range between 70 and 75, which reflect the composition of illite, smectite, and muscovite, and represent a source that is moderately weathered. On the other hand, higher CIA values up to 100 indicate an intense weathering, which eventually produces clayey residues with high content of kaolinite and Al-Oxy-hydroxides (Mongelli et al., 2006).

In order to reconstruct the paleoweathering history of the studied sediments CIA and Plagioclase Index of Alteration (PIA) values, A-CN-K and A-CNK-FM ternary plots are used. The calculated CIA values for the Lambir and Sibuti sediments were higher (77, 79, 81, 81, 70, and 71) for quartz arenite, litharenites, arkoses, wackes, Fe-sand, and wackes (SF), respectively. Higher CIA values (>75) in LF, except the Fe-sand, indicates intensive chemical weathering, whereas <75 in SF and Fe-sands of LF indicates moderate weathering. Low CIA values (<70) are recorded in Ca-rich wackes (SF) and Fe-sand (LF). The PIA values of the SF and LF range between 71 and 98. The highest value is represented by litharenite while Fe-sand records the lowest value. The high PIA values infer an intense weathering in the source region.

By plotting A-CN-K ternary diagram, the weathering history of recent and ancient sediments can be estimated as the molecular ratio of Al_2O_3 (A), $CaO^* + Na_2O$ (CN), and K_2O (K) (Nesbitt and Young, 1984; Nesbitt, 2003). On the A-CN-K plot (Fig. 7.6A; Nesbitt and Young, 1982) all Lambir sediments plot away from the average shale composition and are clustered near illite, whereas Sibuti sediments and Fe-sands plot nearer to average shale properties, which indicates that Lambir sediments have more illite than Sibuti sediments. Similarly, the samples are clustered toward A-K line, which indicates that a considerable

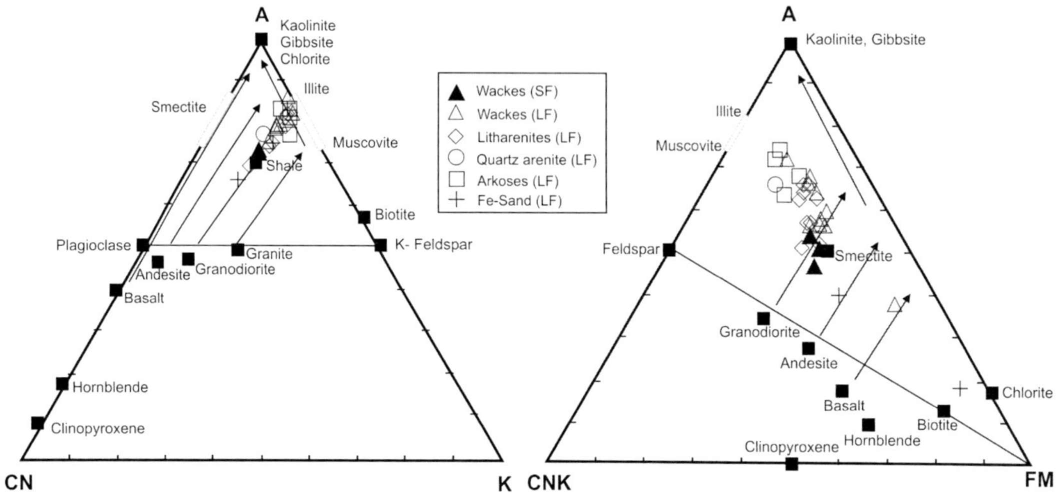

FIGURE 7.6 (A) A-CN-K and (B) A-CNK-FM (after Nesbitt and Young, 1984; Nesbitt and Wilson, 1992) plots showing the weathering trend for the siliciclastic sediments of the Lambir and Sibuti Formations.

amount of Na_2O and CaO has been removed from the bulk composition due to chemical weathering (Khan and Khan, 2015). The corresponding high CIA values signify an intense weathering. The samples are plotted on a weathering trend derived from UCC/granodiorite, which indicates that these sediments are derived from a felsic source rock. This interpretation is further supported by an A-CNK-FM plot (Fig. 7.6B), where Lambir sediments plot away from smectite toward illite, whereas Sibuti sediments plot nearer to smectite and follow the predicted weathering trend of granodiorite than mafic source rocks.

Fe-sands and a wacke of LF are plotted toward FM apex indicating the presence of ferromagnesian minerals. In particular, one of the Fe-sands and a wacke of LF are enriched in MgO (5.52 and 3.04 wt%, respectively) and CaO contents (12.26 and 28.43 wt%, respectively), which should be related to the presence of dolomite. Overall, the stronger chemical weathering indicates that the source area has experienced warm and humid climate conditions (i.e., Nesbitt and Young, 1982). Compared to the wacke (SF), the quartz arenite, litharenites, and arkoses of the LF exhibit higher CIA and PIA values, as well as low ICV and values. This reflects the cumulative effect of multiple cycles of sedimentary recycling and the prolonged chemical weathering history in the northern part of Borneo.

5.3 Sediment Sorting and Recycling

Sorting is the degree of mineral separation according to grain size and the degree of sorting increases as the mineral density increases, which enables an accurate measurement of the mineral grain size distribution. The distribution of Th, U, Zr, Hf, and Nb in clastic sediments is controlled by hydraulic sorting of minerals, which may affect the bulk composition of sedimentary rocks (Armstrong-Altrin, 2009). The relationship between the composition of the source rock and sedimentary processes were analyzed by a Th/Sc versus Zr/Sc plot (McLennan et al., 1993). Zr/Sc ratio can be used to investigate the enrichment of zircon since

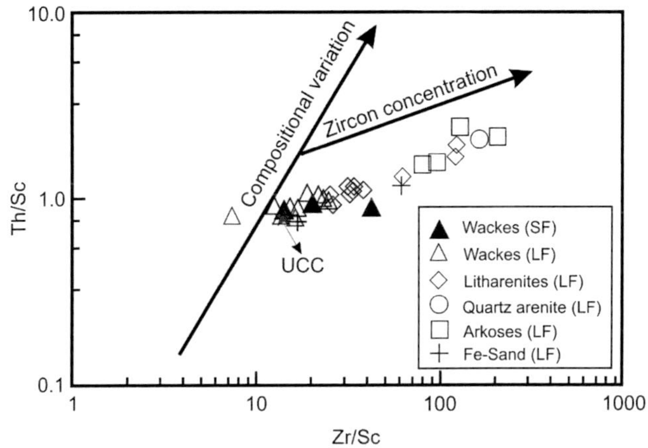

FIGURE 7.7 Th/Sc versus Zr/Sc diagram for the siliciclastic sediments of the Lambir and Sibuti Formations (after McLennan et al., 1993). The concentration of zircon due to sediment sorting and recycling can be seen along Trend 2.

Zr is strongly enriched in zircon, whereas Sc preserves the provenance signature like REEs. Zr and Hf are abundant in zircon and are predominantly associated with felsic igneous rocks, whereas mafic components have high concentrations of Sc. Hence, a Zr/Sc versus Th/Sc bivariate plot (McLennan, 1993) can be used to evaluate the Zr enrichment during sediment sorting and also to differentiate between felsic and mafic compositions. On this plot (Fig. 7.7), the samples deviate from the compositional trend and plot along the zircon concentration trend, indicating zircon concentration during sorting and recycling processes. In particular, arkoses, quartz arenites, and some litharenites of LF show high enrichment of zircon compared to other rock types. Some samples from the LF lie on the compositional variation trend, which was not subjected to sorting and weathering processes. The Al_2O_3-TiO_2-Zr ternary plot (Garcia et al., 1994) eliminates weathering effects and illustrates the effect of sorting-related fractionation based on the proportion of these elements. On the Al_2O_3-TiO_2-Zr ternary diagram (Fig. 7.8), the samples reveal variation in the Al_2O_3/Zr ratio due to the recycling effect. Thus, the recycling process significantly controlled the provenance signature of the clastic sediments of the LF and SF.

The degree of modification by physical and chemical weathering of clastic sediment is defined by the term maturity. Compositional maturity signifies the level of the chemical features in approaching the end product (Ingersoll et al., 2003). The degree of compositional maturity of the clastic sediments is calculated by using the Index of Compositional Variability ($ICV = (Fe_2O_3 + K_2O + Na_2O + CaO + MgO + TiO_2)/Al_2O_3$; Cox et al., 1995) and K_2O/Al_2O_3 ratio (Armstrong-Altrin et al., 2015). Generally nonclay detrital minerals have higher ICV values than clay minerals. Typical rock forming minerals (i.e., feldspars, amphiboles, and pyroxenes) show ICV values > 0.84, whereas alteration products such as kaolinite, illite, and muscovite show <0.84 (Cox et al., 1995). Thus, ICV values decrease with increasing intensity of weathering and/or maturity of the clastic sediments. The ICVs of the studied samples are generally <1 in LF (quartz arenite: 0.56; litharenites: 0.68; arkoses: 0.48), suggesting that the samples are compositionally mature and were likely subjected to recycling during transportation and deposition. The high ICV values (>1) recorded in wackes (5.24, 4.42 in

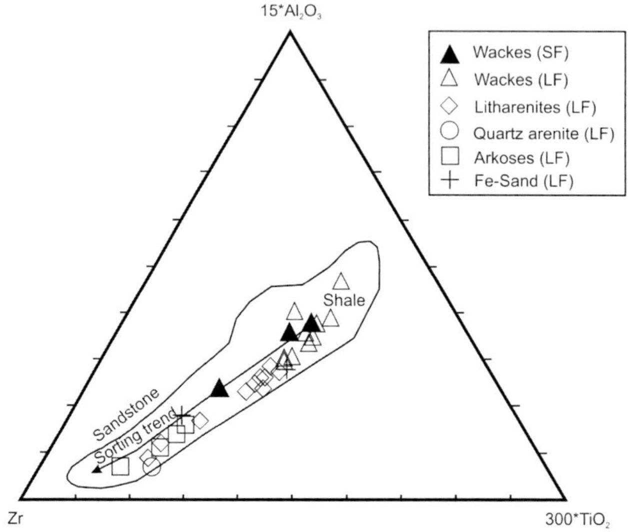

FIGURE 7.8 Al_2O_3-Zr-TiO_2 plot showing the sorting trend for the clastic sediments of the Lambir and Sibuti Formations.

LF and SF) and Fe-sand (1.59, 5.97 in LF) suggest chemical immaturity. In addition, the wackes (SF and LF) and Fe-sands (LF) are enriched in CaO content (Table 7.2). Particularly in the CIA versus ICV plot (Fig. 7.9), Fe-sand and wackes (LF and SF) fall above the ICV values of PAAS, which indicate their chemically immature nature. The amount of alkali feldspar, plagioclase, and clay in the clastic sediments can be deduced by using the ratio between K_2O and Al_2O_3. Clay minerals show a value approaching zero in the K_2O/Al_2O_3 ratio while alkali feldspars range from 0.4 to 1.0 (Cox et al., 1995). The average K_2O/Al_2O_3 ratio in the studied samples is <0.2 (Table 7.2) and is comparable with illite (~ 0.3). This observation is also supported by the mineralogy data and A-CN-K plot (Fig. 7.6A), where the samples are plotted between the average shale and illite compositions. A-CNK-FM plot also supports the abundance of illites that are enriched in smectite/chlorite (Fig. 7.6B), indicating moderate to high chemical maturity.

5.4 Provenance

The high content of quartz (75.79–98.81%) and textural features such as medium- to fine-grained, moderately sorted, and subrounded to rounded shapes, indicate a long transport distance with extensive reworking of the sandstones (arkose, quartz arenite, litharenite, and Fe-sand) and also reveals a cratonic or a recycled source for the Miocene LF (e.g., Zaid and Gahtani, 2015). It is well known that trace elements HFSE, REE, Th, and some transitional elements (Sc and Cr) are useful in constraining the average provenance composition of siliciclastic sediments (Taylor and McLennan, 1985; Cullers, 2000; Basu et al., 2016). The Al_2O_3/TiO_2 ratio is widely used to determine source rock composition (Hayashi et al., 1997; Nagarajan et al., 2015) since these compounds retain their source rock values. The Al_2O_3/TiO_2 ratio

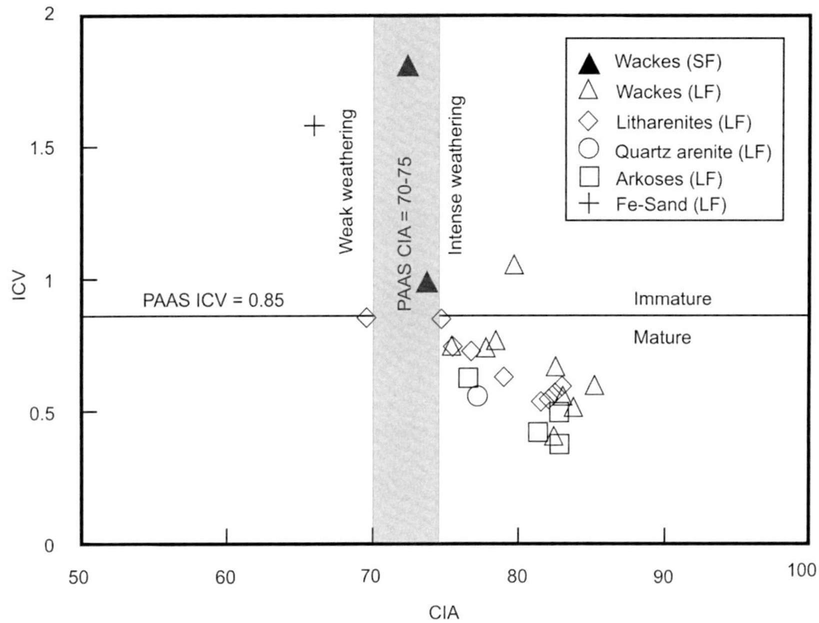

FIGURE 7.9 CIA versus ICV plot shows the intensity of weathering and maturity of the siliciclastic sediments of the Lambir and Sibuti Formations (after Long et al., 2012).

varies among igneous source rocks from 3–8 in mafic igneous rocks, 8–21 for intermediate igneous rocks, to 21–70 for felsic igneous rocks (Hayashi et al., 1997). The Al_2O_3/TiO_2 ratio ranges between 6.25 in quartz arenites (LF), 10.10–18.07 in litharenites (LF), 9.55–13.86 in arkoses (LF), 16.73–25.87 in wacke (LF), 15.88–16.52 in Fe-sand (LF), and 18.35–21.67 in wacke (SF), which are comparable with the Al_2O_3/TiO_2 ratio of intermediate to felsic igneous source rocks. One sample (A19) recorded a very low Al_2O_3/TiO_2 ratio value (6.25). This might be due to the high content of SiO_2 and TiO_2 and low content of Al_2O_3, which is also confirmed petrographically and geochemically; therefore A19 is classified as a quartz arenite.

The provenance discriminant function diagram of major elements proposed by Roser and Korsch (1988) is used to predict the provenance of terrigenous sediments of LF (Fig. 7.10), which comprised four provenance groups (P1: felsic igneous; P2: intermediate igneous; P3: mafic igneous; and P4: quartzose sedimentary). Wacke and litharenites from SF plot in the intermediate igneous provenance field, whereas sandstones from LF plot in the quartzose sedimentary provenance field, which suggests that Lambir sediments are recycled from existing orogenic Rajang/Crocker Formations, as well as indicating provenance with a high sediment maturity. Roser and Korsch (1988) stated that sediments derived from passive continental margins, intracratonic sedimentary basins, and recycled orogenic provinces will fall in the quartzose sedimentary provenance field. The uncertainty of provenance discrimination due to multiple cycles of sediment reworking cannot be excluded when only major element concentrations are used for a provenance study. Therefore, the La/Th versus Hf diagram is used to address the provenance of the study area based on immobile trace

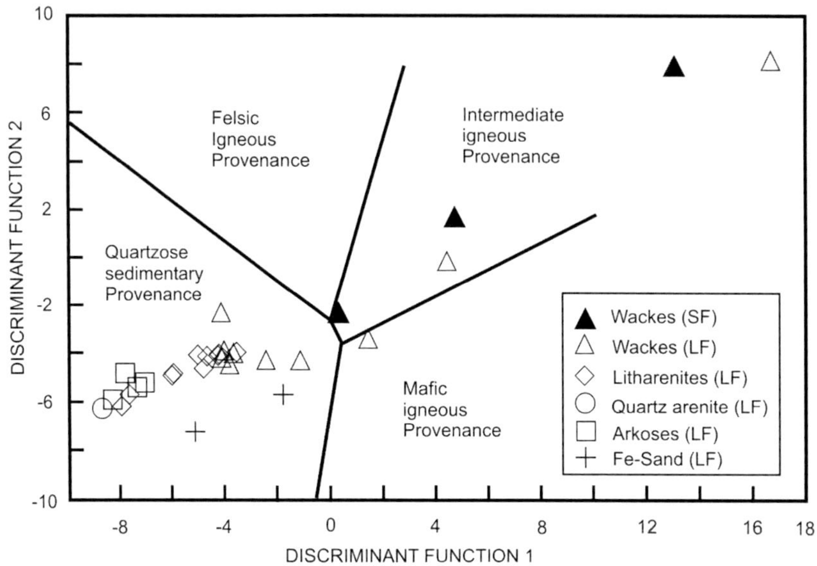

FIGURE 7.10 Discriminant function diagram for the provenance signatures of the Sibuti sandstones using major element (Roser and Korsch, 1988). The discriminant functions are: Discriminant Function 1 = $(-1.773.TiO_2) + (0.607.Al_2O_3) + (0.760.Fe_2O_3) + (-1.500.MgO) + (0.616.CaO) + (0.509.Na_2O) + (-1.224.K_2O) + (-9.090)$; Discriminant Function 2 = $(0.445.TiO_2) + (0.070.Al_2O_3) + (-0.250.Fe_2O_3) + (-1.142.MgO) + (0.438.CaO) + (1.475.Na_2O) + (-1.426.K_2O) + (-6.861)$.

elements. On the Hf versus La/Th diagram, samples plot within the felsic source region and extend toward a passive margin source, which further reaffirms that the sediments were recycled from the preexisting sedimentary and/or metasedimentary source area (Fig. 7.11). The recycled nature of LF sediments is also supported by the enrichment of Zr and Hf contents, and further by abundance of sedimentary to metasedimentary lithic fragments.

Abundances of REE and Th are higher in felsic rocks, while mafic rocks are enriched with Co, Sc, and Cr based on their incompatible and compatible behavior, respectively (McLennan, 1989). Cr/V ratio is a good index to trace Cr enrichment over other ferromagnesian trace elements input from mafic and ultramafic rocks (McLennan et al., 1993). The lower Cr/V ratio in the studied samples (avg. Cr/V ratio $= < 1$) excludes the sediment input from ultramafic and mafic rocks, and suggests a possible sediment source input area dominated by felsic rocks. A Th/Sc versus Sc plot (Fig. 7.12) was constructed for the sandstones and compared with average granite, andesite, basalt (Condie, 1993), UCC, and PAAS (McLennan, 2001). Samples that plot between UCC and granite compositions suggest a felsic-dominated provenance with significant recycling. Recycled sedimentary rocks show Eu/Eu* between 0.60 and 0.70 and Th/Sc ratio >1.0 and are often associated with fractionation and enrichment of heavy minerals, notably zircon (McLennan et al., 1993). The average Eu/Eu* values are recorded as 0.59, 0.61, 0.61, 0.68, 0.68 for quartz arenite, litharenites, arkoses, wackes, and Fe-sand of LF and 0.71 in wackes of SF. The Eu/Eu* values and Th/Sc ratios of the studied samples are comparable with the average ratio values of recycled sedimentary rocks (McLennan, 1993).

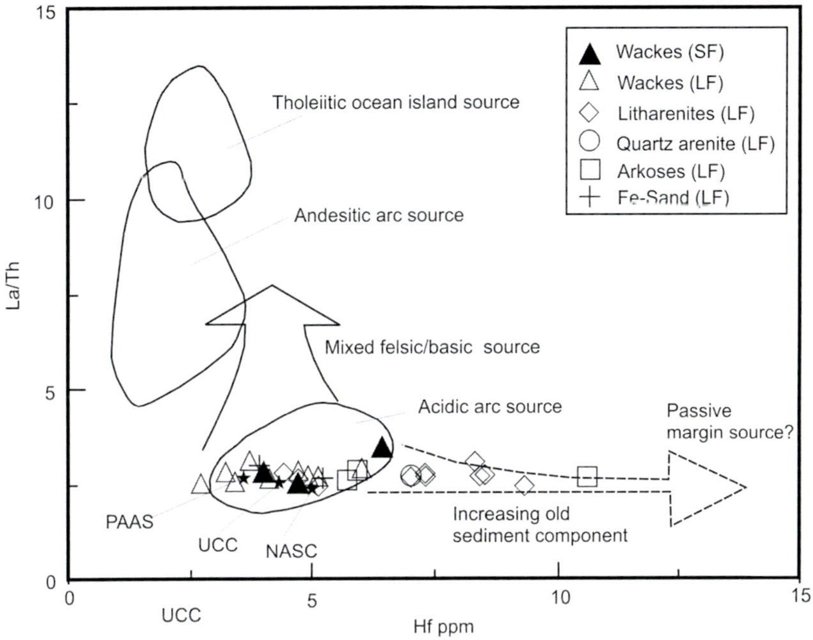

FIGURE 7.11 La/Th versus Hf bivariate diagram for the Lambir and Sibuti sandstones (after Floyd and Leveridge, 1987). *Open symbols*, Lambir Formation; *Filled symbols*, Sibuti Formation; PAAS, NASC, and UCC are from Taylor and McLennan (1985) and McLennan (2001).

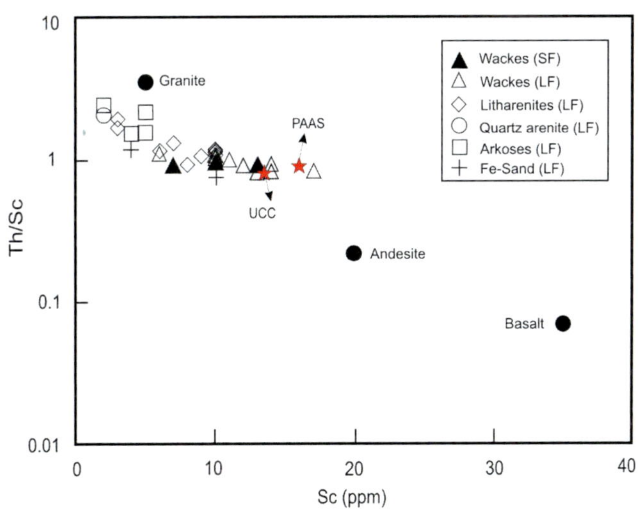

FIGURE 7.12 Th/Sc versus Sc bivariate plot for siliciclastic sediments of the Lambir and the Sibuti Formations. PAAS and UCC (Taylor and McLennan, 1985; McLennan, 2001), and granite, andesite, and basalt (Condie, 1993).

Felsic rock is characterized by high LREE/HREE ratios and negative Eu anomalies, while mafic rocks are low in LREE/HREE ratio and with little or no Eu anomaly (Cullers, 1994; Nagarajan et al., 2007a,b). On the chondrite normalized spider diagram (Fig. 7.5), an enrichment of LREE and flat HREE pattern with negative Eu anomaly is observed for the LF and SF samples. Furthermore, trace element ratios such as La/Sc, Th/Sc, Th/Cr, La/Co, Th/Co, Eu/Eu*, and $(La/Lu)_{cn}$ were calculated and compared with the average sediments derived from felsic and mafic rocks, UCC, and PAAS. These ratios show significant variation between mafic and felsic rocks and so provide useful information on the provenance of the sedimentary rocks (Cullers et al., 1988; Cullers, 1994, 2000; Cullers and Podkovyrov, 2000; Armstrong-Altrin et al., 2004). Using this approach, it was confirmed that the trace element ratios of this study fall within the range of felsic source rock (Table 7.3).

Setiawan et al. (2013) conducted petrography, geochemical characteristics, and LA-ICP-MS U-Pb zircon dating of the Late Triassic metatonalite from the Schwaner Mountains in West Kalimantan and commented on its contribution to sedimentary provenance in Sundaland. The authors found that the metatonalites consist of plagioclase, biotite, quartz, apatite, muscovite, and titanite with relict clinopyroxene surrounded by hornblende. The geochemical characteristics show that the rocks have calc-alkaline affinities and were derived from a subduction-related arc tectonic environment. The result of LA-ICP-MS U-Pb zircon dating reveals that the metatonalite has a magmatic age of 233 ± 3 Ma (Late Triassic), which is the oldest magmatic age in the Schwaner Mountains. This strongly suggests that the Schwaner Mountains has potential to be an important sediment source in Sundaland not only from the Cretaceous but also from the Triassic. The regional implication points to the Schwaner Mountains as not only the sediment provenance of NW Borneo basins, but the mountains might also have contributed to the sedimentary provenance of West Java, together with Tin Belt granites from the Malay Peninsula (Van Huttum et al., 2006).

5.5 Tectonic Setting

In sedimentary geochemistry, the tectonic discrimination diagrams proposed by Bhatia (1983), Bhatia and Crook (1986), and Roser and Korsch (1986) have been widely used to identify the tectonic setting of unknown basins (Xie and Chi, 2016; El-Enen et al., 2016). Although a few studies (Valloni and Maynard, 1981; Ryan and Williams, 2007) identified that the tectonic settings inferred from these diagrams are inconsistent with those inferred from the geology of ancient terranes. In addition, few authors (Armstrong-Altrin, 2015; Armstrong-Altrin and Verma, 2005; Verma and Armstrong-Altrin, 2013) evaluated the functioning of these major and trace element–based discrimination diagrams using Neogene sediments and showed a low success rate for the Bhatia (1983) and Roser and Korsch (1986) diagrams.

Recently, Verma and Armstrong-Altrin (2016) proposed two new discriminant function-based multidimensional diagrams for the discrimination of active and passive margin settings from isometric log-ratio transformations of major and major-trace element concentrations. These two diagrams were constructed based on worldwide examples of Neogene-Quaternary siliciclastic sediments from known tectonic settings. The active margin field includes the sediments from arc and collision settings, while the passive margin field includes sediments from the rift setting. Verma and Armstrong-Altrin (2016) also showed the excellent performance of these two diagrams through the testing of 11 case studies, of Quaternary to

TABLE 7.3 Range of Elemental Ratios of Sandstones of the Sibuti and Lambir Formations, Compared With Sediments From Felsic and Mafic Rocks, Upper Continental Crust, and Post-Archaean Australian Shale

	Sibuti Formation[a]	Lambir Formation[a]					Range of sediments[b]			
Element Ratio	Wacke[1] (n = 3)	Wacke[3] (n = 10)	Litharenite[4] (n = 10)	Arkose[5] (n = 4)	Fe-Sand[6] (n = 2)	Quartz arenite[7] (n = 1)	Felsic rocks	Mafic rocks	UCC[c]	PAAS[c]
La/Sc	2.42–3.14	2.09–2.93	2.59–5.37	4.22–6.40	2.26–3.18	5.70	2.50–16.3	0.43–0.86	2.21	2.40
	2.69 ± 0.40	2.56 ± 0.24	3.43 ± 1.01	5.13 ± 1.02	2.72 ± 0.65					
Th/Sc	0.90–0.96	0.82–1.10	0.94–1.97	1.55–2.45	0.76–1.20	2.10	0.84–20.5	0.05–0.22	0.79	0.90
	0.92 ± 0.03	0.95 ± 0.10	1.27 ± 0.32	1.94 ± 0.45	0.98 ± 0.31					
Th/Cr	0.16–0.20	0.13–0.19	0.17–0.23	0.14–0.20	0.16–0.19	—	0.13–2.7	0.018–0.046	0.13	0.13
	0.17 ± 0.02	0.18 ± 0.02	0.19 ± 0.02	0.17 ± 0.04	0.18 ± 0.02					
La/Co	2.42–4.40	1.18–10.90	1.74–15.80	5.28–9.17	3.18–3.23	5.70	1.80–13.8	0.14–0.38	1.76	1.66
	3.17 ± 1.07	3.41 ± 2.71	5.58 ± 4.13	7.75 ± 2.15	3.20 ± 0.04					
Th/Co	0.96–1.28	0.45–3.53	0.63–5.95	1.98–3.63	1.09–1.20	2.10	0.67–19.4	0.04–1.40	0.63	0.63
	1.07 ± 0.18	1.22 ± 0.85	2.06 ± 1.53	2.90 ± 0.85	1.14 ± 0.08					
Eu/Eu*	0.66–0.76	0.59–0.76	0.56–0.71	0.55–0.72	0.64–0.72	0.59	0.40–0.94	0.71–0.95	0.63	—
	0.71 ± 0.05	0.68 ± 0.04	0.61 ± 0.04	0.61 ± 0.07	0.68 ± 0.06					
$(La/Lu)_{cn}$	8.16–8.62	7.19–10.49	5.75–7.46	4.68–7.94	6.70–7.32	6.23	3.00–27	1.10–7	9.73	—
	8.48 ± 0.28	8.20 ± 1.09	6.67 ± 0.54	6.22 ± 1.35	7.01 ± 0.44					

—, Not determined.
[a]*(1, 2, 3, 4, 5, 6, 7), This Study.*
[b]*Cullers (1994, 2000); Cullers and Podkovyrov (2000); Cullers et al. (1988).*
[c]*Taylor and McLennan (1985, 2001).*

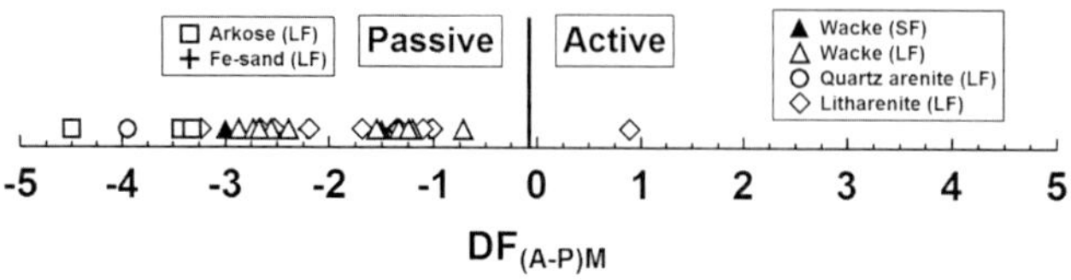

FIGURE 7.13A Discrimination diagrams based on major element (oxides) (after Verma and Armstrong-Altrin, 2016).

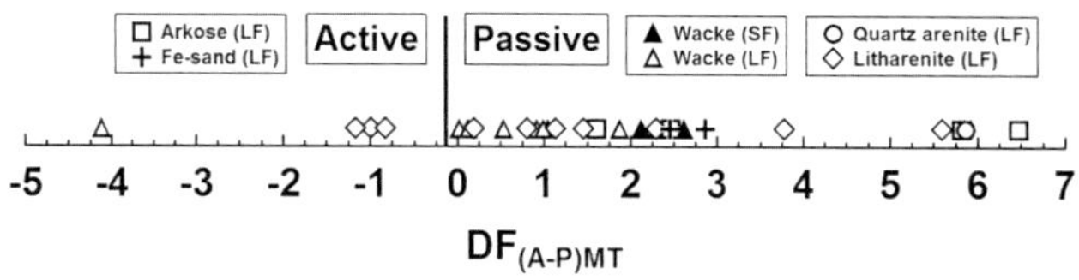

FIGURE 7.13B Discrimination diagrams based on major and trace elements (after Verma and Armstrong-Altrin (2016).

Holocene siliciclastic sediments from known tectonic margins. They recommended that these diagrams should be considered as a tool for successfully discriminating the tectonic setting of older sedimentary basins. All but four samples from this study of SF and LF clastic sediments plot in the passive margin field on these major (Fig. 7.13A) and major-trace (Fig. 7.13B) element discriminant function diagrams. The other four samples plotted in the active margin field may be influenced by collision, which started during the Middle Miocene.

5.5.1 Interpreted Tectonic Setting in the Context of Regional Tectonic Development

The tectonic settings inferred from the discrimination diagrams based on geochemistry are compared to the regional understanding of the tectonic evolution of NW Borneo. During the Late Eocene, the major Rajang Group turbidite flysch, located to the east of the Miri Zone, has been folded, thrust, and uplifted. It continues toward the southwest as the Sibu Zone of Sarawak. This tectonic episode, known as the Sarawak Orogeny (Hutchison, 2007), caused a transition in sedimentation, from flysch to molasse. Arguably, the active margin signature of these older and later recycled sediments could possibly still be preserved in the geochemical fingerprint.

Rifting in the South China Sea occurred during the Early Miocene, which coincided with the depositional period of the SF. The sediments supplied from the Rajang Group after the Sarawak Orogeny event were continuously deposited until the Middle Miocene. By that time, rifting and a potential subduction of a proto-South China Sea had slowed down, which resulted in another phase of uplift of the Borneo landmass. This led to the deposition of the LF sediments, dominated by sandstone compared to mudstone-marl-dominated SF.

The transition from a mud-dominated, low-energy deeper water shelf environment in the Early Miocene started with the deposition of the Setap Shale and SF (Fig. 7.1), and led to the sand-prone, post-MMU/DRU (Mid-Miocene Unconformity/Deep Regional Unconformity) larger sandy shelfal margin, with deposits of the LF (and the Belait, Miri, Tukau, and Seria Formations; Fig. 7.1). The relationship between sedimentation and uplift of the hinterland has been observed and documented by Kessler and Jong (2015b). The authors portray the development of the Miocene shelf from the standpoints of stratigraphy, sea-level fluctuations, hinterland uplift, and sediment recycling; mobile clay tectonics; and the impact of the monsoon climate. Balancing the different viewpoints, the transition from a muddy Mid-Miocene shelf to an unusually sandy one can be attributed to the rise of the Borneo part of Sundaland in the Middle to Late Miocene, caused by tectonic compression, in combination with the influence of the monsoon climate; and the availability, through erosion of the Rajang/Crocker system, with massive amounts of sand delivered to the basin in geologically short time intervals.

6. CONCLUSIONS

Sandstones of the LF predominantly consist of quartz and illite/muscovite with some heavy minerals such as zircon, rutile, and opaque grains based on petrography and XRD studies. Sandstones are chemically much more mature than the mudstones and show a high content of Si and low content of Rb and Sr with the exception of some trace elements. High CaO content in the mudstones of SF and LF (avg. 13.23 and 3.73 wt%) indicates the presence of fossils. High CIA (66–85) and PIA (71–98) values imply that the source area underwent moderate to intensive chemical weathering. Statistical analysis demarcates the effects of weathering, sorting, and recycling. Petrography and geochemistry (LREE enrichment, negative Eu/Eu* value, and flat HREE pattern) and discriminant diagrams illustrate that these sediments were derived from recycled sedimentary/metasedimentary sources and deposited in an evolving passive to active continental margin setting during the Miocene.

Acknowledgments

First author (RN) would like to thank the bachelor degree students (2014) for their help during field work and sample processing. This study was financially supported by Faculty of Engineering Science, Curtin University, and Research Performance fund awarded to RN. JSA is grateful to the Instituto de Ciencias del Mar y Limnología, UNAM, Institutional Project (no. 616). This chapter has benefited greatly from the reviewers (Prof. Abhijit Basu and Prof. R. Nagendra), and editor (Dr. Rajat Mazumder), who contributed significantly to the improvement of this chapter. We also acknowledge Dr. Doug Gillies for improving the language and presentation of the chapter.

References

Ali, S., Stattegger, K., Garbe-Schönberg, D., Frank, M., Kraft, S., Kuhnt, W., 2014. The provenance of Cretaceous to Quaternary sediments in the Tarfaya basin, SW Morocco: evidence from trace element geochemistry and radiogenic Nd–Sr isotopes. Journal of African Earth Sciences 90, 64–76.

Armstrong-Altrin, J.S., Verma, S.P., 2005. Critical evaluation of six tectonic setting discrimination diagrams using geochemical data of Neogene sediments from known tectonic settings. Sedimentary Geology 177 (1–2), 115–129.

Armstrong-Altrin, J.S., Lee, Y.I., Verma, S.P., Ramasamy, S., 2004. Geochemistry of sandstones from the upper Miocene Kudankulam formation, Southern India: implications for provenance, weathering, and tectonic setting. Journal of Sedimentary Research 74 (2), 285–297.

Armstrong-Altrin, J.S., Lee, Y.I., Kasper-Zubillaga, J.J., Carranza-Edwards, A., Garcia, D., Eby, G.N., Balaram, V., Cruz-Ortiz, N.L., 2012. Geochemistry of beach sands along the western Gulf of Mexico, Mexico: implication for provenance. Chemie der Erde-Geochemistry 72, 345–362.

Armstrong-Altrin, J.S., Nagarajan, R., Madhavaraju, J., Rosalez-Hoz, L., Lee, Y.I., Balaram, V., Cruz-Martinez, A., Avila-Ramirez, G., 2013. Geochemistry of Jurassic and upper Cretaceous shales from the Molango Region, Hidalgo, eastern Mexico: implications for source-area weathering, provenance, and tectonic setting. Comptes Rendus Geoscience 345 (4), 185–202.

Armstrong-Altrin, J.S., Nagarajan, R., Lee, Y.I., Kasper-Zubillaga, J.J., Córdoba-Saldaña, L.P., 2014. Geochemistry of sands along the San Nicolás and San Carlos beaches, Gulf of California, Mexico: implications for provenance and tectonic setting. Turkish Journal of Earth Sciences 23, 533–558.

Armstrong-Altrin, J.S., Machain-Castillo, M.L., Rosales-Hoz, L., Carranza-Edwards, A., Sanchez-Cabeza, J.A., Ruíz-Fernández, A.C., 2015. Provenance and depositional history of continental slope sediments in the southwestern Gulf of Mexico unraveled by geochemical analysis. Continental Shelf Research 95, 15–26.

Armstrong-Altrin, J.S., Lee, Y.I., Kasper-Zubillaga, J.J., Trejo-Ramírez, E., 2016. Mineralogy and geochemistry of sands along the Manzanillo and El Carrizal beaches, southern Mexico: implications for provenance and tectonic setting. Geological Journal. http://dx.doi.org/10.1002/gj.2792.

Armstrong-Altrin, J.S., 2009. Provenance of sands from Cazones, Acapulco, and Bahía Kino beaches, Mexico. Revista Mexicana de Ciencias Geológicas 26, 764–782.

Armstrong-Altrin, J.S., 2015. Evaluation of two multi-dimensional discrimination diagrams from beach and deep sea sediments from the Gulf of Mexico and their application to Precambrian clastic sedimentary rocks. International Geology Review 57, 1446–1461.

Bakkiaraj, D., Nagendra, R., Nagarajan, R., Armstrong-Altrin, J.S., 2010. Geochemistry of sandstones from the upper Cretaceous Sillakkudi formation, Cauvery basin, southern India: implication for provenance. Journal of the Geological Society of India 76, 453–467.

Banda, R.M., Honza, E., 1997. Miocene stratigraphy of Northwest Borneo basin. Bulletin of the Geological Society of Malaysia 40, 1–11.

Banda, R.M., 1998. The Geology and Planktic Foraminiferal Biostratigraphy of the NW Borneo Basin Sarawak, Malaysia (Ph.D thesis). University of Tsukuba, Japan.

Basu, A., Bickford, M.E., Deasy, R., 2016. Inferring tectonic provenance of siliciclastic rocks from their chemical compositions: a dissent. Sedimentary Geology 336, 26–35.

Bhatia, M.R., Crook, K.A.W., 1986. Trace element characteristics of graywackes and tectonic setting discrimination of sedimentary basins. Contributions to Mineralogy and Petrology 92 (2), 181–193.

Bhatia, M.R., 1983. Plate tectonics and geochemical composition of sandstones. Journal of Geology 91, 611–627.

Condie, K.C., Lee, D., Farmer, L., 2001. Tectonic setting and provenance of the Neoproterozoic Uinta Mountain and Big Cootonwood groups, northern Utah: constraints from geochemistry, Nd isotopes, and detrital modes. Sedimentary Geology 141–142, 443–464.

Condie, K.C., 1993. Chemical composition and evolution of the upper continental crust: contrasting results from surface samples and shales. Chemical Geology 104, 1–37.

Cox, R., Lowe, D.R., Cullers, R.L., 1995. The influence of sediment recycling and basement composition on evolution of mudrock chemistry in the southwestern United States. Geochimica et Cosmochimica Acta 59 (14), 2919–2940.

Cullers, R.L., Podkovyrov, V.N., 2000. Geochemistry of the Mesoproterozoic Lakhanda shales in southeastern Yakutia, Russia: implications for mineralogical and provenance control, and recycling. Precambrian Research 104 (1–2), 77–93.

Cullers, R.L., Basu, A., Suttner, L.J., 1988. Geochemical signature of provenance in sand-size material in soils and stream sediments near the Tobacco Root batholith, Montana, U.S.A. Chemical Geology 70, 335–348.

Cullers, R.L., 1994. The controls on the major and trace element variation of shales, siltstones and sandstones of Pennsylvanian – Permian age from uplifted continental blocks in Colorado to platform sediment in Kansas, U.S.A. Geochimica et Cosmochimica Acta 58, 4955–4972.

Cullers, R.L., 1995. The controls on the major and trace element evolution of shales, siltstones and sandstones of Ordovician to Tertiary age in the Wet Mountain region, Colorado, U.S.A. Chemical Geology 123 (1–4), 107–131.

Cullers, R.L., 2000. The geochemistry of shales, siltstones and sandstones of Pennsylvanian-Permian age, Colorado, U.S.A.: implications for provenance and metamorphic studies. Lithos 51, 305–327.

Dickinson, W.R., 1970. Interpreting detrital modes of graywacke and arkose. Journal of Sedimentary Petrology 40 (2), 695–707.

El-Enen, M.M.A., Abu-Alam, T.S., Whitehouse, M.J., Ali, K.A., Okrusch, M., 2016. P–T path and timing of crustal thickening during amalgamation of East and West Gondwana: a case study from the Hafafit Metamorphic Complex, Eastern Desert of Egypt. Lithos. http://dx.doi.org/10.1016/j.lithos.2016.01.001.

Floyd, P.A., Leveridge, B.E., 1987. Tectonic environment of the Devonian Gramscatho basin, south Cornwall: framework mode and geochemical evidence from turbiditic sandstones. Journal of the Geological Society 144 (4), 531–542.

Folk, R.L., 1974. Petrology of Sedimentary Rocks. Hemphill Publishing Company, Austin, Texas, 190p.

Garcia, D., Fonteilles, M., Moutte, J., 1994. Sedimentary fractionation between Al, Ti, and Zr and genesis of strongly peraluminous granites. Journal of Geology 102, 411–422.

Gazzi, P., 1966. Le arenarie del flysch sopracretaceo dell'Appennino modensese: Correlazioni con il flysch di Monghidoro. Mineralogica et Petrographica Acta 16, 69–97.

Gotelli, N.J., Ellison, A.M., 2004. A Primer of Ecological Statistics. Sinauer Associates, Sunderland, Massachusetts.

Hall, R., Nichols, G.J., 2002. Cenozoic sedimentation and tectonics in Borneo: climatic influences on orogenesis. In: Jones, S.J., Frostick, L. (Eds.), Sediment Flux to Basins: Causes, Controls and Consequences, 191. Geological Society of London Special Publications, pp. 5–22.

Hall, R., Marco, W.A., van Hattum, M.W.A., Spakman, W., 2008. Impact of India-Asia collision on SE Asia: the record in Borneo. Tectonophysics 451 (1–4), 366–389.

Hall, R., 2013. Contraction and extension in northern Borneo driven by subduction rollback. Journal of Asian Earth Sciences 76, 399–411.

Hayashi, K., Fujisawa, H., Holland, H.D., Ohmoto, H., 1997. Geochemistry of 1.9 Ga sedimentary rocks from northeastern Labrador, Canada. Geochimica et Cosmochimica Acta 61, 4115–4137.

Herron, M.M., 1988. Geochemical classification of terrigenous sands and shales from core or log data. Journal of Sediment Petrology 58 (5), 820–829.

Hutchison, C.S., 2005. Geology of North-West Borneo; Sarawak, Brunei and Sabah. Elsevier, Amsterdam, 421p.

Hutchison, C.S., 2007. Geological Evolution of South-East Asia, second ed. Geological Society of Malaysia, Kuala Lumpur. 433p.

Ingersoll, R.V., Dickinson, W.R., Graham, S.A., 2003. Remnant-ocean submarine fans: largest sedimentary systems on Earth. In: Chan, M.A., Archer, A.W. (Eds.), Extreme Depositional Environments: Mega End Members in Geologic Time, 370. Geological Society of America Special Papers, pp. 191–208.

Jong, J., Barker, S., Kessler, F.L., Tan, T.Q., 2014. The Sarawak Bunguran fold Belt - structural development in the Context of south China sea tectonics. In: 8th International Petroleum Technology Conference, Kuala Lumpur. 18197-MS.

Jong, J., Barker, S., Kessler, F.L., 2015. A comparison of fold-thrust belts in eastern Sundaland: structural commonalities and differences on the circum-Borneo margin. In: Proceedings Indonesian Petroleum Association, 39th Annual Convention & Exhibition, Jakarta. IPA15-G-138. CDROM, 21p.

Jong, J., Kessler, F.L., Noon, S., Tran, Q.T., 2016. Structural development, depositional model and petroleum system of Palaeogene carbonate of the Engkabang-Karap Anticline, onshore Sarawak. Berita Sedimentologi 34, 5–25.

Kessler, F.L., Jong, J., 2014a. Habitat and C-14 ages of lignitic terrace deposits along the northern Sarawak coastline. Bulletin of the Geological Society of Malaysia 60, 27–34.

Kessler, F.L., Jong, J., 2014b. The origin of Canada Hill – a result of strike-slip deformation and hydraulically powered uplift at the Pleistocene/Holocene border? Bulletin of the Geological Society of Malaysia 60, 35–44.

Kessler, F.L., Jong, J., 2015a. Northwest Sarawak: a complete geologic profile from the lower Miocene to the Pliocene covering the upper Setap shale, Lambir and Tukau formations. Persatuan Geologi Malaysia Warta Geologi 41 (3–4), 45–51.

Kessler, F.L., Jong, J., 2015b. Tertiary uplift and the Miocene evolution of the NW Borneo shelf margin. Berita Sedimentologi 33, 21–46.

Kessler, F.L., 2009. Observations on sediments and deformation characteristics, Sarawak Foreland, Borneo Island. Persatuan Geologi Malaysia Warta Geologi 35 (1), 1–10.

Khan, T., Khan, M.S., 2015. Clastic rock geochemistry of Punagarh basin, trans-Aravalli region, NW Indian shield: implications for paleoweathering, provenance, and tectonic setting. Arabian Journal of Geosciences 8 (6), 3621–3644.

Long, X., Yuan, C., Sun, M., Xiao, W., Wang, Y., Cai, K., Jiang, Y., 2012. Geochemistry and Nd isotopic composition of the Early Paleozoic flysch sequence in the Chinese Altai, Central Asia: evidence for a northward derived mafic source and insight into Nd model ages in accretionary orogen. Gondwana Research 22, 554–566.

McLennan, S.M., Nance, W.B., Taylor, S.R., 1980. Rare earth element-thorium correlations in sedimentary rocks, and the composition of the continental crust. Geochimica et Cosmochimica Acta 44 (11), 1833–1839.

McLennan, S.M., Hemming, D.K., Hanson, G.N., 1993. Geochemical approaches to sedimentation, provenance and tectonics. Geological Society of America Special Papers 284, 21–40.

McLennan, S.M., 1989. Rare earth elements in sedimentary rocks; influence of provenance and sedimentary processes. In: Lipin, B.R., McKay, G.A. (Eds.), Geochemistry and Mineralogy of Rare Earth Elements. Reviews in Mineralogy and Geochemistry, 21, pp. 169–200.

McLennan, S.M., 1993. Weathering and global denudation. Journal of Geology 101, 295–303.

McLennan, S.M., 2001. Relationships between the trace element composition of sedimentary rocks and upper continental crust. Geochemistry, Geophysics, Geosystems 2 (4), 1021, 102. http://dx.doi.org/10.1029/2000GC000109.

Mongelli, G., Critelli, S., Perri, F., Sonnino, M., Perrone, V., 2006. Sedimentary recycling, provenance and paleoweathering from chemistry and mineralogy of Mesozoic continental redbed mudrocks, Peloritani Mountains, southern Italy. Geochemical Journal 40, 197–209.

Morley, C.K., Tingay, M., Hillis, R., King, R., 2008. Relationship between structural style, overpressures, and modern tress, Baram Delta Province, NW Borneo. Journal of Geophysical Research 113, B09410. http://dx.doi.org/10.1029/2007JB005324.

Nagarajan, R., Madhavaraju, J., Nagendra, R., Armstrong-Altrin, J.S., Moutte, J., 2007a. Geochemistry of Neoproterozoic shales of Rabanpalli formation, Bhima Basin, Northern Karnataka, southern India: implications for provenance and paleoredox conditions. Revista Mexicana de Ciencias Geologicas 24 (2), 150–160.

Nagarajan, R., Armstrong-Altrin, J.S., Nagendra, R., Madhavaraju, J., Moutte, J., 2007b. Petrography and geochemistry of terrigenous sedimentary rocks in the Neoproterozoic Rabanpalli formation, Bhima basin, southern India: implications for paleoweathering conditions, provenance and source rock composition. Journal of the Geological Society of India 70 (2), 297–312.

Nagarajan, R., Roy, P.D., Jonathan, M.P., Lozano-Santacruz, R., Kessler, F.L., Prasanna, M.V., 2014. Geochemistry of Neogene sedimentary rocks from Borneo basin, East Malaysia: paleo-weathering, provenance and tectonic setting. Chemie der Erde-Geochemistry 74 (1), 139–146.

Nagarajan, R., Armstrong-Altrin, J.S., Kessler, F.L., Hidalgo-Moral, E.L., Dodge-Wan, D., Taib, N.I., 2015. Provenance and tectonic setting of Miocene siliciclastic sediments, Sibuti Formation, northwestern Borneo. Arabian Journal of Geosciences 8, 8549–8565.

Nesbitt, H.W., Wilson, R.E., 1992. Recent chemical weathering of basalts. American Journal of Science 292 (10), 740–777.

Nesbitt, H.W., Young, G.M., 1982. Early Proterozoic climates and plate motions inferred from major element chemistry of lutites. Nature 299 (5885), 715–717.

Nesbitt, H.W., Young, G.M., 1984. Prediction of some weathering trends of plutonic and volcanic rocks based on thermodynamic and kinetic consideration. Geochimica et Cosmo chimica Acta 48 (7), 1523–1534.

Nesbitt, H.W., 2003. Petrogenesis of siliciclastic sediments and sedimentary rocks. In: Lentz, D.R. (Ed.), Geochemistry of Sediments and Sedimentary Rocks: Evolutionary Considerations to Mineral Deposit-forming Environments, 4. Geological Association of Canada, Newfoundland, pp. 39–51. Geo Text.

Osman, M., 1996. Recent to Quaternary River Nile Sediments: A Sedimentological Characterization on Samples from Aswan to Naga Hammadi, Egypt (Unpublished doctoral dissertation). University of Vienna, Vienna.

Potter, P.E., Maynard, J.B., Depetris, P.J., 2005. Mud and Mudstones: Introduction and Overview. Springer Berlin, Heidelberg, New York, 297p. ISBN 3-540-22157-3.

Rollinson, H.R., 1993. Using Geochemical Data: Evaluation, Presentation, Interpretation. Routledge, United Kingdom, 352p.

Roser, B.P., Korsch, R.J., 1986. Determination of tectonic setting of sandstone-mudstone suites using SiO_2 content and K_2O/Na_2O ratio. Journal of Geology 94, 635–650.

Roser, B.P., Korsch, R.J., 1988. Provenance signatures of sandstone–mudstone suites determined using discrimination function analysis of major-element data. Chemical Geology 67, 119–139.

Ryan, K.M., Williams, D.M., 2007. Testing the reliability of discrimination diagram for determining the tectonic depositional environment of ancient sedimentary basins. Chemical Geology 242, 103–125.

Sandal, T. (Ed.), 1996. The Geology and Hydrocarbon Resources of Negara Brunei Darussalam. Brunei Shell Petroleum Co. Sdn. Bhd., and Brunei Museum, 243p.

Setiawan, N.I., Osanai, Y., Nakano, N., Adachi, T., Setiadji, L.D., Wahyudiono, J., 2013. Late Triassic metatonalite from the Schwaner mountains in West Kalimantan and its contribution to sedimentary provenance in the Sundaland. Berita Sedimentologi 28, 4–12.

Simon, K., Bin Amir Hassen, M.H., Barbeito, M.P.J., 2014. Sedimentology and stratigraphy of the Miocene Kampung Opak limestone (Sibuti Formation), Bekenu, Sarawak. Bulletin of the Geological Society of Malaysia 60, 45–53.

Taylor, S.R., McLennan, S.M., 1985. The Continental Crust: Its Composition and Evolution. Blackwell Scientific Publications, Oxford.

Togunwa, O.S., Abdullah, W.H., Hakimi, M.H., Barbeito, P.J., 2015. Organic geochemical and petrographic characteristics of Neogene organic-rich sediments from the onshore West Baram Delta Province, Sarawak Basin: implications for source rocks and hydrocarbon generation potential. Marine and Petroleum Geology 63, 115–126.

Valloni, R., Maynard, J.B., 1981. Detrital modes of recent deep-sea sands and their relation to tectonic setting: a first approximation. Sedimentology 28 (1), 75–83.

van Hattum, M.W.A., Hall, R., Pickard, A.L., Nichols, G.J., 2006. Southeast Asian sediments not from Asia: provenance and geochronology of North Borneo sandstones. Geology 34, 589–592.

van Hattum, M.W.A., Hall, R., Pickard, A.L., Nichols, G.J., 2013. Provenance and geochronology of Cenozoic sandstones of northern Borneo. Journal of Asian Earth Sciences 76 (25), 266–282.

Verma, S.P., Armstrong-Altrin, J.S., 2013. New multi-dimensional diagrams for tectonic discrimination of siliciclastics sediments and their application to Precambrian basins. Chemical Geology 355, 117–133.

Verma, S.P., Armstrong-Altrin, J.S., 2016. Geochemical discrimination of siliciclastic sediments from active and passive margin settings. Sedimentary Geology 332, 1–12.

Viet, N.H., 2014. Sedimentological and Petrophysical Properties of Sandstone Facies Belonging to Lambir Formation at Tusan Beach, Miri, Sarawak. Final year Project Report, Universiti Teknologi Petronas, Malaysia. http://utpedia.utp.edu.my/14097/.

Xie, Y., Chi, Y., 2016. Geochemical investigation of dry- and wet-deposited dust during the same dust-storm event in Harbin, China: constraint on provenance and implications for formation of aeolian loess. Journal of Asian Earth Sciences 120, 43–61.

Zaid, S.M., Gahtani, F.A., 2015. Provenance, diagenesis, tectonic setting and geochemistry of Hawkesbury sandstone (Middle Triassic), southern Sydney Basin, Australia. Turkish Journal of Earth Sciences 24, 72–98.

Zin, I.C.M., 1996. Tertiary Tectonics and Sedimentation History of the Sarawak Basin, East Malaysia (Durham theses). Durham University. http://etheses.dur.ac.uk/5198/.

CHAPTER

8

What Are the Origins of V-Shaped (Chevron) Dunes in Madagascar? The Case for Their Deposition by a Holocene Megatsunami

D. Abbott[1,2], *V. Gusiakov*[3], *G. Rambolamanana*[4], *D. Breger*[2,5], *R. Mazumder*[6], *K. Galinskaya*[7]

[1]City College of New York, New York, NY, United States; [2]Lamont-Doherty Earth Observatory of Columbia University, Palisades, NY, United States; [3]Tsunami Laboratory, ICMMG SD RAS, Novosibirsk, Russia; [4]University of Antananarivo, Antananarivo, Madagascar; [5]Micrographic Arts, Saratoga Springs, NY, United States; [6]Curtin University, Sarawak, Malaysia; [7]Brooklyn College, New York, NY, United States

OUTLINE

Sediment Provenance
http://dx.doi.org/10.1016/B978-0-12-803386-9.00008-3

1. INTRODUCTION

Chevrons are elongated dunes with a V-shaped pattern in map view. In some exposures, smaller Vs are nested within the larger Vs. The term chevron was first used to describe desert dunes (Maxwell and Haynes, 1989) based on their similarity to the nested chevrons used on military uniforms or in heraldry. Chevrons later were identified in coastal regions and proposed to represent megastorm deposits (Hearty et al., 1998; Kindler and Strasser, 2000). Subsequently, other workers suggested that some coastal chevron dunes were tsunami deposits (Bryant and Nott, 2001; Scheffers and Kelletat, 2003; Scheffers et al., 2008).

The proposal that some coastal chevron dune complexes represent tsunami deposits is based on three sets of observations (Scheffers and Kelletat, 2003). The first is that the long axes of many coastal chevron complexes are not oriented parallel to the direction of the prevailing wind. The second is that some chevron complexes extend several kilometers (km) inland and rise to over 100 meters (m) above sea level. Some of these chevron complexes are located on shorelines that lack beaches. In these particular cases, it is difficult to understand how either megastorms or wind could have formed the chevrons. Megastorms cannot move subaerial rock and sediment over km-scale distances with elevation gains of hundreds of meters (Cox et al., 2012; Erdmann et al., 2015). Wind cannot move sediment inland if there are no subaerial, poorly consolidated sediments on the coast.

In this chapter, we describe three chevron complexes, V-shaped, elongated dune complexes on the southern coast of Madagascar. Their origin is disputed because individual dunes are elongated along an azimuth that is close to the direction of the prevailing winds (Abbott et al., 2008; Pinter and Ishman, 2008), although their low angle of deposition generally is inconsistent with aeolian dunes. However, other characteristics preclude their derivation from modern beach deposits, although we do not discount later aeolian reworking of some chevron deposits. In particular, the Madagascar chevrons contain significant proportions of early Holocene carbonate tests resembling shells of marine foraminifera, including some that are partially dolomitized, and some that are infilled with mud. These observations suggest that marine carbonate tests in the chevrons were eroded from the continental shelf, and not from modern beaches. Furthermore, despite having lateral extents of tens of km, characteristics of the chevrons (degree of sediment sorting, carbonate content, and marine microfossil concentrations) do not change significantly along strike, as might be expected for aeolian deposits.

2. BACKGROUND

We previously hypothesized that the Madagascar chevrons were generated by a megatsunami (Gusiakov et al., 2010), and that either a submarine impact in the vicinity of the Burckle Crater candidate (Abbott et al., 2007) or a caldera collapse of Reunion Island (Oehler et al., 2004) could have produced the postulated megatsunami wave. However, despite our previous assertions (Abbott et al., 2008; Gusiakov et al., 2010), no unequivocal impact component has been identified within the Madagascar chevron sands, and the source of a putative megatsunami wave is presently unknown.

The megatsunami origin of Madagascar chevrons is disputed by others, who favor an aeolian origin (Pinter and Ishman, 2008; Bourgeois and Weiss, 2009). In this chapter, we present further information suggesting that the three dune complexes in Madagascar had a megatsunami origin.

On the southern coast of Madagascar, there are marine fossil-bearing chevron dunes that extend over 40 km along-strike, rising to over 175 m above sea level (Figs. 8.1 and 8.2). During a three-week field reconnaissance in 2006, we examined the three most obvious chevrons having the greatest sand thicknesses: Faux Cap, Fenambosy, and Ampalaza. Our group collected sediment samples and marine shells from the surface of the three chevron complexes, in the subsoil, and nearby, along the southern coast (Fig. 8.1; Table 8.1). As we will show, certain characteristics of the chevrons strongly suggest a megatsunami origin.

The Fenambosy Chevron is the most spectacular of the three we sampled. It extends at least 28 km along-strike and has a maximum width of 6 km. It encompasses a steep fault scarp approximately 175 m high (Fig. 8.2). On the elevated portion of the chevron that lies landward of the fault scarp, the edge of the chevron is 6–12 km from the ocean.

3. SAMPLE SELECTION AND PROCESSING

Given that this study provided for initial and rapid reconnaissance of the area, traverse locations largely were constrained to those that were accessible by road. Additionally, the two traverses on the Fenambosy Chevron were located so that people who conducted sampling could safely negotiate the fault scarp cliff on foot. Because local roads are impassable during the wet season, we also timed our trip to coincide with the dry season.

As the weather was dry during our trip, we could not discern sedimentary structures on exposed dune interiors. In most cases, we used a trowel to dig into the surface so that samples represent a mixture of sediment derived from the surface down to several centimeters (cm). At one site on the Ampalaza chevron (S27), we used a shovel to sample at a depth of half a meter to provide for a comparison to near-surface samples.

In the lab, samples were wet-sieved first to remove dust and fine organic carbon. Wastewater from wet sieving was sterilized using bleach to kill potentially dangerous microorganisms. Samples were then dry-sieved using sieves with mesh sizes of 38, 63, 125, 250, and 500 micrometers (μm). If there were significant numbers of particles >500 μm, we used larger sieves to characterize the size distribution of those particles further. Sorting and other sedimentologic parameters were calculated using a statistical package (Blott and Pye, 2001).

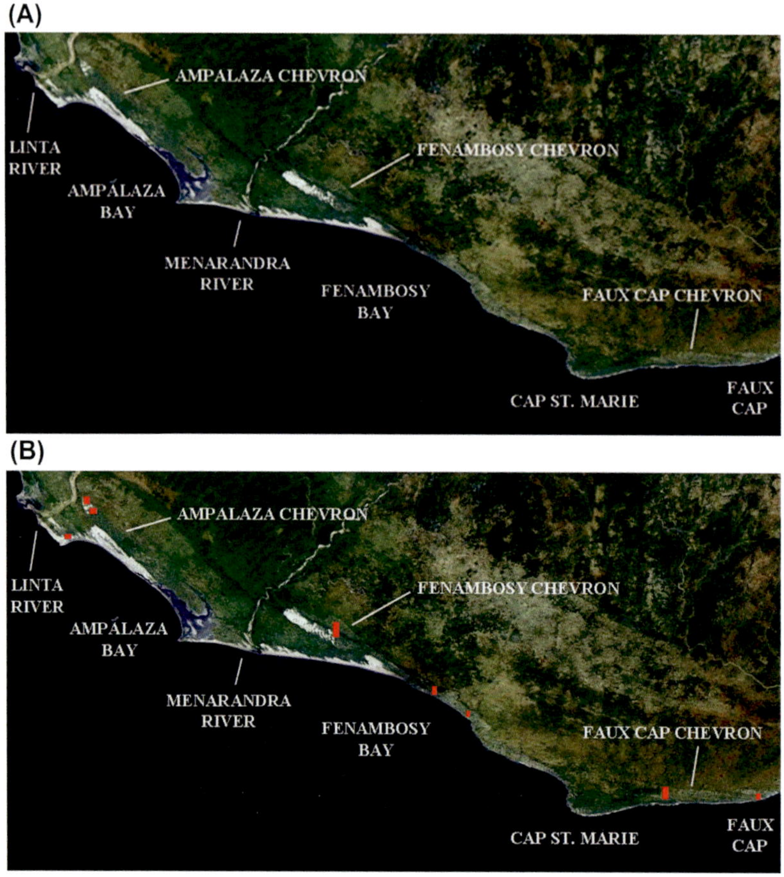

FIGURE 8.1 Satellite image showing the three Madagascar chevron dune complexes. Due north is up. This image was taken during a time of comparatively high rainfall, thereby maximizing the visual contrast between the chevrons and the surrounding terrain. The reworked parts of the chevrons are white and the vegetated portions are a uniform green that stands out against the brown to greenish-gray color of the surrounding bedrock. (A) Annotated image of chevrons and surroundings. (B) Same image as (A) with sampling sites in red. Both images from 2007 Europa Technologies, 2007 Digital Globe, and 2007 TerraMetrics.

Due to our relatively small sample sizes, we could not extend our size analyses to grains <38 μm in diameter, therefore our estimates of the degree of sorting are maximum values, as including the silt and clay fractions in our analyses would reduce the estimated degree of sorting.

Our samples of the Madagascar chevrons were taken as far as possible from local farmers' fields. Many samples are, however, only km distant from active agricultural sites. We interviewed the farmers and they indicated that they were not importing marine carbonate into their fields, nor were they eating shellfish. Larger seashells (3–6 cm wide) found on the surface of the dunes in all chevron complexes have modern ^{14}C ages, suggesting that they were collected and brought inland as souvenirs, or that they are residues from subsistence practices several generations in the past. We are skeptical that their ages accurately date the chevron

FIGURE 8.2 Photo (Image © Dallas Abbott) looking northwest showing the fault scarp separating the elevated plateau portion of the Fenambosy Chevron from the lower coastal plain. This photo clearly shows the scarp cliff in two places. The near cliff encompasses chevron deposits that are continuous on the coastal plain and discontinuous along 9 km of the uplifted wall of the fault scarp (cf., the middle red rectangle on Fenambosy Chevron in Fig. 8.1B). These were sampled on one traverse. The far cliff has ubiquitous chevron deposits extending from the coastal plain to the top of the elevated plateau. They cover a 21 km-long section of the fault. The far cliff includes the westernmost red rectangle (sampling sites on the N–S traverse) of the Fenambosy Chevron shown in Fig. 8.1B.

deposits themselves. Consequently, we use the ^{14}C ages of marine microfossils in the deposits to provide a maximum age for chevron deposition, as discussed later in the chapter.

For carbonate analyses, half a gram of unsieved sample was ground and homogenized with a mortar and pestle. Carbonate was assessed in replicate samples using a CO_2 coulometer.

We used a Zeiss Supra 50 scanning electron microscope (SEM) fitted with an EDAX energy-dispersive X-ray microanalyzer (EDS) located at City College in New York City to evaluate individual sediment grains. We looked at both mounted marine microfossils and sediment lithogenic clasts, as well as clasts and microfossils in thin sections.

The high carbonate content of the chevron sands meant that even with relatively small sample sizes (half liter bags), we had enough material for ^{14}C AMS dating. We used the 125–250 μm size fraction of the chevron sand, which contains well-preserved marine microfossil tests and quartz grains. The maximum grain size below which sand grains are transported by continuous suspension in air is 3.5 ϕ or 88 μm (Visher, 1969; Skoček and Saadallah, 1972). Therefore, our dated sands are unlikely to have been transported by the wind except through saltation or grain creep, both of which would destroy the carbonate tests after transport over a few km (Sharp, 1966). All samples were wet-sieved and appeared very clean. Nevertheless, we carefully examined each sample under a microscope and picked out any material that might bias our dating results. We found a few minor pieces of flat carbonate that could either be fragments of marine bivalves or of land snails. These were removed from

TABLE 8.1 Locations, Names, and Elevations of Sampling Sites

Location	Latitude	Longitude	Site Name	Elevation, m
Ambondro (not on chevron)	−25.23500	45.79700	S2	204
Fenambosy (not on chevron—on subsoil)	−25.35588	44.86937	S4	79
Fenambosy chevron	−25.36143	44.86712	S5	66
Fenambosy chevron	−25.24112	44.69767	S9	153
Fenambosy chevron	−25.25203	44.69060	S12	186
Fenambosy chevron	−25.26695	44.68449	S13	53
Cap St. Marie (not on chevron)	−25.55711	45.15937	S14	140
Cap St. Marie (not on chevron)	−25.58095	45.16374	S17	194
Ampalaza chevron	−24.98868	44.16132	S19	63
Ampalaza chevron	−25.00429	44.16482	S20	88
Ampalaza chevron	−25.05840	44.12059	S22	6
Ampalaza chevron	−25.00907	44.19189	S25	64
Ampalaza chevron	−25.01003	44.19458	S26	68
Ampalaza chevron (65 cm depth)	−25.01337	44.19415	S27	53
Ampalaza chevron (15 cm depth)	−25.01325	44.19415	S28	55
Ampalaza chevron	−25.01299	44.19082	S30	66
Menarandra River	−25.05565	44.67977	S32A	66
Faux Cap chevron field	−25.554567	45.51968	S32B	10
Faux Cap chevron field	−25.55795	45.51803	S33	17
Faux Cap chevron field	−25.57435	45.29113	S35	61
Faux Cap chevron field	−25.56093	45.28285	S36	150
Faux Cap chevron field	−25.54261	45.27870	S37	205

our samples, as were any pieces of possible terrestrial organic matter. The final samples sent out for dating consisted of a mixture of clean quartz sand and clean marine microfossil tests.

4. SEDIMENTARY CHARACTERISTICS OF CHEVRON SANDS

We compared the fossil content (marine fossils per gram in the 250–500 μm size range) and grain size distribution of surrounding areas to that of the chevrons (Tables 8.2 and 8.3). Land snails are not abundant and constitute <1% of the fossils per gram. Land snails and questionable marine fossils are excluded from the count. All four of the samples from off chevron contain 10% $CaCO_3$ or less and all but one (S14) contain no fossils. The samples from

TABLE 8.2 Sedimentologic Parameters of Coarse Silts and Sands From Madagascar[1]

Site#	Mean, ϕ	Sorting, ϕ	Skewedness, ϕ	Kurtosis, ϕ	Sorting
S2 (off)	0.93	0.64	1.64	6.83	MWS
S4 (off)	2.10	1.02	−0.60	4.26	PS
S5 (FC)	1.72	0.57	0.12	7.99	MWS
S9 (FC)	1.81	1.07	−0.04	1.80	PS
S12 (FC)	2.45	0.79	−2.60	14.50	MS
S13 (FC)	1.89	0.88	−0.42	3.08	MS
S14 (off)	1.26	0.77	0.29	4.57	MS
S17 (off)	0.89	0.66	0.55	4.83	MWS
S19 (AC)	2.65	0.48	0.41	3.72	WS
S20 (AC)	2.35	0.60	−0.04	3.05	MWS
S22 (AC)	2.91	0.53	−0.01	2.35	MWS
S25 (AC)	2.25	0.68	−0.42	3.99	MWS
S26 (AC)	1.81	0.86	−0.28	2.05	MS
S27 (AC)	2.16	0.80	−0.41	3.07	MS
S28 (AC)	1.79	0.78	0.12	2.90	MS
S30 (AC)	1.91	0.74	−0.72	4.06	MS
S32A (off)	2.34	0.73	−1.11	5.51	MS
S32B (FCC)	2.25	0.66	−0.70	4.19	MWS
S33 (FCC)	2.18	0.74	−1.62	8.05	MS
S35 (FCC)	1.33	0.66	0.59	5.22	MWS
S36 (FCC)	1.32	0.87	0.46	3.85	MS
S37 (FCC)	1.17	1.13	0.79	2.97	PS

[1]*These are the results of calculations from the data in Appendix 8.1 using the logarithmic method of moments (Krumbein and Pettijohn, 1938; Blott and Pye, 2001). Degree of sorting ranges from poorly sorted (PS: sorting is 1.0–2.0 ϕ), to moderately sorted (MS: sorting is 0.7–1.0 ϕ), to moderately well sorted (MWS: sorting is 0.5–0.7 ϕ), to well sorted (WS: sorting is 0.35–0.5 ϕ).*
Samples from Fenambosy Chevron (FC), from Ampalaza Chevron (AC), and from Faux Cap Chevron (FCC).

the chevrons all contain fossils and carbonate, typically between 40% and 60% $CaCO_3$ and between 21 and 5750 marine fossils per gram. The three samples with counts of 993–5750 fossils per gram are all from sites within 2 km of the ocean. In each sample, we must sort through tens to hundreds of grains to find a single fossil. For the sample with the most abundant fossils (5750 per gram at S22), we calculate that there are 12 mineral grains per fossil (De Villiers, 2005). For the sample with the next highest abundance of fossils (2236 per gram at S33), we calculate that there are 30 mineral grains per fossil. Therefore, the grain size distribution of these samples is primarily determined by the size distribution of the mineral grains

TABLE 8.3 Bulk Carbonate Content Versus Marine Fossil Content >250 μm Size Fraction

Site# (Off Chevron Locations Noted)	Replicate 1, % $CaCO_3$	Replicate 2, % $CaCO_3$	Average % $CaCO_3$, Rounded to Nearest %	Fossils/Gram Sediment
S2 (off)	0.0	0.0	0	0
S4 (off)	10.0	9.9	10	0
S5 (FC)	38.6	38.9	39	46
S9 (FC)	49.2	48.7	49	65
S12 (FC)	52.5	54.0	53	179
S13 (FC)	58.3	57.8	58	215
S14 (off)	2.8	3.5	3	0
S17 (off)	8.5	8.8	9	10
S19 (AC)	40.8	40.8	41	95
S20 (AC)	48.9	48.3	49	100
S22 (AC)	53.8	54.2	54	5750
S25 (AC)	39.4	39.0	39	94
S26 (AC)	44.6	45.1	45	205
S27 (AC)	42.5	42.4	42	140
S28 (AC)	36.0	36.1	36	75
S30 (AC)	35.1	36.5	36	442
S32A (off)	0.0	0.0	0	0
S32B (FCC)	40.4	41.2	41	993
S33 (FCC)	52.0	53.4	53	2237
S35 (FCC)	22.6	23.4	23	38
S36 (FCC)	40.0	37.3	39	56
S37 (FCC)	7.4	7.2	7	21

Samples from Fenambosy Chevron (FC), from Ampalaza Chevron (AC), and from Faux Cap Chevron (FCC).

and not by the size distribution of the fossils. Interestingly the three most fossil-rich samples are all moderately well sorted, not well sorted, as would be expected if the bulk sediment were primarily transported by the wind.

We attribute many of the differences in fossil count to the difficulty of accurately counting fossils whose surfaces have been ablated by later aeolian activity. This hypothesis is consistent with two observations. The first is that the carbonate contents of the fossil-bearing samples vary much less than the fossil counts (Table 8.3). The second is the common occurrence of marine microfossils that are easily identifiable only on one side. At many sites, we observed

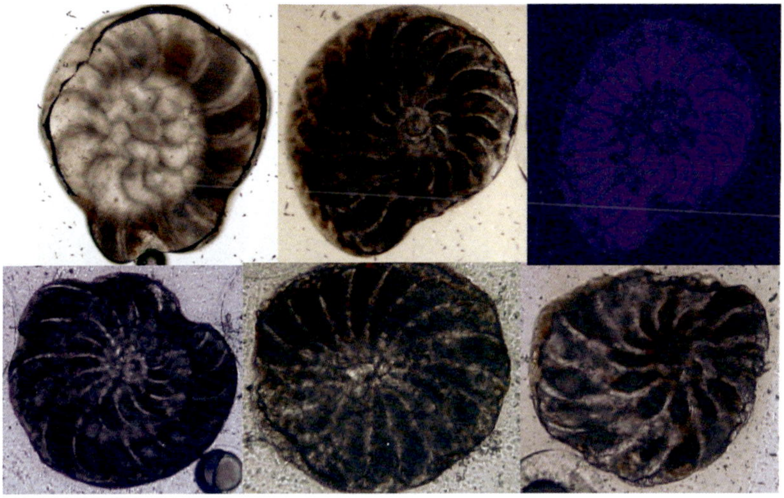

FIGURE 8.3 Transmitted light images and Mg element map (top right) of marine microfossils in thin section from Madagascar chevrons. All microfossils were picked from the 250–500 μm size fraction. Top: Marine microfossils from northwest distal end of Ampalaza Chevron. Bottom: Marine microfossils from Fenambosy Chevron.

marine microfossils with significant differences in preservation between their top and bottom surfaces. This pattern may occur because the fossils are too big to be moved by the wind. As a result, saltating sand grains would tend to preferentially erode upper, exposed surfaces. Lower surfaces facing downward would better preserve the distinctive surface morphology of marine benthic foraminifera.

Because the two mineral forms of calcium carbonate, calcite and aragonite, both have cleavage, individual grains are broken into smaller and smaller pieces as they are transported by saltation. In contrast, quartz has no cleavage and individual grains become more rounded as they are transported by saltation. In an aeolian depositional environment, attrition (collision between grains) is significant and very effective in imparting roundness because the viscosity of air is much lower than that of water (Allen, 1985). As a result, mature aeolian sands and silts consist of nearly pure, rounded quartz grains with minor proportions of heavy minerals and calcium carbonate. If transported by saltation induced by the wind, initially angular quartz and other mineral grains become well-rounded and well-sorted over a relatively short transport distance, about 10–12 km (Sharp, 1966).

In our samples, individual marine carbonate microfossils had ablated surfaces but did not appear broken (Fig. 8.3). Individual sediment clasts that lack cleavage, such as quartz and garnet, were typically angular and irregular, rather than spherical.

5. CHARACTERISTICS OF INDIVIDUAL CHEVRONS

5.1 Fenambosy Chevron

The sand dunes making up the Fenambosy Chevron have two distinct morphologies depending on whether they are vegetated or not (Fig. 8.4: top). In unvegetated areas with

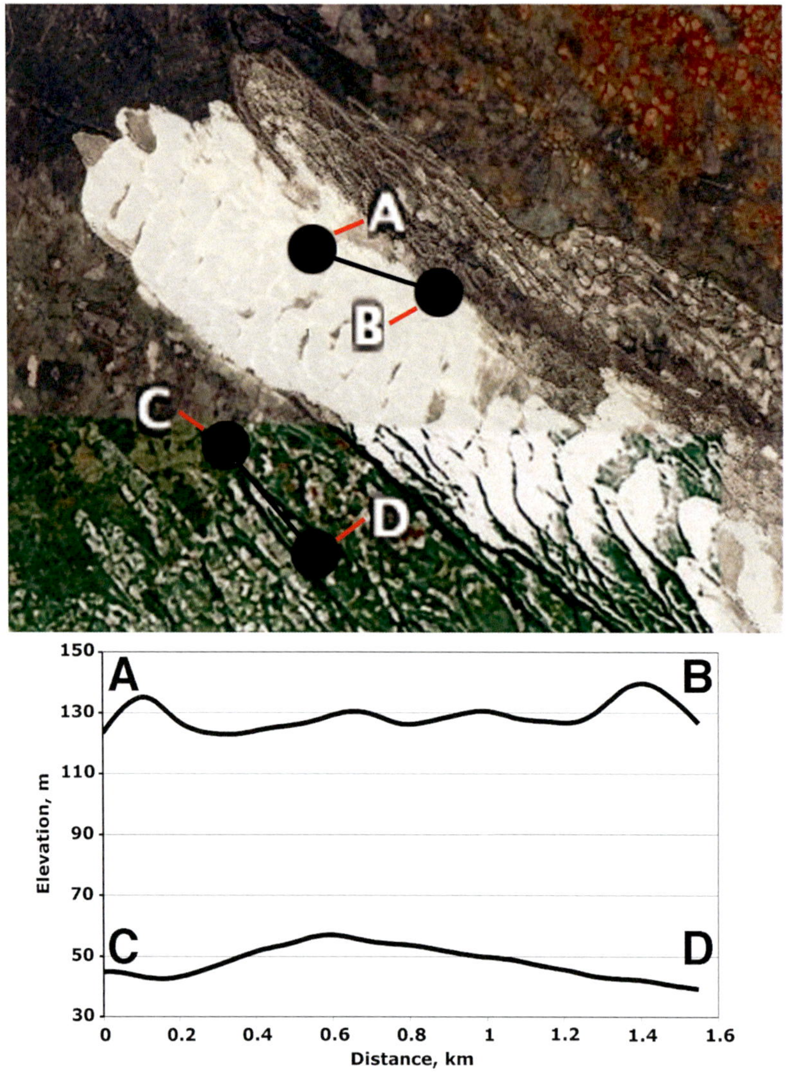

FIGURE 8.4 Top: Satellite image of far end of Fenambosy Chevron from Google Earth (2007) Europa Technologies (2007), Digital Globe (2007) and TerraMetrics (2007). This image was taken during a time of high rainfall. *Red lines* connect letters to *black circles* at ends of cross-sections. The fault scarp cliff runs NW to SE in the gray area that appears near the base of the letter A. Based on the relative whiteness of the image, the sand layer north of the fault appears thinner than the sand layer south of the fault. (In the field, both areas were covered with sand whose total thickness was difficult to determine.) Bottom: Cross-sections—A–B is from an area of active ablation of the dune surfaces; C–D is from a vegetated area that is not being farmed.

sand at the surface (appearing white in Fig. 8.4: top), chevrons are being reworked by the wind, and the dunes are elongated perpendicular to the wind direction. In the areas with the maximum erosion from deflation, the slip faces of the dunes have angles up to 30 degrees and the underlying substrate is exposed.

Conversely, chevrons are covered by vegetation in areas that appear light green in satellite photographs (Fig. 8.4: top). In some places, these chevrons are being farmed, producing a patchy appearance derived from the boundaries of farmers' fields. In other greenish areas, the chevrons are covered by undisturbed vegetation and comprise V-shaped structures of varying size and extent. The regions with V-shaped structures have lower relief and smaller maximum surface slopes (about 10 degrees) than the (white) areas of active erosion. Two cross-sections, the first along the long axis of a white area, and the second along the long axis of a green, nonfarmland area, show the contrast in wavelength and surface slope of the dunes (Fig. 8.4: bottom).

The Fenambosy Chevron has a minimum along-strike length of 28 km (Fig. 8.5). If the chevron was exclusively of aeolian origin, we would expect that the quartz grains in the western half of the chevron would be more rounded than those in the eastern half (Sharp, 1966). The western half of the chevron would contain only finely pulverized carbonate grains and would not contain whole, recognizable marine microfossils. Instead we find large numbers of marine microfossils per gram in samples from a distance of >12 km along the strike of the chevron (Table 8.3). The sorting and marine fossil content of the sands in the Fenambosy Chevron vary locally with no significant along-strike trends. This pattern could occur with

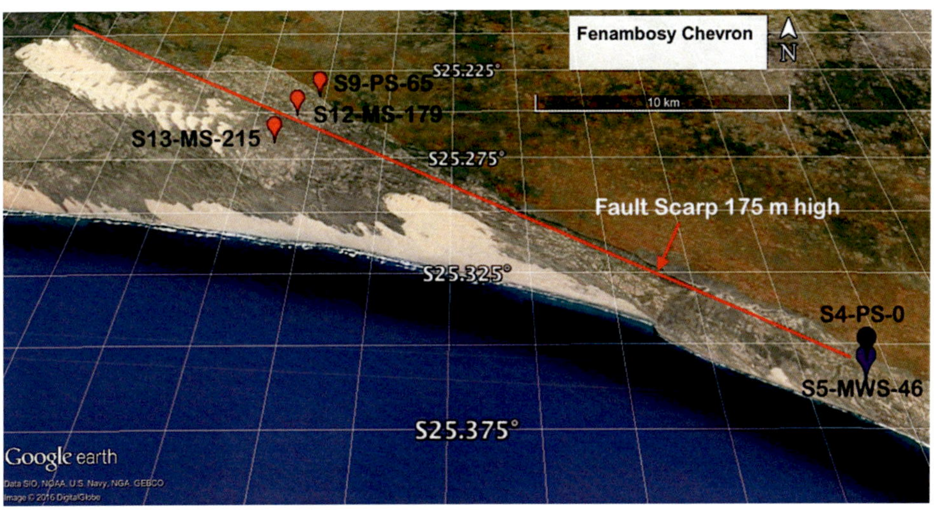

FIGURE 8.5 Google Earth image of Fenambosy Chevron. Image © 2016 DigitalGlobe. Image © 2016 CNES/Astrium. Colored symbols: Sampling locations color-coded by percentage of $CaCO_3$ in the substrate. Black: 0–20%, Blue: 20–40%, Red: 40–60%. Black lettering: Site#-degree of sorting-number of marine fossils per gram. The degree of sorting is *PS*, poorly sorted; *MS*, moderately sorted. *Black line*-trace of 175-m high fault scarp. Note on right side the two closely spaced sites, S4 and S5. At S4, the substrate to the chevron was exposed. It contains neither $CaCO_3$ nor fossils. Site S5, immediately adjacent, contains 46 fossils per gram of sediment and a significant amount of $CaCO_3$.

subaqueous transport but is inconsistent with aeolian transport of the bulk of the material in the chevrons.

In contrast to the well-sorted character of mature aeolian deposits, none of the samples from the Fenambosy Chevron are well sorted (Tables 8.1, 8.2, and 8.3; Fig. 8.6). The black curve with black dots represents the grain size distribution from a Gaussian (normal) curve with the standard deviation of typical well-sorted aeolian sand. The Gaussian curve was calculated using the excel function NORMDIST with the mean set to the mean grain size of each sample in ϕ units and the standard deviation set to 0.425 ϕ: a typical value for well-sorted sediment (Blott and Pye, 2001). The red curves show the grain size distributions derived from sieving bulk samples at 1 ϕ intervals. Most samples contain nearly 100% of material coarser than 38 μm, and grain size distribution results for those samples were not impacted significantly by ignoring the finer than 38 μm component. Only three samples, all located close to the fault scarp, contain large amounts of fine material. As discussed earlier, results for these samples record the maximum degree of sorting since finer-grained materials were discounted.

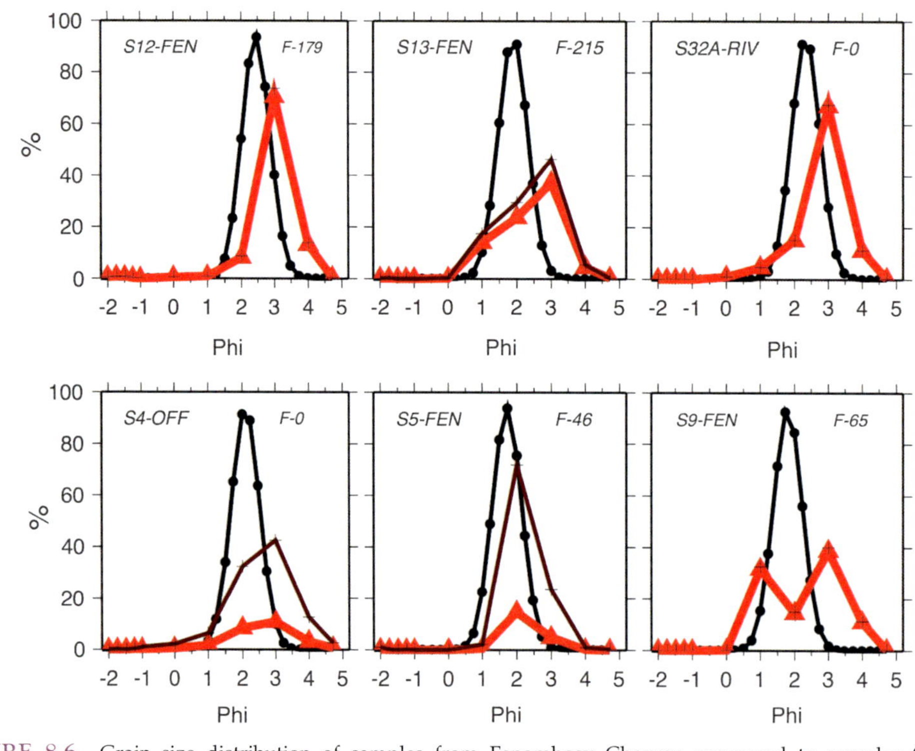

FIGURE 8.6 Grain size distribution of samples from Fenambosy Chevron compared to samples from the Menarandra River (upper right; S32A-RIV) and to samples from off chevron (lower left; S4-OFF). *Black line* with *black dots*: Theoretical model of the grain size distribution of a well-sorted sand—a normal distribution with the same mean grain size as the sample. *Red line* with *triangles*: Grain size distribution uncorrected for unsieved fine material <38 μm in size. *Brown line* with *crosses*: Grain size distribution corrected to 100%, accounting for material washed through the 38 μm sieve, so not applicable to all samples. *F*, fossils per gram.

We also show a reference sample taken from the Menarandra River (Figs. 8.1 and 8.6), recovered while traveling from the Fenambosy to the Ampalaza Chevron. The sample from the Menarandra River was taken upstream of the western end of the Fenambosy Chevron (Fig. 8.7), thus its primary sediment source is weathered material from the basement and its primary mode of transport is aquatic. The grain size distributions of two of the sediment samples from the Fenambosy Chevron, S12 and S5 (the latter corrected to 100%), closely resemble the grain size distribution of fluvial sediments from the Menarandra River (Fig. 8.6). The sorting of the remaining sediments from the Fenambosy Chevron (S13 = moderately sorted and S9 = poorly sorted) more closely resembles the moderately well-sorted river sediments than well-sorted aeolian sediments.

6. AMPALAZA CHEVRON

The Ampalaza Chevron is the longest of those investigated in this study (Fig. 8.7). It extends at least 45 km along-strike and varies from approximately 4 to 6 km wide. Our observations of white, unvegetated areas recorded in satellite photographs demonstrate they are experiencing enough wind erosion to expose the substrate locally, and to transport some sand. The only well-sorted sample on the Ampalaza Chevron is from its northwestern end in one of the unvegetated, white areas (Sample S19; Figs. 8.7 and 8.8). Although the sample is well sorted, it contains about 95 fossils per gram of sample. It was collected at 68 m elevation, well above the 3–5 m maximum rise of Holocene sea level (Camoin et al., 2004; Woodroffe and Horton, 2005) and of the Linta River to the north (Fig. 8.1).

The areas of the Ampalaza Chevron covered by vegetation appear dark gray and gray in Fig. 8.7. There are numerous, low-relief, V-shaped hills within the vegetated areas of the chevron. The Vs point toward the uphill, distal end of the chevron. As within the Fenambosy Chevron, aeolian reworking is destroying the V-shaped hills (Fig. 8.7).

The marine fossil content of the Ampalaza Chevron shows no significant trends along-strike. Individual samples are highly variable, with between 75 and 445 fossils per gram in a small area (Fig. 8.7: bottom).

None of the grain size distributions of the sediments in the Ampalaza Chevron match the theoretical distribution of a well-sorted aeolian sediment (Fig. 8.8, *black curves* with *dots*), even the sample characterized as well sorted (S19). All distributions show a substantial proportion of very fine-grained sand-sized sediment (63–125 μm, $\phi = 3.0–4.0$). The proportion of very fine-grained sand-sized material typically increases with marine fossil content. Coarse-grained sand-sized material (500–1000 μm, $\phi = 0.0–1.0$) occurs in the sample of river sediment, and in all but one of the samples taken from vegetated areas (S26, S27, S28, S30). Two samples are from the same location, S27 from 65 cm depth and S28 from 15 cm depth. These samples demonstrate that the grain size distribution of the dune sand may vary with depth. Overall, the grain size distributions in the Ampalaza Chevron show a closer match to the sediment from the Menarandra River (Figs. 8.7 and 8.8) than to the theoretical grain size distribution of aeolian sediment. The sample from the Menarandra River is from a location well upstream of the Fenambosy Chevron, so it probably represents primary material eroded from the basement. This material is most likely the source of mineral grains in modern, nearby beaches and in early Holocene beach sand incorporated into the Ampalaza Chevron.

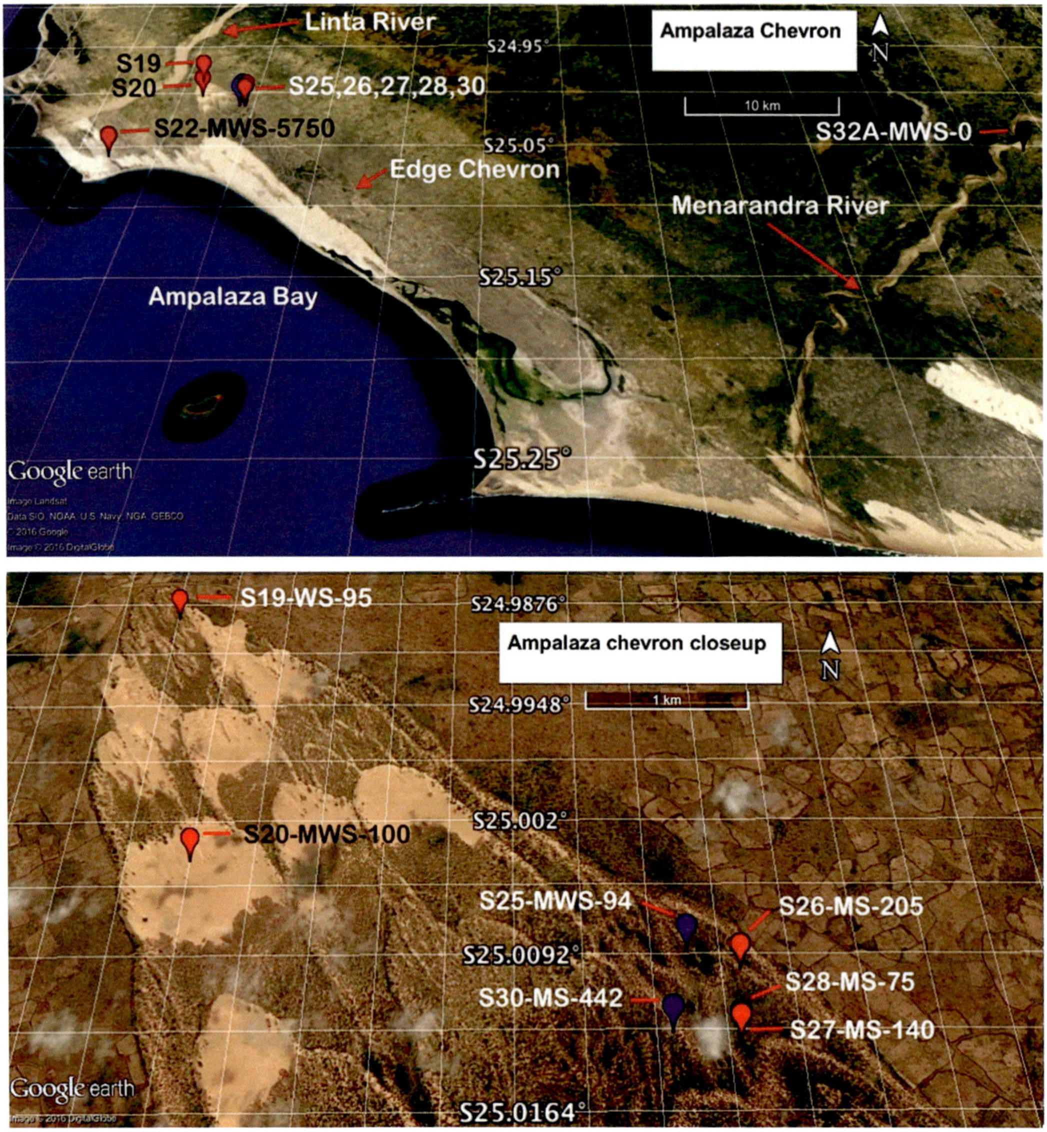

FIGURE 8.7 Satellite images of the Ampalaza Chevron from Google Earth. Images © 2016 DigitalGlobe. Image © 2016 CNES/Astrium. Colored symbols: Sampling locations color-coded by percentage of $CaCO_3$ in the substrate. Black: 0–20%, Blue: 20–40%, Red: 40–60%. Black and white lettering: Site#-degree of sorting-number of marine fossils per gram. The degree of sorting is *MS*, moderately sorted; *MWS*, moderately well sorted. Top: View of entire chevron. Bottom: Enlargement of detailed sampling area on upper left of top image. The grayish-brown colored, vegetated portion of the chevron is covered with V-shaped sand waves with structure at different scales. White lettering: degree of sorting-number of marine fossils per gram. The sands are *MS*, moderately sorted; *MWS*, moderately well sorted; *WS*, well sorted.

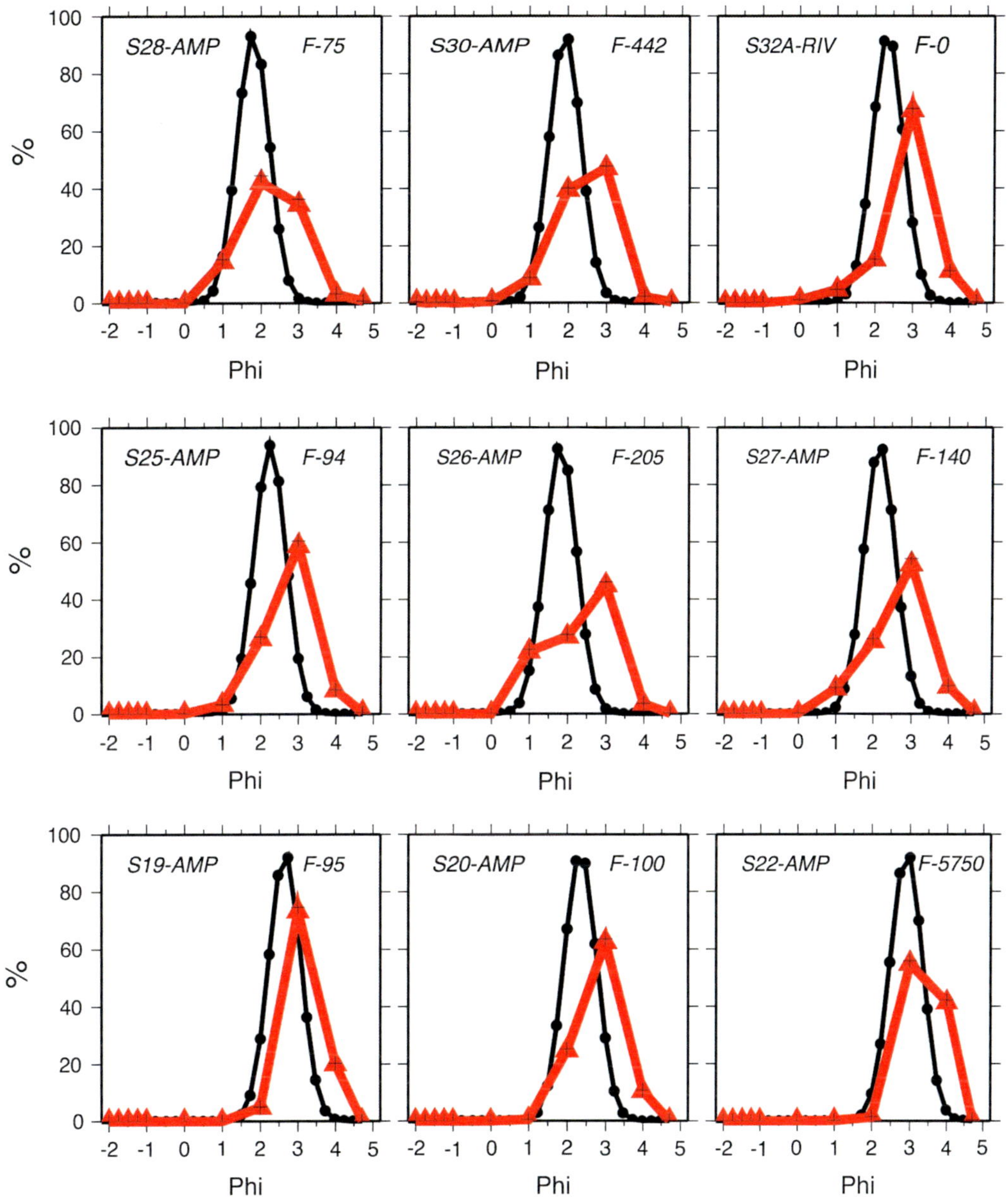

FIGURE 8.8 Grain size distribution of samples from the Ampalaza Chevron compared to samples from the Menarandra River (S32A, upper right). *Black line* with *black dots*: Theoretical model of the grain size distribution of a well-sorted sand-a normal (Gaussian) distribution with the same mean grain size as the sample. *Red line* with *triangles*: Grain size distribution uncorrected for unsieved fine material below 38 μm in size. *F*, fossils per gram.

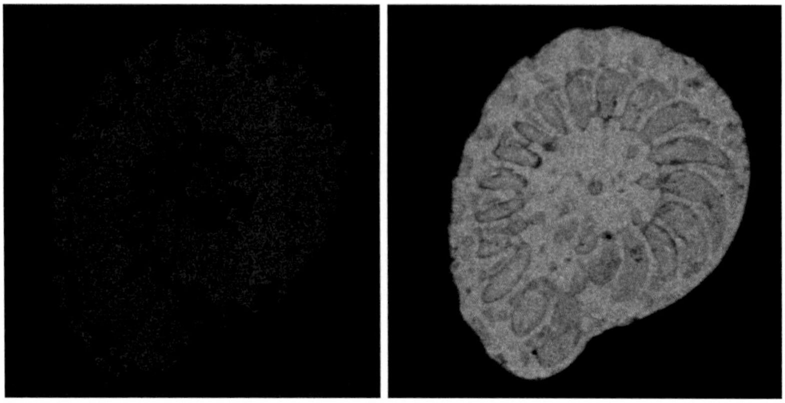

FIGURE 8.9 Element maps of a thin section of a marine microfossil from the distal end of the Ampalaza Chevron. Unaltered calcium carbonate tests of marine microorganisms typically contain too little Mg for the Mg to show up on an element map. (An element needs to be present above the 1% level to be visible in an element map.) The chambers in the microfossil were probably filled with Mg-rich mud after the organism died. Note that some parts of the test also appear Mg-rich, consistent with partial dolomitization of the test.

The fossils within the chevron have ablated surfaces but are still recognizable as marine microfossils, most likely benthic foraminifers. The interiors of the fossils from the chevron sands are often filled, sometimes with Mg-rich material (Fig. 8.9). Outside the chevron, near the ocean, there are sandy, modern beach deposits. The beach sands have a higher fossil content, with thousands of fossils per gram of sediment. S22 is an example. S22 is somewhat landward of the beach but is within reach of coastal flooding from large storms. The marine microfossils in the beach deposits are hollow, lacking an interior filling. Their surfaces are fresh and are not ablated, unlike the marine microfossils within the chevron sands. These differences suggest that the fossils within the Ampalaza Chevron, although geologically young, are not aeolian deposits derived from modern beaches. Instead they represent marine microfossils that were buried, filled, and altered in situ in the marine environment, perhaps on the continental shelf or below the water table on beaches, and were later excavated and deposited within the chevrons.

An EDS element map of a thin section of a marine microfossil from the distal end of the Ampalaza Chevron (sample labeled MWS-100 in Fig. 8.8) shows the outline of a carbonate-rich microfossil test and its interior, the latter filled with Mg-rich material (Fig. 8.9). Note that the shape of the test appears normal; that is, the test does not appear to be significantly ablated or broken.

The Faux Cap Chevron field differs from the Ampalaza and Fenambosy Chevrons, and is more like the chevrons in the rest of Madagascar. Sand deposits are thinner than in the Ampalaza and Fenambosy Chevrons. This may reflect absence of a significant fluvial sand source, and the more nearly perpendicular orientation of the coastline relative to the direction of sediment transport during a putative megatsunami event. Except for very close to the coast, most of the sand contains very few marine microfossils. Due to extensive farming,

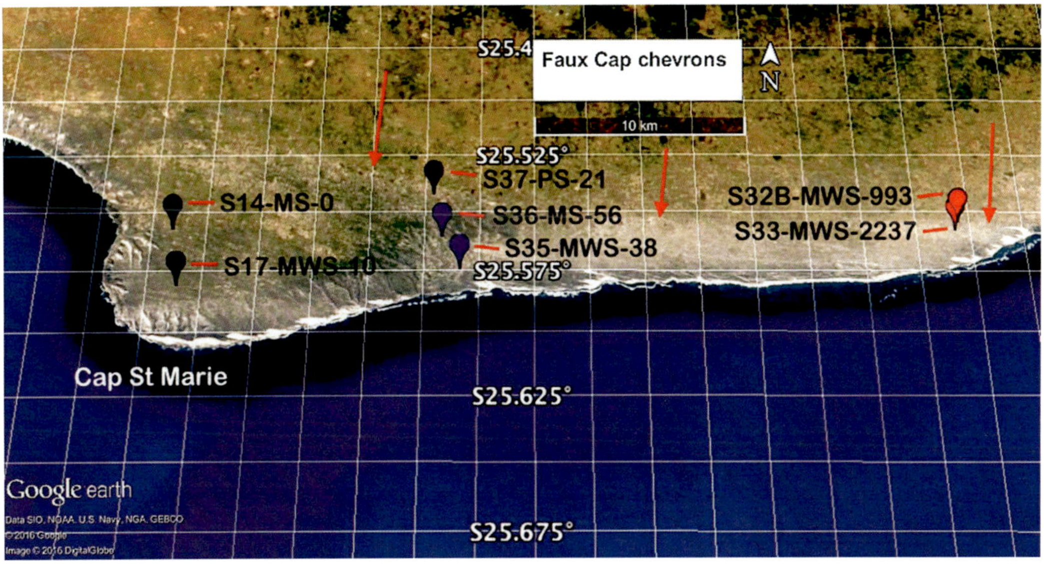

FIGURE 8.10 Image of the Faux Cap Chevron from Google Earth. Image © 2016 DigitalGlobe. Image © 2016 CNES/Astrium. Regional view. Colored symbols: Sampling locations color-coded by percentage of $CaCO_3$ in the substrate. Black: 0–20%, Blue: 20–40%, Red: 40–60%. Black lettering: Site#-degree of sorting-number of marine fossils per gram. The degree of sorting is *PS*, poorly sorted; *MS*, moderately sorted; *MWS*, moderately well sorted.

the Faux Cap Chevrons preserve little of the internal V-shaped structures observed in the Fenambosy and Ampalaza Chevrons (Fig. 8.10). V-shapes are faintly visible in the outlines of the Faux Cap Chevrons.

The sands lacking marine microfossils contain <10% calcium carbonate. These include sand without $CaCO_3$ from the Menarandra River (Fig. 8.1), sampled in a location landward of the chevrons (Fig. 8.7: top). This observation implies that the calcium carbonate in the chevrons is not derived from nearby basement outcrops. Most likely, all the carbonate in the chevrons was originally eroded from local beach and shallow water sediments. The bulk of the fossils would have come from buried fossils within the beach and shelf sand, and would be expected to have some sediment infilling.

The grain size distributions of the sediments from the Faux Cap chevrons (Fig. 8.11) more closely resemble the fluvial material than well-sorted aeolian sand. Thus, the grain size distributions of the chevron sediment are consistent with their transport by water. The samples from the Faux Cap region (Figs. 8.10 and 8.11) also include two examples of sediments from locations close to the ocean but not within a chevron. The samples contain only a small component of carbonate (S14: 3% $CaCO_3$ and S17: 9% $CaCO_3$). The samples contain no (S14) or very few (S17: 10 fossils/gm) marine microfossils. These samples are located above a small beach, a potential source for windblown sediment. Despite the nearshore location and a potential aeolian source area, the sediments from these two sites are poorly to moderately sorted.

One site from the Faux Cap region located well away from the coast (S2, Fig. 8.11) and not on a chevron (Fig. 8.12: *green arrow*) has sand that contains no fossils and is moderately well

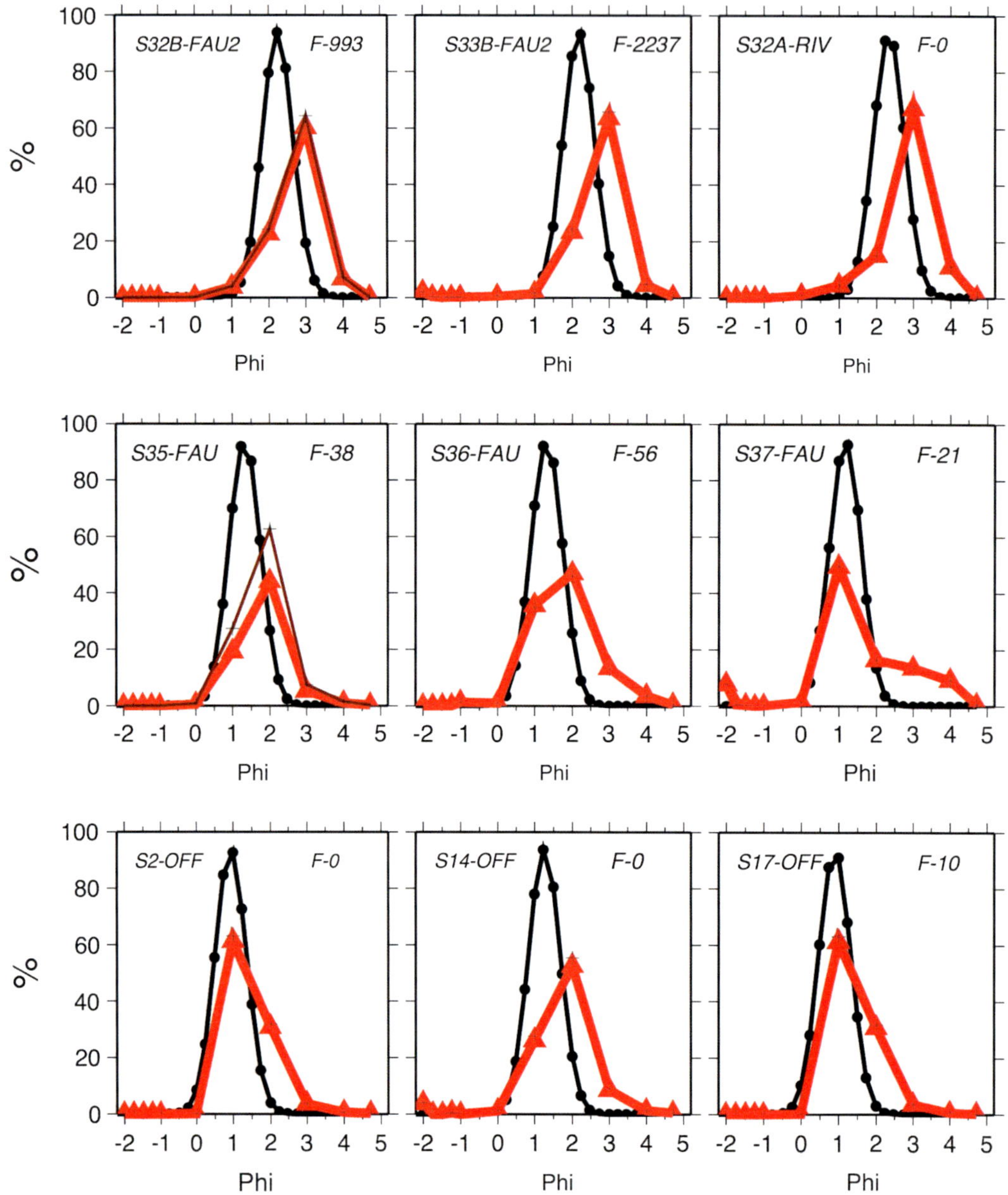

FIGURE 8.11 Grain size distribution of samples from the Faux Cap chevrons compared to samples from the Menarandra River (upper right) and from off chevron areas near Cap St. Marie and to the north (lower row of plots). *Black line* with *black dots*: Theoretical model of the grain size distribution of a well-sorted sand-a normal distribution with the same mean grain size as the sample. *Red line* with *triangles*: Grain size distribution uncorrected for material less than 38 μm in size. *Brown line* with *crosses*: Grain size distribution corrected to 100%, accounting for material washed through the 38 μm sieve, so not applicable to all samples. *F*, fossils per gram.

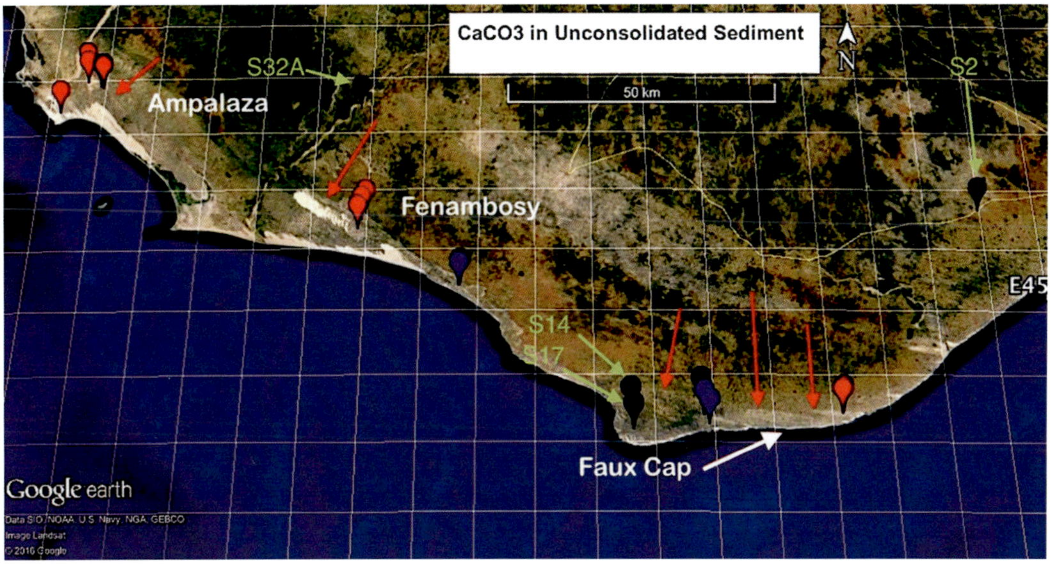

FIGURE 8.12 Summary of $CaCO_3$ content of chevrons and surrounding area. Image from Google Earth is contrast enhanced to show the chevrons. Image © 2016 DigitalGlobe. Image © 2016 CNES/Astrium. Colored symbols: Sampling locations color-coded by percentage of $CaCO_3$ in the sediment. Black: 0–20% $CaCO_3$, Blue: 20–40% $CaCO_3$: Red 40–60% $CaCO_3$. *Red arrows*: Landward edges of chevrons. *Green arrows*: Off chevron sampling sites S2, S14, S17, and S32A.

sorted. It is one of the few samples with a mean grain size that lies within the mean grain size of well-sorted material. The other similar sample is S17, which is also off chevron and contains very few marine microfossils (Fig. 8.12). Samples S2 and S17 contain no material coarser than 0 ϕ (1000 μm), as might be expected of samples that are entirely (S2) or nearly entirely (S17) windblown material. There is some fine sand in S2 and S17 that keeps them from perfectly matching the size distribution of aeolian sand.

Within the chevrons, there is a direct relationship between calcium carbonate content and marine fossil content (Fig. 8.13). As fossil content increases, the average carbonate content increases. As the fossils become more ablated by the wind, they are more difficult to recognize and count. Thus, the fossil counts per gram represent a lower bound in some sediment.

7. DISCUSSION OF MADAGASCAR CHEVRONS

If the chevron sands were derived from the substrate, we would expect their fossil content and grain size distributions to be similar. If the chevron sands were transported inland from beaches by the wind, we would expect them to contain little to no calcium carbonate beyond a few km from the ocean, and no identifiable marine microfossils. We would also expect the associated mineral grains to be well rounded and well sorted (Sharp, 1966). This is not the case. Instead, sediments from locations that are tens of km along-strike in the chevrons contain marine microfossils (Figs. 8.5, 8.7, and 8.10) and significant $CaCO_3$ (summary in Fig. 8.12).

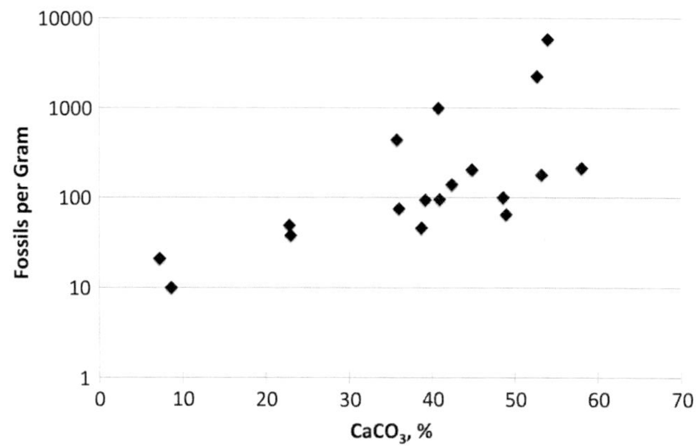

FIGURE 8.13 Bulk calcium carbonate content versus number of marine fossils per gram.

The surrounding sedimentary grains are angular and typically moderately sorted to moderately well sorted (Figs. 8.6, 8.8, and 8.11).

The chevron sands are typically classified as either moderately well sorted or moderately sorted. This may be because the total weight of individual sieved samples is relatively small, about 20–30 g. The largest rocks are not abundant. There are often only one or two per sample. The presence of these larger rocks decreases estimates of the degree of sorting. Thus, the degree of sorting we estimate could be systematically too high. This putative sample size effect may explain why we did not find a large number of sites with poorly sorted sediment within the chevrons.

The presence of Mg-rich fills within the microfossils from the Ampalaza Chevron is complemented by microprobe analyses of individual fossil tests from this site (manuscript in preparation). Microprobe analyses show that MgO is present within the marine microfossil tests at the level of a few percent. This is significantly higher than the MgO content of modern marine foraminiferal tests, which is at most <1% (Nürnberg et al., 1996; Lear et al., 2000, 2002; Reichart et al., 2003). This is not enough MgO for the carbonate in the tests to be called dolomite, but it is enough to suggest that the tests experienced some diagenetic replacement. In dolomite-rich sediments, most of the dolomite is estimated to form within a few tens of m of the sediment–water interface (Baker and Burns, 1985). It is likely that the tests were buried in an environment where calcium carbonate was being replaced by dolomite, with burial long enough for the interiors of the tests to be filled with semilithified material, some with a high Mg content (Fig. 8.9).

8. GEOCHRONOLOGY

We obtained AMS ^{14}C dates of the carbonate microfossils in the chevron sand in three widely separated locations (Figs. 8.14 and 8.15, Table 8.4). The ages of the carbonate range from 13,835 ± 40 to 11,415 ± 35 year BP. The ages of recent marine carbonates in the southwest tropical Indian Ocean vary from 418 ± 57 to 800 ± 59 year BP (Southon et al., 2002),

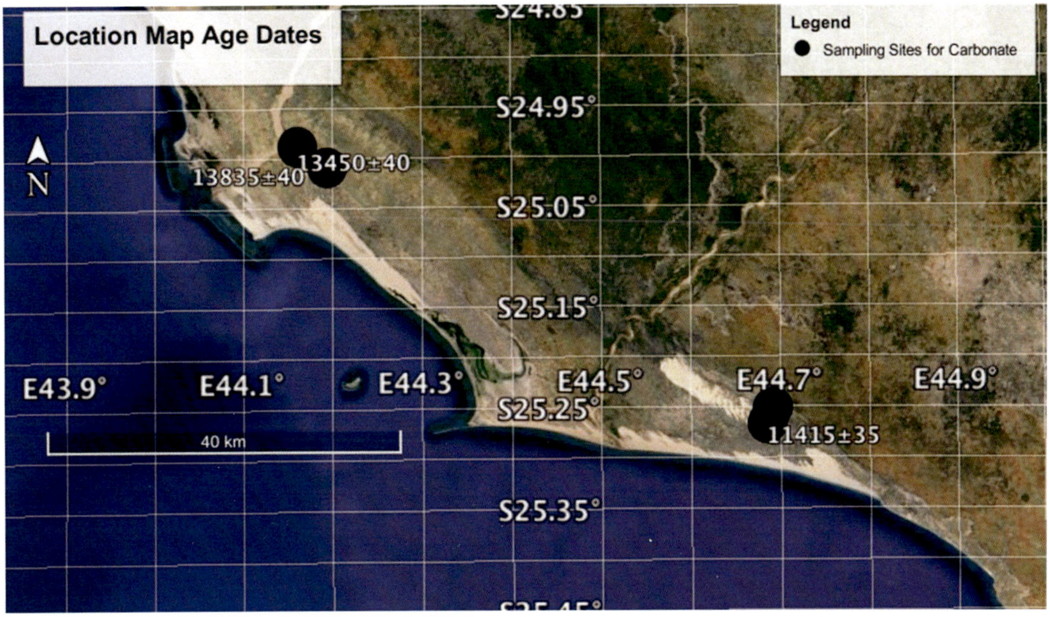

FIGURE 8.14 Location map of ^{14}C sampling sites (*black circles*). Annotations are uncorrected ages with errors. The Ampalaza Chevron on the left (west) has two similar ages. The Fenambosy Chevron on the right (east) has one final age determination and a second age in progress. Because the sand in the chevrons could have been derived from differing levels of erosion of preexisting sediments, all ages are maximum ages and do not preclude the same age of formation of both chevrons.

much younger than the ages of the carbonate microfossils in the chevrons. These ages are not zero because older carbon is incorporated into modern marine calcium carbonate. The correction for this effect is called the marine carbonate reservoir correction.

9. ORIGIN OF THE MADAGASCAR CHEVRONS INVESTIGATED HERE

The age of the chevrons is unknown but it must be geologically young. None of the chevrons are lithified and all the fossils appear as individual tests. Five sets of sedimentologic observations strongly suggest a water-laid, premodern origin for the chevrons. The first is their unusual V-shaped appearance, and the low maximum slopes of the triangular sediment waves in the chevrons, on the order of 5–10 degrees. All of these sediment waves are covered by vegetation. Maximum slopes of water-laid sediment waves are typically <10 degrees, although they are occasionally as high as 20 degrees (Ashley, 1990). The second observation suggesting chevrons were deposited by a megatsunami is the absence of a trend in the degree of sorting along-strike of the chevrons. The sand in the Ampalaza and Fenambosy Chevrons is several m thick and would be expected to become well-sorted and well-rounded within 10–12 km of along-strike transport by the wind (Sharp, 1966). Instead the sediment in the

FIGURE 8.15 Top: White boulder and smaller rocks in farmer's field on top of 175 m high cliff. Middle: Foreground: Typical country rock. Background, middle right: Possible displaced boulders. Some appear white. Farmers seeking to maximize their crop yields may have cleared the smaller boulders from the fields. The red soil in the bank to the left contains no visible rocks and is an unlikely source for boulders of this size. Bottom: View of the ocean, 7 km from edge of cliff and from farmer's field shown above. Boulders here appear gray rather than white.

TABLE 8.4 Uncorrected Radiocarbon Ages of Marine Carbonate

CAMS#	Site Name	Fraction Modern	±	$\delta^{14}C$	±	^{14}C Age	±
172725	MAD 19	0.1874	0.0009	−812.6	0.9	13,450	40
172726	MAD 26	0.1787	0.0009	−821.3	0.9	13,835	40
172727	MAD 13	0.2414	0.0010	−758.6	1.0	11,415	35

chevrons varies in sorting but is typically either moderately well sorted or moderately sorted. The third observation is the difference in maximum slope and shape of the sand waves between the vegetated and unvegetated areas of the chevrons. The sand waves in the white areas form long dunes oriented at right angles to the wind direction. The maximum slopes of the white sand waves on their dip-slopes are 30 degrees, the expected slope for dunes of windblown origin. Therefore, the sand waves in the white, unvegetated areas of the chevrons have experienced a different recent history from the sand waves in the vegetated areas of the chevrons. The fourth is the form, composition, and excellent preservation of the microfossils in the chevrons. The microfossils are abundant and are often filled with Mg-rich material. Their abundance does not decrease along-strike of the chevrons. If the chevrons were entirely of windblown origin, we infer that well-preserved microfossils would be absent beyond 10−12 km along-strike. Instead we see abundant, well-preserved microfossils at distances of tens of km along-strike, including at the most distal end of the Ampalaza Chevron. The sandy substrate of the chevrons contains little carbonate and no fossils (Figs. 8.5, 8.7, and 8.10). There is no or less than 10% $CaCO_3$ and no or sparse marine fossils in sands from the areas surrounding the chevrons (Fig. 8.12). The fifth important observation is that the Ampalaza Chevron is buried on its eastern end by floodplain sediments from the Menarandra River. If the chevrons were modern features that were actively forming through aeolian transport of beach and fluvial sand, we would expect to see well-preserved V-shaped chevrons immediately west of the bank of the Menarandra River. There are possibly a few relict chevrons being farmed in this area (Fig. 8.7: top) but they appear to be partially reworked by the wind.

These sedimentologic observations are consistent with a megatsunami origin for these chevrons; geochronological data help to pinpoint the timing of such an event. Because we infer that carbonate was most likely eroded from a preexisting, fossil-bearing sediment, the ages of the carbonate particles in the sediment are roughly the same as the age of the depositional event and older. That is, the ^{14}C ages are maximum ages for the megatsunami. Considering that these are maximum ages, the range of ages is small. Further, these ages constitute additional evidence that the chevrons are not modern aeolian deposits derived from local beaches. If the carbonate microfossils in the chevrons were blown in from nearby beaches, the ages of the carbonate in the chevrons would be close to or within the range of the reservoir correction for the Indian Ocean. While it is conceivable that nondegraded or lightly degraded fossil shell is present in the littoral margin of southern Madagascar, and could be incorporated into aeolian deposits, and while our radiocarbon sample size is quite small, our overall project data suggest that the chevron deposits are water-laid, and date prior to the late Holocene or modern era.

The ^{14}C age results also appear to preclude another hypothesis: that the individual marine carbonate tests are being eroded from poorly lithified carbonate-rich rocks in the basement. If the carbonate tests were derived from the basement, they would be radiocarbon dead, giving inconclusive ages of >45,000 years.

10. OTHER MODERN TSUNAMI DEPOSITS: MIXTURES OF CARBONATE-RICH SAND AND LARGE ROCKS

Studies of tsunami deposits in modern environments show that mixtures of calcium carbonate-rich sand and large rocks, often of megaboulder size, are common. There are numerous well-documented examples of boulders transported inland by tsunamis (Bryant et al., 1992; Scheffers and Kelletat, 2003; Scicchitano et al., 2007; Paris et al., 2010; Bryant, 2014). Where other sediments are present, the boulders are often in stratigraphic contact with carbonate- and/or silicate-rich sand.

Large boulders and sand sheets deposited from a megatsunami have been documented in the Cape Verde islands (Ramalho et al., 2015). The sand layers contain dispersed rounded and angular clasts. The maximum run-up heights are in excess of 270 m. The megatsunami was produced from a caldera collapse of Fogo volcano about 73,000 years ago.

In Madagascar, there are white boulders within the sands at the top of the 175-m high cliff within the Fenambosy Chevron. Because they are too large for subsistence farmers to move without mechanical assistance, large boulders lie within a field that is being actively farmed (Fig. 8.15; top). Other rocks appear in piles within depressions in the country rock. The whiter rocks lack the dark coating of the basement rock (Fig. 8.15; middle). The white boulders might represent more recent deposits with insufficient time to develop a dark weathering rind. These boulders need to be investigated in more detail, to determine if they were torn from the edge of the cliff or if they were transported over a longer distance (Fig. 8.15; bottom).

11. SUGGESTIONS FOR FURTHER WORK

Many workers have documented the presence of foraminifera within tsunami deposits (Mamo et al., 2009; Sugawara et al., 2009; Pilarczyk and Reinhardt, 2012). We know of no other cases where the foraminifera within tsunami deposits are filled in and partially dolomitized; however, most researchers do not make thin sections of foraminifera or examine them with a scanning electron microscope. In the future, it would be helpful to know if the dolomitization of marine microfossils is nonuniform within tsunamigenic sequences. In areas where dolomitization of offshore sequences is favored, the maximum amount of dolomitization might correlate well with the degree of erosion of offshore sediments by the tsunami, a likely marker for tsunami size.

The high carbonate contents of the Madagascar chevrons suggest that multispectral remote sensing data should be able to evaluate the carbonate content of other V-shaped dune complexes. In tropical regions, coastal chevron dunes with high carbonate contents should be identified and studied to determine their age and origin.

Once modern shoreline tsunami deposits are more fully documented, it will be possible to identify more of their counterparts within Precambrian and Phanerozoic sediments. Although Precambrian carbonates were not precipitated as shells, their precipitation was modulated by a low rate of deposition of clastic material. In areas with a high rate of clastic deposition, carbonate precipitation is overwhelmed and the sediment subsequently becomes a shale or sandstone. Thus, the admixture of large amounts of fine-grained carbonate or dolomite with significant amounts of sand, gravel, or boulders suggests some special circumstances. If the sedimentary structures are also appropriate, such mixed carbonate-coarse clastic sequences could represent ancient megatsunami deposits.

The Madagascar chevrons show that the stratigraphic expression of ancient tsunami deposits on shorelines is likely to be complicated. The water-laid sand waves of low amplitude have been preserved by vegetation in some parts of the chevrons but in other areas the sand is being actively eroded and reworked by the wind. Because there was little to no vegetation during Precambrian time (Long, 2011; Eriksson et al., 2013; Mazumder and Van Kranendonk, 2013), this mixed stratigraphic expression of tsunami deposits could be present in Precambrian sequences. In addition, the high carbonate content of the Madagascar chevrons, typically 30–50%, suggests that abundant carbonate sand mixed with sand, gravel, or boulders that are not carbonate and/or of differing carbonate/dolomite content could be a marker for ancient tsunami deposits (Lowe and Byerly, 1988; Hassler et al., 2000; Hassler and Simonson, 2001; Glikson, 2004; Glass and Simonson, 2012; Lowe et al., 2014). We find that the average bulk carbonate content in the Madagascar sediments is directly related to the fossil content per unit weight (Fig. 8.14). This implies that similar mixed carbonate-sand-gravel-boulder sequences in ancient sediments should be assessed for current structures and sediment waves, characteristic of aqueous deposition.

APPENDIX 8.1: WEIGHT PERCENTAGE DATA OF DIFFERENT GRAIN SIZES USED TO CALCULATE GRAIN SIZE PARAMETERS IN TABLE 8.2

Station Number	>4000	>3360	>2800	>2360	>2000	>1000	>500	>250	>125	>63	>38	Sum
S2	0.0	0.0	0.0	0.0	0.0	0.2	61.3	30.9	3.7	0.7	0.4	97.2
S4	0.0	0.2	0.0	0.4	0.8	2.0	5.7	28.8	38.0	11.3	2.3	89.4
S5	1.0	0.2	0.0	0.1	0.0	0.0	2.1	70.5	23.0	1.0	0.3	98.1
S9	0.0	0.1	0.0	0.0	0.0	0.2	31.6	14.6	38.8	11.0	0.4	96.7
S12	0.5	0.6	0.7	0.2	0.1	0.5	0.8	8.2	70.5	13.2	0.4	95.6
S13	0.2	0.2	0.0	0.0	0.0	0.2	14.0	23.6	37.1	4.4	0.3	80.1
S14	3.8	0.0	0.0	0.5	0.0	1.2	26.2	52.7	8.7	1.3	0.5	94.9
S17	0.0	0.1	0.0	0.1	0.0	2.7	57.3	34.0	3.7	0.3	0.1	98.4

(Continued)

Station Number	>4000	>3360	>2800	>2360	>2000	>1000	>500	>250	>125	>63	>38	Sum
S19	0.0	0.0	0.0	0.0	0.0	0.0	0.0	4.8	73.3	19.9	0.1	98.1
S20	0.0	0.0	0.0	0.0	0.0	0.0	0.5	24.7	62.4	10.6	0.1	98.3
S22	0.0	0.0	0.0	0.0	0.0	0.0	0.1	1.4	54.8	41.4	0.4	98.0
S25	0.0	0.0	0.0	0.0	0.0	0.2	3.3	26.3	58.9	8.3	0.4	97.5
S26	0.0	0.0	0.0	0.0	0.0	0.1	21.8	27.0	44.9	3.5	0.1	97.4
S27	0.0	0.0	0.0	0.0	0.0	0.1	8.9	25.3	52.2	9.6	0.5	96.5
S28	0.0	0.0	0.0	0.0	0.0	0.0	14.5	42.1	34.5	2.9	0.7	94.7
S30	0.3	0.0	0.1	0.1	0.0	0.6	8.7	39.5	47.0	2.2	0.1	98.7
S32A	0.0	0.0	0.0	0.0	0.0	1.0	4.7	15.1	67.1	11.1	0.3	99.3
S32B	0.0	0.0	0.0	0.0	0.0	0.2	3.8	22.7	60.4	6.8	0.1	94.0
S33	2.0	0.3	0.1	0.1	0.2	0.6	1.5	23.4	63.6	4.7	0.3	96.6
S35	0.0	0.0	0.0	0.0	0.0	0.5	19.2	44.1	5.3	1.0	0.3	70.5
S36	0.0	0.0	0.0	0.0	1.3	0.9	34.5	45.4	13.0	3.6	0.5	99.2
S37	7.8	0.5	0.3	0.0	0.1	1.5	49.6	16.5	13.6	9.3	0.6	99.8

Sizes of grains are in μm.

Acknowledgments

We thank WAPMERR for funding our 2006 Expedition to Madagascar. We thank Andriamiranto Raveloson and Hoby Razafinodrakoto for their invaluable assistance in the field. We thank W. Bruce Masse, Ann Isley and A.J. van Loon for helpful suggestions that improved the text. We thank the Museum of Natural History in New York City for access to their microprobe facility. We thank the City College of New York for access to their scanning electron microscope facility. We are grateful to the late Jeff Steiner who was responsible for creating this wonderful facility and for sponsoring our initial access. We thank Tom Guilderson and the Center for AMS dating at Lawrence Livermore National Laboratory for their careful analyses of our carbonate-rich sand. We thank Kara Dennis for help with calcium carbonate analyses. We thank Leanne Darson for grain size analyses and for picking fossils for thin sections. We thank the Earth Institute for support of salary costs for Leanne. We thank NSF grant OCE-13-59194 for support of Karina Galinskaya, the AMS ^{14}C dating, and carbonate analyses.

References

Abbott, D.H., Bryant, E.F., Gusiakov, V., Masse, W., Breger, D., 2008. Impacts, mega-tsunami, and other extraordinary claims: COMMENT. GSA Today 18 (6), e12.

Abbott, D.H., Masse, W.B., Burckle, L.H., Breger, D., Gerard-Little, P., 2007. Burckle abyssal impact crater: did this impact produce a global deluge? In: Papmarinopoulous, S.P. (Ed.), The Atlantis Hypothesis: Searching for a Lost Land. Heliotopos Publications, Greece, pp. 179–190.

Allen, J.R.L., 1985. Principles of Physical Sedimentology. Chapman and Hall.

Ashley, G.M., 1990. Classification of large-scale subaqueous bedforms: a new look at an old problem-SEPM bedforms and bedding structures. Journal of Sedimentary Petrology 60 (1), 160–172.

Baker, P.A., Burns, S.J., 1985. Occurrence and formation of dolomite in organic-rich continental margin sediments. AAPG Bulletin 69 (11), 1917–1930.

Blott, S.J., Pye, K., 2001. GRADISTAT: a grain size distribution and statistics package for the analysis of unconsolidated sediments. Earth Surface Processes and Landforms 26, 1237–1248.

Bourgeois, J., Weiss, R., 2009. "Chevrons" are not mega-tsunami deposits—A sedimentologic assessment. Geology 37 (5), 403–406.

Bryant, E., 2014. Tsunami: The Underrated Hazard. Springer.

Bryant, E.A., Nott, J., 2001. Geological indicators of large tsunami in Australia. Natural Hazards 24 (3), 231–249.

Bryant, E.A., Young, R.W., Price, D.M., 1992. Evidence of tsunami sedimentation on the southeastern coast of Australia. The Journal of Geology 1, 753–765.

Camoin, G.F., Montaggioni, L.F., Braithwaite, C.J.R., 2004. Late glacial to post glacial sea levels in the Western Indian Ocean. Marine Geology 206 (1), 119–146.

Cox, R., Zentner, D.B., Kirchner, B.J., Cook, M.S., 2012. Boulder ridges on the Aran Islands (Ireland): recent movements caused by storm waves, not tsunamis. The Journal of Geology 120 (3), 249–272.

De Villiers, S., 2005. Foraminiferal shell-weight evidence for sedimentary calcite dissolution above the lysocline. Deep Sea Research Part I: Oceanographic Research Papers 52 (5), 671–680.

Erdmann, W., Kelletat, D., Scheffers, A.M., Haslett, S., 2015. Origin and Formation of Coastal Boulder Deposits at Galway Bay and the Aran Islands. Heidelberg Springer Briefs in Geography, Western Ireland.

Eriksson, P.G., Banerjee, S., Catuneanu, O., Corcoran, P.L., Eriksson, K.A., Hiatt, E.E., Laflamme, M., Lenhardt, N., Long, D.G.F., Miall, A.D., Mints, M.V., Pufahl, P.K., Sarkar, S., Simpson, E.L., Williams, G.E., 2013. Secular changes in sedimentation systems and sequence stratigraphy. Gondwana Research 24, 468–489.

Glass, B.P., Simonson, B.M., 2012. Distal impact ejecta layers: Spherules and more. Elements 8, 43–48.

Glikson, A.Y., 2004. Early Precambrian asteroid impact-triggered tsunami: excavated seabed, debris flows, exotic boulders, and turbulence features associated with 3.47–2.47 Ga-old asteroid impact fallout units, Pilbara craton, western Australia. Astrobiology 4 (1), 19–50.

Gusiakov, V., Abbott, D.H., Bryant, E.A., Masse, W.B., Breger, D., 2010. Mega tsunami of the world oceans: chevron dune formation, micro-ejecta, and rapid climate change as the evidence of recent oceanic bolide impacts. In: Beer, T. (Ed.), Geophysical Hazards, Minimizing Risk, Maximizing Awareness Netherlands. Springer, pp. 197–227.

Hassler, S.W., Robey, H.F., Simonson, B.M., 2000. Bedforms produced by impact-generated tsunami, ~2.6 Ga Hamersley basin, Western Australia. Sedimentary Geology 135 (1).

Hassler, S.W., Simonson, B.M., 2001. The sedimentary record of extraterrestrial impacts in deep-shelf environments: evidence from the early Precambrian. The Journal of Geology 109, 1–19.

Hearty, P.J., Neumann, A.C., Kaufman, D.S., 1998. Chevron ridges and runup deposits in the Bahamas from storms late in oxygen-isotope substage 5e. Quaternary Research 50 (3), 309–322.

Kindler, P., Strasser, A., 2000. Palaeoclimatic significance of co-occurring wind-and water-induced sedimentary structures in the last-interglacial coastal deposits from Bermuda and the Bahamas. Sedimentary Geology 131 (1), 1–7.

Krumbein, W.C., Pettijohn, F.J., 1938. Manual of Sedimentary Petrography. Appleton-Century-Crofts, New York.

Lear, C.H., Elderfield, H., Wilson, P.A., 2000. Cenozoic deep-sea temperatures and global ice volumes from Mg/Ca in benthic foraminiferal calcite. Science 287 (5451), 269–272.

Lear, C.H., Rosenthal, Y., Slowey, N., 2002. Benthic foraminiferal Mg/Ca-paleothermometry: a revised core-top calibration. Geochimica et Cosmochimca Acta 66 (19), 3375–3387.

Long, D.G.F., 2011. Architecture and depositional style of fluvial systems before land plants: a comparison of Precambrian, early Paleozoic and modern river deposits. In: Davidson, S.K., Leleu, S., North, C.P. (Eds.), From River to Rock Record: The Preservation of Fluvial Sediments and Their Subsequent Interpretation, vol. 97. SEPM Special Publication, pp. 37–61.

Lowe, D.R., Byerly, G.R., 1988. Identification and effects of large, early Archean, terrestrial meteorite impacts: a geological perspective on late accretion. In: Proceedings Lunar and Planetary Science Conference, vol. 19. Lunar and Planetary Institute, Houston, TX, p. 693.

Lowe, D.R., Byerly, G.R., Kyte, F.T., 2014. Recently discovered 3.42–3.23 Ga impact layers, Barberton Belt, South Africa: 3.8 Ga detrital zircons, Archean impact history, and tectonic implications. Geology 42 (9), 747–750.

Mamo, B., Strotz, L., Dominey-Howes, D., 2009. Tsunami sediments and their foraminiferal assemblages. Earth-Science Reviews 96 (4), 263–278.

Maxwell, T.A., Haynes, C.V., 1989. Large-scale, low-amplitude bedforms (chevrons) in the selima sand sheet, Egypt. Science 243 (4895), 1179–1182.

Mazumder, R., Van Kranendonk, M.J., 2013. Paleoproterozoic terrestrial sedimentation in the Beasley river quartzite, lower Wyloo Group, western Australia. Precambrian Research 231, 98–105.

Nürnberg, D., Bijma, J., Hemleben, C., 1996. Assessing the reliability of magnesium in foraminiferal calcite as a proxy for water mass temperatures. Geochimica et Cosmochimica Acta 60 (5), 803–814.

Oehler, J.F., Labazuy, P., Lénat, J.F., 2004. Recurrence of major flank landslides during the last 2-Ma-history of Reunion Island. Bulletin of Volcanology 66 (7), 585–598.

Paris, R., Fournier, J., Poizot, E., Etienne, S., Morin, J., Lavigne, F., Wassmer, P., 2010. Boulder and fine sediment transport and deposition by the 2004 tsunami in Lhok Nga (western Banda Aceh, Sumatra, Indonesia): a coupled offshore–onshore model. Marine Geology 268, 43–54.

Pilarczyk, J.E., Reinhardt, E.G., 2012. Testing foraminiferal taphonomy as a tsunami indicator in a shallow arid system lagoon: Sur, Sultanate of Oman. Marine Geology 295, 128–136.

Pinter, N., Ishman, S.E., 2008. Impacts, mega-tsunami, and other extraordinary claims. GSA Today 18 (1), 37–38.

Ramalho, R.S., Winckler, G., Madeira, J., Helffrich, G.R., Hipólito, A., Quartau, R., Adena, K., Schaefer, J.M., 2015. Hazard potential of volcanic flank collapses raised by new megatsunami evidence. Science Advances 1 (9).

Reichart, G.J., Jorissen, F., Anschutz, P., Mason, P.R., 2003. Single foraminiferal test chemistry records the marine environment. Geology 31 (4), 355–358.

Scheffers, A., Kelletat, D., 2003. Sedimentologic and geomorphologic tsunami imprints worldwide—A review. Earth-Science Reviews 63, 83–92.

Scheffers, A., Kelletat, D., Scheffers, S.R., Abbott, D.H., Bryant, E.A., 2008. Chevrons–enigmatic sedimentary coastal features. Zeitschrift für Geomorphologie 52 (3), 375–402.

Scicchitano, G., Monaco, C., Tortorici, L., 2007. Large boulder deposits by tsunami waves along the Ionian coast of south-eastern Sicily (Italy). Marine Geology 238, 75–91.

Sharp, R.P., 1966. Kelso dunes, Mojave desert, California. Geological Society of America Bulletin 77, 1045–1074.

Skoček, V., Saadallah, A.A., 1972. Grain-size distribution, carbonate content and heavy minerals in eolian sands, southern desert, Iraq. Sedimentary Geology 8 (1), 29–46.

Southon, J., Kashgarian, M., Fontugne, M., Metivier, B., Yim, W.W., 2002. Marine reservoir corrections for the Indian Ocean and Southeast Asia. Radiocarbon 44 (1), 167–180.

Sugawara, D., Minoura, K., Nemoto, N., Tsukawaki, S., Goto, K., Imamura, F., 2009. Foraminiferal evidence of submarine sediment transport and deposition by backwash during the 2004 Indian Ocean tsunami. Island Arc 18 (3), 513–525.

Visher, G.S., 1969. Grain size distributions and depositional processes. Journal of Sedimentary Research 39 (3), 1074–1106.

Woodroffe, S.A., Horton, B.P., 2005. Holocene sea-level changes in the Indo-Pacific. Journal of Asian Earth Sciences 25 (1), 29–43.

CHAPTER

9

The Contourite Problem

G. Shanmugam
The University of Texas at Arlington, Arlington, TX, United States

OUTLINE

http://dx.doi.org/10.1016/B978-0-12-803386-9.00009-5

1. INTRODUCTION

1.1 Contourite Research

G. Wust (1933), B. C. Heezen (1959), and C. D. Hollister (1967) were the three pioneers of contourite research in the 20th Century. The domain of contourite research has a long history of contributions on both physical oceanography and process sedimentology (Wüst, 1933; Stommel, 1958; Heezen and Hollister, 1964; Hubert, 1964; Hsü, 1964; Heezen et al., 1966; Klein, 1966; Hollister, 1967; Hollister and Heezen, 1972; Bouma and Hollister, 1973; Stow and Lovell, 1979; Lovell and Stow, 1981; Hollister and McCave, 1984; Faugères et al., 1984; Gonthier et al., 1984; Broecker, 1991; Faugères and Stow, 1993; Shanmugam et al., 1993a,b; Viana and Rebesco, 2007; Hunter et al., 2007; McCave, 2008; Rebesco and Camerlenghi, 2008; Rebesco et al., 2008, 2014; Shanmugam, 2008a; Zenk, 2008; Faugères and Mulder, 2011; Stow et al., 2013; Talley, 2013; Hernández-Molina et al., 2013, 2014, 2016; Mulder et al., 2013; Mutti et al., 2014; Pérez et al., 2015; Valencia et al., 2015; among others).

From a sedimentological point of view, the contourite research can be broadly grouped into an early North American phase (1960s–1970s), and the subsequent European phase (1980s–present). Heezen (1959) was the first one to link the ripple bedforms (i.e., traction structures) observed on the deep-sea floor to global ocean currents that are certainly not turbidity currents. Following this novel idea, Hollister's (1967) PhD work at Columbia University (New York) under the supervision of Bruce Heezen, on the western North Atlantic, is truly the initiation of the contourite research. Hollister (1993) provided a succinct history of this early phase. In the midst of the emerging turbidite paradigm in the early 1960s, Hollister presented his pioneering idea, which advocated that strong near-bottom deep-sea contour currents moving in response to thermohaline circulation (THC) patterns were indeed responsible for generating cross-laminated sands with low mud matrix and with heavy mineral concentrations, at the 1963 International Union of Geodesy and Geophysics conference in Berkeley, California (Heezen and Hollister, 1963). He attributed these traction structures to reworking by bottom currents and to winnowing away of fines. His revolutionary ideas were severely criticized. However, Hollister and his students have prevailed in establishing the founding principles of deep-sea contourites (Hollister, 1993), which included the High Energy Benthic Boundary Layer Experiment (Hollister et al., 1980). As a turbidophobe (i.e., one who questions the orthodoxy that all deep-sea sands are turbidites; see Hsü, 2008, p. 14, for a fascinating tale on the turbidite mindset in the 1960s), Hsü (1964) argued that traction structures in deep-marine sands were more meaningful as deposits of contour currents than of turbidity currents. His pioneering idea was met with angry opposition from the academics at the 1963 SEPM meeting in Houston.

Stow (1977), based on his PhD work at Dalhousie University (Canada) under the supervision of David Piper, on the Nova Scotian Continental Margin, proposed some of the early concepts of contourites that prevail today (Rebesco et al., 2014). These concepts are that (1) a continuum may exist between turbidity currents, contour currents, and hemipelagic settling, (2) there are two types of contourites, muddy and sandy, (3) bioturbation is common in both muddy and sandy types, (4) sandy contourites contain traction structures, and (5) sandy contourites may serve as deep-water petroleum reservoirs (Stow and Lovell, 1979).

Subsequently, the influence of the European research community on contourite research is evident in three Geological Society of London publications (Stow and Piper, 1984; Stow et al., 2002; Viana and Rebesco, 2007). The European influence is even more striking in the thematic volume *Contourites*, edited by Rebesco and Camerlenghi (2008). Of the 25 chapters in the volume, 22 (88%) are from the European research community (Table 9.1); only three are from non-European authorships (United States, China, and Brazil). Integrated Ocean Drilling Program (IODP) Expedition 339 (November 2011 to January 2012) in the Gulf of Cadiz and off the West Iberian margin, which drilled contourites, is dominated by the European research community (Hernández-Molina et al., 2013). The two cochief scientists of Expedition 339 were from Spain (F.J. Hernández-Molina) and the United Kingdom (D.A.V. Stow). After nearly four decades, the contribution from global research communities is balanced as reflected by IODP Expedition 342, which drilled Paleogene sediment drifts off Newfoundland (Expedition 342 Scientists, 2012). These comments are not a criticism, but rather an observation from a historical perspective.

In general, most of what we know about modern-day contourites is based primarily on large-scale features observed on seismic and bathymetric data (Table 9.1), with some sediment core data. On the other hand, the literature on ancient contourites offers ample details on small-scale sedimentary features based on outcrop and conventional core data (Natland, 1967; Bouma and Hollister, 1973; Bein and Weiler, 1976; Mutti, 1992; Shanmugam et al., 1993a; Martın-Chivelet et al., 2008; Mutti and Carminatti, 2011; Shanmugam, 2012a), but with only limited information on paleo-oceanography and on large-scale depositional features. This disparity in conjunction with other issues normally associated with deep-water processes and facies (Shanmugam, 2012a) have resulted in a multitude of challenges in interpreting ancient contourites (Hüeneke and Stow, 2008).

1.2 Description of the Problem

Faugères and Stow (1993) presented an overview of the selected problems concerning deep bottom-current-controlled deposits. In a review, Rebesco et al. (2014) provide a useful catalog of contributions on contourite research through the decades. Although Rebesco et al. have acknowledged a few of the contourite problems (e.g., facies model), they have overlooked some fundamental issues. For example, Rebesco et al. (2014, their Fig. 18) promote the bedform-velocity matrix without acknowledging its flaws (see Section 4.10).

In advancing contourite research, a rigorous scrutiny of all basic problems is imperative. Otherwise, the reader is left with a false impression that the science of contourites is mostly settled. In reality, contourite research is at a crisis stage (Shanmugam, 2006a, 2008a,b, 2012a), if we consider Kuhn's (1996) five stages of scientific revolutions, which comprise (1) random observations, (2) first paradigm, (3) crisis, (4) revolution, and (5) normal science. The contourite problem, somewhat analogous to the turbidite problem (Van der Lingen, 1969; Shanmugam, 2000), the tsunamite problem (Shanmugam, 2006b), the landslide problem (Shanmugam, 2015) and the seismite problem (Shanmugam, 2016c), has implications for both process sedimentology and petroleum geology. A total

TABLE 9.1 Contourite Research Contributions by Country for the 25 Chapters in the Edited Volume "Contourites" (Rebesco and Camerlenghi, 2008)

Chapter	Contribution (Chapter Title)	Authorship	First Author's Affiliated Institution or Residence by Country
1	Contourite research: A field in full development	M. Rebesco, A. Camerlenghi, and A.J. Van Loon	Italy
2	Personal reminiscences on the history of contourites	K.J. Hsü	United Kingdom
3	Methods for contourite research	J.A. Howe	United Kingdom
4	Abyssal and contour currents	W. Zenk	Germany
5	Deep-water bottom currents and their deposits	G. Shanmugam	United States
6	Dynamics of the bottom boundary layer	S. Salon, A. Crise, and A.J. Van Loon	Italy
7	Sediment entrainment	Y. He, T. Duan, and Z. Gao	China
8	Size sorting during transport and deposition of fine sediments: sortable silt and flow speed	I.N. McCave	United Kingdom
9	The nature of contourite deposition	D.A.V. Stow, S. Hunter, D. Wilkinson, and F.J. Hernández-Molina	United Kingdom
10	Traction structures in contourites	J. Martín-Chivelet, M.A. Fregenal-Martínez, and B. Chacón	Spain
11	Bioturbation and biogenic sedimentary structures in contourites	A. Wetzel, F. Werner, and D.A.V. Stow	Switzerland
12	Some aspects of diagenesis in contourites	P. Giresse	France
13	Contourite facies and the facies model	D.A.V. Stow and J.-C. Faugères	United Kingdom
14	Contourite drifts: nature, evolution, and controls	J.-C. Faugères and D.A.V. Stow	France
15	Sediment waves and bedforms	R.B. Wynn and D.G. Masson	United Kingdom

TABLE 9.1 Contourite Research Contributions by Country for the 25 Chapters in the Edited Volume "Contourites" (Rebesco and Camerlenghi, 2008)—cont'd

Chapter	Contribution (Chapter Title)	Authorship	First Author's Affiliated Institution or Residence by Country
16	Seismic expression of contourite depositional systems	T. Nielsen, P.C. Knutz, and A. Kuijpers	Denmark
17	Identification of ancient contourites: problems and paleoceanographic significance	H. Hüneke and D.A.V. Stow	Germany
18	Abyssal plain contourites	F.J. Hernández-Molina, A. Maldonado, and D.A.V. Stow	Spain
19	Continental slope contourites	F.J. Hernández-Molina, E. Llave, and D.A.V. Stow	Spain
20	Shallow-water contourites	G. Verdicchio and F. Trincardi	Italy
21	Mixed turbidite–contourite systems	T. Mulder, J.-C. Faugères, and E. Gonthier	France
22	High-latitude contourites	T. van Weering, M. Stoker, and M. Rebesco	The Netherlands
23	Economic relevance of contourites	A.R. Viana	Brazil
24	Paleoceanographic significance of contourite drifts	P.C. Knutz	Denmark
25	The significance of contourites for submarine slope stability	J.S. Laberg, and A. Camerlenghi	Norway

Note that 22 chapters (88%) represent contributions from the European Research Community, and only three chapters (5, 7, and 23) are from non-European countries (United States, China, and Brazil).

of 46 self-citations is included in illustrating my relentless endeavors against the orthodoxy of deep-water facies models.

In the petroleum industry, the concept of contourites is muddled. Based on a detailed core study of Cretaceous and Tertiary deep-water sandstones in the Campos Basin, offshore Brazil, Mutti and Carminatti (2011) have reinterpreted "turbidite" sands as "turbidite-contourite" sands with emphasis on bottom-current reworking by tidal currents. The problem with the Brazilian core study is that deposits of tidal currents have been classified as contourites. The confusion here is that tidalites are not contourites. Furthermore, the reservoir quality of bottom-current reworked sands, which include contourites, has become the

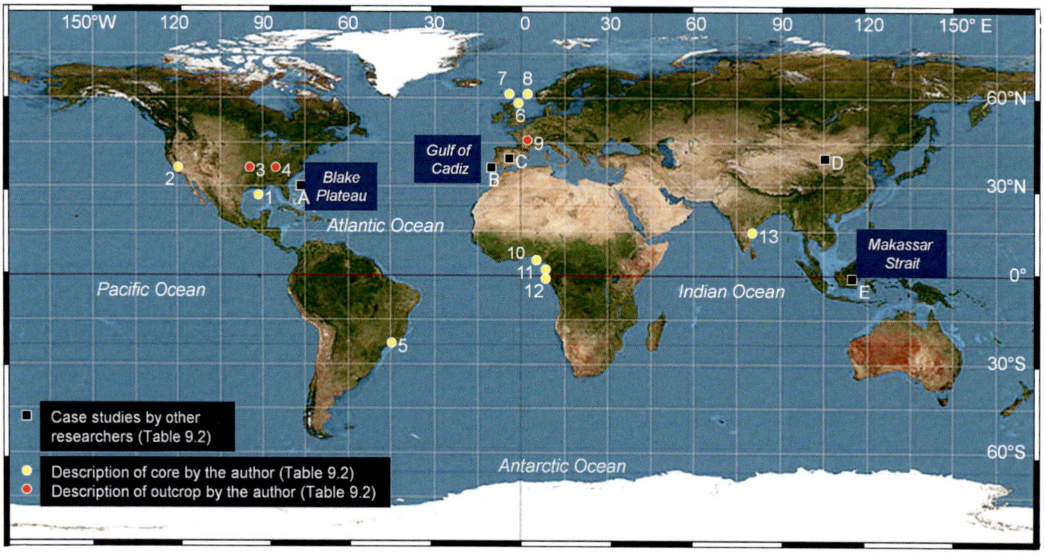

FIGURE 9.1 Map showing the locations of case studies used in this chapter, which include critical case studies by other researchers (locations A, B), and locations of studies by other researchers that resulted in recent debates on deep-water processes (locations C, D, and E). Note 35 locations of core and outcrop descriptions of deep-water sandstones with traction structures that were interpreted by the present author as products of bottom-current reworking (Table 9.2). Blank world map credit: http://upload.wikimedia.org/wikipedia/commons/8/83/Equirectangular_projection_SW.jpg.

center of a lively debate in the AAPG Bulletin (Dunham and Saller, 2014). Dunham and Saller (2014) argued that the reservoir quality of contourites is poor in comparison to turbidites in the Kutei Basin in the Makassar Strait (Fig. 9.1, location E). Their notion is based on a false premise that only turbidites form good-quality reservoir sands (see reply by Shanmugam, 2014a).

During the past three decades, the founding principles of contourites have been gradually eroded away as discussed in this chapter. This attrition has led to a lack of conceptual clarity. Therefore, the primary purpose of this critical review is to identify and explain the basic contourite problems and to offer possible solutions or suggestions in selected cases. In this review, 15 fundamental issues have been identified: (1) the dual forcing of global ocean circulation; (2) continuum between turbidity currents and contour currents; (3) the founding principle of contour currents; (4) regional hiatuses; (5) origin of erosional features; (6) Gulf of Cadiz as the type locality; (7) the contourite facies model; (8) grain-size data and related issues; (9) traction structures and shale clasts; (10) the bedform-velocity matrix; (11) interpretation of seismic profiles, side-scan sonar images, and submarine photographs; (12) oceanic waves (internal, cyclonic, and tsunami); (13) reservoir quality; (14) sediment provenance and (15) abyssal plain contourites.

In representing global examples, critical case studies of modern systems by other researchers (Fig. 9.1, locations A, B), and debates on case studies of ancient systems by the author are included (Fig. 9.1, locations C, D, and E). In addressing the economic significance of bottom-current reworked sands (e.g., Shanmugam et al., 1993a; Viana, 2008), descriptions of deep-water strata from 35 case studies worldwide that include 7832 m of conventional cores from 123 wells, representing 32 petroleum fields are considered (Fig. 9.1, Table 9.2). Hopefully, this collective effort will motivate others in acknowledging and in resolving the contourite problem.

2. GLOBAL THERMOHALINE CIRCULATION

Historically, the concept of contour currents has been attributed to global THC (Heezen et al., 1966). Aspects of THC are discussed by Zenk (2008) and Talley (2013). The THC and related deep-marine bottom currents in modern oceans became popular when Heezen et al. (1966) reported deep-water masses and related contour currents along the continental rise in the US Atlantic margin. An example of such deep-water mass is the Antarctic Bottom Water (AABW). AABW was first identified by Brennecke (1921) in the northwest corner of the Weddell Sea in the Antarctic region (Fig. 9.2).

The deep-water masses in the world's oceans are caused by differences in temperature and salinity. When sea ice forms in the polar regions due to freezing of shelf waters, seawater experiences a concurrent increase in salinity due to salt rejection and a decrease in temperature. The increase in the density of cold saline (i.e., thermohaline) water directly beneath the ice triggers the sinking of the water mass down the continental slope (Fig. 9.2) and the spreading of the water masses to other parts of the ocean. These are called thermohaline water masses.

Stommel (1958) first developed the concept of the global circulation of thermohaline water masses and the vertical transformation of light surface waters into heavy deep-water masses in the oceans. Broecker (1991) presented a unifying concept of the global oceanic "conveyor belt" by linking the wind-driven surface circulation with the thermohaline-driven deep circulation regimes (Fig. 9.2). The large-scale horizontal transport of water masses, which also sink and rise at select locations, are known as the thermohaline circulation (THC). The term THC, which refers to a driving mechanism by high-latitude cooling, is a physical concept and not a measurable quantity (Rahmstorf, 2006). The global conveyor belt system in the North Atlantic originates near Greenland and Iceland where the sea-ice formation produces cold and salty North Atlantic Deep Water (NADW). The NADW sinks and flows southward along the continental slope of North and South America toward Antarctica where the water mass then flows eastward around the Antarctic continent.

According to Talley (2013, p. 81), "Description of the pathways and energetics of the global overturning circulation (GOC) is central to understanding the interaction of different ocean basins and layers as well as the interplay of external forcings." Aspects of the global overturning circulation (GOC) have been discussed in some detail (Gordon, 1986; Schmitz, 1996; Lumpkin and Speer, 2007; Richardson, 2008; Talley, 2013). For example, Schmitz (1996) illustrated meridional sections of interbasin flow with their global linkages among the Indian, Pacific, and Atlantic Oceans using Antarctica as the core of global circulation (Fig. 9.3).

Talley (2013) has shown that the overturning pathways for the surface-ventilated NADW and AABW and the diffusively-formed Indian Deep Water and Pacific Deep Water are intertwined (Fig. 9.4). According to Talley (2013), the GOC includes both large wind-driven

TABLE 9.2 Summary of Deep-Water Published Case Studies by Other Researchers That Are Used in This Article (Locations: A, B, C, D, and E, Filled Squares; see Fig. 9.1)

Location Symbol and Number in Fig. 9.1	Case Studies	Thickness of Core and Outcrop Described by the Author (Not Applicable to Studies by Other Researchers)[a]	Comment (This Chapter)
A. Case study: Blake Plateau and Blake-Bahama outer ridge (Heezen et al., 1966)	Modern Contour currents	Echo sounding, bottom photographs, sediment cores	Introduction of basic concept of contour currents
B. Case study: Gulf of Cadiz (Faugères et al., 1984; Gonthier et al., 1984; Stow and Faugères, 2008)	Modern Faro contourite drift	3.5 kHz seismic profiles, sediment cores	Discussion of problematic contourite facies model (discussed in this chapter)
B. Case study: Gulf of Cadiz (Hernández-Molina et al., 2006; García et al., 2009)	Modern Faro contourite drift	Seismic profiles, bottom photographs, sediment cores	Discussion of complex origin of erosional features (discussed in this chapter)
B. Case study: Gulf of Cadiz (Mulder et al., 2013)	Modern Faro contourite drift	Sediment cores, grain-size analysis, thin-section studies	Discussion of problematic contourite facies model in terms of velocity (discussed in this chapter)
B. Case study: Gulf of Cadiz (Stow et al., 2013)	Modern Cadiz Channel	2 gravity cores and over 3000 submarine photographs (Stow et al., 2013)	Discussion of problematic origin contourite sands (discussed in this chapter)
C. Case study: NE Spain (Pomar et al., 2012)	Ricla Section, Upper Jurassic	1 outcrop section (Bádenas et al., 2012; Pomar et al., 2012)	Discussion of problematic internal-wave and internal-tide deposits (Shanmugam, 2013a,b,c)
D. Case study: China (He et al., 2011)	Ningxia, Middle Ordovician	Several outcrop sections (He et al., 2011)	Discussion of problematic internal-wave and internal-tide deposits (Shanmugam, 2012b, 2014b)
E. Case study: Makassar Strait (Saller et al., 2006)	Kutei Basin, Miocene	2 wells (Saller et al., 2006, 2008a,b)	Discussion of deep tidal currents (Shanmugam, 2008a)
E. Case study: Makassar Strait (Dunham and Saller, 2014)	Kutei Basin, Miocene	2 wells (Saller et al., 2006, 2008a,b)	Reply to a discussion on the reservoir quality of bottom-current reworked sands (Shanmugam, 2014a)

TABLE 9.2 Summary of Deep-Water Published Case Studies by Other Researchers That Are Used in This Article (Locations: A, B, C, D, and E, Filled Squares; see Fig. 9.1)—cont'd

Location Symbol and Number in Fig. 9.1	Case Studies	Thickness of Core and Outcrop Described by the Author (Not Applicable to Studies by Other Researchers)[a]	Comment (This Chapter)
1. Gulf of Mexico, US (Shanmugam et al., 1988)	**1.** Mississippi Fan, Quaternary, DSDP Leg 96	~500 m DSDP core (selected intervals described)	Modern submarine fan
1. Gulf of Mexico, US (Shanmugam et al., 1993a, b; Shanmugam and Zimbrick, 1996)	**2.** Green Canyon, late Pliocene **3.** Garden Banks, middle Pleistocene **4.** Ewing Bank 826, Pliocene-Pleistocene **5.** South Marsh Island, late Pliocene **6.** South Timbalier, middle Pleistocene **7.** High Island, late Pliocene **8.** East Breaks, late Pliocene-Holocene	1067 m Conventional core And piston core 25 wells	Sandy mass-transport deposits and bottom-current reworked sands common
2. California (Shanmugam and Clayton, 1989; Shanmugam, 2006a, 2012a)	**9.** Midway Sunset Field, upper Miocene, onshore	650 m Conventional core 3 wells	Sandy mass-transport deposits and bottom-current reworked sands
3. Ouachita Mountains, Arkansas and Oklahoma, US (Shanmugam and Moiola, 1995)	**10.** Jackfork Group, Pennsylvanian	369 m 2 outcrop sections	Sandy mass-transport deposits and bottom-current reworked sands common
4. Southern Appalachians, Tennessee, US (Shanmugam, 1978; Shanmugam and Benedict, 1978; Shanmugam and Walker, 1978, 1980)	**11.** Sevier Basin, Middle Ordovician	2152 m 5 outcrop sections	Ancient submarine fan
5. Brazil (Shanmugam, 2006a, 2012a)	**12.** Lagoa Parda Field, lower Eocene, Espirito Santo Basin, onshore **13.** Fazenda Alegre Field, upper Cretaceous, Espirito Santo Basin, onshore	200 m Conventional core 10 wells	Sandy mass-transport deposits and bottom-current reworked sands common

(Continued)

TABLE 9.2 Summary of Deep-Water Published Case Studies by Other Researchers That Are Used in This Article (Locations: A, B, C, D, and E, Filled Squares; see Fig. 9.1)—cont'd

Location Symbol and Number in Fig. 9.1	Case Studies	Thickness of Core and Outcrop Described by the Author (Not Applicable to Studies by Other Researchers)[a]	Comment (This Chapter)
	14. Cangoa Field, upper Eocene, Espirito Santo Basin, offshore **15.** Peroá Field, lower Eocene to upper Oligocene, Espirito Santo Basin, offshore **16.** Marlim Field, Oligocene, Campos Basin, offshore **17.** Marimba Field, upper Cretaceous, Campos Basin, offshore **18.** Roncador Field, upper Cretaceous, Campos Basin, offshore		
6. North Sea (Shanmugam et al., 1995a)	**19.** Frigg Field, lower Eocene, Norwegian North Sea **20.** Harding Field (formerly Forth Field), lower Eocene, UK North Sea **21.** Alba Field, Eocene, UK, North Sea **22.** Fyne Field, Eocene, UK, North Sea **23.** Gannet Field, Paleocene, UK, North Sea **24.** Andrew Field, Paleocene, UK, North Sea **25.** Gryphon Field, upper Paleocene-lower Eocene, UK, North Sea	3658 m Conventional core 50 wells	Sandy mass-transport deposits and bottom-current reworked sands common
7. UK Atlantic margin (Shanmugam et al., 1995)	**26.** Faeroe area, Paleocene, west of the Shetland Islands **27.** Foinaven Field, Paleocene, West of the Shetland Islands	Thickness included in the North sea count 1 well Conventional core 1 well	Sandy mass-transport deposits and bottom-current reworked sands common; contourites have been reported (Damuth and Olson, 2001)

TABLE 9.2 Summary of Deep-Water Published Case Studies by Other Researchers That Are Used in This Article (Locations: A, B, C, D, and E, Filled Squares; see Fig. 9.1)—cont'd

Location Symbol and Number in Fig. 9.1	Case Studies	Thickness of Core and Outcrop Described by the Author (Not Applicable to Studies by Other Researchers)[a]	Comment (This Chapter)
8. Norwegian sea and vicinity (Shanmugam et al., 1994)	**28.** Mid-Norway region, Cretaceous, Norwegian Sea **29.** Agat region, Cretaceous, Norwegian North Sea	500 m Conventional core 14 wells	Sandy mass-transport deposits and bottom-current reworked sands common
9. French Maritime Alps, Southeastern France (Shanmugam, 2002a, 2003)	**30.** Annot Sandstone, Eocene-Oligocene	610 [b] 1 outcrop section (12 units described)	Sandy mass-transport deposits and bottom-current reworked sands common (deep tidal currents)
10. Nigeria (Shanmugam, 1997a; Shanmugam, 2006a, 2012a)	**31.** Edop Field, Pliocene, offshore	875 m Conventional core 6 wells	Sandy mass-transport deposits and bottom-current reworked sands common (deep tidal currents)
11. Equatorial Guinea (Famakinwa et al., 1996; Shanmugam, 2006a, 2012a)	**32.** Zafiro Field, Pliocene, offshore **33.** Opalo Field, Pliocene, offshore	294 m Conventional core 2 wells	Sandy mass-transport deposits and bottom-current reworked sands common
12. Gabon (Shanmugam, 2006a, 2012a)	**34.** Melania Formation, lower Cretaceous, offshore (includes four fields)	275 m Conventional core 8 wells	Sandy mass-transport deposits and bottom-current reworked sands common
13. Bay of Bengal, India (Shanmugam et al., 2009)	**35.** Krishna-Godavari Basin, Pliocene	313 m Conventional core 3 wells	Sandy debrites and tidalites common
Total thickness of rocks described by the author		11,463 m	

Note: conventional core and outcrop description carried out by the author worldwide (locations: 1–13, filled circles, see Fig. 9.1). Traction structures of bottom-current origin are common in All 35 case studies carried out by the author.

[a] *The rock description of 35 case studies of deep-water systems comprises 32 petroleum-producing massive sands worldwide. Description of core and outcrop was carried out at a scale of 1:20 to 1:50, totaling 11,463 m, during 1974–2011, by G. Shanmugam as a PhD student (1974–1978), as an employee of Mobil Oil Corporation (1978–2000), and as a consultant (2000–2011). Global studies of cores and outcrops include a total of 7832 m of conventional cores from 123 wells, representing 32 petroleum fields worldwide (Shanmugam, 2013c,d). These modern and ancient deep-water systems include both marine and lacustrine settings.*

[b] *The Peira Cava outcrop section was originally described by Bouma (1962), and later by Pickering and Hilton (1998, their Fig. 62), among others.*

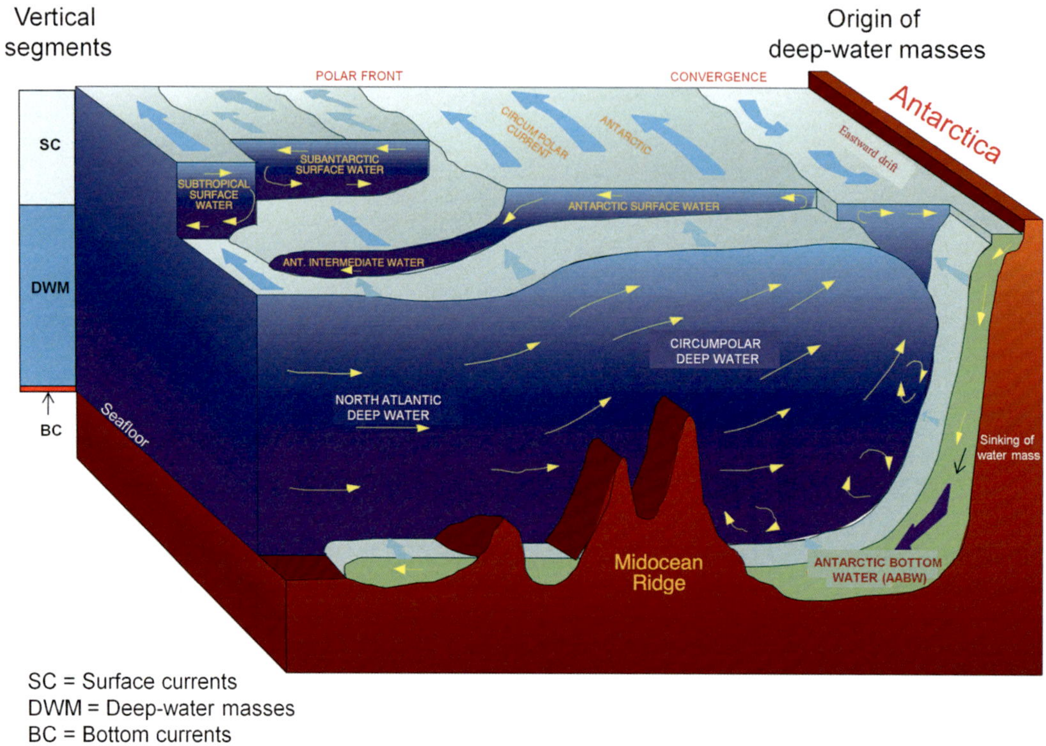

FIGURE 9.2 A conceptual model of the Southern Ocean showing three vertical segments, composed of the upper surface currents, the middle deep-water masses, and the lower bottom currents, forming a vertical continuum (left). Note the origin of AABW by freezing of shelf waters (right). As a consequence, the increase in the density of cold saline (i.e., thermohaline) water triggers the sinking of the water mass down the continental slope and the spreading of the water masses to other parts of the ocean. Modified after Hannes Grobe, April 7, 2000, http://en.wikipedia.org/wiki/File:Antarctic_bottom_water_hg.png. Figure from Shanmugam (2012a). Publication: Elsevier Books. Handbook of Petroleum Exploration and Production, Volume 9 (2012). With permission from Elsevier. Copyright Clearance Center's RightsLink: Licensee: G. Shanmugam. License Number: 3739590300706. License Date: October 31, 2015.

upwelling in the Southern Ocean and important internal diapycnal transformation in the deep Indian and Pacific Oceans (Fig. 9.4).

3. DEEP-WATER BOTTOM CURRENTS

Southard and Stanley (1976) recognized five types of bottom currents at the shelf break based on their origin. These currents are generated by (1) thermohaline differences, (2) wind forces, (3) tidal forces, (4) internal waves, and (5) surface waves. In addition, tsunami-related traction currents have been speculated to occur in bathyal waters (Yamazaki et al., 1989). Also, cyclone-related bottom currents are common (Shanmugam, 2008c).

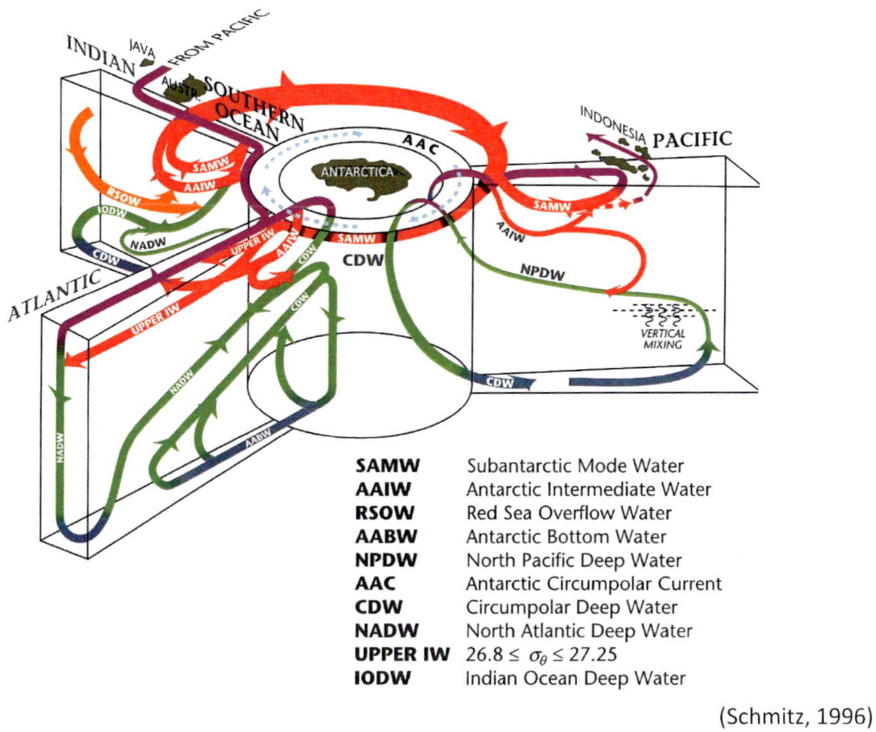

FIGURE 9.3 Schematic meridional sections of interbasin flow with their global linkages among the Indian, Pacific, and Atlantic Oceans using Antarctica as the core of global circulation. Note that the acronym AAC for the Antarctic Circumpolar Current is synonymous with ACC in other studies (e.g., Richardson, 2008). Diagram modified after Schmitz (1996). Similar concepts and diagrams of the overturning circulation from a Southern Ocean perspective were published by Gordon (1986), Lumpkin and Speer (2007), Richardson (2008), and Talley (2013).

However, the mechanics of such currents are not yet well understood (Shanmugam, 2008c, 2012c). In this review, I have selected four major types of deep-water bottom currents (Shanmugam, 2008a), namely (1) thermohaline-induced geostrophic contour currents (Heezen et al., 1966); (2) wind-driven bottom currents (Pequegnat, 1972); (3) tide-driven bottom currents, mostly in submarine canyons (Shepard et al., 1979; Shanmugam, 2003); and (4) internal wave/tide-driven baroclinic currents (Lonsdale et al., 1972; Cacchione et al., 2002; Shanmugam, 2013a). Studies have shown that all four types of bottom currents (i.e., thermohaline-induced contour currents, wind-driven bottom currents, deep-marine tidal currents, and baroclinic tidal currents) have produced similar bedforms and traction structures (Fig. 9.5) (Hsü, 1964; Hubert, 1964; Hollister, 1967; Lonsdale et al., 1972; Pequegnat, 1972; Klein, 1975; Mutti, 1992; Shanmugam et al., 1993a; Mutti and Carminatti, 2011; Shanmugam, 2008a, 2013a). This similarity in sedimentary structures stresses the need for a better understanding of all four processes and their depositional mechanics in order to develop criteria for distinguishing their respective deposits. In the context of understanding sediment provenance, it is worth noting that the four types of bottom currents are reworking agents, and as such they are generally not involved in transporting large volumes of

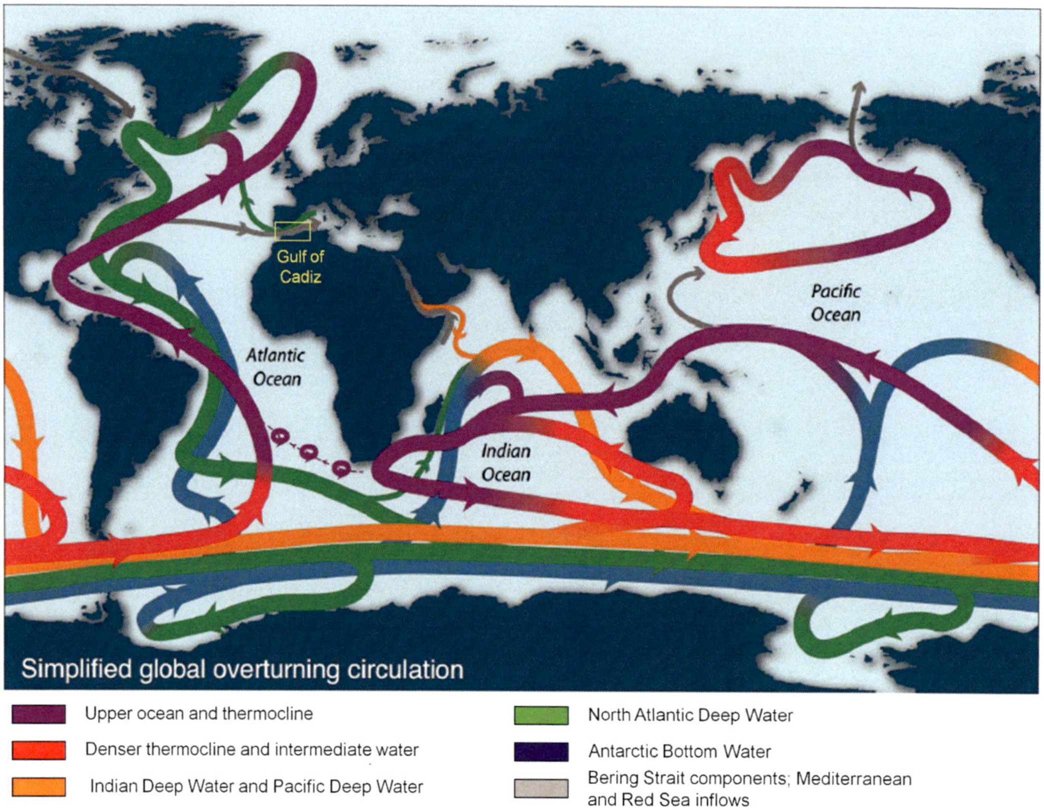

FIGURE 9.4 Map showing the global overturning circulation (GOC). The location of Gulf of Cadiz is added in this article. This site served as the type locality for the contourite facies model (see Section 4.7 in the text). The global circulation is not important in interpreting the primary sediment provenance at a given site. Modified after Talley (2013), with permission from the Oceanography Society. A simpler version of thermohaline circulation (THC) pattern was first published by Broecker (1991); it was later simplified by Rahmstorf (2002, 2006).

coarse detrital sediment (e.g., gravel, coarse sand, and medium sand) from the source to sites of deposition.

3.1 Thermohaline-Driven Geostrophic Contour Currents

Thermohaline-driven bottom currents tend to winnow, rework, and deposit sediment on the seafloor for a sustained period of time. They are popularly known as contour currents because they follow bathymetric contours (Heezen et al., 1966). Maximum current velocities of bottom currents in different parts of the world's oceans are summarized in Table 9.3. Measured current velocities usually range from 1 to 20 cm s^{-1} (Hollister and Heezen, 1972); however, exceptionally strong, near-bottom currents with maximum velocities of up to 300 cm s^{-1} were recorded in the Strait of Gibraltar (Gonthier et al., 1984). Bottom-current velocities of 73 cm s^{-1} were measured at a water depth of 5 km on the lower continental rise off Nova Scotia (Richardson et al., 1981). Heezen and Hollister

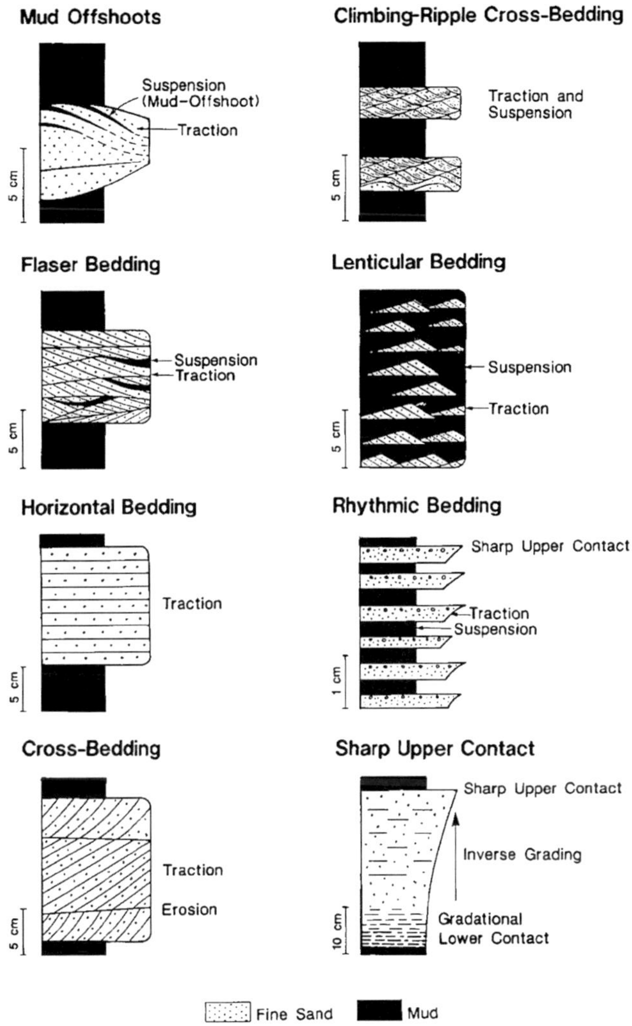

FIGURE 9.5 Summary of traction features interpreted as indicative of deep-water bottom-current reworking by all types of bottom currents. Each feature occurs randomly and should not be considered as part of a vertical facies model. From Shanmugam et al. (1993a), with permission from AAPG.

(1971) suggested that at extremely high bottom velocities of over 100 cm s^{-1}, relict pockets of sand and gravel may occur on the barren seafloor. According to Bulfinch and Ledbetter (1983/1984), the Deep Western Boundary Undercurrent (DWBUC) flows southward along the North American continental slope and rise between 1000 and 5000 m. The DWBUC has a 300-km wide high-velocity zone, with a maximum measured velocity of 73 cm s^{-1}, which winnows both fine and very fine silt, and results in deposition of medium and coarse silt. Traction structures are common in contour-current deposits (Fig. 9.6) (Hollister, 1967; Bouma and Hollister, 1973).

TABLE 9.3 Maximum Current Velocities of Bottom Currents in Different Parts of the World's Oceans

Study Area	Depth (m) (Dominant Driving Mechanism, This Chapter)	Maximum Current Velocity (cm s^{-1})
Straits of Gibraltar, Mediterranean Outflow Water (Gonthier et al., 1984; see also Hernández-Molina et al., 2013)	400–1400 (Thermohaline)	300
Upper slope. Offshore Brazil, Equatorial Atlantic (Viana et al., 1998)	200 (Thermohaline)	300
Gulf of Mexico, Loop Current (Cooper et al., 1990)	100 (Wind-driven)	204
Green Canyon 166 area, Gulf of Mexico. Drilling operations were temporarily suspended in August of 1989 because of high current velocities that reached 153 cm s^{-1} (Koch et al., 1991).	45 (Wind-driven)	153
Faeroe Bank Channel, North Atlantic (Crease, 1965)	760 (Thermohaline)	109
Rise, Off Nova Scotia, North Atlantic (Richardson et al., 1981)	5000 (Thermohaline)	73
Base of North American continental rise (Bulfinch and Ledbetter, 1983/84)	5022 (thermohaline)	73
Trench, Ryukyu Trench, Japan (Tsuji, 1993)	340 (tidal)	51
Samoan Passage, Western South Pacific (Hollister et al., 1974)	?	50
Hebrides Slope, North Atlantic (Howe and Humphrey, 1995)	403–468 (Thermohaline)	48
Faeroe-Shetland Channel, North Atlantic (Akhurst, 1991)	900 (Thermohaline)	33
Rise, near Hatteras Canyon, North Atlantic (Rowe, 1971)	(Thermohaline)	33
Carnegie Ridge, Eastern Equatorial Pacific (Lonsdale and Malfait, 1974)	1000–2000 (?)	>30
SE of Iceland, North Atlantic (Steele et al., 1962)	2100 slope (Thermohaline)	30
Argentine Basin, Western South Atlantic (Ewing et al., 1971)	(Thermohaline)	30
Amirante Passage, Western Indian Ocean (Johnson and Damuth, 1979)	4000–4600 (Thermohaline)	30
Rise, Off New England, North Atlantic (Zimmerman, 1971)	3000–5000 (Thermohaline)	26.5
Blake Bahama Outer Ridge, North Atlantic (Amos et al., 1971)	4300–5200 (Thermohaline)	26

TABLE 9.3 Maximum Current Velocities of Bottom Currents in Different Parts of the World's Oceans—cont'd

Study Area	Depth (m) (Dominant Driving Mechanism, This Chapter)	Maximum Current Velocity (cm s^{-1})
Off North Carolina, North Atlantic (Rowe and Menzies, 1968)	1500–4000 (Thermohaline)	25
Off Cape Cod, North Atlantic (Volkman, 1962)	10–3200 (Thermohaline)	21.5
Off Cape Hatteras, North Atlantic (Barrett, 1965)	(Thermohaline)	21
Greater Antilles Outer Ridge, North Atlantic (Tuholke et al., 1973)	5300–5800 (Thermohaline)	20
Off Blake Plateau, North Atlantic (Swallow and Worthington, 1961)	3300–3500 (Thermohaline)	20
Tonga Trench and vicinity, Western South Pacific (Reid, 1969)	>4800 (?)	19
Western North Atlantic (Wüst, 1950)	2000–3000 (Thermohaline)	17
West Bermuda Rise, North Atlantic (Knauss, 1965)	5200 (Thermohaline)	17
[a]Scotia Ridge, Antarctic Circumpolar Current, Antarctica (Zenk, 1981)	3008 (Wind-driven) (Howe et al., 1997)	17[b]
Greenland-Iceland-Faeroes Ridge, North Atlantic (Worthington and Volkman, 1965)	2000–3000 (Thermohaline)	12
Antillean-Caribbean Basin (outer), North Atlantic (Wust, 1963)	4000–8000 (Thermohaline)	10

[a]*Antarctic Circumpolar Current has both wind-driven and thermohaline-driven components (CIMAS, 2015).*
[b]*1-year vector averaged speed.*
? indicates that the precise origin is unknown.

3.2 Wind-Driven Bottom Currents

The wind-driven bottom current, a product of wind stress (i.e., atmospheric forcing) exerted at the sea surface that causes flows to extend all the way to the sea floor thousands of meters below, is well documented in the world's oceans. For example, the Gulf Stream is a powerful, warm, and swift Atlantic Ocean current that originates at the tip of Florida (Fig. 9.7A), and follows the eastern coastlines of the United States and Newfoundland before crossing the Atlantic Ocean. The Gulf Stream proper is a western-intensified current, largely driven by wind stress (Wunsch, 2002). The Loop Current in the eastern Gulf of Mexico is a wind-driven surface current (Pequegnat, 1972) (Fig. 9.7A), and it is genetically linked to the Gulf Stream in the Atlantic Ocean (Mullins et al., 1987). Velocities in eddies that have detached from the Loop Current have been recorded as high as 200 cm s^{-1} at a depth of 100 m (Cooper et al., 1990). Computed flow velocities of the

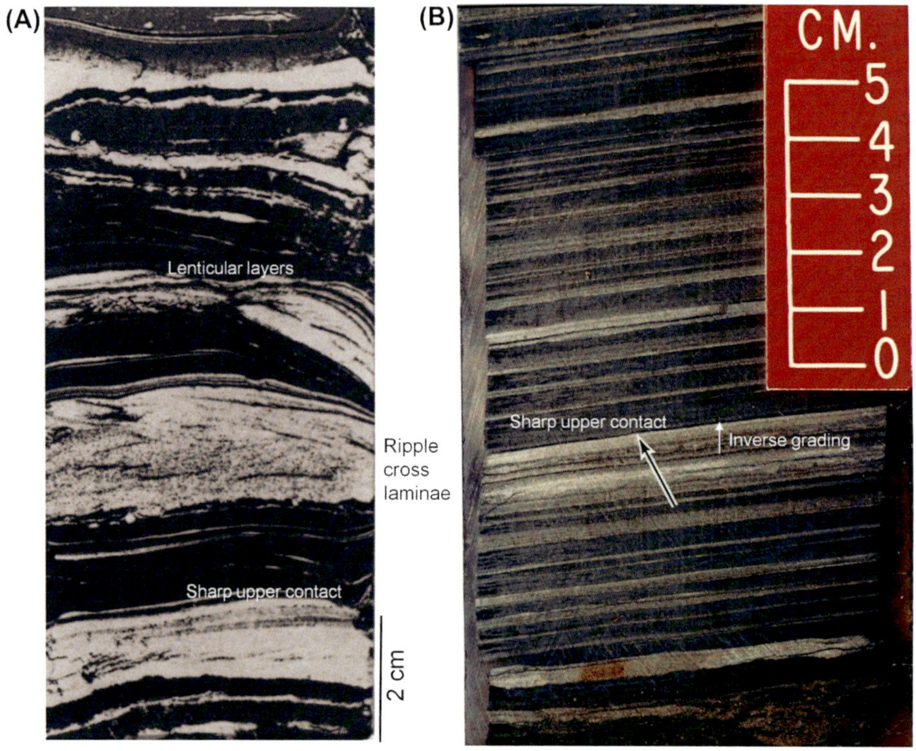

FIGURE 9.6 (A) Core photograph showing well-sorted fine-grained sand and silt layers (light gray) with interbedded mud layers (dark gray). Note sand layers with sharp upper contacts, internal ripple cross-laminae, and mud offshoots. Also note lenticular nature of some sand layers. Pleistocene, continental rise off Georges Bank, Vema 18-374, 710 cm, water depth 4756 m. After Hollister (1967, his Figure VI-1, p. 208) and Bouma and Hollister (1973), reproduced with permission from SEPM. (B) Core photograph showing rhythmic layers of sand and mud, inverse grading, and sharp upper contacts of sand layers (*arrow*), interpreted as bottom-current reworked sands. Paleocene, North Sea. Figure from Shanmugam (2008a). Publication: Elsevier Books. Developments in Sedimentology, Volume 60, "Contourites" (2008). With permission from Elsevier. Copyright Clearance Center's RightsLink: Licensee: G. Shanmugam. License Number: 3747140683175. License Date: October 31, 2015.

Loop Current vary from nearly 100 cm s^{-1} at the sea surface to more than 25 cm s^{-1} at 500 m water depth (Nowlin and Hubert, 1972). Kenyon et al. (2002) reported 25 cm s^{-1} current velocity measured 25 m above the seafloor. Such currents are capable of reworking fine-grained sand on the seafloor. Current ripples, composed of sand at a depth of 3091 m on the seafloor (Fig. 9.8), are the most convincing empirical evidence of wind-driven bottom-current activity in the Gulf of Mexico today (Pequegnat, 1972). Another example of a wind-driven bottom current is the eastward-flowing Antarctic Circumpolar Current (ACC), which has influenced sedimentation on the slope and floor of the western Falkland Trough, where the axis of the current is topographically constrained (Howe et al., 1997). This deep-water flow (below 3000 m) has produced a symmetrical sediment drift on the trough floor, with nondepositional margins indicating higher current velocities at the base of slope.

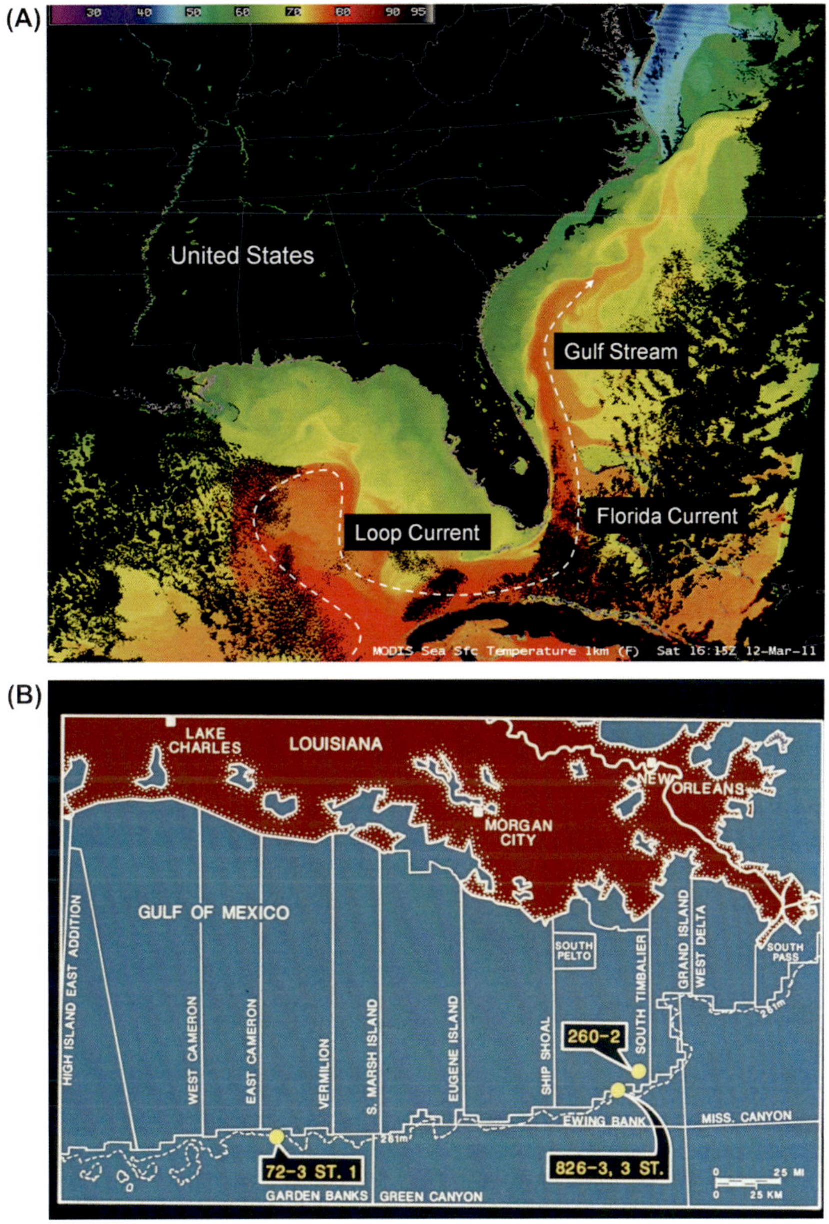

FIGURE 9.7 (A) Sea surface temperature (SST) image showing the Loop Current in the Gulf of Mexico and the axis of the Gulf Stream in the Atlantic Ocean along the US Continental margin on March 12, 2011. *Darker orange to red color enhancement* represents temperatures in the upper 70°s F (upper 20°s C). Image credit: NOAA's Cooperative Institute for Meteorological Satellite Studies, University of Wisconsin Madison, US, http://cimss.ssec.wisc.edu/goes/blog/wpcontent/uploads/2011/03/MODIS_SST_20110312_1615_largescale.png. Figure from Shanmugam (2012a). Publication: Elsevier Books. Handbook of Petroleum Exploration and Production, Volume 9 (2012). With permission from Elsevier. Copyright Clearance Center's RightsLink: Licensee: G. Shanmugam. License Number: 3739570762326. License Date: October 31, 2015. (B) Location map of the Ewing Bank and adjacent areas in the Northern Gulf of Mexico. Solid dots show locations of cores. After Shanmugam et al. (1993a), with permission from AAPG.

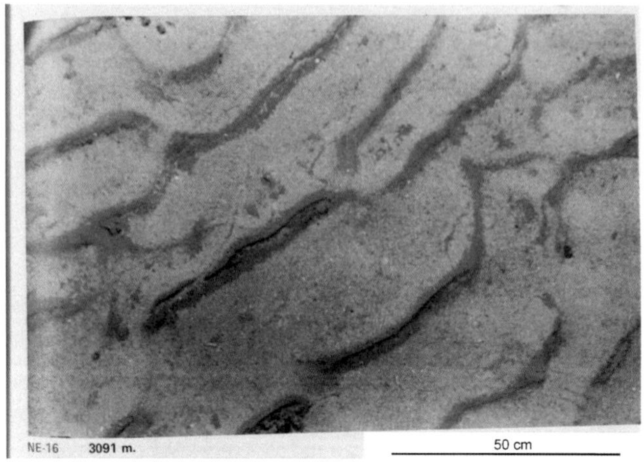

FIGURE 9.8 Undersea photograph showing possible mud-draped (*arrow*) current ripples at 3091 m water depth in the Gulf of Mexico. Similar mud drapes may explain the origin of mud offshoots observed in the core (see Fig. 9.5). A current measuring nearly 18 cm s^{-1} was recorded on the day this photograph was taken. Current flow was from upper left to lower right. Bar scale is 50 cm. Alaminos Cruise 69-A-13, St. 35. Photograph originally published by Pequegnat (1972). Figure from Shanmugam (2012a). Publication: Elsevier Books. Handbook of Petroleum Exploration and Production, Volume 9 (2012). With permission from Elsevier. Copyright Clearance Center's RightsLink: Licensee: G. Shanmugam. License Number: 3739570762326. License Date: October 31, 2015.

Deposits of the Loop Current have been interpreted in the cores from the Ewing Bank 826 Field, Plio-Pleistocene, Gulf of Mexico. The Ewing Bank Block 826 Field is located nearly 100 km off the Louisiana coast in the northern Gulf of Mexico (Fig. 9.7B). It contains hydrocarbon-producing clastic reservoir sands that have been interpreted as bottom-current-reworked sands (Shanmugam et al., 1993a,b). Cores from the Ewing Bank and adjacent areas exhibit traction structures (Fig. 9.9) such as horizontal layers, low-angle cross-laminae, ripple cross-laminae, flaser bedding in ripples, mud offshoots in ripples, eroded and preserved ripples, and inverse grading (see Shanmugam et al., 1993a,b for additional core photographs).

3.3 Tide-Driven Bottom Currents

In understanding tide-induced bottom currents, Shepard et al. (1979) measured current velocities in 25 submarine canyons worldwide at water depths ranging from 46 to 4200 m by suspended current meters, usually 3 m above the sea bottom (Fig. 9.10A). Shepard et al. (1979) also documented systematically that up- and down-canyon currents closely correlated with timing of tides (Fig. 9.10B). These submarine canyons include the Hydrographer, Hudson, Wilmington, and Zaire in the Atlantic Ocean; and the Monterey, Hueneme, Redondo, La Jolla/Scripps, and Hawaii canyons in the Pacific Ocean. Maximum velocities of up- and down-canyon currents commonly ranged from 25 to 50 cm s^{-1} (Shepard et al., 1979). Keller and Shepard (1978) reported velocities as high as 70–75 cm s^{-1}, velocities sufficient to transport even coarse-grained sand, from the Hydrographer Canyon.

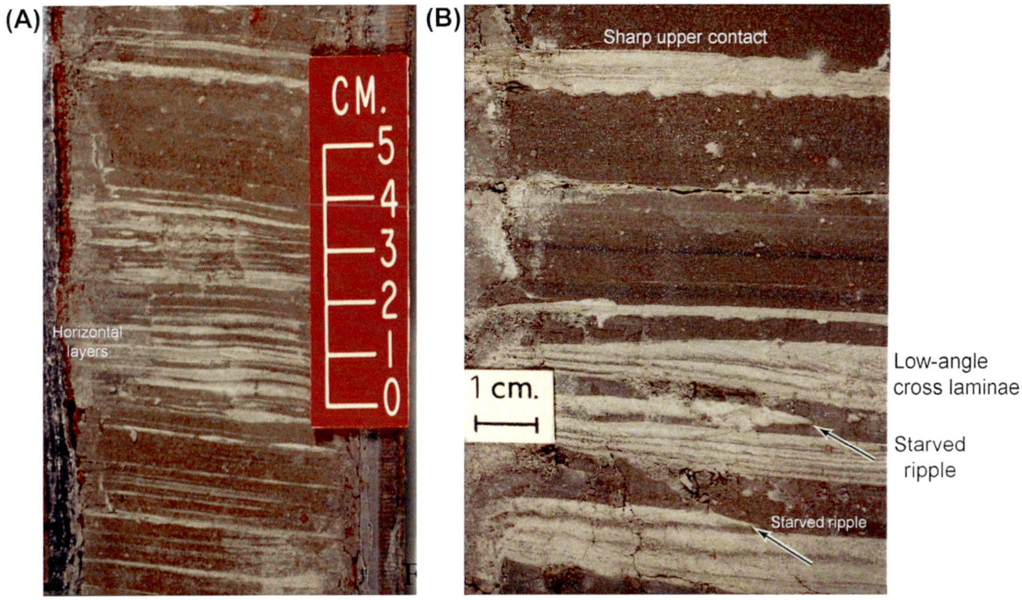

FIGURE 9.9 (A) Core photograph showing rhythmic layers of sand and mud. Middle Pleistocene, Gulf of Mexico. Figure from Shanmugam (2012a). Publication: Elsevier Books. Handbook of Petroleum Exploration and Production, Volume 9 (2012). With permission from Elsevier. Copyright Clearance Center's RightsLink: Licensee: G. Shanmugam. License Number: 3739570762326. License Date: October 31, 2015. (B) Core photograph showing discrete thin sand layers with sharp upper contacts (*top arrow*). Traction structures include horizontal laminae, low-angle cross-laminae, and starved ripples. Dip of cross-laminae to the right suggests current from left to right. Note rhythmic occurrence of sand and mud layers. Middle Pleistocene, Gulf of Mexico. After Shanmugam et al. (1993a), with permission from AAPG.

Deep-water petroleum reservoirs exhibit parallel laminae and double mud layers in offshore Nigeria (Fig. 9.11A) and in the Bay of Bengal (Fig. 9.11B). Double mud layers are unique to deposition from tidal currents in both shallow-water (Visser, 1980; Shanmugam et al., 2000) and deep-water environments (Shanmugam, 2003; Shanmugam et al., 2009; Mutti and Carminatti, 2011; Mazumder and Arima, 2013). However, such parallel laminae are commonly mislabeled as Bouma Tb divisions and misinterpreted as turbidites (Saller et al., 2006; see critique by Shanmugam, 1997b, 2008b, 2014a).

3.4 Internal Wave- and Tide-Driven Baroclinic Currents

Apel (2002), Apel et al. (2006), and Jackson (2004a) documented internal waves and tides worldwide (Fig. 9.12). A sedimentologic and oceanographic review of baroclinic currents associated with internal waves and tides is provided by Shanmugam (2013a). Internal waves are gravity waves that oscillate along oceanic pycnoclines (Fig. 9.13A). In a stratified ocean, internal tides are generated commonly above an area of steep bathymetric variation, such as the shelf break, seamount, and so on. Empirical data on physical properties of internal solitary waves and tides, which include wave speed, have been compiled for 51 locations in the world's oceans (Shanmugam, 2013a; his Table 2). Turnewitsch et al. (2008) discussed internal tides and

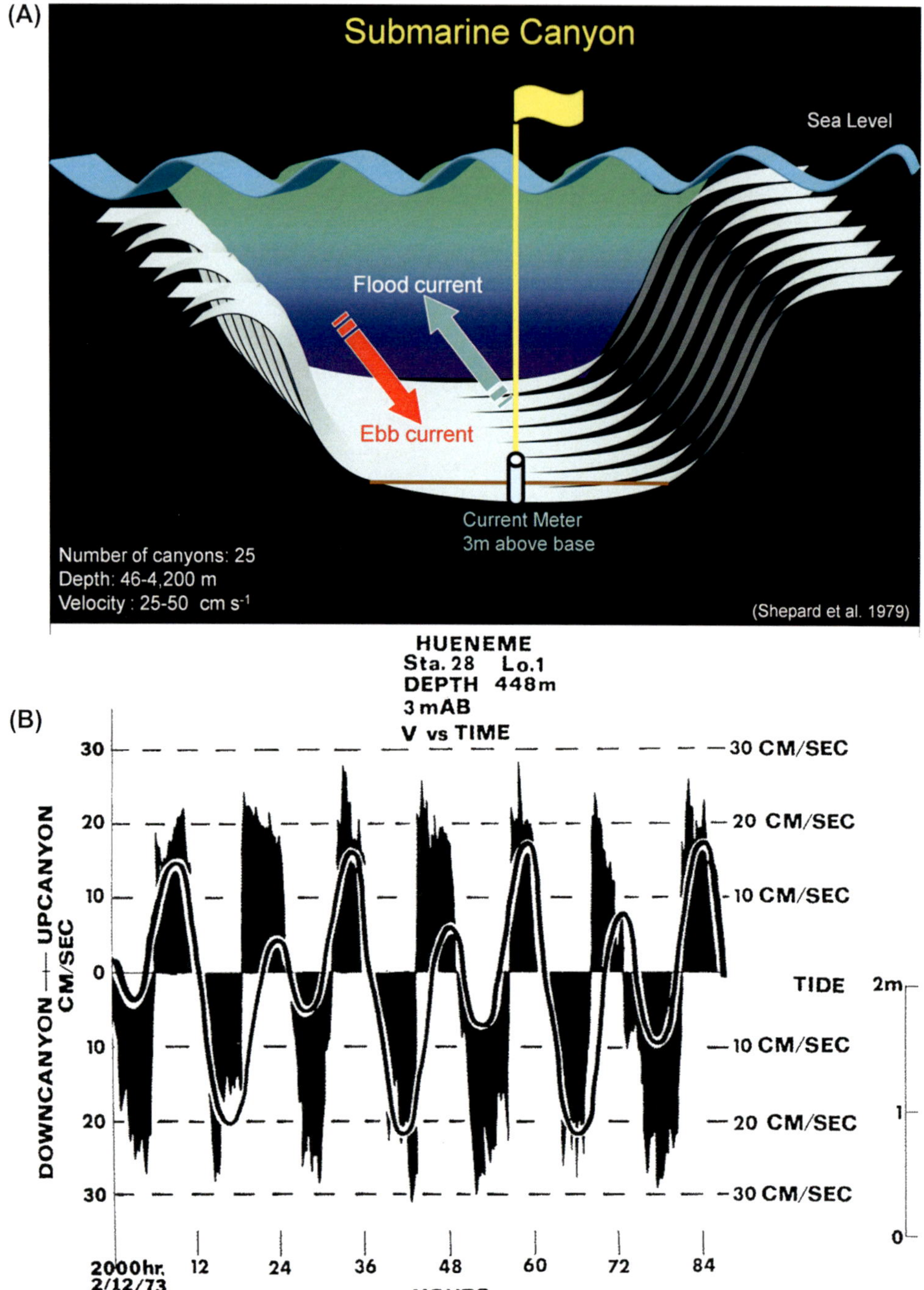

FIGURE 9.10 (A) Conceptual diagram showing a cross-section of a submarine canyon with ebb and flood tidal currents (*opposing arrows*). Shepard et al. (1979) measured current velocities in 25 submarine canyons at water depths ranging from 46 to 4200 m by suspending current meters commonly 3 m above the sea bottom. Measured maximum velocities commonly range from 25 to 50 cm s^{-1}. Figure from Shanmugam (2003). Publication: Marine and Petroleum Geology. With permission from Elsevier. Copyright Clearance Center's RightsLink: Licensee: G. Shanmugam. License Number: 3739551194950. License Date: October 31, 2015. (B) Time-velocity plot from data obtained at 448 m in the Hueneme Canyon, California, showing excellent correlation between the timing of up- and down-canyon currents and the timing of tides obtained from tide tables (*solid curve*). 3 mAB = Velocity measurements were made 3 m above sea bottom. Modified after Shepard et al. (1979), with permission from AAPG.

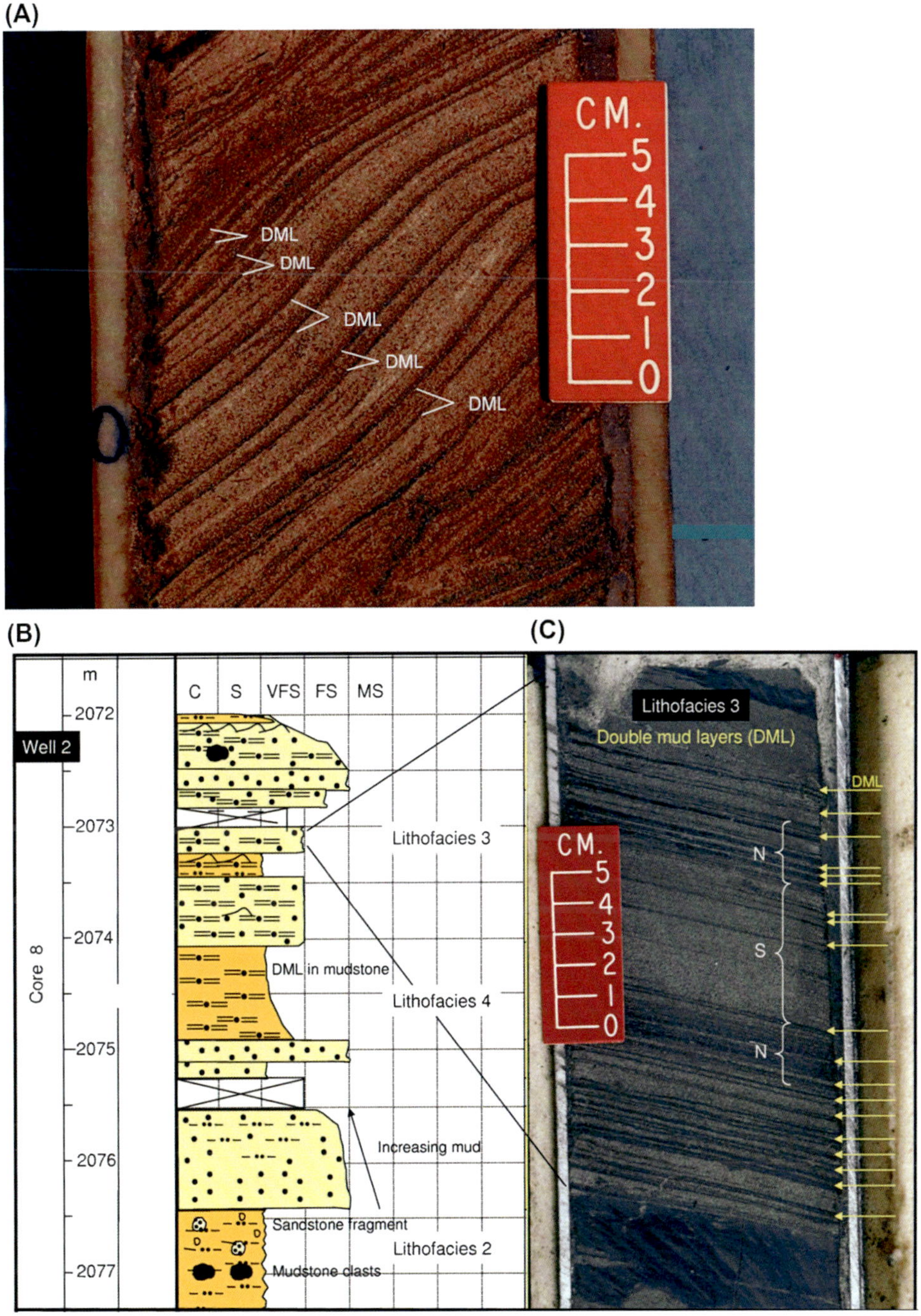

FIGURE 9.11 (A) Core photograph showing double mud layers (DML), indicative of deposition by deep-marine tidal currents, in a submarine-canyon setting. Pliocene strata, Edop Field, offshore Nigeria. Figure from Shanmugam (2003). Publication: Marine and Petroleum Geology. With permission from Elsevier. Copyright Clearance Center's RightsLink: Licensee: G. Shanmugam. License Number: 3739551194950. License Date: October 31, 2015. (B) Sedimentological log showing alternation of sand (lithofacies 3) and mudstone (lithofacies 4) intervals with continuous presence of double mud layers (DML). Note muddy debrite facies (Lithofacies 2) near the bottom. Wentworth grain-size classes: *C*, clay; *S*, silt; *VFS*, very fine sand; *FS*, fine sand; *MS*, medium sand. (C) Lithofacies 3 core photograph showing rhythmic bedding (rhythmites) and DML (*arrows*) in sand. *N*, Neap (*thin*) bundle; *S*, Spring (*thick*) bundle. Note that we could designate the DML intervals as Tb and the massive sand unit (between scale divisions 2 and 4 cm) as Ta using the Bouma Sequence; however, Shanmugam et al. (2009) did not. Core photograph from Shanmugam et al. (2009), with permission from SEPM.

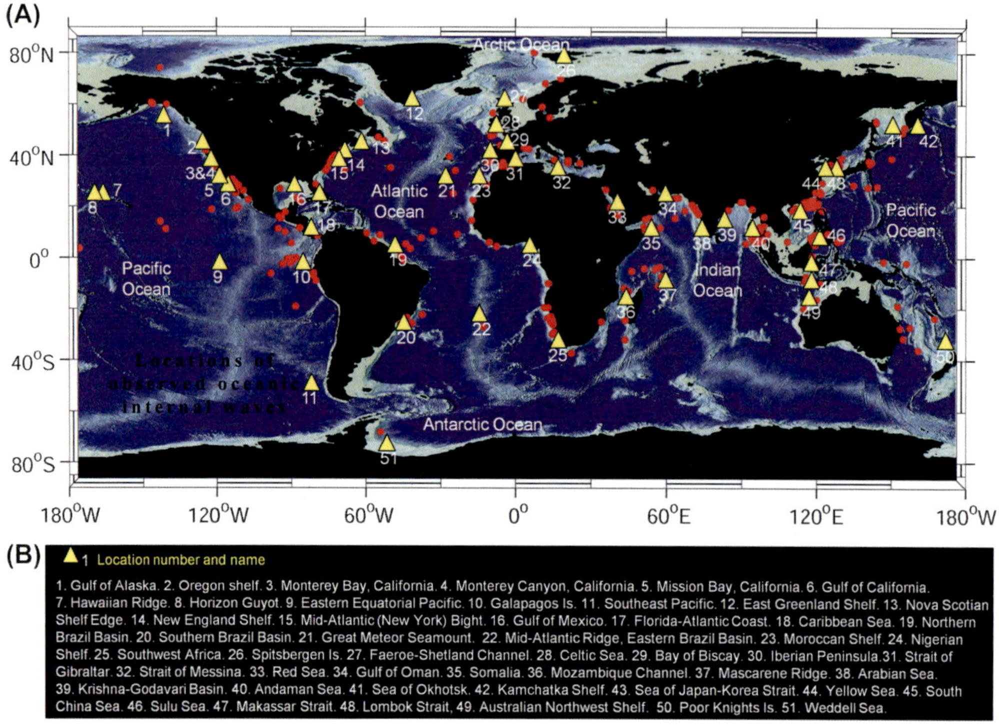

FIGURE 9.12 (A) Map showing 51 locations (*red dots*) of observed oceanographic internal waves and tides in coastal seas and in the open ocean (after Apel, 2002; Jackson, 2004a). (B) Explanation of symbols and numbers. *Yellow triangles* and numbers represent locations of internal waves used for physical properties in a study by Shanmugam (2013a, his Table 2). Base map courtesy of C.R. Jackson, Global Ocean Associates. From Shanmugam (2013a), with permission from AAPG.

sediment dynamics in the deep sea using evidence from radioactive $^{234}Th/^{238}U$ disequilibria. Brandt et al. (2002) reported results of high-resolution velocity measurements carried out by means of a vessel-mounted acoustic Doppler current profiler on the November 12, 2000 in the equatorial Atlantic, at 44°W between 4.5°N and 6°N. The data showed the presence of three large-amplitude internal solitary waves. The pulse-like intense solitary disturbances propagated perpendicular to the Brazilian Shelf, toward the north-northeast. These internal waves were characterized by maximum horizontal velocities of about 200 cm s^{-1} and maximum vertical velocities of about 20 cm s^{-1}. Shepard (1975) suggested that internal waves, which occur in canyon depths of up to 3500 m, were mostly tidal in origin (i.e., internal tides).

In the Suruga Trough in Japan, semidiurnal tidal fluctuations are evident in the current with the total amplitude reaching 50 cm s^{-1} at a depth of 1370 m. These currents have been associated with internal tides (Matsuyama et al., 1993). Velocity measurements associated with internal tides in the Gaoping Submarine Canyon off southwestern Taiwan have revealed maximum velocities of more than 100 cm s^{-1} (Lee et al., 2009). At these velocities, even gravel-grade grains can be eroded, transported, and deposited by baroclinic tidal currents. In fact, Lonsdale et al. (1972) documented asymmetrical dunes and asymmetrical

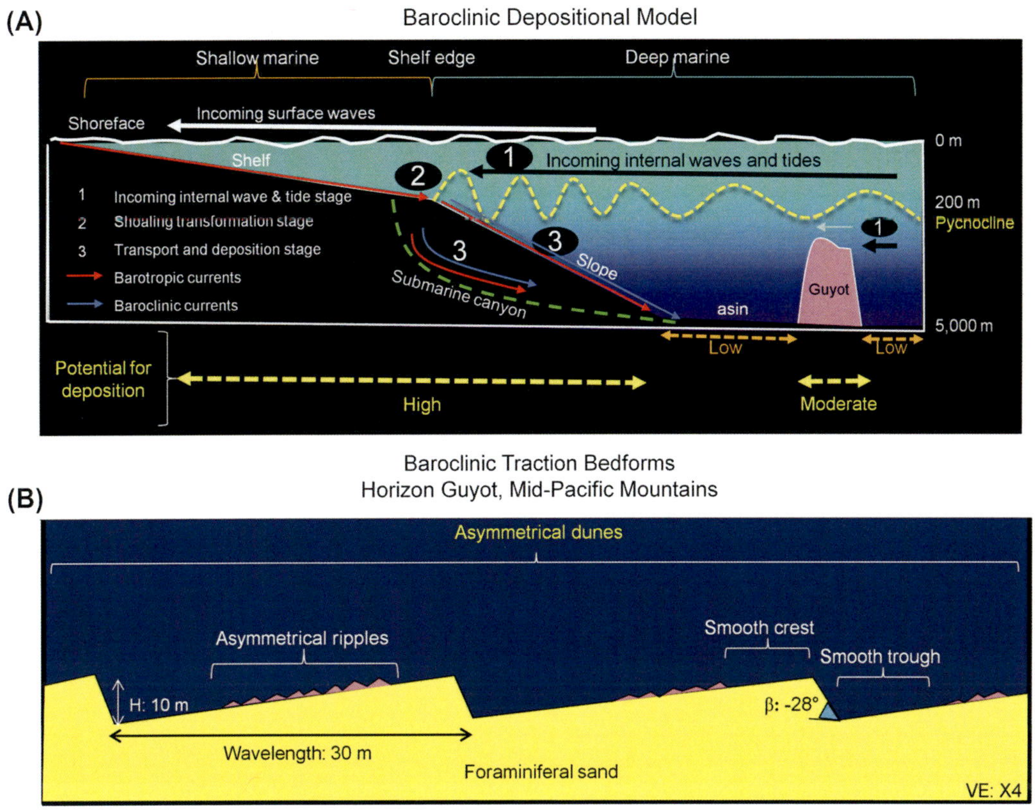

FIGURE 9.13 (A) Conceptual oceanographic and sedimentologic framework showing deposition from baroclinic currents on continental slopes, in submarine canyons, and on guyots. On continental slopes and in submarine canyons, deposition occurs in three progressive stages: (1) incoming internal wave and tide stage, (2) shoaling transformation stage, and (3) sediment transport and deposition stage. Continental slopes and submarine canyons are considered to be environments with high potential for deposition from baroclinic currents. In the open ocean, baroclinic currents can rework sediments on flat tops of towering guyot terraces, without the need for three stages required for deposition on continental slopes. In this model, basin plains are considered unsuitable environments for deposition of baroclinic sands. Not to scale. From Shanmugam (2013a), with permission from AAPG. (B) Cross-profile showing asymmetrical dunes and asymmetrical ripples observed from side-looking sonar and photographic evidence obtained from the terrace of the Horizon Guyot, Mid-Pacific Mountains. Bathymetry of bedforms: 1630–1632 m. Dune heights (H) were estimated from the length of acoustic shadows. Redrawn from Lonsdale et al. (1972, their Figure 10), with permission from the Geological Society of America.

ripples observed using side-looking sonar and photographic evidence obtained from the terrace of the Horizon Guyot, Mid-Pacific Mountains at a depth of 1630–1632 m (Fig. 9.13B).

Despite a great wealth of oceanographic information published on internal waves and tides (Apel et al., 2006), there is a clear lack of published sedimentological details of baroclinic currents (Shanmugam, 2013a). This knowledge gap hinders distinguishing baroclinites (i.e., deposits of baroclinic currents) from contourites in the ancient stratigraphic record.

4. FUNDAMENTAL CONTOURITE PROBLEMS

Although there are numerous contourite problems, the following 15 fundamental issues have been selected for discussion. This is somewhat analogous to the author's previous review of "Ten turbidite myths" in identifying fundamental problems (Shanmugam (2002a). For example, the concept of high-density turbidity currents in explaining gravelly and sandy turbidites (Shanmugam, 1996), somewhat analogous to the concept of irrational numbers in mathematics (Havil, 2014), is incommensurable (Shanmugam, 2016a,b). Furthermore, despite the constant promotion of the turbidite-fan link (Grotzinger et al., 2007), the number of documented cases of the existence of gravelly and sandy turbidity currents in modern deep-water environments is zero!

4.1 Dual Forcing of Global Ocean Circulation

The paradigm of global ocean circulation has been the thermohaline forcing of two independent water masses, namely the NADW or the "great ocean conveyor" (Broecker, 1991) and the AABW (Gordon, 1986). The global ocean circulation is initiated in the Southern Ocean (Antarctica) as the cumulative result of (1) wind-driven (adiabatic) upwelling, (2) surface buoyancy flux, and (3) deep-water formation by cooling and saline rejection (i.e., thermohaline) (Fig. 9.14). Both atmospheric forcing (i.e., surface-wind stress) and thermohaline forcing (i.e., bottom-water formation) are necessary to induce and maintain

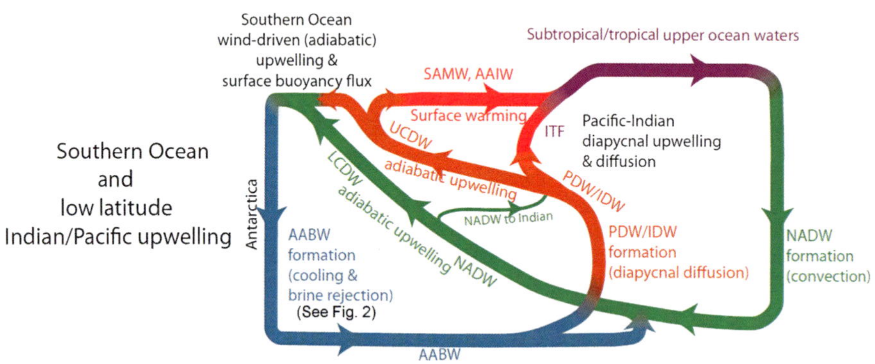

Acronyms

SAMW:	Subantarctic Mode Water
AAIW:	Antarctic Intermediate Water
ITF:	Indonesian Throughflow
UCDW:	Upper Circumpolar Deep Water
LCDW:	Lower Circumpolar Deep Water
PDW:	Pacific Deep Water
IDW:	Indian Deep Water
NADW:	North Atlantic Deep Water
AABW:	Antarctic Bottom Water

FIGURE 9.14 Schematic diagram showing the wind-driven and thermohaline-driven mechanisms in the Southern Ocean (Antarctica) in initiating global ocean circulation. From Talley (2013), with permission from the Oceanography Society.

global ocean circulation (Talley, 2013). For example, the ACC is widely accepted as being dominantly a wind-driven current (Howe et al., 1997). Therefore, a sound knowledge of global ocean surface currents is critical for understanding ocean bottom currents (Gill, 1982; Apel, 1987; Stewart, 2008; CIMAS, 2015). In light of the dual forcing of most water masses, it is inappropriate to classify an ancient layer as a contourite routinely, with a skewed emphasis on thermohaline forcing and with a total avoidance of the role of atmospheric forcing.

The term contourite drift is used commonly in the geologic literature (see book chapters by Faugères and Stow, 2008; Faugères and Mulder, 2011). The ACC produces drifts at great depths of over 3000 m (Howe et al., 1997; Pudsey and Howe, 1998). These drift sediments are products of currents that follow bathymetric contours, and therefore they could be classified as contourites. However, these drift sediments are not genuine contourites because they are products of mostly wind-driven currents, not thermohaline-driven currents. In other words, contourites could be generated by more than one type of bottom currents. The problem here is that there are no sedimentological criteria for distinguishing deposits of purely wind-driven bottom currents from those of thermohaline-driven bottom currents. Therefore, the application of the term contourites to the ancient stratigraphic record, with little information on forcing mechanisms, should proceed with caution. A solution is to replace the genetic term "contourite drift" with the nongenetic term "sediment drift."

4.2 Continuum Between Turbidity Currents and Contour Currents

Rebesco et al. (2014, their Fig. 9.1) begin their review with a ternary diagram with three end members composed of contourites, turbidites, and pelagites. The ternary diagram is based on the continuum principle of these three basic deep-sea sediment types that was advocated nearly 35 years ago by Stow and Lovell (1979). It is difficult, however, to reconcile a process continuum between turbidity currents and contour currents. By definition, the term "continuum" refers to a gradual transition from one end member to the other, without any abrupt changes. The continuum principle is unsustainable for the following reasons:

- Downslope-flowing turbidity currents and along-slope flowing contour currents are almost at right angles with each other (Fig. 9.15). Even if the two interact with each other, the interaction would be ephemeral and is of no sedimentological significance.
- Turbidity currents are local or regional in transport, whereas most contour currents are global in scale.
- Turbidity currents are episodic (Kuenen and Migliorini, 1950) or surge-type events that fail to develop equilibrium conditions (Allen, 1985), whereas contour currents persist for long periods of time and can develop equilibrium conditions.

In addition, the ternary diagram totally ignores the importance of mass-transport deposits (Mosher et al., 2010), tidal currents in submarine canyons (Shepard et al., 1979), baroclinic currents (Shanmugam, 2013a), and bottom currents associated with cyclones and tsunamis in the deep ocean (Shanmugam, 2008c).

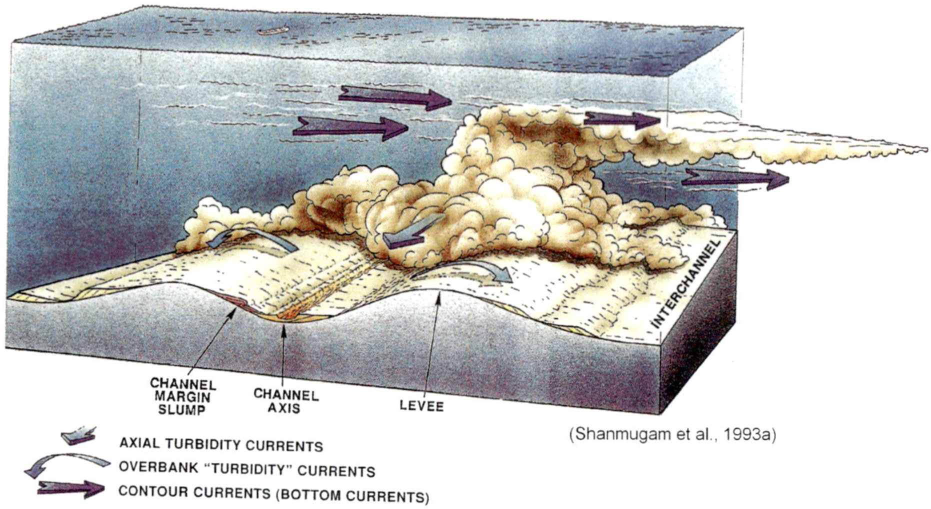

FIGURE 9.15 Conceptual model showing the spatial relationship between downslope turbidity currents and along-slope contour currents. This is an unlikely scenario for developing a process continuum between the two types. Note that turbidity currents transport sediment downslope from the primary sediment source to basin along with mass-transport processes, whereas contour currents are reworking agents and as such they are unrelated to the primary sediment provenance. After Shanmugam et al. (1993a), with permission from AAPG.

4.3 Revision of the Basic Principle of Contour Currents

The basic principle of contour currents was introduced first by Heezen and Hollister (1964) to the marine geologic community and later by Heezen et al. (1966) to the general scientific community. Their studies were based on a regional study of the continental rise off eastern United States in the Atlantic Ocean, covering the Blake Plateau and Blake-Bahama Outer Ridge (Fig. 9.1, location A). Their seminal study was based on a robust dataset composed of echo sounding, bottom photographs, and sediment cores. Therefore, it is useful to revisit the following three fundamental points from Heezen et al. (1966):

- Page 502: "Geostrophic contour-following bottom currents involved in the deep thermohaline circulation of the world ocean appear to be the principal agents which control the shape of the continental rise and other sediment bodies."
- Page 504: "Pressure gradients indicated by the inclined isopycnals must be opposed by an opposite and equal force which would seem to be provided by a current in which the Coriolis forces are acting normal to the direction of motion (to the right in the northern hemisphere). These currents flow along isopycnals which are approximately parallel to the bathymetric contours. We refer to these currents as contour currents." (See Fig. 9.4.)
- Page 507: "In marked contrast to the steady, low velocity (2 to 20 cm/sec) contour-following geostrophic currents which never flow downslope, turbidity currents are intermittent, high-velocity (up to 2500 cm/sec) downslope movements..."

As envisioned by Heezen et al. (1966), the basic principle of contour currents was scientifically sound, and there is no need to revise it. Nevertheless, other authors have broadened the meaning. For example:

- Johnson et al. (1980) applied the term contourites to sediments in Lake Superior, in the United States.
- Lovell and Stow (1981, p. 349) conclude that "Contourite: a bed deposited significantly reworked by a current that is persistent in time and space and flows along slope in relatively deep water (certainly below wave base). The water may be fresh or salt; *the cause of the current is not necessarily critical to the application of the term*." I have used Italics for the last phrase to emphasize their point that contourites can be produced by any kind of bottom current (Fig. 9.16), irrespective of their origin (i.e., thermohaline, wind, tide, or baroclinic).
- The last phrase in Lovell and Stow (1981) (see preceding) has served as the foundation and a continuum for a subsequent paper by Stow et al. (2008), in which they expanded the meaning of the term contourite. For example, Stow et al. (2008, p. 144) explicitly state that "Bottom (contour) currents are those currents that operate as part of either the normal thermohaline circulation or wind-driven circulation systems..."

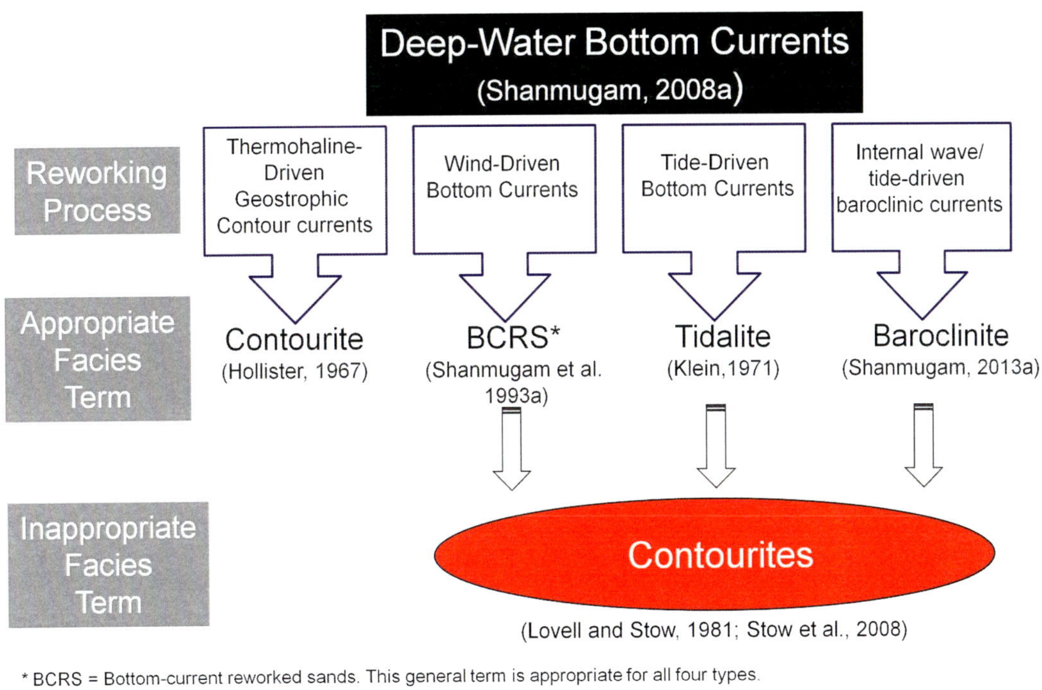

FIGURE 9.16 Four types of bottom currents and their depositional facies. The facies term "contourites" is appropriate only for deposits of thermohaline-driven geostrophic contour currents in deep-water environments, but not for deposits of other three types of bottom currents (i.e., wind, tide, or baroclinic). Note that BCRS represent only sandy lithofacies, but may also be applicable to silty lithofacies. Figure from Shanmugam (2016a), with permission from Elsevier.

- Furthermore, Stow et al. (2008, p. 145) state that "Bottom currents are highly variable in location, direction and velocity over relatively short time scales (from hours to months). Velocity increase, decrease and flow reversal occur as a result of deep tidal effects (e.g. Shanmugam, 2008)" (i.e., Shanmugam, 2008a in this chapter).

Although Stow et al. (2008) were justified in searching for a broad term to represent all bottom currents, their choice of the term "contour currents" for all types is inappropriate. As noted earlier, there are four basic types of bottom currents, namely (1) thermohaline-driven contour currents, (2) wind-driven bottom currents, (3) tidal bottom currents, and (4) baroclinic currents. Major problems associated with broadening the meaning of the term contour currents are as follows:

- Unlike thermohaline-driven contour currents, the other three types do not originate due to thermohaline forcing. The Loop Current in the Gulf of Mexico, for example, is strictly a wind-driven current (Pequegnat, 1972; CIMAS, 2015); no thermohaline forcing is involved. It would be incorrect to classify deposits of the Loop Current as contourites.
- Unlike thermohaline-driven contour currents, the other three types commonly do not follow bathymetric contours. The wind-driven Loop Current in the Gulf of Mexico, for example, does not follow bathymetric contours (Pequegnat, 1972; Mullins et al., 1987; Shanmugam et al., 1993a). The Loop Current also triggers eddies that fail to follow bathymetric contours.
- Deep-marine tidal currents flow up and down submarine canyons (Shepard et al., 1979).
- In some cases, baroclinic tidal currents flow across the canyon and in a direction parallel to the shelf break (Allen and Durrieu de Madron, 2009).

Rebesco et al. (2008, p. 6) argued that a strict adaptation of the basic definition of Heezen et al. (1966) would prevent the application of the contour-current concept to ancient deposits, where both depth and direction of the currents can rarely be precisely reconstructed. Although interpretation of ancient deep-water strata will always remain a challenge, we should not compromise the basic principles of contour currents for the sake of convenience and simplicity. A solution is to adopt the general term "bottom currents" for all four types.

As a continuation of this problem, the original meaning of the term contourite has been broadened. The tradition of genetic nomenclature in sedimentary geology began with the introduction of the term turbidite for a deposit of a turbidity current in deep-water environment (Kuenen, 1957). Shanmugam (2006b) presented a detailed review of the problems associated with genetic nomenclatures. The term contourite was first introduced in a publication for deposits of contour currents by Hollister and Heezen (1972), although Hollister (1967) discussed contourites earlier in his unpublished PhD dissertation. In these early contributions, contourites solely referred to deposits of contour currents. But other researchers have widened the definition to include deposits of a variety of bottom currents that include wind-driven currents and tidal currents (Stow et al., 2008). Such a broad application of the term contourite undermines the very basic tenet of process sedimentology, which is to distinguish deposit of one specific process from that of the other. In acknowledging this conceptual-nomenclatural problem, Rebesco et al. (2008, p. 7) state, "This implies the risk of an excessively wide application of the term 'contourite', and consequently of a loss of significance." Although the original contourite concept was designed solely for deep-water deposits (Hollister and Heezen, 1972), it has been expanded to include shallow-water deposits (e.g., Verdicchio and Trincardi, 2008), causing additional confusion.

These problems can be alleviated by simply being faithful to the original definition of the term contourite as envisaged by the founding fathers of the concept: the late B.C. Heezen and the late C.D. Hollister. In discussing gravity-driven downslope processes, Middleton and Hampton (1973) proposed four types of sediment-gravity flows, namely grain flow, fluidized flow, debris flow, and turbidity current, based on sediment-support mechanisms. No one would classify deposits of all four types of sediment-gravity flows as turbidites! Similarly, we should not classify all four types of bottom currents as contourites.

4.4 Hiatuses in Contourites

In nonmarine and shallow-marine clastic environments, hiatuses (breaks in sedimentation) are ubiquitous. For example, Miall (2014) reported that only 10% of elapsed time is represented by sediment in these environments; the remainder (90%) is nothing but hiatuses. In deep-marine environments, regional erosion throughout thousands of square kilometers of seafloor has been attributed to bottom currents (Berggren and Hollister, 1977; Tucholke and Embley, 1984). In the Gulf of Cadiz, the lower core of the Mediterranean Outflow Water (MOW) tends to cause more erosion (Hernández-Molina et al., 2014). In the Rockall Trough region, bottom currents associated with the NADW have caused an erosive area extending over 8500 km^2 in water depths of 500–2000 m (Howe et al., 2001). This erosive phase, which eroded approximately 150 m of sediment and lasted nearly 35 Ma (Early Oligocene-Holocene), existed through four supercycles (second order) and 23 cycles (third order) of sea-level rise and fall in the global chronostratigraphic chart of Haq et al. (1988). Viana (2008) cautioned on the potential dangers of misinterpreting regional unconformities at the base of contourites as "sequence boundaries" on seismic profiles using examples from the Santos Drift, offshore Brazil (Duarte and Viana, 2007). Clearly, there is no simple correlation between current-induced erosional surfaces (unconformities) and eustasy. These practical challenges exist because there are no objective criteria to recognize erosional surfaces, caused by deep-marine bottom currents versus other processes, on seismic profiles (Shanmugam, 1988, 2007).

4.5 Origin of Erosional Features

Pérez et al. (2015) discussed erosional and depositional features associated with contourites on seismic data. However, there are conceptual and sedimentological problems.

- In defining the contourite depositional system (CDS), Hernández-Molina et al. (2008, p. 350) state, "An association of various drifts and related erosional features has been termed a 'contourite depositional system' (CDS)..." This inclusion of erosional features under the term "contourite depositional system" is conceptually confusing. It is useful to maintain a distinction between erosion and deposition. A solution is simply to group both erosion and deposition under "contourite system" instead.
- Following Hernández-Molina et al. (2006, 2008), García et al. (2009) attributed the origin of four types of erosive features, including contourite channel, to erosion exclusively by the MOW in the Gulf of Cadiz. However, these authors did not consider the alternative possibility of erosion by baroclinic currents in the Gulf of Cadiz, where internal waves and internal tides are active oceanic phenomena

(Cairns, 1980; Armi and Farmer, 1988; LaViolette and Lacombe, 1988; Apel, 2000; Morozov et al., 2002; Vargas-Yañez et al., 2002; Chérubin et al., 2003, 2007; Serra, 2004; Pavec et al., 2005; Ambar et al., 2008; Huthnance et al., 2008; Sánchez-Román et al., 2008; Vsemirnova et al., 2009; León et al., 2014). The other problem is that there are no detailed measurements and observations on the velocities and erosive power of baroclinic currents on the deep seafloor. This is a potential topic for future research.

- Stow et al. (2013, p. 112) state, "In this paper, we have detailed the development and characteristics of a contourite channel, which is as long, wide and deep as many turbidity current channels, but which has been cut and shaped by bottom currents, and by their interaction with a bottom topography influenced by neotectonics. In places it is floored by contourite sands and gravel." If the channel was cut and shaped by bottom currents that include four types (Shanmugam, 2008a), it is misleading to classify any channel a contourite channel with a skewed emphasis on contour currents, ignoring the other three bottom currents.
- There are no sedimentological criteria to distinguish deep-sea channels cut by turbidity currents from those cut by contour currents. This problem is further complicated when similar depositional features, such as mud drapes, are associated with channels of different origins. For example, mud drapes have been reported from (1) turbidite channels (Miocene) exposed at the San Clemente State Beach, California (Walker, 1975) and from (2) estuarine tidalite channels (Cretaceous) in the subsurface conventional cores, Oriente Basin, Ecuador (Shanmugam et al., 2000).
- Erosion by strong bottom currents tends to cause lag deposits in submarine environments. Various aspects of contourite lag deposits were discussed by other authors (Hüeneke and Stow, 2008; Martın-Chivelet et al., 2008; Stow and Faugères, 2008; Wetzel et al., 2008). The grain size of the lag deposits merely indicates which grain-size fractions could not be transported. Besides, a lag represents a gap in the sedimentary record, which may cause problems with the construction of high-resolution age models of sediment cores.

By nature, erosion does not leave behind any clue in the rock record for establishing the type of process that caused the erosion. Furthermore, modern unfilled submarine channels and canyons are a testimony to the fact that the processes that created these erosional features in the past are probably not the same processes that will fill them in the future. Therefore, there is a need to develop criteria for distinguishing erosional features cut by contour currents from those cut by other processes, such as turbidity currents.

4.6 Gulf of Cadiz as the Type Locality

Hernández-Molina et al. (2013) characterized the Gulf of Cadiz as "the world's premier contourite laboratory." The modern Gulf of Cadiz has served as the center for contourite research activities since the 1970s (Fig. 9.1, location B). For example:

- The Gulf Cadiz is the birthplace of the first contourite facies model (Faugères et al., 1984; Gonthier et al., 1984).
- The MOW (Fig. 9.17) and related properties have been well studied (Zenk, 1975; Ambar and Howe, 1979; Zenk and Armi, 1990; Pinardi and Masetti, 2000; Criado-Aldeanueva

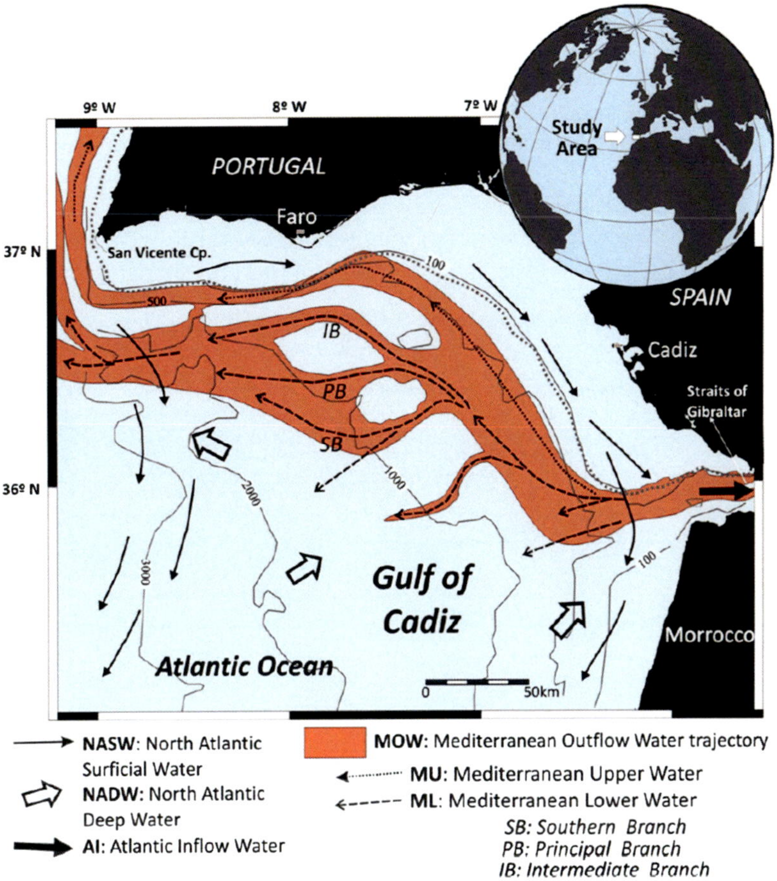

FIGURE 9.17 Map showing the main water-mass circulation in the Gulf of Cadiz. Note the trajectory of the Mediterranean Outflow Water (MOW) flowing westward in the gulf and turning northward as it enters the Atlantic Ocean at Cape São Vicente (San Vicente cp.). The initial black-and-white version was published by Hernandez-Molina et al. (2003); modified by Llave et al. (2011) and Stow et al. (2013). With permission from the Geological Society of America.

et al., 2006; Hernández-Molina et al., 2003, 2006, 2014; García et al., 2009; Alves et al., 2011; Mulder et al., 2013; Stow et al., 2013), with salinity, temperature, and velocity measurements (Price et al., 1993; Baringer and Price, 1999).

- Internal waves and internal tides have been documented in the Gulf of Cadiz (Cairns, 1980; Armi and Farmer, 1988; LaViolette and Lacombe, 1988; Apel, 2000; Bruno et al., 2006; Alvarado-Bustos, 2011; Sanchez-Garrido et al., 2011; Quaresma and Pichon, 2013).
- Sedimentary bedforms on the seafloor were documented using side-scan sonar images (Kenyon and Belderson, 1973) and submarine photographs (Stow et al., 2013).
- The Gulf of Cadiz was the site of the IODP Expedition 339 (Hernández-Molina et al., 2013).

The Gulf of Cadiz, despite its popularity, has its limitations. Although the MOW in the Gulf of Cadiz is a thermohaline-driven water mass (Alves et al., 2011), it is not a genuine contour current. For example, Zenk (2008, p. 45) characterizes the behavior of MOW as follows: "The warm and salty Mediterranean outflow water (MOW) in the Gulf of Cadiz of the eastern North Atlantic represents *an excellent example for the transition* (*italics* for emphasis) between a purely bottom-following current to a genuine contour current…" Empirical data indeed support the transition of the MOW in the Gulf of Cadiz. The MOW undergoes three progressive stages of evolution during its journey from the Strait of Gibraltar where it enters the Gulf of Cadiz to Cape São Vicente where it exits the gulf before entering the Atlantic Ocean (Fig. 9.18).

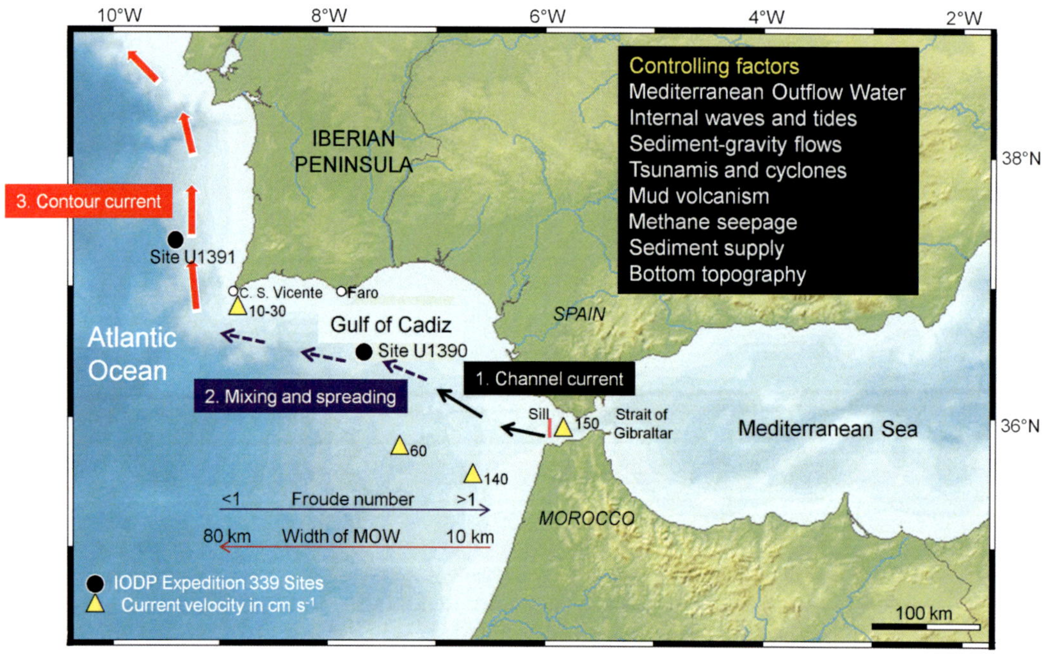

FIGURE 9.18 Schematic diagram showing the location of Gulf of Cadiz and complex transport nature of the Mediterranean Outflow Water (MOW), involving three stages of evolution: (1) channel-current stage, (2) mixing and spreading (i.e., transition) stage, and (3) genuine contour-current stage (see Zenk, 2008, his Fig. 4.10). Velocity at the Strait of Gibraltar is from Heezen and Johnson (1969). Velocity near Cape São Vicente is from Prater and Sanford (1994) and Baringer and Price (1999). Other velocity values, Froude numbers, and MOW widths are from Baringer and Price (1997, 1999). Details on IODP Expedition 339 cores are discussed by Hernández-Molina et al. (2013), who reported 300 cm s^{-1} (118.11 in. s^{-1}) velocity at the Strait of Gibraltar (see also Gonthier et al., 1984) and ~80–100 cm s^{-1} near Cape São Vicente. The popular Faro contourite drift (Faugères et al. (1984) is located just south of the town of Faro offshore. *C.S. Vicente*, Cape São Vicente, Cape St. Vincent (in some publications); *Sill*, Camarinal Sill (Sánchez-Román et al., 2008). Blank base map credit: http://search.aol.com/aol/imageDetails?s_it=imageDetails&q=gulf+of+cadiz&v_t=wscreen50-bb&b=image%3Fenabled_terms%3D%26s_it%3Dwscreen50-bb%26q%3Dgulf%2Bof%2Bcadiz%2B%2B%26oreq%3D24c0082b9d3f4e468816812f471e3793&img=http%3A%2F%2Fupload.wikimedia.org%2Fwikipedia%2Fcommons%2Fthumb%2F8%2F8f%2FAlboran_Sea_map.png%2F220px-Alboran_Sea_map.png&host=http%3A%2F%2Fen.wikipedia.org%2Fwiki%2FGulf_of_C%25C3%25A1diz&width=80&height=82&thumbUrl=http%3A%2F%2Fimages-partners-tbn.google.com%2Fimages%3Fq%3Dtbn%3AANd9GcTrt3F8dPieSVqdoqx7k_zjHT2FTU1uxbncPOgNzNk7T_h_RX3IJDMxkX0&imgWidth=220&imgHeight=225&imgSize=32693&imgTitle=gulf+of+cadiz. Figure from Shanmugam (2016a), with permission from Elsevier.

4.6.1 Channel-Current Stage

Price et al. (1993), based on the 1988 Gulf of Cadiz Expedition that included 99 full–depth profiles of temperature and salinity and 56 horizontal current profiles, characterized the MOW in the Gulf of Cadiz as a "steady channel flow" near the Strait of Gibraltar (Fig. 9.18). At this first stage, the current was highly turbulent and the Froude number was above 1. The transport was downslope from east to west (Fig. 9.18); however, the descent was asymmetric and occurred in two preferred modes or cores (Baringer and Price, 1997).

4.6.2 Mixing and Spreading Stage

Mixing and spreading of MOW represents the second transition stage (Fig. 9.18). Within 100 km downstream from the Strait of Gibraltar, the MOW was affected by the Coriolis force. Due to mixing, the MOW lost its density and increased its transport volume westward. The velocity progressively decreased westward from 150 cm s^{-1} at the strait to 10–30 cm s^{-1} near Cape São Vicente (Fig. 9.18). At this turning point, the MOW became neutrally buoyant in the lower portion of the North Atlantic thermocline (Baringer and Price, 1999). In the western Gulf of Cadiz, where the entrainment was much weaker, Froude numbers were consistently below 1 (Baringer and Price, 1997).

4.6.3 Contour-Current Stage

After making a 90° turn to the right (north) in the open Atlantic Ocean due to the full effect of the Coriolis force, the MOW attains total geostrophic balance and flows northward nearly parallel to the bottom topography of the Atlantic Ocean, off the western Iberian margin (Zenk, 2008, his Fig. 4.10; Hernández-Molina et al., 2011, their Fig. 4). At this final stage, the MOW is considered a genuine contour current (Fig. 9.18). In summary, the Gulf of Cadiz is a highly complex oceanographic location for studying depositional and erosional aspects of genuine contour currents because the deep-sea sediments in this gulf are controlled by the following factors (Fig. 9.18):

- Transitory MOW (Zenk, 2008)
- Internal waves and tides (Apel, 2000; Alvarado-Bustos, 2011)
- Sediment-gravity flows (Hernández-Molina et al., 2013)
- Pelagic and hemipelagic settling
- Tsunamis (Lario et al., 2010)
- Cyclones (Lario et al., 2010)
- Mud volcanism (Pinheiro et al., 2003)
- Methane seepage (Magalhães et al., 2012)
- Sediment supply (Mulder et al., 2013)
- Pore-water venting and hydraulic pumping (León et al., 2014)
- Channels and ridges (Stow et al., 2013)
- The Camarinal Sill (Gómez-Enri et al., 2007)

Complex localities like the Gulf of Cadiz requires an understanding of all processes in concert with each other because deep-water processes are tightly intertwined with shallow-water processes by oceanic wave phenomena, such as internal waves and tsunamis. Therefore, the archaic notion of dealing with a particular deep-water process (e.g., contour currents) in a vacuum is over. The 21st century necessitates the rigor of holistic process sedimentology.

4.7 The Contourite Facies Model

Faugères et al. (1984) explained the role of MOW in developing the first muddy contourite facies model from the Gulf of Cadiz (Fig. 9.19). Students (Brackenridge, 2014; Lathrop, 2015) and researchers (Rebesco et al., 2014) use this model routinely. Nevertheless, the vertical facies model suffers for the following reasons.

4.7.1 *Five Internal Divisions*

Faugères et al. (1984) developed the original facies model without internal divisions. Stow and Faugères (2008, their Fig. 13.9), however, revised the original model with five internal

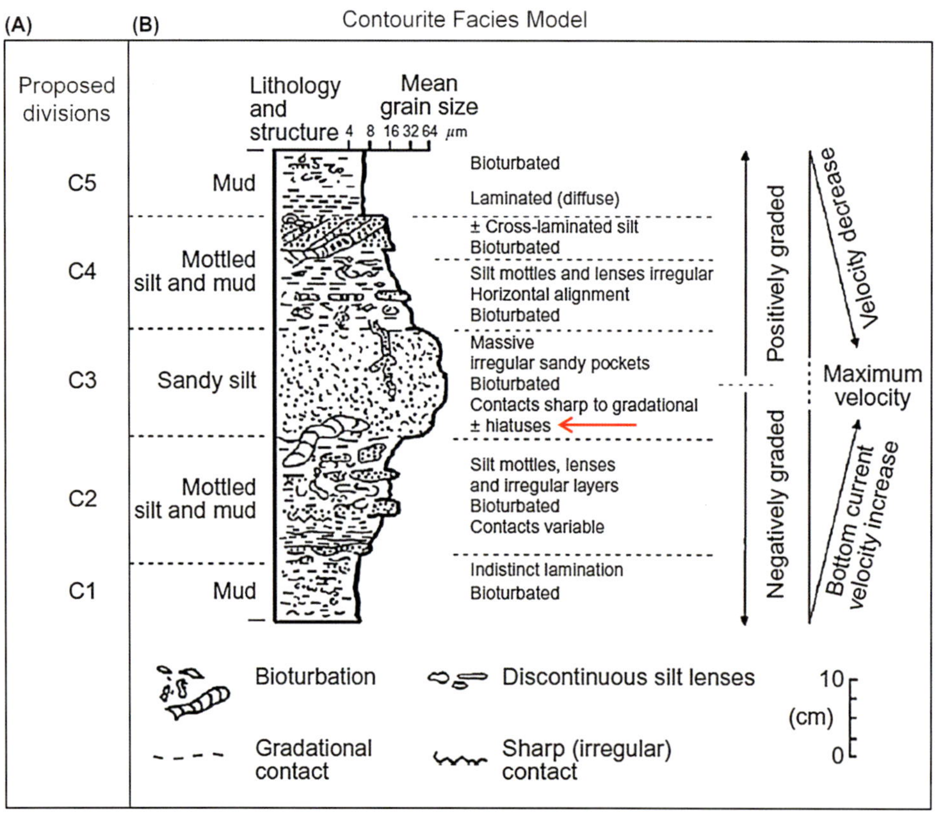

FIGURE 9.19 (A) Revised contourite facies model with five divisions proposed by Stow and Faugères (2008). (B) Original contourite facies model by Faugères et al. (1984). Note that the original authors of this model did not include the five internal divisions (Faugères et al., 1984). The version of this model by Faugères and Mulder (2011) contains neither the five internal divisions nor the hiatuses in the C3 division (*red arrow* inserted in this article). Originally from Faugères et al. (1984), with permission from the Geological Society of America.

divisions (C1, C2, C3, C4, and C5) (Fig. 9.19A) analogous to the Bouma turbidite model (Bouma, 1962). In their most recent version, Faugères and Mulder (2011, their Fig. 3.18) have reverted back to the 1984 version, without the five internal divisions. Reasons for such back-and-forth fundamental changes to the facies model, by the same group of authors, need to be explained in the literature for the benefit of the international research community. If recognized in the ancient rock record, these five divisions would reveal nothing about deposition from thermohaline-driven geostrophic contour currents in deep-water environments.

4.7.2 *Current Velocities*

The vertical facies model, composed of a basal upward-coarsening interval followed by an upward-fining interval (Fig. 9.19B), has been attributed to a successive increase and decrease in contour-current velocity and competency (Faugères et al., 1984). However, Mulder et al. (2013) suggest that the origin of this vertical sequence is much more complex than due to a simple velocity variation. Mulder et al. (2013, p. 357) state that "… the contourite sequence is only in part related to changes in bottom current velocity and flow competency, but may also be related to the supply of a coarser terrigeneous particle stock, provided by either increased erosion of indurated mud along the flanks of confined contourite channels (mud clasts), or by increased sediment supply by rivers (quartz grains) and downslope mass transport on the continental shelf and upper slope. The classical contourite facies association may therefore not be solely controlled by current velocity, but may be the product of a variety of depositional histories." No further explanation is necessary.

4.7.3 *Internal Hiatuses*

In the original contourite facies model, Faugères et al. (1984, their Fig. 4) did not include internal hiatuses. However, Stow and Faugères (2008, their Fig. 13.9) included hiatuses in the middle C3 division of their revised contourite facies model (Fig. 9.19A; see *horizontal red arrow*). In the most recent (2011) version of the model (Faugères and Mulder, 2011, their Fig. 3. 18), the hiatuses are absent once again. How can a natural, observed, sedimentary feature (i.e., hiatus) simply vanish? The authors need to explain this puzzle.

Wetzel et al. (2008, p. 189) state, "When bottom currents prevent deposition for a considerable time span, and/or erode sediments, submarine hiatuses develop, represented by semi-consolidated firm- or hard grounds or stable cohesive partially dewatered muddy substrates." Because hiatuses occur in the C3 division (Fig. 9.19B), the lower and upper intervals must represent two different depositional events. Conventionally, a genetic facies model is designed for a single depositional event, without internal hiatuses (e.g., the turbidite facies model, Bouma, 1962). In fact, Walther's Law (Middleton, 1973) is not meaningful for sequences with internal hiatuses. This is because a hiatus can represent a considerable span of time (spanning millions of years) that is missing in the sedimentary record (Howe et al., 2001).

4.7.4 *Bioturbation*

A characteristic feature of the contourite facies model is the bioturbation (Fig. 9.19B), which has generated debates (Shanmugam, 2002b; Mulder et al., 2002). Conventionally, a genetic facies model (e.g., the turbidite facies model, Bouma, 1962) is based on vertical disposition of primary physical sedimentary structures. This is because physical structures can be used to interpret a particular physical process in the rock record. But bioturbation cannot be used as a criterion for interpreting deposit of a single process (i.e., contour currents). The bioturbation criterion is defective because ancient deep-water turbidites (e.g., in the Late Cretaceous Point Loma Formation near San Diego, California) are also extensively bioturbated and even contain the trace fossil *Ophiomorpha* (Nilsen and Abbott, 1979). Furthermore, convincing cases of contourites without bioturbation have been documented in the rock record (Dalrymple and Narbonne, 1996). In describing the Canterbury Drifts from SW Pacific Ocean, *Carter (2007*, p. 129) state that *"Bioturbation is moderate and rarely destroys the pervasive background, centimetre-scale, planar or wispy alternation of muddy and sandy silts displayed by Formation Micro-Scanner imagery.* The muddy contourite facies model with emphasis on bioturbation defies the very first principle of process sedimentology, which is to interpret the fluid mechanics of depositional processes using primary physical sedimentary structures (Sanders, 1963).

4.7.5 Multiple Interactive Processes

The muddy contourite facies model was based on the notion that a single process, namely deposition from contour currents, was solely responsible for the deposit (Faugères et al., 1984). But Stow et al. (2013) have demonstrated that multiple interactive processes are operating in the Gulf of Cadiz. In 1984, prior to detailed velocity measurements of MOW (Price et al., 1993) and numerous other investigations of internal waves and internal tides in the Gulf of Cadiz, it was reasonable for Faugères et al. (1984) to propose a contourite facies model at a time when we were grappling with complex deep-water processes, without much data. But today, a great wealth of empirical data (see references in Stow et al., 2013) is available. The Gulf of Cadiz is an extremely complex setting in terms of physical oceanography with multiple processes (e.g., MOW, internal waves, and internal tides) and bottom topography with channels, ridges, and sills. The physical, chemical, and sedimentological aspects of the MOW are equally complex (Ambar et al., 2002; Criado-Aldeanueva et al., 2006). Rebesco et al. (2014, p. 139) acknowledge that "Regardless, the previous research on this issue holds two important lessons: firstly, that there is no unique facies sequence for contourites; and secondly, that traction sedimentary structures are also common within contourites..." Deep-water depositional processes are variable in time and space. Furthermore, extensive bioturbation caused by influx of prolific oxygen in deep-sea currents obliterates physical structures. From a practical viewpoint of interpreting ancient deposits as contourites on land, there is no way of knowing the contours of the paleo-seafloor (Stow et al., 1998). In summary, the global applicability of the contourite facies model is dubious.

4.8 Grain-Size Data and Related Issues

A fundamental aspect of many sedimentological studies is the documentation of detailed vertical grain-size variation that is plotted on a sedimentological log. It is so vital that the present author has allotted the maximum space for grain size (i.e., expanded column widths for silt, very fine sand, medium sand, etc.) in sedimentological logs (see Fig. 9.11B). But such sedimentological logs illustrating vertical grain-size variations and other sedimentological details for sandy contourite intervals are absent in publications by Stow and Faugères (2008) and by Stow et al. (2008). In fact, none of the 19 core photographs (six from the Gulf of Cadiz, eight from the Brazilian margin, and five from the UK margin) has associated sedimentological logs in Stow and Faugères (2008). Consequently, the reader is left with core photographs of sandy contourites without the fundamental grain-size data.

During the IODP Expedition 339, five sites were drilled in the Gulf of Cádiz and two sites off the West Iberian margin (Hernández-Molina et al., 2013). The total length of recovered core is 5447 m, with an average recovery of 86.4% (Expedition 339 Scientists, 2012). Published results of the IODP 339 core studies, although preliminary, are useful in testing the contourite facies model.

- A key element of the contourite facies model is the vertical grain-size variations (Fig. 9.19B). However, none of the published lithologic columns of drilled intervals contains Wentworth grain-size class on the abscissa (Fig. 9.20). Even the detailed lithologic logs for individual sites lack the Wentworth scale (Figs. 9.21 and 9.22).

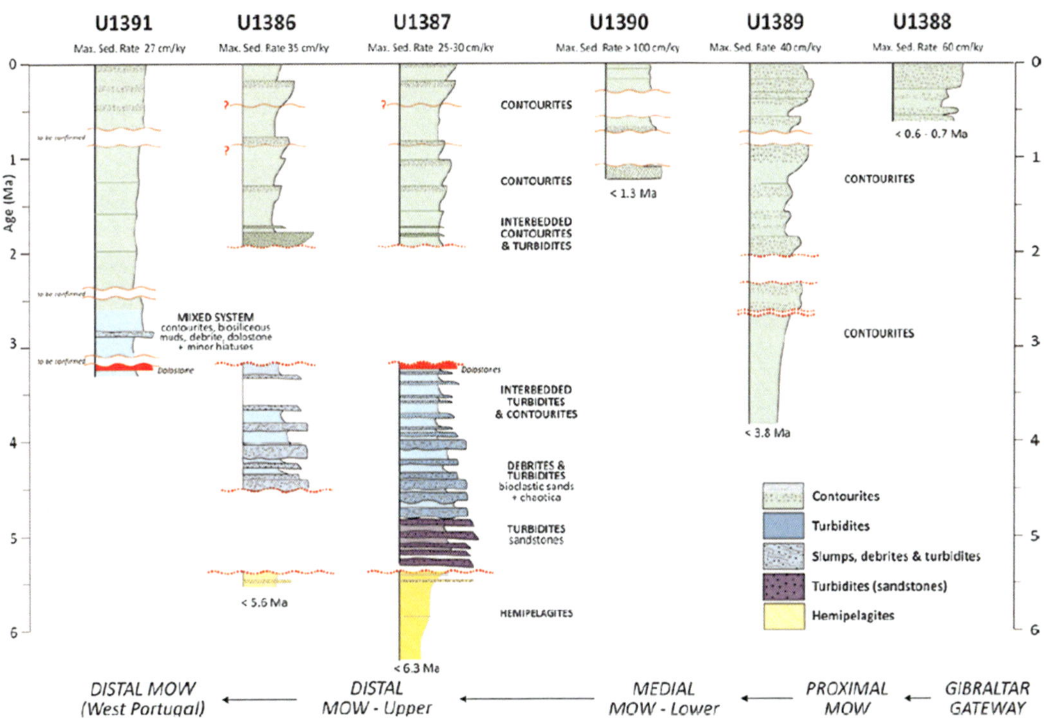

FIGURE 9.20 Lithologic summary for the sites drilled during IODP Expedition 339 in the Contourite Depositional System of the Gulf of Cadiz and west off Portugal. A general interpretation, including the position of principal hiatuses, is indicated. Age models are based on biostratigraphic datums and magnetostratigraphy. Sedimentation rates for the Pliocene = 15–25 cm $(ka)^{-1}$ and for the Quaternary = ~30 to >100 cm $(ka)^{-1}$. Note locations of sites U1390 within the Gulf of Cadiz and U1391 outside the Gulf of Cadiz. Also note the absence of Wentworth grain-size class on the abscissa on each log. From Hernández-Molina et al. (2013), with permission from IODP Expedition 339 Scientific Drilling.

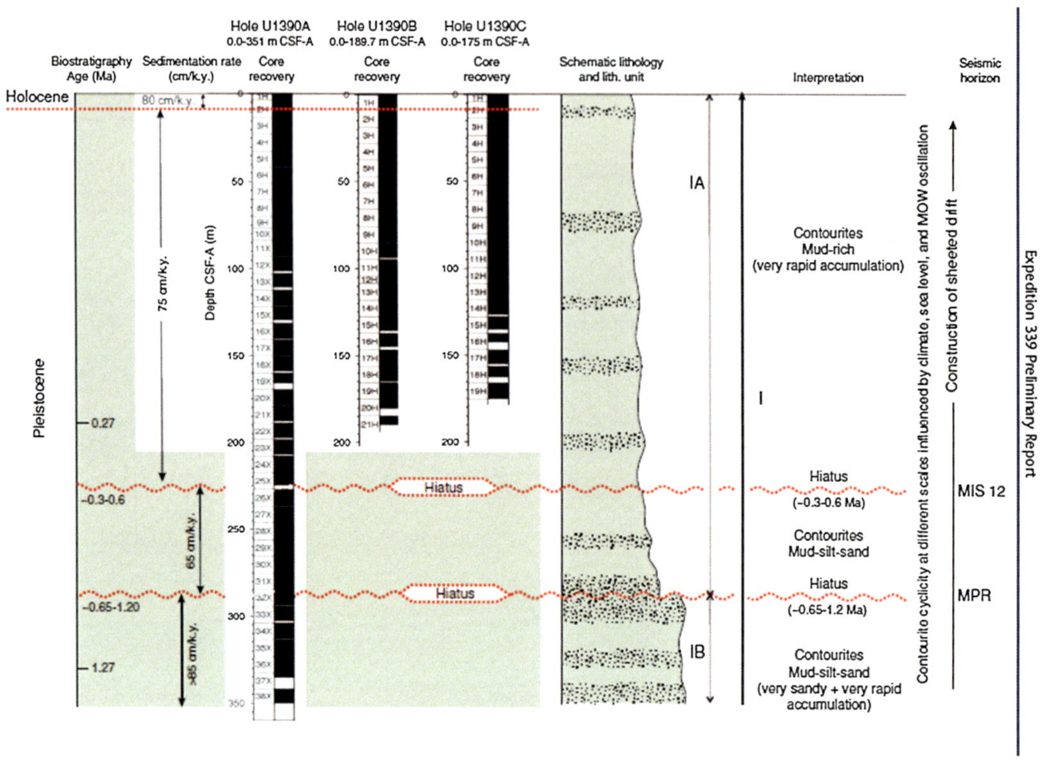

FIGURE 9.21 Lithologic summary for the Site U1390 located within the Gulf of Cadiz. *MOW*, Mediterranean Outflow Water; *MPR*, mid-Pleistocene revolution discontinuity; *BQD*, base Quaternary discontinuity. Note the absence of Wentworth grain-size class on the abscissa. From Expedition 339 Scientists (2012).

- Core photographs labeled as bigradational sequences (Fig. 9.23A) and sandy contourite (Fig. 9.23C) do not show vertical grain-size variations based on measurements.
- Specific sedimentological criteria used for distinguishing base cut-out contourites with normal grading (Fig. 9.23B) from turbidites with normal grading (Fig. 9.23D) are not discussed.
- The five internal divisions of the contourite facies model are not evident in any of the published core intervals. Even in the core interval U1390A-8H-6A, labeled Bigradational grading, which presumably represents the entire contourite sequence, the five internal divisions are not evident (Fig. 9.23A).
- The Expedition 339 Scientists (2012) reported hiatuses in contourites (Figs. 9.19 and 9.20). It is unclear as to how these hiatuses fit into the contourite facies model. Do these hiatuses represent the C3 division in the model (Fig. 9.19B)?
- Unlike turbidites with a sharp or an erosional contact at the base, contourites with gradational bases do not have a precise point of origin (Fig. 9.23). As a consequence, the starting point of a basal inversely graded contourite sequence is purely subjective.

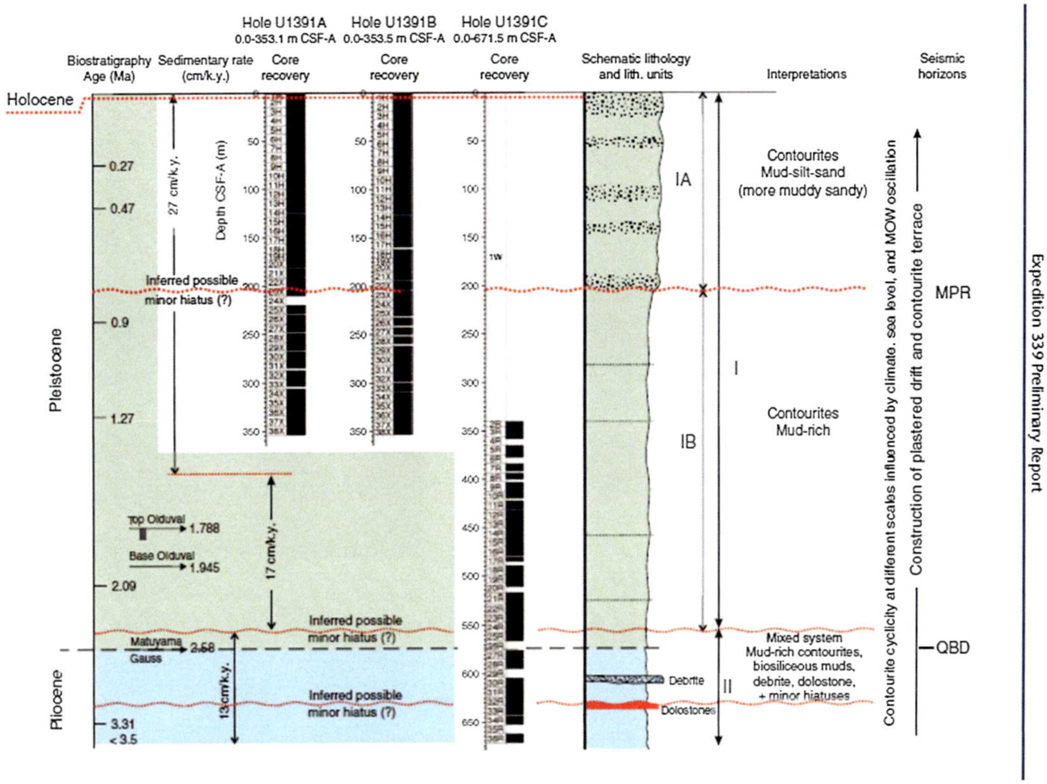

FIGURE 9.22 Lithologic summary for the Site U1391 located outside the Gulf of Cadiz. *MOW*, Mediterranean Outflow Water; *MIS*, marine isotope stage; *MPR*, mid-Pleistocene revolution discontinuity. Note the absence of Wentworth grain-size class on the abscissa. From Expedition 339 Scientists (2012).

- The Expedition 339 Scientists (2012) report that cored intervals at both sites of U1390 and U1391 show similar features, such as bigradational trends, a lack of five internal divisions, and internal hiatuses. The problem is that Site U1390 is located within the Gulf of Cadiz (36°19.110′N; 7°43.078′W) (Fig. 9.21), whereas Site U1391 is located outside the Gulf of Cadiz (Fig. 9.22), on the southwest Iberian Margin (37°21.532′N; 9°24.656′W). Therefore, the true significance of MOW in developing unique properties of contourite deposits within the Gulf of Cadiz (touted as the premier contourite site) is unconvincing.

In summarizing the results of IODP 339 cores, Stow et al. (2014) reported the following characteristics:

- The uniformity in sedimentation of muddy contourites
- The dominance of greenish-gray color
- The general absence of primary sedimentary structures
- The sediment homogenization by bioturbational mottling

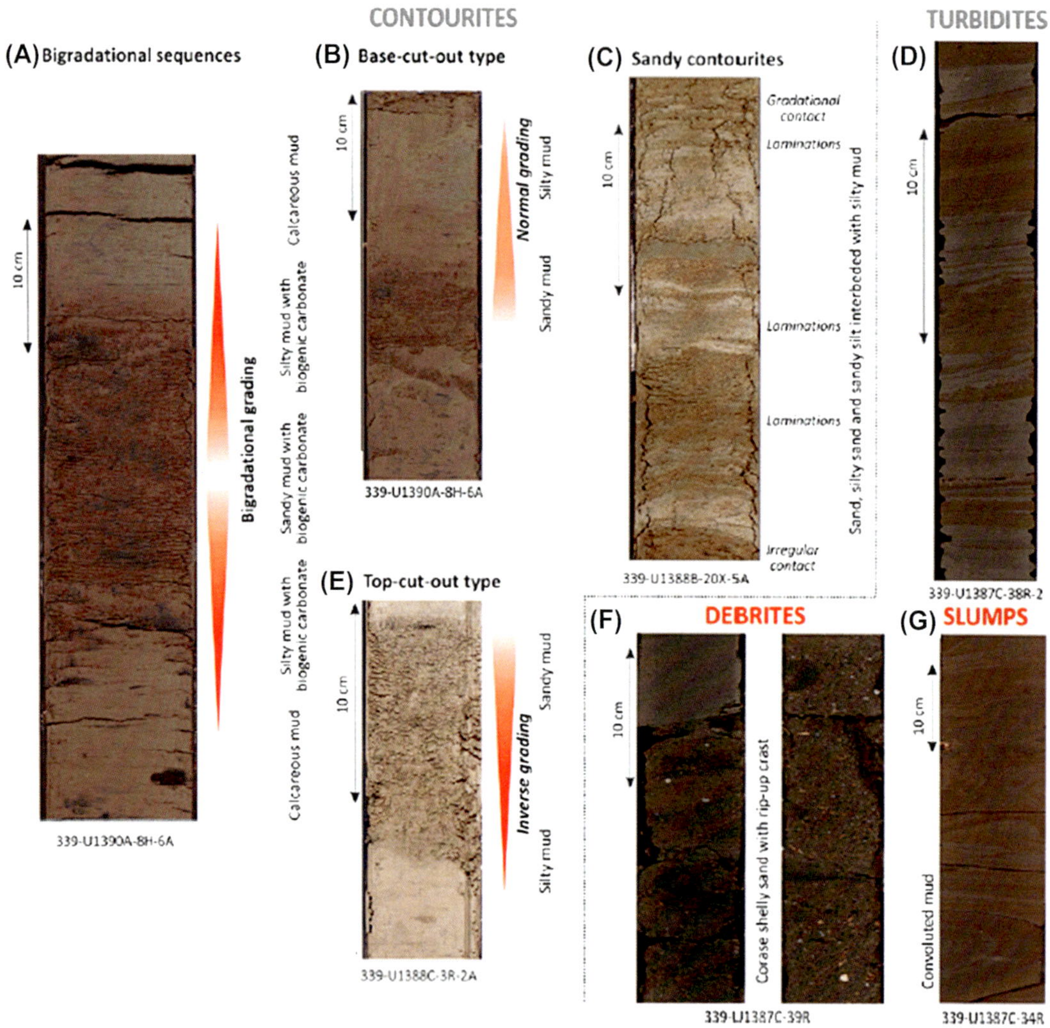

FIGURE 9.23 Core photographs showing sedimentary facies of contourites (A–C, E), turbidites (B), debrites (F), and slumps (G) recovered during IODP Expedition 339. Note that vertical grain-size variations showing grading are schematic (*red arrows*), not factual using the Wentworth grain-size class on the abscissa. From Hernández-Molina et al. (2013), with permission from IODP Expedition 339 Scientific Drilling.

- The uniformly mixed biogenic-terrigenous composition
- The consistent cyclicity of facies
- The grain size in bigradational units

Two fundamental problems are evident from the IODP 339 cores: (1) the absence of primary sedimentary structures, which renders it impossible to interpret depositional processes (e.g., Sanders, 1963); and (2) thin, bigradational muddy units, the underpinning

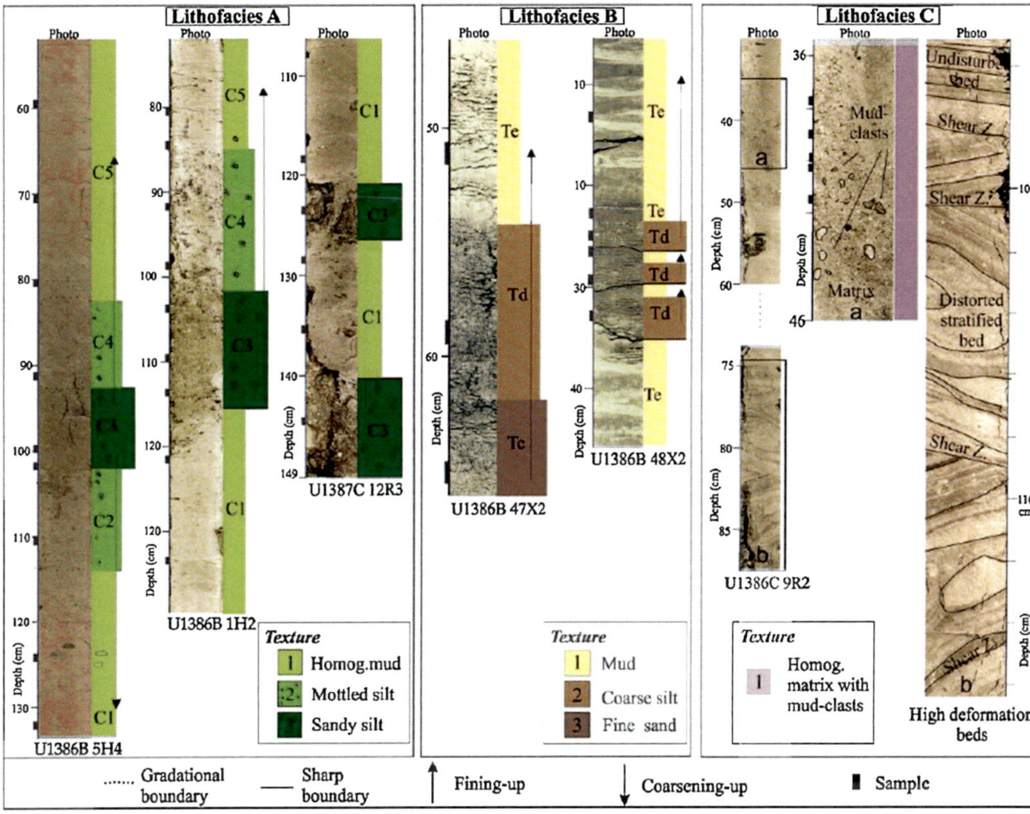

FIGURE 9.24 Core photographs showing the main sedimentary sequences of the Pleistocene Faro Drift deposits as interpreted by Alonso et al. (2016). The sequences of lithofacies A display complete contourite sequence with five divisions (C1 to C5) and truncated sequences (C3 to C5, and C3); the sequences of lithofacies B show fining-up sequence; and the sequences of lithofacies C display a matrix with mud-clasts (a) and highly deformed beds (b). Legend: C1 to C5 refer to the contourite divisions of Stow and Faugères (2008); Tc, Td, and Te are the turbidite divisions of the Bouma sequence (see Fig. 9.25); *Homog*, Homogeneous. Note the absence of Wentworth grain-size class on the abscissa. Photographs from Alonso et al. (2016). Publication: Marine Geology. With permission from Elsevier. Copyright Clearance Center's RightsLink: Licensee: G. Shanmugam. License Number: 3822650497346. License Date: March 5, 2016.

characteristic of the model, are impossible to recognize in the compacted mudstone of the ancient record.

In a study of IOPD 339 cores, Alonso et al. (2016) have identified all five divisions of the contourite facies model, namely C1, C2, C3, C4, and C5, in core photographs (Fig. 9.24), but failed to provide corresponding vertical grain-size variation using Wentworth scale. Instead, each contourite division is shown to exhibit a grain-size trend in a vertical column without any scale on the abscissa, which makes it practically impossible to evaluate the true vertical variation in grain size. Even if there are subtle differences in grain size among the five divisions, it would be impossible to recognize these massive contourite divisions without primary sedimentary structures (e.g., ripple cross-laminae) in the ancient rock record due to compaction. The ultimate goal of studying modern analogs, such as the Gulf of Cadiz, is to gain knowledge in interpreting ancient deposits as contourites for which the information on paleocurrent circulation is absent. But the sedimentological features observed in the cores of IOPD 339 sites yet failed to provide that basic knowledge for interpreting ancient strata as contourites.

Alonso et al. (2016) have also recognized internal divisions, composed of Tc, Td, and Te (Fig. 9.24) of the now defunct turbidite facies model known as the Bouma Sequence (Shanmugam, 1997b). The problem is that Tc, Td, and Te turbidite divisions can also be formed by bottom-current reworking, composed of contour currents (Fig. 9.25). For example, in areas in which both downslope sandy debris flows and along-slope-bottom currents operate concurrently (Fig. 9.25A), the reworking of the tops of sandy debris flows by bottom currents may be expected. Such a scenario could generate a basal massive sand division and an upper reworked division, mimicking a partial Bouma Sequence (Fig. 9.25B). The reworking of deep-water sands by bottom currents has been suggested by other researchers as well (e.g., Stanley, 1993; Ito, 2002; Strzeboński, 2015). But Alonso et al. (2016) ignored this alternative possibility in their interpretation. Genetic facies models are nothing more than a "groupthink" (Shanmugam, 2012a, p. 153) that tends to thrive more on custom and complacency than on intellect and innovation.

4.9 Traction Structures and Shale Clasts

The presence of traction structures in cores and outcrops (Fig. 9.5) have long been recognized as evidence for bottom-current reworked sands by contour currents, wind-driven currents, and tidal currents in deep-water strata (Hsü, 1964, 2008; Hubert, 1964; Klein, 1966; Hollister, 1967; Natland, 1967; Piper and Brisco, 1975; Shanmugam et al., 1993a,b; Shanmugam, 2008a; Martın-Chivelet et al., 2008; Mutti and Carminatti, 2011). As noted earlier, ripples and dunes have been associated with internal tidal currents (Lonsdale and Malfait, 1974). In other words, traction structures and bedforms have been associated with all four types of bottom currents. The challenge is how to distinguish a traction structure (e.g., ripple or parallel laminae) formed by contour currents from those formed by wind-driven bottom currents in the ancient stratigraphic record.

In discussing the origin of shale clasts in muddy and sandy contourites, Stow and Faugères (2008, p. 231) state, "The shale clasts are generally millimetric in size, and occur with long axes sub-parallel to bedding and, presumably, also sub-parallel to the current direction." Alternatively, the planar clast fabric (i.e., alignment of long axis of clasts parallel to the bedding surface) could be interpreted as evidence for laminar debris flow (Fisher, 1971; Enos, 1977; Shanmugam and Benedict, 1978). In short, there are no reliable sedimentological criteria that we can apply in interpreting the ancient rock record as sandy contourites.

4.10 Bedform-Velocity Matrix

Van Rooij (2013) used the bedform-velocity matrix (Fig. 9.25) of Stow et al. (2009) in discussing the challenges associated with processes and products of deep-water bottom currents. Problems associated with the bedform-velocity matrix are as follows:

- Stow et al. (2009) proposed a bedform-velocity matrix (Fig. 9.25) for deep-water bottom currents. This matrix diagram is a slightly modified version of Figs. 3.1 and 3.2 in

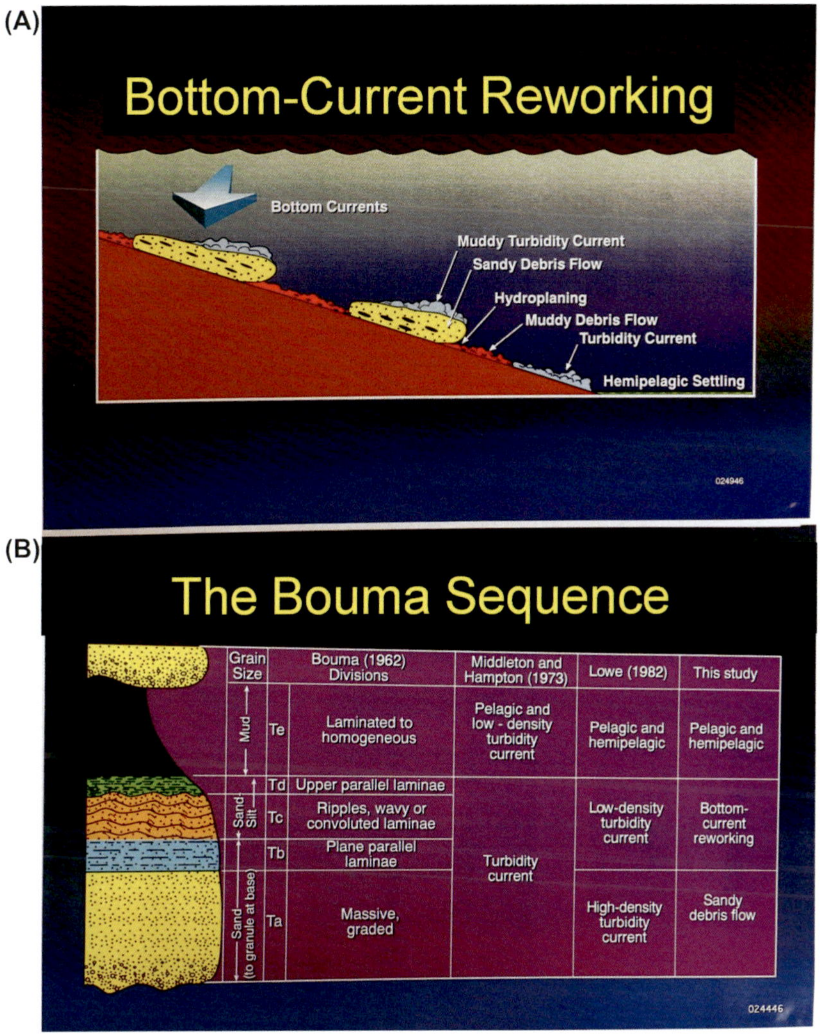

FIGURE 9.25 (A) Conceptual model showing reworking the tops of downslope sandy debris flows by along-slope bottom currents. Such complex deposits would generate a sandy unit with a basal massive division and upper reworked divisions with traction structures (ripple laminae), mimicking the Bouma sequence. Figure from Shanmugam (2006a). Publication: Elsevier Books. Handbook of Petroleum Exploration and Production, Volume 5 (2006). With permission from Elsevier. Copyright Clearance Center's RightsLink: Licensee: G. Shanmugam. License Number: 3795990304484. License Date: January 25, 2016. (B) The turbidite facies model (i.e., the Bouma Sequence) showing Ta, Tb, Tc, Td, and Te divisions. Conventional interpretation is that the entire sequence is a product of a turbidity current (Bouma, 1962; Walker, 1965; Middleton and Hampton, 1973). According to Lowe (1982), the Ta division is a product of a high-density turbidity current and Tb, Tc, and Td divisions are deposits of low-density turbidity currents. In this article, the Ta division is considered to be a product of a turbidity current only if it is normally graded, otherwise it is a product of a sandy debris flow; the Tb, Tc, and Td divisions are considered to be deposits of bottom-current reworking. Figure from Shanmugam (1997b). Publication: Earth-Science Reviews. With permission from Elsevier. Copyright Clearance Center's RightsLink: Licensee: G. Shanmugam. License Number: 3795980450687. License Date: January 25, 2016.

Belderson et al. (1982). Stow et al. (2009) applied the bedform-velocity matrix, developed by Belderson et al. (1982) for shelf tidal currents, to all types of deep-water bottom currents. But shallow-water tidal currents and deep-water bottom currents are not one and the same hydrodynamically. As mentioned earlier, at least four different types of deep-water bottom currents exist (Shanmugam, 2008a). The underpinning assumption of the matrix, which is that all four deep-water bottom currents hydrodynamically behave the same as the shallow-water tidal currents, is incongruous.

- Stow et al. (2009) acknowledged that (1) although the velocity data presented by them were for near-bottom flow, they did not define the exact height above seafloor; (2) they did not address the variable nature of the benthic boundary layer that will also complicate how flow velocity affects seafloor bedform; (3) for most of their data sets it was impossible to know the precise flow velocity (mean or peak) that created the observed bedform; (4) they rarely had the opportunity of witnessing the development of deep-water bedforms in situ; and (5) they did not consider the effects of sediment supply and bed roughness on bedform development. In other words, the matrix was built without the necessary empirical data.
- The concept of bedform-velocity matrix became popular in the 1960s with the advent of matrix diagrams of alluvial sedimentary structures based on empirical data derived from flume experiments (Simons et al., 1965). However, the matrix diagram proposed by Stow et al. (2009) is not based on experiments; meaning that their data are neither verifiable nor reproducible independently.
- In commenting on the problems with the bedform-velocity diagram of Stow et al. (2009) and Dykstra (2012, his Fig. 14.2 caption) states, "Note that this Fig. does not take into account either the duration of a current or sediment availability, both of which are important controls on the development of bedforms…."

Given these uncertainties, it is unreliable to estimate current velocities for modern bedforms using the bedform-velocity matrix.

4.11 Seismic Profiles, Sonar Images, and Submarine Photographs

Nelson et al. (1993) interpreted sandy contourites in the Gulf of Cadiz based on seismic data, but without critical sedimentological data. Well-developed wave geometries seen on seismic profiles, interpreted as megasediment waves formed by the MOW off Southwest Portugal, have been reported (Nielsen et al., 2008). However, seismic wave geometry has also been associated with sand dunes formed by internal solitary waves (Reeder et al., 2011). Furthermore, no objective criteria exist to distinguish wave geometry created by contour currents from wave geometry created by tidal currents or by turbidity currents on seismic profiles (see review by Shanmugam, 2012a). In their comprehensive review of seismic expression of contourite depositional systems, Nielsen et al. (2008) state that "… because the reflections result from changes in the physical parameters through the sedimentary succession, there is no unequivocal correlation between seismic facies and sedimentary structures within the facies. A seismic facies characterized by a parallel reflection configuration, for example, need not necessarily indicate the existence of fine parallel banding or stratification of the sediments." Clearly, there are fundamental problems in using seismic facies for interpreting bottom-current deposits. For example, bottom-current-reworked sands are difficult to recognize even from the direct examination of the rocks because of the presence of traction structures in deposits of all four types of bottom currents.

Sedimentary bedforms on the seafloor have been documented using side-scan sonar images (Kenyon and Belderson, 1973). Stow et al. (2013, p. 101) state, "Examination of bottom photographs is one of the principal methods by which we can determine the nature of processes operating at the present day in deep water environments." Although they have used over 3000 submarine photographs, interpreting a specific process from a bird's-eye view of the submarine photograph is problematical. Photographic images of ripples and other bedforms on the seafloor are useful for inferring current directions, but not current types (i.e., hydrodynamic behaviors). Identical ripple types can be formed by more than one type of bottom current in the deep sea. In the deep Pacific Ocean, for example, ripples and dunes were attributed to internal tidal currents (Lonsdale and Malfait, 1974) (Fig. 9.13B). But in the deep Gulf of Mexico, ripples were related to the wind-driven Loop Current (Pequegnat, 1972) at a depth of 3091 m (Shanmugam, 2012a, see Fig. 9.8). The problem is that there are no objective criteria to distinguish ripple types associated with contour currents from those associated with wind-driven bottom currents. In the modern Gulf of Cadiz, where both MOW and internal tides are active, we cannot distinguish the type of ripples formed by MOW-related bottom currents from those formed by baroclinic tidal currents.

Turbidity currents and debris flows can develop normal grading and inverse grading, respectively. But such *internal* features cannot be resolved on submarine photographs of *external* bedform-surfaces. Internal sedimentary structures are best studied using core and outcrop, which are the key to interpreting fluid mechanics of depositional processes (Sanders, 1963).

4.12 Oceanic Waves

Oceanic waves are composed of three main types, namely internal waves and tides, cyclonic waves, and tsunami waves. All three waves have associated bottom currents. A common depositional attribute of these three types is traction structures. Because traction structures are also common in deposits of contour currents (Hollister, 1967), wind-driven currents (Pequegnat, 1972), and tidal currents (Klein, 1975), it is necessary to discuss the types of oceanic waves here.

4.12.1 Internal Waves and Tides

Depositional aspects of oceanic waves (e.g., internal waves, cyclonic waves, and tsunami waves) and their bottom currents are still a poorly understood entity (Shanmugam, 2008c, 2012c, 2013a). In particular, the topic of internal waves and internal tides has generated lively debates with direct implications for turbidite and contourite research (Table 9.2). Internal waves and internal tides are active oceanic phenomena in the Gulf of Cadiz at various depths (Armi and Farmer, 1988; Apel, 2000; Serra, 2004; Pavec et al., 2005; Ambar et al., 2008; Huthnance et al., 2008; Magalhães et al., 2010; Alvarado-Bustos, 2011; Sanchez-Garrido et al., 2011; Quaresma and Pichon, 2013; León et al., 2014). Of particular significance is the study by Alvarado-Bustos (2011), who states, "Semi-diurnal internal tides and a continuous MOW flow are observed on the slope. The MOW flow is persistent reaching >0.40 m s^{-1}, but varies in strength with the tides. The Internal wave field in the Gulf of Cadiz can play an important role affecting the MOW signal over the continental slope; MOW can be displaced by the internal tide. Internal waves are generated by tides and MOW flow interacting with the bottom, the two most energetic sources locally." In this complex environment, it would be a challenge to distinguish sands deposited by MOW-related transitional currents from deep-water sands deposited by baroclinic currents associated with internal waves and internal tides. This is because there are at present not yet objective sedimentological criteria to recognize baroclinic sands (Shanmugam, 2012b, 2013a,b, 2014a).

According to Stow et al. (2013), the sandy bedforms in the Gulf of Cadiz are a product of both MOW-related bottom currents and deep tidal currents. It illustrates the problems with conducting contourite research in a complex oceanographic setting, such as the Gulf of Cadiz, with multiple interactive processes. Even if the influence of baroclinic currents is minimal in depositing the sandy deposits in the Gulf of Cadiz, the fact that the MOW is in transition undermines the legitimacy of the contourite story.

Based on swath bathymetric data and on chirp and 2D seismic data, León et al. (2014) proposed that "...pockmark formation on either side of the Strait of Gibraltar resulted from gas and/or sediment porewater venting from overpressured shallow gas reservoirs entrapped in coarse-grained contourites of levee deposits and Pleistocene palaeochannel infillings. Venting was either triggered or promoted by hydraulic pumping associated with topographically forced internal waves. This mechanism is analogous to the long-known effect of tidal pumping on the dynamics of unit pockmarks observed along the Norwegian continental margin." Given that the origin of contourites in the Gulf of Cadiz is already a problematic issue, the origin of pockmarks in contourites associated with complex factors, such as possible porewater venting and hydraulic pumping attributed to internal waves, further complicates the problem.

In distinguishing deposits of internal waves and internal tides in the ancient stratigraphic record, bidirectional cross-bedding has been used (Gao and Eriksson, 1991). This is based on the notion that up- and down-currents in channel environments develop bidirectional cross-bedding. However, satellite images of modern internal waves reveal that the directions of propagation of internal waves are highly variable with respect to the shoreline, the shelf edge, and the channel axis (Fig. 9.26). Furthermore, no systematic linking exists of wave-propagation directions seen as the sea-surface manifestations on satellite images (Fig. 9.26) with their respective influence on internal sedimentary structures (i.e., dip directions) in the depositional bedforms on the modern seafloor. This is further compounded by the presence of local sills on the seafloor because sills invariably control the direction of wave propagation (Fig. 9.27D–F), which include the Camarinal Sill (Fig. 9.27E). Since the first publication on vertical facies models of internal-tide deposits by Gao and Eriksson (1991), there has not been any systematic process-sedimentological research on baroclinic currents in establishing their vertical disposition of sedimentary structures either by using sediment cores from modern marine settings, or by conducting laboratory experiments in validating vertical facies trends. The stalled status of research on internal-tide deposits is evident in a review article by Gao et al. (2013), which has resulted in a discussion (Shanmugam, 2014b) and reply (Gao et al., 2014).

On a positive note, the study by Stow et al. (2013) offers some hope in advancing research on bottom currents in the Gulf of Cadiz because, for the first time, the authors acknowledge

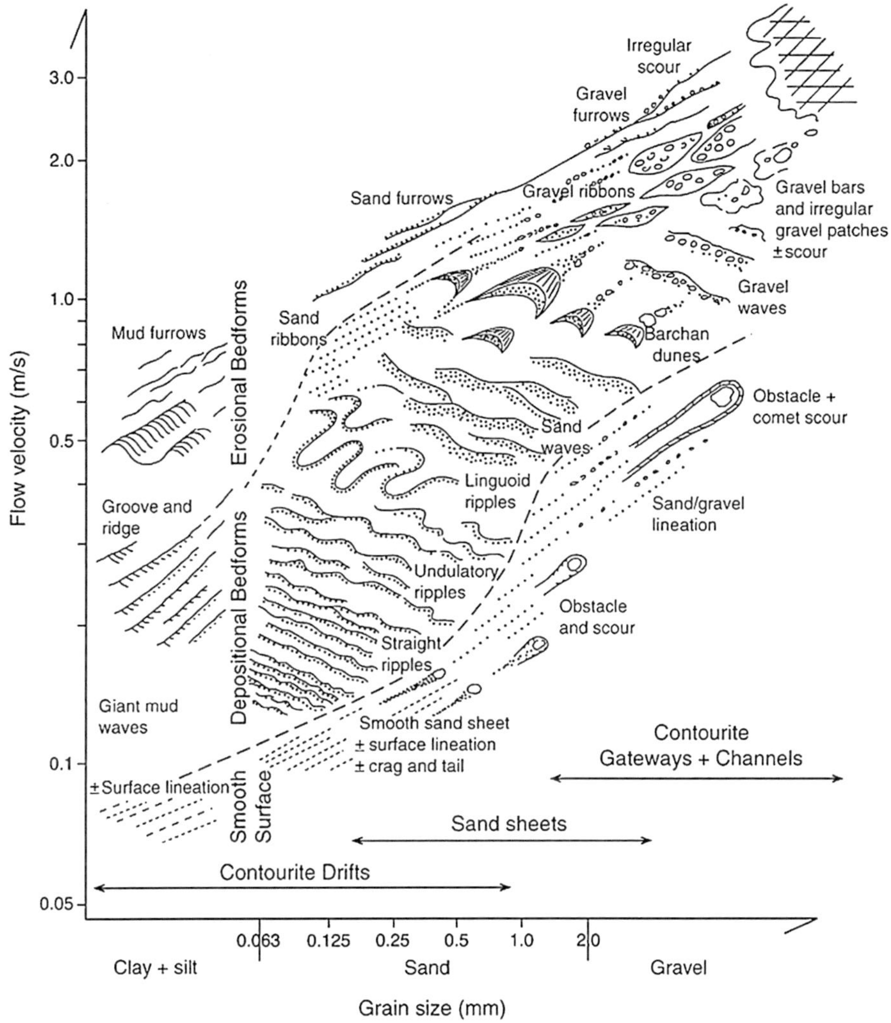

FIGURE 9.26 Bedform-velocity matrix for deep-water bottom currents. From Stow et al. (2009), with permission from the Geological Society of America.

the sedimentological significance of internal waves and internal tides in the Gulf of Cadiz, although the oceanographic significance of internal waves has been well known in the Gulf of Cadiz.

4.12.2 *Cyclonic Waves*

In the Gulf of Mexico, the propagation of tropical cyclones over the wind-driven Loop Current was investigated by Jaimes (2009), Oey and Wang (2009), and Jaimes and Shay (2010), among others. In the northern Gulf of Mexico, empirical data show that the wind speeds of Hurricane Katrina increased dramatically as it passed through the warm waters of the

Loop Current toward the Gulf Coast in late August in 2005. The increased wind velocity of hurricanes has implications for increasing velocities of bottom currents associated with cyclones (Shanmugam, 2008c, 2012a). It is worth noting that although both tropical cyclones and the Loop Current are wind-driven phenomena, the Loop Current can penetrate the entire water column and affect the seafloor (Pequegnat, 1972).

Cyclonic waves can erode and transport sediment in deeper shelf environments at 200 m (Komar et al., 1972) because cyclone-induced *combined flows*, a combination of unidirectional currents and oscillatory motion driven by waves, are powerful agents of sediment transport on the shelf (Swift et al., 1986). Such combined forces can increase shear stress in the current direction up to 10 times more than the shear stress exerted by the unidirectional current alone (Silvester, 1974). Measured velocities of cyclone-induced bottom flows in various submarine settings (e.g., shelf, slope, canyon, reentrant, and trough) are given in Table 9.4. Maximum velocities of cyclone-triggered bottom flows are in the range of 100−300 cm s^{-1} on the shelf and 200−7000 cm s^{-1} in submarine canyons and troughs (Table 9.4). At these high bottom velocities, even gravel-size grains would be eroded and transported.

In the Gulf of Mexico, south of Mobile Bay (Alabama), Teague et al. (2006) have estimated that extensive bottom scouring along the outer continental shelf under Hurricane Ivan resulted in the displacement of more than 100 million m^3 of sediment from a 35 km × 15 km region directly under Ivan's path. Sediment resuspension was accomplished by the extreme waves generated by Ivan and transported by strong near-bottom wind-driven currents. Bottom scouring results from a combination of wave-driven sediment resuspension and current-driven transport of the resuspended sediment (Keen and Glenn, 2002). Hurricane Ivan produced the largest wave field ever measured under a hurricane with maximum and significant wave heights about 28 and 18 m, respectively, near the locations under maximum wind stress (Wang et al., 2005). Near-bottom currents ranged from 40 to 120 cm s^{-1} at all six moorings during Hurricane Ivan's passage (Mitchell et al., 2005) while scouring occurred. The Gulf of Cadiz has also been subjected to cyclones (Lario et al., 2010). The implication is that there are no criteria to distinguish erosional and depositional features associated with contour currents from those associated with cyclonic bottom currents.

4.12.3 *Tsunami Waves*

The Gulf of Cadiz has also been subjected to tsunamis (Lario et al., 2010). Tsunami waves not only cause erosion and deposition during inundation of coastlines in subaerial environments, but also trigger backwash flows in submarine environments. These incoming waves and outgoing flows emplace sediment in a wide range of environments, which include coastal lake, beach, marsh, lagoon, bay, open shelf, slope, and basin. Holocene deposits of tsunami-related processes from these environments exhibit a multitude of physical, biological, and geochemical features (Shanmugam, 2012c, his Fig. 3). These features include horizontal planar laminae, cross-stratification, and hummocky cross-stratification. In the context of the present review on contourites, tsunami-related traction structures are of relevance because they represent both landward- and seaward-dipping cross-stratification (Fig. 9.27) (Shanmugam, 2012c). In interpreting sediment provenance of deep-water sediments with bottom-current deposits, such opposing current directions need to be evaluated with the possibility of tsunamis that affect virtually all marine basins (Fig. 9.28).

TABLE 9.4 Measured Velocity Values of Cyclone-Induced Bottom Flows in Various Submarine Settings. Updated after Shanmugam (2008c)

Meteorological Event (Date)	Submarine Setting (Bathymetry)	Velocity in cm s^{-1} (References)
Category 2 Hurricane Isabel (September 18, 2003)	Shelf (Onslow Bay), North Carolina 30 m	>50 (Wren and Leonard, 2005)
Tropical Storm Delta (September, 1973)	Shelf, Gulf of Mexico 21 m	50–75 (Forristall et al., 1977)
Unnamed cyclone (December 13, 1995)	Shelf (Eel), northern California 50 m	80 (Cacchione et al., 1999)
Unnamed cyclone (October 28, 1999)	Shelf (Eel), northern California 60 m	88 (Puig et al., 2003)
Category 5 Hurricane Allen (August, 1980)	Shelf (Texas), Gulf of Mexico 70 m	80–90 (Snedden et al., 1988)
Category 5 Hurricane Katrina (August, 2005)	Shelf, Gulf of Mexico 73–100 m	>100 (Welsh et al., 2009)
[a]Tropical Storm Floyd (September 18, 1999)	Shelf, New Jersey 12 m	80–100 (Kohut et al., 2006)
Category 3 hurricane Diana (September 11–13, 1984)	Shelf (Onslow bay), North Carolina 24–33 m	125 (Mearns et al., 1988)
Category 4 hurricane Lili (October 3, 2002)	Shelf (Atchafalaya), Gulf of Mexico 4.5 m	140 (Allison et al., 2005)
Category 5 hurricane Ivan (September 16, 2004)	Shelf (Alabama), Gulf of Mexico 89 m	150 (Stone et al., 2005)
Category 2 Unnamed hurricane (March 3, 1999)	Shelf (Columbia river Mouth), Oregon 35 m	>150 (Moritz, 2004)
Category 5 hurricane Camille (August, 1969)	Shelf, gulf of Mexico 10 m	160 (Murray, 1970)
Category 5 hurricane Rita (September, 2005)	Outer continental shelf, Gulf of Mexico 40 m	250–400 (Gearhart et al., 2011, their Figure 21)
Category 3 hurricane Joy (December, 1990)	Shelf, Great Barrier Reef, Australia 12 m	140 >300 (Larcombe and Carter, 2004)
Unnamed cyclone (January 7–11, 1989)	Slope, Middle Atlantic Bight 500 m	40 (Brunner and Biscaye, 1997)
Category 5 hurricane Ivan (September, 2004)	Upper continental slope Gulf of Mexico 500–1000 m	>200 (Teague et al., 2007)
Category 2 hurricane Georges (September 24–28, 1998)	Canyon (Mississippi), Gulf of Mexico 300 m	68 (Burden, 2000)
Unnamed cyclone (October 28, 1999)	Canyon (Eel), northern California 120 m	78 (Puig et al., 2003)
Unnamed cyclone (February, 2004)	Canyon (Cap de Creus), Gulf of Lions 300 m	80 (Palanques et al., 2006).

(Continued)

TABLE 9.4 Measured Velocity Values of Cyclone-Induced Bottom Flows in Various Submarine Settings. Updated after Shanmugam (2008c)—cont'd

Meteorological Event (Date)	Submarine Setting (Bathymetry)	Velocity in cm s^{-1} (References)
Unnamed cyclone (December 17–19, 2002)	Canyon (Monterey), northern California 1300 m	150–500+ (MBARI, 2003)
Unnamed cyclone (November 24, 1968)	Canyon (Scripps), southern California 44 m	190 (Inman et al., 1976).
Category 3 hurricane Hugo (September, 1989)	Canyon (salt river), St. Croix, V.I. >100 m	200–400 (Hubbard, 1992)
Category 1 hurricane Iwa (November, 1982)	Reentrant (Kahe point), Oahu, Hawaii 220 m	300 (Dengler et al., 1984)
Unnamed cyclone (August, 1990)	Trough (Suruga), Japan >500 m	7000 (Mitsuzawa et al., 1993)

[a]*Category 4 Hurricane Floyd weakened to a Tropical Storm strength offshore New Jersey.*

4.13 Reservoir Quality

Perhaps the first application of the contourite concept to a major petroleum reservoir was in the Frigg Field, North Sea (Heritier et al., 1979). These authors interpreted a wavy surface, between wells 25/1–1 and 25/1–5, on a seismic profile as evidence for contour currents. The Frigg field was considered one of the largest gas fields in the world in the 1970s. Despite numerous published contourite reservoirs (Shanmugam et al., 1993a, 1995a; Moraes et al., 2007; Viana, 2008; Mutti and Carminatti, 2011; Shanmugam, 2006a, 2012a, 2014a; Maslin, 2015), some petroleum geologists still believe that reservoir quality of bottom-current reworked sands, which include contourites, is poor in comparison to that of turbidites. In discussing the reservoir quality of deep-water Miocene sands in the Kutei Basin, Makassar Strait (Fig. 9.1, location E), Dunham and Saller (2014) claim that "The key point from the perspective of the Exploration-Geologist is that bottom currents did not transport or redistribute these Kutei basin reservoir-sands from their original-depositional locations. If significant redistribution of sand had occurred, our exploration-model would have failed, and we would not have found thick high-quality reservoir sands in our prospects. We based our interpretations (Saller et al., 2006, 2008b) on evidence from seismic data, cores, and exploration discoveries."

Contrary to this claim, published data do show that bottom-current reworked sands have good porosity and permeability. Selected examples include the following:

- Off the Great Bahama Bank, sandy calciclastic contourites (Middle Miocene to Pleistocene) have a measured maximum porosity of 40% and a maximum permeability of 9880 mD (Mullins et al., 1980). The high permeability has been attributed to the winnowing away of muds from the intergranular primary pores by vigorous contour currents. These carbonate sandy contourite drifts are hemiconical-shaped bodies that are up to 600 m in thickness and nearly 60 km in length.
- In the Ewing Bank Block 826 area (Fig. 9.7B), bottom-current reworked sands (Plio-Pleistocene) show 25–40% measured porosity and 100–1800 mD permeability

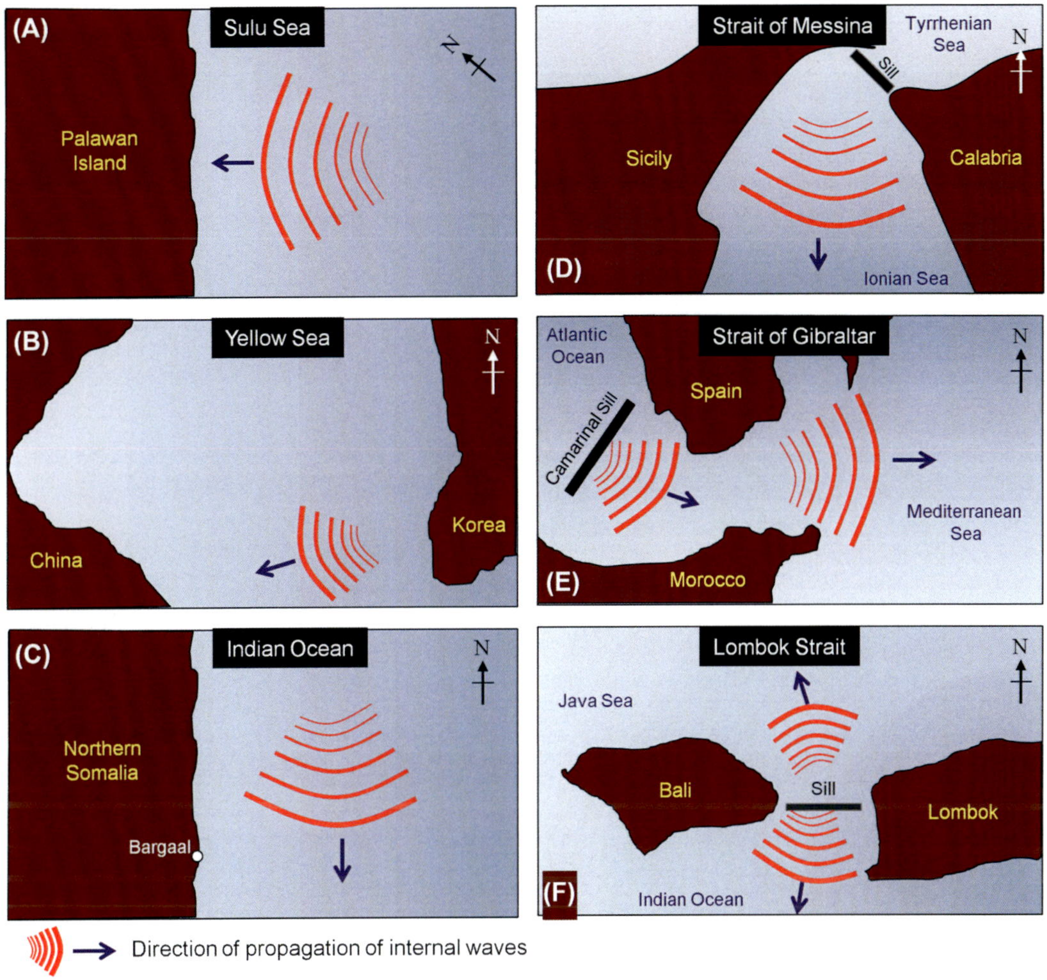

FIGURE 9.27 Maps showing the variable directions of propagation of internal waves with respect to shoreline or shelf edge seen as surface manifestations on satellite images. (A) Internal waves propagating toward the shoreline of Palawan Island in the Sulu Sea. (B) Internal waves propagating away from the shoreline or shelf edge in the Yellow Sea (Hsu et al., 2000, their Fig. 8). (C) Internal waves propagating nearly parallel to the shoreline of northern Somalia in the Indian Ocean (Jackson, 2004b, his Fig. 3). (D) Internal waves propagating parallel to the strait or channel axis in the Strait of Messina. (E) Internal waves propagating in the same direction on both sides of the Strait of Gibraltar. Note the position of the Camarinal Sill at the point of origin of internal waves (Gómez-Enri et al., 2007). (F) Internal waves propagating in opposite directions from the point of origin, which is a sill in the Lombok Strait (Susanto et al., 2005). Baroclinic currents, associated with internal waves and tides, are reworking agents and as such they are unrelated to the primary sediment provenance. Features shown are schematic and not to scale. From Shanmugam (2013a), with permission from AAPG.

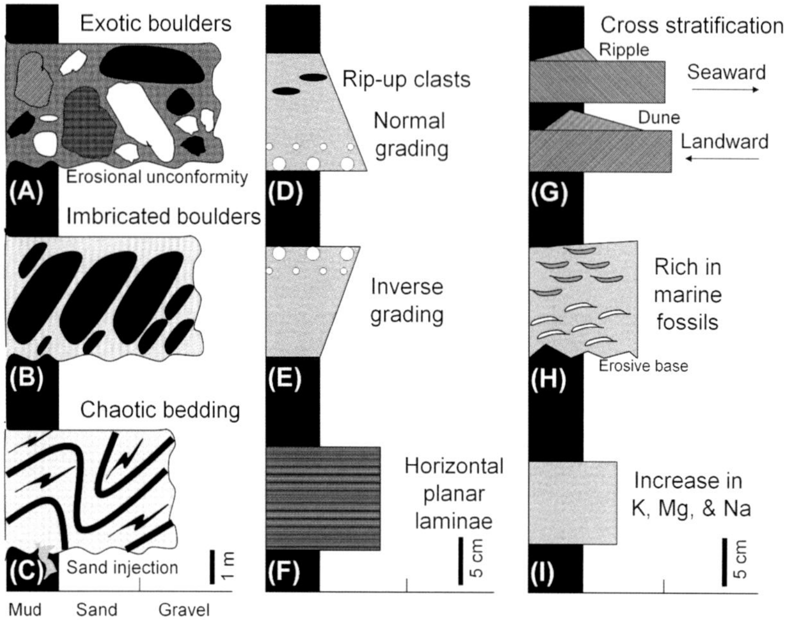

FIGURE 9.28 Published sedimentological features claim to be associated with tsunami-related deposits by other authors. These features are also associated with cyclone-related deposits. Note both landward- and seaward-dipping cross-stratification (g). Figure from Shanmugam (2012c). Publication: Natural Hazards. With permission from Springer. Copyright Clearance Center's RightsLink: License: G. Shanmugam. License Number: 3739560425172. License Date: October 31, 2015.

(Shanmugam et al., 1993a, their Table 1). Individual reworked sand layers commonly range in thickness from 5 to 10 cm, but the entire unit reached up to 6 m in total thickness.

- In the Bay of Bengal (Fig. 9.1, core description Fig. 9.11), high-quality Pliocene petroleum-producing reservoir sands formed by deep-marine sandy debris flows and tidal currents have been documented in the Krishna-Godavari Basin. Tidalite sands show measured porosity values of 34–41% and permeability values of 525–5977 mD (Shanmugam et al., 2009, their Table 4). Individual tidalite units vary from a few centimeters to nearly a meter in thickness (Fig. 9.11B).
- In the Gulf of Cadiz, a 10-m thick sheet sand has been interpreted as contourites (Stow et al., 2011).
- In southeastern South Africa, Fleming (1980, p. 179) studied bedforms formed by reworking by the Agulhas Current near the shelf edge. He documented a variety of bedform types, which include gravel pavements, sand ribbons, comet marks, sand streamers, dunes, and smooth sand sheets. The implication is that siliciclastic sandy and gravelly contourites near the shelf edge can develop important reservoirs with high porosity and permeability. If preserved, these sandy and gravelly contourites may occupy areas covering 10s of km in length (i.e., parallel to the shelf edge) and about 5 km in width (i.e., perpendicular to the shelf edge).

In summary, bottom-current reworked sands have better reservoir quality than turbidites in many cases (Shanmugam, 2012a, 2014a).

4.14 Sediment Provenance

4.14.1 Current Directions

Commonly, primary sedimentary structures and related current directions are used in deciphering sediment provenance (Pettijohn, 1975; Potter and Pettijohn, 1977; Zuffa, 1985). However, complex current directions associated with all four types of bottom currents pose immense challenges in inferring the primary sediment source. For example:

- Contour currents are global in circulation pattern and flow parallel to the strike of the regional slope (Figs. 9.3 and 9.14).
- Wind-driven bottom currents are complex in circulation pattern in the Gulf of Mexico (Fig. 9.7A), which include circular motions (gyres) unrelated to the slope. Such bottom currents have been reported beneath the Gulf Stream Gyre at a depth of nearly 4 km in the northern Bermuda Rise (Laine, 1978). Laine and Hollister (1981) suggest that the Deep Gulf Stream Return Flow entrains suspended sediment in a deep gyre and may be responsible for the deposition at the base of the continental rise.
- Deep-marine tidal currents are bidirectional in nature and they flow up and down submarine canyons (Fig. 9.10A).
- Baroclinic currents are extremely variable in propagation directions with respect to sediment source (Fig. 9.26).
- Because bottom currents are strictly a reworking agent, their sedimentary structures do not reflect the true direction of the primary sediment source (Fig. 9.29). Therefore, the conventional approach of inferring source directions (i.e., sediment provenance) using current ripples and cross-beddings is unreliable when dealing with deep-marine bottom currents and their deposits (Fig. 9.29).

4.14.2 Detrital Composition

The other important criterion in interpreting sediment provenance is the detrital composition (Zuffa, 1985; Arribas et al., 2007). However, reworking by bottom currents may not alter the original composition of the sediment derived from the primary provenance. For example, in understanding the compositional difference between contourites and turbidites in the Bounty Submarine Fan, New Zealand, cored intervals from the Ocean Drilling Program Site 1122 on Leg 181 have been studied. In discussing the results, Shapiro et al. (2007, p. 277) state that "… there are no significant trends among thickness, grain size, composition, and depth of Site 1122 sand samples, except that thicker beds tend to contain slightly more metamorphic rock fragments. The generally homogeneous composition of Site 1122 sand indicates that it may have had a relatively uniform source back into the early Miocene. Thus, the up-section change from sandy contourite to turbidite deposits at Site 1122 is not reflected in sand composition. This suggests that the sand provenance remained constant while the depositional processes of sand at Site 1122 changed." Distinguishing compositional variations caused by variations in deep-sea depositional processes is a potential area of future research on sediment provenance.

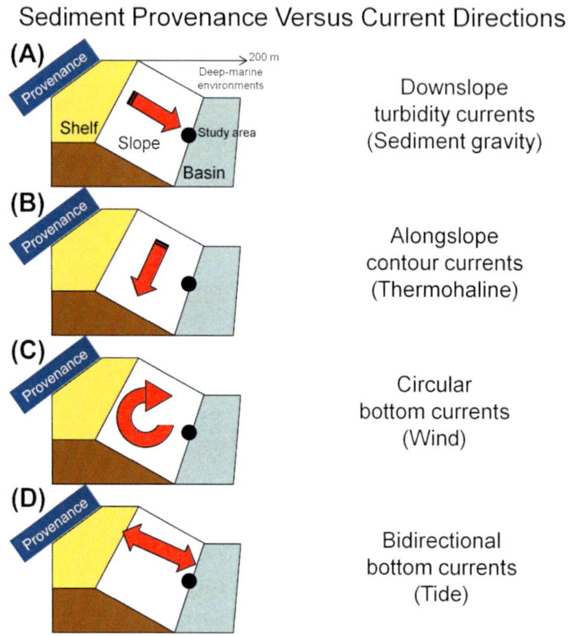

FIGURE 9.29 Four conceptual models showing the physical relationship between primary sediment provenance and current directions (red arrows) in deep-marine environments. (A) Downslope, unidirectional, turbidity currents. Current ripples in turbidites are reliable indicators of sediment provenance. (B) Along-slope, thermohaline-driven contour currents. Current ripples and cross-beddings in contourites are not reliable indicators of sediment provenance. (C) Circular, wind-driven bottom currents. Current ripples in these deposits are not reliable indicators of sediment provenance. (D) Bidirectional, tide-driven bottom currents are common in submarine canyons (Fig. 9.10A) (Shepard et al., 1979). Current ripples in deep-marine tidalites are also not reliable indicators of sediment provenance. Some sites, such as the Gulf of Cadiz (Fig. 9.18) that served as the type locality for the contourite facies model (Fig. 9.19), are also affected by bottom currents associated with internal waves, cyclones, and tsunamis, causing complex current directions.

4.15 Abyssal Plain Contourites

Hernández-Molina et al. (2008) discussed "abyssal plain contourites". Conventionally, the term "abyssal plain" refers to a flat region of the ocean floor, usually at the base of a continental rise, where slope is less than 1:1,000 (Heezen et al., 1959). It represents the deepest and flat part of the ocean floor that occupies between 4,000 and 6,500 m in the U.S. Atlantic Margin. A more general term "basin plain" is commonly used in referring to ancient examples (Shanmugam, 2016d). However, Hernández-Molina et al. (2008) consider abyssal plains or basin plains to include up to 10 distinct morphological elements: (1) continental rise; (2) abyssal plains; (3) oceanic rises, (4) distal fans and their distributary channels; (5) sediments drifts; (6) abyssal hills; (7) seamounts; (8) transfer fracture zones; (9) mid-ocean channels; and (10) oceanic trenches. This reclassification of abyssal plains, ignoring the basic principles of classification of continental shelf, slope, rise, and plain based on the position of seafloor depths, is confusing and unnecessary. This reclassification defies the basic concept of "contour currents" that was introduced for contour-following bottom currents along continental slope and rise, not for bottom currents flowing over flat abyssal plains.

5. CONCLUDING REMARKS

The contourite problems, composed of conceptual, nomenclatural, empirical, and methodological issues, have effectively hindered progress on contourite research during the past six decades. Failure to acknowledge and rectify these issues will only further muddle the problem. Because the real-world oceans are ubiquitously affected by multiple processes concurrently, the grand ingrained principle of "one deposit for one flow type" is nothing more than a misplaced optimism. The contourite problem is not just incidental, it is fundamental to the basic understanding of all deep-water sediments.

Acknowledgments

I thank Rajat Mazumder, the volume editor, for encouraging me to contribute this iconoclastic review of contourites. I also thank both Tasha Frank and Marisa LaFleur, Associate Acquisition Editors (Elsevier), for their enthusiastic help with various issues. I am deeply indebted to George Devries Klein, a sedimentologic pioneer on contourites and tidalites, for his total endorsement of science in this chapter and for his helpful editorial comments. I also thank A.J. (Tom) van Loon, who served as the Series Editor for Elsevier's Developments in Sedimentology 60 on "Contourites" (Rebesco and Camerlenghi, 2008) for his meticulous editing of the manuscript. As always, I am grateful to my wife Jean for her general comments on this manuscript and on all my other publications since 1976.

I acknowledge with gratitude the following organizations and colleagues involved in various academic activities that are of relevance in this chapter:

- My interest on provenance began with my research on sandstone reservoirs at Mobil Oil Company in 1978. As a consequence, I was an invited speaker at the NATO Advanced Study Institute Conference on "Reading Provenance from Arenites" held in Calabria, Italy (1984) by G.G. Zuffa. In a related conference volume edited by Zuffa (1985), my contribution dealt with "Types of porosity in sandstones and their significance in interpreting provenance" (Shanmugam, 1985).
- My sedimentological research on deep-water bottom currents began in 1974 as part of my PhD work on the Middle Ordovician of the Southern Appalachians in the United States (Shanmugam, 1978; Shanmugam and Walker, 1978, 1980) and has continued through my employment with Mobil Oil Company (Shanmugam and Moiola, 1982, 1984; Shanmugam, 1990; Shanmugam et al., 1993a,b) to the present as an adjunct professor and as a consultant (Shanmugam, 2006a, 2008a, 2012a, 2013a, 2014a).
- As my manager and coresearcher, R.J. Moiola provided enthusiastic support for my contourite research throughout my employment with Mobil (1978–2000). As a Mobil colleague, J.E. "Jed" Damuth provided me historical information on contourite research at Lamont-Doherty Earth Observatory of Columbia University (New York) where he received his PhD under Bruce Heezen. I am indebted to numerous colleagues at Mobil and other oil companies, petroleum-related service companies, academic institutions, and government agencies for assisting me in core and outcrop descriptions worldwide during the past 40 years (Table 9.2).
- As an invited lecturer in the SEPM Pacific Section Short Course held in San Francisco, as part of the 1990 AAPG Convention, I presented a lecture (Shanmugam, 1990) entitled "Deep-marine facies models and the interrelationship of depositional components in time and space." This lecture included emphasis on deep-water bottom currents. SEPM Course organizers: G.C. Brown, D.S. Gorsline, W.J. Schweller.
- My first major paper on process sedimentology and reservoir quality of sandy contourites, which focused on the significance of traction structures in contourites following Heezen's (1959) pioneering concept, was peer-reviewed by Charles Hollister for the AAPG Bulletin (Shanmugam, 1993a). I dedicate this paper to the late Charles Davis Hollister (1936–1999), considered to be "the father of contourites" (McCave, 2002), who died in a climbing

accident while on vacation in Wyoming with his family at an untimely age of 63. His pioneering publications have greatly influenced my research during the past 40 years.

- In response to an invitation from R.D. Winn Jr. and J.M. Armentrout, I (Shanmugam et al., 1995b) participated in the 1995 SEPM Core Workshop held in Houston, Texas. This study dealt with core examination of traction sedimentary structures indicating bottom-current reworking in the Gulf of Mexico.
- In response to an invitation from the UK Department of Trade and Industry, I organized a deep-water sandstone workshop in Edinburgh, Scotland, for petroleum geoscientists from various countries in Europe in 1995 (October). This workshop utilized cores from the UK Atlantic Margin (Table 9.2, Item 7) that contain deposits of sandy mass-transport deposits and bottom-current reworked sands (Shanmugam et al., 1995a).
- In response to an invitation from M. Rebesco, I contributed Chapter 5 (Shanmugam, 2008a), entitled, "Deep-water bottom currents and their deposits," to the thematic volume on contourites (Rebesco and Camerlenghi, 2008).
- In response to an invitation from A.J. (Tom) van Loon, I reviewed a book (Shanmugam, 2008d) entitled, *Economic and Palaeoceanographic Significance of Contourite Deposits*, edited by Viana and Rebesco (2007), for *Geologos* (republished in *Journal of Sedimentary Research*).
- I also reviewed a book (Shanmugam, 2011) entitled, *Deep-Sea Sediments*, edited by Hüneke and Mulder (2011), with Chapter 3 on "Contourites," for *Geologos* (republished in *Journal of Sedimentary Research*).

References

Akhurst, M.A., 1991. Aspects of Late Quaternary Sedimentation in the Faeroe-Shetland Channel, Northwest U.K. Continental Margin. British Geological Survey Technical Report, WB/91/2.

Allen, J.R.L., 1985. Loose-boundary hydraulics and fluid mechanics: selected advances since 1961. In: Brenchley, P.J., Williams, P.J. (Eds.), Sedimentology: Recent Developments and Applied Aspects. Blackwell Scientific, Oxford, pp. 7–28.

Allen, S.E., Durrieu de Madron, X., 2009. A review of the role of submarine canyons in deep-ocean exchange with the shelf. Ocean Science 5, 607–620.

Allison, M.A., Sheremet, A., Goni, M.A., Stone, G.W., 2005. Storm layer deposition on the Mississippi–Atchafalaya subaqueous delta generated by Hurricane Lili in 2002. Continental Shelf Research 25, 2213–2232.

Alonso, B., Ercilla, G., Casas, D., Stow, D.A.V., Rodŕıguez-Tovar, F.J., Dorador, J., Hernández-Molina, F.J., 2016. Contourite vs gravity-flow deposits of the Faro Drift (Gulf of Cadiz) during the Pleistocene: sedimentological and mineralogical approaches. Marine Geology 377, 77–94. http://dx.doi.org/10.1016/j.margeo.2015.12.016.

Alvarado-Bustos, R., 2011. Mixing in the Continental Slope: Study Case Gulf of Cadiz (Ph.D. thesis). University of Liverpool, Liverpool, UK, 138 p.

Alves, J., Carton, X., Ambar, I., 2011. Hydrological structure, circulation and water mass transport in the gulf of cadiz. International Journal of Geosciences 2, 432–456.

Ambar, I., Howe, M.R., 1979. Observations of the Mediterranean outflow I. Mixing in the Mediterranean outflow. Deep-Sea Research 26A, 535–554.

Ambar, I., Serra, N., Brogueira, M.J., Cabecadas, G., Abrantes, F., Freitas, P., Goncalves, C., Gonzalez, N., 2002. Physical, chemical and sedimentological aspects of the Mediterranean outflow off Iberia. Deep-Sea Research Part II 49, 4163–4177. http://dx.doi.org/10.1016/S0967-0645(02)00148-0.

Ambar, I., Alvarado-Bustos, R., Hobbs, R., Huthnance, J., Krahmann, G., Moate, B., Silva, P., Quentel, E., 2008. Gulf of Cadiz oceanography for comparison with seismic imaging. In: ALSO/AUG/TOS Ocean Sciences Meeting 2008: From the Watershed to the Global Ocean, 7th March 2008, Orlando, Florida, USA. American Society of Limnology and Oceanography.

Amos, A.F., Gordon, A.L., Schneider, E.D., 1971. Water masses and circulation patterns in the region of the Blake-Bahama Outer Ridge. Deep-Sea Research 18, 145–165.

Apel, J.R., 1987. Principles of Ocean Physics. In: International Geophysics, vol. 38. Academic Press, London, 63 p.

Apel, J.R., 2000. Solitons Near Gibraltar: Views from the European Remote Sensing Satellites. Report GOA 2000–1. Global Ocean Associates, Silver Spring, MD.

Apel, J.R., 2002. Oceanic internal waves and solitons. In: Jackson, C.R., Apel, J.R. (Eds.), An Atlas of Oceanic Internal Solitary-Like Waves and Their Properties (May 2002) by Global Ocean Associates Prepared for Office of Naval

Research – Code 322PO, pp. 1–40. http://www.internalwaveatlas.com/Atlas_PDF/IWAtlas_Pg001_Introduction.PDF.

Apel, J.R., Ostrovsky, L.A., Stepanyants, Y.A., Lynch, J.F., 2006. Internal Solitons in the Ocean. Woods Hole Oceanographic Institution Technical Report WHOI-2006-04. Woods Hole Oceanographic Institution, Woods Hole, MA 02543, 109 p.

Armi, L., Farmer, D.M., 1988. The flow of Mediterranean water through the Strait of Gibraltar. Progress in Oceanography 21, 1–105.

Arribas, J., Johnsson, M.J., Critelli, S. (Eds.), 2007. Sedimentary Provenance and Petrogenesis: Perspectives from Petrography and Geochemistry. GSA Special Paper 379.

Bádenas, B., Pomar, L., Aurell, M., Morsilli, B., 2012. A facies model for internalites (internal wave deposits) on a gently sloping carbonate ramp (Upper Jurassic, Ricla, NE Spain). Sedimentary Geology. http://dx.doi.org/10.1016/j.sedgeo.2012.05.020.

Baringer, M.O., Price, J.F., 1997. Mixing and spreading of the Mediterranean outflow. Journal of Physical Oceanography 27, 1654–1677.

Baringer, M.O., Price, J.F., 1999. A review of the physical oceanography of the Mediterranean outflow. Marine Geology 155, 63–82.

Barrett, J.R., 1965. Subsurface currents off Cape Hatteras. Deep-Sea Research 12, 173–184.

Bein, A., Weiler, Y., 1976. The Cretaceous Talme Yafe Formation: a contour current shaped sedimentary prism of calcareous detritus at the continental margin of the Arabian craton. Sedimentology 23, 511–532.

Belderson, R.H., Johnson, M.A., Kenyon, N.H., 1982. Bedforms. In: Stride, A.H. (Ed.), Offshore Tidal Sands. Chapman & Hall, London, pp. 27–57.

Berggren, W.A., Hollister, C.D., 1977. Plate tectonics and paleocirculation – Commotion in the ocean. Tectonophysics 38, 11–48.

Bouma, A.H., 1962. Sedimentology of Some Flysch Deposits: A Graphic Approach to Facies Interpretation. Elsevier, Amsterdam, 168 p.

Bouma, A.H., Hollister, C.D., 1973. Deep ocean basin sedimentation. In: Emiddleton, G.V., Bouma, A.H. (Eds.), Turbidites and Deep-Water Sedimentation. SEPM, Anaheim, CA, SEPM Pacific Section Short Course, pp. 79–118.

Brackenridge, R.E., 2014. Contourite Sands in the Gulf of Cadiz: Characterisation, Controls and Wider Implications for Hydrocarbon Exploration (Ph.D. dissertation). Heriot-Watt University, Edinburgh, Scotland, 265 p.

Brandt, P., Rubino, A., Fischer, J., 2002. Large-amplitude internal solitary waves in the north equatorial countercurrent. Journal of Physical Oceanography 32, 1567–1573.

Brennecke, W., 1921. Die ozeanographischen Arbeiten der deutschen antarktischen Expedition 1911–1912. Archiv d. Deutschen Seewarte 39, 1–216.

Broecker, W.S., 1991. The great ocean conveyor. Oceanography 4, 79–89.

Brunner, C.A., Biscaye, P.E., 1997. Storm-driven transport of foraminifers from the shelf to the upper slope, southern Middle Atlantic Bight. Continental Shelf Research 17 (5), 491–508.

Bruno, M., Vázquez, A., Gómez-Enri, J., Vargas, J.M., García-Lafuente, J.M., Ruiz-Cañavate, A., Mariscal, L., Vidal, J.M., 2006. Observations of internal waves and associated mixing phenomena in the Portimao Canyon area. Deep Sea Research Part II: Topical Studies in Oceanography 53 (11–13), 1219–1240.

Bulfinch, D.L., Ledbetter, M.T., 1983/1984. Deep western boundary undercurrent delineated by sediment texture at base of North American continental rise. Geo-Marine Letters 3, 31–36.

Burden, C.A., 2000. Sediment Transport in the Mississippi Canyon: The Role of Currents and Storm Events on Optical Variability. (December 1999). http://www-ocean.tamu.edu/~pdgroup/abstracts/burden/bur_ab_01.html.

Cacchione, D.A., Pratson, L.F., Ogston, A.S., 2002. The shaping of continental slopes by internal tides. Science 296, 724–727.

Cacchione, D.A., Wiberg, P.L., Lynch, J., Irish, J., Traykovski, P., 1999. Estimates of suspended-sediment flux and bedform activity on the inner portion of the Eel continental shelf. Marine Geology 154, 83–97.

Cairns, J.L., 1980. Variability in the Gulf of Cadiz: internal waves and globs. Journal of Physical Oceanography 10, 579–595.

Carter, R.M., 2007. The role of intermediate-depth currents in continental shelf-slope accretion: Canterbury Drifts, SW Pacific Ocean. In: Viana, A.R., Rebesco, M. (Eds.), Economic and Palaeoceanographic Significance of Contourite Deposits. Geol. Soc. London Spec. Publ., pp. 129–154.

Chérubin, L., Serra, N., Ambar, I., 2003. Low-frequency variability of the Mediterranean undercurrent downstream of Portimao Canyon. Journal of Geophysical Research 108, 3058. http://dx.doi.org/10.1029/2001JC001229.

Chérubin, L., Carton, X., Dritschel, D.G., 2007. Vortex dipole formation by baroclinic instability of boundary currents. Journal of Physical Oceanography 37, 1661–1677.

CIMAS (The Cooperative Institute for Marine and Atmospheric Studies), 2015. Ocean Surface Currents. http://oceancurrents.rsmas.miami.edu/.

Cooper, C., Forristall, G.Z., Joyce, T.M., 1990. Velocity and hydrographic structure of two Gulf of Mexico warm-core rings. Journal of Geophysical Research 95 (C2), 1663–1679.

Crease, J., 1965. The flow of Norwegian sea water through the Faeroe bank channel. Deep-Sea Research 12, 143–150.

Criado-Aldeanueva, F., García-Lafuente, J., Vargas, J.M., Río, J.D., Vázquez, A., Reul, A., Sánchez, A., 2006. Distribution and circulation of water masses in the Gulf of Cadiz from in situ observations. Deep-Sea Research II 53, 1144–1160.

Dalrymple, R.W., Narbonne, G.M., 1996. Continental slope sedimentation in the Sheepbed Formation (Neoproterozoic, Windermere Supergroup), Mackenzie Mountains, N.W.T. Cananadian Journal of Earth Sci 33, 848–862.

Damuth, J.E., Olson, H.C., 2001. Neogene-Quaternary contourite and related deposition on the West Shetland Slope and Faeroe-Shetland Channel revealed by high-resolution seismic studies. Marine Geophysical Researches 22, 363–398.

Dengler, A.T., Wilde, P., Noda, E.K., Normark, W.R., 1984. Turbidity currents generated by Hurricane Iwa. Geo-Marine Letters 4, 5–11.

Duarte, C.S.L., Viana, A.R., 2007. Santos drift system: stratigraphic organization and implications for late Cenozoic palaeocirculation in the Santos Basin. In: Viana, A.R., Rebesco, M. (Eds.), Economic and Palaeoceanographic Significance of Contourite Deposits. Geological Society London Special Publications 276, pp. 171–198.

Dykstra, M., 2012. Deep-water tidal sedimentology. In: Davis Jr., R.A., Dalrymple, R.W. (Eds.), Principles of Tidal Sedimentology. Springer, Berlin, Germany, pp. 371–396.

Dunham, J., Saller, A.H., 2014. Modern internal waves and internal tides along oceanic pycnoclines: challenges and implications for ancient deep-marine baroclinic sands: Discussion. AAPG Bulletin 98, 851–857.

Enos, P., 1977. Flow regimes in debris flow. Sedimentology 24, 133–142.

Expedition 339 Scientists, 2012. Mediterranean outflow: environmental significance of the Mediterranean Outflow Water and its global implications. Integrated Ocean Drilling Program (IODP) Preliminary Report 339. http://dx.doi.org/10.2204/iodp.pr.339.2012.

Expedition 342 Scientists, 2012. Paleogene Newfoundland sediment drifts. Integrated Ocean Drilling Program (IODP) Preliminary Report 342. http://dx.doi.org/10.2204/iodp.pr.342.2012.

Ewing, M., Ettrium, S.L., Ewing, J.L., Le Pichon, X., 1971. Sediment transport and distribution in the Argentine Basin: part 3, Nepheloid layer and process of sedimentation. In: Ahrens, L.A., Press, F., Runcorn, S.K., Urey, H.C. (Eds.), Physics and Chemistry of the Earth. Pergamon Press, London, pp. 55–77.

Famakinwa, S.B., Shanmugam, G., Hodgkinson, R.J., Blundell, L.C., 1996. Deep-water slump and debris flow dominated reservoirs of the Zafiro Field area, offshore Equatorial Guinea. In: Offshore West Africa Conference and Exhibition, Libreville, Gabon, November 5–7, pp. 1–14.

Faugères, J.-C., Stow, D.A.V., 1993. Bottom-current-controlled sedimentation: a synthesis of the contourite problem. Sediment. Geology 82, 287–297.

Faugères, J.-C., Stow, D.A.V., 2008. Contourite drifts: nature, evolution and controls. In: Rebesco, M., Camerlenghi, A. (Eds.), Contourites, Developments in Sedimentology, vol. 60. Elsevier, Amsterdam, pp. 259–288 (Chapter 14).

Faugères, J.-C., Mulder, T., 2011. Contour currents and contourite drifts. In: Hüneke, H., Mulder, T. (Eds.), Deep-Sea Sediments, Developments in Sedimentology, vol. 63. Elsevier, Amsterdam, pp. 149–214 (Chapter 3).

Faugères, J.-C., Gonthier, E., Stow, D.A.V., 1984. Contourite drift moulded by deep Mediterranean outflow. Geology 12, 296–300.

Fisher, R.V., 1971. Features of coarse-grained, high-concentration fluids and their deposits. Journal of Sedimentary Petrology 41, 916–927.

Flemming, B.W., 1980. Sand transport and bedform patterns on the continental shelf between Durban and Port Elizabeth (southeast African continental margin). Sediment. Geol. 26, 179–205.

Forristall, G.Z., Hamilton, R.C., Vardone, V.J., 1977. Continental shelf currents in Tropical Storm Delta: observations and theory. Journal of Physical Oceanography 87, 532–546.

Gao, Z., Eriksson, K.A., 1991. Internal-tide deposits in an Ordovician submarine channel: previously unrecognized facies? Geology 19, 734–737.
Gao, Z., He, Y., Li, X., Duan, T., 2013. Review of research in internal-wave and internal-tide deposits of China. Journal of Palaeogeography 2 (1), 56–65.
Gao, Z., He, Y., Li, X., Duan, T., 2014. Reply to Shanmugam, G. "Review of research in internal-wave and internal-tide deposits of China: discussion". Journal of Palaeogeography 3 (4), 351–358.
García, M., Hernández-Molina, F.J., Llave, E., Stow, D.A.V., León, R., Fernández-Puga, M.C., Díaz del Río, V., Somoza, L., 2009. Contourite erosive features caused by the Mediterranean Outflow Water in the Gulf of Cádiz: Quaternary tectonic and oceanographic implications. Marine Geology 257, 24–40.
Gearhart II, R., Jones, D., Borgens, A., Laurence, S., DeMunda, T., Shipp, J., 2011. Impacts of Recent Hurricane Activity on Historic Shipwrecks in the Gulf of Mexico Outer Continental Shelf. U.S. Dept. of the Interior, Bureau of Ocean Energy Management, Regulation and Enforcement, Gulf of Mexico OCS Region, New Orleans, LA. OCS Study BOEMRE 2011-003. 202 p.
Gill, A.E., 1982. Atmosphere-Ocean Dynamics. In: International Geophysics Series, vol. 30. Academic Press, An Imprint of Elsevier, San Diego, 662 p.
Gómez-Enri, J., Vázquez, A., Bruno, M., Mariscal, L., Villares, P., 2007. Characterization of internal waves in the Strait of Gibraltar, using SAR and in-situ measurements. In: Proceedings of the 'Envisat Symposium 2007', Montreux, Switzerland April 23–27, 2007: ESA SP-636, July 2007). In: http://earth.esa.int/workshops/envisatsymposium/proceedings/posters/4P14/464335go.pdf.
Gonthier, E.G., Faugères, J.-C., Stow, D.A.V., 1984. Contourite facies of the Faro Drift, Gulf of Cadiz. In: Stow, D.A.V., Piper, D.J.W. (Eds.), Fine-Grained Sediments: Deep-Water Processes and Facies. Geological Society of London, pp. 275–292. Geological Society of London Special Publication 15.
Gordon, A.L., 1986. Is there a global scale ocean circulation? EOS. Transactions of the American Geophysical Union 67, 109–110.
Grotzinger, J., Jordan, T.H., Press, F., Siever, R., 2007. Understanding Earth, fifth ed. W. H. Freeman & Company, New York. 579 p.
Haq, B.U., Hardenbol, J., Vail, P.R., 1988. Mesozoic and Cenozoic chronostratigraphy and cycles of sea-level change. In: Wilgus, C.K., Hastings, B.S., Kendall, C.G.StC., Posamentier, H.W., Ross, C.A., Van Wagoner, J.C. (Eds.), Sea-Level Changes: An Integrated Approach. SEPM, Tulsa, OK, pp. 71–108. Special Publication 42.
Havil, J., 2014. The Irrationals: A Story of the Numbers You Can't Count on. Princeton University Press, Princeton, 312 p.
He, Y.-B., Luo, J.-X., Li, X.-D., Gao, Z.-Z., Wen, Z., 2011. Evidence of internal-wave and internal-tide deposits in the Xujiajuan Formation of the Xiangshan Group, Ningxia, China. Geo-Marine Letters 31, 509–523.
Heezen, B.C., 1959. Dynamic processes of abyssal sedimentation; erosion, transportation, and re-deposition on the deep-sea floor. Geophysical Journal of the Royal Astronomical Society 2, 142–163.
Heezen, B.C., Hollister, C.D., 1963. Evidence of deep-sea bottom currents from abyssal sediments. Abstracts of papers. In: International Association of Physical Oceanography, 13th General Assembly, International Union Geodesy and Geophysics, vol. 6, p. 111.
Heezen, B.C., Hollister, C.D., 1964. Deep sea current evidence from abyssal sediments. Marine Geology 1, 141–174.
Heezen, B.C., Hollister, C.D., 1971. The Face of the Deep. Oxford University Press, New York, 659 p.
Heezen, B.C., Johnson, G.L., 1969. Mediterranean under-current and microphysiography west of Gibraltar. Bulletin de l'Institut Océanographique de Monaco 67 (1382), 1–97.
Heezen, B.C., Tharp, M., Ewing, M., 1959. The floors of the oceans: I. The North Atlantic. Geological Society of America Special Paper, 65, 122 p.
Heezen, B.C., Hollister, C.D., Ruddiman, W.F., 1966. Shaping of the continental rise by deep geostrophic contour currents. Science 152, 502–508.
Heritier, F.E., Lossel, P., Wathne, E., 1979. Frigg field: large submarine-fan trap in Lower Eocene rocks of the North Sea Viking graben. AAPG Bulletin 63, 1999–2020.
Hernández-Molina, F.J., Llave, E., Somoza, L., Fernández-Pug, M.C., Maestro, A., León, R., Barnolas, A., Medialdea, T., García, M., Vázquez, J.T., Díaz del Río, V., Fernández-Salas, L.M., Lobo, F., Alveirinho Dias, J.M., Rodero, J.Y., Gardner, J., 2003. Looking for clues to paleoceanographic imprints: a diagnosis of the Gulf of Cadiz contourite depositional systems. Geology 31 (1), 19–22.

Hernández-Molina, F.J., Llave, E., Stow, D.A.V., García, M., Somoza, L., Vázquez, J.T., Lobo, F., Maestro, A., Díaz del Río, V., León, R., Medialdea, T., Gardner, J., 2006. The contourite depositional system of the Gulf of Cadiz: a sedimentary model related to the bottom current activity of the Mediterranean Outflow Water and the continental margin characteristics. Deep-Sea Research II 53, 1420–1463. http://dx.doi.org/10.1016/j.dsr2.2006.04.016.

Hernández-Molina, F.J., Maldonado, A., Stow, D.A.V., 2008. Abyssal plain contourites. In: Rebesco, M., Camerlenghi, A. (Eds.), Contourites, Developments in Sedimentology, vol. 60. Elsevier, Amsterdam, pp. 345–378 (Chapter 18).

Hernández-Molina, F.J., Serra, N., Stow, D.A.V., Ercilla, G., Llave, E., Van Rooij, D., 2011. Along-slope oceanographic processes and sedimentary products around Iberia. Geo-Marine Letters 31, 315–341.

Hernández-Molina, F.J., Stow, D.A.V., Alvarez-Zarikian, C., 2013. IODP Expedition 339 in the Gulf of Cadiz and off West Iberia: decoding the environmental significance of the Mediterranean outflow water and its global influence. Scientific Drilling 16, 1–11.

Hernández-Molina, F.J., Llave, E., Preu, B., Ercilla, G., Fontan, A., Bruno, M., Serra, N., Gomiz, J.J., Brackenridge, R.E., Sierro, F.J., Stow, D.A.V., García, M., Juan, C., Sandoval, N., Arnaiz, A., 2014. Contourite processes associated with the Mediterranean Outflow Water after its exit from the Strait of Gibraltar: global and conceptual implication. Geology 42 (3), 227–230.

Hernández-Molina, F.J., Wåhlin, A., Bruno, M., Ercilla, G., Llave, E., Serra, N., Roson, G., Puig, P., Rebesco, M., Van Rooij, D., Roque, D., González-Pola, C., Sánchez, F., Gómez, M., Preu, B., Schwenk, T., Hanebuth, T.J.J., Sánchez Leal, R.F., García-Lafuente, J., Brackenridge, R.E., Juan, C., Stow, D.A.V., S'anchez-González, J.M., 2016. Oceanographic processes and morphosedimentary products along the Iberian margins: a new multidisciplinary approach. Marine Geology 378, 127–156. http://dx.doi.org/10.1016/j.margeo.2015.12.008.

Hollister, C.D., 1967. Sediment Distribution and Deep Circulation in the Western North Atlantic (Ph.D. dissertation). Columbia University, New York, 467 p.

Hollister, C.D., 1993. The concept of deep-sea contourites. Sedimentary Geology 82, 5–11.

Hollister, C.D., Heezen, B.C., 1972. Geological effects of ocean bottom currents: western North Atlantic. In: Gordon, A.L. (Ed.), Studies in Physical Oceanography, vol. 2. Gordon and Breach, New York, pp. 37–66.

Hollister, C.D., McCave, I.N., 1984. Sedimentation under deep-sea storms. Nature 309, 220–225.

Hollister, C.D., Johnson, D.A., Lonsdale, P.F., 1974. Current-controlled abyssal sedimentation: Samoan Passage, equatorial west Pacific. Journal of Geology 82, 275–300.

Hollister, C.D., Nowell, A.R.M., Smith, J.D., 1980. The Third Annual Report of the High Energy Benthic Boundary Layer Experiment. WHOI Tech. Report 80-32, 48.

Howe, J.A., Humphrey, J.D., 1995. Photographic evidence for slope current activity on the hebrides slope, North-East Atlantic Ocean. Scott. Journal of Geology 30, 107–115.

Howe, J.A., Pudsey, C.J., Cunningham, A.P., 1997. Pliocene-Holocene contourite deposition under the Antarctic circumpolar current, western Falkland trough, South Atlantic Ocean. Marine Geology 138 (1–2), 27–50.

Howe, J.A., Stoker, M.S., Woolfe, K.J., 2001. Deep-marine seabed erosion and gravel lags in the northwestern Rockall Trough, North Atlantic ocean. Journal of the Geological Society, London 158, 427–438.

Hsü, K.J., 1964. Cross-laminated sequence in graded bed sequence. Journal of Sedimentary Petrology 34, 379–388.

Hsü, K.J., 2008. Personal reminiscences on the history of contourites. In: Rebesco, M., Camerlenghi, A. (Eds.), Contourites, Developments in Sedimentology, vol. 60. Elsevier, Amsterdam, pp. 11–15 (Chapter 2).

Hsu, M., Liu, A.K., Liu, C., 2000. A study of internal waves in the China Seas and Yellow Sea using SAR. Continental Shelf Research 20, 389–410. http://dx.doi.org/10.1016/S0278-4343(99)00078-3.

Hubbard, D.K., 1992. Hurricane-induced sediment transport in open shelf tropical systems_Anexample from St. Croix, U.S. Virgin Islands. Journal of Sedimentary Petrology 62, 946–960.

Hubert, J.F., 1964. Textural evidence for deposition of many western North Atlantic deep-sea sands by ocean-bottom currents rather than turbidity currents. Journal of Geology 72, 757–785. http://dx.doi.org/10.1086/627031.

Hüeneke, H., Stow, D.A.V., 2008. Identification of ancient contourites: problems and palaeoceanographic significance. In: Rebesco, M., Camerlenghi, A. (Eds.), Contourites, Developments in Sedimentology, vol. 60. Elsevier, Amsterdam, pp. 323–346 (Chapter 17).

Hüneke, H., Mulder, T., 2011. In: Deep-Sea Sediments. Developments in Sedimentology, vol. 63. Elsevier, Amsterdam, 849 p.

Hunter, S.E., Wilkinson, D., Louarn, E., McCave, I.N., Rohling, E., Stow, D.A.V., Bacon, S., 2007. Deep western boundary current dynamics and associated sedimentation on the Eirik Drift, Southern Greenland Margin. Deep-Sea Research I 54, 2036–2066.

Huthnance, J., Alvarado, R., Ambar, I., Hobbs, R., Krahmann, G., Quentel, E., Silva, P., 2008. Gulf of Cadiz oceanography for comparison with seismic imaging. Geophysical Research Abstracts 10. EGU2008-A-03530, 2008, SRef-ID: 1607–7962/gra/EGU2008-A-03530.

Inman, D.L., Nordstrom, C.E., Flick, R.E., 1976. Currents in submarine canyons: an air-sea-land interaction. Annual Review of Fluid Mechanics 8, 275–310.

Ito, M., 2002. Kuroshio current-influenced sandy contourites from the Plio-Pleistocene Kazusa forearc basin, Boso Peninsula, Japan. In: Stow, D.A.V., Pudsey, C.J., Howe, J.A., Faugères, J.-C., Viana, A.R. (Eds.), Deep-Water Contourite Systems, Modern Drifts and Ancient Series, Seismic and Sedimentary Characteristics, vol. 22. Geological Society, London, Memoirs, pp. 421–432.

Jackson, C.R. (Ed.), February 2004a. An Atlas of Oceanic Internal Solitary-like Waves and Their Properties, second ed. Global Ocean Associates (Prepared for Office of Naval Research) – Code 322PO http://www.internalwaveatlas.com/Atlas2_index.html.

Jackson, C.R., 2004b. Northeast Africa. In: Jackson, C.R. (Ed.), An Atlas of Oceanic Internal Solitary-like Waves and Their Properties, second ed. Global Ocean Associates (Prepared for the Office of Naval Research) http://www.internalwaveatlas.com/Atlas2_PDF/IWAtlas2_Pg535_NEAfrica.pdf.

Jaimes, B., 2009. On the Response to Tropical Cyclones in Mesoscale Oceanic Eddies (Ph.D. dissertation). University of Miami, Miami, Florida, 145 p.

Jaimes, B., Shay, L.K., 2010. Near-inertial wave wake of Hurricanes Katrina and Rita over mesoscale oceanic eddies. Journal of Physical Oceanography 40, 1320–1337.

Johnson, T.C., Carlson, T.W., Evans, J.E., 1980. Contourites in Lake Superior. Geology 8, 437–441.

Johnson, D.A., Damuth, J.E., 1979. Deep thermohaline flow and current controlled sedimentation in the Amirante Passage: western Indian Ocean. Marine Geology 33, 1–44.

Keen, T.R., Glenn, S.,M., 2002. Predicting bed scour on the continental shelf during Hurricane Andrew. Journal of Waterway Port Coastal Ocean Engineering 128, 249–257.

Keller, G.H., Shepard, F.P., 1978. Currents and sedimentary processes in submarine canyons off the northeast United States. In: Stanley, D.J., Kelling, G.K. (Eds.), Sedimentation in Submarine Canyons, Fans, and Trenches. Dowden, Hutchinson & Ross, Stroudsburg, PA, pp. 15–32.

Kenyon, N.H., Belderson, R.H., 1973. Bed forms of the Mediterranean undercurrent observed with side-scan sonar. Sedimentary Geology 9, 77–99.

Kenyon, N.H., Akhmetzhanov, A.M., Twichell, D.C., 2002. Sand wave fields beneath the Loop Current, Gulf of Mexico: reworking of fan sands. Marine Geology 192, 297–307.

Klein, G.D., 1966. Dispersal and petrology of sandstones of Stanley-Jackfork boundary, Ouachita fold belt, Arkansas and Oklahoma. AAPG Bulletin 50, 308–326.

Klein, G.D., 1975. Resedimented pelagic carbonate and volcaniclastic sediments and sedimentary structures in Leg 30 DSDP cores from the western equatorial Pacific. Geology 3, 39–42.

Knauss, J., 1965. A technique for measuring deep ocean currents close to the bottom with an unattached current meter and some preliminary results. Journal of Marine Research 23, 237–245.

Koch, S.P., Barker, J.W., Vermersch, J.A., 1991. The gulf of mexico loop current and deepwater drilling. Journal of Petroleum Technology 43, 1046–1050, 1118–1119.

Kohut, J.T., Glenn, S.M., Paduan, J.D., 2006. Inner shelf response to Tropical Storm Floyd. Journal of Geophysical Research 111. http://dx.doi.org/10.1029/2003JC002173. C09S91. http://www.agu.org/pubs/crossref/2006/2003JC002173.shtml.

Komar, P.D., Neudeck, R.H., Kulm, L.D., 1972. Origin and significance of deep-water oscillatory ripple marks on the Oregon continental shelf. In: Swift, D.J.P., Duane, D.B., Pilkey, O.H. (Eds.), Shelf Sediment Transport Processes and Pattern. Dowden, Hutchinson, and Ross, Stroudsburg, pp. 601–619.

Kuenen, Ph H., 1957. Sole markings of graded greywacke beds. Journal of Geology 65, 231–258.

Kuenen, Ph H., Migliorini, C.I., 1950. Turbidity currents as a cause of graded bedding. Journal of Geology 58, 91–127.

Kuhn, T.S., 1996. The Structure of Scientific Revolutions, third ed. The University of Chicago Press, Chicago. 212 p.

Laine, E.P., 1978. Geological Effects of the Gulf Stream in the North American Basin. Ph.D. dissertation. Woods Hole Oceanographic Institution/Massachusetts Institute of Technology joint program in oceanography and ocean engineering, Woods hole, MA, 164 p.

Laine, E.P., Hollister, C.D., 1981. Geological effects of the gulf Stream system on the northern Bermuda Rise. Marine Geology 39, 277–310.

Larcombe, P., Carter, R.M., 2004. Cyclone pumping, sediment partitioning and the development of the Great Barrier Reef shelf system: a review. Quaternary Science Reviews 23, 107–135.

Lathrop, E., 2015. Sediment Composition in the Gulf of Cádiz Contourites during the Pleistocene (Undergraduate research thesis). The Ohio State University, Columbus, Ohio, 61 p.

Lario, J., Luque, L., Zazo, C., Goy, J.L., Spencer, C., Cabero, A., Bardají, T., Borja, F., Dabrio, C.J., Civis, J., González-Delgado, J.Á., Borja, C., Alonso-Azcárate, J., 2010. Tsunami vs. storm surge deposits: a review of the sedimentological and geomorphological records of extreme wave events (EWE) during the Holocene in the Gulf of Cadiz, Spain. Zeitschrift für Geomorphologie 54 (Suppl. 3), 301–316. Stuttgart, Juli 2010.

LaViolette, P.E., Lacombe, H., 1988. Tidal-induced pulses in the flow through the strait of Gibraltar. In: Minas, H.J., Nival, P. (Eds.), Ocdanographie Pdlagique Mdditerrandenne: Oceanologica Acta, pp. 13–27.

Lee, I.H., Lien, R.C., Liu, J.T., Chuang, W.S., 2009. Turbulent mixing and internal tides in Gaoping (Kaoping) Submarine Canyon, Taiwan. Journal of Marine Systems 76, 383–396. http://dx.doi.org/10.1016/j.jmarsys.2007.08.005.

León, R., Somoza, L., Medialdea, T., Javier González, F., Gimenez-Moreno, C.J., Pérez-López, R., 2014. Pockmarks on either side of the Strait of Gibraltar: formation from overpressured shallow contourite gas reservoirs and internal wave action during the last glacial sea-level lowstand? Geo-Marine Letters 34, 131–151.

Llave, E., Matias, H., Hernández-Molina, F.J., Ercilla, G., Stow, D.A.V., Medialdea, T., 2011. Pliocene–Quaternary contourites along the northern Gulf of Cadiz margin: sedimentary stacking pattern and regional distribution. Geo-Marine Letters 31, 377–390.

Lonsdale, P., Malfait, B., 1974. Abyssal dunes of foraminiferal sand on the Carnegie Ridge. GSA Bulletin 85, 1697–1712.

Lonsdale, P., Nornaark, W.R., Newman, W.A., 1972. Sedimentation and erosion on Horizon Guyot. Geological Society of America Bulletin 83, 289–316.

Lovell, J.P.B., Stow, D.A.V., 1981. Identification of ancient sandy contourites. Geology 9, 347–349.

Lowe, D.R., 1982. Sediment gravity flows, II. depositional models with special reference to the deposits of high-density turbidity currents. Journal of Sedimentary Petrology 52, 279–297.

Lumpkin, R., Speer, K., 2007. Global ocean meridional overturning. Journal of Physical Oceanography 37, 2,550–2,562. http://dx.doi.org/10.1175/JPO31.

Magalhães, J., da Silva, J.C., Grimshaw, R., Griffiths, S., New, A., 2010. Internal solitary waves off the western Iberian coast and in the Gulf of Cadiz: main hotspots and some generation mechanisms. In: SEASAR 2010: The 3rd International Workshop on Advances in SAR Oceanography from Envisat, ERS and ESA Third Party Missions. 25–29 January, 2010, Rome, Italy.

Magalhães, V.H., Pinheiro, L.M., Ivanov, M.K., Kozlova, E., Blinova, V., Kolganova, J., Vasconcelos, C., McKenzie, J.A., Bernasconi Kopf, A.J., Díaz-del-Río, V., González, F.J., Somoza, L., 2012. Formation processes of methane-derived authigenic carbonates from the Gulf of Cadiz. Sedimentary Geology 243–244, 155–168.

Martın-Chivelet, J., Fregenal-Martnez, M.A., Chacón, B., 2008. Traction structures in contourites. In: Rebesco, M., Camerlenghi, A. (Eds.), Contourites, Developments in Sedimentology, vol. 60. Elsevier, Amsterdam, pp. 159–182 (Chapter 10).

Maslin, E., 2015. Scientific Drilling. Offshore Engineer, OE. http://www.oedigital.com/geoscience/item/10495-scientific-drilling.

Matsuyama, M., Ohta, S., Hibiya, T., Yamada, H., 1993. Strong tidal currents observed near the bottom in the Suruga Trough, central Japan. Journal of Oceanography 49, 683–696. http://dx.doi.org/10.1007/BF02276752.

Mazumder, R., Arima, M., 2013. Tidal rhythmites in a deep sea environment: An example from Mio-Pliocene Misaki Formation. Miura Peninsula, Japan. Marine and Petroleum Geology 43, 320–325.

MBARI (Monterey Bay Aquarium Research Institute), 2003. Monterey Canyon Turbidity Flow: Caught in the Act…again!, December 17–19, 2002. http://www.mbari.org/benthic/Turbidity%20Event%202002/canyon%20events.html.

McCave, I.N., 2002. Charles Davis Hollister, 1936-1999: a personal scientific appreciation of the father of 'contourites'. In: Stow, D.A.V., Pudsey, C.J., Howe, J.A., Faugères, J.-C., Viana, A.R. (Eds.), Deep-water Contourite Systems, Modern Drifts and Ancient Series, Seismic and Sedimentary Characteristics, vol. 22. Geological Society, London, Memoirs, pp. 1–6.

McCave, I.N., 2008. Size sorting during transport and deposition of fine sediments: sortable silt and flow speed. In: Rebesco, M., Camerlenghi, A. (Eds.), Contourites, Developments in Sedimentology, vol. 60. Elsevier, Amsterdam, pp. 121–142 (Chapter 8).

Mearns, D.L., Hine, A.C., Riggs, S.R., 1988. Comparison of sonographs taken before and after Hurricane Diana; Onslow Bay, North Carolina. Geology 16, 267–270.

Miall, A.D., 2014. The emptiness of the stratigraphic record: a preliminary evaluation of missing time in the Mesaverde Group, Book cCiffs, Utah, U.S.A. Journal of Sedimentary Research 84, 457–469.

Middleton, G.V., 1973. Johannes Walther's Law of the correlation of fades. Geological Society of America Bulletin 84, 979–988.

Middleton, G.V., Hampton, M.A., 1973. Sediment gravity flows: mechanics of flow and deposition. In: Middleton, G.V., Bouma, A.H. (Eds.), Turbidites and Deep-Water Sedimentation. SEPM, Anaheim, CA, pp. 1–38. SEPM Pacific section Short Course.

Mitchell, D.A., Teague, W.J., Jarosz, E., Wang, D.W., 2005. Observed currents over the outer continental shelf during Hurricane Ivan. Geophysical Research Letters 32, L11610. http://dx.doi.org/10.1029/2005GL023014.

Mitsuzawa, K., Momma, H., Fukasawa, M., Hotta, H., 1993. Observation of deep sea current and change of bottom shapes in the Suruga Trough. In: OCEANS apos 93, Engineering in Harmony with Oceanapos, Proceedings, vol. 3, Issue 18–21, October 1993, pp. III149–III154. Digital Object Identifier: 10.1109/OCEANS.1993.326176. http://ieeexplore.ieee.org/Xplore/login.jsp?url=/iel2/1081/7739/00326176.pdf?isnumber=7739&prod=CNF&arnumber=326176&arSt=III149&ared=III154+vol.3&arAuthor=Mitsuzawa%2C+K.%3B+Momma%2C+H.%3B+Fukasawa%2C+M.%3B+Hotta%2C+H.

Moraes, M.A.S., Maciel, W.B., Braga, M.S.S., Viana, A.R., 2007. Bottom-current reworked Palaeocene- Eocene deep-water reservoirs of the Campos Basin, Brazil. In: Viana, A.R., Rebesco, M. (Eds.), Economic and Palaeoceanographic Significance of Contourite Deposits. Geological Society London Special Publications 276, pp. 81–94.

Moritz, H.R., 2004. The effects of infragravity energy and storm-induced current on short waves beyond the surf zone. In: 8th International Workshop on Wave Hindcasting and Forecasting: North Shore, Oahu, Hawaii, November 14–19, 2004. http://www.waveworkshop.org/8thWaves/Papers/O2.pdf.

Morozov, E.G., Trulsen, K., Velarde, M.G., Vlasenko, V.I., 2002. Internal tides in the strait of Gibraltar. Journal of Physical Oceanography 32, 3193–3206.

Mosher, D.C., Moscardelli, L., Shipp, R.C., Chaytor, J.D., Baxter, C.D.P., Lee, H.J., Urgeles, R., 2010. Submarine mass movements and their consequences. In: Mosher, D.C., et al. (Eds.), Submarine Mass Movements and Their Consequences, Advances in Natural and Technological Hazards Research, vol. 28, pp. 1–8.

Mulder, T., Migeon, S., Savoye, B., Faugeres, J.-C., 2002. Reply to discussion by Shanmugam on Mulder et al. (2001, Geo-Marine Letters, 21, 86–93) Inversely graded turbidite sequences in the deep Mediterranean. A record of deposits from flood-generated turbidity currents? Geo-Marine Letters 22, 112–120.

Mulder, T., Hassan, R., Ducassou, E., Zaragosi, S., Gonthier, E., Hanquiez, V., Marchès, E., Toucanne, S., 2013. Contourites in the Gulf of Cadiz: a cautionary note on potentially ambiguous indicators of bottom current velocity. Geo-Marine Letters 33, 357–367.

Mullins, H.T., Neumann, A.C., Wilber, R.J., Hine, A.C., Chinburg, S.J., 1980. Carbonate sediment drifts in the northern Straits of Florida. AAPG Bulletin 64, 1701–1717.

Mullins, H.T., Gardulski, A.F., Wise Jr., S.W., Applegate, J., 1987. Middle Miocene oceanographic event in the eastern Gulf of Mexico: implications for seismic stratigraphic succession and Loop Current/Gulf Stream circulation. GSA Bulletin 98, 702–713.

Murray, S.P., 1970. Bottom currents near the coast during Hurricane Camille. Journal of Geophysical Research 75, 4579–4582.

Mutti, E., 1992. Turbidite Sandstones. Special Publication. Agip, Milan, p. 275.

Mutti, E., Carminatti, M., 2011. Deep-Water Sands in the Brazilian Offshore Basins: AAPG Search and Discovery Article 30219. http://www.searchanddiscovery.com/documents/2012/30219mutti/ndx_mutti.pdf.

Mutti, E., Cunha, R.S., Bulhoes, E.M., Arienti, L.M., Viana, A.R., 2014. Contourites and turbidites of the Brazilian marginal basins. Search and Discovery Article #51069. Adapted from oral presentation at AAPG Annual Convention & Exhibition. Houston, USA, April 6–9, 2014.

Natland, M.L., 1967. New classification of water-laid clastic sediments. AAPG Bulletin 51, 476.

Nelson, C.H., Baraza, J., Maldonado, A., 1993. Mediterranean undercurrent sandy contourites, Gulf of Cadiz, Spain. Sedimentary Geology 82, 103–131.

Nielsen, T., Knutz, P.C., Kuijpers, A., 2008. Seismic expression of contourite depositional systems. In: Rebesco, M., Camerlenghi, A. (Eds.), Contourites, Developments in Sedimentology, vol. 60. Elsevier, Amsterdam, pp. 301–321 (Chapter 16).

Nilsen, T.H., Abbott, P.L., 1979. Introduction. In: Nilsen, T.H., Arthur, M.A. (Eds.), Upper Cretaceous Deep-Sea Fan Deposits. Annual Meeting of Geological Society of America, Fieldtrip 11, San Diego, CA, pp. 137–166.

Nowlin Jr., W.D., Hubert, J.M., 1972. Contrasting summer circulation patterns for the eastern Gulf. In: Capurro, L.R.A., Reid, J.L. (Eds.), Contributions on the Physical Oceanography of the Gulf of Mexico, Texas A&M University Oceanographic Studies 2. Gulf Publishing, Houston, TX, pp. 119–137.

Oey, L.-Y., Wang, D.-P., 2009. Modeling Waves and Currents Produced by Hurricanes Katrina, Rita, and Wilma. U.S. Dept. of the Interior, Minerals Management Service, Herndon, Virginia. OCS Study MMS 2009-060. xviii + 135 p.

Palanques, A., Durieu de Madron, X., Puig, P., Fabres, J., Guillén, J., Calafat, A.M., Canals, M., Bonnin, J., 2006. Suspended sediment fluxes and transport processes in the Gulf of Lions submarine canyons. The role of storms and dense water cascading. Marine Geology 234, 43–61.

Pavec, M., Carton, X., Swaters, G., 2005. Baroclinic instability of frontal geostrophic currents over a slope. Journal of Physical Oceanography 35, 911–918.

Pequegnat, W.E., 1972. A deep bottom-current on the Mississippi Cone. In: Capurro, L.R.A., Reid, J.L. (Eds.), Contribution on the Physical Oceanography of the Gulf of Mexico, Texas A&M University Oceanographic Studies 2. Gulf Publishing, Houston, TX, pp. 65–87.

Pérez, L.F., Hernández-Molina, F.J., Esteban, F.D., Tassone, A., Piola, A.R., Maldonado, A., Preu, B., Violante, R.A., Lodolo, E., 2015. Erosional and depositional contourite features at the transition between the western Scotia Sea and southern South Atlantic Ocean: links with regional water-mass circulation since the Middle Miocene. Geo-Marine Letters 35, 271–288.

Pettijohn, F.J., 1975. Sedimentary Rocks, third ed. Harper & Row, Publishers, New York, p. 628.

Pickering, K.T., Hilton, V., 1998. Turbidite Systems of Southeast France. Vallis Press, London, 229 p.

Pinardi, N., Masetti, E., 2000. Variability of the large scale general circulation of the Mediterranean Sea from observations and modelling: a review. Palaeogeography, Palaeoclimatology, Palaeoecology 158, 153–173. http://dx.doi.org/10.1016/S0031-0182(00) 00048–1.

Pinheiro, L.M., Ivanov, M.K., Sautkin, A., Akhmanov, G., Magalhaes, V.H., Volkonskaya, A., Monteiro, J.H., Somoza, L., Gardner, J., Hamouni, N., Cunha, M.R., 2003. Mud volcanism in the Gulf of Cadiz: results from the TTR-10 cruise. Marine Geology 195, 131–151.

Piper, D.J.W., Brisco, C.D., 1975. Deep-water continental-margin sedimentation, DSDP Leg 28, Antarctica. In: Hayes, D.E., et al. (Eds.), Initial Reports of the Deep Sea Drilling Project. U.S. Govt. Printing Office, Washington, D.C, pp. 727–755.

Pomar, L., Morsilli, M., Hallock, P., Bádenas, B., 2012. Internal waves, an under-explored source of turbulence events in the sedimentary record. Earth-Science Reviews 111, 56–81.

Potter, P.E., Pettijohn, F.J., 1977. Paleocurrents and Basin Analysis. Springer, 425 p.

Prater, M.D., Sanford, T.B., 1994. A meddy off Cape St. Vincent. Part 1: description. Journal of Physical Oceanography 24 (7), 1572–1586.

Price, J.F., Baringer, M.O., Lueck, R.G., Johnson, G.C., Ambar, I., Parrilla, G., Cantos, A., Kennelly, M.A., Sanford, T.B., 1993. Mediterranean outflow mixing dynamics. Science 259, 1277–1282.

Pudsey, C.J., Howe, J.A., 1998. Quaternary history of the Antarctic Circumpolar Current: evidence from the Scotia Sea. Marine Geology 148, 83–112.

Puig, P., Ogston, A.S., Mullenbach, B.L., Nittrouer, C.A., Sternberg, R.W., 2003. Shelf-to-canyon sediment-transport processes on the Eel continental margin (northern California). Marine Geology 193, 129–149.

Quaresma, L.S., Pichon, A., 2013. Modelling the barotropic tide along the West-Iberian margin. Journal of Marine Systems 109–110, S3–S25.

Rahmstorf, S., 2002. Ocean circulation and climate during the past 120,000 years. Nature 419, 207–214.

Rahmstorf, S., 2006. Thermohaline ocean circulation. In: Elias, S.A. (Ed.), Encyclopedia of Quaternary Sciences. Elsevier, Amsterdam, pp. 1–10.

Rebesco, M., Camerlenghi, A. (Eds.), 2008. Contourites, Developments in Sedimentology, vol. 60. Elsevier, Amsterdam, 663 p.

Rebesco, M., Camerlenghi, A., Van Loon, A.J., 2008. Contourite research: a field in full development. In: Rebesco, M., Camerlenghi, A. (Eds.), Contourites, Developments in Sedimentology, vol. 60. Elsevier, Amsterdam, pp. 3–10 (Chapter 1).

Rebesco, M., Hernández-Molina, F.J., Van Rooij, D., Wåhlin, A., 2014. Contourites and associated sediments controlled by deep-water circulation processes: state-of-the-art and future considerations. Marine Geology 352, 111–154.

Reeder, D.B., Ma, B.B., Yang, Y.J., 2011. Very large subaqueous sand dunes on the upper continental slope in the South China Sea generated by episodic, shoaling deep-water internal solitary waves. Marine Geology 279, 12–18. http://dx.doi.org/10.1016/j.margeo.2010.10.009.

Reid, J.L., 1969. Preliminary results of measurements of deep currents in the Pacific Ocean. Nature 221, 848.

Richardson, P.L., 2008. On the history of meridional overturning circulation schematic diagrams. Progress in Oceanography 76, 466–486. http://dx.doi.org/10.1016/j.pocean.2008.01.005.

Richardson, M.J., Wimbush, M., Mayer, L., 1981. Exceptionally strong near-bottom flows on the continental rise of Nova Scotia. Science 213, 887–888.

Rowe, G.T., 1971. Observations on bottom currents and epibenthonic populations in Hatteras submarine canyon. Deep-Sea Research 18, 569–581.

Rowe, G.T., Menzies, R.T., 1968. Deep bottom currents off the coast of North Carolina. Deep-Sea Research 15, 711–720.

Saller, A.H., Lin, R., Dunham, J., 2006. Leaves in turbidite sands: the main source of oil and gas in the deep-water Kutei Basin, Indonesia. AAPG Bulletin 90, 1585–1608.

Saller, A.H., Dunham, J., Lin, R., 2008a. Leaves in turbidite sands: the main source of oil and gas in the deep-water Kutei Basin, Indonesia: Reply. AAPG Bulletin 92, 139–141.

Saller, A.H., Werner, K., Sugiaman, F., Cebastiant, A., May, R., Glenn, D., Barker, C., 2008b. Characteristics of Pleistocene deep-water fan lobes and their application to an upper Miocene reservoir model, offshore East Kalimantan, Indonesia. AAPG Bulletin 92, 919–949.

Sanchez-Garrido, J.C., Sannino, G., Liberti, L., 2011. Generation and Evolution of Internal Waves in the Strait of Gibraltar. EAI (Energia, Ambiente, Innovazione) 4–5, pp. 74–79.

Sánchez-Román, A., Criado-Aldeanueva, F., García-Lafuente, J., Sánchez, J.C., 2008. Vertical structure of tidal currents over Espartel and Camarinal sills, Strait of Gibraltar. Journal of Marine Systems 74, 120–133.

Sanders, J.E., 1963. Concepts of fluid mechanics provided by primary sedimentary structures. Journal of Sedimentary Petrology 33, 173–179.

Schmitz, W.J., 1996. On the World Ocean Circulation: Volume I. Some Global Features/North Atlantic Circulation. Woods Hole, MA. Woods Hole Oceanographic Institution Technical Report. WHOI-96-03, 141 p.

Serra, N., 2004. Observations and Numerical Modeling of the Mediterranean Outflow (Ph.D. thesis). Faculdade de Ciencias, Universidade de Lisboa, Portugal, p. 234 (unpublished).

Shanmugam, G., 1978. The Stratigraphy, Sedimentology, and Tectonics of the Middle Ordovician Sevier Shale Basin in East Tennessee (Ph.D. dissertation). The University of Tennessee, Knoxville, Tennessee, 222 p.

Shanmugam, G., 1985. Types of porosity in sandstones and their significance in interpreting provenance. In: Zuffa, G.G. (Ed.), Provenance of Arenites. D. Reidel Publishing Company, Dordrecht, pp. 115–137.

Shanmugam, G., 1988. Origin, recognition and importance of erosional unconformities in sedimentary basins. In: Kleinspehn, K.L., Paola, C. (Eds.), New Perspectives in Basin Analysis. Springer-Verlag, New York, pp. 83–108.

Shanmugam, G., 1990. Deep-marine facies models and the interrelationship of depositional components in time and space. In: Brown, G.C., Gorsline, D.S., Schweller, W.J. (Eds.), Applied Deep-Marine Sedimentation: SEPM Short Course, San Francisco, pp. 199–246.

Shanmugam, G., 1996. High-density turbidity currents, are they sandy debris flows? Journal of Sedimentary Research 66, 2–10.
Shanmugam, G., 1997a. Deep-water exploration: conceptual models and their uncertainties. NAPE (Nigerian Association of Petroleum Explorationists) Bulletin 12 (01), 11–28.
Shanmugam, G., 1997b. The Bouma Sequence and the turbidite mind set. Earth-Science Reviews 42, 201–229.
Shanmugam, G., 2000. 50 years of the turbidite paradigm (1950s–1990â€™): deep-water processes and facies models-a critical perspective. Marine and Petroleum Geology 17, 285–342.
Shanmugam, G., 2002a. Ten turbidite myths. Earth-Science Reviews 58, 311–341.
Shanmugam, G., 2002b. Discussion on Mulder et al. 2001, Geo-Marine Letters, 21, 86-93. Inversely graded turbidite sequences in the deep Mediterranean. A record of deposits from flood-generated turbidity currents? Geo-Marine Letters 22, 108–111.
Shanmugam, G., 2003. Deep-marine tidal bottom currents and their reworked sands in modern and ancient submarine canyons. Marine and Petroleum Geology 20, 471–491. http://dx.doi.org/10.1016/S0264-8172(03)00063-1.
Shanmugam, G., 2006a. Deep-Water Processes and Facies Models: Implications for Sandstone Petroleum Reservoirs. Elsevier, Amsterdam, 476 p.
Shanmugam, G., 2006b. The tsunamite problem. Journal of Sedimentary Research 76, 718–730.
Shanmugam, G., 2007. The obsolescence of deep-water sequence stratigraphy in petroleum geology. Indian Journal of Petroleum Geology 16 (1), 1–45.
Shanmugam, G., 2008a. Deep-water bottom currents and their deposits. In: Rebesco, M., Camerlenghi, A. (Eds.), Contourites, Developments in Sedimentology, vol. 60. Elsevier, Amsterdam, pp. 59–81 (Chapter 5).
Shanmugam, G., 2008b. Leaves in turbidite sandthethe main source of oil and gas in the deep-water Kutei Basin, Indonesia: Discussion. AAPG Bulletin 92, 127–137.
Shanmugam, G., 2008c. The constructive functions of tropical cyclones and tsunamis on deepwater sand deposition during sea level highstand: implications for petroleum exploration. AAPG Bulletin 92, 443–471.
Shanmugam, G., 2008d. In: Viana, A.R., Rebesco, M. (Eds.), Book Review: "Economic and Palaeoceanographic Significance of Contourite Deposits". Geological Society (London) Special Publication 276, 2007. Hardcover, 350 p. Journal of Sedimentary Research (Online).
Shanmugam, G., 2011. Book review: "Deep-sea sediments". In: Hüneke, H., Mulder, T. (Eds.), Developments in Sedimentology, vol. 63. Elsevier, Amsterdam. Hardbound, 849 p. Geologos (Online).
Shanmugam, G., 2012a. New Perspectives on Deep-water Sandstones: Origin, Recognition, Initiation and Reservoir Quality. In: Handbook of Petroleum Exploration and Production, vol. 9. Elsevier, Amsterdam, 524 p.
Shanmugam, G., 2012b. Discussion of He et al. 2011, Geo-Marine Letters) Evidence of internalwaveand internal-tide deposits in the Middle Ordovician Xujiajuan Formation of the Xiangshan Group, Ningxia, China. Geo-Marine Letters 32, 359–366. http://dx.doi.org/10.1007/s00367-011-0264-9.
Shanmugam, G., 2012c. Process-sedimentological challenges in distinguishing paleo-tsunamis deposits. In: Kumar, A., Nister, I. (Eds.), Paleo-tsunamis: Natural Hazards, vol. 63, pp. 5–30.
Shanmugam, G., 2013a. Modern internal waves and internal tides along oceanic pycnoclines: challenges and implications for ancient deep-marine baroclinic sands. AAPG Bulletin 97 (5), 767–811.
Shanmugam, G., 2013b. Comment on "Internal waves, an underexplored source of turbulence events in the sedimentary record" by L. Pomar, M. Morsilli, P. Hallock, and B. Bádenas [Earth-Science Reviews, 111 (2012), 56–81]. Earth-Science Reviews 116, 195–205.
Shanmugam, G., 2013c. New perspectives on deep-water sandstones: Implications. Petroleum Exploration and Development 40 (3), 216–324.
Shanmugam, 2013d. Deep-Water Processes. Blog. http://g-shanmugam.blogspot.com/.
Shanmugam, G., 2014a. Modern internal waves and internal tides along oceanic pycnoclines: challenges and implications for ancient deep-marine baroclinic sands: Reply. AAPG Bulletin 98, 858–879.
Shanmugam, G., 2014b. Review of research in internal-wave and internal-tide deposits of China: Discussion. Journal of Palaeogeography 3 (4), 332–350.
Shanmugam, G., 2015. The landslide problem. Journal of Palaeogeography 4 (2), 109–166.
Shanmugam, G., 2016a. Slides, slumps, debris flows, turbidity currents, and bottom currents. In: Reference Module in Earth Systems and Environmental Sciences. Elsevier, 2016 (Online).

Shanmugam, G., 2016b. Submarine fans: a critical retrospective (1950–2015). Journal of Palaeogeography 5 (2), 110–184.

Shanmugam, G., 2016c. The seismite problem. Journal of Palaeogeography 5 (4) in press.

Shanmugam, G., 2016d. Glossary: A supplement to "Submarine fans: A critical retrospective (1950-2015)" in the Journal of Palaeogeography (2016, 5[2]). Journal of Palaeogeography 5 (3), 258–277.

Shanmugam, G., Benedict, G.L., 1978. Fine-grained carbonate debris flow, Ordovician basin margin, Southern Appalachians. Journal of Sedimentary Petrology 48, 1233–1240.

Shanmugam, G., Moiola, R.J., 1982. Eustatic control of turbidites and winnowed turbidites. Geology 10, 231–235.

Shanmugam, G., Moiola, R.J., 1984. Eustatic control of calciclastic turbidites. Marine Geology 56, 273–278.

Shanmugam, G., Clayton, C.A., 1989. Reservoir description of a sand rich submarine fan complex for a steamflood project: Upper Miocene Potter Sandstone, North Midway Sunset Field, California. AAPG Bulletin 73, 411.

Shanmugam, G., Moiola, R.J., 1995. Reinterpretation of depositional processes in a classic flysch sequence in the Pennsylvanian Jackfork Group, Ouachita Mountains. AAPG Bulletin 79, 672–695.

Shanmugam, G., Walker, K.R., 1978. Tectonic significance of distal turbidites in the Middle Ordovician Blockhouse and lower Sevier formations in east Tennessee. American Journal of Science 278, 551–578.

Shanmugam, G., Walker, K.R., 1980. Sedimentation, subsidence, and evolution of a foredeep Basin in the Middle Ordovician, Southern Appalachians. American Journal of Science 280, 479–496.

Shanmugam, G., Zimbrick, G., 1996. Sandy slump and sandy debris flow facies in the Pliocene and Pleistocene of the Gulf of Mexico: implications for submarine fan models. In: AAPG International Congress and Exhibition, Caracas, Venezuela, Official Program, A45.

Shanmugam, G., Moiola, R.J., McPherson, J.G., O'Connell, S., 1988. Comparison of turbidite facies associations in modern passive-margin Mississippi Fan with ancient active-margin fans. Sedimentary Geology 58, 63–77.

Shanmugam, G., Lehtonen, L.R., Straume, T., Syversten, S.E., Hodgkinson, R.J., Skibeli, M., 1994. Slump and debris flow dominated upper slope facies in the Cretaceous of the Norwegian and Northern North Seas (61°–67°N): implications for sand distribution. AAPG Bulletin 78, 910–937.

Shanmugam, G., Bloch, R.B., Mitchell, S.M., Beamish, G.W.J., Hodgkinson, R.J., Damuth, J.E., Straume, T., Syvertsen, S.E., Shields, K.E., 1995a. Basin-floor fans in the North Sea: sequence stratigraphic models vs. sedimentary facies. AAPG Bulletin 79, 477–512.

Shanmugam, G., Spalding, T.D., Rofheart, D.H., 1995b. Deep-marine bottom-current reworked sand (Pliocene-Pleistocene) Ewing Bank 826 field, gulf of Mexico. In: Winn Jr., R.D., Armentrout, J.M. (Eds.), Turbidites and Associted Deep-Water Facies: SEPM Core Workshop No. 20, Houston, Texas, pp. 25–54.

Shanmugam, G., Poffenberger, M., Toro Alava, J., 2000. Tide-dominated estuarine facies in the Hollin and Napo ("T" and "U") Formations (Cretaceous), Sacha Field, Oriente Basin, Ecuador. AAPG Bulletin 84, 652–682.

Shanmugam, G., Shrivastava, S.K., Das, B., 2009. Sandy debrites and tidalites of Pliocene reservoir sands in upper-slope canyon environments, offshore Krishna-Godavari Basin (India): implications. Journal of Sedimentary Research 79, 736–756.

Shanmugam, G., Spalding, T.D., Rofheart, D.H., 1993a. Process sedimentology and reservoir quality of deep-marine bottom-current reworked sands (sandy contourites): an example from the Gulf of Mexico. AAPG Bulletin 77, 1241–1259.

Shanmugam, G., Spalding, T.D., Rofheart, D.H., 1993b. Traction structures in deep-marine bottom current–reworked sands in the Pliocene and Pleistocene, Gulf of Mexico. Geology 21, 929–932.

Shapiro, S.A., Marsaglia, K.M., Carter, L., 2007. The petrology and provenance of sand in the Bounty Submarine Fan, New Zealand. In: Arribas, J., Johnsson, M.J., Critelli, S. (Eds.), Sedimentary Provenance and Petrogenesis: Perspectives from Petrography and Geochemistry. GSA Special Paper 420, pp. 277–296.

Shepard, F.P., 1975. Progress of internal waves along submarine canyons. Marine Geology 19, 131–138.

Shepard, F.P., Marshall, N.F., McLoughlin, P.A., Sullivan, G.G., 1979. Currents in submarine canyons and other sea valleys. AAPG Studies in Geology 8, 173.

Silvester, R. (Ed.), 1974. Coastal Engineering. Elsevier, Amsterdam, 338 p.

Simons, D.B., Richardson, E.V., Nordin Jr., C.F., 1965. Sedimentary structures generated by flow in alluvial channels. In: Middleton, G.V. (Ed.), Primary Sedimentary Structures and Their Hydrodynamic Interpretation. SEPM, Tulsa, OK, pp. 34–52. SEPM Special Publication 12.

Snedden, J.W., Nummedal, D., Amos, A.F., 1988. Storm- and fair-weather combined flow on the Central Texas continental shelf. Journal Sedimentary Petrology 58, 580–595.

Southard, J.B., Stanley, D.J., 1976. Shelf-break processes and sedimentation. In: Stanley, D.J., Swift, J.P. (Eds.), Marine Sediment Transport and Environmental Management. John Wiley & Sons, New York, pp. 351–377.
Stanley, D.J., 1993. Model for turbidite-to-contourite continuum and multiple process transport in deep marine settings, example in the rock record. Sedimentary Geology 82, 241–255.
Steele, T.H., Barrett, T.R., Worthington, L.V., 1962. Deep currents south of Iceland. Deep-Sea Research 9, 465–474.
Stewart, R.H., 2008. Introduction to Physical Oceanography. E-book. Texas A&M University, College Station, TX, 353 p. http://oceanworld.tamu.edu/resources/ocng_textbook/PDF_files/book.pdf.
Stommel, H., 1958. The abyssal circulation. Deep-Sea Research 5, 80–82.
Stone, G.W., Walker, N.D., Hsu, S.A., Babin, A., Liu, B., Keim, B.D., Teague, W., Mitchell, D., Leben, R., 2005. Hurricane Ivan's impact along the northern Gulf of Mexico. EOS, Transactions, American Geophysical Union 86 (48), 497–508.
Stow, D.A.V., 1977. Late Quaternary Stratigraphy and Sedimentation on the Nova Scotian Outer Continental Margin (Ph.D. dissertation). Dalhousie University, Halifax, Canada, 361 p.
Stow, D.A.V., Faugères, J.C., 2008. Contourite facies and the facies model. In: Rebesco, M., Camerlenghi, A. (Eds.), Contourites, Developments in Sedimentology, vol. 60,. Elsevier, Amsterdam, pp. 223–256 (Chapter 13).
Stow, D.A.V., Lovell, J.P.B., 1979. Contourites: their recognition in modern and ancient sediments. Earth-Science Reviews 14, 251–291.
Stow, D.A.V., Piper, D.J.W. (Eds.), 1984. Fine-Grained Sediments: Deep-Water Processes and Facies: London. Geological Society of London Special Publication 15, p. 659.
Stow, D.A.V., Faugéres, J.-C., Viana, A.R., Gonthier, E., 1998. Fossil contourites: a critical review. Sedimentary Geology 115, 3–31.
Stow, D.A.V., Hunter, S., Wilkinson, D., Hernández-Molina, F.J., 2008. The nature of contourite deposition. In: Rebesco, M., Camerlenghi, A. (Eds.), Contourites, Developments in Sedimentology, vol. 60. Elsevier, Amsterdam, pp. 143–156 (Chapter 9).
Stow, D.A.V., Pudsey, C.J., Howe, J.A., Faugères, J.-C., Viana, A.R. (Eds.), 2002. Deep-Water Contourite Systems: Modern Drifts and Ancient Series, Seismic and Sedimentary Characteristics. Geological Society, London, Memoirs 22.
Stow, D.A.V., Hernández-Molina, F.J., Llave, E., Sayago, M., Díaz del Río, V., Branson, A., 2009. Bedform-velocity matrix: the estimation of bottom current velocity from bedform observations. Geology 37 (4), 327–330.
Stow, D.A.V., Brackenridge, R., Hernandez-Molina, F.J., 2011. Contourite sheet sands: new deepwater exploration target. In: Abstract American Association of Petroleum Geologists Annual Conference, Houston, 2011.
Stow, D.A.V., Hernández-Molina, F.J., Llave, E., Bruno, M., García, M., Díaz del Rio, V., Somoza, L., Brackenridge, R.E., 2013. The Cadiz Contourite Channel: sandy contourites, bedforms and dynamic current interaction. Marine Geology 343, 99–114.
Stow, D.A.V., Hernández-Molina, F.J., Alvarez-Zarikian, c., the Expedition 339 Shipboard Scientis, 2014. New advances in the contourite Paradigm: IODP Expedition 339, Gulf of Cadiz. In: 2nd Deep-Water Circulation Congress, 10–12 September 2014, Ghent, Belgium.
Strzeboński, P., 2015. Late Cretaceous – early Paleogene sandy-to-gravelly debris flows and their sediments in the Silesian Basin of the Alpine Tethys (western outer Carpathians, Istebna Formation). Geological Quarterly 59 (1), 195–214. http://dx.doi.org/10.7306/gq.1183.
Susanto, R.D., Mitnik, L., Zheng, Q., 2005. Ocean internal waves observed in the Lombok Strait. Oceanography 18, 80–87. http://dx.doi.org/10.5670/oceanog.2005.08.
Swallow, J.C., Worthington, L.V., 1961. An observation of a deep counter current in the western North Atlantic. Deep-Sea Research 8, 1–9.
Swift, D.J.P., Han, G., Vincent, C.E., 1986. Fluid processes and sea floor response on a modern storm-dominated shelf; middle Atlantic shelf of North America. Part 1: the storm-current regime. In: Knight, R.J., McLean, J.R. (Eds.), Shelf Sands and Sandstones, Canadian Society of Petroleum Geologists Memoir, vol. 11, pp. 99–119.
Talley, L.D., 2013. Closure of the global overturning circulation through the Indian, Pacific, and Southern Oceans: schematics and transports. Oceanography 26 (1), 80–97.
Teague, W.J., Jarosz, E., Keen, T.R., Wang, D.W., Hulbert, M.S., 2006. Bottom scour observed under Hurricane Ivan. Geophysical Research Letters 33. http://dx.doi.org/10.1029/2005GL025281. L07607.

Teague, W.J., Jarosz, E., Wang, D.W., Mitchell, D.A., 2007. Observed oceanic response over the upper continental slope and outer shelf during Hurricane Ivan. Journal of Physical Oceanography 37, 2181–2206.

Tsuji, Y., 1993. Tide influenced high energy environments and rhodolith-associated carbonate deposition on the outer shelf and slope off the Miyako Islands, southern Ryukyu Island Arc, Japan. Marine Geology 113, 255–271.

Tucholke, B.E., Embley, R.W., 1984. Cenozoic regional erosion of the abyssal sea floor off South Africa. In: Schlee, J.S.M. (Ed.), Interregional Unconformities and Hydrocarbon Accumulation, Memoir, vol. 36. AAPG, Tulsa, OK, pp. 145–164.

Tuholke, B.E., Wright, W.R., Hollister, C.D., 1973. Abyssal circulation over the greater Antilles Outer Ridge. Deep-Sea Research 20, 973–995.

Turnewitsch, R., Reyss, J.L., Nycander, J., Waniek, J.J., Lampitt, R.S., 2008. Internal tides and sediment dynamics in the deep sea—evidence from radioactive $^{234}Th/^{238}U$ disequilibria. Deep-Sea Research I 55, 1727–1747.

Valencia, J., Ercilla, G., Hernández-Molina, F.J., Casas, D., 2015. Oceanographic mower cruise. In: The International Archives of the Photogrammetry, Remote Sensing and Spatial Information Sciences, Volume XL-5/W5, 2015 Underwater 3D Recording and Modeling, 16–17 April 2015, Piano di Sorrento, Italy.

Van der Lingen, G.J., 1969. The turbidite problem. New Zealand Journal of Geology and Geophysics 12, 7–50.

Van Rooij, D., 2013. Deep-water bottom current dynamics: processes, products & challenges. In: Marine and River Dune Dynamics – MARID IV – 15 & 16 April 2013-Bruges, Belgium, pp. 297–300.

Vargas-Yañez, M., Viola, T.S., Jorge, F.P., Rubin, J.P., Garc!ıa-Martínez, M.C., 2002. The influence of tide-topography interaction on low-frequency heat and nutrient fluxes. Application to Cape Trafalgar. Continental Shelf Research 22, 115–139.

Verdicchio, G., Trincardi, F., 2008. Shallow-water contourites. In: Rebesco, M., Camerlenghi, A. (Eds.), Contourites, Developments in Sedimentology, vol. 60. Elsevier, Amsterdam, pp. 409–434 (Chapter 20).

Viana, A.R., 2008. Economic relevance of contourites. In: Rebesco, M., Camerlenghi, A. (Eds.), Contourites, Developments in Sedimentology, vol. 60. Elsevier, Amsterdam, pp. 493–510 (Chapter 23).

Viana, A.R., Rebesco, M. (Eds.), 2007. Economic and Palaeoceanographic Significance of Contourite Deposits: London. Geological Society Special Publication 276, 350 p.

Viana, A.R., Faugeres, J.C., Kowsmann, R.O., Lima, J.A.M., Caddah, L.F.G., Rizzo, J.G., 1998. Hydrology, morphology, and sedimentology of the Campos continental margin, offshore Brazil. Sedimentary Geology 115, 133–157.

Visser, M.J., 1980. Neap-spring cycles reflected in Holocene subtidal large-scale bedform deposits: a preliminary note. Geology 8, 543–546.

Volkman, C., 1962. Deep current observations in the western North Atlantic. Deep-Sea Research 9, 493–500.

Vsemirnova, E., Hobbs, R., Serra, N., Klaeschen, D., Quentel, E., 2009. Estimating internal wave spectra using constrained models of the dynamic ocean. Geophysical Research Letters 36. http://dx.doi.org/10.1029/2009GL039598. L00D07.

Walker, R.G., 1965. The origin and significance of the internal sedimentary structures of turbidites. Proceedings of the Yorkshire Geological Society 35, 1–32.

Walker, R.G., 1975. Nested submarine-fan channels in the Capistrano Formation, San Clemente California. Geological Society of America Bulletin 86, 915–924.

Wang, D.W., Mitchell, D.A., Teague, W.J., Jarosz, E., Hulbert, M.S., 2005. Extreme waves under Hurricane Ivan. Science 309, 896.

Welsh, S.E., Inoue, M., Rouse Jr., L.J., Weeks, E., 2009. Observation of the Deepwater Manifestation of the Loop Current and Loop Current Rings in the Eastern Gulf of Mexico. U.S. Dept. of the Interior, Minerals Management Service, Gulf of Mexico OCS Region, New Orleans, LA. OCS Study MMS 2009-050. 110 p.

Wetzel, A., Werner, F., Stow, D.A.V., 2008. Bioturbation and biogenic sedimentary structures in contourites. In: Rebesco, M., Camerlenghi, A. (Eds.), Contourites, Developments in Sedimentology, vol. 60. Elsevier, Amsterdam, pp. 183–202 (Chapter 11).

Worthington, L.V., Volkman, G.H., 1965. The volume transport of the Norwegian Sea overflow in the North Atlantic. Deep-Sea Research 12, 667–676.

Wren, P.A., Leonard, L.A., 2005. Sediment transport on the mid-continental shelf in Onslow Bay, North Carolina during Hurricane Isabel. Estuarine. Coastal and Shelf Science 63, 43–56.

Wunsch, C., 2002. What is the thermohaline circulation? Science 298 (5596), 1179–1181. http://dx.doi.org/10.1126/science.1079329. PMID: 12424356.

Wüst, G., 1933. Schichtung und Zirkulation des Atlantischen Ozeans. Das Bodenwasser und die Gliederung der Atlantischen Tiefsee. Wiss. Erg. Dt. Atl. Exp. "Mete" (1925–1927) 6(1), 106 pp. In: Olson, B.E. (Ed.), Bottom Water and the Distribution of the Deep Water of the Atlantic, Slessers, M. (translator). US Naval Oceanographic Office, Washington, DC, 145 p.

Wüst, G., 1950. Block diagramme der Atlantischhen Zirkulation auf Grand der Meteor Ergibribe. Kider Meeresforsch 7, 24–34.

Wüst, G., 1963. On the stratification and circulation in the cold water sphere of the Antillean-Caribbean basins. Deep-Sea Research 10, 165–167.

Yamazaki, T., Yamaoka, M., Shiki, T., 1989. Miocene offshore tractive current-worked conglomerates— Tsubute-gaura, Chita Peninsula, central Japan. In: Taira, A., Masuda, F. (Eds.), Sedimentary Features in the Active Plate Margin. Terra Scientific, Tokyo, Japan, pp. 483–494.

Zenk, W., 1975. On the Mediterranean outflow west of Gibraltar. Meteor Forscir Ergebuisse 16, 23–34.

Zenk, W.O., 1981. Detection of overflow events in the Shag Rocks Passage, Scotia Ridge. Science 213, 1113–1114.

Zenk, W., 2008. Abyssal and contour currents. In: Rebesco, M., Camerlenghi, A. (Eds.), Contourites, Developments in Sedimentology, vol. 60. Elsevier, Amsterdam, pp. 37–57 (Chapter 4).

Zenk, W., Armi, L., 1990. The complex spreading patterns of Mediterranean water off the Portuguese continental slope. Deep-Sea Research 37, 1805–1823.

Zimmerman, H.B., 1971. Bottom currents on the New England continental rise. Journal of Geophysical Research 76, 5865–5876.

Zuffa, G.G. (Ed.), 1985. Provenance of Arenites. D. Reidel Publishing, Dordrecht, 407 p.

CHAPTER

10

Fluvial Systems, Provenance, and Reservoir Development in the Eocene Brennan Basin Member of the Duchesne River Formation, Northern Uinta Basin, Utah

T. Sato[1], *M.A. Chan*[2]

[1]INPEX Corporation, Tokyo, Japan; [2]University of Utah, Salt Lake City, UT, United States

OUTLINE

Sediment Provenance
http://dx.doi.org/10.1016/B978-0-12-803386-9.00010-1

1. INTRODUCTION

Fluvial sandstone reservoirs are one of the most important hydrocarbon exploration targets in continental basins (e.g., Bohacs, 2012). However, the prediction of high-quality and large-volume reservoir facies distribution can be problematic as fluvial deposits are generally influenced by many complex allogenic controls, such as accommodation, topography, and climate (Shanley and McCabe, 1994, 1998; Catuneanu, 2006).

The purpose of this paper is to characterize the sandstones and document a comprehensive source-to-sink system in the Brennan Basin member of the Duchesne River Formation, with emphasis on how tectonics (uplifts) and induced climatic feedback mechanisms (Sato and Chan, 2015) influenced sediment sources and the sandstone provenance. The Eocene Duchesne River Formation of the northern Uinta Basin was deposited on an alluvial plain adjacent to the sediment and water sources in the mountain ranges of the Uinta Mountains to the north, and in the Wasatch Range (Sevier Fold Thrust Belt) to the west (Warner, 1965, 1966; Andersen and Picard, 1972, 1974; Bruhn et al., 1986).

2. GEOLOGICAL CONTEXT

The Uinta Basin, an intermontane foreland basin in northeastern Utah, is an asymmetric basin, bounded to the north by a high-angle (basement-involved) reverse fault (e.g., Fouch, 1975; Bruhn et al., 1983, 1986). This basin was developed as a part of the Laramide Lake Basin system in the present-day Rocky Mountain region during the latest Cretaceous to early Tertiary (Dickinson et al., 1986, 1988) (Fig. 10.1). Paleogene deposits in the basin include, in ascending order, the Wasatch Formation (fluvial), Green River Formation (lacustrine), Uinta Formation (fluvial—lacustrine transition), and Duchesne River Formation (fluvial) (Fig. 10.1). This Paleogene package exhibits a typical upward coarsening lacustrine basin-fill succession (Visher, 1965; Picard and High, 1972; Lambiase, 1990), from lacustrine mudstone-dominated Green River Formation to fluvial sandstone-dominated Duchesne River Formation, overlying a coarse-grained lowermost unit of the Wasatch Formation. Organic-rich shales of the lacustrine Green River Formation are renowned as hydrocarbon source rocks (Fouch et al., 1994). The stratigraphic relationship of hydrocarbon source rocks overlain by fluvial sandstone reservoirs constitutes a common, favorable petroleum system in worldwide lacustrine basins; e.g., Cretaceous rift basins in Sudan (Schull, 1988), presalt rift basins of the West Africa Atlantic margin (Beglinger et al., 2012), and Oligocene strata in the Indonesia Natuna Basin (Phillips et al., 1997). In the Uinta Basin, most past, regional stratigraphic studies focused on the Green River Formation (e.g., Keighley et al., 2003). In contrast, the overlying Uinta and Duchesne River Formations have received much less attention despite their good exposures, probably due to their lesser known economic significance.

The Duchesne River Formation includes four stratigraphic units. In ascending order these are the Brennan Basin member, Dry Gulch Creek member, Lapoint member, and Starr Flat member (Fig. 10.2). These units were originally defined as lithostratigraphic units by Andersen and Picard (1972), and later regionally mapped by Bryant et al. (1989). Sato and Chan (2015) followed these studies, and proposed a regional stratigraphic framework with

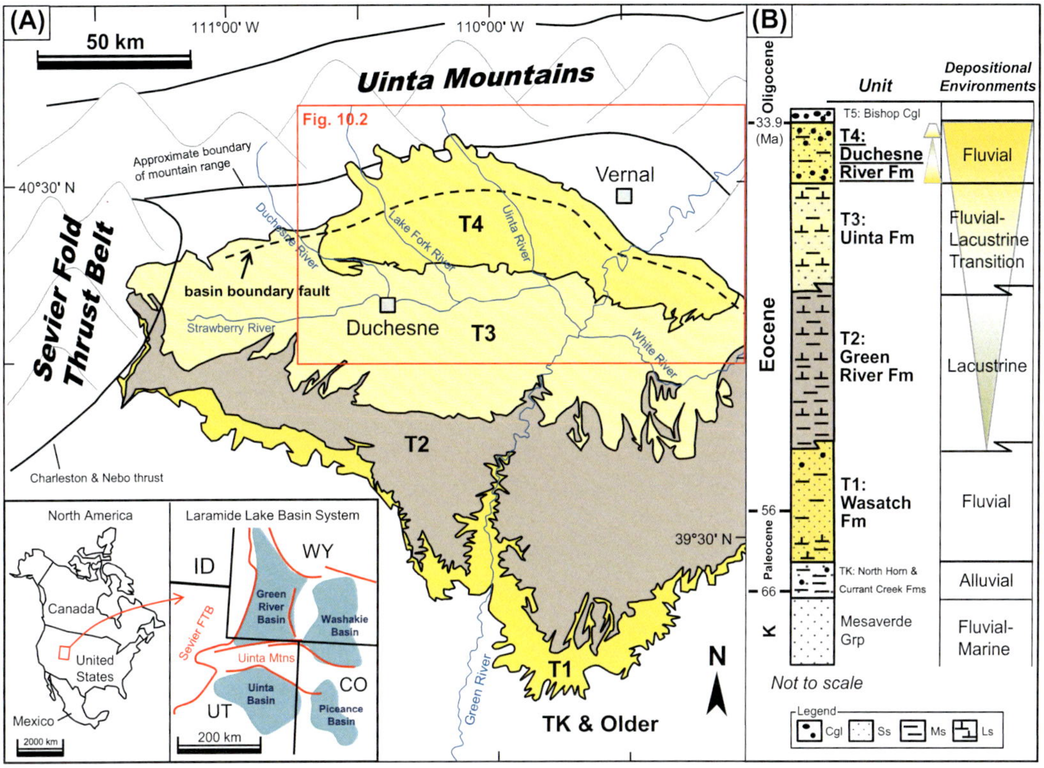

FIGURE 10.1 (A) Geological map of the Uinta Basin. Regional dip is to the north and formations are progressively younger toward the Uinta Mountains. The basin is currently surrounded by high mountain ranges of the Uinta Mountains and Sevier Fold Thrust Belt (FTB). (The map of Laramide lake basin system is from Dickinson et al., 1988. The geological map is modified from Andersen and Picard (1974), Bryant et al. (1989), Bryant (1992), Hintze et al. (2000), Sprinkel (2006, 2007), and Bryant (2010)). (B) Schematic geologic column showing the Paleogene sequence of the Uinta Basin (modified from Hintze et al., 2000). T2 to T4 exhibit a typical upward-coarsening/shallowing lacustrine basin-fill succession.

a sequence stratigraphic context. The study demonstrated that the base of the Duchesne River Formation is a sequence boundary that is a visible time-marker. The base of the Lapoint member, Dl (Fig. 10.2), is defined by the first occurrence of tuff, or tuffaceous beds. These tuffs are ~40 Ma in age (McDowell et al., 1973; Andersen and Picard, 1974; Prothero and Swisher, 1992; Kelly et al., 2012; Sprinkel, 2013), representing a consistent time-marker. The base of Dl is used as a stratigraphic datum for generating regional geological cross-sections (Fig. 10.3). The basal member of the Duchesne River Formation (Brennan Basin member: Db) marks the initial stage of an upward-fining fluvial sequence or cycle, that corresponds to the progradation of an alluvial plain environment following the cessation of a lake environment in the Uinta Formation (Fig. 10.3). The base of this member is characterized by abrupt depositional facies changes, here interpreted as a sequence boundary, related to uplift of surrounding mountain ranges in the Uinta Mountains and possibly Sevier Fold Thrust Belt (Wasatch Range) (Sato and Chan, 2015).

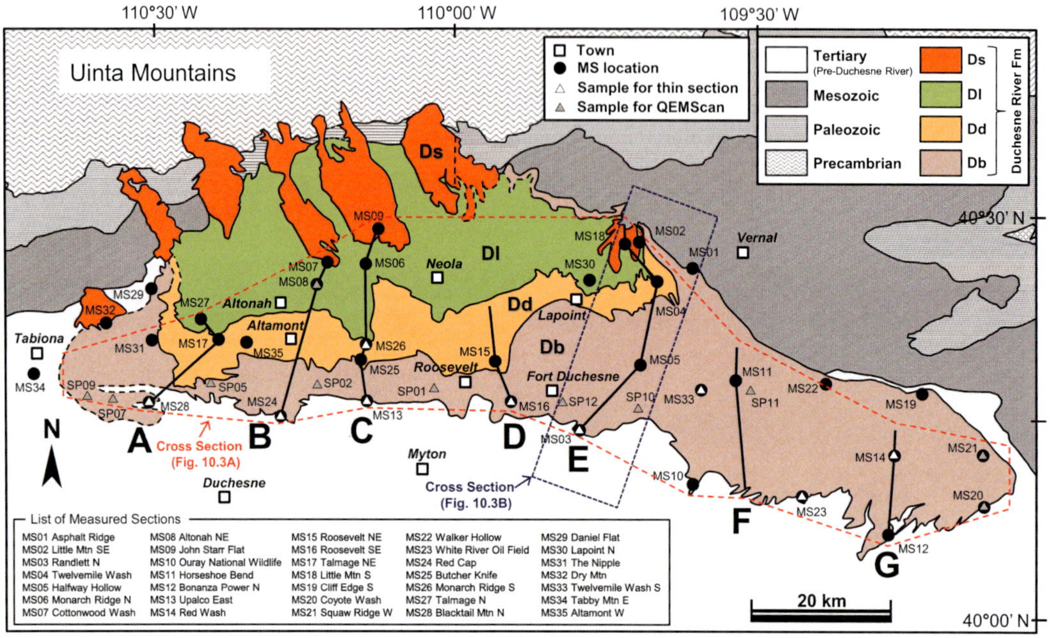

FIGURE 10.2 Geological map of the four members of the Duchesne River Formation, with sandstone sample locations. Regional dip is to the north, and the Duchesne River members (Db, Brennan Basin member; Dd, Dry Gulch Creek member; Dl, Lapoint member; Ds, Starr Flat member) get progressively younger toward the Uinta Mountains. The locations of 35 measured sections (MS), sandstone samples for thin section (*white triangles*) and for QEMScan (*gray triangles*), and composite sections A to G (*black lines*) are shown on the map. The map is modified after Andersen and Picard (1974), Rowley et al. (1985), Bryant et al. (1989), and Sprinkel (2006, 2007).

3. REGIONAL SEDIMENTARY FACIES OF THE BRENNAN BASIN MEMBER

Sato and Chan (2015) show detailed stratigraphic and facies analyses of the Duchesne River Formation based on 35 measured field sections (a total of 2750 m of strata). Their study defined four facies associations (FA1, 2, 3, and 4) within the basal member (Brennan Basin member: Db) of the Duchesne River Formation (Table 10.1). Here, we briefly describe characteristics and occurrences of each facies association and its internal lithofacies.

FA1 (amalgamated braided fluvial channels) is dominated by amalgamated channelized sandstones with widely connected (>1000 m) sandbodies (lithofacies Sc1), accompanying relatively minor red mudstones (lithofacies Mr) and thin-layered sandstones and siltstones (lithofacies Sth) (see the detailed lithological descriptions in Table 10.1). This facies association occurs in the western part of the basin within Db and exhibits a high net-sandstone-to-gross-thickness ratio (NTG) (0.75 at MS28) (Fig. 10.4). Trace fossils in FA1 are commonly observed but less abundant than in FA2. FA1 is interpreted to represent a fluvial style of widespread multiple interweaving fluvial channels (i.e., braided channels of Sc1) and narrow dry floodplain environments (Sth and Mr). The decreased abundance of trace fossils compared to FA2

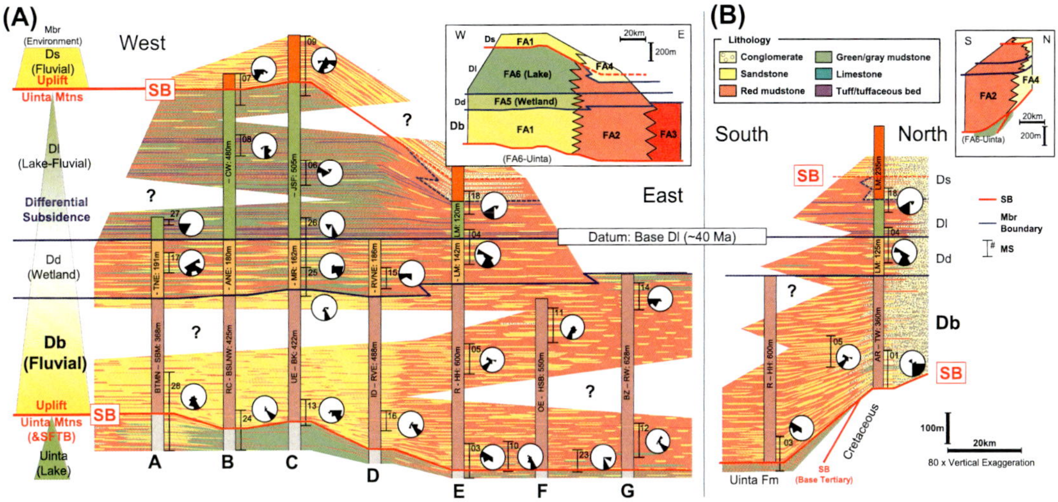

FIGURE 10.3 Regional geological cross-sections showing detailed basin-scale facies architecture, paleocurrent data at each measured section location, and the sequence stratigraphic framework of the uppermost Uinta and Duchesne River Formations (modified from Sato and Chan, 2015). The stratigraphic datum is set at the base (basal tuffs) of Dl, which represents a nearly isochronous boundary. (A) E–W regional correlations of composite sections A to G (location of cross-section in Fig. 10.2). The succession of the uppermost Uinta and the Duchesne River Formation is characterized by upward-fining continental cycles. The architecture of facies associations is shown in the upper right inset panel. Note the significant contrast of facies (facies association) between the western and eastern portions in the Brennan Basin member (Db). (B) N–S cross-section (location of cross-section in Fig. 10.2). The architecture of facies associations is shown in the upper right inset panel. FA4 (alluvial–fan complex) occurs in the northern part (i.e., foothills of the Uinta Mountains) of the Brennan Basin member (Db) distribution. Note that Db is juxtaposed with the Cretaceous Mesaverde Group in the north where the Tertiary Uinta and Green River formations are completely eroded out by the unconformity (sequence boundary) related to the uplift of the Uinta Mountains.

might reflect destructions of bioturbated substrates due to repetitive cut-and-fill patterns of the amalgamated fluvial channels (Sc1).

FA2 (extensive flood plain and stacked fluvial channels) is dominated by stacked channelized sandstones (lithofacies Sc2) and red mudstones (Mr) (Table 10.1). Overall these channelized sandstones (Sc2) are less laterally connected (with apparent connected bodies of >100 m) than the amalgamated channelized sandstones (Sc1) of FA1 and occasionally exhibit lateral-accretion features. This facies association occurs in the central-eastern part of the basin within Db, and shows a moderate NTG (0.5 at MS33) (Fig. 10.4). Abundant trace fossils observed in FA2 include root structures (rhizoliths) in mudstones and a variety of meniscate backfill burrows and nesting structures both in mudstones and sandstones. FA2 is interpreted to represent a depositional environment of extensive dry flood plains (Mr) with mixed braided, meandering, and isolated small river systems (Sc2 and Sc3). Occasional lateral bar-accretion features of Sc2 indicate some rivers were at least more sinuous than those of FA1. The abundance of trace fossils in this facies association indicates prosperous organic communities under moderately prolonged stable conditions and high preservation potential of organic traces due to the aggradational stacking pattern.

TABLE 10.1 Summary of Facies Associations in the Brennan Basin Member of the Duchesne River Formation.

FA#	Facies Association	Occurrence in Db	Lithofacies Components: Code	Lithofacies Components: Description	Lithofacies Components: Interpretation	Trace Fossils (Rare / Sparse / Common / Abundant)	Ss/Ms Ratio	Apparent Sandbody Dimensions
FA1	Amalgamated Braided Fluvial Channels	Western part of basin	Sc1	Fine- to coarse-grained, yellowish and reddish gray, poorly to well-sorted, channelized and trough cross-stratified sandstones with strongly amalgamated bodies (with apparent connected bodies over lateral distances of > 1,000 m) and common downstream (bar) accretion features.	Amalgamated braided fluvial channels	Common	75/25 (MS28)	> 1,000 m (MS28)
			Sth	Silt, fine to medium, red, grayish white, greenish gray, light gray, yellowish gray, poorly to well sorted, thin-layered (commonly < 1 m thick), massive or trough cross-stratified sandstone and siltstone with common intensive bioturbation	Overbank deposit, typically pedogenically altered			
			Mr	Clay- to silt-size, red, massive or mottled mudstone with occasional slickensides and common vertical and semi-vertical burrows	Well-drained flood-plain paleosol			
FA2	Extensive Flood Plain and Stacked Broad Fluvial Channels	Central-eastern part of basin	Sc2	Fine- to coarse-grained, light and yellowish gray, channelized and trough cross-stratified sandstones with stacked bodies (with apparent connected bodies over lateral distances of > 100 m) and uncommon lateral-accretion features	Braided and sinuous fluvial channels	Abundant	50/50 (MS33)	> 100 m (MS33)
			Sc3	Fine- to coarse-grained, light gray and grayish and yellowish white, channelized and trough cross-stratified sandstones with isolated narrow bodies (with apparent connected bodies under lateral distances of < 100 m)	Isolated small stream channel			
			Sth	See the above description and interpretation for Sth				
			Mr	See the above description and interpretation for Mr				
			My	Clay- to silt-size, yellow to brown, mottled mudstone with common relict bedding	Moderately drained flood-plain paleosol			

FA3	Extensive Flood Plain and Isolated Small Steams	Eastern part of basin	Sc3	See the above description and interpretation for Sc3						15/85 (MS14)	< 100 m (MS14)
			Sth	See the above description and interpretation for Sth							
			Mr	See the above description and interpretation for Mr							
			My	See the above description and interpretation for My							
FA4	Alluvial Fan Complex	Northern margin of basin	Cc	Granule- to boulder-size (max 1 m), poorly sorted, structureless or imbricate conglomerates with thick), and very fine- to very coarse-grained, trough cross-stratified sandstones with channelized or lenticular-shaped bodies	Alluvial-fan channel and lobe					70/30 (cgl+ss/ms) (MS01)	n/a
			Mr	See the above description and interpretation for Mr							
			Mg1	Clay- to silt-size, dominantly green and gray to partly yellow, purple and red, mottled mudstone with thin carbonaceous (e.g., fossil plants) mudstone layers and intensive gypsum veins	Playa or wetland deposit in the distal fan						

Abbreviations: FA = Facies Association, MS = Measured Section. Ss = Sandstone, Ms = Mudstone

Note: Paleosol moisture (drainage) interpretations are based on Kraus (2002), Atchley et al. (2004) and Kraus and Hasiotis (2006).

Modified from Sato and Chan (2015)

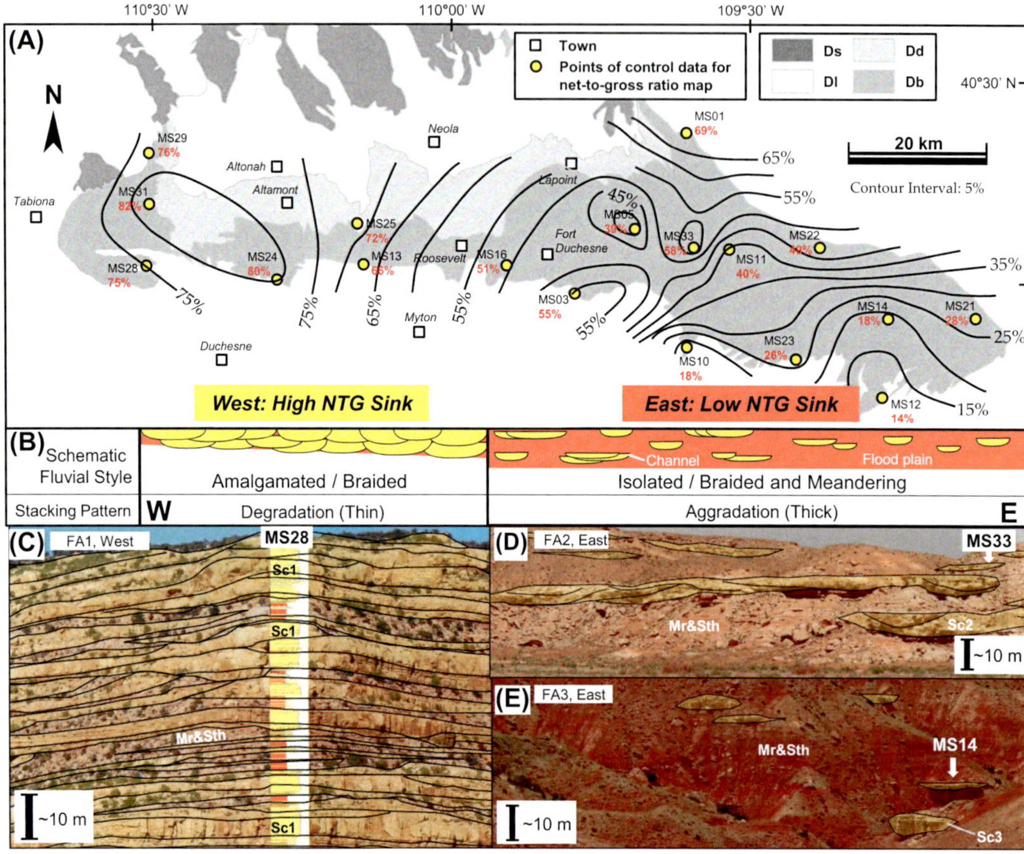

FIGURE 10.4 Contrasting facies in fluvial deposits of the Brennan Basin member (Db) between the western and eastern sinks. (A) Net-sandstone-to-gross-thickness ratio (NTG) map over the basin. Points of control data for net-to-gross ratio map are highlighted by circles (accompanied with numbers/percentages of NTG used for contouring). The western part of the basin (from Tabiona to Roosevelt) has a high NTG (over 60%), whereas the eastern part of the basin (Roosevelt to the eastern margin of Db distribution) has a lower NTG (60–14%). (B) Schematic fluvial styles and stacking patterns of Db, showing amalgamated braided channels with a degradational stacking pattern in the west and relatively isolated channels with aggradational stacking pattern in the east. (C) Outcrop (MS28) photo of representative high NTG facies with laterally continuous fluvial channels in the west. Resistant fluvial channel sand bodies are highlighted in yellow. (D) Outcrop (MS33) photo of representative moderate NTG facies with relatively isolated fluvial channels in the east. (E) Outcrop (MS14) photo of representative low NTG facies with very narrow isolated small fluvial channels in the east.

FA3 (extensive flood plain and isolated small streams) is dominated by red mudstones of Mr with some narrow (<100 m width) and isolated channelized sandstones (lithofacies Sc3) (Table 10.1), and tends to form very muddy, slope-forming "badlands" outcrops. This facies association occurs in the eastern part of the basin within Db, and exhibits a low NTG (0.15 at MS14) (Fig. 10.4). Trace fossils in this facies association are common although less abundant than FA2. FA3 is interpreted to represent a depositional environment of extensive dry flood plains (Mr) with only isolated small streams (Sc3). The low abundance of trace fossils in FA3 compared to FA2 could be resulted from a sampling bias due to poorly exposed (covered) outcrop conditions of this muddy facies association.

FA4 (alluvial–fan complex) is dominated by the conglomeratic lithofacies Cc including granule to boulder size (max 1 m), massive to imbricate conglomerates with channelized or lenticular-shaped bodies (max 10 m thick), and very fine- to very coarse-grained, trough cross-stratified sandstones with channelized or lenticular-shaped bodies (Table 10.1). This facies association occurs only in the northern margin of the basin (i.e., foothills of the Uinta Mountains) within Db and exhibits a high percentage of coarse-grained deposits (e.g., the ratio of conglomerate/sandstone to mudstone is 70:30 at MS01) (Fig. 10.3). Invertebrate trace fossils are scarce in this facies association, although there are intensive large rhizoliths at one locality (MS01). FA4 is interpreted to represent an alluvial fan complex including interchannel and playa or wetland environments. Lithofacies Cc contains both structureless and imbricated conglomerates indicating debris flows and traction transport, respectively (Nemec and Steel, 1984). These mixed transportation mechanisms and radiated paleocurrents at MS01 (Fig. 10.3) suggest very high-energy seasonal to perennial gravel-bed-river processes, and episodic and repetitive avulsions and lobe switching (e.g., Crews and Ethridge, 1993).

It should be noted that there are significant changes in sedimentary facies (fluvial style) and thickness (stacking pattern) in Db along the E–W basin-wide cross-section (Figs. 10.3 and 10.4). The western Db facies (FA1) represents an amalgamated braided fluvial channel system and a degradational (erosional cut and fill) stacking pattern with a high NTG (Fig. 10.4). In contrast, the eastern Db facies (FA2 and FA3) represent a mixed braided, meandering, and isolated river system and an aggradational stacking pattern with a moderate NTG or an isolated small channel system with a low NTG (Fig. 10.4). This distinct basin-scale west–east contrast has a spatially abrupt facies transition around the town Myton.

4. METHOD

In this study, sandstone sampling and subsequent compositional and petrographic analyses focused specifically on the Brennan Basin member (Db) because this is the only Duchesne River unit suitable for evaluating compositional changes within a basin-wide, regional fluvial (drainage) systems context (Fig. 10.2). The petrographic studies here integrated sandstone composition data by Andersen and Picard (1974), who examined the composition of 121 clastic rocks (~70 sandstone and ~50 conglomerate and siltstone samples) from the Duchesne River Formation and reported geographical differences (but not member-level stratigraphic differences) in composition between the western and eastern parts of the basin. The latest field-based sedimentological studies and new petrographic studies show the additionally detailed relationships between fluvial sedimentary facies associations, sandstone composition, and reservoir properties (porosity) within the Brennan Basin member (Db).

Two different approaches were used to evaluate sandstone compositional data: (1) conventional thin-section petrography, and (2) QEMScan (Quantitative Evaluation of Materials by Scanning electron microscopy) automated disaggregate counts (Allen et al., 2012). A total of 20 representative fine- and medium-grained sandstone samples were collected mainly from the Brennan Basin member (Db) (sample locations in Fig. 10.2). Nine samples (8 samples from Db and one sample from Dd) were used for conventional thin-section examinations, following the Gazzi–Dickinson point-counting method (Gazzi, 1966; Dickinson, 1970) updated by Ingersoll et al. (1984) (Table 10.2). In this method, any mineral >0.0625 mm is counted as an individual grain component such as quartz and feldspar, even if this mineral forms a part of volcanic or sedimentary rock fragments. In contrast, previous petrographic work by Andersen and Picard (1974) followed the classification of Folk (1968) in rock fragments to represent individual sandstone framework components. In this study, there appears to be no significant difference in the resultant compositions between the two approaches in sandstone classification, as rock fragments with large (>0.0625 mm) minerals/grains are minimal and the majority of rock fragments are carbonates and cherts, as described later.

The compositions of all 20 samples were examined by QEMScan, an effective tool to provide quantitative analysis and mapping of minerals in solid materials. Although several different QEMScan analytical approaches are available (e.g., mineralogic mapping on thin section), the QEMScan automated disaggregate count technique of Allen et al. (2012) was used in this study. In this method each sandstone sample is ground (mechanically abraded) into its component grains, and cement and matrix materials are removed to scan only sand particles. We used 150- and 600-mm mesh sieves to remove large blocky samples and small materials such as cement and matrix. Nine of the 20 samples were the same sandstone samples as those used for conventional thin-section analysis (Table 10.2), and were included to ensure the consistency of the results between the two different methods. Thus 11 of 20 samples (10 samples from Db, one sample from Dl) were used to supplement or fill in gaps between thin-section data points.

Here we summarize pros and cons of QEMScan automated disaggregated count analysis (Table 10.3). The major advantage of this method is to shorten the analytical time, although the duration depends both on the resolution and the number of particles to be counted. In this study, a 16-μm resolution (pixel size) was used, with a scan duration of 1.5 h, using QEMScan in particle-counting mode. For all 20 samples, the effective numbers of particles scanned (minimum 1344, maximum 2118 grains, average 1818 grains) allowed comparison with results from the Gazzi–Dickinson point counts. The major disadvantages of QEMScan automated disaggregated count are lack of some textural information (e.g., matrix, epimatrix -derived matrix of Dickinson, 1970; cementation, pore system) by disaggregation process and nondifferentiation of some minerals (e.g., monocrystalline, polycrystalline quartz, and chert).

Several different ways of postprocessing in QEMScan automated disaggregated counts were tested to investigate the best fit with mineral compositions compared with thin-section point counts: (1) bulk mineral area proportion calculation without filter, (2) bulk mineral area calculation with grain size filter, and (3) particle count with grain size and mineral identification filters. The results of bulk mineral area proportion calculation (method 1) showed the best correlations with thin-section-derived data in terms of major components, such as quartz (plus chert) and carbonate, although the other two methods also kept good correlations and consistencies (Fig. 10.5). The estimated K-feldspar tends to be slightly lower than the ideal correlation (1:1) line. This is possibly due to the sample preparation issue, as some highly weathered K-feldspars could wash away in sample grinding process. However,

TABLE 10.2 List of Sandstone Samples and Results of Thin-Section and QEMScan Analysis

				Thin Section								QEMScan (Bulk Mineral Area%)		
							R% (breakdown)							
ID	Mbr	Locality	Coordinates (NAD1927)	Q%	K%	R%	Clastics	Carb	Chert	Cement /Matrix %	Porosity %	Q%	K%	R (Ls + Dol) %
#1	Db	MS28, Blacktail Mtn north	N40.27920, W110.50679	90.0	3.6	6.4	1.7	4.3	0.4	5.3	14.7	89.2	2.3	4.3
#2	Db	MS24, red Cap	N40.25983, W110.28766	96.0	0.6	3.4	2.1	0.0	1.3	1.9	17.6	97.7	1.0	0.0
#3	Dd	MS26, Monarch Ridge south	N40.34881, W110.14320	97.4	0.2	2.4	2.2	0.2	0.0	6.7	19.0	97.9	0.7	0.0
#4	Db	MS13, Upalco east	N40.27922, W110.15182	97.2	0.6	2.2	2.2	0.0	0.0	4.2	17.4	97.8	1.0	0.0
#5	Db	MS16, Roosevelt SE	N40.27232, W109.90966	62.0	1.0	37.0	8.3	27.2	1.4	16.9	9.7	68.7	1.4	24.0
#6	Db	MS03, Randlett north	N40.24093, W109.80736	79.6	1.2	19.2	4.9	10.8	3.5	10.4	12.1	82.7	1.2	12.5
#7	Db	MS33, Twelvemiles Wash south	N40.29030, W109.60100	67.8	3.0	29.2	4.1	18.9	6.3	16.3	6.3	82.5	1.4	12.4
#8	Db	MS23, white river Oil field	N40.15940, W109.44488	79.4	2.6	18.0	1.6	13.8	2.6	17.5	12.3	76.4	1.7	14.7
#9	Db	MS14, red Wash	N40.20829, W109.28806	89.6	0.4	10.0	1.2	7.0	1.8	25.0	6.1	95.9	1.2	1.9
#10	Db	SP09, Blacktail Mtn west	N40.27206, W110.58815	n/a	n/a	n/a	n/a	n/a	n/a	n/a	n/a	95.5	2.9	0.0
#11	Db	SP07, Blacktail Mtn west	N40.27932, W110.54595	n/a	n/a	n/a	n/a	n/a	n/a	n/a	n/a	89.6	3.2	4.0
#12	Db	SP05, Duchesne north	N40.28607, W110.36244	n/a	n/a	n/a	n/a	n/a	n/a	n/a	n/a	96.8	1.7	0.0
#13	Dl	MS08, NE Altonah	N40.43758, W110.21249	n/a	n/a	n/a	n/a	n/a	n/a	n/a	n/a	90.8	3.4	2.6
#14	Db	SP02, Big sand Wash	N40.29228, W110.21679	n/a	n/a	n/a	n/a	n/a	n/a	n/a	n/a	98.4	0.6	0.0
#15	Db	SP01, Roosevelt west along I-40	N40.28993, W109.99861	n/a	n/a	n/a	n/a	n/a	n/a	n/a	n/a	74.3	2.4	14.8
#16	Db	SP12, Fort Duchesne east	N40.28402, W109.84258	n/a	n/a	n/a	n/a	n/a	n/a	n/a	n/a	68.0	1.1	26.9
#17	Db	SP10, Pelican Lake north	N40.24449, W109.70727	n/a	n/a	n/a	n/a	n/a	n/a	n/a	n/a	91.0	0.6	6.4
#18	Db	SP11, Horseshoe Bend east	N40.28274, W109.53285	n/a	n/a	n/a	n/a	n/a	n/a	n/a	n/a	92.7	1.2	4.0
#19	Db	MS21, Squaw Ridge west	N40.21300, W109.13760	n/a	n/a	n/a	n/a	n/a	n/a	n/a	n/a	70.6	2.7	18.1
#20	Db	MS20, Coyote Wash	N40.12432, W109.17005	n/a	n/a	n/a	n/a	n/a	n/a	n/a	n/a	88.8	1.2	8.4

Carb, carbonate; *Dol*, dolomite; *K*, feldspar; *Ls*, limestone; *Q*, quartz; *R*, rock fragments.

TABLE 10.3 Comparison Between Thin-Section Point Counts and QEMScan Automated Disaggregated Counts

Method	Pros	Cons
Conventional thin-section point counts	• Accurate and detailed information on grains if operator is skillful • Visible original texture (e.g., sorting, cementation, pore geometry, etc.)	• Possible human error (depending on operator's skill) • Possible inconsistent results between operators/researchers
QEMScan automated disaggregated Counts	• Fast and easy • Automated • Repeatable • Consistency	• Losing some information by disaggregation process (e.g., matrix, epimatrix, cementation, weak grains, pore system) • No porosity data • Does not differentiate some minerals (e.g., monocrystalline, polycrystalline quartz and chert)

this loss would only be a minor component, and a reasonable trend is apparent (Fig. 10.5). Consequently QEMScan mineral composition data can be used with confidence to supplement the thin-section data.

5. SANDSTONE COMPOSITION AND PALEOCURRENTS

The newly acquired sandstone composition data by thin-section examinations of the Brennan Basin member samples can be categorized as quartzose, sublithic, and lithic arenites based on Folk's (1968) classification (Fig. 10.6A), as do the earlier point counts of Andersen and Picard (1974). Also most composition data are plotted in the area of "recycled orogenic provenance" in the Dickinson et al. (1983) classification (Fig. 10.6B). These plots confirm that feldspar is only a minor component of sandstones across and along the basin. The ratio of lithic grains in sublithic and lithic sandstones is shown in Fig. 10.6C. Although the data in this plot are highly scattered, carbonate grains are the most common, and siliciclastic grains (siltstone/mudstone) are relatively minor. Chert grains are also minor in the new data from the Brennan Basin member, and therefore would not have a critical influence on the following provenance interpretations integrating composition data by the QEMScan method, which is not able to differentiate between quartz and chert.

The regional trend of sandstone compositions was investigated using a cross-plot of longitude versus percent rock fragments of grains, in comparison with paleocurrent data from Sato and Chan (2015) (Fig. 10.7). This plot shows a significant difference in sandstone composition between the western and eastern parts of the basin; i.e., a low proportion of rock fragments in the west (high NTG sink) and high in the east (low NTG sink). This means sandstones in the west are rich in quartz (over 90% of total grains), and those in the east contain less quartz and more carbonate and siliciclastic rock fragments (Table 10.2).

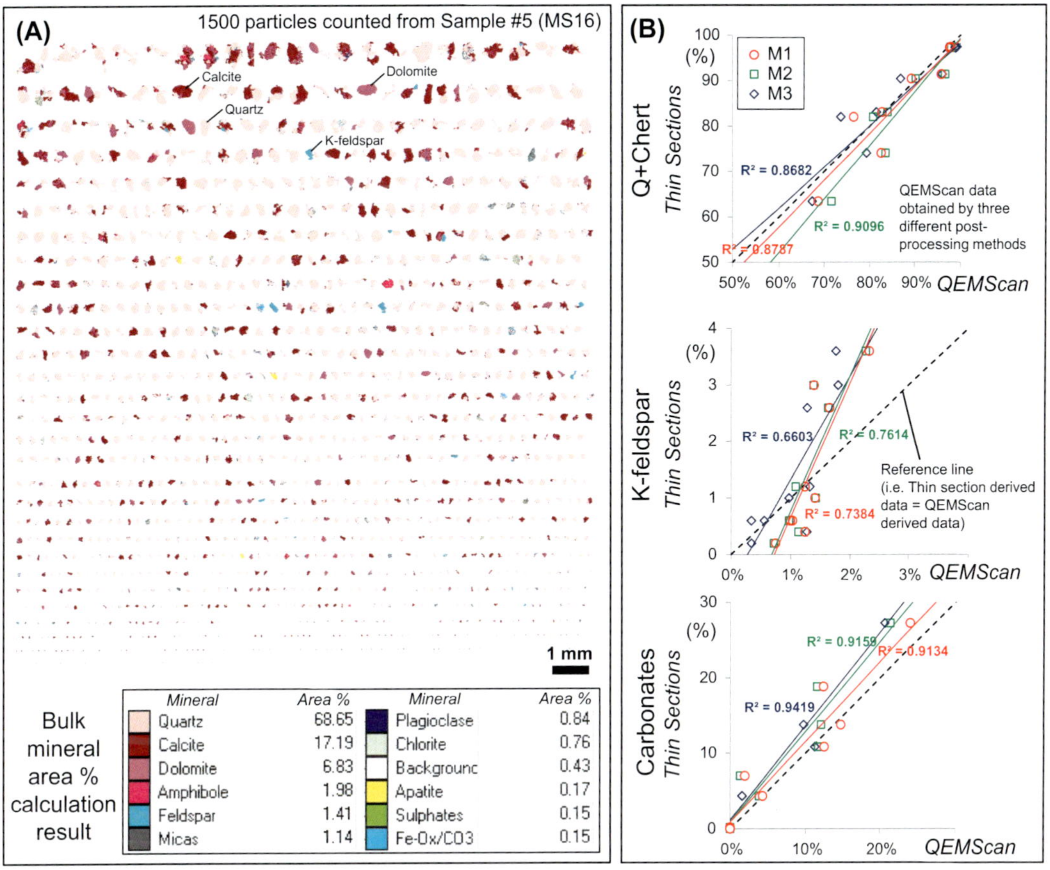

FIGURE 10.5 (A) An example of QEMScan automated disaggregated count data (scanned 1500 particles from sample 5 and postprocessing output by method 1 (M1), bulk mineral area% calculation method without filter. (B) A series of cross-plots of grain-type proportions from QEMScan on the X-axis and proportions from thin-section examination on the Y-axis. QEMScan data by three different postprocessing methods—method 1 (M1); method 2 (M2), bulk mineral area% calculation with grain size filter; and method 3 (M3), particle count with grain size and mineral identification filters—are shown in different symbols. Major components such as quartz and carbonate exhibit strong correlations with thin-section data. A minor component of K-feldspar shows minor deviation from the ideal (1:1) correlation line. Nevertheless, the overall trend is still reasonable.

A total of 264 paleocurrent measurements were acquired throughout the fluvial-channel-dominated Brennan Basin member (Fig. 10.7). The majority of paleocurrent data were measured based on trough cross-bedding structures in fluvial sandstones, although some data were derived from clast imbrications in conglomeratic facies (FA4) at some locations (MS01 and MS22). Although paleocurrents show overall southward transport that confirms earlier reports by Warner (1965, 1966) and Andersen and Picard (1974), more detailed basin-wide examination indicates significant features that assist in interpreting the paleodrainage patterns. The western part of the basin exhibits dominantly eastward and southeastward flows, whereas the central-eastern part of the basin shows south-southwestward flows,

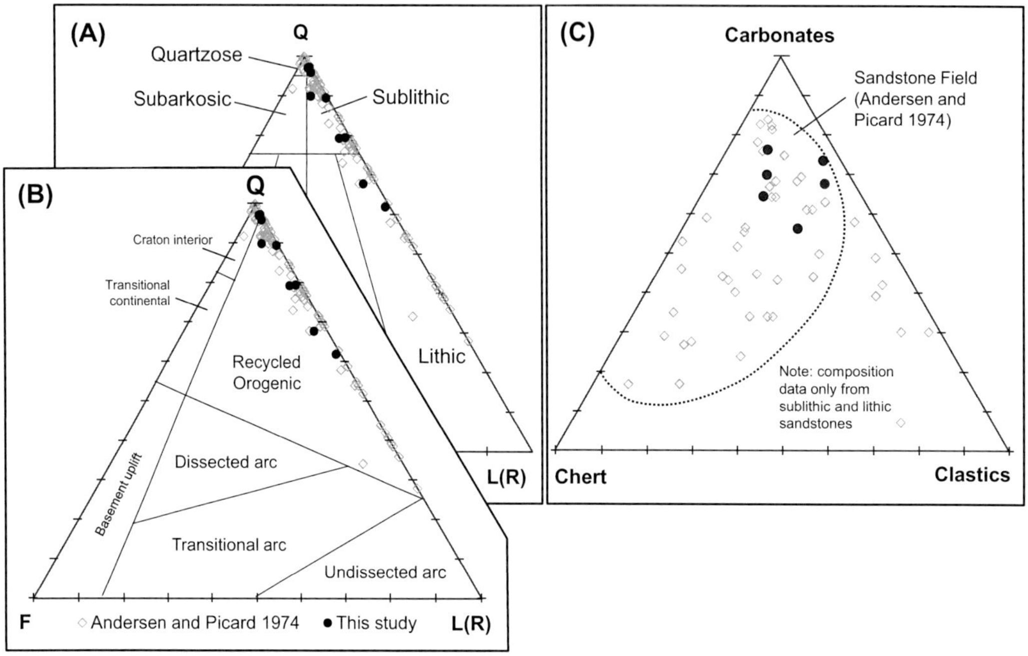

FIGURE 10.6 Ternary QFL(R) plots showing sandstone composition of the Duchesne River Formation samples (data from Andersen and Picard, 1974, and this study). Plots (A) Folk's (1968) classification and (B) Dickinson et al. (1983) classification both indicate that feldspar is a very minor component of sandstones over the basin. Plot (C) shows the relative abundance of rock fragment grains and indicates that carbonate grains are the most common, and siliciclastics are relatively minor.

or variably directed flows, suggesting some differences in drainage pattern between the west and east. Correspondingly, there is a significant contrast in fluvial styles (facies association) and sandstone compositions between the western and eastern parts of the basin: amalgamated channel dominated FA1 with quartz-rich sandstones in the west and relatively isolated channel and mudstone dominated FA2 and FA3 with rock fragment-rich sandstones in the east (Figs. 10.4 and 10.7).

6. PETROGRAPHY

Thin-section petrography provides both compositional data and visual information on sandstone textures and reservoir characterization properties such as porosity. Overall petrographic trends suggest a distinct difference between the western and eastern parts of the basin. The western quartz-rich samples (samples 1, 2, and 4) exhibit low matrix and/or cement materials (1.9–5.3% of total counts) and high porosity (point count porosities ranging from 14.7% to 17.6%). In contrast, the eastern rock fragment-rich samples (samples 5 to 9) tend to have high matrix and/or cement materials (10.4–25.0% of total counts) and lower porosity

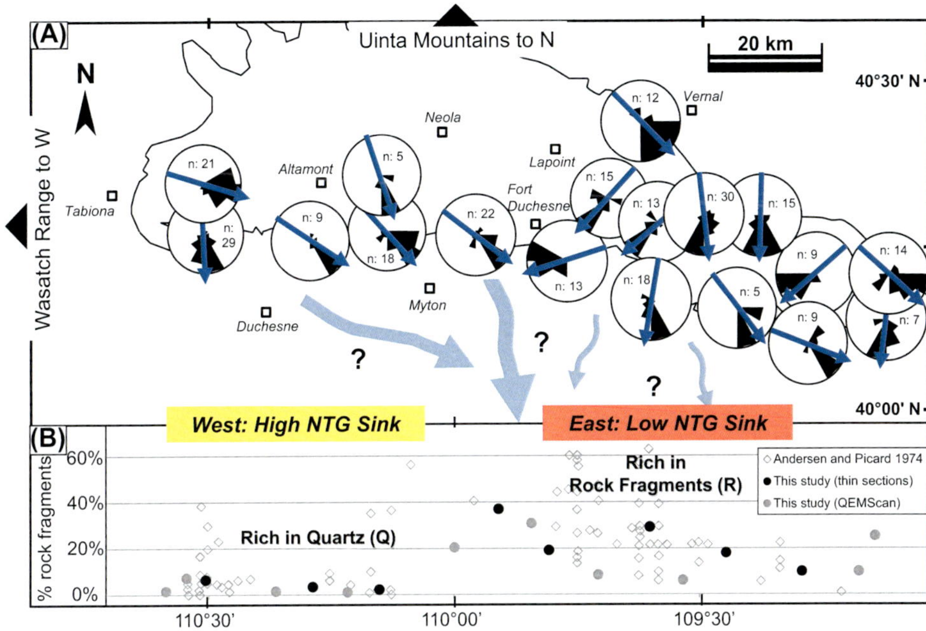

FIGURE 10.7 Paleocurrents from the Brennan Basin member (Db) and longitude versus percent rock fragments of grains. (A) Paleocurrent data (rose diagram with the average direction shown as a long arrow, total paleocurrents $n = 264$) exhibit an overall southerly transport from the Uinta Mountains. However, the western part of the basin shows evidence of more eastward and southeastward flows, whereas the eastern part of the basin is characterized by south-southwestward or variably directed flows. The map is modified from Sato and Chan (2015). (B) Plot of percent rock fragments across the basin provides an indication of relative abundance of quartz grains, as feldspar is only a minor component of all sandstones. Note the relative abundance of quartz in the west, and rock fragments in the east.

(6.1–12.3%) (Fig. 10.8). These characteristics of textural immaturity indicate that some lithic grains were deformed (pseudomatrix of Dickinson, 1970) or dissolved and migrated or precipitated into the original pore space during diagenesis. In other words, compositionally and texturally mature quartz-rich sandstones in the west were favorable in maintaining the original pore space without significant porosity occlusion by cementation or compaction/grain deformation.

Samples from the western part of the basin, notably samples 2 (MS24) and 4 (MS13), are remarkably better sorted, more porous, and richer in quartz than sample 1 (MS28), which was taken from the western limit of the Brennan Basin member distribution (Fig. 10.8). This trend suggests MS28 was located in the upstream (proximal) part, and MS24 and MS13, which contain more texturally mature sediments, were located in the downstream (distal) part of the drainage system in the western part of the basin. Collectively, in combination with paleocurrent data, the texturally mature and porous sandstones in the central-eastern part of the basin are interpreted to reflect a long transport distance from the source terrains in the Uinta Mountains in the north, and probably the Sevier Fold Thrust Belt (FTB) in the west.

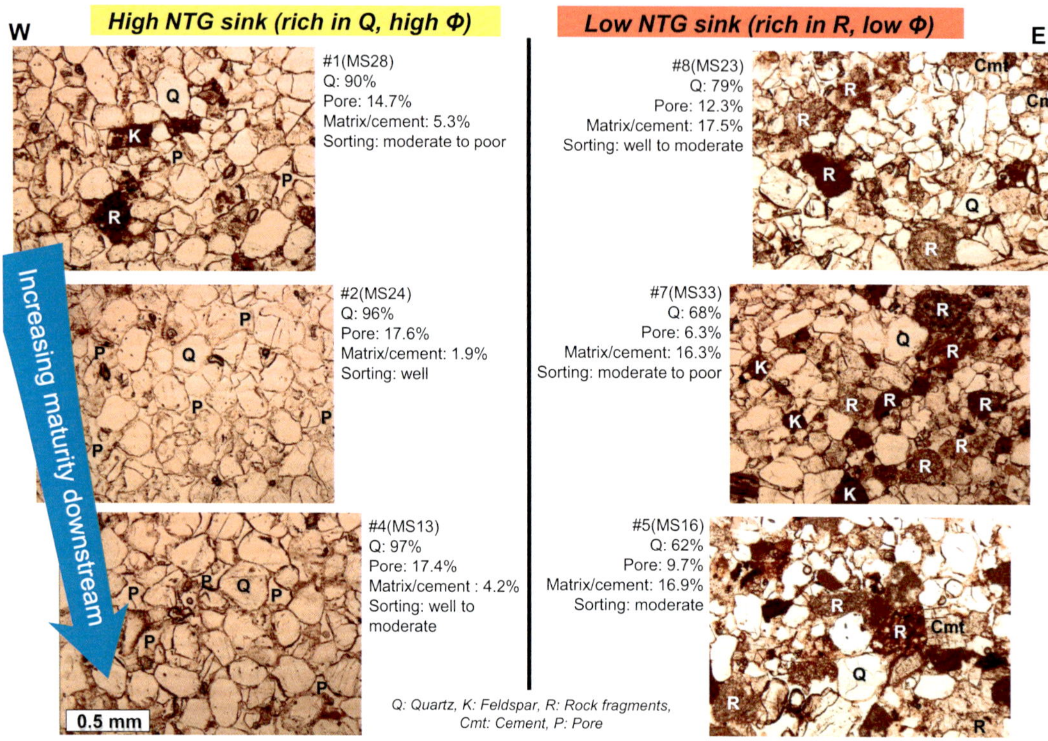

FIGURE 10.8 Thin-section petrography of sandstone samples from the Brennan Basin member (Db). Note there are distinct compositional and porosity differences between the west (rich in quartz, higher porosity) and east (rich in rock fragments, lower porosity). A trend observed in samples 1 (moderately sorted sandstone with 90% quartz) to 4 (well to moderately sorted sandstone with 97% quartz) indicates sandstones become texturally more mature downstream. All thin-section figures are at the same scale.

7. SYNTHESIS OF PALEODRAINAGE MODEL

Here the source-to-sink (paleodrainage) interpretation on the Brennan Basin member (Db) is synthesized by integrating stratigraphic, sedimentological and petrologic data. In this intermontane lacustrine basin, the ultimate allogenic control on fluvial sedimentation of the Duchesne River Formation is tectonic uplift(s) of the sediment and water source in the mountain range(s) of the Uinta Mountains in the north, and possibly the Sevier FTB in the west, which is marked by an unconformable sequence boundary at the base of Db (Sato and Chan, 2015) (Fig. 10.9). However, the marked lateral facies changes of Db within the basin record previously undocumented local and specific allogenic controls stemming from regional tectonic uplift(s).

The western high-NTG braided fluvial channel system suggests high discharge and/or topographic gradient, relative to the eastern low-NTG mixed (relatively narrow and isolated) channel system, based on the classic fluvial channel style/width concept explained by discharge and slope relationships (e.g., Leopold and Wolman, 1957; Bridge, 2001). Paleocurrents and geographical change in sandstone composition indicate two distinct drainage

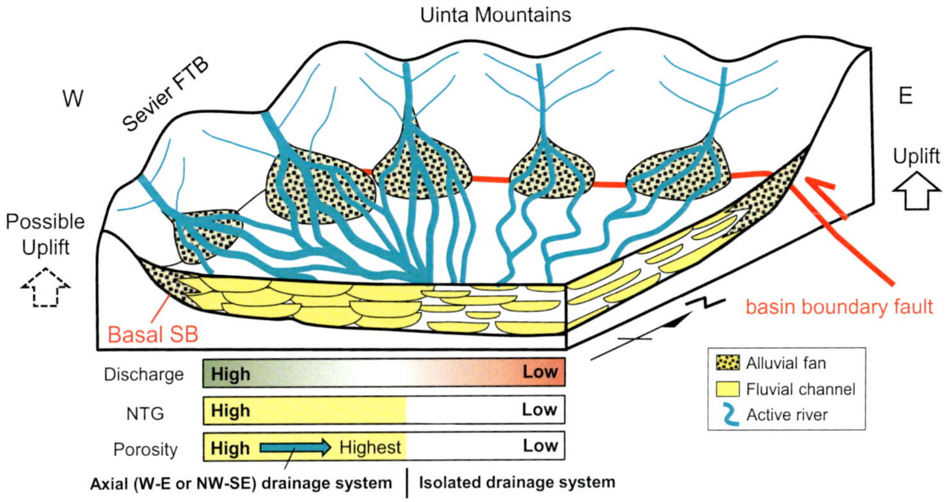

FIGURE 10.9 Schematic paleoenvironmental model for deposition of the Brennan Basin member (Db) (modified from Sato and Chan, 2015). Db reflects high-energy fluvial deposition after uplifts in the Uinta Mountains and Sevier FTB. The W–E or NW–SE axial drainage system with high NTG and high porosity in the western part of the basin mainly reflects high discharge from two source terrains in the north and west. In contrast, a relatively isolated drainage system with the low NTG and low porosity in the eastern part of the basin indicates low discharge from a single source terrain in the north.

systems in the western and eastern parts of the basin. Specifically, texturally and compositionally mature (quartz-rich) sandstones and eastward and southeastward paleocurrents indicate a long transportation along the E–W or NW–SE basin axis (axial drainage system), with the contributions both from the Uinta Mountains to the north and from the Sevier FTB to the west (Fig. 10.9). In this system, the best sandstone reservoirs, in terms of quantity and quality (porosity), were found in the central-western part of the basin (around MS13 and MS24), where porous and amalgamated braided channel sandstones developed (Fig. 10.10). In contrast, rock fragment-rich, poor quality (low porosity) sandstone reservoirs

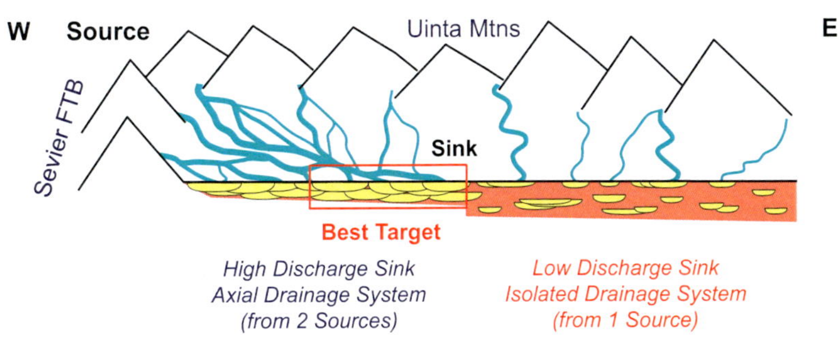

FIGURE 10.10 Schematic source to sink model for the Eocene Brennan Basin member of the Duchesne River Formation in the Uinta Basin. Note two different drainage systems in the sink: high discharge in the west and low discharge in the east.

with southward and variably-oriented paleocurrents occur in the relatively isolated drainage systems in the east. This suggests contribution from a single source terrain of the Uinta Mountains in a relatively short distance to the north. The provenance difference in source terrain contributions (i.e., two sources in the west and a single source in the east) influenced both water and sediment discharge into the basin, resulting in the contrasting fluvial style and sedimentary facies.

8. DISCUSSION

The observations just made indicate that discharge, which is related to the local source climate and the number of source terrains, strongly controlled fluvial styles of the Brennan Basin member in this basin. A high discharge fluvial system in the west implies a wet climate and multiple source terrains, whereas the low discharge fluvial system in the east indicates a drier climate and a single source terrain. It should be noted that a climatic contrast is observed even in the present-day Uinta Basin and surrounding ranges (Fig. 10.11A); a wetter climate and higher precipitation in the western area and a drier climate and lower precipitation in the eastern area (Greer, 1981; Jensen et al., 1990; Gillies and Ramsey, 2009). Although the modern Green River flowing across the eastern Uinta Mountains gives a significant amount of discharge into the eastern dry Uinta Basin at present (Fig. 10.1), this large drainage system opened in the late Miocene or early Pliocene (Hansen, 1986), and did not exist in the Eocene.

Studies of modern fluvial environments and drainage patterns have been greatly enhanced in recent years due to easily accessible satellite image data (e.g., Google Earth). The modern Himalayan foreland province provides a possible analogous setting to the basin at the time of deposition of the Brennan Basin member (Db), where various drainage patterns develop adjacent to source mountain ranges (e.g., Leier et al., 2005; Weissmann et al., 2010; Hartley et al., 2010). The modern upper Brahmaputra River exhibits a drainage pattern of a distributive fluvial system (DFS) terminating in an axial system (Hartley et al., 2010). In this area, the modern alluvial plain (basin area) is surrounded by multiple source mountains in the north and east and receives multiple water and sediment inputs from these source terrains (Fig. 10.11B). The uppermost streams form small-scale (narrow and elongated) fluvial fans, and terminate in an axial drainage of large-scale (wide) braided fluvial channel belt character. The western high-discharge (high-NTG) river system of the Brennan Basin member (Db) in the Uinta Basin exhibits some common features: multiple source inputs from surrounding high mountain ranges and widespread braided channel belts with axial drainage patterns. In addition, the size of this ancient western Db drainage system (approximately 50 $\times$ 50 km area exposed) is possibly comparable to the modern upper Brahmaputra River example (Fig. 10.11C). Although the geometric parameters of this modern upstream Brahmaputra River channel or channel belt (e.g., width, depth, bankfull discharge) need to be investigated in detail in the future, this appears to be a reasonable candidate as a modern upstream analog of a high-discharge fluvial system where multiple source inputs terminate in an axial system at the uppermost streams of the intermontane foreland basin.

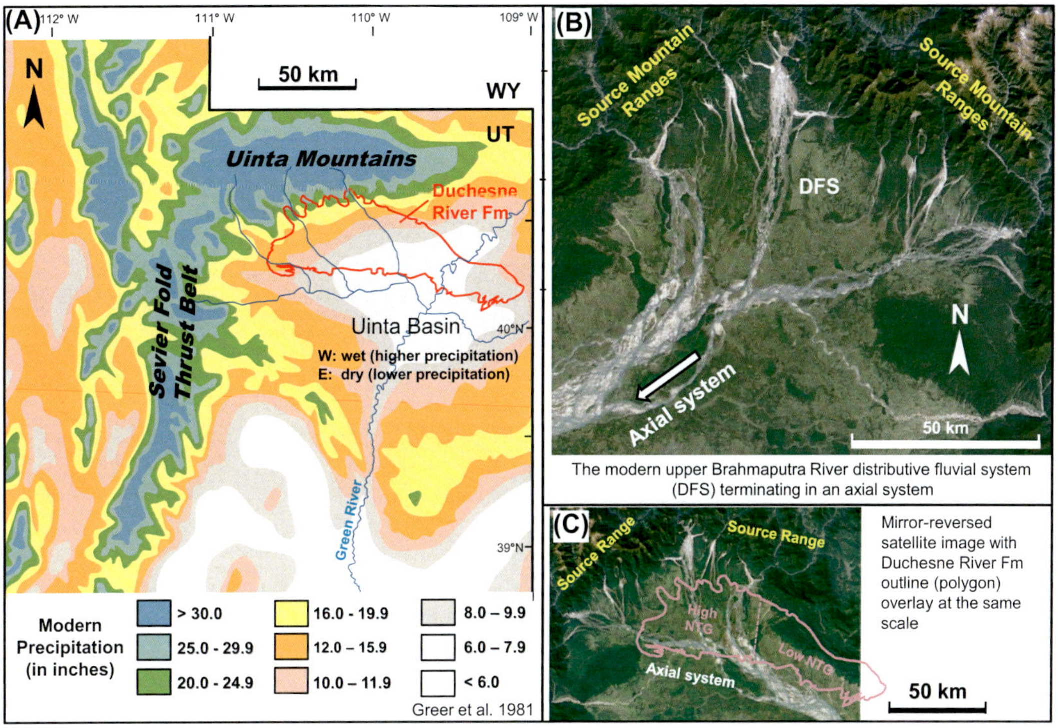

FIGURE 10.11 (A) Modern precipitation in and around the Uinta Basin (after Greer, 1981). Note that a wetter climate and higher precipitation characterize the western part of the basin and adjacent mountain ranges (Uinta Mountains to the north and Sevier FTB to the west), and a drier climate and lower precipitation exist in the eastern part of the basin. This modern example implies local source tectonics have a great influence on both discharge and distribution of fluvial systems in the basin (sink) area. (B) Google Earth satellite image of the modern upper Brahmaputra River in the Himalayan foreland basin, India. This basin includes multiple distributive fluvial systems (DFS) in proximal settings to the north and east (Hartley et al., 2010). These are surrounded by highlands, providing water and sediment sources. These DFS terminate in a northeast to southwest axial drainage system, marked by high-discharge, well-developed braided channels. (C) Mirror-reversed satellite image of the modern upper Brahmaputra River with the Duchesne River Formation outline (polygon) overlay at the same scale for basin size comparison. This modern upper Brahmaputra River is a possible modern analog for the axial paleodrainage (fluvial) system of the Eocene Brennan Basin member in the Uinta Basin.

9. CONCLUSIONS

The source-to-sink depositional system of the Brennan Basin member of the Duchesne River Formation in the Uinta Basin was evaluated by integrating regional sedimentology and sandstone petrology determined from both point counts and QEMScan methods. This study reveals the importance of source terrain control on fluvial reservoir facies in the basin (sink) to provide an important analog example for hydrocarbon exploration in similar continental basins as summarized below.

- The Brennan Basin member of the Duchesne River Formation exhibits a distinct facies change in fluvial style between the western and eastern parts of the basin (sink): i.e., a high-NTG, wide braided fluvial channel system with degradational (cut and fill) stacking patterns in the west, and a low-NTG, mixed braided, meandering, and isolated small fluvial channel system with aggradational stacking patterns in the east.
- Fluvial sandstone composition and texture show a distinct change between the west and east: i.e., quartz-rich sandstones with high porosity in the west and rock fragment-rich sandstones with low porosity (high matrix or cement contents) in the east.
- Compositional and fabric contrasts are shown to be strongly affected by different discharge from distinct source terrains: i.e., high discharge from two source terrains in the Uinta Mountains and the Sevier FTB to the western sink, and low discharge from a single source terrain in the Uinta Mountains to the eastern sink.
- The best fluvial sandstone reservoirs in terms of quantity (sandstone thickness and connectivity) and quality (porosity) occur in the central-western part of the basin: i.e., the distal parts of the high-discharge west-to-east drainage system in the western sink.

Acknowledgments

We thank Allan Ekdale, Erich Petersen, Cari Johnson, Lisa Stright, and Lauren Birgenheier at the University of Utah for their useful comments on this project. Wil Mace and Quintin Sahratian helped with thin-section and QEMScan analyses. Douglas Sprinkel at the Utah Geological Survey was a great supporter and provided valuable insight on the Uinta Basin geology. We gratefully acknowledge the input of reviewers. We acknowledge the Ute Indian Tribe and the Bureau of Land Management in Vernal and Ouray National Wildlife Refuge who generously provided essential permissions for the field work.

References

Allen, J.L., Johnson, C.L., Heumann, M.J., Gooley, J., Gallin, W., 2012. New technology and methodology for assessing sandstone composition: a preliminary case study using a quantitative electron microscope scanner (QEMScan). Geological Society of America 487, 177–194 special publication.

Andersen, D.W., Picard, M.D., 1972. Stratigraphy of the Duchesne River Formation (Eocene-Oligocene?), northern Uinta Basin, Northeastern Utah. Utah Geological and Mineral Survey Bulletin 97, 29 p.

Andersen, D.W., Picard, M.D., 1974. Evolution of synorogenic clastic deposits in the intermontane Uinta Basin of Utah. In: Dickinson, W.R. (Ed.), Tectonics and Sedimentation, vol. 22. SEPM Special Publication, pp. 167–189.

Atchley, S.C., Nordt, L.C., Dworkin, S.I., 2004. Eustatic control on alluvial sequence stratigraphy: a possible example from the Cretaceous—Tertiary transition of the Tornillo Basin, Big Bend National Park, West Texas, USA. Journal of Sedimentary Research 74, 391–404.

Beglinger, S.E., Doust, H., Cloetingh, S., 2012. Relating petroleum system and play development to basin evolution; West African South Atlantic basins. Marine and Petroleum Geology 30, 1–25.

Bohacs, K.M., 2012. Relation of hydrocarbon reservoir potential to Lake-Basin type: an integrated approach to unraveling complex genetic relations among Fluvial, Lake-Plain, Lake Margin, and Lake Center Strata. In: Baganz, O.W., Bartov, Y., Bohacs, K., Nummedal, D. (Eds.), Lacustrine Sandstone Reservoirs and Hydrocarbon Systems, vol. 95. American Association of Petroleum Geologists Memoir, pp. 13–56.

Bridge, J.S., 2001. Characterization of fluvial hydrocarbon reservoirs and aquifers: problems and solutions. AAS Revista, Revista de la Asociación de Sedimentología 8, 87–114.

Bruhn, R.L., Picard, M.D., Beck, S.L., 1983. Mesozoic and early Tertiary structure and sedimentology of the central Wasatch Mountains, Uinta Mountains, and Uinta Basin. Utah Geological and Mineralogical Survey Special Studies 59, 63–105. Salt Lake City.

Bruhn, R.L., Picard, M.D., Isby, J.S., 1986. Tectonics and sedimentation of Uinta Arch, Western Uinta Mountains and Uinta Basin. In: Peterson, J.A. (Ed.), Paleotectonics and Sedimentation in Rocky Mountain Region, United States, vol. 41. American Association of Petroleum Geologists Memoir, pp. 333–352.

Bryant, B., 1992. Geologic and structure maps of the Salt Lake City 1 degree × 2 degrees quadrangle, Utah and Wyoming. U.S. Geological Survey Miscellaneous Investigations Series Map I-1997.

Bryant, B., 2010. Geologic map of the east half of the Salt Lake City 1 degree × 2 degrees (Duchesne and Kings Peak 30 × 60 quadrangles), Duchesne, Summit, and Wasatch Counties, Utah, and Uinta County, Wyoming (digitized and modified from U.S. Geological Survey Miscellaneous Investigations Series Map I-1997). Utah Geological Survey.

Bryant, B., Naeser, C.W., Marvin, R.F., Mehnert, H.H., 1989. Upper Cretaceous and Paleogene sedimentary rocks and isotopic ages of Paleogene tuffs, Uinta Basin, Utah. U.S. Geological Survey Bulletin 1787-J 22.

Catuneanu, O., 2006. Principles of Sequence Stratigraphy. Elsevier, Amsterdam, 375 p.

Crews, S.G., Ethridge, F.G., 1993. Laramide tectonics and humid alluvial fan sedimentation: NE Uinta Uplift, Utah and Wyoming. Journal of Sedimentary Petrology 63, 420–436.

Dickinson, W.R., 1970. Interpreting detrital modes of graywacke and arkose. Journal of Sedimentary Petrology 40, 695–707.

Dickinson, W.R., Beard, L.S., Brakenridge, G.R., Erjavec, J.L., Ferguson, R.C., Inman, K.F., Knepp, R.A., Lindberg, F.A., Ryberg, P.T., 1983. Provenance of North American Phanerozoic sandstones in relation to tectonic setting. Geological Society of America Bulletin 94, 222–235.

Dickinson, W.R., Klute, M.A., Hayes, M.J., Janecke, S.U., Lundin, E.R., McKittrick, M.A., Olivares, M.D., 1988. Paleogeographic and paleotectonic setting of Laramide sedimentary basins in the central Rocky Mountain region. Geological Society of America Bulletin 100, 1023–1039.

Dickinson, W.R., Lawton, T.F., Inman, K.F., 1986. Sandstone detrital modes, central Utah foreland region: stratigraphic record of Cretaceous–Paleogene tectonic evolution. Journal of Sedimentary Petrology 56, 276–293.

Folk, R.L., 1968. Petrology of Sedimentary Rocks. Hemphills's, Austin, 170 p.

Fouch, T.D., 1975. Lithofacies and related hydrocarbon accumulations in Tertiary strata of the western and central Uinta Basin, Utah. In: Bolyard, D.W. (Ed.), Deep Drilling Frontiers of the Central Rocky Mountains, Rocky Mountain Association of Geologists Symposium, pp. 163–173. Denver, Colorado.

Fouch, T.D., Nuccio, V.F., Anders, D.E., Rice, D.D., Pitman, J.K., Mast, R.F., 1994. The Green River petroleum system, Uinta Basin, Utah, U.S.A. In: Magoon, L.B., Dow, W.C. (Eds.), The Petroleum System—From Source to Trap, vol. 60. American Association of Petroleum Geologists Memoir, pp. 399–421.

Gazzi, P., 1966. Le arenarie del flysch sopracretaceo dell'Appennino modenese; correlazioni con il flysch di Monghidoro. Mineraologica e Petrografica Acta 12, 69–97.

Gillies, R.R., Ramsey, R.D., 2009. Climate of Utah. In: Banner, R.E., Baldwin, B.D., McGinty, E.L. (Eds.), Rangeland Resources of Utah (Revised), Utah State University, Cooperative Extension Service, pp. 39–45.

Greer, D.C., Director, 1981. Atlas of Utah. Brigham Young University Press, Provo, Utah, 300 p.

Hansen, W.R., 1986. Neogene tectonics and geomorphology of the eastern Uinta Mountains in Utah, Colorado, and Wyoming. U.S. Geological Survey Professional Paper 1356, 78.

Hartley, A.J., Weissmann, G.S., Nichols, G.J., Warwick, G.L., 2010. Large distributive fluvial systems; characteristics, distribution, and controls on development. Journal of Sedimentary Research 80, 167–183.

Hintze, L.F., Willis, G.C., Laes, D.Y.M., Sprinkel, D.A., Brown, K.D., 2000. Digital Geologic Map of Utah, vol. 1, p. 500000. Utah Geological Survey.

Ingersoll, R.V., Bullard, T.F., Ford, R.L., Grimm, J.P., Pickle, J.D., Sares, S.W., 1984. The effects of grain size on detrital modes: a test of the Gazzi-Dickinson point-counting method. Journal of Sedimentary Petrology 54, 103–116.

Jensen, D.T., Bingham, G.E., Ashcroft, G.L., Malek, E., McCurdy, G.D., McDougal, W.K., 1990. Precipitation Pattern Analysis Uinta Basin-Wasatch Front, Report to Division of Water Resources, State of Utah under Contract Number 90–3078. Office of the State Climatologist, Utah State University, Logan, Utah, 41 p.

Keighley, D., Flint, S., Howell, J., Moscariello, A., 2003. Sequence stratigraphy in lacustrine basins: a model for part of the Green River formation (Eocene), southwest Uinta Basin, Utah. Journal of Sedimentary Research 73, 987–1006.

Kelly, T.S., Murphey, P.C., Walsh, S.L., 2012. New records of small mammals from the middle Eocene Duchesne River Formation, Utah, and their implications for the Uintan-Duchesnean North American Land Mammal Age transition. Paludicola 8, 208–251.

Kraus, M.J., 2002. Basin-scale change in flood plain paleosols: implications for interpreting alluvial architecture. Journal of Sedimentary Research 72, 500–509.

Kraus, M.J., Hasiotis, S.T., 2006. Significance of different modes of rhizolith preservation to interpreting paleoenvironmental and paleohydrologic settings: example from Paleogene paleosols, Bighorn Basin, Wyoming, U.S.A. Journal of Sedimentary Research 76, 633–646.

Lambiase, J.J., 1990. A model for tectonic control of lacustrine stratigraphic sequences *in* continental rift basins. In: Katz, B.J. (Ed.), Lacustrine Basin Exploration—Case Studies and Modern Analogs, vol. 50. American Association of Petroleum Geologists Memoir, pp. 265–276.

Leier, A.L., DeCelles, P.G., Pelletier, J.D., 2005. Mountains, Monsoons and Megafans: Geology 33, 289–292.

Leopold, L.B., Wolman, M.G., 1957. River Channel Patterns: Braided, Meandering and Straight. U.S. Geological Survey Professional. Paper 282-B, 51 p.

McDowell, F.W., Wilson, J.A., Clark, J., 1973. K-Ar dates for biotite from two paleontologically significant localities: Duchesne River Formation, Utah, and Chadron Formation, South Dakota. Isochron/West 7, 11–12.

Nemec, W., Steel, R.J., 1984. Alluvial and coastal conglomerates: their significant features and some comments on gravelly mass-flow deposits. In: Koster, E.H., Steel, R.J. (Eds.), Sedimentology of Gravels and Conglomerates, vol. 10. Canadian Society of Petroleum Geologists, Memoir, pp. 1–31.

Picard, M.D., High, L.R., 1972. Criteria for recognizing lacustrine rocks. In: Rigby, J.K., Hamblin, W.K. (Eds.), Recognition of Ancient Sedimentary Environments, vol. 16. SEPM Special Publication, pp. 108–145.

Phillips, S., Little, L., Michael, E., Odell, V., 1997. Sequence stratigraphy of tertiary petroleum systems in the West Natuna Basin. In: Indonesian Petroleum Association, Petroleum Systems of SE Asia and Australasia, Conference Proceedings, pp. 381–390.

Prothero, D.R., Swisher, C.C., 1992. Magnetostratigraphy and geochronology of the terrestrial Eocene-Oligocene transition in North America. In: Prothero, D.R., Berggren, W.A. (Eds.), Eocene-Oligocene Climatic and Biotic Evolution. Princeton University Press, pp. 46–74.

Rowley, P.D., Hansen, W.R., Tweto, O., Carrara, P.E., 1985. Geologic map of the Vernal 1ø × 2ø quadrangle, Colorado, Utah, and Wyoming. U.S. Geological Survey Miscellaneous Investigations Series I-1526.

Sato, T., Chan, M.A., 2015. Fluvial facies architecture and sequence stratigraphy of the Tertiary Duchesne River Formation, Uinta Basin, Utah. Journal of Sedimentary Research 85, 1438–1454.

Shanley, K.W., McCabe, P.J., 1994. Perspectives on the sequence stratigraphy of continental strata. American Association of Petroleum Geologists Bulletin 78, 544–568.

Shanley, K.W., McCabe, P.J., 1998. Relative Role of Eustasy, Climate, and Tectonism in Continental Rocks, vol. 59. SEPM Sepcial Publication, p. 234.

Schull, T.J., 1988. Rift basins of interior Sudan: petroleum exploration and discovery. American Association of Petroleum Geologists Bulletin 72, 1128–1142.

Sprinkel, D.A., 2006. Interim Geologic Map of the Dutch John 30′ X 60′ Quadrangle, Daggett and Uintah Counties Utah, Moffat County, Colorado, and Sweetwater County. Utah Geological Survey, Wyoming.

Sprinkel, D.A., 2007. Interim Geologic Map of the Vernal 30′ X 60′ Quadrangle, Uintah and Duchesne Counties, Utah, and Moffat and Rio Blanco Counties. Utah Geological Survey, Colorado.

Sprinkel, D.A., 2013. Interim geologic map of the eastern part of the Duchesne 30′ × 60′ quadrangle, Duchesne and Wasatch Counties, Utah, year 1 of 6. Utah Geological Survey Open-File Report 625, 19, 1 plate, scale 1:50,000.

Visher, G.S., 1965. Use of vertical profile in environmental reconstruction. American Association of Petroleum Geologists Bulletin 49, 41–61.

Warner, M.M., 1965. Cementation as a clue to structure, drainage patterns, permeability, and other factors. Journal of Sedimentary Petrology 35, 797–804.

Warner, M.M., 1966. Sedimentational analysis of the Duchesne River formation, Uinta Basin, Utah. Geological Society of America Bulletin 77, 945–957.

Weissmann, G.S., Hartley, A.J., Nichols, G.J., Scuderi, L.A., Olson, M., Buehler, H., Banteah, R., 2010. Fluvial form in modern continental sedimentary basins. Distributive Fluvial Systems: Geology 39, 39–42.

CHAPTER

11

Changes in the Shape of Breccia Lenses Sliding From Source to Sink in the Cambrian Epeiric Sea of the North China Platform

A.J. (Tom) Van Loon[1], *Z. Han*[2], *Y. Han*[3]

[1]Geocom Consultants, Benitachell, Spain; [2]Shandong University of Science and Technology, Qingdao, China; [3]China University of Geosciences, Beijing, China

OUTLINE

http://dx.doi.org/10.1016/B978-0-12-803386-9.00011-3

1. INTRODUCTION

Transport of sedimentary particles can take place in two ways: by either grain-by-grain transport or mass transport. During the first decades of sedimentological research, there was hardly any attention for mass-flow transport, whereas grain-by-grain transport received much attention. Consequently, this mode of transport became increasingly well understood, and the various processes involved are now understood in great detail. It is therefore not surprising that much sedimentological research nowadays focuses on mass transport. This has resulted in a wealth of data regarding the various modes of mass transport, and it has become obvious that mass-transport processes are much more varied (and in many respects much more complicated) than the processes that are responsible for grain-by-grain transport. Moreover, new studies indicate that not all features of mass-transported sediments are known yet, so that it must be deduced that at least part of the processes involved in this type of transport are still insufficiently understood, if understood at all.

Here we deal with a feature—semiconsolidated breccia lenses that slid down a gently inclined slope—that has been described only twice before (Van Loon et al., 2012; Su et al., 2016), and that was found in a succession that accumulated in an environment (an epeiric sea) from which mass-transported sediments are hardly known. We describe the shape of the breccia lenses with the objective to interpret which processes were involved in this shaping.

2. GEOLOGICAL SETTING

The limestone lenses under study form part of the Late Cambrian (Furongian) Chaomidian Formation in eastern China (Chen et al., 2009a, 2011). This formation accumulated in an epeiric sea (Meng et al., 1997). The depositional conditions in epeiric seas are commonly quiet, as expressed by the presence of lime-mud layers, but regionally tides and wind-induced waves can affect the bottom, as expressed by the presence of oolites. Moreover, occasional storms can be so strong that they may lead to waves that can break up the bottom of the shallow sea, as expressed by the presence of breccia layers. The sea bottom has commonly a relatively low relief. These conditions tend to result during most of the time to low-energy accumulation of a sedimentary succession in which mass-transported event layers are commonly absent or rare. The North China Platform, where the sediments under study are located, is a representative example (Chen et al., 2009a, 2010). It formed on a stable craton, the Sino-Korean Block (Meng et al., 1997) (Fig. 11.1A), and the Cambrian-Ordovician of the platform consists of a thick (1800 m) succession of mixed siliciclastic and carbonate deposits (Meng et al., 1997; Chough et al., 2010) (Fig. 11.1B).

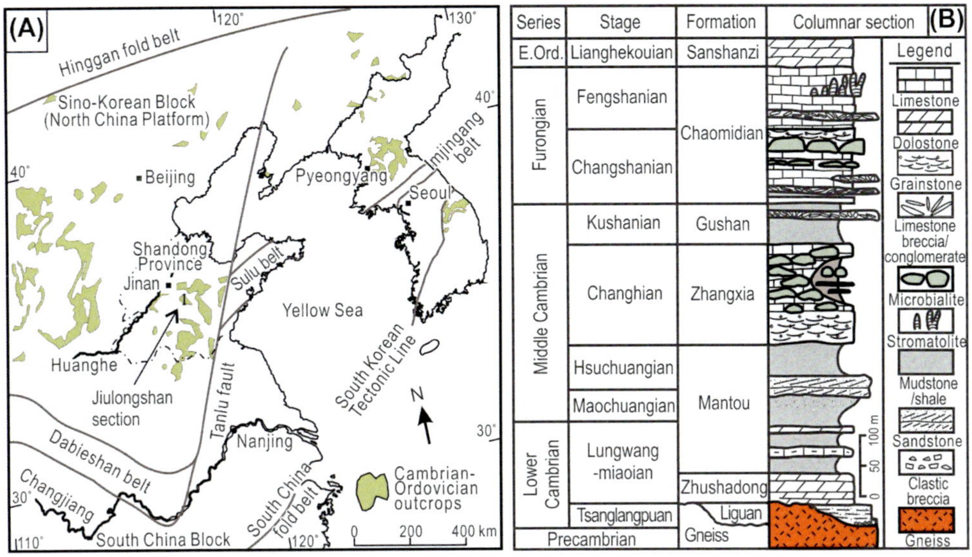

FIGURE 11.1 Characteristics of the study. (A) location map of the study area (modified after Chen et al., 2009a); (B) schematic Cambrian lithostratigraphy in Shandong Province, China.

The upper part of the Cambrian succession (Chaomidian Formation) consists mainly of various carbonate deposits (e.g., lime-mudstones, a spectrum of all compositions from wackestones to grainstones, and microbialites), and, as common in carbonate successions of epeiric seas, especially a number of limestone breccias and conglomerates (Chen et al., 2009a,b, 2011). Most breccias in the Chaomidian Formation are due to storm waves that affected the bottom of the epeiric sea, which was situated below fair-weather base. For more information about the depositional environment of the Chaomidian Formation and the carbonate platform on which it was deposited, we refer to Meng et al. (1997), Chough et al. (2001), Mei and Ma (2001), Kwon et al. (2002) and Chen et al. (2009a,b, 2010, 2011).

3. THE BRECCIA LENSES

We found eight breccia lenses, all in the same oolite layer, within an east–west (E–W) (085–265 degrees) trending wall of the Jiulongshan section (Fig. 11.2); this section is situated in Shandong Province. All these lenses have an identical composition and fabric. The lenses range in visible length from a few decimeters to several meters. They are mound-shaped to teardrop-shaped, have a tail, and are commonly underlain by a shear zone; the genesis of the lenses has been detailed by Van Loon et al. (2012); let it suffice here to mention that they were formed by the breaking up of a breccia layer. The thus formed lenses slid down the locally gently sloping sea floor (Fig. 11.3).

We should mention here that the term breccia is not entirely correct, because in addition to truly angular clasts, clasts also (and even more commonly) occur with rounded edges. The

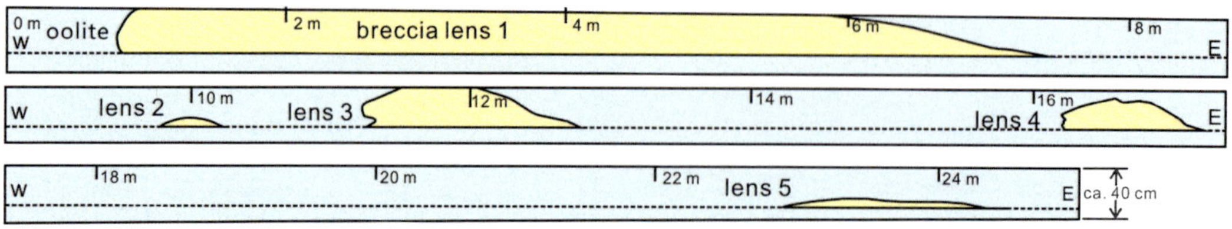

FIGURE 11.2 Overview of the oolite (light blue) with the five easternmost breccia lenses (1–5) (yellowish brown). The uppermost section shows the western part of the section under discussion, whereas the lowermost section shows the easternmost part (see the 2-m interval indications). Three more comparable lenses occur to the west of the upper diagram, the largest one being approximately 12 m long. These three westernmost lenses are not depicted here because they show characteristics that are identical to the five depicted here and described in the text.

common occurrence of angular fragments makes the term conglomerate also incorrect, however. The term breccias will be used here for the sake of simplicity. The type of breccia that constitutes the lenses in the oolite layer is oligomictic, clast-supported, with a marlstone matrix.

Although the majority of the clasts are flat, suggesting a consolidated or even lithified nature during the fragmentation process that resulted in a breccia, clasts of the same composition occur that show bending (Figs. 11.4 and 11.5), proving that at least part of the clasts were not lithified during fragmentation. We therefore conclude that the layer from which the clasts were derived was not in a lithified state when the brecciation occurred, but rather in a semiconsolidated to consolidated state.

All eight lenses have a flat, slightly irregular base and a roughly mound- to teardrop-shaped geometry (Fig. 11.6). Most lenses show truncation of their top parts (Fig. 11.7). The base of all eight lenses is situated at the same level in the oolite. This level can be traced from one breccia lens to another within the oolite because of the presence of a horizon (Fig. 11.8) with mainly angular fragments of the same composition as the clasts and matrix of the breccias and the oolite in which the breccia lenses are hosted.

3.1 Shapes and Characteristics of the Lenses

The shape of the breccia lenses is uncommon: they are not irregular (as might be expected if a layer is broken up), but rather tend to have a broadly rounded outline at their western ends and a tail (detailed later) on their eastern side (Fig. 11.9). They are thickest in their middle part.

The three westernmost lenses have characteristics that are identical to those of the five eastern lenses, apart from the fact that they originally must have had, like lenses 1 and 3 (see Fig. 11.2 for their position), a height that surpassed the present thickness of the oolite layer in which they are embedded: their top parts were abraded to the same level as this oolite layer. Because of the general resemblance, these three westernmost lenses will not be dealt with here in detail; let it suffice to mention that the westernmost lens is the largest of all, with a visible width (or length) of some 12 m. The five lenses dealt with in more detail here are numbered 1–5 (from west to east) in the following (see Fig. 11.2).

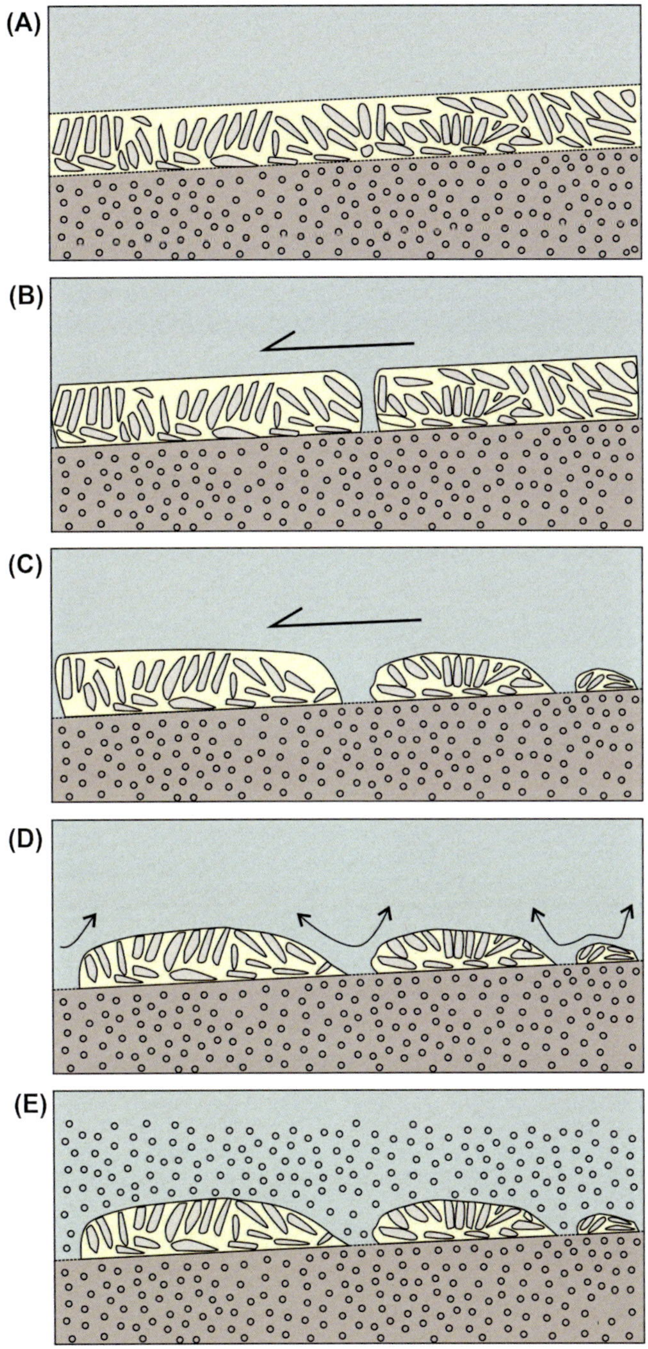

FIGURE 11.3 Genetic model of the breccia lenses (yellowish brown) that slid down over a slightly inclined sea floor during ongoing accumulation of oolite (light purple). (A) Formation of a breccia layer. (B) Initial break-up and subsequent sliding along a very shallow inclined surface of oolite. (C) Further break-up during sliding. (D) Abrasion by waves after sliding of the breccia lenses stopped. (E) Burial by ongoing accumulation of the host oolite.

FIGURE 11.4 Slightly bent platy clasts prove a semiconsolidated stage during the diagenetic stage when the breccias originated, but before the breccia layer started sliding and became fragmented.

FIGURE 11.5 Many of the breccias in the Jiulongshan section show (sub)horizontal platy clasts at the bottom, slightly to steeply inclined clasts in their middle part, and (sub)vertical clasts in their upper part. This occurred before the not-yet consolidated breccia became fluidized due to upward water/sediment escape under high pressure (see Van Loon et al., 2013). Before the breccia layer became fragmented, it had become semiconsolidated.

FIGURE 11.6 Breccia lens with the mound-like shape that characterizes most of the lenses. The internal fabric is chaotic, with numerous more or less vertical limestone clasts. The matrix is colored orange-brown. Note the flat lower boundary (sliding plane) with the underlying oolite; the boundary between the oolite and the underlying breccia is also plane and may represent an abrasion level. The top of the breccia lens is abraded.

FIGURE 11.7 Truncation of clasts at the top of lens 3 interpreted as abrasion by waves. The matrix between the large vertical clasts consists mainly of micrite. Several clasts are well rounded, indicating that they underwent individual transport (or movement by waves) before they became embedded in the breccia.

FIGURE 11.8 A thin horizon characterized by fragments of the breccia lenses and the underlying oolite connects the bases of the breccia lenses.

FIGURE 11.9 The tail (the well visible part is 35 cm long) of lens 3, which gradually thins to the east (= right). Note that the light gray fragments, because of their color, form only the best visible part of the tail. Above and below the gray fragments, the tail consists of darker fragments. The lowermost tail fragments are positioned at the same level as the base of the breccia lens.

3.1.1 Lens 1

The exposed part of the lens 1 is 6.58 m wide and has a maximum thickness of 33 cm. Its westernmost part is covered by the host oolite, but its thickness increases rapidly toward the east so that no oolite cover is present. In its middle part it has been abraded to the same level as the oolite in which it is embedded. At its eastern end, the breccia lens thins again toward its tail and is overlain by the oolite. All over its upper boundary, most of the clasts are truncated due to abrasion; wave activity must be held responsible.

At the base of the lens, locally a relatively strongly weathered zone is present with—in contrast to most of the lens—exclusively horizontal to subhorizontal clasts that have been broken into small fragments (Fig. 11.10). Moreover, this weathered zone under the lens contains not only breccia clasts (both grayish wackestone and brownish dolomitized lime mudstone) but also oolite clasts; the characteristics of this weathered zone indicate that it is a shear zone.

3.1.2 Lens 2

Lens 2 (Fig. 11.2) is the smallest lens, 45 cm wide and maximally 5 cm thick, showing a flat mound-shaped geometry. This small lens has neither a tail nor a shear zone underneath. Its top is not truncated by abrasion because it is below the level affected by the wave activity that removed the top part of lens 1.

3.1.3 Lens 3

The mound-shaped lens 3 (Figs. 11.4, 11.5, and 11.7) has a flat bottom. It is 132 cm wide and maximally 28 cm thick. Because of its relative thickness, it has been abraded in the middle of its upper part. At its eastern end, a tail is present.

The most obvious irregularity is at its western boundary, where oolite seems to penetrate the otherwise rounded breccia lens. At this place, the overall horizontally stratified oolite penetrating the lens shows a vague lamination dipping eastward, resembling the wedge-shaped structures that are formed in front of an advancing glacier that "bulldozes" the sediments forward (cf. Morawski, 2009).

FIGURE 11.10 The shear zone underneath lens 1 (diameter of pen 8 mm). The clasts consist of oolite with the same composition as the underlying oolite, and of grayish wackestones and brownish dolomitized lime mudstones that have the same composition as the clasts in the breccia lenses.

A shear zone, consisting of clasts with horizontal or eastward-dipping orientations, underlies the breccia body, suggesting movement with a component from east to west.

3.1.4 Lens 4

Lens 4 is only 91 cm wide and maximally 21 cm thick. It resembles lens 3 in almost each detail; since its height is less than that of the host oolite, its top has not been abraded, however. This lens shows, like lens 3, a westernmost rounded boundary that is somewhat irregular. A shear zone is visible under the breccia, and a tail is present, again on the eastern side.

An interesting aspect of this lens is that, some 30 cm west of its western end, a small north–south (N–S) trending section is exposed perpendicular to the general E–W trending wall. Part of the breccia lens is visible in this N–S trending section (Fig. 11.11). It starts 5 cm north of the E–W trending exposure and shows part of the breccia lens. A width of 35 cm is visible, and a maximum height of 28 cm at a place where the lens is still covered with 10 cm of oolite, but where the thickness decreases rapidly to the north, strongly suggesting only a small northward extent. This configuration suggests an orientation of the breccia lens of ~115–295 degrees (ESE–WNW).

3.1.5 Lens 5

Lens 5 is the last lens that we describe from this series (see earlier). It has a much lower height/width ratio than the other lenses, at 142 cm long and maximally 7 cm thick. The topmost clasts of the breccia are truncated. The western boundary is irregular and the eastern side features a tail. No shear zone is visible.

3.2 Shapes and Characteristics of the Tails

Four of the five lenses show a tail (only lens 2 does not). They are all present at the eastern end of the respective lenses (Fig. 11.12). They resemble the tails that have frequently been described from slump masses: they consist of fragments with the same composition as the

FIGURE 11.11 Western end of lens 4, where a roughly N–S orientated exposure is present perpendicular to the main E–W trending exposure, allowing a rough estimate of the elongation direction of the lens.

FIGURE 11.12 Detail of the tail at the eastern end of lens 3. Note that the individual fragments that constitute the tail are predominantly built of the matrix material of the breccia lens. Apparently the matrix could be destroyed more easily than the breccia clasts.

lens—with a very small fraction consisting of the material from the sediment over which the downslope movement occurred, here the oolite—and each of them forms an eastwards gradually thinning body.

It appears that the individual fragments in the tails decrease in average size from the lens body to the end of the tail. Moreover, the farther away from the lens body, the less fragments tend to form the tail (Fig. 11.8), until only small isolated fragments are left at the base level of the lenses. Eventually, only a horizon with some particles remains; almost all of these have a lens composition, but some rare particles with a composition that is similar to that the oolite may also be present. This horizon can be traced from lens to lens, thus proving that the bases of all lenses are situated in the host oolite at the same level.

3.3 Shapes and Characteristics of the Shear Planes

A fairly chaotic level occurs below lenses 1, 3, and 4 (Fig. 11.10). It consists of clast-supported angular fragments that represent both the material from the underlying oolite layer and that of the breccia lenses; in between these clasts, a poorly sorted matrix is present that is derived—as far as can be observed in the field— from the same sources. The clasts show a chaotic fabric (though almost all fragments have a horizontal to subhorizontal position) and at some places concentrations exist of clasts derived from the lenses, whereas clasts derived from the material below seem to predominate at other places.

This zone thus shows, though at a much smaller scale, the same characteristics as the shear zones found under large overthrusts (nappes), for instance, like in the Alps. We did not find any other feasible explanation for the feature under the lenses.

The shear planes have a somewhat irregular thickness, but form roughly tabular lithosomes. The maximum thickness is some 3 cm, but most commonly the thickness varies between 1.5 and 2.5 cm. The irregular thickness must be ascribed to the lithology of the breccia lenses: the brecciated components and the matrix had different degrees of consolidation (and perhaps some clasts were already lithified), so that the base of the lenses not only became differentially eroded, but also itself differentially eroded the oolitic substratum.

4. GENETIC INTERPRETATION OF THE LENSES AND THEIR SHAPES

The origin of the lenses has been discussed in detail by Van Loon et al. (2012). Let it suffice here to mention that a large block broke off from a breccia layer and started sliding over a gentle slope. During the sliding, this block broke into several pieces—ranging from a few dm to over 10 m (Fig. 11.3). The fragments came to rest in an area where oolites accumulated (Fig. 11.1). For more details, we refer to Van Loon et al.

5. CHANGE OF SHAPE DURING AND AFTER SLIDING

When a layer breaks up into pieces, the fragments tend to have a more or less platy shape with irregular side planes. The limestone lenses under study here, however, have shapes that are entirely different: (1) their top parts are either flat by abrasion or concave, (2) several lenses are underlain by a horizon with roughly horizontal clasts, (3) most of the lenses have a tail at their western end, and (4) the front (eastern) parts of the lenses tend to show a tear shape. These characteristics must have been obtained during or after the sliding process. The postdepositional processes are most easy to reconstruct, and therefore will be dealt with first.

5.1 Postdepositional Changes in Architecture

Two postdepositional processes played a role in the final shaping of the lenses: (1) the shaping of the upper surface of the lenses and (2) the formation and shaping of the tails.

5.1.1 Shaping of the Upper Surface

The lenses are embedded in an oolite. The oolite only extremely rarely shows any signs of current activity, so it may be deduced that their formation was classical: limestone precipitation around nuclei that were moved to and fro by wave activity. This indicates that the sedimentary surface was within reach of waves (possibly storm waves) during the more or less continuous oolite accumulation. During this accumulation the lenses became emplaced on the sedimentary surface, and it is only logical that they, too, were occasionally affected by storm wave activity. This is proven by the truncation of the relatively large lenses, and by the truncation of clasts at the top of the smaller lenses (Fig. 11.7). This explains the fairly flat upper surfaces of the lenses, particularly in the middle top parts.

Wave activity must also be held responsible for the convex upper surface of some of the smaller lenses: the side parts of the lenses were exposed most to the water movement (see Fig. 11.3), and consequently were affected most (comparable with the weathering of granites that are exposed to water percolating through their joints, resulting eventually in a concentration of ball-shaped granite blocks). The consequence is that the lenses obtained a more or less convex upper surface.

5.1.2 Formation and Shaping of the Tails

As explained in Section 3.2, the back sides of the lenses tend to have a tail. The clasts found in the tails are, like in shear zones, mainly fragments of the breccia lenses with—in the very lowest part of the tails—a small addition of oolite fragments. The bases of all tails are situated at the same level, which corresponds with the bases of the breccia lenses, and which can also be traced in the oolite between the lenses because of the presence of small breccia fragments (Fig. 11.8). This is proof that these parts of the tails consist of particles left by the sliding blocks due to friction at the sedimentary surface.

Some tails show, however, a feature that requires special attention: although they gradually thin out quickly, some tails, particularly close to the lens, can be so thick that they almost reach to the top of the lens. Friction at the sedimentary surface obviously cannot be an explanation. A second remarkable feature is that, in two cases, two tails are present behind a lens, starting at different heights, with normal oolite in between the two tails (Fig. 11.13).

Analysis of the components in these tails shows that all clasts situated more than about 1 cm above the sliding plane consist of material derived from the breccia lenses. This indicates that solely the lenses must be the source of the tails as far as above the sliding planes. The responsible process must therefore have been able to place fragments of a lens at its back side at a level that may be a decimeter or even somewhat more above the sliding plane. No structures have been found that indicate the presence of processes (e.g., upward escape of a water/sediment mixture, as a result of overpressurized pore water) that may lift fragments (sometimes with sizes of a few centimeters) from the sedimentary surface to such heights. It must therefore be concluded that the fragments came from above, and indeed, a

FIGURE 11.13 Detail of the succession just behind a lens. Two distinct tails are present, with oolites below, between, and above them. Note that some clasts are also present in the intermediate oolite level, suggesting that earlier eroded fragments of the lens were in an unstable position and fell off the lens during a phase of oolite accumulation.

process was present that was able to do so: the wave activity that eroded the upper (and side) parts of the lenses, giving them their convex upper surface. The wave-eroded fragments formed rubble that tumbled down from the higher parts of the lenses, thus forming some kind of talus, just like the talus alongside a reef. This genesis also explains why the tails become thinner with increasing distance from the lenses.

This formation as lens-derived abrasion fragments also explains the occasional presence of two tails (Fig. 11.13) separated by normal oolite at the back end of two lenses: wave abrasion (probably during a storm) must have produced rubble that formed a tail; then normal oolite accumulation took over, followed by a new phase of wave abrasion producing a tail on top of the oolite, followed eventually again by normal oolite accumulation.

It thus must be concluded that the tails represent material eroded by wave activity from the top and side parts of the lenses after these had come to rest at their final depositional site.

5.2 Changes in Shape During Sliding

The shapes and characteristics of the basis, the back side, and the front side of the lenses must be ascribed to processes that took place during their sliding. These processes are dealt with in Sections 5.2.1 and 5.2.2.

5.2.1 *Formation of the Breccia Level Under the Lenses*

As explained in Section 3.3, the level immediately under the lenses—but above the oolite—with mostly platy and horizontally positioned fragments of the breccia lenses (with some additional fragments of the oolite) must be explained as a shear zone that originated due to the friction between the sliding lenses and the oolitic substratum; the shearing was, by definition, a process that took place during the sliding of the blocks.

Shearing occurred because the friction at the sedimentary surface caused fragments of both the lenses and the autochthonous sediment to break off. This was facilitated by the lithology of the lens: the clasts must have had a resistance against attrition that differed from that of the matrix. The fragments that were set free were pushed into a roughly horizontal position by the moving breccia lens (Fig. 11.10). The breaking off of fragments from the base of the lens caused the commonly somewhat irregular lower surface of the lenses. This, in turn, may have increased the friction during sliding, and may thus have facilitated the breaking off of more fragments.

5.2.2 *The Tear Shape of the Frontal Part*

The remarkable tear-shaped frontal part of some of the breccia fragments is uncommon for limestone. The fronts resemble in several aspects the head parts of slumped masses, which owe their shape to two main processes: (1) the rotational movement of the highly vicious mass and (2) the resistance posed by the water to the moving mass. In the case of the limestone fragments under study, a rotational movement can be excluded, as the limestone lenses were, in spite of their semiconsolidated state, certainly not viscous but rigid. This leaves resistance met by the sliding blocks as the only possibility. Apparently the limestone masses were insufficiently consolidated during the sliding to remain underformed. The shearing at the sedimentary surface thus caused the vertically middle part of the limestone to move slightly faster than the bottom part.

FIGURE 11.14 Flattened and rounded nose of a lens, with clasts that seem to have become reorientated due to the stress field resulting from the friction between the sliding semiconsolidated breccia lens and the ambient water.

It is not yet completely clear whether the frontal top part of the slid-down lenses reached less far than the middle part only because of postdepositional abrasion [which certainly played a role (see Section 5.1.1)] or whether the friction with the water mass also slowed down the movement of the breccia mass, just as friction with the sedimentary stratum did for the lower part. It is interesting in this context that the fabric in the head parts of some of the lenses seems to have adapted to the resistance met: the elongated (platy) fragments in the breccia seem to have become reorientated (which was possible because of the semiconsolidated state of the breccias) according to the flattened and rounded nose of the lens (Fig. 11.14).

6. DISCUSSION

The Late Cambrian North China Platform was a typical epeiric platform (Meng et al., 1997), thus representing a low-relief environment where sliding is not typically expected. Sliding seems nevertheless the only satisfactory explanation for the features shown by the breccia lenses, because all other known geological processes and mechanisms that might explain the shape, position, and characteristics of the lenses appear inadequate. We discussed this earlier in detail (Van Loon, 2012) and will therefore only do so shortly in the following sections.

6.1 Possible Other Genetic Mechanisms

Concave sediment lenses can have different origins. Examples are (1) channel fills, (2) megaripples, (3) slumps, (4) microbialite structures (see Kiessling, 2003), (5) abraded lifted soft-sediment blocks, and (6) sedimentary sills. The following reasons indicate why these modes of genesis cannot be applied here.

1. The limestone lenses cannot be channels because the base of the lenses is horizontal whereas the top is convex, which cannot be ascribed to differential compaction, because the oolite does not show more compaction than the breccias. Moreover, the vertical position of a large number of clasts is not consistent with the fabric of clasts in a channel fill.
2. The lenses do not represent megaripples because in that case the clasts should preferentially be orientated according to the lee-side inclination. Moreover, megaripples with a length of some 12 m are difficult to explain in the shallow environment of an epeiric sea where the bottom is almost continuously affected by wave action. It must also be noted that no megaripples (pointing at a high-energy flow regime) are distinguishable in this section of the Chaomidian Formation where even current ripples (pointing at a lower flow regime) are extremely rare.
3. Slumps undergo rotational movement, whereas the clasts in the lenses were not reorientated according to the flowage pattern that characterizes slumps (cf. Van Loon, 1983).
4. The lenses are neither metazoan reefs nor microbial mounds, although some minor microbial infilling between some of the clasts is present. Moreover, talus from a microbial mound is not built of mainly vertically oriented fragments.
5. It is known that blocks of water-saturated sediment with a mass comparable to that of water (e.g., peat) can be uplifted by waves, transported over some distance, and deposited elsewhere (Van Loon and Wiggers, 1976), but semiconsolidated limestone breccias are too heavy. Moreover, no tails are formed in this case, and there would not be a distinct horizon with breccia clasts connecting the bases of the breccia lenses; finally, such a genesis cannot explain the shear zone under most breccia lenses.
6. A sedimentary sill (dyke) cannot explain the lenses because the clasts would show a preferential orientation according to the flow lines like crystals carried along in an intruding magmatic vein (Hiscott, 1979; Parize and Fries, 2003) whereas in the lenses, the clasts form clusters with other orientations; moreover, an intrusion would neither explain the flat lower boundaries of the lenses nor the tails consisting of isolated clasts. In addition, only the clasts in the upper (convex) margin of breccia lenses are truncated, which indicates that the breccia lenses were abraded after emplacement. This truncation implies that the lenses either were exposed at the sedimentary surface or positioned at such a shallow depth that wave action could abrade their topmost parts.

From these reasons it is clear that only a slump origin cannot be fully excluded. However, a slump requires plastic behavior and would have most likely resulted in deformed (rotated, contorted, or overturned) bedding inside the slump mass (Martinsen, 2003). The clasts, however, do not show any evidence of mutual displacement during slumping, but have retained a fabric similar to that in non-slumped breccias of the Chaomidian Formation which resulted from reorientation by upward escape of pore water and fluidized sediment (Chen et al., 2009a; Van Loon et al., 2012); it must therefore be concluded that the breccia lenses were at least semiconsolidated (this can also be deduced from the rounding of many of the clasts). If the mechanisms just described were not responsible for the formation of the breccia lenses inside the oolite, sliding of semiconsolidated blocks is the only process that can explain the various characteristics detailed in a satisfactory way.

6.2 Required Inclination of the Slide Plane

It has been reported before (Pedley et al., 1992) that sliding can occur on a carbonate ramp, but this sliding was ascribed to seismic shocks. No indications for seismic activity, however, have been found in the section under study here. Therefore, specific conditions must, for at least once during the long depositional history of the Chaomidian Formation, have initiated the breaking off of a breccia layer that then started to slide down along a sedimentary surface with a very low gradient.

Such a very gentle inclination does not pose a problem because even a very gentle slope is sufficient for mass transport (Gibert et al., 2005; Moretti and Sabato, 2007; Alsop and Marco, 2011); examples of slumping and sliding over nearly horizontal sedimentary surfaces (in other environments) have been described several times, also for inclinations of less than 1 degrees (e.g., García-Tortosa et al., 2011) and even less than 0.25 degrees (Field et al., 1982). Owen (1996) demonstrated very low-angle movement experimentally.

6.3 Fragmentation of the Breccia Layer

The occurrence of broken-up limestone layers is a rare phenomenon. Therefore an important question is which mechanism(s) triggered the initial break-up and subsequent sliding of the breccia fragments. Several mechanisms can do so in principle, including overloading, tsunamis, earthquakes, storms, and sea-level fluctuations (Spence and Tucker, 1997; Kullberg et al., 2001; Moretti and Sabato, 2007; Spalluto et al., 2007; Van Loon, 2009). As mentioned earlier, no evidence of seismic activity is present in the Chaomidian Formation in this region, which excludes an earthquake. No signs of a tsunami are present either, and the sediments below and above the host oolite of the breccia lenses do not indicate rapid sea-level fluctuations. This leaves cyclic wave overloading by storm-induced waves as the most likely trigger [cf. Bouchette et al., 2001; this is consistent with the restricted lateral occurrence (probably a few km^2)] of the breccia lenses.

It is not self-evident either why only one layer hosts a number of slid-down breccia blocks (we found only one badly exposed layer a few meters higher with just one such block, and one isolated slid-down block a few hundred meters further at a not precisely correlatable level in the formation). However we also may ask why descriptions of comparable features are extremely rare (Pedley et al., 1992), and why several blocks might slide down over the bottom of an epeiric sea. Since there are no clues from the field, an answer can be only speculative. Considering the many layers in the succession that have been broken up and now form breccia layers, storms must have been relatively frequent. Present-day observations indicate that storms with a specific power/magnitude decrease exponentially in number with increasing power/magnitude. Since the Chaomidian Formation covers more or less the entire Furongian (which lasted approximately 12 million years), numerous extremely heavy storms must have occurred (cf. Meng et al., 1986). Having no other clues, we consider the sliding of the breccia block (and its subsequent fragmentation) therefore as the probable result of cyclic wave overloading due to extremely heavy storms.

6.4 Source Area

In a study aimed at reconstruction of sediment transport from source to sink, attention should be paid to the source, in this case the original position of the breccia layer that has

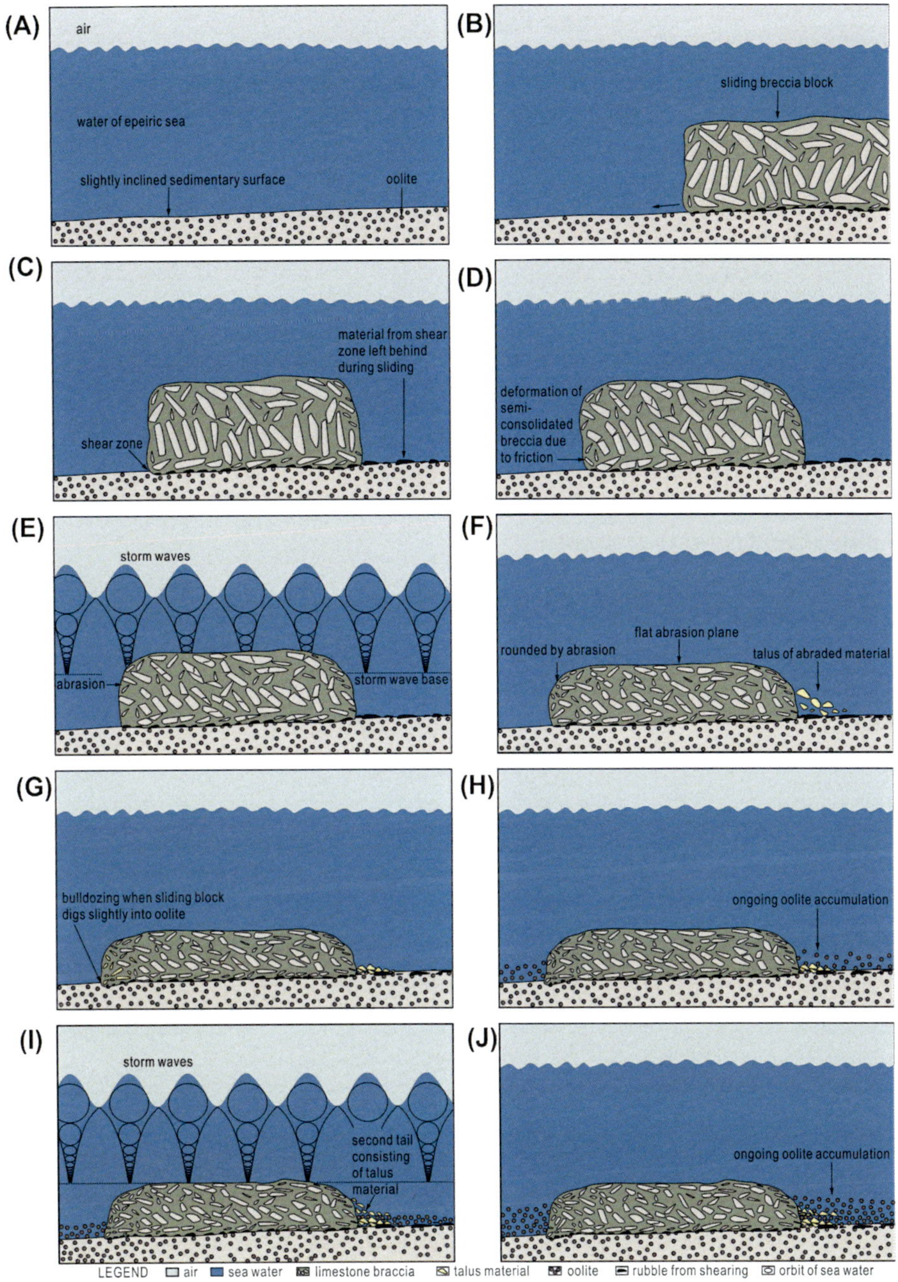

FIGURE 11.15 Model (not to scale) showing the presumed development of the reshaping of the lenses during sliding and after deposition. Note: Some of the phases described here may have taken place simultaneously. (A) Situation before emplacement of a breccia lens. (B) A breccia block arrives sliding over a gently inclined substratum consisting of oolite. (C) A shear zone, consisting of more or less horizontally lying angular fragments of both breccia and oolite develops between the sliding block and the autochthonous oolite. Some of this shear breccia is left on the sedimentary surface after the sliding block has passed. (D) The friction between the sliding block and the substratum results in deformation of the semiconsolidated breccia, resulting in a basal part that remains a bit behind the middle part, just like in slumps. (E) Storm waves reach the breccia lens, resulting in abrasion of the top part and rounding of the upper outer edges. (F) A talus of eroded breccia fragments if formed behind the lens. (G) Locally, the semiconsolidated breccia may, probably due to a shearing-induced irregular basis, dig slightly into the unconsolidated oolite substratum, causing some bulldozing of the oolite that partly is pressed into the breccia. (H) Oolite accumulation continues, covering the breccia talus. (I) A new storm results in further abrasion, resulting in a second talus that is deposited on the fresh oolite surface. (J) Oolite accumulation continues above the breccia fragment, thus embedding it completely.

the characteristics mentioned earlier for the various lenses. Since such breccias are common (dozens in this section), the type of source rock does not pose any problem: numerous breccias of this type (Fig. 11.13) are present in the Chaomidian Formation, both above and below the layer with the lenses. Unfortunately it cannot be checked whether a layer exists from which a large piece was broken off; the reason is that the transport direction, as deduced from the only 3D outcrop, indicates that the source area must have been located in a direction (probably ~115 degrees) where no hills are present that reach to this stratigraphic level. Considering the shear zone below several of the blocks, the sliding blocks must have met fairly much resistance. It is therefore likely that the transport distance was small, probably at most a few hundred meters.

An emplacement model for the breaking up and sliding of the lenses was already shown (Fig. 11.3); how the various fragments may have received their peculiar shape is shown in Fig. 11.15.

Acknowledgments

We gratefully acknowledge the financial support by the National Natural Science Foundation of China (40972043 and 41040018), the PhD Programs Foundation of the Ministry of Education of China (20093718110001), and the SDUST Research Fund (2015TDJH101).

References

Alsop, G.I., Marco, S., 2011. Soft-sediment deformation within seismogenic slumps of the Dead Sea Basin. Journal of Structural Geology 33, 433–457.

Bouchette, F., Seguret, M., Moussine-Pouchkine, A., 2001. Coarse carbonate breccias as a result of water-wave cyclic loading (uppermost Jurassic–South-East Basin, France). Sedimentology 48, 767–789.

Chen, J., Chough, S.K., Chun, S.S., Han, Z., 2009a. Limestone pseudoconglomerates in the Late Cambrian Gushan and Chaomidian Formations (Shandong Province, China): soft-sediment deformation induced by storm-wave loading. Sedimentology 56, 1174–1195.

Chen, J., Van Loon, A.J., Han, Z., Chough, S.K., 2009b. Funnel-shaped, breccia-filled clastic dykes in the Late Cambrian Chaomidian Formation (Shandong Province, China). Sedimentary Geology 221, 1–6.

Chen, J., Han, Z., Zhang, X., Fan, A., Yang, R., 2010. Early diagenetic deformation structures of the Furongian ribbon rocks in Shandong Province of China – a new perspective of the genesis of limestone conglomerates. Science China, Earth Sciences 53, 241–252.

Chen, J., Chough, S.K., Han, Z., Lee, J.H., 2011. An extensive erosion surface of a strongly deformed limestone bed in the Gushan and Chaomidian Formations (late Middle Cambrian to Furongian), Shandong Province, China: sequence-stratigraphic implications. Sedimentary Geology 233, 129–149.

Chough, S.K., Kwon, Y.K., Choi, D.K., Lee, D.J., 2001. Autoconglomeration of limestone. Geosciences Journal 5, 159–164.

Chough, S.K., Lee, H.S., Woo, J., Chen, J., Choi, D.K., Lee, S.-B., Kang, I., Park, T.-Y., Han, Z., 2010. Cambrian stratigraphy of the North China Platform: revisiting principal sections in Shandong Province, China. Geosciences Journal 14, 235–268.

Field, M.E., Gardner, V., Jennings, A.E., Edwards, B.D., 1982. Earthquake-induced sediment failures on a 0.25° slope, Klamath river delta, California. Geology 10, 542–546.

García-Tortosa, F.J., Pedro Alfaro, P., Gibert, L., Scott, G., 2011. Seismically induced slump on an extremely gentle slope (<1°) of the Pleistocene Tecopa paleolake (California). Geology 39, 1055–1058.

Gibert, L., Sanz De Galdeano, C., Alfaro, P., Scott, G., Lopez Garrido, A.C., 2005. Seismic induced slump in Early Pleistocene deltaic deposits of the Baza Basin (SE Spain). Sedimentary Geology 179, 279–294.

Hiscott, R.N., 1979. Clastic sills and dikes associated with deep-water sandstones, Tourelle Formation, Ordovician, Quebec. Journal of Sedimentary Petrology 49, 1–10.

Kiessling, W., 2003. Reefs. In: Middleton, G.V. (Ed.), Encyclopedia of Sediments and Sedimentary Rocks. Kluwer Academic Publishers, Dordrecht, pp. 557–560.

Kullberg, J.C., Oloriz, F., Marques, B., Caetano, P.S., Rocha, R.B., 2001. Flat-pebble conglomerates: a local marker for Early Jurassic seismicity related to syn-rift tectonics in the Sesimbra area (Lusitanian Basin, Portugal). Sedimentary Geology 139, 49–70.

Kwon, Y.K., Chough, S.K., Choi, D.K., Lee, D.J., 2002. Origin of limestone conglomerates in the Choson Supergroup (Cambro-Ordovician), mid-east Korea. Sedimentary Geology 146, 265–283.

Martinsen, O.J., 2003. Slide and slump structures. In: Middleton, G.V. (Ed.), Encyclopedia of Sediments and Sedimentary Rocks. Kluwer Academic Publishers, Dordrecht, pp. 666–668.

Mei, M.X., Ma, Y.S., 2001. Study on sequence Stratigraphy and sea-level changes of Late Cambrian in northern part of North China – discussion on the correlation of sea-level change with that of North America. Journal of Stratigraphy 25, 201–206 (in Chinese, with English abstract).

Meng, X.H., Qiao, X.F., Ge, M., 1986. Study on ancient shallow sea carbonate storm deposits (tempestite) in North China and Dingjiatan – model of facies sequences. Acta Sedimentologica Sinica 5, 1–18 (in Chinese, with English abstract).

Meng, X., Ge, M., Tucker, M.E., 1997. Sequence stratigraphy, sea-level changes and depositional systems in the Cambro-Ordovician of the North China carbonate platform. Sedimentary Geology 114, 189–222.

Morawski, W., 2009. Neotectonics induced by ice-sheet advances in NE Poland. Geologos 15, 199–217.

Moretti, M., Sabato, L., 2007. Recognition of trigger mechanisms for soft-sediment deformation in the Pleistocene lacustrine deposits of the Sant'Arcangelo Basin (Southern Italy): seismic shock vs. overloading. Sedimentary Geology 196, 31–45.

Owen, G., 1996. Experimental soft-sediment deformation: structures formed by the liquefaction of unconsolidated sands and some ancient examples. Sedimentology 43, 279–293.

Parize, O., Fries, G., 2003. The Vocontian clastic dykes and sills; a geometric model. In: van Resenbergen, P., Hillis, A.J., Morley, C.K. (Eds.), Subsurface Sediment Mobilization. Geological Society, London, pp. 51–72. Special Publications 216.

Pedley, H.M., Cugno, C., grasso, M., 1992. Gravity slide and resewdimentation processes in a Miocene carbonate ramp, hyblean Plateau, southeastern Sicily. Sedimentary Geology 79, 189–202.

Spalluto, L., Moretti, M., Festa, V., Tropeano, M., 2007. Seismically-induced slumps in Lower-Maastrichtian peritidal carbonates of the Apulian Platform (southern Italy). Sedimentary Geology 196, 81–98.

Spence, G.H., Tucker, M.E., 1997. Genesis of limestone megabreccias and their significance in carbonate sequence stratigraphic models: a review. Sedimentary Geology 112, 163–193.

Su, D.-C., Van Loon, A.J., Sun, A.-P., 2016. How quiet was the epeiric sea when the Middle Cambrian Zhangxia Formation was deposited in SW Beijing, China? Marine and Petroleum Geology 72, 209–217.

Van Loon, A.J., Wiggers, A.J., 1976. Primary and secondary synsedimentary structures in the lagoonal Almere Member (Groningen Formation, Holocene, The Netherlands). Sedimentary Geology 16, 89–97.

Van Loon, A.J., Han, Z., Han, Y., 2012. Slide origin of breccia lenses in the Cambrian of the North China Platform: new insight into mass transport in an epeiric sea. Geologos 18, 223–235.

Van Loon, A.J., Han, Z., Han, Y., 2013. Origin of the vertically orientated clasts in brecciated shallow-marine limestones of the Chaomidian Formation (Furongian, Shandong Province, China). Sedimentology 60, 1059–1070.

Van Loon, A.J., 1983. The stress system in mud flows during deposition, as revealed by the fabric of some Carboniferous pebbly mudstones in Spain. In: van den Berg, M.W., Felix, R. (Eds.), Geologie en Mijnbouw, vol. 62, pp. 493–498. Special issue in honour of J.D. de Jong.

Van Loon, A.J., 2009. Soft-sediment deformation structures in siliciclastic sediments: an overview. Geologos 15, 3–55.

CHAPTER

12

Provenance of Chert Rudites and Arenites in the Northern Canadian Cordillera

D.G.F. Long

Laurentian University, Sudbury, ON, Canada

OUTLINE

1. INTRODUCTION

Chert is a common, and often dominant component of sandstones and conglomerates in the Cretaceous of the Canadian Cordillera. It is abundant in the terrestrial components of retroarc foreland basins in Alberta and Northwest Territories (Eisbacher et al., 1974; Eisbacher, 1981; Ross et al., 2005; Raines et al., 2013), as well as in peripheral foreland basins and piggyback basins (*sensu* Busby and Ingersoll, 1995) within the interior of the Cordilleran orogen

Sediment Provenance
http://dx.doi.org/10.1016/B978-0-12-803386-9.00012-5

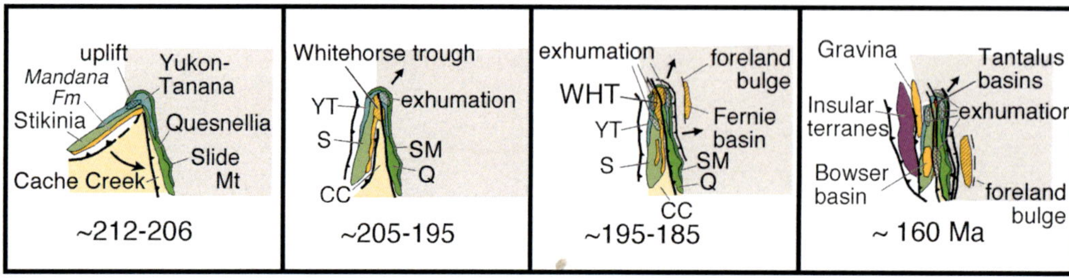

FIGURE 12.1 Model for the oroclinal closure of the Whitehorse trough based on Colpron et al. (2015). The 212–206 Ma paleogeography corresponds to deposition of the Lewes River Group in forearc settings. The 205–195 Ma interval reflects closure of the Cache Creek ocean, and deposition of the lower part of the Laberge Group. The 195–185 Ma interval reflects final closure and deposition of the upper part of the Laberge Group in a piggyback basin, fed largely from the north. The final post 160 Ma paleogeography reflects basin geometry during and after deposition of the Tantalus formation.

in British Columbia and Yukon (Eisbacher, 1981; Ricketts et al., 1992; Long, 2005; Evenchick et al., 2010). Where microfossils are well preserved in the cherts it is possible to identify a potential carbonate source (Cordey, 1992a,b), hence should provide a good record of chert-bearing strata (mostly carbonates) in the local provenance area.

Current models of Cordilleran amalgamation suggest that the Stikinia and Quesnellia terranes once formed a continuous arc system adjacent to the Yukon-Tanana terrane that became wrapped around a remnant ocean basin represented by the Cache Creek terrane (Fig. 12.1; Colpron et al., 2015). These combined terranes began to collide with the North American plate in Upper Triassic to Lower Jurassic times (Mihalynuk et al., 1994; Nelson and Colpron, 2007). Counterclockwise rotation of Stikinia around a flexural hinge north of the Whitehorse trough led to both subduction and obduction (Gordey and Stevens, 1994; Bickerton et al., 2013) of parts of the Cache Creek terrane, with subsequent enclosure of remnants of the Cache Creek ocean. Rotational collision, with some northward translation, continued until closure with Quesnellia in the Middle Jurassic. This allowed strata of the Laberge Group (Fig. 12.2 right) in the Whitehorse trough to accumulate initially in a forearc setting within a marginal ocean basin that was progressively transformed into piggyback basins, beginning at the north end of the embayment (White et al., 2012; Colpron et al., 2015), at the same time that arc-related rocks of the Hazelton Group accumulated to the south in northern British Columbia.

Strata of the Tantalus formation accumulated in a further series of strike-slip influenced piggyback basins that developed above strata of the Whitehorse trough during the Late Kimmeridgian to Valangian (Fig. 12.2). Most of the Tantalus formation consists of chert pebble conglomerate, and chert arenite of fluvial origin, with only minor feldspathic and volcanic components (Long, 1986, 2005, 2015; Long and Lowey, 2006). The abundance of chert clasts in the formation is problematic as underlying strata of the Laberge and Lewes River groups contain very little chert: the obvious source is the Cache Creek terrane, which currently lies to the south (Fig. 12.2 left), however both the observed paleocurrent information and maximum clast size trends indicate sources to the north of the Whitehorse trough (Long, 2015).

The object of this chapter is to evaluate how systematic petrographic observations and zircon geochronology can be used in combination with routine sedimentological observations to better evaluate potential source areas in a highly complex geotectonic setting.

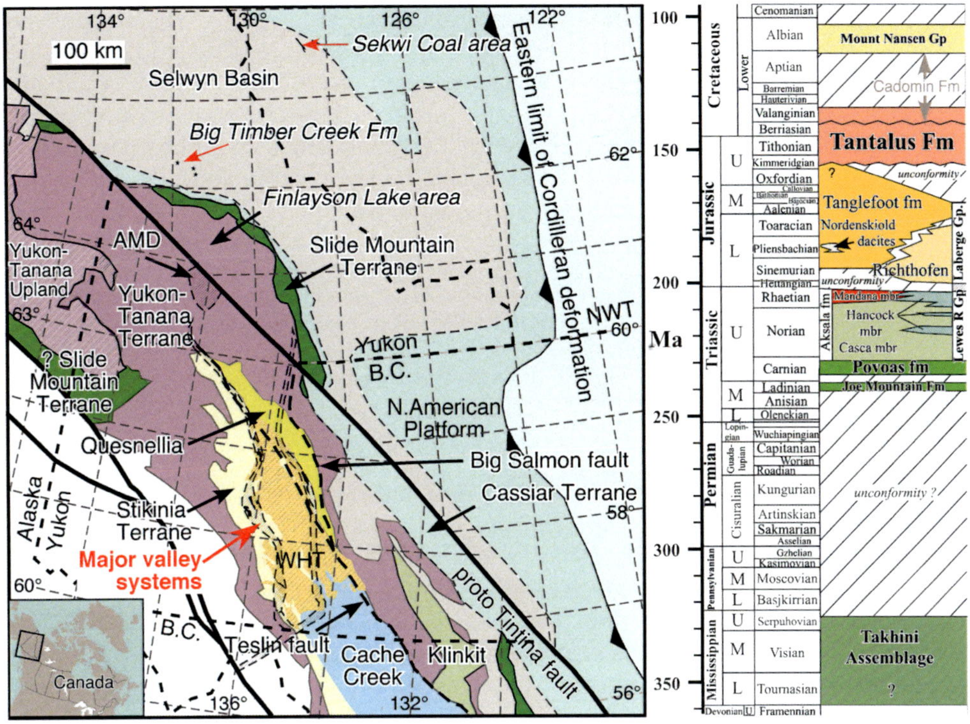

FIGURE 12.2 Left: Tectonic framework of the Whitehorse trough, adjusted for 430 km Eocene dextral stike-slip along the Tintina fault in the Eocene (Gabrielse et al., 2006), based on Colpron (2011) and Long (2015). The Whitehorse trough (Laberge Group) overlaps the Stikinia (west), and Stikinia (east), and may onlap the Cache Creek terrane to the south. Right: Stratigraphy of Mississippian to Early Cretaceous strata of Stikinia, based on Long (2015). Ages based on Cohen et al. (2013). Lower Cretaceous strata of the Big Timber Creek formation (Gordey, 2013) may represent early proximal foreland basin deposits developed during emplacement of the Slide Mountain terrane onto cratonic North America.

2. LITHOLOGY AND SEDIMENTOLOGY

Strata of the Tantalus formation (Bostock, 1936) occur in a number of narrow elongate basins that run parallel to major structures in the Whitehorse trough, and overlap strata within the Whitehorse trough, Stikinia, Quesnellia, and locally the Yukon-Tanana terrane (Fig. 12.2). In the type area, at Tantalus Butte, at the north end of the trough, the formation is at least 370 m thick. In the west central part of the trough the maximum preserved thickness is at least 1273 m. The minimum preserved thickness is less than 100 m at the southwest edge of the basin (Long, 2015).

Conglomerate forms the bulk of the exposed parts of the Tantalus formation (86% of exposed strata in all measured sections), with sandstones forming about 10%, and mudstones only 4%. Coal forms less than half a percent of the exposed parts of measured sections. The true abundance of mudstone and coal may be slightly underestimated, as these are more readily obscured by slope debris in some sections.

Long (2005, 2015) interpreted most of the extensive chert pebble conglomerate in the formation as the deposits of shallow (<3 m) and deep (>3 m) gravel-bed braided rivers, with local development of deep meandering gravel-bed rivers (comparable to models B, C, and F of Miall, 1996, p. 203). Locally gravel dominated Gilbert-type deltas developed where rivers debouched into floodplain ponds or lakes (Long and Lowey, 2006). Sandstone and mudstone are predominantly of floodplain origin, including levee, splay, marsh, swamp, and pond deposits. Coal deposits developed locally on abandoned segments of floodplains within confined river valleys, in places associated with high-constructive single- or multichannel (anastomosed) fluvial systems, with sand-filled channels (analogous to model J of Miall, 1996).

3. PETROGRAPHY

3.1 Conglomerates

Conglomerates of the Tantalus formation are dominated by well-sorted to moderately well-sorted, medium and large pebble conglomerate, with well-rounded clasts consisting predominantly of varicolored black, gray, white, and rare red and light green chert, with minor sandstone, igneous, and metamorphic clasts. Maximum grain size (intermediate diameter) is 260 mm (large cobble grade) in the northeast of the basin at Claire Creek (61°56′53″N, 135°22′23″W). Elsewhere maximum grain size is from 23 to 130 mm (medium pebble to small cobble grade). Petrographic analysis of 80 thin sections of conglomerate were made using the Gazzi-Dickinson point-counting method (Ingersoll et al., 1984) to avoid grain size bias. The average framework composition includes 13.4% quartz, 2.4% feldspar, and 84.2% lithic fragments (Fig. 12.3). The quartz is predominantly strained (72.7%), with almost equal abundance of monocrystalline (13.3%) and polycrystalline (12.7%) varieties. The lithic component is dominated by chert (91.3%), with minor igneous (4.6%), sedimentary (3.4%), and metamorphic rock fragments (<1%). In thin sections 48.5% (average) of the chert is white (range 10–100%), 35.5% is yellow to gray-brown (range 0–79%), 10.4% is black (range 0–61%), and 5.6% is gray (range 0–39%). When considered in terms of textural varieties, thin sections average 37% chert with spheres (range 0–87%) presumably representing casts of radiolarians, 35.7% have uniform massive textures (range 5–88%), and 28.1% have brecciated textures (range 2–95%) with multiple phases of chert cementation. Voids form an average of 1.1% of the 80 conglomerate samples examined. Cement-filled pore space makes up an average of 3.6% of the samples examined. Of this 2.7% is ferruginous cement (hematite and iron-sesquioxide), 0.7% calcite, and 0.2% quartz. Clay cement was observed in only one sample (average = 0.01%).

3.2 Sandstones

Sandstones form less than 10% of exposed strata in the Tantalus formation. Units associated with the conglomerates are typically moderately well sorted, medium to very coarse sand grade. Pebbly sandstones (i.e., with less than 30% gravel) are common. Finer grained sandstones are more commonly associated with floodplain facies. All the sandstone units in the Tantalus formation have a speckled "salt-and-pepper" appearance due to the abundance of chert grains. Petrographic analysis of 26 thin sections indicates an average

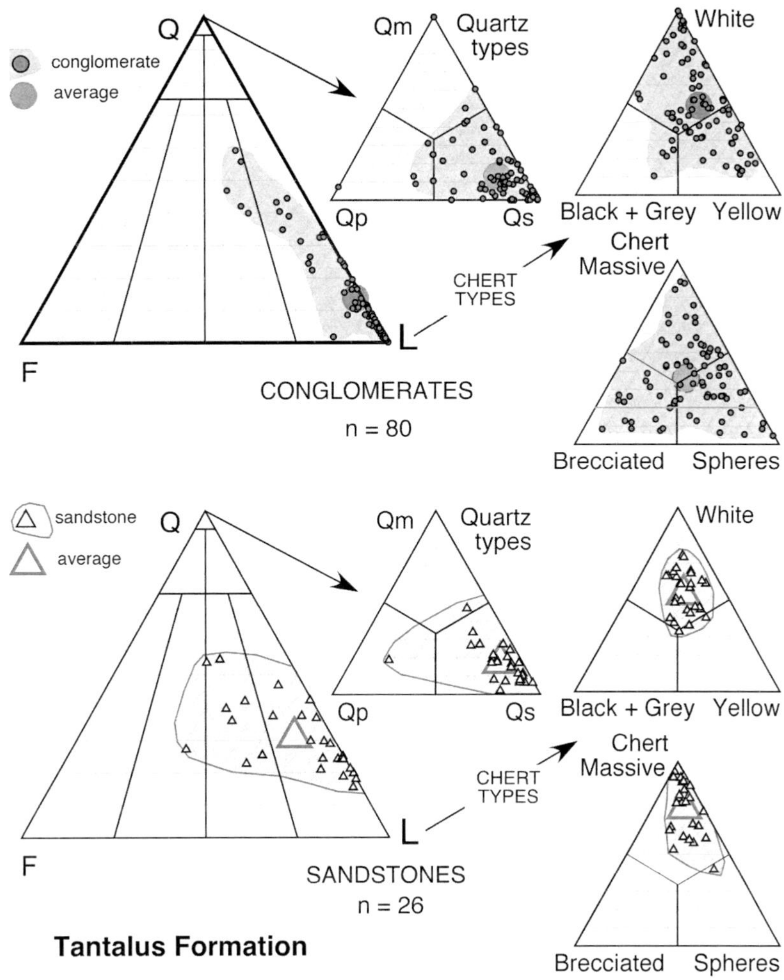

FIGURE 12.3 Framework composition of conglomerates (top) and sandstones (bottom) of the Tantalus Formation (*Q*, quartz, excluding chert; *F*, feldspars, including epimatrix; *L*, lithic fragments). Daughter triangles show ratios of quartz types (*Qm*, massive; *Qp*, polycrystalline; *Qs*, strained). *Data from Long, D.G.F., 2015. Provenance and Depositional Framework of Braided and Meandering Gravel-bed River Deposits and Associated Coal Deposits in Active Intermontane Piggyback Basins: The Upper Jurassic to Lower Cretaceous Tantalus Formation, Yukon, Canada. Yukon Geological Survey, Open File Report 2015-23. http://www.geology.gov.yk.ca*

composition of 31% quartz (range 16–55%), 9.7% feldspar (range 0–42%), and 58.9% lithic fragments (range 24–82%), of which 95.4% are chert, with minor igneous (2.9%), sedimentary (1.0%), and metamorphic (0.7%) rock fragments (Fig. 12.3). As in the conglomerates, the quartz is predominantly strained (average 71.3%). Monocrystalline grains appear to be slightly more abundant (average 17.8%) and polycrystalline quartz slightly less abundant (average 10.9%), although the overall range is similar (Fig. 12.3). In thin sections 53.5% of the chert is white (range 33–54%), 26.4% is yellow (range 13–44%), 10.3% is black (range

3–21%), and 9.4% is gray (range 3–27%). The overall distribution of chert color is more tightly clustered than in the conglomerates. Chert textures are likewise more tightly clustered, with 75.3% massive textures (range 42–75%), 16.8% containing spheres (range 2–47%), and 7.8% with brecciated textures (range 1–22%). Voids (empty pores) form an average of 4% of the samples examined. Filled pores form a further 7.2% (average) of the thin sections, including 4.6% ferruginous cement, 1.7% carbonate, and 0.9% quartz. Chlorite cement was detected in trace amounts in only one sample.

4. ZIRCON GEOCHRONOLOGY

Four samples of the Tantalus formation were collected as part of an extensive regional study of rocks in and around the Whitehorse trough (Colpron et al., 2015). Samples TB1 and TB2 are both from the open pit mine at Tantalus Butte (62°08′32″N, 136°15′59″W). Sample TB3 is from a stratigraphically lower level, exposed further south at Tantalus Butte. Sample C1 is from the upper half of the exposed section at the north end of Corduroy Mountain (61°20′33″N, 135°58′45″W). The Tantalus samples were analyzed by Dr. George Gehrels at the University of Arizona LaserChron Centre. Relative age probability plots and primary geochemical information for these and older detrital zircon samples shown in Fig. 12.4 are provided in Colpron et al. (2015), along with analytical methods. Interpreted ages are based on $^{206}Pb/^{238}U$ for <800 Ma grains and on $^{206}Pb/^{207}Pb$ for >800 Ma grains. Analyses that were >30% discordant (by comparison of $^{206}Pb/^{238}U$ and $^{206}Pb/^{207}Pb$ ages) or >5% reverse discordant were excluded. Interpreted ages are shown on relative age–probability diagrams (Fig. 12.4, left, following Ludwig, 2001) and as cumulative histograms (Fig. 12.4, right). The relative age probability diagrams show each age and its uncertainty (for measurement error only) as a normal distribution, and sum all ages from a sample into a single curve. Grains over 400 Ma have not been plotted as these represent less than 2% of the total population. Sample T1 contained two grains (1925 and 2261 Ma), sample TB2 one grain (2080 Ma), sample TB3 three grains (2039, 2059, and 2292 Ma), and sample C1 contained no grains over 400 Ma.

5. INTERPRETATION

5.1 Detrital Provenance

5.1.1 Quartz

Quartz makes up 13.4% of framework grains in conglomerates, and 31.4% in the sandstones. It consists mainly of strained varieties (Fig. 12.3), indicating that it may have been derived predominantly from highly deformed metamorphic sources, possibly within the Yukon-Tanana terrane (Fig. 12.1) or was deformed in situ by later tectonic processes. As conglomerates and sandstones in the underlying Tanglefoot formation (informal) also contain a similar ratio of strained quartz, much of the detrital quartz in the Tantalus could have been derived from underlying strata of the Laberge Group, or from a common source in deformed Permian and older plutonic rocks within adjacent strata of Stikinia, Quesnellia, and/or Yukon-Tanana.

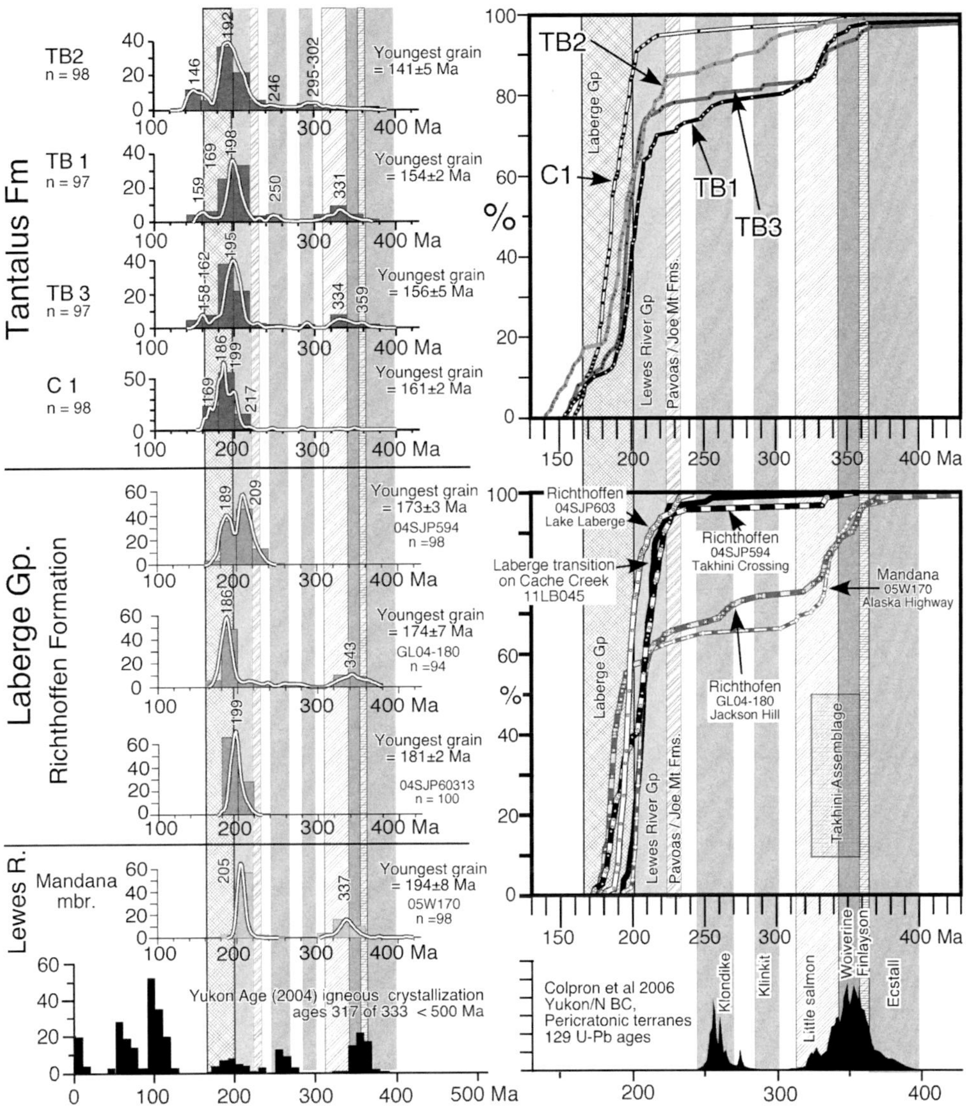

FIGURE 12.4 Age distribution of zircons from strata of the Whitehorse trough (Colpron et al., 2015) compared with crystallization ages from Yukon age (Breitsprecher and Mortensen, 2004), and U-Pb ages and events in Pericratonic terranes from Colpron et al. (2006). Data are presented as histograms and normalized age probability plots (left) and cumulative histograms (right). Ages over 430 Ma are not plotted.

Monocrystalline quartz grains are considered by Folk (1974) and Basu (1985), to be derived from undeformed igneous sources. They form 13.3% and 17.8% of the quartz content of the Tantalus formation conglomerates and sandstones, compared with 18.7% and 25.3% of conglomerates and sandstones in the underlying Laberge Group (Long, 2015). In both cases the most obvious primary source of igneous quartz would be Upper Triassic and early Lower

Jurassic intrusions in the roots of Stikinia, and adjacent parts of Yukon-Tanana and Quesnellia. In the case of the Tantalus formation some of this monocrystalline quartz was probably reworked from underlying strata of the Laberge and Lewes River groups.

The relative abundance of polycrystalline quartz is more than twice as abundant in the Tantalus formation (12.7% in conglomerates and 10.9% in sandstones) than in the Tanglefoot formation (4.7% in conglomerates and 4.5% in sandstones). Polycrystalline quartz is typically ascribed to annealing of metamorphic quartz during retrograde metamorphism: this could indicate contributions from Yukon-Tanana, or from metasedimentary rocks in the core of Stikinia or Quesnellia.

5.1.2 Feldspar

Feldspar, along with patches of epimatrix (patches of clay replacing feldspar: Dickinson, 1970), form an average of 2.4% of framework of conglomerate and 9.7% of grains in sandstone from the Tantalus formation. This is significantly lower than in the Tanglefoot formation where they form 25% of framework in the conglomerate and 33.4% of the framework in the sandstones. Microcline is present in only minor quantities, representing 0.5% of the conglomerate and 0.2% of the sandstone. Plagioclase (predominantly andesine) forms 2.2% of the conglomerate and 4.1% of the sandstone. Nontwinned potassium feldspars are the most abundant recognizable feldspar in both the conglomerate (30.4%) and sandstone (38.7%). It is notable that only 14% of the Tantalus formation conglomerate samples and 62% of the sandstone samples examined contained recognizable feldspars: this is increased to 48% and 84% when epimatrix is included. In contrast, conglomerates and sandstones in the Tanglefoot formation almost all contain some feldspar and/or epimatrix.

The similarity of feldspar types in the Tantalus and Tanglefoot formations indicates that they were probably derived from a common source (Upper Triassic and early Lower Jurassic plutons within Stikinia, Quesnellia, or Yukon-Tanana), or that the feldspars preserved in strata of the Tantalus formation were produced largely by reworking of the Tanglefoot formation. As microcline tends to be more resistant to weathering than orthoclase and plagioclase (Folk, 1974), the general paucity of microcline suggests that pegmatite was not especially abundant in the source area, even though they are locally abundant within the Aishihik batholith and younger plutons of the Long Lake plutonic suite, to the west of the Whitehorse trough (Johnston et al., 1996). Although identifiable orthoclase is far more abundant in the thin sections than is plagioclase, this may not reflect the primary detrital composition. In the Tantalus formation, 67% of the feldspar in the conglomerate and 57% of the feldspar in the sandstones have been converted to clay to form patches of epimatrix. This is much higher than in the underlying Tanglefoot formation, where only 16% of the feldspar in the conglomerate and 20% of the feldspar in the sandstones is represented by epimatrix, indicating more intense in situ weathering by groundwater following deposition and shallow burial. Feldspars probably converted to clays in a humid temperate setting (c.f. Bauluz et al., 2014). Folk (1974) and Nesbitt et al. (1997) suggest that microcline and orthoclase are more resistant to both weathering and in situ diagenetic alteration than plagioclase, consequently the relative abundance of plagioclase plus epimatrix in the Tantalus formation tends to support a dominantly volcanic-plutonic source area within Stikinia or Quesnellia, with more aggressive diagenetic alteration of feldspars in situ than in the Tanglefoot formation.

5.1.3 Nonchert Lithic Grains

Petrographic analysis indicates that nonchert lithic grains form 7.3% of the framework grains in conglomerates, and 2.7% of framework grains in the sandstones of the Tantalus formation (Table 12.1). They are most abundant in coarser grained conglomerates on the northeast side of the Whitehorse trough (Claire Creek in Table 12.1).

Nonchert lithic fragments in the Tantalus formation include sandstone, silicified mudstone, felsic igneous, and metamorphic rock fragments. Fine-grained igneous (plutonic and subvolcanic) fragments are by far the most abundant nonchert rock type (Table 12.1) and may have been derived from erosion of local strata within Stikinia and/or Quesnellia. Extrusive volcanic rock fragments are comparatively rare in the Tantalus formation, reflecting a tendency to break down during weathering and transport prior to burial, and may reflect deeper erosion of older arc sources in Stikinia and Quesnellia. Sedimentary rock fragments appear to be slightly more abundant in the conglomerate than the sandstone, suggesting that these were also labile (weakly cemented) and broke down readily during transport. Metamorphic rock fragments are significantly less abundant in the Tantalus formation than in the underlying Tanglefoot formation (Table 12.1), but may indicate a minor source in Yukon-Tanana.

5.1.4 Chert

Chert is by far the most abundant lithic fragment preserved in the Tantalus formation, making up 77% of the framework of the conglomerate and 56% of the sandstone. This is in marked contrast with the underlying Tanglefoot formation, in which only 16% of the framework grains in conglomerates and 5.5% in sandstones consist of chert. Chert is even less abundant in pre-Bajocian parts of the Laberge Group (Colpron et al., 2015). This means that most of the chert must be extrabasinal, and was not a significant component of the detrital supply prior to 170 Ma. Although 110 varieties of chert were observed during this study (Long, 2015), it is difficult to locate specific sources (Table 12.2). When chert types are grouped based on color (white, black and gray, and yellow) or texture (massive, brecciated, sphere bearing), there is little difference between different parts of the basin. In the conglomerates, brecciated and sphere-bearing chert types are slightly more abundant than in the associated sandstones (Fig. 12.3). Black, gray, and yellow brecciated and sphere-bearing chert varieties are notably less abundant in the sandstones than the conglomerates.

Wheeler (1961) suggested that the Cache Creek terrane, south of the Whitehorse trough, may have been the main source of chert in the Tantalus formation, as did Hart and Radloff (1990), who specifically indicated the Kedahda formation as a probable source (Table 12.2). Cordey (1992a) reported Permian radiolaria in a chert pebble from the Whitehorse Coal area on the southeast margin of the Whitehorse trough, but found only Middle to Late Triassic radiolaria in pebbles from the Carmacks area in the north. He suggested that these were most likely derived from the northeastern belt of the Cache Creek terrane in Southern Yukon and northern British Columbia, where both Triassic and Jurassic chert is present (Cordey et al., 1991), but Pennsylvanian and Permian radiolarian chert is more abundant (Monger, 1975). This implied southern provenance presents major problems, as none of the pebbles examined by Cordey (1992a) from Tantalus Butte contained any Pennsylvanian or Permian radiolaria. In addition, both paleocurrent trends and a southerly decrease in

TABLE 12.1 Absolute and Relative Abundance of Minor Grain Types: Igneous Includes Volcanic Plus Fine-Grained Plutonic Grains

Area/Grain Type	**Abundance**		**Sedimentary**		**Metamorphic**		**Igneous**	
Tantalus Fm	**Sst.**	**Cong.**	**Sst.**	**Cong.**	**Sst.**	**Cong.**	**Sst.**	**Cong.**
Mt Granger-Whitehorse coal	6.1 (2.4–14)	6.4 (0–20.4)	6.7 (0–19.4)	21.4 (0–79.4)	7.9 (0–15.7)	18.1 (0–97.4)	85.4 (80.4–92.3)	60.5 (0–100)
Braeburn-Kynocks Vowel Mountain	1.1 (0.2–3.6)	2.7 (0–40.0)	0.3 (0–0.9)	0.1 (0–2.0)	1.3 (0–3.8)	1.0 (0–10.9)	1.3 (0–3.8)	98.9 (88.2–100)
Carmacks	1.8 (0.2–3.2)	5.5 (0–18.4)	0.3 (0–0.6)	0.9 (0–6.4)	2.6 (0–14.0)	0.1 (0–2.0)	97.0 (85.2–100)	99.0 (97.3–100)
Hootalinqua Claire Creek	1.9 (1.0–3.0)	22.1 (1.0–71)	0.4 (0–1.0)	0.6 (0–4.9)	0 (0–0)	0 (0–0)	99.6 (99.4–100)	99.4 (95.1–100)
TANGLEFOOT FM								
Mt Granger	21.3 (5.4–55.8)	16.8 (15.4–18.8)	1.9 (0–11.8)	2.4 (0–7.4)	1.9 (0–9.1)	2.4 (0–7.4)	96.3 (86.3–100)	95.2 (85.2–100)
Braeburn-Kynocks Vowel Mountain	4.4 (0–8.0)	5.5 (0–19.2)	0.5 (0–1.9)	0.4 (0–0.9)	21.8 (10.2–27.3)	20.3 (0–25.5)	77.7 (72.0–89.8)	79.3 (73.6–85.6)
Carmacks	13.5 (5.6–21.4)	21.3 (0–74.5)	0 (0–0)	0.6 (0–3.3)	17.4 (9.1–24.4)	18.5 (0–35.5)	82.6 (75.6–90.9)	81.0 (64.5–96.7)
Hootalinqua Claire Creek	1.2	NA	0	—	10.2	—	89.8	—

Range indicated in brackets.

TABLE 12.2 Potential Chert Sources

Terrane/Age (L = Lower; M = Middle; U = Upper)	Unit	Red	Green	Gray	Black	White/ Clear	Yellow- Brown	Clay/ Mica	Radiolaria	References
NORTH AMERICAN PLATFORM (MACKENZIE MOUNTAINS)										
Permian	Fantasque Fm			X	X		X		X	MacNaughton (2002) and Martel et al. (2011)
Permian	Jungle Creek Fm					X				Bamber and Waterhouse (1971)
U Mississippian	Hart river Fm						X			Bamber and Waterhouse (1971) and Dixon (1992)
L Mississippian	Ford Lake Fm			X	X					Richards et al. (1997)
L Mississippian	Tischu group					X				Martel et al. (2011)
Devonian	Imperial Fm			X	X	X	X	X		Norris (1968) and Pyle and Jones (2009)
Devonian	Canol Fm			X	X		X	X	X	Mackenzie (1974) and Martel et al. (2011)
L-M Devonian	Sombre Fm				X					Martel et al. (2011)
U Silurian–L Devonian	Tsetso Fm (Delorme Gp)					X				Martel et al. (2011)
U Ordovician–Silurian	Mt Kindle Fm			X	X	X	X		X	Pyle and Jones (2009), and Martel et al. (2011)
Ordovician–Silurian	Bouvette Fm			X	X	X	X	X		Morrow (1999)
U Cambrian–L Ordovician	Franklin Mountain Fm			X		X	X			Pyle and Jones (2009) and Martel et al. (2011)
Cambrian	Slats Creek Fm					X				Morrow (1999)
Cambrian	Illtyd Fm			X	X			X		Gordey and Makepeace (2001)
Neoproterozoic	Coates Lake Gp.		X	X	X					Martel et al. (2011)
Mesoproterozoic	Mackenzie Mts. SGp.			X	X	X				Martel et al. (2011)
Paleoproterozoic	Wernecke SGp.		X	X	X					Delaney (1981)

(Continued)

TABLE 12.2 Potential Chert Sources—cont'd

Terrane/Age (L = Lower; M = Middle; U = Upper)	Unit	Red	Green	Gray	Black	White/ Clear	Yellow-Brown	Clay/ Mica	Radiolaria	References
NORTH AMERICAN SLOPE (SELWYN BASIN)										
L. Permian	Mount Christie F	Tr	X	X	X					Martel et al. (2011) and Gordey (2013)
Mississippian	Tay Fm				X					Gordey (2013)
U Devonian–L Mississippian	Prevost Fm			X	X					Martel et al. (2011)
Devonian	Portrait Lake Fm			X		Tr				Martel et al (2011) and Gordey (2013)
L Silurian	Steel Fm-Road river Gp		X	X	Tr	Tr				Gordey (2013)
M Ordovician–L Silurian	Duo Lake Fm-R. R. Gp			X	X					Martel et al. (2011), and Gordey (2013)
Late Cambrian–L Devonian	Rabbitkettle Fm		X	X	X					Gordey (2013)
Neoproterozoic	Yusezyu Fm				Tr					Gordey (2013)
SLIDE MOUNTAIN TERRANE										
Mississippian–L Permian	Campbell range formation	X	X						X	Plint and Gordon (1997), Pigage (2004), and Murphy et al. (2006)
U Pennsylvanian–L Permian	Fortin Creek Fm		X	X					X	Pigage (2004)
Mississippian–L Permian	Rose Mountain Fm	X	X	X	X	X	X		X	Pigage (2004)
Devonian–L Mississippian	Mount Aho Fm	X	X		X	X	X			Pigage (2004)
Southwest Yukon and Alaska										
Triassic	Wellesley Lake Fm	0058	X		X		X			Murphy et al. (2008)

YUKON-TANANA TERRANE								
U Triassic?	Faro peak Fm			X	X			Pigage (2004)
L Permian	Gatehouse Fm	X	X	X		X		Murphy et al. (2006)
L Permian	Money Creek Fm	X	X	X			X	Murphy et al. (2006)
Carboniferous	White Lake/ King Arctic fms	X	X			X	X	Murphy et al. (2006)
L Mississippian	Tuchitua river Fm							Murphy et al. (2006)
U Devonian	Waters Creek Fm							Murphy et al. (2006)
U Devonian–L Mississippian	Cleaver Lake Fm							Murphy et al. (2006)
Southern Yukon (Including Klinkit Assemblage)								
Triassic	Teh and Logjam Fms				X			Roots et al. (2006)
U Mississippian–M Permian	Little Salmon Fm	X	X	X				Colpron et al. (2006)
U Mississippian	Screw Creek Lst – Klinkit Gp	X	X					Roots et al. (2006)
Devonian–Mississippian?	Swift river group		X	X	X			Roots et al. (2006)
U Mississippian	Big Salmon complex	X						Mihalynuk and Peter (2001)
L Mississippian	Ram Creek complex		X	X		X		Roots et al. (2006)
Mississippian	Little Kalzas Fm	X	X	X				Colpron et al. (2006)

(Continued)

TABLE 12.2 Potential Chert Sources—cont'd

Terrane/Age (L = Lower; M = Middle; U = Upper)	Unit	Red	Green	Gray	Black	White/ Clear	Yellow-Brown	Clay/ Mica	Radiolaria	References
STIKINE TERRANE										
U Triassic	Hancock mbr, Aksala Fm			X		X				This study
M Triassic	Joe Mountain Fm									Hart and Orchard (1996)
L-M Pennsylvanian	Boswell + Semminof fms		X		X	X				Simard and Devine (2003)
L Permian	Ambition Fm, Astika Gp			X	X					Gunning et al. (1994)
Devonian–Permian	Stikine assemblage									Evenchick and Thorkelson (2005)
CACHE CREEK TERRANE										
M Triassic–L Jurassic	Teenah Lake assemblage		X	X	X	X			X	Jackson (1992)
Jurassic	Unnamed									Cordey (1991)
Triassic	Kedahda Fm	X	X		X	X	X		X	Cordey (1991) and Mihalynuk (1999)
Permian	Teslin Fm				X				X	Monger (1975)
Pennsylvanian-Permian	Horsefeed Fm			X	X				X	Monger (1975)
Pennsylvanian-Permian	Kedahda Fm		X	X		X	X		X	Monger (1975) and Bickerton et al. (2013)

maximum grain size of clasts within the Tantalus formation (Long, 2015) indicate that potential sources should be located north or northwest of the Whitehorse trough.

The exact source of individual chert varieties is difficult to identify based on petrography, using either color or texture. Most populations include clasts with spherules of microcrystalline chert, presumably representing recrystallized radiolaria. Others are uniform, or have patchy replacement textures, or are brecciated (Fig. 12.5). Chert is most abundant in platformal and proximal slope facies of cratonic North America, especially in slope facies of the Middle Ordovician to Silurian Road River Group (Gordey, 2013; Table 12.2). Radiolaria have been identified in the Permian Fantasque formation, as well as the Devonian Canol formation, and the Upper Ordovician to Silurian Mount Kindle formation (Bamber and Waterhouse, 1971; Beauchamp and Baud, 2002; Martel et al., 2011). These sources appear to have contributed to isolated bodies of chert pebble conglomerate in the Lower Cretaceous Big Timber Creek formation (Gordey, 2008, 2013). They may not have been available to the Tantalus rivers

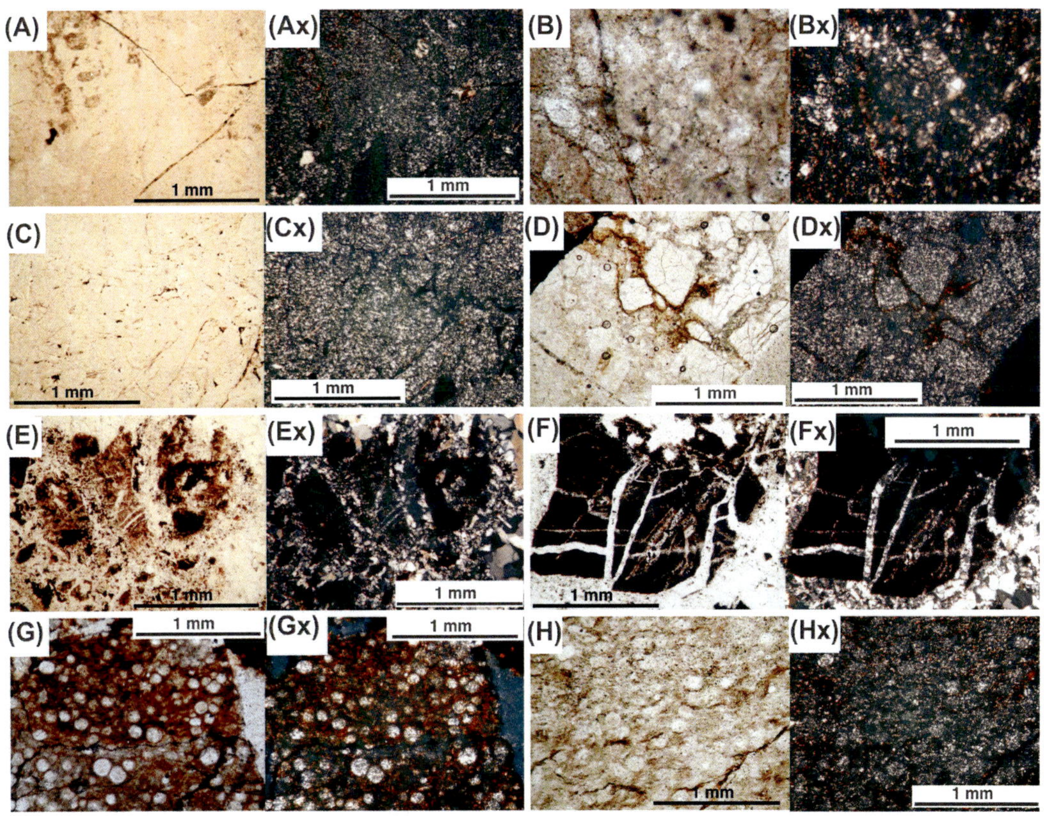

FIGURE 12.5 Representative chert types from the Tantalus Formation in plane-polarized and cross-polarized light. (A) Uniform nonstructured microcrystalline chert. (B) Uniform chert with minor clusters of quartz reflecting partial recrystallization. (C) Weakly brecciated uniform chert. (D–F) Chert breccias. (G) Chert with spherical microlites of chalcedony (possibly pseudomorphs after radiolarians). (H) Recrystallized sphere-bearing chert. *X* after letter indicates view in cross-polarized light.

during the late Upper Jurassic and early Lower Cretaceous as they would have been on the far side of the mountain ranges generated by thrusting of the Slide Mountain and Yukon-Tanana terranes onto North America (Fig. 12.1).

Varicolored radiolarian chert is a conspicuous component of the Slide Mountain terrane in Devonian to Lower Permian units, including the Mount Aho, Rose Mountain, Fortin Creek, and Campbell Range formations (Table 12.2; Plint and Gordon, 1997; Pigage, 2004; Colpron et al., 2006; Murphy et al., 2006). This would have been to the north of the Whitehorse trough at the time of deposition of the Tantalus formation, but does not contain any known Triassic chert. Remnants of the Slide Mountain terrane located west of the Yukon-Tanana terrane, about 100 km west of the north end of the Whitehorse trough (Israel et al., 2014), do contain varicolored Triassic chert in the Wellesley Lake formation (Murphy et al., 2008) and are potentially a source of sphere-bearing chert grains.

The Yukon-Tanana terrane, which surrounds the northern Whitehorse trough, contains minor varicolored chert of Devonian to Triassic age (Table 12.2). Unfortunately none of these has been shown to contain radiolaria. The Stikine terrane, which underlies and flanks the Whitehorse trough, contains minor chert-bearing units in the Devonian–Permian Stikine Assemblage (including the Permian Ambition formation in northern British Columbia; Gunning et al., 1994; Evenchick and Thorkelson, 2005), as do the Pennsylvanian Boswell and Semenof formations in Quesnellia, east of the trough, and Middle Triassic intervals within the Joe Mountain formation, within and west of the trough (Hart and Orchard, 1996; Simard and Devine, 2003). Within the Whitehorse trough, Triassic strata of the Hancock member of the Aksala formation (informal) contain minor gray and white chert, but in insufficient quantities to have been a major supplier to the Tantalus formation. Red (or pink) chert is present in 8% of the samples from the Tantalus formation, and makes up less than 1% of the total chert population (Long, 2015). Pink manganiferous chert is present locally in the Yukon-Tanana terrane to the east of the Whitehorse trough, in the Little Salmon area (Colpron and Reinecke, 2000; Colpron et al., 2006), and southeast of the trough in northern British Columbia (Big Salmon complex: Mihalynuk and Peter, 2001).

The Cache Creek terrane, located south of the northern Whitehorse trough, contains abundant varicolored cherts of Pennsylvanian to Jurassic age (Monger, 1975; Cordey, 1991; Jackson, 1992; Bickerton et al., 2013). It is considered to be the main supplier of chert to the Bowser Basin, which overlaps and lies above Stikinia in northern British Columbia, and developed following obduction of some of the Cache Creek terrane onto Stikinia in the late Lower to early Middle Jurassic (Evenchick and Thorkelson, 2005). For the currently exposed remnants of the Cache Creek terrane (south of the Whitehorse trough) to have provided significant quantities of chert to the northern Whitehorse trough, major north-flowing rivers would have had to develop east of Stikinia, within major orogen parallel intermountain valleys to link with rivers at the north end of the trough. This is considered unlikely, although the modern Columbia River in southern British Columbia, which follows the southern Rocky Mountain trench northward for about 375 km from Columbia Lake before diverting to the southwest toward Revelstoke, has a similar pattern.

The Wellesley Lake formation, located in the Slide Mountain terrane, northwest of the Whitehorse trough (Fig. 12.1), contains some varicolored Triassic chert (Murphy et al., 2008), although radiolaria have yet to be documented. Given that lithic fragments recovered from the Claire Creek exposures of the Tantalus formation at the north end of the Whitehorse trough contain pebbles and cobbles of schistose Yukon-Tanana material, it is possible that the Wellesley Lake formation could have been a source of at least some of the chert within the formation.

5.2 Zircon Provenance

Sixty-nine percent of the detrital zircon population in the four Tantalus formation samples fall between 220 and 170 Ma, consistent with reworking of older strata of the Laberge and Lewes River groups within the Whitehorse trough, supplemented by common sources in adjacent parts of Quesnellia and Stikinia (Figs. 12.1–12.4). Between 10% and 26% of the zircon populations in individual samples from the Tantalus formation correspond to parts of the Lewes River Group, and 36–80% correspond to depositional ages of the Laberge Group (Fig. 12.4). Using the Overlap-Similarity program described in Gehrels (2000) it is clear that the Tantalus Butte samples have peak positions that closely match (>0.8 similarity) the majority of samples from the Richthoffen and Tanglefoot formations (Table 12.3). Sample C1 from Corduroy Mountain has slightly weaker peak-similarity, except for one sample of the Tanglefoot formation from near Tantalus Butte (primary data from Colpron et al., 2015; Long, 2015). Peak overlap is less well defined, with some values over 0.7.

TABLE 12.3 Comparison of Zircon Populations in the Tantalus Formation and Underlying Rocks in the Whitehorse Trough, Using the Overlap-Similarity Program Described in Gehrels (2000)

SIMILARITY (location of peaks, 1 = 100% overlap)

Tantalus Fm (TB2)	TB2	0.85	0.87	0.78	0.65	0.77	**0.80**	0.76	**0.85**	**0.84**
Tantalus Fm (TB1)	0.57	TB1	0.91	0.74	**0.82**	0.76	0.79	**0.81**	**0.85**	**0.83**
Tantalus Fm (TB3)	0.63	0.72	TB3	0.79	0.76	0.78	0.79	**0.83**	**0.83**	**0.86**
Tantalus Fm C1	0.61	*0.47*	*0.41*	C1	*0.49*	0.79	0.78	0.74	0.75	**0.88**
Mandana mbr (05W170)	0.72	0.52	0.53	0.64	W170	0.60	0.71	0.73	0.76	0.68
Richthoffen mbr. (GL04-108b)	0.53	0.61	*0.42*	*0.41*	0.51	108b	0.76	0.68	0.74	**0.81**
Richthoffen mbr. (04SJP594)	0.61	*0.46*	*0.40*	0.79	0.68	*0.41*	P594	**0.80**	**0.91**	**0.87**
Richthoffen mbr. (04SJP603)	0.54	*0.40*	*0.35*	0.69	0.67	*0.36*	0.79	P603	**0.88**	**0.89**
Tanglefoot fm. Eagles Nest	0.79	0.62	0.54	0.76	0.82	0.56	0.68	0.72	Tang. E.N.	**0.90**
Tanglefoot fm, Tantalus Butte	0.70	0.54	0.51	0.68	0.70	0.52	0.52	0.54	0.79	Tang. T.B.

OVERLAP (abundance of peaks, 1 = 100% overlap of peaks)

Program available at http://www.geo.arizona.edu/alc, Higher numbers (*gray background*) indicate a higher degree of similarity in the position of peaks (values below 0.5 are not considered significant).

A minor peak at ~250 Ma in samples TB1 and TB2 coincides with ages from the Klondike schist, which occurs as remnants within the Yukon-Tanana terrane, north, east, and west of the trough (Fig. 12.2; Colpron et al., 2006), or alternatively represents a contribution from strata in the Cache Creek terrane, analogous to the Kutcho assemblage in northern British Columbia, which has yielded ages of 246 to 242 Ma (Childe and Thompson, 1997; Childe et al., 1998; Schiarizza, 2012). The minor peak in sample TB2 between 208 and 303 Ma could have been derived from rocks of the Klinkit assemblage, which represents part of Yulon-Tanana, preserved within the roots of Quesnellia, east of the trough (Simard et al., 2003; Colpron et al., 2006; Beranek and Mortensen, 2011), or from some as yet undated source in the Slide Mountain terrane further to the northwest (Murphy et al., 2008). The prominent secondary peak (~7%) at 342 to 211 Ma in samples TB 1 and TB3 may likewise have an eastern source, in rocks of the Little Salmon complex, also part of Yukon-Tanana trapped in the roots of Quesnellia (Simard et al., 2003). This older peak is also evident in some samples from the Richthoffen formation (informal) and Mandana member of the Aksala formation (Fig. 12.4), so may in part have been derived from rocks within the trough. The absence of peaks between 208 and 342 Ma in sample C1 indicates that these eastern sources did not contribute significantly to strata on the far western side of the trough, at least during the early phases of deposition of the Tantalus formation.

The apparent absence of Archean zircons in all the Tantalus samples suggests that proximal cratonic strata from North America did not contribute large volumes of detritus to the Tantalus rivers. Only six Paleoproterozoic zircons (2292–1925 Ma) were recorded from the Tantalus formation (Fig. 12.6), all of which were from samples collected at Tantalus butte. This limited distribution of Rhyacian and Orosirian grains may indicate a source in pericratonic strata within Yukon-Tanana (Nelson and Gehrels, 2007; Beranek and Mortensen, 2011). Comparison of grains over 1 Ga in the Tantalus formation (Table 12.4) with pericratonic strata indicate a weak similarity of peaks with the Swift River assemblage, and weak overlap of abundance with rocks from the Dorsey complex in British Columbia (Ross and Harms, 1998). Alternatively the six Paleoproterozoic grains may represent inherited cores from Paleozoic intrusions within Yukon-Tanana, or from Upper Triassic to Lower Jurassic plutons that intrude Stikinia, Quesnellia, and the Yukon-Tanana terrane (Mortensen, 1990; Colpron et al., 2006).

Eleven percent of all zircons recovered from the Tantalus formation have ages younger than 170 Ma (8–17% of individual samples), with peaks at 169 to 146 Ma (Fig. 12.4). The youngest grains in individual samples are from 161 ± 2 to 141 ± 5 Ma, and directly overlap the suspected time of deposition of the unit (Fig. 12.2). The Yukon Age database (Breitsprecher and Mortensen, 2004) contains no record of igneous crystallization ages within the time span of 156 ± 6 to 141 ± 5 Ma in the Yukon (see histogram on lower left of Fig. 12.4). Undated rocks of this age may be present west of the Denali fault in the Insular terranes of Wrangellia and Alexandria (c.f. Gehrels et al., 2009). Rocks of this age are known from the southern part of Stikinia in central British Columbia (Evenchick et al., 2007, 2010). Cretaceous strata within the Bowser Basin in northern British Columbia are also known to contain zircons of the same age as the host strata: for example, strata in the upper part of the Todagin formation contain an ash horizon dated at 158 ± 1 Ma, and strata of the Devils Claw Formation contain an ash horizon dated at 141 ± 1 Ma (McNicoll, personal Communication, 2007), hence a common volcanic source, in the Skeena arch, south of the Bowser Basin, is possible (McNicoll et al., 2005;

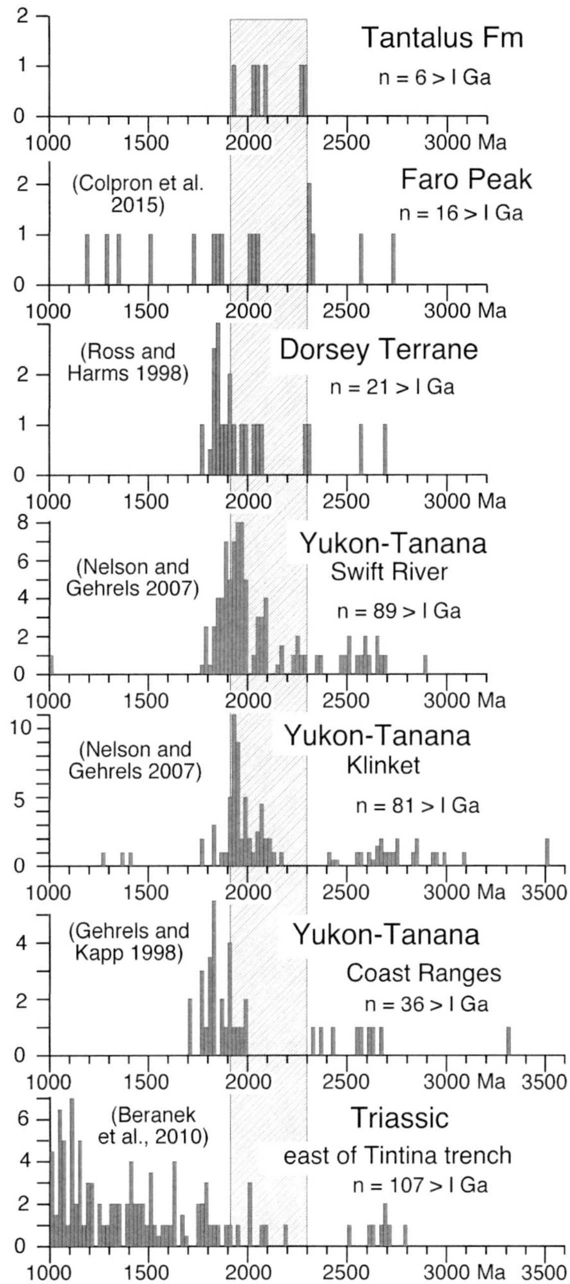

FIGURE 12.6 Histograms of >1 Ga zircons (20 Ma bins) from the Tantalus Formation, compared with strata west of the Tintina Trench, and in the Dorsey and Yukon-Tanana terranes.

TABLE 12.4 Comparison of Zircon Populations >1 Ga in the Tantalus Formation and Selected Pericratonic Strata West of the Tintina Trench, and in the Dorsey and Yukon-Tanana Terranes, Using the Overlap-Similarity Program Described in Gehrels (2000)

Ages over 1 GA	SIMILARITY (location of peaks, 1 = 100% overlap)						
All Tantalus at Tantalus Butte	Tantalus	*0.41*	0.52	0.61	*0.48*	*0.19*	*0.19*
Faro Peak	*0.46*	Faro Peak	0.55	0.71	0.65	*0.46*	0.56
Dorsey	0.64	0.52	Dorsey	0.70	0.59	0.58	*0.30*
Yukon Tanana Swift River	*0.43*	0.78	*0.48*	Swift R	0.82	0.59	*0.45*
YukonTanana-Klinket Assemblage	*0.25*	0.74	*0.34*	0.65	Klinket	0.52	*0.46*
YukonTanana-Coast Ranges	*0.27*	*0.38*	0.54	*0.39*	*0.34*	Coast	*0.31*
Triassic E of Tintina trench	*0.32*	0.74	*0.41*	0.73	0.64	0.33	Triassic
	OVERLAP (abundance within peaks, 1 = 100%)						

Program available at http://www.geo.arizona.edu/alc, Higher numbers indicate a greater degree of similarity in the position of peaks (values below 0.5 are not considered significant).

Evenchick et al., 2007, 2010). The younger grains in the Tantalus formation could likewise have been transported to the drainage basin as wind-borne volcanic ash from a westerly point source within the Coast Ranges, or from southerly point sources in central British Columbia.

6. DISCUSSION

Strata of the Tantalus formation were deposited in a series of isolated terrestrial inter-mountain successor basins during collision-induced uplift and deformation of underlying strata within the Whitehorse trough (Long, 2015). As such, they represent piggyback basins (Mihalynuk et al., 1994, 2004; White et al., 2012; Bickerton et al., 2013; Colpron et al., 2015). Parallelism of the preserved basins with major northwest oriented faults may imply limited strike slip activity within and along the local basin margins (Tempelman-Kluit, 2009). As erosional remnants of the Tantalus formation are confined to small, elongate basins, predominantly along the western and eastern sides of the Whitehorse trough, these may have developed as two separate axial trunk rivers, whose positions were influenced by right lateral strike slip during late stages of collision (Fig. 12.2). Tectonic discrimination diagrams (Dickinson and Suczek, 1979; Dickinson et al., 1983; Dickinson, 1985) demonstrate the hybrid nature of the basins, showing most data straddling several tectonic fields, depending on where chert is included in the plots. The scatter on the standard QFL plot is somewhat reduced by including chert at the quartz pole (Fig. 12.7A and B) such that the average value for the Tantalus formation lies within the continental interior field. Plots of monocrystalline quartz (Qm) against total feldspars (including epimatrix) and noncarbonate rock fragments

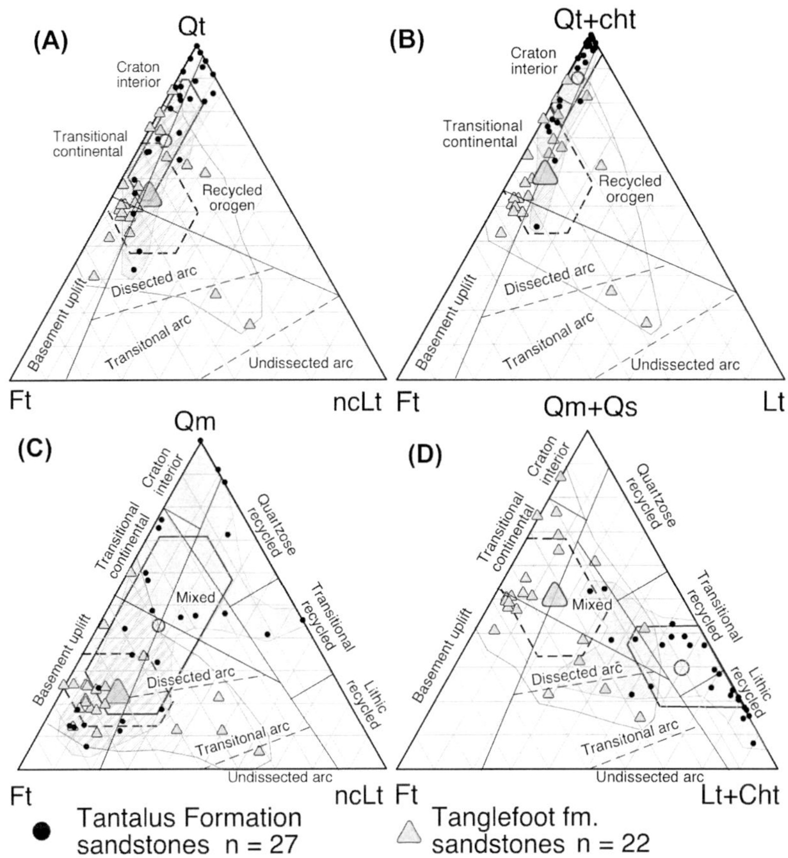

FIGURE 12.7 Tectonic discrimination diagrams for sandstones of the Tantalus Formation (dots, $n = 27$) and Tanglefoot formation (triangles, $n = 22$). *Qt*, total quartz (excluding chert); *Ft*, total feldspar (including epimatrix); *ncLt*, total noncarbonate rock fragments (excluding chert); *Lt* + *Cht* = total rock fragments (including chert). *Large symbols* indicate averages: *polygon* indicates one standard deviation from average.

excluding chert (ncLt), show a marked difference from plots in which the strained quartz is included at the Qm pole and chert is included at the Lithic pole (Fig. 12.7C and D), where the average value for the Tantalus formation falls in the transitional, recycled field.

The dominance of quartz and chert in all samples of the Tantalus formation, combined with in situ decomposition of feldspar during diagenesis, indicates that both the drainage basin and depositional basin were sites of intense chemical weathering in a humid setting (Amorosi and Zuffa, 2011). As labile (chemically unstable) grains would have been rapidly broken down within the soil in the catchment area, and in alluvium in the trunk streams, the preserved (nonchert) lithic fragments are probably more representative of the petrography of local sources than the upper reaches of the drainage basin. This is evident from the dominance of fine-grained igneous rock fragments in the Tantalus formation (Table 12.1) that are assumed to have a very local provenance both within the Whitehorse trough and in adjacent arc terranes. Limestone fragments were not present in any of the samples examined, despite the abundance of chert. A more distant, extrabasinal component is potentially represented by the presence of both sedimentary and metamorphic rock fragments, although these are not especially diagnostic of any specific terrane. The relatively high abundance of these grain types in strata in the southwestern part of the trough (Table 12.1: Mt Granger–Whitehorse Coal area) may indicate local provenance in rocks within Stikinia and/or Yukon-Tanana to the north or northwest of the trough.

Given the southerly trend of paleocurrents and a southerly decrease in maximum grain size in strata of the Tantalus Formation documented by Long (2015), it is considered unlikely that any of the known (preserved) chert sources discussed earlier and in Table 12.2 contributed more than minor amounts of lithic material to the Tantalus basins. The most viable alternative is that much of the chert was derived from a segment of the Cache Creek terrane that had been emplaced onto the Yukon-Tanana terrane north of the Whitehorse trough during accretion of Stikinia and Quesnellia onto the North American craton, and has since been completely eroded. Colpron et al. (2015) suggest that counterclockwise rotation of Stikinia during closure of the Whitehorse trough may have led to obduction of parts of the Cache Creek terrane, with major exhumation of terranes surrounding the northern end of the trough beginning in the late Norian to Rhaetian. This continued through closure of the Whitehorse trough, with obduction of the combined terranes onto North America, and eventual deposition of the Tantalus Formation (Fig. 12.1). Evidence for crustal stacking in the area north of the trough comes from the presence of high-grade metamorphic rocks in Yukon-Tanana, both to the north in the Stewart River area (Berman et al., 2007), and to the north-northeast on the far side of the Tintina trench, in the Finlayson Lake area (Staples et al., 2013, 2014). The Finlayson Lake area is currently located to the east of the Whitehorse trough, but was originally located to the north, prior to ~430 km of dextral strike slip movement along the Tintina fault in the Eocene (Gabrielse et al., 2006).

Most of the Yukon-Tanana west of the Tintina fault was metamorphosed in the Upper Permian (Longpingian) and Lower to Middle Triassic (260–239 Ma: Berman et al., 2007), with a further phase of burial between 195 and 187 Ma, followed by rapid exhumation. To the west of the trough Johnston et al. (1996) suggested rapid uplift following emplacement of the Aishihik batholiths during the Pleinsbachian at ~186 Ma, and intrusion of the Long Lake plutonic suite at ~187 Ma. They also suggested that cooling ages of 160–165 Ma might indicate a second burial event associated with obduction of part of the Cache Creek terrane in this area. This could conceivably have provided a source of chert to the northwest of the Tantalus basins. Staples et al. (2013, 2014) identified a broad area of Middle Jurassic to early Lower Cretaceous (Berrisian; 169–142 Ma) prograde metamorphism within Yukon-Tanana in the Finlayson Lake area, east of the Tintina fault, that may indicate burial to ~25 km. West of the Tintina trench, Staples et al. (2013, 2014) identified a possible core complex in the Australian Mountain domain where prograde metamorphism occurred at ~30 km depth during the Lower Cretaceous (Berrisian to Aptian: 146–118 Ma). The presence of these high-grade metamorphic rocks in the area north of the Whitehorse trough (AMD in Fig. 12.2) may indicate that the Yukon-Tanana terrane in this area could also have been overridden by a plate of Cache Creek terrane during closure of the Whitehorse trough, and hence provide a viable source for the abundant chert in the Tantalus formation.

Although the chert component within the Tantalus formation can be explained by a northern source in a now-eroded remnant of the Cache Creek terrane, the zircon data are dominated by local sources within the Whitehorse trough and adjacent arc terranes of Stikinia and Quesnellia. As the existing remnants of the Cache Creek did not contain abundant felsic strata it is to be expected that a distinct zircon signature of this terrane may not have been preserved. The presence of six Paleoproterozoic zircons in analyzed samples of the formation indicates that at least some of the Yukon-Tanana terrane was exposed in the catchment basin(s) of the Tantalus rivers.

7. CONCLUSIONS

Fluvial chert pebble conglomerate and chert arenite of the Tantalus formation accumulated within confined orogen-parallel intermountain river valleys, in a humid temperate setting, during the late stages of oroclinal closure of the Canadian Cordilleran margin in the Upper Jurassic and Lower Cretaceous. Intense weathering in the drainage basin led to a dominance of resistate (quartz, chert) grains surviving transport from the headwaters of the river system(s), with minor surviving nonchert lithic fragments coming from more local basement uplifts adjacent to the depositional basin(s). Paleocurrent distributions, and trends of maximum clast size in the Tantalus formation indicate a source, or sources to the north of the Whitehorse trough, excluding a source in rocks of the Cache Creek terrane now located south of the trough as a potential chert source. Age profiles of detrital zircon assemblages are dominated by local contributions from reworking of strata within the trough, with lesser contributions from uplifted fragments of the Stikinia and Quesnellia terranes, which wrapped around the northern end of the trough. In addition, more distal northerly sources in the Yukon-Tanana and adjacent terranes are indicated. The absence of Archean zircons indicates that proximal North American cratonic sources to the east were isolated by continued uplift of the Cordilleran fold belt. The youngest grains in the Tantalus formation are not represented by any known intrusions in the vicinity of the Whitehorse trough, and appear to have come from airborne volcanic ash, possibly derived from either the Skeena arch, 6–800 km to the south, or the coast ranges (Wrangellia) of northern British Columbia or western Yukon, which are at least 160 km to the southwest, confirming that collision of the insular terranes was occurring at the same time as deposition of the Tantalus formation. As no radiolarians have been reported from chert in Stikinia, Quesnellia, or Yukon-Tanana, a northerly source, in a now-eroded klippen of Cache Creek material, is required to explain the abundance of spherical quartz clusters in chert from the Tantalus formation. This study demonstrates that provenance studies should always involve more than one approach, and should not be undertaken in isolation from basic sedimentological studies of grain size, paleocurrents, and facies distributions, especially in areas with a complicated geological history.

Acknowledgments

I thank Rajat Mazumder for suggesting I write this chapter, and Pat Eriksson and Abhijit Basu for critically reviewing the manuscript. I thank Dirk Tempelman-Kluit, Maurice Colpron, Grant Lowey, and many others for stimulating discussions, and sharing their insight of Yukon geology.

References

Amorosi, A., Zuffa, G.G., 2011. Sand composition changes across key boundaries of siliciclastic and hybrid depositional sequences. Sedimentary Geology 236, 153–163.

Bamber, E.W., Waterhouse, J.B., 1971. Carboniferous and permian stratigraphy and paleontology, Northern Yukon Territory, Canada. Bulletin of Canadian Petroleum Geology 19, 29–250.

Basu, A., 1985. Reading provenance from detrital quartz. In: Zuffa, G.G. (Ed.), Provenance of Arenites. D. Reidel Publishing Company, pp. 231–247.

Bauluz, B., Yusye, A., Mayayo, M.J., Canudo, J.I., 2014. Early kaolinization of detrital Weald facies in Galve sub-basin (Central Iberian Chain, northeast Spain) and its relationship to paleoclimate. Cretaceous Research 50, 214–227.

Beauchamp, B., Baud, A., 2002. Growth and demise of Permian biogenic chert along northwest Pangea: evidence for end-Permian collapse of thermohaline circulation. Palaeogeography, Palaeoclimatology, Palaeoecology 184, 37–64.

Beranek, L.P., Mortensen, J.K., 2011. The timing and provenance of the Late Permian Klondike orogeny in northwestern Canada and arc-continent collision along western North America. Tectonics 30. http://dx.doi.org/10.1029/2010TC002849 cit. no. TC5017. http://www.agu.org.

Berman, R.G., Ryan, J.J., Gordey, S.P., Villeneuve, M., 2007. Permian to Cretaceous polymetamorphic evolution of the Stewart River region, Yukon-Tanana terrane, Yukon, Canada: *P-T* evolution linked with in situ SHRIMP monazite geochronology. Journal of Metamorphic Geology 25, 803–827.

Bickerton, L., Colpron, M., Gibson, D., 2013. Cache Creek terrane, Stikinia, and overlap assemblages of the eastern Whitehorse (NTS 105D) and western Teslin (NTS 105C) map areas. In: MacFarlane, K.E., Nordling, M.G., Sack, P.J. (Eds.), Yukon Exploration and Geology 2012. Yukon Geological Survey, pp. 1–17.

Bostock, H.S., 1936. Carmacks district, Yukon. Canada, department of mines, geological survey. Memoir 187, p. 67.

Breitsprecher, K., Mortensen, J.K., 2004. YukonAge 2004: A Database of Isotopic Age Determinations for Rock Units From Yukon Territory (compilers). Yukon Geological Survey, CD-ROM.

Busby, C.J., Ingersoll, R.V., 1995. Tectonics of Sedimentary Basins. Blackwell Science, pp. 579.

Childe, F.C., Thompson, J.F.H., 1997. Geological setting, U-Pb geochronology, and radiogenic isotope characteristics of the Permo-Triassic Kutcho assemblage north-central British Columbia. Canadian Journal of Earth Sciences 34, 1310–1324.

Childe, F.C., Thompson, J.F.H., Mortensen, J.K., Friedman, R.M., Schiarizza, P., Bellefontaine, K., Marr, J.M., 1998. Primative Permo-Triassic volcanism in the Canadian Cordillera: tectonic and metallogenic implications. Economic Geology 93, 224–231.

Cohen, K.M., Finney, S.C., Gibbard, P.L., Fan, J.-X., 2013. The ICS International Chronostratigraphic Chart. Episodes 36, 199–204.

Colpron, M., Reinecke, M., 2000. Glenlyon project: coherent stratigraphic succession of Yukon-Tanana terrane in the Little Salmon Range, and its potential for volcanic-hosted massive sulphide deposits, central Yukon. In: Emond, D.S., Weston, L.H. (Eds.), Yukon Exploration and Geology 1999, Exploration and Geological Services Division, Yukon, Indian and Northern Affairs Canada, pp. 87–100.

Colpron, M., Mortensen, J.K., Gehrels, G.E., Villeneuve, M., 2006. Basement complex, Carboniferous magnetism and Paleozoic deformation in Yukon-Tanana terrane of central Yukon: field, geochemical and geochronological constraints from Glenlyon map area. In: Colpron, M., Nelson, J.L. (Eds.), Paleozoic Evolution and Metallogeny of Pericratonic Terranes at the Ancient Pacific Margin of North America, Canadian and Alaskan Cordillera. Geological Association of Canada, pp. 131–151. Special Paper, No. 45.

Colpron, M., Crowley, J.L., Gehrels, G., Long, D.G.F., Murphy, D.C., Beranek, L., Bickerton, L., 2015. Birth of the northern Cordilleran orogen, as recorded by detrital zircons in Jurassic synorogenic strata and regional exhumation in Yukon. Lithosphere. http://dx.doi.org/10.1130/L451.1.

Colpron, M., 2011. Geological Compilation of the Whitehorse Trough – Whitehorse (105D), Lake Laberge (105E), and Parts of Carmacks (115I) and Teslin (105C) Yukon Geological Survey, Geoscience Map 2011–1, 3 Maps, Legend and Appendices (compiler).

Cordey, F., Gordey, S.P., Orchard, M.J., 1991. New biostratigraphic data for the northern Cache Creek terrane, Teslin map area, southern Yukon. In: Current Research, Part E. Geological Survey of Canada, pp. 67–76. Paper 91-1E.

Cordey, F., 1991. New Biostratigraphic Data for the Northern Cache Creek Terrane, Teslin Map Area, Southern Yukon. Geological Survey of Canada, pp. 67–76. Paper 91-1E.

Cordey, F., 1992a. Radiolarian Ages From Chert Pebbles of the Tantalus Formation, Carmacks Area, Yukon Territory. Geological Survey of Canada, pp. 53–59. Paper 92-1E.

Cordey, F., 1992b. Radiolarians and terrane analysis in the Canadian Cordillera: the "clastic approach". Palaeogeography, Palaeoclimatology, Palaeoecology 96, 155–159.

Delaney, G.D., 1981. The Mid-Proterozoic Wernecke Supergroup, Wernecke Mountains, Yukon Territory. Geological Survey of Canada, pp. 1–23. Paper 81-10.

Dickinson, W.R., Suczek, C.A., 1979. Plate tectonics and sandstone compositions. American Association of Petroleum Geologists, Bulletin 63, 2164–2182.

Dickinson, W.R., Beard, L.S., Brakenridge, G.R., Erjavec, J.L., Ferguson, R.C., Inman, K.F., Knepp, R.A., Lindberg, F.A., Ryberg, P.T., 1983. Provenance of North American Phanerozoic sandstones in relation to tectonic setting. Geological Society of America, Bulletin 94, 222–235.

Dickinson, W.R., 1970. Interpreting detrital modes of greywacke and arkose. Journal of Sedimentary Petrology 40, 695–707.

Dickinson, W.R., 1985. Interpreting provenance relations from detrital modes of sandstones. In: Zuffa, G.G. (Ed.), Provenance of Arenites. D. Reidel Publishing Company, pp. 333–361.

Dixon, J., 1992. Stratigraphy of mesozoic strata, eagle plain area, northern Yukon. Geological Survey of Canada, Bulletin 408, 58.

Eisbacher, G.H., Carrigy, M., Camppbell, R.B., 1974. Paleodrainage pattern and late-orogenic basins of the Canadian Cordillera. In: Dickinson, W.R. (Ed.), Tectonics and Sedimentation. Society of Economic Paleontologists and Mineralogists, pp. 143–166. Special Publication 22.

Eisbacher, G.H., 1981. Late mesozoic – paleogene bowser basin molasse and Cordilleran tectonics, Western Canada. In: Miall, A.D. (Ed.), Sedimentation and Tectonics in Alluvial Basins. Geological Association of Canada, pp. 125–151. Special Paper 23.

Evenchick, C.A., Thorkelson, D.J., 2005. Geology of the Spatsizi river map area, north-central British Columbia. Geological Survey of Canada, Bulletin 577, 276.

Evenchick, C.A., McMechan, M.E., McNicoll, V.J., Carr, S.D., 2007. A synthesis of the Jurassic-Cretaceous tectonic evolution of the central and southeastern Canadian Cordillera: exploring links across the orogeny. In: Sears, J.W., Harms, T.A., Evenchick, C.A. (Eds.), Whence the Mountains? Inquiries into the Evolution of Orogenic Systems: A Volume in Honor of Raymond a Price. Geological Society of America, pp. 117–145. Special Paper, No. 433.

Evenchick, C.A., Polton, T.P., McNicoll, V.J., 2010. Nature and significance of the diachronous contact between the Hazelton and Bowser Lake groups (Jurassic), north-central British Columbia. Bulletin of Canadian Petroleum Geology 58, 235–237.

Folk, R.L., 1974. Petrology of Sedimentary Rocks. Hemphill Publishing Co., Austin, Texas, USA.

Gabrielse, H., Murphy, D.C., Mortensen, J.K., 2006. Cretaceous and Cenozoic dextral orogen-parallel displacements, magmatism and paleogeography, north-central Canadian Cordillera. In: Haggart, J.W., Monger, J.W.H., Enkin, R.J. (Eds.), Paleogeography of the North American Cordillera: Evidence for and Against Large-scale Displacements. Geological Association of Canada, pp. 255–276. Special Paper, No. 46.

Gehrels, G., Rusmore, M., Woodsworth, G., Crawford, M., Andronicos, C., Hollister, L., Patchett, J., Ducea, M., Butler, R., Klepeis, K., Davidson, C., Friedman, R., Haggart, J., Mahoney, B., Crawford, W., Pearson, D., Girardi, J., 2009. U-Th-Pb geochronology of the Coast Mountains batholith in north-coastal British Columbia: constraints on age and tectonic evolution. Geological Society of America, Bulletin 121, 1341–1361.

Gehrels, G., 2000. Introduction to detrital zircon studies of Paleozoic and Triassic strata in western Nevada and northern California. In: Soreghan, M., Gehrels, G. (Eds.), Paleozoic and Triassic Paleogeography and Tectonics of Western Nevada and Northern California. Geological Society of America, pp. 1–18. Special Paper 347.

Gordey, S.P., Makepeace, A.J., 2001. Bedrock Geology, Yukon Territory. Geological Survey of Canada. Open-File 3754.

Gordey, S.P., Stevens, R.A., 1994. Tectonic framework of the Teslin region, southern Yukon Territory. In: Current Research 1994-A. Geological Survey of Canada, pp. 11–18.

Gordey, S.P., 2008. Bedrock Geology, Whitehorse (105D), Yukon. Geological Survey of Canada. Open File 5640, scale 1: 250 000.

Gordey, S.P., 2013. Evolution of the Selwyn Basin region, Sheldon Lake and Tay River map areas, central Yukon. Geological Survey of Canada, Bulletin 599, 176.

Gunning, M.H., Bamber, E.W., Brown, D.A., Rui, L., Mamet, B.L., Orchard, M.J., 1994. The Permian Ambition Formation of northwestern Stikinia, British Columbia. In: Embry, A.F., Beauchamp, B., Glass, D.J. (Eds.), Pangea: Global Environments and Resources. Canadian Society of Petroleum Resources, Memoir 17, pp. 589–619.
Hart, C.J.R., Orchard, M.J., 1996. Middle Triassic (Ladinian) volcanic strata in southern Yukon Territory, and their Cordilleran correlatives. Geological Survey of Canada, Current Research 1996-A 11–18.
Hart, C.J.R., Radloff, J.K., 1990. Geology of Whitehorse, Alligator Lake, Fenwick Creek, Carcross and Part of Robinson Map Areas (105D/11, 6, 3, 2, and 7). Indian and Northern Affairs Canada, Northern Affairs, Yukon Region. Open File 1990-4. p. 113.
Ingersoll, R.V., Bullard, T.F., Ford, R.L., Grimm, J.P., Pickle, J.D., Sares, S.W., 1984. The effect of grain size on detrital modes: a test of the Gazzi-Dickinson point-counting method. Journal of Sedimentary Petrology 54, 103–116.
Israel, S., Beranek, L., Friedman, R.M., Crowley, J.L., 2014. New ties between the Alexander terrane and Wrangellia and implications for North American Cordilleran evolution. Lithosphere 6, 270–276.
Jackson, J.L., 1992. Tectonic Analysis of the Nisling, Northern Stikine and Northern Cache Creek Terranes, Yukon and British Columbia (Ph.D. thesis). University of Arizona, p. 200.
Johnston, S.T., Mortenson, J.K., Erdmer, P., 1996. Igneous and metaigneous age constraints for the Aishihik metamorphic suite, southwest Yukon. Canadian Journal of Earth Sciences 33, 1543–1555.
Long, D.G.F., Lowey, G.M., 2006. Anatomy of a Late Jurassic Gilbert-type delta in basal strata of the Tantalus formation, Whitehorse Trough, Yukon, Canada. In: Emond, D.S., Bradshaw, G.D., Lewis, L.L., Weston, L.H. (Eds.), Yukon Exploration and Geology 2005, pp. 195–205.
Long, D.G.F., 1986. Coal in Yukon. In: Morin, J.D. (Ed.), Mineral Deposits of the Northern Cordillera. Canadian Institute of Mining and Metallurgy, pp. 311–318. Special Paper, No. 37.
Long, D.G.F., 2005. Sedimentology and hydrocarbon potential of fluvial strata in the Tantalus and Aksala formations, northern Whitehorse Trough, Yukon. In: Emond, D.S., Lewis, L.L., Bradshaw, G.D. (Eds.), Yukon Exploration and Geology 2004, pp. 165–174.
Long, D.G.F., 2015. Provenance and Depositional Framework of Braided and Meandering Gravel-bed River Deposits and Associated Coal Deposits in Active Intermontane Piggyback Basins: The Upper Jurassic to Lower Cretaceous Tantalus Formation. Yukon Geological Survey, Yukon, Canada. Open File Report 2015-23. http://www.geology.gov.yk.ca.
Ludwig, K.R., 2001. Isoplot/Ex, rev.2.49. Berkeley Geochronology Center. Special Publication, No. 1a, p. 56.
Mackenzie, W.S., 1974. Radiolaria from the Canol Formation, Northwest Territories. Geological Survey of Canada. Paper 74-1, p. 319.
MacNaughton, R.B., 2002. Sedimentology of Triassic siliciclastic strata in the Mount Martin and Mount Merrill map areas, Yukon Territory. In: Geological Survey of Canada, Current Research 2002–A4, p. 10.
Martel, E., Turner, E.C., Fischer, B.J., 2011. Geology of the Central Mackenzie Mountains of the Northern Canadian Cordillera, Sekwi Mountain (106P), Mount Eduni (106A), and Northern Wigley Lake (95M) Map-areas, Northwest Territories. Northwest Territories Geoscience Office. NWT Special Volume, No. 1, p. 423.
McNicoll, V.J., Evenchick, C.A., Mustard, P.S., 2005. Provenance studies on the depositional histories of the Bowser and Sustut basins and their implications for tectonic evolution of the northern Canadian Cordillera. In: Geological Association of Canada, Annual Meeting Abstracts, 30, p. 133.
Miall, A.D., 1996. The Geology of Fluvial Deposits. Springer-Verlag, Berlin, p. 582.
Mihalynuk, M.G., Peter, J.M., 2001. A hydrothermal origin for "crinkle chert" of the Big Salmon Complex. In: Geological Fieldwork 2000, British Columbia Geological Survey, Paper 2001-1, pp. 83–84.
Mihalynuk, M.G., Nelson, J., Diakow, L., 1994. Cache Creek Terrane entrapment: oroclinal paradox within the Canadian Cordillera. Tectonics 13, 575–595.
Mihalynuk, M.G., Erdmer, P., Ghent, E.D., Cordey, F., Archibald, D.A., Friedman, R.M., Johannson, G.G., 2004. Coherent French Range blueschist: subduction to exhumation in <2.5 m.y.? Geological Society of America, Bulletin 116, 910–922.
Mihalynuk, M.G., 1999. Geology and Mineral Resources of the Tagish Lake Area (NTS 104M/8, 9, 10E, 15 and 104N/12W) Northwestern British Columbia. BC Ministry of Energy and Mines, Energy and Minerals Division, Geological Survey Branch. Bulletin, no. 105.
Monger, J.W.H., 1975. Upper Paleozoic rocks of the Atlin terrane, northwestern British Columbia and south-central Yukon. Geological Survey of Canada. Paper 74-47, p. 63.
Morrow, D.W., 1999. Lower Paleozoic stratigraphy of Northern Yukon Territory and northwestern district of Mackenzie. Geological Survey of Canada, Bulletin 538, 202.

Mortensen, J.K., 1990. Geology and U-Pb geochronology of the Klondike district, west-central Yukon Territory. Canadian Journal of Earth Sciences 27, 903–914.

Murphy, D.C., Mortensen, J.K., Piercey, S.J., Orchard, M.J., Gehrels, G.E., 2006. Mid-Paleozoic to early Mesozoic tectonostratigraphic evolution of Yukon-Tanana and Slide Mountain terranes and affiliated overlap assemblages, Finlayson Lake massive sulphide district, southeastern Yukon. In: Colpron, M., Nelson, J.L. (Eds.), Paleozoic Evolution and Metallogeny of Pericratonic Terranes at the Ancient Pacific Margin of North America, Canadian and Alaskan Cordillera, 45. Geological Association of Canada, pp. 75–105. Special Paper.

Murphy, D.C., van Stall, C., Mortensen, J.K., 2008. Windy McKinley terrane, Stevenson Ridge area (115JK), western Yukon: composition and proposed correlations, with implications for mineral potential. In: Emond, D.S., Blackburn, L.R., Hill, R.P., Weston, L.W. (Eds.), Yukon Exploration and Geology 2007. Yukon Geological Survey, pp. 225–235.

Nelson, J., Colpron, M., 2007. Tectonics and metallogeny of the British Columbia, Yukon and Alaskan Cordillera, 1.8 Ga to the present. In: Goodfellow, W.D. (Ed.), Mineral Deposits of Canada: A Synthesis of Major Deposit-Types, District Metallogeny, the Evolution of Geological Provinces, and Exploration Methods. Geological Association of Canada, Mineral Deposits Division, pp. 755–791. Special Publication No. 5.

Nelson, J., Gehrels, G., 2007. Detrital zircon geochronology and provenance of the southeastern Yukon-Tanana terrane. Canadian Journal of Earth Sciences 44, 297–316.

Nesbitt, H.W., Fedo, C.M., Young, G.M., 1997. Quartz and feldspar stability, steady and non-steady state weathering and petrogenesis of siliciclastic sands and muds. Journal of Geology 105, 173–192.

Norris, A.W., 1968. Reconnaissance Devonian Stratigraphy of Northern Yukon Territory and Northwestern District of MacKenzie. Geological Survey of Canada. Paper 67–53.

Pigage, L.C., 2004. Bedrock Geology Compilation of the Anvil District (Parts of NTS 105K/2, 3, 5, 6, 7 and 11), Central Yukon. Yukon Geological Survey. Bulletin, No. 15, p. 103.

Plint, H.E., Gordon, T.M., 1997. The Slide mountain terrane and structural evolution of the Finlayson Lake Fault Zone, southeastern Yukon. Canadian Journal of Earth Sciences 34, 105–126.

Pyle, L.J., Jones, A.L., 2009. Regional Geoscience Studies and Petroleum Potential, Peel Plateau and Plain, Northwest Territories and Yukon: Project Volume. Northwest Territories Geoscience Office, and Yukon Geological Survey. NWT Open File 2009-02/YGS Open File 2009-25, p. 540.

Raines, M.K., Hubbard, S.M., Kukulski, R.B., Leir, A.L., Gehrels, G.E., 2013. Sediment Dispersal in an Evolving Foreland: Detrital Zircon Geochronology From Upper Jurassic and Lowermost Cretaceous Strata, Alberta Basin, Canada, vol. 125. Geological Society of America, Bulletin, pp. 741–755.

Richards, B.C., Bamber, E.W., Utting, J., 1997. The geology, mineral and hydrocarbon potential of northern Yukon Territory and northwestern District of Mackenzie. Geological Survey of Canada, Bulletin 422, 201–251.

Ricketts, B.D., Evenchick, C.A., Anderson, R.G., Murphy, D.C., 1992. Bowser basin, northern British Columbia: constraints on the timing of initial subsidence and Stikinia – North America terrane interactions. Geology 20, 1119–1122.

Roots, C.F., Nelson, J.L., Simard, R.-L., Harms, T.A., 2006. Continental fragments, mid-Paleozoic arcs and overlapping late Paleozoic arc and Triassic sedimentary strata in the Yukon-Tanana terrane of northern British Columbia and southern Yukon. In: Colpron, M., Nelson, J.L. (Eds.), Paleozoic Evolution and Metallogeny of Pericratonic Terranes at the Ancient Pacific Margin of North America, Canadian and Alaskan Cordillera. Geological Association of Canada, pp. 153–177. Special Paper, No. 45.

Ross, G.M., Harms, T.A., 1998. Detrital zircon geochemistry of sequence "C" grits, Dorsey Terrane (Thirtymile Range, southern Yukon): provenance and stratigraphic correlation. In: Radiogenic Age and Isotope Studies, Report 11, Geological Survey of Canada, Current Research, 1998-F, pp. 107–115.

Ross, G.M., Patchett, P.J., Hamilton, M., Heaman, L., DeCelles, P.G., Rosenberg, E., Giovanni, M.K., 2005. Evolution of the Cordilleran Orogen (Southwestern Alberta Canada) Inferred From Detrital Mineral Geochronology, Geochemistry, and Nd Isotpes in the Foreland Basin, vol. 117. Geological Society of America, Bulletin, pp. 747–763.

Schiarizza, P., 2012. Geology of the Kutcho Assemblage between Kehlechoa and Tucho Rivers, Northern British Columbia. British Columbia Geological Survey, pp. 75–98. Geological Fieldwork 2011, Paper 2012-1.

Simard, R.-L., Devine, F., 2003. Preliminary geology of the southern Semenof Hills, central Yukon (105E/1,7,8). In: Emond, D.S., Lewis, L.L. (Eds.), Yukon Exploration and Geology 2002. Exploration and Geological Services Division, Yukon Region, Indian and Northern Affairs Canada, pp. 213–222.

Simard, R.-L., Dorstal, J., Roots, C.F., 2003. Development of late Paleozoic volcanic arcs in the Canadian Cordillera: an example from the Klinkit Group, northern British Columbia and southern Yukon. Canadian Journal of Earth Sciences 40, 907–924.

Staples, R.D., Gibson, H.D., Berman, R.G., Ryan, J.J., Colpron, M., 2013. A window into the Early to mid-Cretaceous infrastructure of the Yukon-Tanana terrane recorded in multi-stage garnet of west-central Yukon. Journal of Metamorphic Geology 31, 729–753.

Staples, R.D., Murphy, D.C., Gibson, H.D., Colpron, M., Berman, R.G., Ryan, J.J., 2014. Middle Jurassic to Earliest Cretaceous Mid-crustal Tectono-metamorphism in the Northern Canadian Cordillera: Recording Foreland-directed Migration of an Orogenic Front, vol. 126. Geological Society of America, Bulletin, pp. 1511–1530.

Tempelman-Kluit, D.J., 2009. Geology of Carmacks and Laberge Map Areas, Central Yukon: Incomplete Draft Manuscript on Stratigraphy, Structure and its Early Interpretation (Ca. 1986). Geological Survey of Canada. Open File, 5982, p. 399.

Wheeler, J.O., 1961. Whitehorse Map-area, Yukon Territory, 105D. Geological Survey of Canada. Memoir 312, p. 156.

White, D., Colpron, M., Buffett, G., 2012. Seismic and geological constraints on the structure and hydrocarbon potential of the northern Whitehorse trough, Yukon Canada. Bulletin of Canadian Petroleum Geology 60 (4), 239–255.

CHAPTER

13

Late Neoproterozoic to Early Mesozoic Sedimentary Rocks of the Tasmanides, Eastern Australia: Provenance Switching Associated With Development of the East Gondwana Active Margin

C.L. Fergusson[1], R.A. Henderson[2], R. Offler[3]

[1]University of Wollongong, Wollongong, NSW, Australia; [2]James Cook University, Townsville, QLD, Australia; [3]University of Newcastle, Callaghan, NSW, Australia

OUTLINE

Sediment Provenance
http://dx.doi.org/10.1016/B978-0-12-803386-9.00013-7

1. INTRODUCTION

The composition of clastic sediments is a function of rock assemblages in the region from which their components were sourced; the climatic influence on weathering of the source region; and the sedimentary processes involved during dispersal, transport, and accumulation at the sink, including factors such as source area relief, transport mechanisms, travel time, and diagenesis (Boggs, 2009). In ancient systems, connections between sources and sinks are commonly no longer preserved and in many situations the sources themselves may have been removed by erosion or hidden by overlying units. Numerous tools are available to identify potential sources, such as petrographic analysis, whole-rock and trace-element geochemistry, whole-rock isotopic methods, and geochronology (Weltje and von Eynatten, 2004; Gehrels, 2014). These are particularly pertinent for the resolution of tectonic settings of sedimentary successions in orogenic belts where reorganization of the upper crust is commonplace such that past plate tectonic arrangements can be difficult to resolve.

Relationships between plate tectonic settings and sandstone compositions were examined following the development of plate tectonics (Dickinson and Suczek, 1979). This approach has been widely applied to orogenic and basinal systems. Its application to the Tasmanide Orogenic Belt (Tasmanides) of eastern Australia (Figs. 13.1 and 13.2; Korsch, 1978, 1981, 1984; Cowan, 1993; Veevers et al., 1994; Colquhoun et al., 1999; Fergusson and Tye, 1999; Leitch et al., 2003) has been to constrain tectonic settings of various orogenic systems, such as those of the Lachlan Orogen, that have been subject to wide debate (Foster et al., 1999; Glen et al., 2009; Quinn et al., 2014). Geochemical and isotopic whole-rock methods have been less commonly used in the Tasmanides (Bhatia and Taylor, 1981; Turner et al., 1993; Gray and Webb, 1995; Haines et al., 2009). Methods involving detrital zircon ages in provenance analysis, based on U-Pb isotopic systematics, have had widespread application (Williams, 1998; Williams and Pulford, 2008; Sircombe, 1999; Fergusson et al., 2001, 2007, 2013; Veevers, 2015). Detrital zircon ages determined by the Sensitive High Resolution Ion MicroProbe and by Laser Ablation Inductively Coupled Plasma Mass Spectrometry (LA-ICP-MS) methods have become increasingly important in understanding the evolution of the Tasmanides.

The recognition of common 600–500 Ma zircon ages, the so-called Pacific-Gondwana signature, identifies igneous rocks of this age bracket as a major sediment source to the Tasmanides. This signature is widespread in quartz-rich units such as the Kanmantoo Group

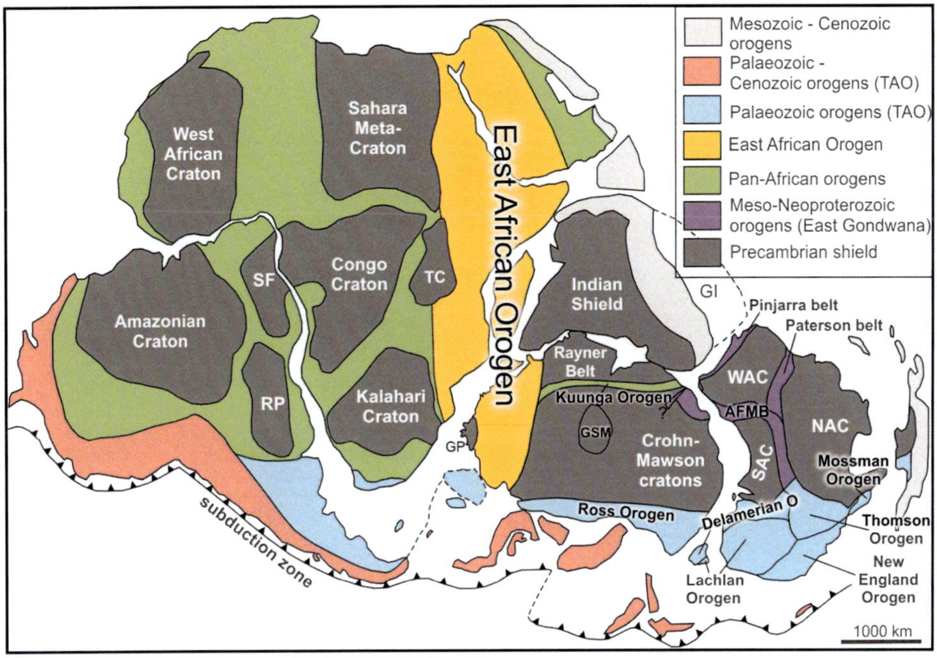

FIGURE 13.1 Gondwana following the reconstruction by de Wit et al. (1988) but modified after Myers et al. (1996), Gray et al. (2008), Boger (2011), and Torsvik and Cocks (2013). Cratons in Australia and Antarctica are from Myers et al. (1996) and Boger (2011), respectively. *AFMB*, Albany–Fraser–Musgrave belt; *Delamerian O*, Delamerian Orogen; *GI*, Greater India; *GSM*, Gamburtev Subglacial Mountains; *GP*, Grunehogna Province; *NAC*, North Australian Craton; *RP*, Río de la Plata Craton; *SAC*, South Australian Craton; *SF*, São Francisco Craton, *TAO*, Terra Australis Orogen, Tanzania, *TC*, Tanzania Craton; *WAC*, West Australian Craton.

of southeastern South Australia, the Ordovician turbidites of southeastern Australia, and the Triassic Hawkesbury Sandstone of the Sydney Basin (Ireland et al., 1998; Sircombe, 1999; Fergusson and Fanning, 2002). It has been extensively passed on, through reworking, to modern beach sands in eastern Australia (Sircombe, 1999; Sircombe and Hazelton, 2004; Boyd et al., 2008; Veevers, 2015). Because zircon grain age spectra are unaffected by the other influences on sandstone composition, particularly that of climate with the removal of more labile grains due to weathering, provenance evaluation based on data sets of this type is particularly robust.

This paper presents a review of the provenance of sedimentary rocks in the Tasmanides of eastern Australia (Fig. 13.2). The Tasmanides, part of the Terra Australis Orogen (Cawood, 2005), are dominantly of Paleozoic age. This belt is the most widely exposed part of the Pacific-facing East Gondwana active margin and developed progressively from about 550 Ma to 220 Ma. Resolving the tectonic evolution of the Tasmanides, therefore, places significant constraints on the tectonic development of the East Gondwana active margin. The provenance of sedimentary rocks in the Lachlan and New England Orogens has played an important role in analysis of their tectonic settings; for example, the quartz-rich nature of the Ordovician turbidites in southeastern Australia has been presented as an argument

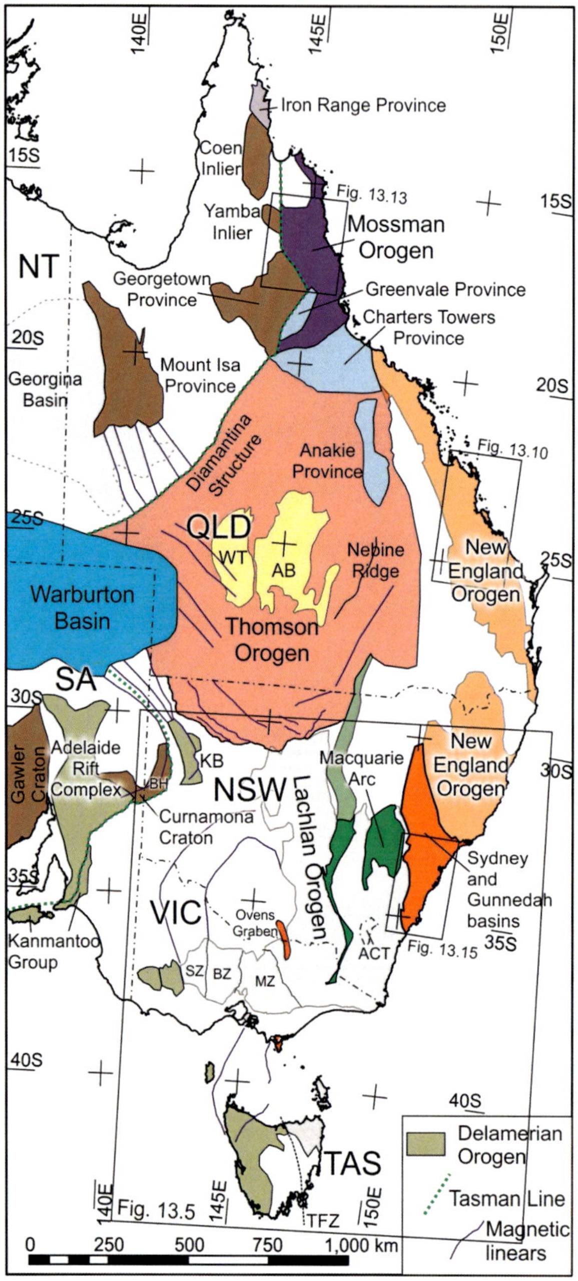

FIGURE 13.2 Orogenic belts and some sedimentary basins of the Tasmanides of eastern Australia. *Blue lines* are major magnetic and gravity lineaments. Tasman Line shown by *dotted green line*. *AB*, Adavale Basin; *ACT*, Australian Capital Territory; *BH*, Broken Hill; *BZ*, Bendigo Zone; *KB*, Koonenberry Belt; *MZ*, Melbourne Zone; *NSW*, New South Wales; *NT*, Northern Territory; *QLD*, Queensland; *SZ*, Stawell Zone; *TAS*, Tasmania; *TFZ*, Tamar Fracture Zone; *VIC*, Victoria; *WT*, Warrabin Trough. Locations of Figs. 13.5, 13.10, 13.13, and 13.15 shown.

against their proposed accretion in subduction complex settings (Aitchison and Buckman, 2012). In contrast, quartz-rich turbidites of the Carboniferous Shoalwater Formation in central Queensland have been recognized as accreted to the Late Paleozoic subduction complex of the New England Orogen (Fergusson et al., 1990; Leitch et al., 2003; Korsch et al., 2009a).

The Tasmanides are a composite orogenic complex containing five orogens and numerous foreland and successor sedimentary basins. We concentrate on the most widely studied aspects of provenance of sedimentary rocks within the Tasmanides and these include: (1) transition from localized Grenvillian sources in Late Neoproterozoic sedimentary and metasedimentary rocks to the widespread Pacific-Gondwana source in the Middle Cambrian of the Thomson and Delamerian Orogens, (2) provenance of the Ordovician turbidites of the Lachlan Orogen and contrast with the volcanic-dominated Macquarie Arc succession, (3) foreland sedimentary successions of the Silurian-Devonian Melbourne Trough in central Victoria and the related upper part of the Mathinna Supergroup in northeast Tasmania, (4) Devonian to Carboniferous provenance switching in the subduction complex of the New England Orogen, (5) local derivation of clastic detritus in the Silurian to Late Devonian subduction complex of the Mossman Orogen, and (6) mixed sources of clastic detritus in the Permian–Triassic Sydney Basin.

2. GEOLOGICAL SETTING AND SUBDIVISIONS OF THE TASMANIDES

Gondwana came into existence in 550–500 Ma with the collision of West Gondwana, East Gondwana, and India (including the Rayner Belt in East Antarctica) along the East African and Kuunga Orogens (Fig. 13.1; Boger and Miller, 2004). In the East Gondwana segment, the Pacific-facing margin had already changed from an older passive margin to an active margin by 580 Ma along the Ross Orogen in East Antarctica (Goodge et al., 2002, 2004a,b) and somewhat later (~550–515 Ma) in the Delamerian Orogen in southeastern Australia (Haines and Flöttmann, 1998; Turner et al., 2009; Gibson et al., 2011). The Tasmanides represent the mainly Paleozoic development of the active East Gondwana margin from the former East Gondwana passive margin that is represented by the Adelaide Rift Complex of the Delamerian Orogen in southeastern South Australia (Preiss, 2000) as well as the exposed part of the Thomson Orogen of northeastern Australia (Fergusson et al., 2007, 2009).

The Tasmanides are divided into five orogens including the inner Delamerian and Thomson Orogens, the Lachlan Orogen in the south, the New England Orogen in the east, and the Mossman Orogen in far northeastern Australia (Fig. 13.2; Glen, 2005, 2013; Withnall and Henderson, 2012). They are divided from the Proterozoic cratons to the west along the Tasman Line. Although the usefulness of this boundary has been questioned for southeastern Australia (Direen and Crawford, 2003), it is exposed as faulted/sheared contacts along the western boundaries of the Mossman and Thomson Orogens in north Queensland (Fergusson and Henderson, 2013; Henderson et al., 2013). The Tasman Line continues beneath Mesozoic cover in western Queensland as the Diamantina Structure, a striking boundary between the Proterozoic Mount Isa Province to the northwest and the Thomson Orogen to the southeast revealed by gravity and magnetic imaging (Withnall and Hutton, 2013). The southern part of the Tasmanides includes inferred subsurface Precambrian rocks known as the Selwyn Block in central Victoria that have been traced southward to western Tasmania (Cayley et al., 2002), although it is unclear if Late Cambrian turbiditic rocks of the Stawell Zone in the western Lachlan Orogen continue southward west of Tasmania (Gibson et al., 2011; Moore et al., 2013).

In central Australia the Tasman Line is under cover, but a connection between the Tasmanides and the East Gondwana interior most likely existed through central Australia, and remained a site of discontinuous sedimentation from the early Neoproterozoic up until the Carboniferous across the juncture of all three Australian Proterozoic cratons (Veevers, 2000). The idea that the Proterozoic North, South, and West Australian cratons (Fig. 13.1) amalgamated prior to formation of the Centralian Basin is widely assumed in the literature (Cawood and Korsch, 2008). Paleomagnetic data indicates a 40 degrees anticlockwise intracratonic rotation of the North Australian Craton relative to an amalgam of the West and South Australian cratons during 650–550 Ma that includes the timing of the Petermann Orogeny (Li and Evans, 2011; Schmidt, 2014). This rotation implies separation of the North and South Australian cratons along the Diamantina Structure, thereby forming accommodation space for the Thomson Orogen as a rifted oceanic basin (Fergusson and Henderson, 2015). It is therefore possible that the Thomson Orogen does not extensively overlie Precambrian basement as argued by Spampinato et al. (2015).

The Delamerian Orogen is developed in southeastern South Australia including the Adelaide Rift Complex, as well as western Victoria and the Koonenberry Belt in northwestern New South Wales (Fig. 13.2). Much of the western two-thirds of Tasmania include Precambrian rocks with many metamorphic units showing Middle Cambrian Delamerian overprint, known locally as the Tyennan Orogeny (Berry et al., 2007). In the Adelaide Rift Complex initial rifting began around 827 Ma, as indicated by intrusion of the Gairdner Dyke Swarm in Proterozoic crust to the west (Wingate et al., 1998). Several episodes of rifting characterized the history of the Adelaide Rift Complex (Preiss, 2000). Breakup during continental separation has been suggested at around 700 Ma by Preiss (2000) and at ~580 Ma by Crawford et al. (1997). According to Foden et al. (1999, 2006), the Delamerian Orogeny began after deposition of the Kanmantoo Group, potentially constrained to 514 ± 4 Ma by the age of the Rathjen Gneiss. In this interpretation, the Kanmantoo Group was considered the fill of an extensional basin that cut across the preexisting basinal geometry (Preiss, 2000; Haines et al., 2009). Alternatively, older Ar/Ar ages on foliation indicate that the Delamerian Orogeny may have begun much earlier, at around 550 Ma, with the Kanmantoo Group deposited in a synorogenic trough rather than an extensional basin (Turner et al., 2009). An earlier start to the Delamerian Orogeny and synorogenic deposition of the Kanmantoo Group has been supported by Gibson et al. (2011, 2015).

In western Tasmania, and western and central Victoria, the Delamerian (Tyennan) Orogeny involved a collision between an island arc and a passive margin, as indicated by the ages of overthrust ophiolitic rocks with a boninitic chemistry in Tasmania (Berry and Crawford, 1988; Turner et al., 1998) and the inferred occurrence of ultramafic rocks in central Victoria (based on prominent magnetic anomalies generated at depth in the eastern Melbourne Zone; McLean et al., 2010). Alternatively, in western Victoria development of an east-dipping subduction zone with accretion of an older (590–580 Ma) hyperextended margin has been argued by Gibson et al. (2011, 2015) for the Glenelg Zone. In western Victoria, an east-facing Andean-style convergent margin resulted in the Mount Stavely

Volcanics, from around 510 Ma, postdating the earlier island arc collision (Cayley, 2011; Taylor et al., 2014). Development of a west-facing continental margin arc (515–505 Ma) and following Delamerian Orogeny at 505–500 Ma has been argued for the Koonenberry Belt in northwestern New South Wales (Greenfield et al., 2011).

Much of the Thomson Orogen occurs under cover in western and central Queensland but is known from widespread basement cores collected during petroleum exploration in the overlying basins (Murray, 1994). It is exposed in the Anakie, Charters Towers, and Greenvale provinces east and southeast of the Georgetown Inlier (Fig. 13.2). Detrital zircon ages from the upper metamorphic rocks of the exposed part of the Thomson Orogen, and from the basement cores, show that much of the orogen consists of a quartz-rich metasedimentary succession with maximum depositional ages of 510 to 495 Ma (i.e., Late Cambrian; Brown et al., 2014; Carr et al., 2014; Fergusson and Henderson, 2015). Age constraints for overlying units and plutonic rocks overlap with the timing of major shortening in the later part of the Delamerian Orogeny (latest Cambrian), and indicate rapid continental growth (Fergusson and Henderson, 2013, 2015). In the northern part of the Thomson Orogen, a late Neoproterozoic succession associated with rifting is inferred from detrital zircon in the lower metamorphic units of the Anakie Province (Fergusson et al., 2009). This rifting is ~100 Ma younger than rifting in the Georgina Basin and continental breakup in the Adelaide Rift Complex proposed at around 700 Ma (Preiss, 2000; Greene, 2010; Fergusson and Henderson, 2015). It is similar to the 580 Ma rifting suggested for the Koonenberry Belt, western Victoria, and Tasmania by Crawford et al. (1997) and Direen and Crawford (2003). In the subsurface, the southwest Thomson Orogen is continuous with the Cambrian-Ordovician succession of the Warburton Basin in northeastern South Australia (Fig. 13.2) that lacks effects from the Delamerian Orogeny (PIRSA, 2007). In contrast to the Lachlan Orogen to the south, the Thomson Orogen is largely devoid of Silurian sedimentary rocks but is overlain by Devonian backarc basinal successions, such as the Adavale Basin (McKillop, 2013).

The Lachlan Orogen is over 600 km wide across Victoria and has a complex structural arrangement that includes the subsurface Selwyn Block in central Victoria and the Macquarie Arc in eastern New South Wales (Fig. 13.2). The orogen has widespread Early to Middle Ordovician turbidites that in some areas have an oceanic basement of forearc and backarc mafic volcanic rocks overlain by deep-marine chert (VandenBerg et al., 2000; Crawford et al., 2003). In western Victoria, the Stawell Zone is dominated by probable Late Cambrian quartz-rich turbidites, whereas the adjoining Bendigo Zone has a well-established Early to Middle Ordovician turbidite succession with Late Ordovician turbidites in the southeastern part (VandenBerg et al., 2000). The Melbourne Zone contains a thick Silurian to mid-Devonian succession of turbidites and shallow marine rocks that overlie Ordovician turbidites and the Selwyn Block, which is the northern continuation of rock assemblages in western Tasmania (Cayley et al., 2002). East of the Melbourne Zone several zones are recognized in the Lachlan Orogen with abundant Ordovician turbidites developed both west, east, and within the Ordovician–Early Silurian Macquarie Arc. The Macquarie Arc consists of calc-alkaline and shoshonitic mafic and intermediate igneous rocks and associated volcaniclastic successions. Its geochemistry is consistent with an island arc setting (Crawford et al., 2007), although an alternative backarc setting has been proposed by Quinn et al. (2014). The paleogeographic relationships between the island arc rocks and the Ordovician quartz-rich turbidites have been a matter of debate, with suggestions ranging from major terrane translation (Glen et al., 2009), overthrusting of the island arc over a passive margin (Aitchison and Buckman, 2012), and arc rotations and subduction reversal (Fergusson, 2009).

Widespread shortening and medium pressure metamorphism occurred in the Late Ordovician to Early Silurian Benambran Orogeny (Offler et al., 1998) that was associated with the development of three major structural zones: the Bendigo Zone, Wagga-Omeo and Tabberabbera Zones, and the Bega-Mallacoota Zone. These zones have been interpreted as subduction complexes (Foster and Gray, 2000; Fergusson, 2014). In the mid-Silurian to mid-Devonian the Lachlan Orogen in New South Wales and eastern Victoria was dominated by a wide zone of extensional basinal development involving abundant magmatic activity in an arc to backarc setting (Collins, 2002; VandenBerg, 2003; Fergusson, 2010). Intermittent shortening and high temperature/low pressure metamorphism occurred during this interval, particularly in the mid-Devonian Tabberabberan Orogeny, and was followed in the Late Devonian to Early Carboniferous by widespread shallow marine to fluvial deposition with development of a major quartzose sand sheet and less common igneous activity in a backarc setting (Powell, 1984; Glen, 2005). A terminal mid-Carboniferous (Kanimblan) contraction affected the Lachlan Orogen with effects dying out westward in the Broken Hill region in far western New South Wales. In the east, contraction was postdated by intrusions of Carboniferous granitic rocks related to the west-dipping subduction zone preserved in the New England Orogen (Powell, 1984; Glen, 2005).

Tasmania has been considered enigmatic in terms of its relationships to the rest of the Tasmanides (Cayley, 2011; Gibson et al., 2011; Moore et al., 2013, 2015). West of the Tamar Fracture Zone (Fig. 13.2), Precambrian metasedimentary and very low-grade sedimentary successions are abundant and have been strongly affected by the Tyennan Orogeny, which is the local equivalent of the Delamerian Orogeny (Berry and Bull, 2012). Magnetic data indicate that these rocks continue northward and form the Selwyn Block that is basement to the Melbourne Zone (Cayley et al., 2002). It has been argued that western Tasmania has been displaced northward along the East Gondwana margin (Cayley, 2011). It must have arrived somewhere near its present location no later than during the Benambran Orogeny (Cayley, 2011; Gibson et al., 2011). In contrast, northeastern Tasmania has a distinctive Ordovician-Devonian stratigraphy and Devonian granites both with an inherited zircon age pattern that is indicative of an association with the eastern Lachlan Orogen (Reed, 2001; Black et al., 2004, 2010).

In northeastern Australia, the Tasmanides are at their narrowest width (140 km, Fig. 13.2) and a single orogen, the Mossman Orogen, is represented (Withnall and Henderson, 2012). To the south, the Mossman Orogen abuts the Thomson Orogen, but further north it is faulted against Mesoproterozoic rocks of the North Australian Craton exposed in the Yamba Inlier (Fig. 13.2). The Mossman Orogen contains two main assemblages, an older unit that, in the south, is a Late Ordovician island arc that has been thrust westward over the Early Paleozoic margin in the Benambran Orogeny (Henderson et al., 2011). This was followed by the development of an east-facing, Silurian to Devonian active margin (Henderson et al., 2013). The eroded roots of a magmatic arc of comparable age occur west of the Tasman Line in Mesoproterozoic inliers. A dismembered forearc basin is mainly preserved in the southwestern part of the orogen. To the east, a broad tract of subduction complex rocks is characterized by widespread imbrication of trench-wedge turbidite units and abundant stratal disruption.

The New England Orogen is the eastern-most component of the Tasmanides (Fig. 13.2); it is dominated by middle to late Paleozoic rocks but also includes limited early Paleozoic rocks (Murray, 1997; Donchak et al., 2013). Cambrian and Ordovician rocks of very limited extent occur mainly in the south and show evidence of similar styles of forearc and island arc settings to those of the Cambrian of the Lachlan Orogen (Glen, 2013). A Late Silurian–Devonian island arc and backarc assemblage (Gamilaroi-Calliope island arc) is widely developed and was accreted to the East Gondwana margin in the Late Devonian (Aitchison and Flood, 1995; Offler and Murray, 2011). This was followed by formation of a Late Devonian to Carboniferous continental convergent margin with magmatic arc, forearc basin and subduction complex that dominates much of the geology of the orogen (Murray et al., 1987). In the Early Permian, a major episode of extension affected the New England Orogen and adjoining regions causing rifting and development of the Sydney-Gunnedah-Bowen Basin followed by establishment of an Andean convergent margin in the Late Permian to Late Triassic (Hunter-Bowen Orogeny) with a foreland basin to the west and a magmatic arc in the New England Orogen (Veevers et al., 1994; Korsch et al., 2009b,c; Cawood et al., 2011).

3. PROVENANCE

A prominent aspect of the provenance of many sedimentary successions in the Tasmanides is the abundance of quartz-rich, siliciclastic sedimentary rocks of various ages that are characterized by the Pacific-Gondwana detrital zircon age signature (600–500 Ma). These rocks contrast with other sedimentary successions that reflect more local provenance including some assemblages of island arc affinity as developed in the Ordovician–Early Silurian Macquarie Arc of the eastern Lachlan Orogen, the Late Silurian–Devonian Gamilaroi-Calliope island arc of the New England Orogen, and other units related to continental provenance of adjacent regions. We highlight these provenance characteristics of the Tasmanides by reference to the following examples.

3.1 Influx of Pacific-Gondwana Sediment

The Tasmanides are associated with substantial deposition of siliciclastic sediments derived and/or recycled from Gondwana as indicated by the predominance of quartz in many sandstones consistent with a continental source. Neoproterozoic development of the East Gondwana margin is best documented for the deformed successions making up the Delamerian Orogen in southeastern South Australia where formation of the Adelaide Rift Complex was accompanied by numerous rifting episodes and the influx of mainly quartz-rich to arkosic siliciclastic detritus from the East Gondwana craton to the west (Preiss, 2000). Sources in the neighboring Gawler Craton are indicated by samples with detrital zircon ages common at 1900–1550 Ma, such as in the Niggly Gap beds near the base of the rift succession and the Mount Terrible Formation at the base of the Cambrian Normanville Group, which also has a dominant peak at 1830 Ma (Fig. 13.3A and B; Gehrels et al., 1996; Ireland et al., 1998; Preiss, 2000). Other samples from the succession, such as the Mitcham Quartzite, Marino Arkose, Bonney Sandstone (Fig. 13.3C), and Heatherdale Shale (Ireland et al., 1998), show mixed sources with zircon ages consistent with derivation from the Gawler

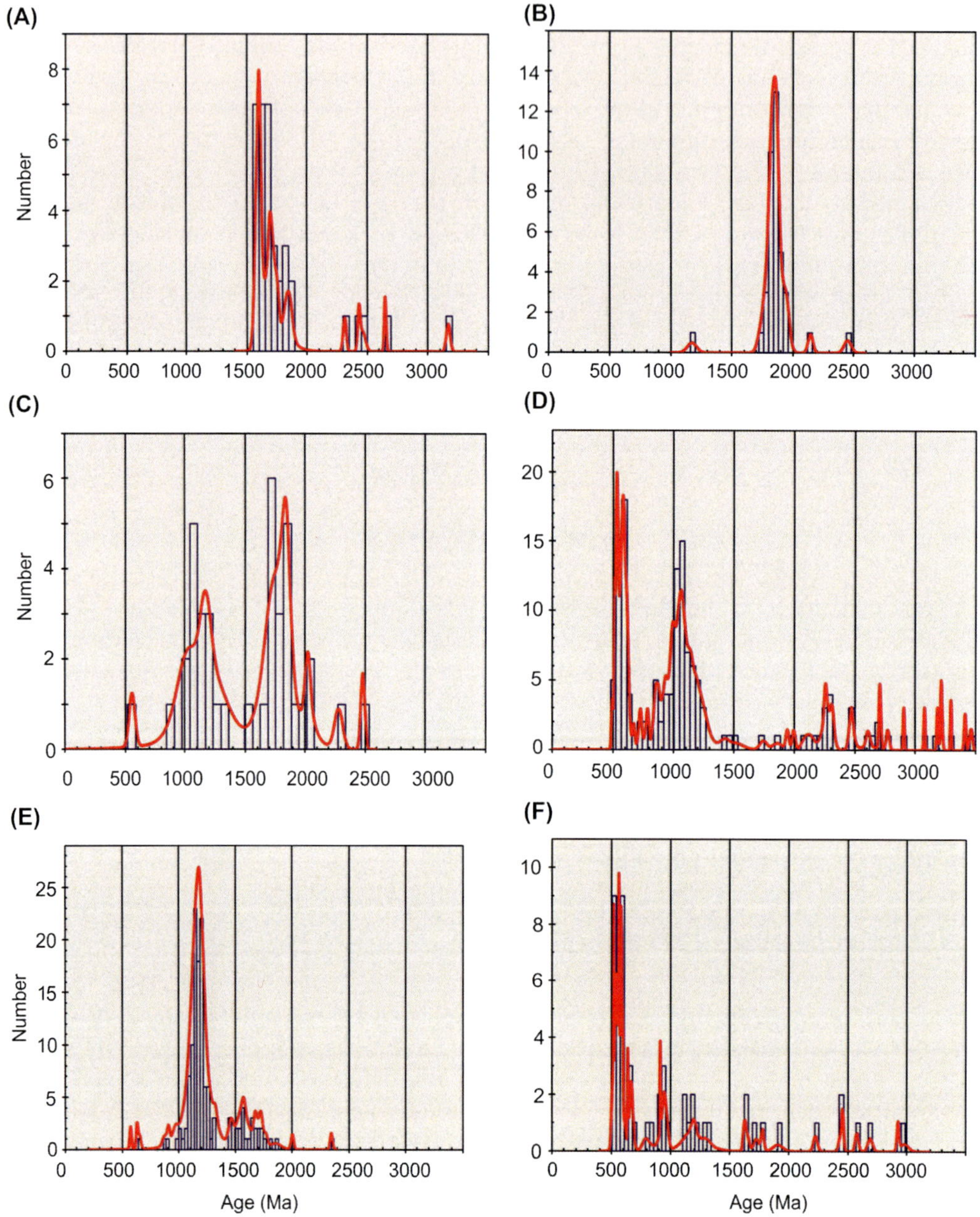

FIGURE 13.3 Selected relative probability plots (*red lines*) and histograms (blue) of detrital zircon ages from the Adelaide Rift Complex, a composite of three samples from the Kanmantoo Group, Delamerian Orogen (data from Ireland et al., 1998; supplementary data) and from the southern Anakie Province (Fergusson et al., 2001; supplementary data). Data replotted using the Isoplot program of Ludwig (2003). In this compilation age estimates for all individual grain analyses are <15% concordant; $^{207}Pb/^{206}Pb$ data are used for age estimates >1 Ga. (A) Niggly Gap Beds (36 analyses). (B) Mt. Terrible Formation (38 analyses). (C) Bonney Sandstone (39 analyses). (D) Kanmantoo Group (3 samples, 143 analyses). (E) Bathampton Metamorphics (3 samples, 134 analyses). (F) Wynyard Metamorphics (59 analyses).

Craton and prominent Grevillian 1200–900 Ma sources, consistent with derivation from the distant Musgrave and Albany–Fraser provinces. Ar/Ar ages of detrital muscovite from these samples are usually overlapping with, and younger than, the zircon ages, reflecting cooling and/or alteration in the source terranes, as shown by younger ages from the margins of detrital muscovites determined by UV laser profiling (Haines et al., 2004). Nd-Sm isotopic data indicate a different and older source for the Cambrian sedimentary successions compared to the Neoproterozoic Adelaide Rift Complex units (Turner et al., 1993), and this change in provenance is supported by the detrital zircon and muscovite ages.

Incoming of the Pacific Gondwana zircons in the Delamerian Orogen is shown by the 650–550 Ma U-Pb zircon ages found in the Kanmantoo Group (Fig. 13.3D), a predominantly quartz-rich turbidite succession deposited in the southern part of the Delamerian Orogen with equivalents in the Arrowie Basin in the north (Ireland et al., 1998; Preiss, 2000). Paleocurrent data from the turbiditic facies of the Kanmantoo Group indicate derivation from the south implying a new source, as is indicated by the detrital zircon ages (Flöttmann et al., 1998; Haines et al., 2009). The new source is also evident from Ar/Ar ages of muscovite, although these are mainly 600–550 Ma and in the older part of the main zircon peak (Haines et al., 2004). Timing of the provenance switch is well constrained by the age of tuff in the underlying Normanville Group at 526 ± 4 Ma (Cooper et al., 1992). An upper limit to the age of the Kanmantoo Group is given from a U-Pb zircon age of the intrusive Rathjen Gneiss at 514 ± 4 Ma (Foden et al., 1999), indicating an interval of 525–510 Ma for rapid filling of the Kanmantoo basin (Haines and Flöttmann, 1998).

The provenance switch to Pacific Gondwana zircon ages in the Delamerian Orogen in South Australia is also reflected in the Koonenberry Belt of northwestern New South Wales, Delamerian Orogen of western Victoria, the Thomson Orogen of central and southern Queensland, and partly in Tasmania. In the Koonenberry Belt, a rifted margin is preserved in the Late Neoproterozoic Grey Range Group with alkaline mafic rocks containing rare silicic volcanic intervals that have a U-Pb zircon age of 586 ± 3 Ma (Greenfield et al., 2011). This package has distinctive detrital U-Pb zircon and rutile ages indicating a Grenville source (Johnson et al., 2012) as evident in parts of the Adelaide Rift Complex (Ireland et al., 1998). It is most likely derived from the Musgrave Province or an eastward extension of it. The Musgrave Province formed a prominent source for sedimentary successions in the western Centralian Basin, including Uluru (Ayers Rock) (Camacho et al., 2002), and in the eastern Amadeus Basin, Harts Range Group, and Georgina Basin during the Petermann Orogeny (Maidment et al., 2007, 2013).

In northeastern Australia, metasedimentary rocks (Bathampton Metamorphics, Cape River Metamorphics, lower Argentine Metamorphics) containing almost unimodal zircon age distributions indicating a Grevillian source are found in the Anakie and Charters Towers provinces of the exposed Thomson Orogen (Fig. 13.3E; Fergusson et al., 2001, 2007). Similar age distributions have also been identified in two basement cores of sedimentary rocks (GSQ Machattie 1, HPP Goleburra 1, Brown et al., 2014) in the Machattie Beds, which are located southeast of the Diamantina Structure in the northwestern Thomson Orogen (Carr et al., 2014; Withnall and Hutton, 2013). These metasedimentary and sedimentary rocks are relatively immature and contain quartz, feldspar, and lithic fragments. The Machattie Beds and the samples from the Anakie Province have maximum depositional ages in the latest Neoproterozoic to Early Cambrian and their deposition has been related to uplift of the

Musgrave Province in the Cambrian Petermann Orogeny (Brown et al., 2014) similar to those of the Amadeus Basin (Camacho et al., 2002). The general immaturity and widespread distribution of these sedimentary and metasedimentary rocks has also been interpreted as evidence for the eastward continuation of the Musgrave Province into northeastern Australia (Fergusson et al., 2007; Fergusson and Henderson, 2015).

Basement cores collected during petroleum exploration in central and southern Queensland provided samples of quartz turbidites (Thomson Beds) of the Thomson Orogen that are typically of low metamorphic grade, steeply dipping, and were considered a northern continuation of the Ordovician turbidites of the Lachlan Orogen (Murray, 1986, 1994). These rocks are dominantly quartzose, and on a QFL plot, straddle the continental block and recycled orogen fields and partly overlap with the Ordovician turbidites of the Lachlan Orogen (Fig. 13.4). Numerous samples, which have been processed for detrital zircon ages, are characterized by the Pacific-Gondwana provenance with a maximum depositional age of 495 Ma (Brown et al., 2014; Carr et al., 2014; Kositcin et al., 2015) and marginally younger than quartz-rich metasandstones that have the same detrital age signature in the Anakie and Charters Towers provinces (Fergusson et al., 2001, 2007; Fergusson and Henderson, 2015). Radiometric ages from associated rocks indicate that the main phase of shortening/metamorphism and crustal development in the Thomson Orogen of Queensland was in the interval 510–480 Ma, overlapping the timing of the Delamerian Orogeny in southeastern Australia (Fergusson and Henderson, 2015). These results were unexpected and indicate that the Pacific-Gondwana sediment influx had a much greater volume and distribution than previously recognized, and have greatly contributed to continental growth of the Tasmanides.

In western Victoria, the Glenelg and Grampians-Stavely zones (Fig. 13.5) contain Early Cambrian quartzose siliciclastic rocks (Moralana Supergroup) equivalent to the Kanmantoo Group and the Middle Cambrian Glenthompson Sandstone (VandenBerg et al., 2000; Morand et al., 2003). These rocks are quartzose with plagioclase, K-feldspar, muscovite, with lithic fragments including silicic and mafic volcanic rock fragments, granite, low-grade schist,

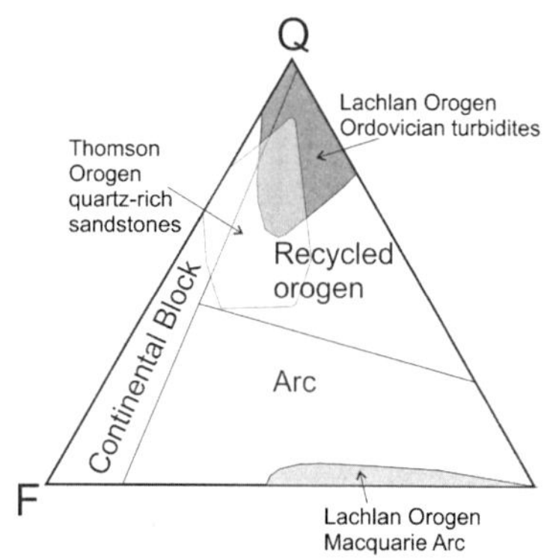

FIGURE 13.4 QFL plot showing provenance discriminating fields from Dickinson et al. (1983) with provenance fields of Late Cambrian sandstones from basement cores of the Thomson Orogen (Murray, 1994), Ordovician turbidite sandstones, and Macquarie Arc sandstones from the Lachlan Orogen (Colquhoun et al., 1999; Fergusson and Tye, 1999).

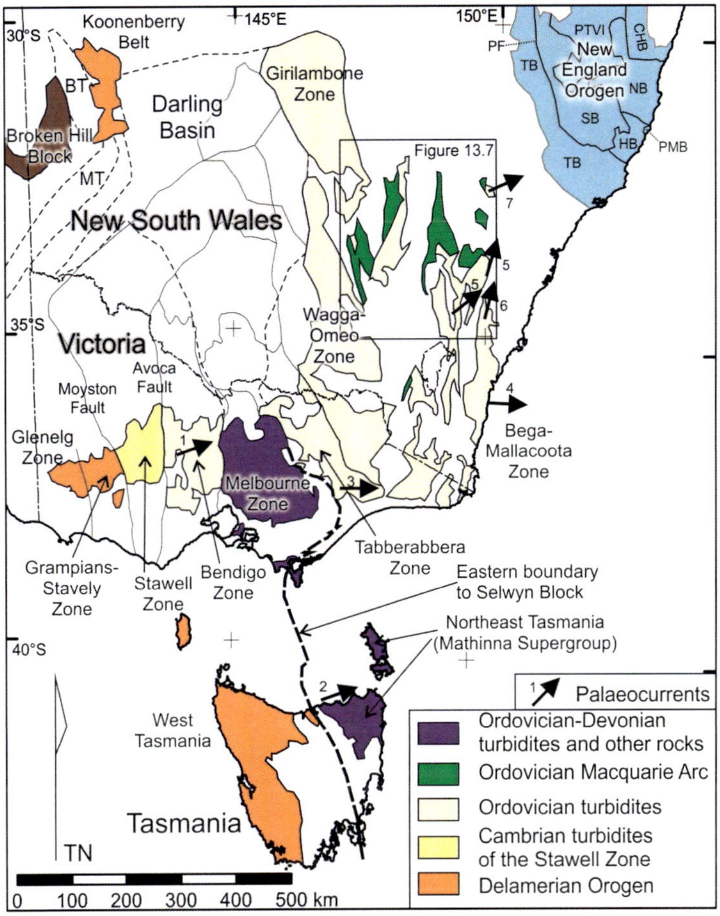

FIGURE 13.5 Map of the Lachlan Orogen in southeast Australia showing the main extent of exposure of the Ordovician quartz turbidite and the Macquarie Arc successions and various structural zones. Eastern boundary of the Selwyn Block is after Moore et al. (2015). Darling Basin (*dashed border*) is mainly exposed in the east and includes the Cobar Basin; further west the Darling Basin is mainly in the subsurface with limited exposures including within the Koonenberry Belt. Full extent shown by *dashed line*. *BT*, Bancannia Trough (*dashed border*, mainly in the subsurface); *MT*, Menindee Trough (*dashed border*, mainly in the subsurface). Subdivisions and labels in the New England Orogen. *CHB*, Coffs Harbor Block; *HB*, Hastings Block; *NB*, Nambucca Block; *PF*, Peel Fault; *PMB*, Port Macquarie Block; *PTVI*, Permian-Triassic volcanic and intrusive rocks; *SB*, subduction complex; *TB*, Tamworth Belt. Details of paleocurrent directions are: 1—Bouma C cross-laminations (Cas and VandenBerg, 1988; 562 measurements, vector mean 069 degrees), 2—lower Mathinna Supergroup Bouma C cross-laminations (Powell et al., 1993; 151 measurements, vector mean 069 degrees), 3—flutes (Fergusson et al., 1989; 131 measurements, vector mean 088 degrees), 4—Bouma C cross-laminations (Powell, 1983, 402 measurements, generalized direction from 9 areas), 5—flutes (Cas et al., 1980; 22 measurements, vector mean 049 degrees and 14 measurements, vector mean 022 degrees), 6—flutes (Jones et al., 1993; 65 measurements, vector mean 012 degrees), 7—flutes (Fergusson and Colquhoun, 1996; 19 measurements, vector mean 064 degrees). Location shown in Fig. 13.2. Location of Fig. 13.7 shown.

and chert (Stuart-Smith and Black, 1999; VandenBerg et al., 2000; Morand et al., 2003). In the Grampians-Stavely Zone, the Glenthompson Sandstone overlies the calc-alkaline, intermediate to silicic Mount Stavely Volcanics (Morand et al., 2003). Detrital zircon ages from these siliciclastic rocks have the typical Pacific-Gondwana age pattern with abundant 600−500 Ma ages and less common ages around 1100 Ma as found in the Kanmantoo Group and elsewhere (Morand et al., 2003; Squire et al., 2006a; Gibson et al., 2011). In the Koonenberry Belt, detrital zircon ages from samples before and after the Delamerian Orogeny, which is locally constrained to 505−498 Ma, have the Pacific-Gondwana signature with several samples having a notable age peak at 580 Ma (Greenfield et al., 2010, 2011; Johnson et al., 2012).

Tasmania is the southernmost part of the Tasmanides and its connection to the rest of the orogenic system has always been enigmatic (Cayley, 2011). Western Tasmania has abundant metasedimentary successions dominated by schist and quartzite with the psammitic rocks showing common detrital zircon ages of 1800−1700 Ma (Berry et al., 2001; Black et al., 2004) and widespread Delamerian (Tyennan) metamorphic ages (Berry et al., 2007). The 1800−1700 Ma detrital ages have been interpreted to reflect a North American provenance (Berry et al., 2001). One sample is dominated by Grenville-age zircons (Wings Sandstone) and several other samples show some Grenville-age zircons in addition to more common zircons with ages of 1900−1400 Ma (Turner et al., 1998; Black et al., 2004). Overall the abundance of distinctive detrital zircon ages in the metasedimentary basement of western Tasmania supports the inference that it must have been derived from further south along the East Gondwana margin (Cayley, 2011; Gibson et al., 2011; Moore et al., 2015). Paleomagnetic data indicate that Tasmania must have been located somewhere near its present position relative to Gondwana in the Late Cambrian to Early Ordovician (Li et al., 1997). This is consistent with the timing of Tyennan metamorphism across the island although some later displacement is required to account for the emplacement of the Selwyn Block in the Melbourne Zone, that prior to the Late Ordovician−Early Silurian Benambran Orogeny, must have lain hundreds of kilometers east of the Stawell and Bendigo zones in western Victoria (Gray et al., 2006; Cayley, 2011). The provenance of the Ordovician-Devonian Mathinna Supergroup in northeastern Tasmania is completely unrelated to that of western Tasmania and is similar to the Lachlan Orogen (see next).

3.2 Ordovician Turbidites and Macquarie Arc in the Lachlan Orogen

Relationships between the widespread Ordovician turbidite succession and the Ordovician to Early Silurian Macquarie Arc have been considered problematic, resulting in numerous tectonic models for the Lachlan Orogen (Quinn et al., 2014). Provenance of these two contrasting successions has been a critical issue as both successions span similar time intervals yet apparently show no evidence of facies interdigitation along numerous contacts between them. In the literature, the same contacts between these two successions have been considered as both stratigraphic and faulted by different authors (Fergusson and Colquhoun, 1996; Meffre et al., 2007; Quinn et al., 2014). This has led to suggestions that the Macquarie Arc is somehow structurally emplaced among the Ordovician turbidites by either overthrusting and/or strike-slip faulting, despite the development of an excellent geophysical database showing the presence of many significant faults but unable to verify the existence of the proposed terrane bounding structures. For example, Quinn et al. (2014) have presented models involving no major fault dislocation between the Macquarie Arc succession and the Ordovician turbidites, whereas these authors had earlier argued for major terrane displacements between these units in the early Paleozoic history of the Lachlan Orogen (Glen et al., 2009).

In the Stawell Zone of western Victoria (Fig. 13.5), the quartz-rich turbidite succession is mainly Late Cambrian, as inferred from scarce fossils (acritarchs), geochronological data, and the lack of graptolites (Squire et al., 2006a,b). In the Bendigo Zone (Fig. 13.5), the age of the succession (Castlemaine Group), Early to Middle Ordovician, is well established from abundant thin beds of graptolitic black shale interbedded with the turbidite succession (VandenBerg et al., 2000). In the southeastern Bendigo Zone, the Sunbury Group consists of quartz-rich turbidites interbedded with graptolitic shales that indicate continuous deposition through the Late Ordovician, which is in contrast to the remainder of the Bendigo Zone (VandenBerg et al., 2000). The Melbourne Zone lacks a widespread Ordovician turbidite succession, except in the southwest in the Mornington Peninsula where the basal part of the Ordovician turbidite succession is similar in thickness to the equivalent succession in the Bendigo Zone but is overlain by a condensed interval (24 m thick) of chert and graptolitic black shale assigned by VandenBerg et al. (2000) as Bendigonian 4 to Castlemainian 1 in age (481–473 Ma, using timescale in Percival et al., 2011). This is interpreted as a result of the turbidites being deposited on the margins of a paleotopographic high formed by the Selwyn Block, whereas in the eastern Melbourne Zone Middle to Late Ordovician black shale reflects sediment starvation over the high (Cayley et al., 2002).

The Ordovician turbidite succession north and east of the Melbourne Zone (Adaminaby, Wagga, and Girilambone groups) is remarkably homogenous. Its stratigraphy is now known in several regions due to interbedded thin chert intervals that contain conodonts, and based on these age data, the succession has been subdivided into a number of packages (VandenBerg and Stewart, 1992; Percival et al., 2011; Percival, 2012). Thin-bedded chert intervals occur in three widespread units of Chewtonian, mid-Darriwilian, and late Darriwilian age (Percival et al., 2011). The upper part of the succession is an interval of black shale (400–500 m thick) of Late Ordovician age (Bendoc Group and equivalents); sandstone is almost completely absent from this unit, indicating starvation of turbidite sediment on the submarine fan (VandenBerg et al., 2000). The cherts have a continental margin geochemical signature shown by high Al_2O_3/Fe_2O_3 ratios, LREE enrichment, and low total REEs (Bruce and Percival, 2014), consistent with their setting interbedded with the terrigenous-derived turbidites.

The Late Cambrian and Ordovician turbidite successions of the Lachlan Orogen are uniformly quartz-rich as shown on the QFL plot (Fig. 13.4) with minor plagioclase, muscovite, and lithic fragments including low-grade metamorphic fragments, fine sedimentary and volcanic rock fragments (Colquhoun et al., 1999; Fergusson and Tye, 1999). The Ordovician turbidites are well known for their Pacific-Gondwana detrital zircon signature with abundant ages at 600–500 Ma usually with a subordinate peak around 1200–1000 Ma as shown by published data (Fig. 13.6A; Ireland et al., 1998; Fergusson and Fanning, 2002; Fergusson et al., 2005, 2013; Squire et al., 2006a; Glen et al., 2013) and a large unpublished database collected by Ian Williams (Williams, 1998, 2001; Williams and Pulford, 2008). Detrital zircon ages with similar patterns to these are given for Bendigonian and Darriwilian sandstone

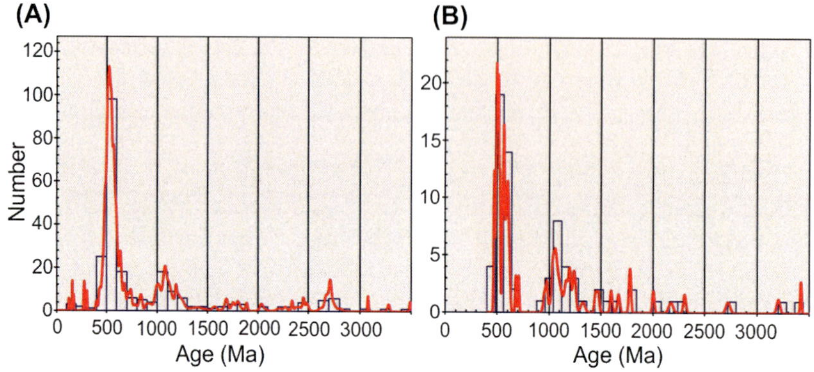

FIGURE 13.6 Selected relative probability plots (*red lines*) and histograms (blue) of detrital zircon ages from the Ordovician turbidites (composite of four samples, data from Fergusson and Fanning, 2002, supplementary data; and Fergusson et al., 2005, supplementary data) and one sample from the Mitchell Formation near the base of the Macquarie Arc succession (data from Glen et al., 2011, supplementary data). Data replotted using the Isoplot program of Ludwig (2003). In this compilation age estimates for all individual grain analyses are <15% concordant; $^{207}Pb/^{206}Pb$ data are used for age estimates >1 Ga. (A) Ordovician turbidites (4 samples, 233 analyses). (B) Mitchell Formation (71 analyses).

samples from mainly west of the Macquarie Arc in figures without accompanying data tables by Glen et al. (2011) and Glen (2013). Detrital muscovite ages are mainly of Delamerian age indicating that the Delamerian and Ross Orogens were a significant source (Turner et al., 1996).

Paleocurrents based on flutes, scour marks, and Bouma C cross-laminations have been measured for some intervals of the Ordovician turbidites especially in the eastern part of the orogen where the rocks are usually better exposed (Powell, 1984). These measurements indicate sediment derivation from the west in Victoria and Tasmania with a swing in trends to northeasterly and northerly sediment flow in eastern New South Wales (Fig. 13.5). However, interpreting these paleocurrent trends is not straightforward as major rotations of some regions may have occurred due to late megakinking (Powell et al., 1985). Moreover, oroclinal folding has been proposed (Cayley, 2012; Musgrave, 2015), thus implying a 90 degrees anticlockwise rotation of the paleocurrent data in the southern Tabberabbera Zone. A simplistic interpretation is that these directions reflect westerly derivation in the more western part of the superfan with a swing to the north in the northeastern part, as the superfan was deflected around the Macquarie Arc.

The Macquarie Arc succession is dominated by mafic to intermediate volcanic rocks, associated volcaniclastic rocks (including breccias, conglomerates, sandstones, and mudstones), and numerous mafic/intermediate and rarer silicic intrusions (Percival and Glen, 2007; Crawford et al., 2007). Parts of the succession contain shallow-marine limestone, whereas bedded chert and black shale occur in deep-marine settings (Percival and Glen, 2007). The succession is divided into several belts that are separated by younger rocks and widespread exposure of the Ordovician quartz-rich turbidites between the western and central belts of the Macquarie Arc in central New South Wales (Fig. 13.7). Two main intervals of activity are inferred with an Early Ordovician phase, separated by a hiatus of 9 Ma from a late Middle

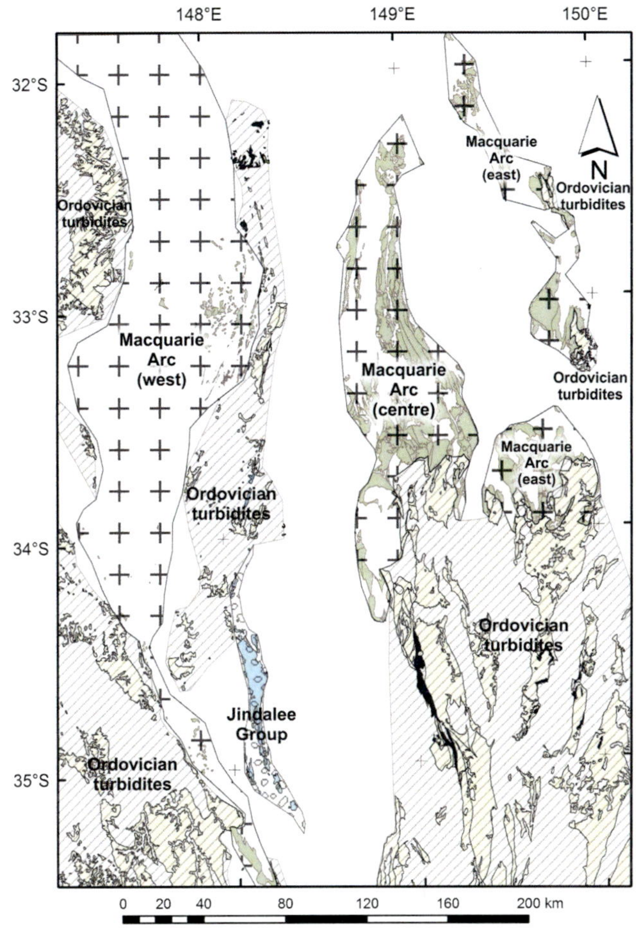

FIGURE 13.7 Outcrop distribution and interpreted extent of the Macquarie Arc (light green = outcrop), Ordovician turbidites (= light yellow, dark gray = black shale), and Jindalee Group (ultramafics, mafic volcanics, and chert of Middle Ordovician age interpreted as formed by rifting, Lyons and Percival, 2002) in central New South Wales. Extent of each unit has been interpreted from magnetic data. Outcrop distribution from Raymond et al. (2012). Location shown in Fig. 13.5.

Ordovician to Early Silurian interval that has been subdivided into three phases by Percival and Glen (2007). Most of the exposed parts of the Macquarie Arc consist of rocks formed in the latter longer interval. The source of sedimentary rocks in the succession is dominated by mafic to intermediate volcanic detritus typical of interlayered and neighboring volcanic successions. In general quartz is mostly absent in the volcaniclastic rocks, but has been found in several thin horizons derived from hydrothermal deposits and uncommon silicic igneous rocks (Packham et al., 2003). The geochemical and isotopic characteristics of the volcanic rocks indicate an intraoceanic arc setting (Crawford et al., 2007). Unexpectedly, it has been found in the Early Ordovician (Phase 1) part of the succession that volcaniclastic rocks contain a detrital zircon signature typical of the Ordovician quartz turbidites with prominent Pacific-Gondwana and Grenville peaks showing a provenance linkage with Gondwana (Fig. 13.6B; Glen et al., 2011).

Detrital zircons from volcaniclastic sedimentary rocks associated with the second magmatic interval have ages that are the same as the associated volcanics. Inherited zircons from the volcanic rocks no older (in general) than 500 Ma have positive ε_{Hf}, indicative of primitive sources rather than being derived from Gondwana (Glen et al., 2011). One plausible explanation is that over time the Macquarie Arc was progressively removed from the vicinity of Gondwana by sea floor spreading in the Wagga Marginal Sea, which separated the arc from the East Gondwana margin (Crawford et al., 2007; Glen et al., 2011; Quinn et al., 2014; see Discussion). A similar conclusion has been made by Bruce and Percival (2014) based on geochemical data from bedded cherts interbedded with the Ordovician turbidites. Additionally, the arc itself has undergone rifting, especially in the late Middle Ordovician, when MORB volcanism and chert deposition (associated with the Jindalee Group) took place (Fig. 13.7; Lyons and Percival, 2002; Quinn et al., 2014).

3.3 Silurian-Devonian Foreland Successions in the Western and Southern Lachlan Orogen

The Melbourne Zone in central Victoria and the Mathinna Supergroup in northeast Tasmania consist of Ordovician to Devonian sedimentary successions that, in the Silurian-Devonian, formed in a foreland setting to the eastern part of the Lachlan Orogen (Powell et al., 1993, 2003). The Darling Basin to the north in central New South Wales was also in a similar tectonic setting (Powell, 1984; Neef, 2012). Sandstones in the Melbourne Zone and Mathinna Supergroup resemble those of the Ordovician turbidites of the Lachlan Orogen and are dominantly quartzose, but on a QFL plot (Fig. 13.8) are slightly less feldspathic and have a greater range in lithic fragment content including common volcanic rock fragments (Powell et al., 1993, 2003). The sandstones from the Darling Basin, including the Bancannia Trough, are also compositionally similar to those of the Melbourne Zone (Neef and Bottrill, 1991, 2001; Neef et al., 1995). In all three basins, paleocurrent directions, based on flutes, scour marks,

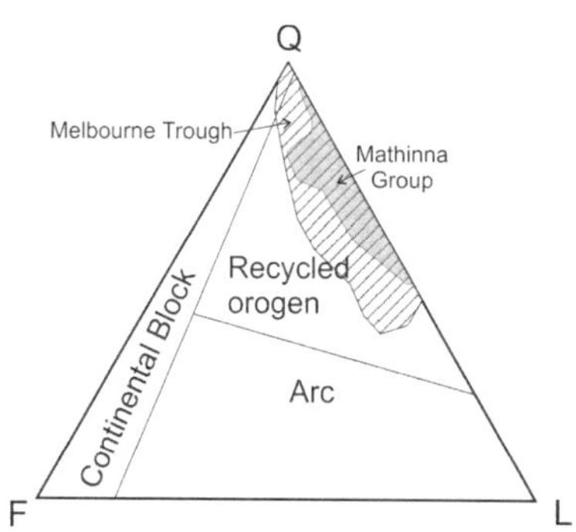

FIGURE 13.8 QFL plot showing provenance discriminating fields Dickinson et al. (1983) with provenance fields of Silurian-Devonian Melbourne trough sandstones (Powell et al., 2003) and Ordovician to Devonian Mathinna Group sandstones (Powell et al., 1993).

and cross-lamination in turbidites and cross-bedding in shallow marine to fluvial units are complicated but show common derivation from the west, with bimodal paleocurrents aligned along the north to northwest-trending basin axes (Powell et al., 1993, 2003; Neef, 2012). For strata of Emsian age in the Melbourne Zone, paleocurrents and sediment provenance show a significant change, with the Norton Gully Sandstone having an eastern source with an increased volcaniclastic component derived from the eastern Lachlan Orogen (Powell et al., 2003). Similar changes occurred in the Givetian in the eastern Darling Basin where an influx of hornfelsed metasedimentary clasts in conglomerates and pebbly sandstones was derived from the east (Powell, 1984).

Detrital zircon ages have been determined for three sandstone samples from the Melbourne Trough with the stratigraphically lowest sample from the Wenlockian Kilmore Siltstone (Fig. 13.9A), and two samples from the Late Silurian to Early Devonian succession

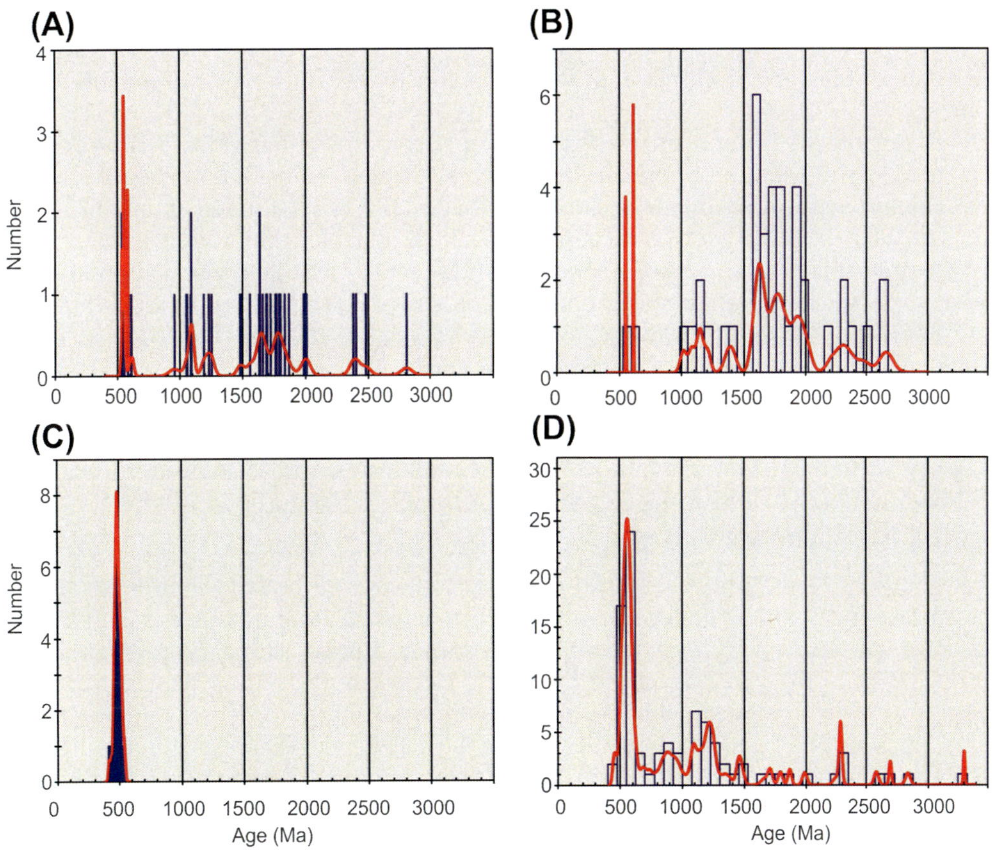

FIGURE 13.9 Selected relative probability plots (*red lines*) and histograms (blue) of detrital zircon ages from the Melbourne Zone, western Lachlan Orogen (data from Squire et al., 2006a, supplementary data) and the Mathinna Group, northeastern Tasmania (Black et al., 2004; data supplied by Simon Bodorkos). Data replotted using the Isoplot program of Ludwig (2003). In this compilation age estimates for all individual grain analyses are <15% concordant; $^{207}Pb/^{206}Pb$ data are used for age estimates >1 Ga. (A) Kilmore Siltstone (35 analyses). (B) Humevale Siltstone (40 analyses). (C) "Glen Creek Sandstone" (46 analyses). (D) Mathinna Group (2 samples, 92 analyses).

(Fig. 13.9B and C) (Squire et al., 2006a). From the Mathinna Supergroup two sandstone samples have had detrital zircon ages determined (Fig. 13.9D); one from the Ordovician Stony Head Sandstone and another from the upper part of the unit (Black et al., 2004). All samples show a prominent Pacific-Gondwana signature but in detail there are minor variations (Fig. 13.9). The sample from the Glen Creek Sandstone (Squire et al., 2006a) has abundant zircons with ages 510–480 Ma, indicating a source from Delamerian igneous activity, as in western Victoria and western Tasmania, and lacks Grenville-age zircons. The sample, which according to Squire et al. (2006a) was taken from the Norton Gully Sandstone, is in fact located within the Humevale Siltstone (according to the coordinates provided by the authors; see supplementary data). This sample has common ages of 2000–1500 Ma, consistent with a local source in adjacent western Tasmania. We suggest that rather than reflecting transport from a distal source (Squire et al., 2006a), the Pacific-Gondwana zircons in these samples have most likely come from reworking of Ordovician and older metasedimentary sources deformed and uplifted in the Delamerian and Benambran orogenies, such as in the Stawell and Bendigo zones (Cayley et al., 2011). None of these five samples has any significant peak younger than 450 Ma, thus indicating that concurrent silicic igneous activity was not a source.

Data available for Tasmania show that detrital zircon ages in the sedimentary successions are matched by the ages of inherited zircons in Paleozoic granites that intrude these successions. Thus in western Tasmania, granites with ages 374–351 Ma have abundant inherited zircons of 1800–1700 Ma ages, as is common in supracrustal Precambrian quartzites in western Tasmania (Black et al., 2010). In northeast Tasmania, granites with ages 400–373 Ma display approximately the same age pattern of inherited zircons as in the Mathinna Supergroup implying that these rocks occur at deeper levels in the crust and were involved in melting and contamination to form the granites (Black et al., 2010).

3.4 Provenance of New England Orogen Sandstones and Conglomerates and Provenance Switching in Subduction Complex Sandstones of the Northern New England Orogen

The southern New England Orogen consists of a western foreland fold-thrust belt with forearc basin and arc flank deposits mainly of Devonian to Carboniferous age with overlying Early Permian sedimentary and volcanic rocks formed during rifting (Veevers et al., 1994; Murray, 1997). In the Late Silurian to mid–Late Devonian, the setting was an intraoceanic arc/backarc, which collided with the active continental margin of the Lachlan Orogen at around 375 Ma (Offler and Murray, 2011). After the collision a new west-dipping subduction zone formed and in the late Late Devonian to Carboniferous the setting was a continental arc and forearc (Murray et al., 1987; Offler and Murray, 2011). Sandstone compositions are well documented in the forearc basin represented by the Tamworth Belt (Fig. 13.5). They are lithic to feldspathic with minor and even extremely rare quartz clasts (Korsch, 1984). From the Devonian to the Late Carboniferous the composition of sandstones changes from dominantly mafic to andesitic detritus lower in the succession to an increasingly dacitic to rhyolitic composition of lithic fragments in the upper part of the succession (Korsch, 1984). These changes are accompanied by detrital pyroxene lower in the succession being replaced by detrital hornblende higher in the succession (Korsch, 1984). Conglomerates of Cambrian age at the base of the succession, as well as Devonian conglomerates, have abundant intermediate volcanic clasts and less common plutonic clasts with a calc-alkaline geochemistry indicative of a magmatic arc with minimal continental crust (Leitch and Willis, 1982; Leitch and Cawood, 1987; Morris, 1988).

Sandstones in the subduction complex in the southern New England Orogen have been studied not only to determine provenance, but also to provide constraints on the ages of these rocks that are poorly known due to their deep-marine depositional setting and lack of macrofossils (Korsch, 1984). They are a highly deformed assemblage with abundant turbidites, tuffaceous rocks, and less common chert and mafic volcanic rocks that represent typical oceanic plate and overlying trench-wedge turbidite successions (Cawood, 1982; Fergusson, 1985; Offler et al., 1988; Aitchison et al., 1992). Ages are poorly constrained in general apart from some limited radiolarian ages for cherts (Aitchison, 1988; Aitchison et al., 1992). Volcaniclastic and feldspathic compositions of subduction complex sandstones reflect derivation from the volcanic arc to the west, and show a similar variation in provenance to sandstones in the forearc basin succession (Korsch, 1978, 1981, 1984). Detrital zircon ages and hornblende ages have been determined by Korsch et al. (2009a) for two samples from the Coramba beds in the subduction complex in the Coffs Harbour Block; zircons and amphiboles give consistent ages of 323−318 Ma. Given the abundance of primary volcanic-derived detritus in these rocks, the ages are considered to reflect the age of deposition. The zircons were selectively dated on the basis of their euhedral shapes to provide an estimate of the age of the host sandstone, which is consistent with an earlier determination by a Rb-Sr isochron at 318 ± 8 Ma, indicating a metamorphic age that provides a minimum constraint on the depositional age based on samples from the southern Coffs Harbour Block (Graham and Korsch, 1985). Thus although these data indicate the importance of the magmatic arc in the source of these sedimentary rocks, consistent with their volcaniclastic nature, they did not include randomly determined ages of rounded zircons and thus were not designed to determine their provenance spectrum (Korsch et al., 2009a).

In the northern New England Orogen, the sedimentary petrography of sandstones is less well established but broadly known from lithological descriptions of mapped rock units, as summarized by Donchak et al. (2013) and data provided by Leitch et al. (2003) for sandstones of the subduction complex (Fig. 13.10). Some detrital zircon age spectra are also available (Korsch et al., 2009a). A dissected Carboniferous magmatic arc is exposed in the west (Connors-Auburn Province) with the forearc basin in the center (Yarrol Province) and the subduction complex in the east (Curtis Island Group and equivalents to the south) (Fig. 13.10). As in the southern New England Orogen, the magmatic history of the arc is well documented from the Yarrol Province forearc, which has a lower succession of Late Silurian to Devonian age dominated by mafic to intermediate volcanic and volcaniclastic rocks with geochemical characteristics of an island arc setting (Offler and Murray, 2011). The Late Devonian to Early Carboniferous succession of the Yarrol Province shows a transition from the underlying island arc to a source from an Andean continental margin arc exposed to the west in the Connors-Auburn Province (Donchak et al., 2013). In contrast to the Tamworth Belt, however, Late Carboniferous units in the Yarrol Province are much less widespread and are largely restricted to the Rockhampton region of central Queensland.

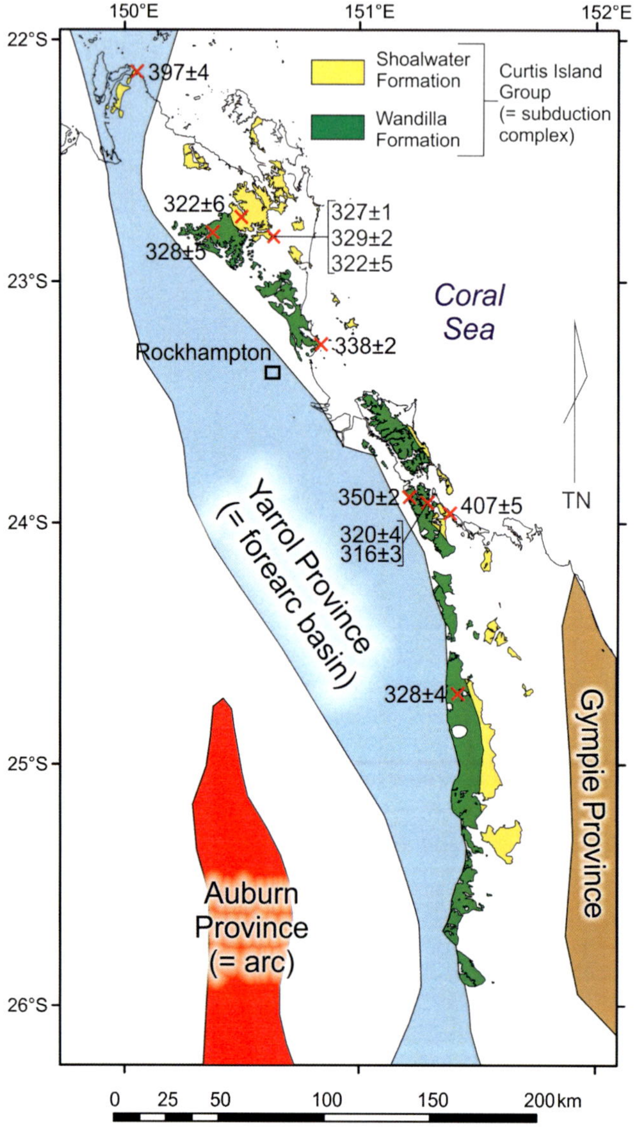

FIGURE 13.10 Map of part of coastal central Queensland showing outcrop of the Curtis Island Group (Wandilla and Shoalwater formations) of the Devonian-Carboniferous subduction complex. Red crosses mark locations of U-Pb zircon samples with ages of the youngest identified zircon peak (from Table 3 in Korsch et al., 2009a). Geology modified from the 1:1 million scale digital geological map of Australia (Raymond et al., 2012). Note the Shoalwater Formation shown in the northern Yarrol Province is of uncertain significance. Location shown in Fig. 13.2.

The subduction complex sandstones in the northern New England Orogen show a remarkable provenance switch from the volcaniclastic detritus in the Wandilla Formation and equivalents to quartz-rich sandstones in the Shoalwater Formation (Fig. 13.11; Leitch et al., 2003). In contrast, sandstones in most of the southern New England Orogen are

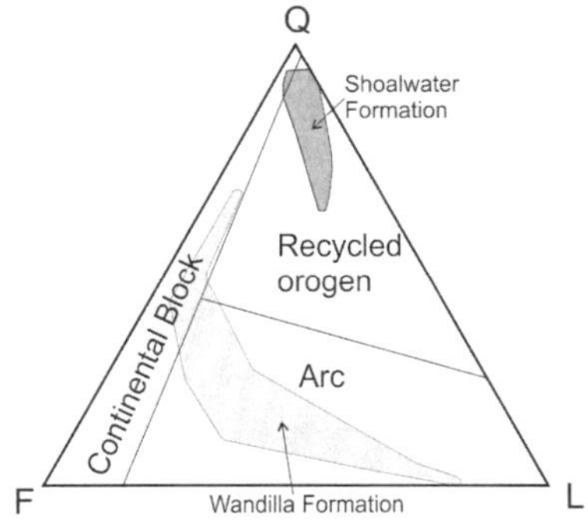

FIGURE 13.11 QFL plot showing provenance discriminating fields from Dickinson et al. (1983) with provenance fields of sandstones from the Carboniferous Wandilla Formation and the Carboniferous Shoalwater Formation in the subduction complex of the northern New England Orogen (Leitch et al., 2003).

volcanolithic and/or feldspathic and have only minor amounts of quartz (Korsch, 1984). The subduction complex rocks lack macrofossils, and as for the southern New England Orogen, their age has been inferred by provenance linkage to the Yarrol Province to the west. The Early Carboniferous forearc basin is characterized by oolitic limestones interbedded with the dominantly clastic succession, and oolites are found widely dispersed among associated volcanically derived sandstones. The Wandilla Formation has been mapped for a distance of nearly 400 km along strike and is notable for oolite-bearing lithic sandstones that have been correlated with Early Carboniferous strata of the Yarrol Province. Similar lithic sandstones have been mapped further southward into the southern New England Orogen where they occur on the limbs of the Texas and Coffs Harbour oroclines (Murray et al., 1987; Murray, 1997; Rosenbaum, 2012). The zircon ages from the Wandilla Formation indicate that the unit is of Early to Late Carboniferous age and of longer duration than previously thought (Murray et al., 1987; Korsch et al., 2009a). This implies uplift in the Late Carboniferous in the forearc basin with reworking of Early Carboniferous oolitic-bearing sands and their redeposition into the Late Carboniferous trench.

Volcaniclastic sandstones of the Wandilla Formation are a poorly sorted mix of volcanic lithic fragments of mafic to silicic composition, plagioclase, quartz, and less common mineral fragments including micas and augite and various other types of lithic fragments such as plutonic, metamorphic, and sedimentary rocks (Leitch et al., 2003). The Shoalwater Formation to the east, by contrast with the Wandilla Formation, largely lacks chert and mafic volcanic units and is dominated by turbidites. Unlike the Wandilla Formation, sandstones of the Shoalwater Formation are dominated by quartz but other clast types are similar albeit with greatly reduced abundance (Leitch et al., 2003). Some lithic sandstones in the Wandilla Formation, east of Rockhampton along the coast, are more quartzose than normally seen in this unit and appear transitional to the higher quartz contents of the Shoalwater Formation, although a complete transition was not achieved (Fig. 13.11). Detrital zircon ages for sandstones from the subduction complex assemblage were provided by Korsch et al. (2009a); Fig. 13.12. Their study involved six samples from five localities of volcanic-lithic sandstones from the Wandilla Formation, five samples from three localities of quartzose lithology from the Shoalwater Formation, and four samples of quartzose and volcanolithic lithologies from

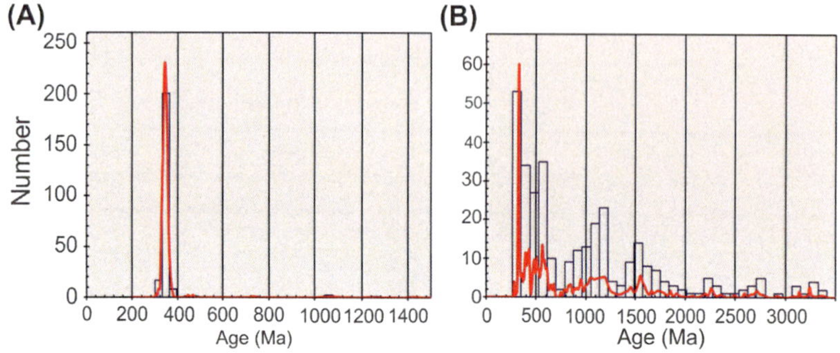

FIGURE 13.12 Relative probability plots (*red lines*) and histograms (blue) of detrital zircon ages from the northern New England Orogen including a composite of six samples from the Wandilla Formation (note only euhedral zoned igneous zircons have been analyzed, see text) and a composite of five samples from the Shoalwater Formation (data from Korsch et al., 2009a; supplementary data). Data replotted using the Isoplot program of Ludwig (2003). In this compilation age estimates for all individual grain analyses are <15% concordant; $^{207}Pb/^{206}Pb$ data are used for age estimates >1 Ga. Locations of samples shown in Fig. 13.10. (A) Wandilla Formation (235 analyses). (B) Shoalwater Formation (324 analyses).

separate localities in the Neranleigh-Fernvale beds of southeastern Queensland. The youngest zircons in these samples varied from concentrations at ~407 Ma to 327–322 Ma consistent with Devonian to mid-Carboniferous ages with the younger ages similar to those of the more outboard lithic sandstones. The quartz-rich sandstone samples with abundant Carboniferous zircons indicate partial derivation from the magmatic arc of this age. All the Shoalwater Formation samples have abundant older zircons with common ages in the pre-Carboniferous with peaks at 400 Ma and 650–500 Ma, in addition to common Precambrian zircons with common ages at 1300–1000 Ma and 1850 Ma (Fig. 13.12B). These ages are consistent with sources that would have been widely exposed in the Late Paleozoic in central to north Queensland, including rocks of the Paleoproterozoic–early Mesoproterozoic inliers and those of the Thomson Orogen (Korsch et al., 2009a).

Early Permian extensional basins developed along the New England Orogen in an interval of tectonic readjustment between the Carboniferous Andean active margin and the site of a new Andean magmatic arc shown by Late Permian to Early Triassic plutonic and volcanic rocks across the former forearc basin and subduction complex (Veevers et al., 1994; Korsch et al., 2009b). Detrital zircon ages have been established from sandstone samples in the Nambucca Block, one of these extensional basins overlying the former subduction complex in the southern New England Orogen (Adams et al., 2013a; Shaanan et al., 2015). These detrital zircon studies have shown that for the Nambucca Block, zircons are mainly Devonian to Carboniferous but include some Early Permian ages and older components such as the Pacific-Gondwana and Grenville ages. The age spectra are consistent with derivation from eastern Australia, in particular its Late Devonian to Carboniferous magmatic arc. Additionally, detrital zircon ages have been determined from samples from the Permian to Triassic succession of the Gympie Province in the northeastern part of the New England Orogen (Li et al., 2015). This province has been considered as either having developed upon an attenuated eastern part of the subduction complex (Holcombe et al., 1997) or to have formed part of an exotic terrane accreted to the eastern part of the orogen (Aitchison and Buckman, 2012). For the Gympie Province, zircon ages are dominantly Carboniferous and Permian and reflecting sources within the New England Orogen to the west, ruling out an exotic origin for this assemblage (Li et al., 2015).

3.5 Local Derivation in the Northern Tasmanides (Mossman Orogen)

Much of the Mossman Orogen is dominated by the disrupted Silurian-Devonian Hodgkinson Formation and similar rocks within the Broken River Province nearer its southern margin. The assemblage consists mainly of turbidites with minor chert and mafic volcanic rocks. Melange, complex folding, and multiple foliation development are widespread in the assemblage and the common consistency of younging directions to the west, in combination with steeply dipping units, indicates either exceptional thicknesses, or more likely, imbrication of the deep-marine succession (Henderson et al., 2013). It has been interpreted as the fill of a backarc basin (Donchak in Glen, 2005), but the widespread disruption indicates accretion in an east-facing subduction complex thought to be synchronous with the development of a coeval magmatic arc for which the eroded roots are exposed as plutonic rocks west of the Tasman Line (Pama Igneous Association, Fig. 13.13; Henderson et al., 2013).

Sandstones within the Hodgkinson Formation are typically of quartz intermediate composition with abundant quartz, altered plagioclase, less common K-feldspar, and minor lithic fragments (Domagala, 1997). Lithic fragments consist of felsic and mafic volcanic rock fragments, sedimentary fragments, and metamorphic rock fragments. From a combination of petrography, geochemical analyses of graywackes on provenance plots and limited U-Pb zircon ages from igneous clasts and detrital zircons, Domagala (1997, p. 233) considered that the main source of the Hodgkinson Formation was the craton to the west with a significant contemporaneous igneous input (in some sandstones). A compilation of detrital zircon ages from six samples, dated by U-Pb LA-ICP-MS techniques (Adams et al., 2013b), confirms the importance of the western cratonic source for half of the samples, which have many ages in the range 1750−1500 Ma (Fig. 13.14A). These ages are consistent with the range of igneous and metamorphic ages found in the adjacent Georgetown, Coen, and Yamba inliers (Fig. 13.2). Three other samples are dominated by zircons with ages in the range 490−400 Ma (Fig. 13.14B), consistent with derivation from igneous rocks of the Early Ordovician Macrossan Igneous Association, the Late Ordovician accreted island arc, Late Ordovician felsic igneous rocks, and the felsic igneous rocks of the older part of the Pama Igneous Association. Overall the assemblage is clearly derived from a variety of sources, which all developed within the cratonic region to the west but not dominated by basement derivation as considered from the petrographic and geochemical data. The sample with the youngest detrital zircon has a cluster of five grains at 360 ± 7 Ma, which is an age within the error of that of cross-cutting granite in the east (U-Pb zircon age of 357 ± 6 Ma; Zucchetto et al., 1999). This suggests a short gap between sedimentation, deformation, and plutonism (Adams et al., 2013b).

An age spectrum for detrital zircon is known from a sample of sandstone from the Carriers Well Formation, a unit with diverse sedimentary and volcanic units that is part of an island arc assemblage located on the southwestern margin of the Mossman Orogen and accreted in the Early Silurian (Henderson et al., 2011). Its youngest cluster at 454 Ma is consistent with

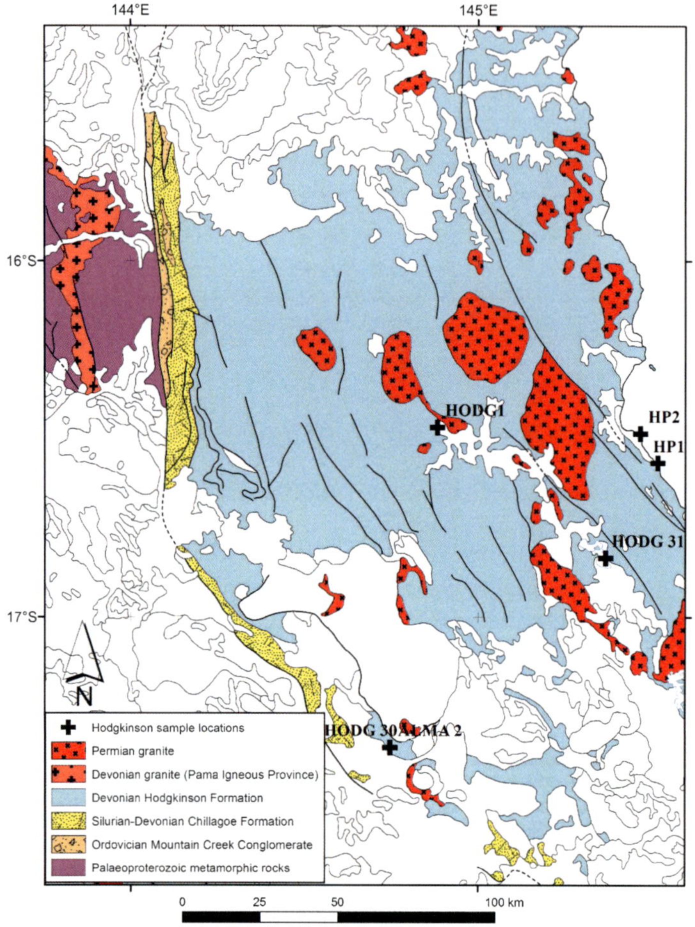

FIGURE 13.13 Geology of the Hodgkinson Province of the northern Mossman Orogen highlighting the main units; adapted from Geological Survey of Queensland (2012) 1:2 million scale geological digital map of Queensland. Location shown in Fig. 13.2.

the age of fossils from this unit and represents arc-derived detritus. However, ages between 3400 and 700 Ma reflect a contribution also from continental sources.

3.6 Orogenic and Cratonic Sources in the Permian–Triassic Sydney Basin

The Permian–Triassic Sydney Basin has a complex history with rifting in the Early Permian followed by thermal subsidence. It then formed part of a foreland basin setting during deformation and igneous activity in the Late Permian–Late Triassic Hunter-Bowen Orogeny that took place in the adjoining New England Orogen (Veevers et al., 1994). A cratonic source of quartz-rich sediment was derived from the southwest and dominates sandstones along the western margin of the basin, with incursions across the basin, such as the

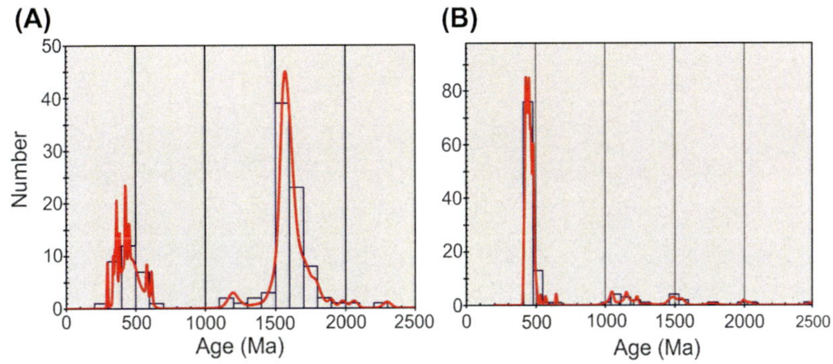

FIGURE 13.14 Relative probability plots (*red lines*) and histograms (blue) for the interval 2500–0 Ma for pooled samples from the western (samples ALMA2, HODG30, and HODG1) and eastern Hodgkinson formation (samples Hodg31, HP1, and HP2) from the northern Mossman Orogen (data from Chris Adams and Bob Henderson for figures showing probability plots for all these samples in Adams et al., 2013b). Data replotted using the Isoplot program of Ludwig (2003). In this compilation age estimates for all individual grain analyses are <15% concordant; $^{207}Pb/^{206}Pb$ data are used for age estimates >1 Ga. See Fig. 13.13 for sample locations. (A) Western Hodgkinson Formation (3 samples, 115 analyses). (B) Eastern Hodgkinson Formation (3 samples, 118 analyses).

Hawkesbury Sandstone (Fig. 13.15) near the top of the succession. Lithic sandstones and chert-bearing conglomerates were derived from the northeast, and consist of volcanic and other lithic detritus sourced from the New England Orogen. At the boundary between the Narrabeen Group and the overlying Hawkesbury Sandstone, mixing of the two provenances has occurred with changes from lithic sandstones to overlying quartz-rich sandstones (Cowan, 1993). This provenance mixing was facilitated by the fluvial environments and consistent with the swing in paleocurrents of both wedges of sediment as they turn from across the basin into the main trunk distributary system that flows along the basin (Conaghan et al., 1982; Veevers et al., 1994; Veevers, 2015).

Detrital zircon ages determined for several samples in the Sydney Basin by Sircombe (1999) show that the provenance pattern is complex and reflects both distal and more proximal cratonic sources. Two samples from the Hawkesbury Sandstone lack zircons of Lachlan Orogen age and are dominated by those of Pacific-Gondwana age (Fig. 13.16C). This implies rejuvenation of the source that supplied the Ordovician turbidites of the Lachlan Orogen, and the older successions of the Kanmantoo Group and Thomson Orogen in Queensland. In contrast, one sample from the Tallong Conglomerate, at the base of the western margin of the Sydney Basin, shows dominant Lachlan Orogen age zircons (Fig. 13.16A), consistent with a local source, whereas the Terrigal Formation sample, from beneath the Hawkesbury Sandstone near the eastern margin of the Sydney Basin, shows zircon ages typical of the adjacent New England Orogen (Fig. 13.16B). This pattern contrasts with Permian and Triassic sedimentary rocks of the subsurface Ovens Graben in northern Victoria and southern New South Wales (Fig. 13.2), and coeval with the Sydney Basin succession. Samples analyzed from these rocks have detrital zircon age patterns indicative of local Lachlan Orogen sources on either side of the rift (Fig. 13.16D; Sircombe and Hazelton, 2004).

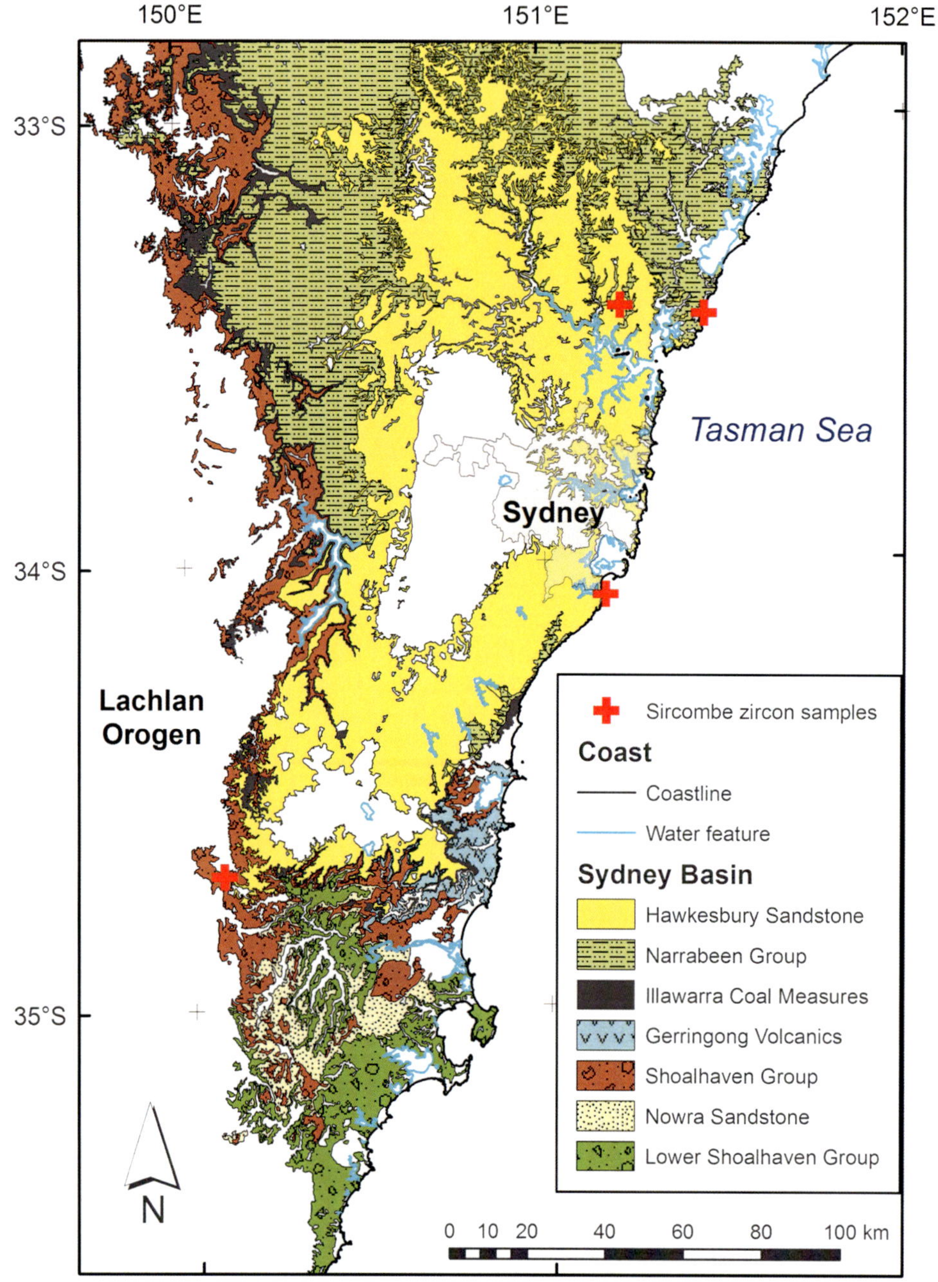

FIGURE 13.15 Map of the central and southern part of the Sydney Basin showing major units discussed in text and the location of the four detrital zircon samples of Sircombe (1999). Map modified from the 1:1 million scale digital geological map of Australia (Raymond et al., 2012). Location shown in Fig. 13.2.

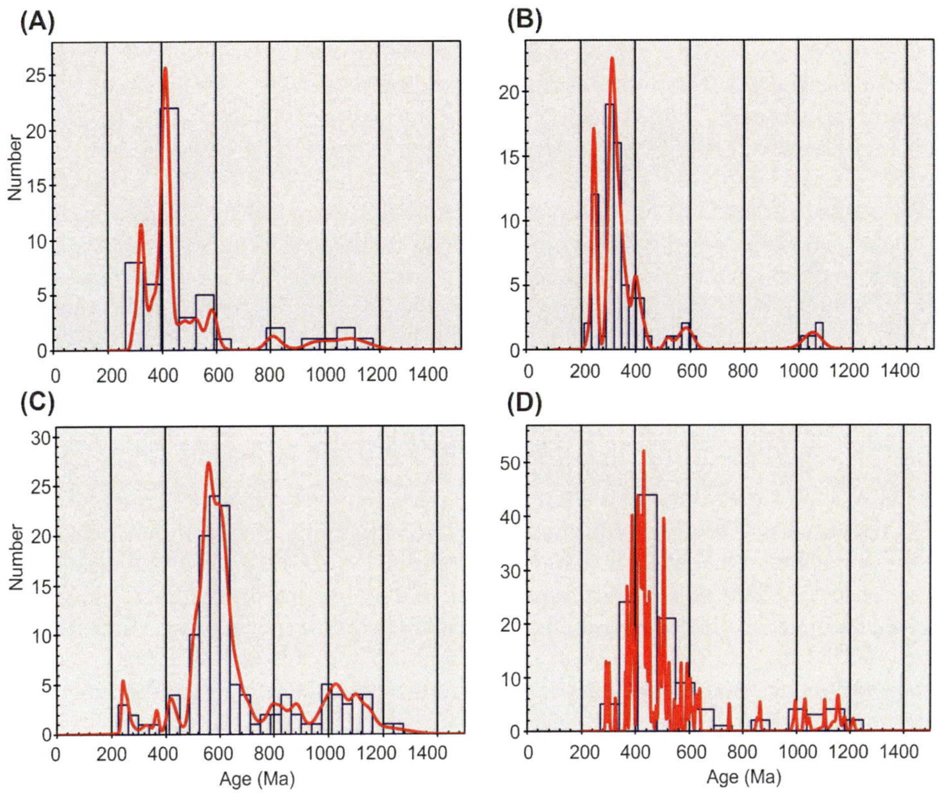

FIGURE 13.16 Relative probability plots (*red lines*) and histograms (blue) for the interval 1500–0 Ma from the Sydney Basin (data from Sircombe, 1999) and the Ovens Graben (data from Sircombe and Hazelton, 2004). Data replotted using the Isoplot program of Ludwig (2003). (A) Tallong Conglomerate (56 analyses). (B) Terrigal Formation (73 analyses). (C) Hawkesbury Sandstone (2 samples, 132 analyses). (D) Ovens Graben (3 samples, 123 analyses).

4. DISCUSSION

4.1 Sources of Sedimentary Rocks in the Tasmanides

The most puzzling and controversial issue about the source of sediment in the Tasmanides is where the Pacific-Gondwana zircons came from. The combination of prominent zircon ages in 700–500 Ma, but mainly skewed to 600–500 Ma, in addition to zircons in 1300–900 Ma, has been considered to be sourced from the East African Orogen, also known as the Trans-gondwanan Supermountains (Squire et al., 2006a; Williams and Pulford, 2008; Veevers, 2015). It has been found that with younger depositional ages from Cambrian to Ordovician samples, the proportion of 600–500 Ma zircons increase and the proportion of 1300–900 Ma zircons decrease, as does the content of pre-1500 Ma zircons indicating the greater prominence of the younger source over time (Adams et al., 2013c). The 700–500 Ma zircons have been widely reported in detrital zircon samples from many parts of Gondwana including north Africa, Arabia, South America, and northern India, and also in peri-Gondwanan terranes in Spain and Germany (Cawood et al., 2007; Rino et al., 2008; Díez Fernández et al., 2010; Voice et al., 2011; Meinhold et al., 2013; Rösel et al., 2014). An exceptional distance of sediment transport is implied along the northern margin of India in the Cambrian to Ordovician because a relatively consistent detrital age signature is found in these rocks along the whole 2000 km length of the Himalayan Orogen (Myrow et al., 2010). The ultimate source of these zircons is inferred to lie within the central East African Orogen (Myrow et al., 2010). The East African Orogen formed by collision of West and East Gondwana and extends from Arabia and adjoining northeastern Africa to southeastern Africa and East Antarctica (Fig. 13.1; Torsvik and Cocks, 2013). For units in the Tasmanides, it is implied that sediment containing the Pacific-Gondwana and less prominent Grenville zircon ages has traveled across and/or alongside the East Antarctic craton from the East African Orogen toward eastern Australia, a distance of at least 4000 km (Fig. 13.17A). Subsequent transport into the undeformed basinal settings of the Tasmanides suggests that an additional 1500−2500 km were traveled by these zircons.

Alternatively, a closer but still distant source from the East Antarctic craton has been proposed (Veevers, 2000) for the sedimentary rocks containing the Pacific-Gondwana zircon signature in the Tasmanides (Crohn-Mawson cratons, Fig. 13.17B). This reduces the distance of transport to less than half of that required for a distal source in the East African Orogen. Most of East Antarctica is covered by a thick ice sheet so that its geology cannot be directly mapped. Numerous studies have been undertaken on sedimentary successions, clasts from moraines, and marine sediments bordering East Antarctica that are considered to reflect sources within the covered East Antarctic geology (Veevers and Saeed, 2008, 2011; Veevers et al., 2008; Goodge and Fanning, 2010; Elliot et al., 2015). These studies indicate that Grenville and Pacific-Gondwana age components are present. Potential sources are in the Gamburtsev Subglacial Mountains and the Ross Orogen that has abundant detritus containing detrital zircons of these ages (Goodge et al., 2004a,b; Gibson et al., 2011; Adams et al., 2013c; Elliot et al., 2015). In contrast with earlier publications, Veevers (2015) favored an East African Orogen source for the Hawkesbury Sandstone and older units in eastern Australia, with 700−500 Ma and Grenville-age zircons, indicating very long distance transport as argued by Squire et al. (2006a) and Williams and Pulford (2008). Veevers (2015) also suggested that the Gamburtsev Subglacial Mountains represented either an additional primary source or a secondary source containing recycled zircons derived from the East African Orogen. Our interpretation is that the Pacific-Gondwana zircons (mainly 600−500 Ma) and the smaller Grenville peak in Cambrian, Ordovician, and Triassic rocks in the Tasmanides reflect derivation from sources in East Antarctica adjacent to and within the interior opposite the Australian margin, rather than derived from the more distant East African Orogen. This is consistent with the interpretation of Adams et al. (2013c) for Cambrian-Ordovician successions in Zealandia and the Ross Orogen and equivalents in the Swanson Formation in West Antarctica. A similar provenance has been suggested based on detrital zircon ages in Devonian and Permian strata in the Beardmore Glacial region of the central Transantarctic Mountains (Elliot et al., 2015).

A source in East Antarctica for Cambrian, Ordovician, and even Triassic siliciclastic units in the Tasmanides contrasts with other quartzose siliciclastic units such as the Carboniferous Shoalwater Formation (New England Orogen) and the Silurian-Devonian Hodgkinson

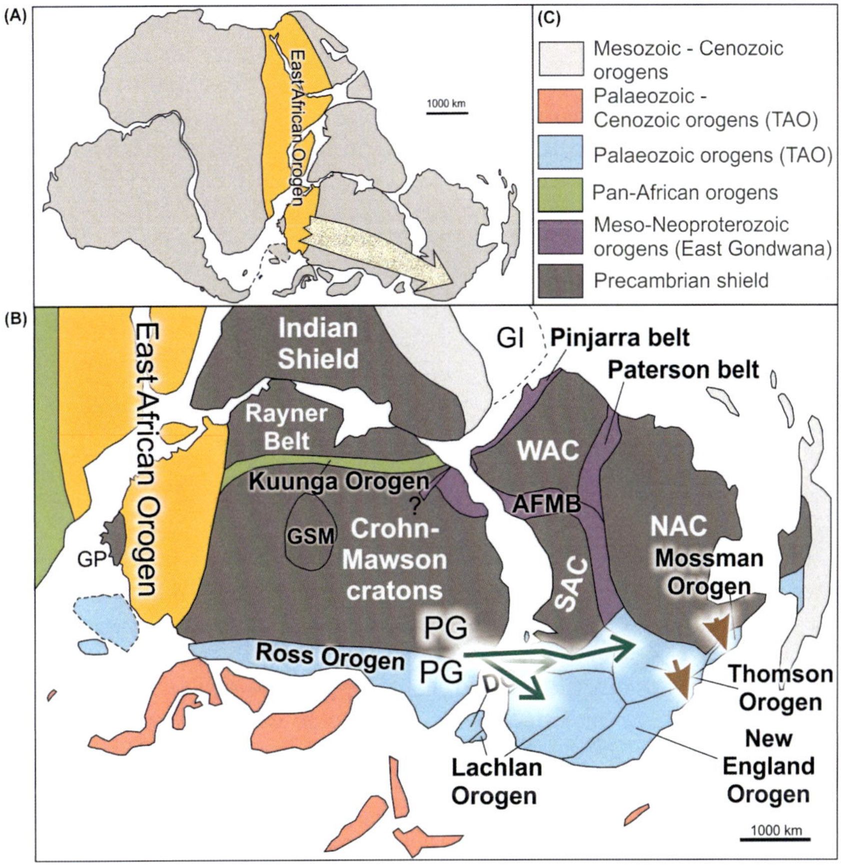

FIGURE 13.17 (A) Gondwana showing direct sediment path (*stippled light yellow arrow*) from the East African Orogen (Transgondwanan Supermountains) to the Tasmanides. (B) Highlighted *green arrows* show sediment paths from Pacific-Gondwana source inferred under East Antarctic ice sheet in the hinterland and inner part of the Ross Orogen. Highlighted brown arrows in eastern Australia show local source directions for sediment in the Hodgkinson formation of the Mossman Orogen and the Shoalwater Formation of the northern New England Orogen. *AFMB*, Albany–Fraser–Musgrave belt; *DO*, Delamerian Orogen; *GI*, Greater India; *GSM*, Gamburtev Subglacial Mountains; *GP*, Grunehogna Province; *NAC*, North Australian Craton; *PG*, Pacific-Gondwana sediment source; *SAC*, South Australian Craton; *TAO*, Terra Australis Orogen; *WAC*, West Australian Craton. (C) Key for map in (B).

Formation (Mossman Orogen), which contains zircon age signatures indicative of local sources rather than reflecting sediment transport over thousands of kilometers. We consider that much of the quartzose siliciclastic units, such as the Thomson Beds of the Thomson Orogen, the Kanmantoo Group of the Delamerian Orogen, and the Ordovician turbidites of the Lachlan Orogen, were derived from a Pacific-Gondwana source developed within, and/or inboard of, the Ross Orogen in East Antarctica (Fig. 13.17). This is consistent with the abundance of plutonic and low-grade metamorphic debris in addition to the zircons. A distal source is not possible for a sample from the latest Cambrian Bilpa Conglomerate in the Koonenberry Belt in northwestern New South Wales. This sample is from a primary post-Delamerian unit deposited in deltaic environments; the unit contains coarse detritus, including clasts over 1 m across and clasts of phyllite, mafic/felsic volcanics, limestone, schist, vein quartz, and granite (Pahl and Sikorska, 2004; Greenfield et al., 2010, p. 139). The unit has rare clasts of mudstone containing Early Cambrian trilobites consistent with derivation from nearby underlying units eroded during the Delamerian Orogeny (Percival et al., 2011, p. 435). Ages of detrital zircons have been determined from the matrix of the conglomerate and show dominant ages in 600–500 Ma and a smaller peak in 1300–1050 Ma (Greenfield et al., 2010, p. 346). Local derivation of the unit is clear, and the detrital zircons reflect recycling from pre-Delamerian units in the Koonenberry Belt that contain the Pacific-Gondwana zircons, as well as from Neoproterozoic units that are dominated by Grenville zircons (Johnson et al., 2012). Recycling of these zircons associated with the erosion of the Delamerian topography is indicated by this conglomerate and has potentially contributed to maintaining the influx of siliciclastic sand in the Ordovician turbidites of the Lachlan Orogen. The abundance of black shale in the Late Ordovician throughout the eastern Lachlan Orogen indicates that the volume of clastic input had significantly receded in this interval (Jones et al., 1993; Fergusson and Tye, 1999). We consider that recycling of uplifted Cambrian units, such as the Kanmantoo Group and its widespread equivalents such as in the Glenelg Zone and Koonenberry Belt, has resulted in the 600–500 Ma and 1300–900 Ma zircon age peaks in samples from the Melbourne Trough and Mathinna Group in northeast Tasmania. The Hawkesbury Sandstone presumably reflects reactivation of these sources in the Ross-Delamerian Orogens and hinterland in East Antarctica, rather than reflecting distal transport from the East African Orogen. The full extent of this far-travelled clastic wedge in the Sydney-Bowen Basin is not documented. It is presently based on detrital zircon age spectra for two samples from the Sydney Basin (Fig. 13.15) and it is not known how much of this signature applies to cratonic derived units in the western Sydney Basin and its equivalents further north. Data are available for just one sample from the Tallong Conglomerate at the base of the succession and indicate local derivation from the underlying Lachlan Orogen (Sircombe, 1999).

Even a sample of volcanic lithic sandstone from the Early Ordovician succession at the base of the Macquarie Arc shows the typical Pacific-Gondwana pattern with most zircon ages of 625–490 Ma and less common ages of 1250–970 Ma (Glen et al., 2011). This sample is enigmatic but sparse zircon ages from two samples of Early Ordovician siltstones confirm the Gondwana provenance signal (Glen et al., 2011). The absence of cratonic detritus other than zircons in these samples, and the predominant mafic volcanic source consistent with abundant volcanic units and shallow intrusions in the Macquarie Arc succession, highlight the problem. As discussed by Glen et al. (2011), it is unclear how these zircons came to be mixed in with volcaniclastic sediment. Undoubtedly they are inherited and must have been separated from the siliciclastic sediments that normally contain them in an intraoceanic arc setting. We consider that the Pacific-Gondwana and older zircons were eroded from Early Ordovician igneous rocks, so that the zircons are ultimately derived from either cratonic-derived metasedimentary rocks within the lower crust of the island arc, or subducted Gondwana-derived siliciclastic sediments as suggested by Glen et al. (2011). Similar complexities have been found in modern island arcs. For example, in east Java, detrital zircons from Early Cenozoic igneous, volcaniclastic, and sedimentary rocks indicate a Gondwana fragment with Archean-Cambrian zircon ages in the lower crust (Smyth et al.,

2007). Also, inherited zircons, with significant age populations in 2800–220 Ma indicative of Australian sources have been found in Eocene-Miocene igneous rocks of the New Hebrides island arc in the Southwest Pacific Ocean (Buys et al., 2014).

4.2 Tectonic Setting and Provenance Switching

The Paleozoic, and in particular the early Paleozoic tectonic history of the Tasmanides, has been a subject of considerable discussion in the literature. Provenance characteristics of clastic successions have been an important constraint in determining past tectonic configurations. For the Late Permian to Early Triassic tectonics of eastern Australia by contrast, most authors agree that a major magmatic arc developed in the New England Orogen with the upper part of the Sydney-Bowen Basin succession formed in a foreland basin setting. A clastic wedge derived from the New England Orogen occurs mainly in the eastern part of the basin and is interlayered with cratonic quartzose sandstones derived from the west and southwest. The pattern of provenance switching from lithic detritus derived from the orogen to the incoming quartzose sheet of the Hawkesbury Sandstone is well illustrated by the detailed study of Cowan (1993), and reflects fluvial mixing of diverse sands at the junction between the incoming clastic sheets. It is also consistent with changing paleocurrents as the ancient streams swing from a high-angle to the basin into longitudinal flow along the basin axis (Conaghan et al., 1982).

Provenance switching has also occurred in the subduction complex of the northern New England Orogen where the volcaniclastic sandstones of the Wandilla Formation change to the quartz-rich sandstones of the Shoalwater Formation across a sharp, probably faulted, contact. Both units formed in deep-marine settings. The Wandilla and Shoalwater formations were originally interpreted by Fergusson et al. (1990) as having being accreted in the subduction complex with the change in composition reflecting a difference in age, with the Wandilla Formation being Early Carboniferous and the Shoalwater Formation Late Carboniferous. The detrital zircon ages provide maximum depositional ages for both units (Korsch et al., 2009a), which show that they stratigraphically overlap each other (Fig. 13.10). Our revised interpretation of the cause of this provenance switch is that the Wandilla Formation formed in the trench derived from the magmatic arc, whereas the Shoalwater Formation represented a turbidite fan that locally flooded the trench and extended well beyond it into an open oceanic setting. The detritus for the fan may well have been derived from a major distributary system, such as the Late Devonian to Carboniferous Drummond Basin draining the Gondwana interior of the Queensland region behind and including the magmatic arc. A modern example of this type of arrangement is the Miocene siliciclastic turbidites derived from the Chinese mainland and deposited in the backarc Shikoku Basin that is being subducted with formation of the Nankai Trough accretionary prism (Clift et al., 2013; Pickering et al., 2013).

The most significant provenance switch in the Tasmanides occurred in the Early Cambrian where paleocurrents indicate southward derivation of the Pacific-Gondwana clastic wedge (Flöttmann et al., 1998). This clastic wedge involved a huge sediment volume that required sediment transport of hundreds to thousands of kilometers from its source in the Ross Orogen and its hinterland in East Antarctica along the length of the Tasmanides to include the Kanmantoo Group and equivalents of the Delamerian Orogen as well as the Thomson

Orogen. Initiation of this provenance switch coincides with the latter part of the Ross Orogeny and the Delamerian Orogeny (Goodge et al., 2004a,b; Cayley, 2011; Gibson et al., 2011, 2015) and presumably reflects increased erosion resulting from a change in circumstances along the East Gondwana active margin. These may have included, singly or in combination, increased plate motions, major climatic variations, and increased uplift rates. The sediment supply continued from the Middle Cambrian into the Ordovician and reappeared as a source in the Triassic. Presumably, the Pacific-Gondwana zircon signature was largely maintained as a sediment characteristic by reworking of the earlier phase of deposition preserved in the Delamerian and Ross Orogens, as shown by the sample of the Bilpa Conglomerate matrix (see earlier).

One of the most controversial issues in the tectonic development of the Tasmanides has been the tectonic setting of the Ordovician–Early Silurian Macquarie Arc and its linkage to the Ordovician turbidites. For example, Aitchison and Buckman (2012) inferred major overthrusting of the Macquarie Arc over the Ordovician turbidites, whereas Quinn et al. (2014) argued for stratigraphic contacts between them. This difference in interpretations has arisen from the strong contrast between the predominantly mafic to intermediate volcanic units and their sedimentary derivatives, which characterizes the Macquarie Arc and the craton-derived, quartz-rich turbidite succession. The quartz turbidite succession has interbedded chert intervals and a thick black shale unit in the Late Ordovician that lack evidence for any contemporaneous igneous activity. Geophysical and other evidence for the Macquarie Arc forming part of a huge allochthon analogous to the Semail Ophiolite in Oman has not been forthcoming. Therefore, we favor the apparent lack of interdigitation of facies between the Ordovician turbidites and the Macquarie Arc as a result of paleogeography at the time of deposition.

The Macquarie Arc succession formed in proximal shallow marine to potentially subaerial environments, with volcanic centers erupting volumes of lava and pyroclastic rocks that fed into surrounding sediment aprons including in widespread deep-marine environments (Simpson et al., 2007). In contrast, the Ordovician turbidites were derived from a distant source in the Ross Orogen, and potentially the interior of the East Antarctic craton, with a potential closer source in the Delamerian mountains from reworking of the uplifted Kanmantoo Group and equivalents. These sediments were deposited in the Wagga Marginal Sea between the East Gondwana margin and the Macquarie Arc. They are also found east of and partially enclosing the Macquarie Arc (Fergusson, 2009). The Ordovician turbidites that were deposited on the distal flanks of the Macquarie Arc would already have crossed a wide marginal sea at least 1000 km in width (Gray et al., 2006), and in the distal part of the basin would have been constrained to topographical lows in the sea bed. It is unlikely, and certainly is not observed, that they were interlayered with volcanic-derived clastic wedges flanking the Macquarie Arc in a similar way that the Hawkesbury Sandstone is interleaved with lithic detritus of the underlying Narrabeen Group and overlying Wianamatta Group in the Sydney Basin (Conaghan et al., 1982). In the second phase of igneous activity associated with the Macquarie Arc, there was a broadening of the arc edifice resulting in mafic to intermediate clastic wedges extending eastward and stratigraphically overlying Ordovician turbidites in the northeastern Lachlan Orogen (Fergusson and Colquhoun, 1996) and also in southeastern New South Wales (Quinn et al., 2014). Thus both elements must have formed adjacent to each other, and the lack interdigitation/sediment mixing reflects restriction of the Ordovician turbidites to deeper settings away from the Macquarie Arc. Unlike subaerial and shallow marine environments, it is difficult to imagine how in deep-ocean environments sediment mixing of the distally derived quartz-rich turbidites could have occurred with the volcanic-derived wedges on the flanks of the Macquarie Arc. In contrast, the Early Silurian Kabadah Formation, which is located between the western and central belts of the Macquarie Arc, shows sediment mixing between the Macquarie Arc, Girilambone Group (deformed Ordovician turbidites), Early Silurian volcanic rocks, and ultramafic sources (Barron et al., 2007). In this case, the diverse provenance of this unit reflects uplift during the Benambran Orogeny that enabled subaerial to shallow marine transport and mixing of detritus with deposition in a shallowing basin.

4.3 Exotic Terranes in the Tasmanides

An issue in orogenic belts is the recognition of exotic terranes, such as the Cache Creek Terrane in the Cordillera of western North America (Johnston and Borel, 2007). Within the Tasmanides, the most likely assemblages that could be classed as exotic terranes are the intraoceanic island arc assemblages including the Ordovician–Early Silurian Macquarie Arc in the Lachlan Orogen, the Lucky Springs assemblage of the Mossman Orogen, the Late Silurian–Devonian Gamilaroi-Calliope Arc and the Permian–Triassic Gympie Province in the New England Orogen. Cambrian island arc assemblages in the Lachlan Orogen and eastern Delamerian Orogen also would have been exotic to the Tasmanides along with the Precambrian basement units of western Tasmania. On the basis of the available zircon age data, we consider that most of the island arc assemblages have formed in the paleo-Pacific Ocean in close proximity to the Gondwana margin. The Cambrian island arcs of the Lachlan Orogen have been covered by the widespread Ordovician turbidites, indicating that they were relatively close to the Gondwanan margin in the Late Cambrian to Early Ordovician, with ophiolite emplacement indicated in Tasmania (Berry and Crawford, 1988; Bruce and Percival, 2014). Similarly, Precambrian rocks of western Tasmania must have been in close proximity to their present location before the end of the Ordovician, and paleomagnetic data indicates close proximity to their present location by the Early Ordovician (Li et al., 1997). The Gympie Province shows characteristics of an intraoceanic island arc in the Permian, but the ages of detrital zircons indicate a connection with the Carboniferous to Early Permian magmatic arc of the New England Orogen (Li et al., 2015). Just how the intraoceanic affinity of mafic volcanic rocks in the Gympie Province relates to a setting in the eastern New England Orogen remains unresolved. The Late Silurian–Devonian Gamilaroi-Calliope island arc and backarc assemblage have been characterized by geochemistry of volcanic rocks, but so far, any provenance linkage with Gondwana has yet to be tested on the basis of detrital zircon ages. The Tasmanides have mostly formed as an assemblage developed relatively proximal to their present setting along the Paleozoic East Gondwana active margin, apart from intraoceanic arcs developed outboard of the margin (Glen, 2013). Additionally, Precambrian units of western Tasmania prior to the Late Cambrian have most likely been transported from a distant location further southward along the East Gondwana margin (Berry et al., 2008; Cayley, 2011; Gibson et al., 2011; Moore et al., 2015).

5. CONCLUSIONS

Sedimentary successions in the Tasmanides show provenance characteristics indicative of numerous sources, the most significant of which has been the major Pacific-Gondwana clastic input containing zircons with many ages of 600–500 Ma and fewer of 1300–1000 Ma. These ages are typical of a Gondwana source and are widely represented in most parts of the supercontinent and some other continental fragments (Rino et al., 2008; Voice et al., 2011). This sedimentary signature was first introduced in the Kanmantoo Group and its equivalents in the Delamerian Orogen, but has been widely recognized in Late Cambrian siliciclastic rocks of the Thomson Orogen. It indicates sedimentary transport of at least 500–2000 km from sources such as the Ross-Delamerian Orogens and the hinterland of the Ross Orogen in East Antarctica, presently covered by the East Antarctic ice sheet. It is particularly well illustrated by the Ordovician turbidites of the Lachlan Orogen. Reworking of these early Paleozoic successions is thought responsible for this distinctive zircon signature occurring in foreland basin deposits in the Silurian to Devonian of the western Lachlan Orogen, such as the Melbourne Trough, and even more recently being widely distributed in modern beach sands along the coast of eastern Australia (Sircombe, 1999; Veevers, 2015). These provenances contrast with locally derived Ordovician–Early Silurian Macquarie Arc and the Late Silurian–Devonian Gamilaroi-Calliope island arc, which are dominated by mafic to intermediate volcanic and volcaniclastic rocks. Other locally derived successions are the Late Devonian, Carboniferous, and Permian sedimentary successions of the New England Orogen derived from the contemporaneous magmatic arc that was located on older Gondwana basement. In the Mossman Orogen of northeastern Australia, the Hodgkinson Formation and its equivalents further south show derivation from the Precambrian basement and contemporaneous igneous assemblages developed west of the Tasman Line. A controversial issue in terms of provenance in the Tasmanides is to how the Gondwana-derived Ordovician turbidites have developed apparently adjacent to and enveloping the Macquarie Arc. We consider that the combination of widespread deep-marine settings inhibiting provenance mixing and a major phase of arc expansion in the Late Ordovician is the best explanation for the dramatic provenance switch that characterizes these elements.

Acknowledgments

We acknowledge past joint work and many discussions with Paul Carr, Gary Colquhoun, Mark Fanning, Brian Jones, Evan Leitch, Allen Nutman, Stuart Tye, and Ian Withnall. We have also discussed aspects of provenance in the Tasmanides with Charlotte Allen, Dominic Brown, Sol Buckman, Patrick Carr, Andrew Cross, John Greenfield, Emma Johnson, and David Purdy. We thank Chris Adams, Simon Bodorkos, Russell Korsch, and Richard Wormald for supplying tables of data of detrital zircon ages. This work was supported by the GeoQuEST Research Center at the University of Wollongong. We thank the reviewers George Gibson and Gideon Rosenbaum for their many suggestions that have significantly improved the final manuscript.

References

Adams, C.J., Korsch, R.J., Griffin, W.L., 2013a. Provenance comparisons between the Nambucca Block, Eastern Australia and the Torlesse composite terrane, New Zealand: connections and implications from detrital zircon age patterns. Australian Journal of Earth Sciences 60, 241–253.

Adams, C.J., Wormald, R., Henderson, R.A., 2013b. Detrital zircons from the Hodgkinson formation: constraints on its maximum depositional age and provenance. In: Jell, P.A. (Ed.), The Geology of Queensland. Queensland Government, pp. 239–241.

Adams, C.J., Bradshaw, J.D., Ireland, T.R., 2013c. Provenance connections between late Neoproterozoic and early Palaeozoic sedimentary basins of the Ross Sea region, Antarctica, south-east Australia and southern Zealandia. Antarctic Science 26, 173–182.

Aitchison, J.C., 1988. Late Paleozoic radiolarian ages from the Gwydir terrane, New England Orogen, eastern Australia. Geology 16, 793–795.

Aitchison, J.C., Buckman, S., 2012. Accordion vs. quantum tectonics: insights into continental growth processes from the Palaeozoic of eastern Gondwana. Gondwana Research 22, 674–680.

Aitchison, J.C., Flood, P.G., 1995. Gamilaroi terrane: a Devonian rifted intra-oceanic island-arc assemblage, NSW, Australia. In: Smellie, J.L. (Ed.), Volcanism Associated with Extension at Consuming Plate MarginsGeological Society of London, pp. 155–168. Special Publication, 81.

Aitchison, J.C., Flood, P.G., Spiller, F.C.P., 1992. Tectonic setting and paleoenvironment of terranes in the southern New England Orogen as constrained by radiolarian biostratigraphy. Palaeogeography, Palaeoclimatology, Palaeoecology 93, 31–54.

Barron, L.M., Meffre, S., Glen, R.A., 2007. Arc and mantle detritus in the post-collisional, lower Silurian Kabadah formation, Lachlan Orogen, New South Wales. Australian Journal of Earth Sciences 54, 353–362.

Berry, R.F., Bull, S.W., 2012. The pre-Carboniferous geology of Tasmania. Episodes 35, 195–204.

Berry, R.F., Chmielowski, R.M., Steele, D.A., Meffre, S., 2007. Chemical U–Th–Pb monazite dating of the Cambrian Tyennan Orogeny, Tasmania. Australian Journal of Earth Sciences 54, 757–771.

Berry, R.F., Crawford, A.R., 1988. The tectonic significance of Cambrian allochthonous mafic–ultramafic complexes in Tasmania. Australian Journal of Earth Sciences 35, 523–533.

Berry, R.F., Jenner, G.A., Meffre, S., Tubrett, M.N., 2001. A North American provenance for Neoproterozoic to Cambrian sandstones in Tasmania? Earth and Planetary Science Letters 192, 207–222.

Berry, R.F., Steele, D.A., Meffre, S., 2008. Proterozoic metamorphism in Tasmania: implications for tectonic reconstructions. Precambrian Research 166, 387–396.

Bhatia, M.R., Taylor, S.R., 1981. Trace element geochemistry and sedimentary provinces: study from the Tasman geosyncline, Australia. Chemical Geology 33, 115–125.

Black, L.P., Calver, C.R., Seymour, D.B., Reed, A., 2004. SHRIMP U–Pb detrital zircon ages from Proterozoic and early Palaeozoic sandstones and their bearing on the early geological evolution of Tasmania. Australian Journal of Earth Sciences 51, 885–900.

Black, L.P., Everard, J.L., McClenaghan, M.P., Korsch, R.J., Calver, C.R., Fioretti, A.M., Brown, A.V., Foudoulis, C., 2010. Controls on Devonian–Carboniferous magmatism in Tasmania, based on inherited zircon age patterns, Sr, Nd and Pb isotopes, and major and trace element geochemistry. Australian Journal of Earth Sciences 57, 933–968.

Boggs, S., 2009. Petrology of Sedimentary Rocks, second ed. Cambridge University Press, Cambridge, UK. 612 p.

Boger, S.D., 2011. Antarctica — before and after Gondwana. Gondwana Research 19, 335–371.

Boger, S.D., Miller, J.M., 2004. Terminal suturing of Gondwana and the onset of the Ross–Delamerian Orogeny: the cause and effect of an early Cambrian reconfiguration of plate motions. Earth and Planetary Science Letters 219, 35–48.

Boyd, R., Ruming, K., Goodwin, I., Sandstrom, M., Schröder-Adams, C., 2008. Highstand transport of coastal sand to the deep ocean: a case study from Fraser Island, southeast Australia. Geology 36, 15–18.

Brown, D., Purdy, D., Carr, P., Cross, A., Kositcin, N., 2014. New Isotopic Data from the Thomson Orogen Basement Cores: A Possible Link with the Centralian Superbasin. Geological Society of Australia, pp. 243–244. Abstracts 110.

Bruce, M.C., Percival, I.G., 2014. Geochemical evidence for provenance of Ordovician cherts in southeastern Australia. Australian Journal of Earth Sciences 61, 927–950.

Buys, J., Spandler, C., Holm, R.J., Richards, S.W., 2014. Remnants of ancient Australia in Vanuatu: implications for crustal evolution in island arcs and tectonic development of the southwest Pacific. Geology 42, 939–942.

Camacho, A., Hensen, B.J., Armstrong, R., 2002. Isotopic test of a thermally driven intraplate orogenic model, Australia. Geology 30, 887–890.

Carr, P., Purdy, D., Brown, D., 2014. Peeking under the covers: undercover geology of the Thomson Orogen. Geological Society of Australia Abstracts 110, 244–245.

Cas, R.A.F., Powell, C.McA., Crook, K.A.W., 1980. Ordovician palaeogeography of the Lachlan Fold Belt: a modern analogue and tectonic constraints. Journal of the Geological Society of Australia 27, 19–31.

Cas, R.A.F., VandenBerg, A.H.M., 1988. Ordovician. In: Douglas, J.G., Ferguson, J.A. (Eds.), Geology of Victoria. Victorian Division. Geological Society of Australia, pp. 63–102.

Cawood, P.A., 1982. Structural relations in the subduction complex of the Paleozoic New England Fold Belt, eastern Australia. Journal of Geology 90, 381–392.

Cawood, P.A., 2005. Terra Australis Orogen: Rodinia breakup and development of the Pacific and Iapetus margins of Gondwana during the Neoproterozoic and Palaeozoic. Earth-Science Reviews 69, 249–279.

Cawood, P.A., Johnson, M.R.W., Nemchin, A.A., 2007. Early Palaeozoic orogenesis along the Indian margin of Gondwana: tectonic response to Gondwana assembly. Earth and Planetary Science Letters 255, 70–84.

Cawood, P.A., Korsch, R.J., 2008. Assembling Australia: Proterozoic building of a continent. Precambrian Research 166, 1–38.

Cawood, P.A., Leitch, E.C., Merle, R.E., Nemchin, A., 2011. Orogenesis without collision: stabilizing the Terra Australis accretionary Orogen, eastern Australia. Geological Society of America Bulletin 123, 2240–2255.

Cayley, R.A., 2011. Exotic crustal block accretion to the eastern Gondwanaland margin in the late Cambrian–Tasmania, the Selwyn block, and implications for the Cambrian–Silurian evolution of the Ross, Delamerian, and Lachlan orogens. Gondwana Research 19, 628–649.

Cayley, R.A., 2012. Oroclinal folding in the Lachlan Fold Belt: consequence of southeast-directed Siluro-Devonian subduction roll-back superimposed on an Ordovician accreted arc assemblage in eastern Australia. In: Selwyn Symposium 2012, vol. 103. GSA Victorian Division, Geological Society of Australia, pp. 34–43. Abstracts.

Cayley, R.A., Taylor, D.H., VandenBerg, A.H.M., Moore, D.H., 2002. Proterozoic – early Palaeozoic rocks and the Tyennan Orogeny in central Victoria: the Selwyn block and its tectonic implications. Australian Journal of Earth Sciences 49, 225–254.

Cayley, R.A., Korsch, R.J., Moore, D.H., Costelloe, R.D., Nakamura, A., Willman, C.E., Rawling, T.J., Morand, V.J., Skladzein, P.B., O'Shea, P.J., 2011. Crustal architecture of central Victoria: results from the 2006 deep crustal reflection seismic survey. Australian Journal of Earth Sciences 58, 113–156.

Clift, P.D., Carter, A., Nicholson, U., Masago, H., 2013. Zircon and apatite thermochronology of the Nankai Trough accretionary prism and trench, Japan: sediment transport in an active and collisional margin setting. Tectonics 32, 377–395.

Collins, W.J., 2002. Nature of extensional orogens. Tectonics 21. http://dx.doi.org/10.1029/2000TC001272.

Colquhoun, G.P., Fergusson, C.L., Tye, S.C., 1999. Provenance of early Palaeozoic sandstones, southeastern Australia, part 2: cratonic to arc switching. Sedimentary Geology 125, 153–163.

Conaghan, P.J., Jones, J.G., McDonnell, K.L., Royce, K., 1982. A dynamic fluvial model for the Sydney Basin. Journal of the Geological Society of Australia 29, 55–70.

Cooper, J.A., Jenkins, R.J.F., Compston, W., Williams, I.S., 1992. Ion probe zircon dating of a mid-Early Cambrian tuff in South Australia. Journal of the Geological Society of London 149, 185–192.

Cowan, E.J., 1993. Longitudinal fluvial drainage patterns within a foreland basin-fill: Permo-Triassic Sydney Basin, Australia. Sedimentary Geology 85, 557–577.

Crawford, A.J., Stevens, B.P.J., Fanning, C.M., 1997. Geochemistry and tectonic setting of some Neoproterozoic and early Cambrian volcanics in western New South Wales. Australian Journal of Earth Sciences 44, 831–852.

Crawford, A.J., Cayley, R.A., Taylor, D.H., Morand, V.J., Gray, C.M., Kemp, A.I.S., Wohlt, K.E., VandenBerg, A.H.M., Moore, D.H., Maher, S., Direen, N.G., Edwards, J., Donaghy, A.G., Anderson, J.A., Black, L.P., 2003. Neoproterozoic and Cambrian. In: Birch, W.D. (Ed.), Geology of Victoria. Geological Society of Australia. Geological Society of Australia (Victoria Division), pp. 73–92. Special Publication, 23.

Crawford, A.J., Meffre, S., Squire, R.J., Barron, L.M., Falloon, T., 2007. Middle and late Ordovician magmatic evolution of the Macquarie Arc, Lachlan Orogen, New South Wales. Australian Journal of Earth Sciences 53, 181–214.

de Wit, M., Jeffery, M., Bergh, H., Nicolaysen, L., 1988. Geological Map of Sectors of Gondwana Reconstructed to Their Positions ~150 Ma, Scale 1:10,000,000. American Association of Petroleum Geologists.

Dickinson, W.R., Suczek, C.A., 1979. Plate tectonics and sandstone compositions. American Association of Petroleum Geologists Bulletin 63, 2164–2182.

Dickinson, W.R., Beard, I.S., Brakenridge, G.R., Erjavec, J.L., Ferguson, R.C., Inman, K.F., Knepp, R.A., Lindberg, F.A., Ryberg, P.T., 1983. Provenance of North American Phanerozoic sandstones in relation to tectonic setting. Geological Society of America Bulletin 93, 222–235.

Díez Fernández, R.D., Martínez Catalán, J.R., Gerdes, A., Abati, J., Arenas, R., Fernández-Suárez, J., 2010. U–Pb ages of detrital zircons from the Basal allochthonous units of NW Iberia: provenance and paleoposition on the northern margin of Gondwana during the Neoproterozoic and Paleozoic. Gondwana Research 18, 385–399.

Direen, N.G., Crawford, A.J., 2003. The Tasman Line: where is it, what is it, and is it Australia's Rodinian breakup boundary? Australian Journal of Earth Sciences 50, 491–502.

Domagala, J., 1997. Hodgkinson formation (Hodgkinson province. In: Bain, J.H.C., Draper, J.J. (Eds.), North Queensland Geology. AGSO Bulletin 240, vol. 9. Queensland Geology, pp. 232–233.

Donchak, P.J.T., Purdy, D.J., Withnall, I.W., Blake, P.R., Jell, P.A., 2013. New England Orogen. In: Jell, P.A. (Ed.), The Geology of Queensland. Queensland Government, pp. 305–472 (Chapter 5).

Elliot, D.H., Fanning, C.M., Hulett, S.R.W., 2015. Age provinces in the Antarctic craton: evidence from detrital zircons in Permian strata from the Beardmore Glacier region, Antarctica. Gondwana Research 28, 152–164.

Fergusson, C.L., 1985. Trench-floor sedimentary sequences in a Palaeozoic subduction complex, eastern Australia. Sedimentary Geology 42, 181–200.

Fergusson, C.L., 2009. Tectonic evolution of the Ordovician Macquarie Arc, central New South Wales: arguments for subduction polarity and anticlockwise rotation. Australian Journal of Earth Sciences 56, 179–193.

Fergusson, C.L., 2010. Plate-driven extension and convergence along the East Gondwana active margin: late Silurian–middle Devonian tectonics of the Lachlan Fold Belt, southeastern Australia. Australian Journal of Earth Sciences 57, 627–649.

Fergusson, C.L., 2014. Late Ordovician to mid-Silurian Benambran subduction zones in the Lachlan Orogen, southeastern Australia. Australian Journal of Earth Sciences 61, 587–606.

Fergusson, C.L., Carr, P.F., Fanning, C.M., Green, T.J., 2001. Proterozoic-Cambrian detrital zircon and monazite ages from the Anakie Inlier, central Queensland: Grenville and Pacific-Gondwana signatures. Australian Journal of Earth Sciences 48, 857–866.

Fergusson, C.L., Cas, R.A.F., Stewart, I.R., 1989. Ordovician turbidites of the Hotham Group, eastern Victoria: sedimentation in deep-marine channel–levee complexes. Australian Journal of Earth Sciences 36, 1–12.

Fergusson, C.L., Colquhoun, G.P., 1996. Early Palaeozoic quartz turbidite fan and volcaniclastic apron, Mudgee district, northeastern Lachlan Fold Belt, New South Wales. Australian Journal of Earth Sciences 43, 497–507.

Fergusson, C.L., Fanning, C.M., 2002. Late Ordovician stratigraphy, zircon provenance and tectonics, Lachlan Fold Belt, southeastern Australia. Australian Journal of Earth Sciences 49, 423–436.

Fergusson, C.L., Fanning, C.M., Phillips, D., Ackerman, B.R., 2005. Structure, detrital zircon U-Pb ages and 40Ar/39Ar geochronology of the early Palaeozoic Girilambone group, central New South Wales: subduction, contraction and extension associated with the Benambran Orogeny. Australian Journal of Earth Sciences 52, 137–159.

Fergusson, C.L., Henderson, R.A., Leitch, E.C., 1990. Structural history and tectonics of the Palaeozoic Shoalwater and Wandilla terranes, northern New England Orogen, Queensland. Australian Journal of Earth Sciences 37, 387–400.

Fergusson, C.L., Henderson, R.A., 2013. Thomson Orogen. In: Jell, P.A. (Ed.), The Geology of Queensland. Queensland Government, pp. 113–224 (Chapter 3).

Fergusson, C.L., Henderson, R.A., 2015. Early Palaeozoic continental growth in the Tasmanides of northeast Gondwana and its implications for Rodinia assembly and rifting. Gondwana Research 28, 933–953.

Fergusson, C.L., Henderson, R.A., Fanning, C.M., Withnall, I.W., 2007. Detrital zircon ages in Neoproterozoic to Ordovician siliciclastic rocks, northeastern Australia: implications for the tectonic history of the East Gondwana continental margin. Journal of the Geological Society, London 164, 215–225.

Fergusson, C.L., Nutman, A.P., Kamiichi, T., Hidaka, H., 2013. Evolution of a Cambrian active continental margin: the Delamerian–Lachlan connection in southeastern Australia from a zircon perspective. Gondwana Research 24, 1051–1066.

Fergusson, C.L., Offler, R., Green, T.J., 2009. Late Neoproterozoic passive margin of East Gondwana: geochemical constraints from the Anakie Inlier, central Queensland, Australia. Precambrian Research 168, 301–312.

Fergusson, C.L., Tye, S.C., 1999. Provenance of Early Palaeozoic sandstones, southeastern Australia, part 1: vertical changes through the Bengal fan-type deposit. Sedimentary Geology 125, 135–151.

Flöttmann, T., Haines, P., James, P., Jago, J., Belperio, A., Gum, J., 1998. Formation and reactivation of the Cambrian Kanmantoo trough, southeast Australia—implications for early Palaeozoic tectonics at eastern Gondwana's plate margin. Journal of the Geological Society of London 155, 525–539.

Foden, J., Sandiford, M., Dougherty-Page, J., Williams, I., 1999. Geochemistry and geochronology of the Rathjen Gneiss: implications for the early tectonic evolution of the Delamerian Orogen. Australian Journal of Earth Sciences 46, 377–389.

Foden, J., Elburg, M.A., Dougherty-Page, J., Burtt, A., 2006. The timing and duration of the Delamerian Orogeny: correlation with the Ross Orogen and implications for Gondwana assembly. Journal of Geology 114, 189–210.

Foster, D.A., Gray, D.R., 2000. Evolution and structure of the Lachlan Fold Belt (Orogen) of eastern Australia. Annual Review of Earth and Planetary Sciences 28, 47–80.

Foster, D.A., Gray, D.R., Bucher, M., 1999. Chronology of deformation within the turbidite-dominated Lachlan Orogen: implications for the tectonic evolution of eastern Australia and Gondwana. Tectonics 18, 452–485.

Gehrels, G., 2014. Detrital zircon U-Pb geochronology applied to tectonics. Annual Review of Earth and Planetary Sciences 42, 127–149.

Gehrels, G.E., Butler, R.F., Bazard, D.R., 1996. Detrital zircon geochronology of the Alexander terrane, southeastern Alaska. Geological Society of America Bulletin 108, 722–734.

Geological Survey of Queensland, 2012. Queensland Geology, Scale 1:2 000 000. Department of Natural Resources and Mines, Brisbane.

Gibson, G.M., Morse, M.P., Ireland, T.R., Nayak, G.K., 2011. Arc–continent collision and orogenesis in western Tasmanides: insights from reactivated basement structures and formation of an ocean–continent transform boundary off western Tasmania. Gondwana Research 19, 608–627.

Gibson, G.M., Champion, D.C., Ireland, T.R., 2015. Preservation of a fragmented late Neoproterozoic–earliest Cambrian hyper-extended continental-margin sequence in the Australian Delamerian Orogen. In: Gibson, G.M., Roure, F., Manatschal, G. (Eds.), Sedimentary Basins and Crustal Processes at Continental Margins: From Modern Hyper-Extended Margins to Deformed Ancient Analogues. Geological Society, London, pp. 269–299. Special Publications 413.

Glen, R.A., 2005. The Tasmanides of eastern Australia. In: Vaughan, A.P.M., Leat, P.T., Pankhurst, R.J. (Eds.), Terrane Processes at the Margins of Gondwana. Geological Society, London, pp. 23–96. Special Publications 246.

Glen, R.A., 2013. Refining accretionary Orogen models for the Tasmanides of eastern Australia. Australian Journal of Earth Sciences 60, 315–370.

Glen, R.A., Korsch, R.J., Hegarty, R., Saeed, A., Poudjom Djomani, Y., Costelloe, R.D., Belousova, E., 2013. Geodynamic significance of the boundary between the Thomson Orogen and the Lachlan Orogen, northwestern New South Wales and implications for Tasmanide tectonics. Australian Journal of Earth Sciences 60, 371–412.

Glen, R.A., Percival, I.G., Quinn, C.D., 2009. Ordovician continental margin terranes in the Lachlan Orogen, Australia: implications for tectonics in an accretionary orogen along the East Gondwana margin. Tectonics 28, TC6012. http://dx.doi.org/10.1029/2009TC002446.

Glen, R.A., Saeed, A., Quinn, C.D., Griffin, W.L., 2011. U–Pb and Hf isotope data from zircons in the Macquarie Arc, Lachlan Orogen: implications for arc evolution and Ordovician palaeogeography along part of the East Gondwana margin. Gondwana Research 19, 670–685.

Goodge, J.W., Fanning, C.M., 2010. Composition and age of the East Antarctic shield in eastern Wilkes Land determined by proxy from Oligocene-Pleistocene glaciomarine sediment and Beacon Supergroup sandstones. Antarctica Geological Society of America Bulletin 122, 1135–1159.

Goodge, J.W., Myrow, P., Williams, I.S., Bowring, S.A., 2002. Age and provenance of the Beardmore Group, Antarctica: constraints on Rodinia supercontinent breakup. Journal of Geology 110, 393–406.

Goodge, J.W., Williams, I.S., Myrow, P., 2004a. Provenance of Neoproterozoic and lower Palaeozoic siliciclastic rocks of the central Ross Orogen, Antarctica: detrital record of rift-, passive- and active-margin sedimentation. Geological Society of America Bulletin 116, 1253–1279.

Goodge, J.W., Myrow, P., Phillips, D., Fanning, C.M., Williams, I.S., 2004b. Siliciclastic record of rapid denudation in response to convergent-margin orogenesis, Ross Orogen, Antarctica. In: Bernet, M., Spiegel, C. (Eds.), Detrital Thermochronology—Provenance Analysis, Exhumation, and Landscape Evolution of Mountain Belts. Geological Society of America, pp. 105–126. Special Paper 378.

Graham, I.G., Korsch, R.J., 1985. Rb–Sr geochronology of coarse grained greywackes and argillites from the Coffs Harbour Block, eastern Australia. Chemical Geology (Isotope Geoscience Section) 58, 45–54.

Gray, C.M., Webb, J.A., 1995. Provenance of Palaeozoic turbidites in the Lachlan orogenic belt: strontium isotopic evidence. Australian Journal of Earth Sciences 42, 95–105.

Gray, D.R., Foster, D.A., Meert, J.G., Goscombe, B.D., Armstrong, R., Trouw, R.A.J., Passchier, C.W., 2008. A Damaran Orogen perspective on the assembly of southwestern Gondwana. In: Pankhurst, R.J., Trouw, R.A.J., Brito Neves, B.B., De Wit, M.J. (Eds.), West Gondwana: Pre-Cenozoic Correlations Across the South Atlantic Region, vol. 294. Geological Society of London, pp. 257–278. Special Publication.

Gray, D.R., Willman, C.E., Foster, D.A., 2006. Crust restoration for the western Lachlan Orogen using the strain-reversal, area balancing technique: implications for crustal components and original thicknesses. Australian Journal of Earth Sciences 53, 329–341.

Greene, D.C., 2010. Neoproterozoic rifting in the southern Georgina Basin, central Australia: implications for reconstructing Australia in Rodinia. Tectonics 29, TC5010. http://dx.doi.org/10.1029/2009TC002543.

Greenfield, J.E., Gilmore, P.J., Mills, K.J., 2010. Explanatory notes for the Koonenberry Belt geological maps. Geological Survey of New South Wales. Bulletin 35, 528.

Greenfield, J.E., Musgrave, R.J., Bruce, M.C., Gilmore, P.J., Mills, K.J., 2011. The Mount Wright Arc: a Cambrian subduction system developed on the continental margin of East Gondwana, Koonenberry Belt, eastern Australia. Gondwana Research 19, 650–669.

Haines, P.W., Flöttmann, T., 1998. The Delamerian Orogeny and potential foreland sedimentation: a review of age and stratigraphic constraints. Australian Journal of Earth Sciences 45, 559–570.

Haines, P.W., Turner, S.P., Foden, J.D., Jago, J.B., 2009. Isotopic and geochemical characterisation of the Cambrian Kanmantoo Group, South Australia: implications for stratigraphy and provenance. Australian Journal of Earth Sciences 56, 1095–1110.

Haines, P.W., Turner, S.P., Kelley, S.P., Wartho, J.-A., Sherlock, S., 2004. $^{40}Ar/^{39}Ar$ dating of detrital muscovite from the Delamerian Orogen, South Australia: implications for provenance, tectonics and crustal evolution. Earth and Planetary Science Letters 227, 297–311.

Henderson, R.A., Innes, B.M., Fergusson, C.L., Crawford, A.J., Withnall, I.W., 2011. Collisional accretion of a late Ordovician oceanic island arc, northern Tasman orogenic zone, Australia. Australian Journal of Earth Sciences 58, 1–19.

Henderson, R.A., Donchak, P.J.T., Withnall, I.W., 2013. Mossman Orogen. In: Jell, P.A. (Ed.), The Geology of Queensland. Queensland Government, pp. 225–304 (Chapter 4).

Holcombe, R.J., Stephens, C.J., Fielding, C.R., Gust, D., Little, T.A., Sliwa, R., McPhie, J., Ewart, A., 1997. Tectonic evolution of the northern New England Fold Belt: Carboniferous to early Permian transition from active accretion to extension. In: Ashley, P.M., Flood, P.G. (Eds.), Tectonics and Metallogenesis of the New England Orogen, pp. 66–79. Special Publication of the Geological Society of Australia 19.

Ireland, T.R., Flöttmann, T., Fanning, C.M., Gibson, G.M., Preiss, W.V., 1998. Development of the early Palaeozoic Pacific margin of Gondwana from detrital-zircon ages across the Delamerian Orogen. Geology 26, 243–246.

Johnson, E.L., Allen, C.M., Phillips, G., 2012. A U–Pb zircon–rutile geochronology study: implications for the Cambrian evolution of the Koonenberry margin. In: 34th International Geological Congress, Brisbane, p. 2061. Abstracts.

Johnston, S.T., Borel, G.D., 2007. The odyssey of the Cache Creek terrane, Canadian Cordillera: implications for accretionary orogens, tectonic setting of Panthalassa, the Pacific superwell, and break-up of Pangea. Earth and Planetary Science Letters 253, 415–428.

Jones, B.G., Fergusson, C.L., Zambelli, P.F., 1993. Ordovician contourites in the Lachlan Fold Belt, eastern Australia. Sedimentary Geology 82, 257–270.

Korsch, R.J., 1978. Petrographic variations within thick turbidite sequences: an example from the Late Paleozoic of eastern Australia. Sedimentology 25, 247–265.

Korsch, R.J., 1981. Some tectonic implications of sandstone petrofacies in the Coffs Harbour association, New England Orogen, New South Wales. Journal of the Geological Society of Australia 28, 261–269.

Korsch, R.J., 1984. Sandstone compositions from the New England Orogen, eastern Australia: implications for tectonic setting. Journal of Sedimentary Petrology 54, 192–211.

Korsch, R.J., Adams, C.J., Black, L.P., Foster, D.A., Fraser, G.L., Murray, C.G., Foudoulis, C., Griffin, W.L., 2009a. Geochronology and provenance of the late Paleozoic accretionary wedge and Gympie terrane, New England Orogen, eastern Australia. Australian Journal of Earth Sciences 56, 655–685.

Korsch, R.J., Totterdell, J.M., Cathro, D.L., Nicoll, M.G., 2009b. Early Permian east Australian rift system. Australian Journal of Earth Sciences 56, 381–400.

Korsch, R.J., Totterdell, J.M., Fomin, T., Nicoll, M.G., 2009c. Contractional structures and deformational events in the Bowen, Gunnedah and Surat basins, eastern Australia. Australian Journal of Earth Sciences 56, 477–499.

Kositcin, N., Purdy, D.J., Brown, D.D., Bultitude, R.J., Carr, P.A., 2015. Summary of Results Joint GSQ-GA Geochronology Project: Thomson Orogen and Hodgkinson Province, 2012–2013. Queensland Geological Record 2015/2, 68 p.

Leitch, E.C., Cawood, P.A., 1987. Provenance determination of volcaniclastic rocks: the nature and tectonic significance of a Cambrian conglomerate from the New England Fold Belt, eastern Australia. Journal of Sedimentary Petrology 57, 630–638.

Leitch, E.C., Fergusson, C.L., Henderson, R.A., 2003. arc to craton provenance switching in a late Palaeozoic subduction complex, Wandilla and Shoalwater terranes, New England Fold Belt, eastern Australia. Australian Journal of Earth Sciences 50, 919–929.

Leitch, E.C., Willis, S.G.A., 1982. Nature and significance of plutonic clasts in Devonian conglomerates of the New England Fold Belt. Journal of the Geological Society of Australia 29, 83–89.

Li, P., Rosenbaum, G., Yang, J.-H., Hoy, D., 2015. Australian-derived detrital zircons in the Permian-Triassic Gympie terrane (eastern Australia): evidence for an autochthonous origin. Tectonics 34. http://dx.doi.org/10.1002/2015TC003829.

Li, Z.X., Baillie, P.W., Powell, C.McA., 1997. Relationship between northwestern Tasmania and East Gondwanaland in the late Cambrian/early Ordovician: paleomagnetic evidence. Tectonics 16, 161–171.

Li, Z.X., Evans, D.A.D., 2011. Late Neoproterozoic 40° intraplate rotation within Australia allows for a tighter-fitting and longer-lasting Rodinia. Geology 39, 39–42.

Ludwig, K.R., 2003. Isoplot 3.0: A Geochronological Toolkit for Microsoft Excel. Berkeley Geochronological Center Special Publication 4. Berkeley Geochronological Center, Berkeley, California, 70 pp.

Lyons, P., Percival, I.G., 2002. Middle to late Ordovician age for the Jindalee group of the Lachlan Fold Belt, New South Wales: conodont evidence and some tectonic implications. Australian Journal of Earth Sciences 49, 801–808.

Maidment, D.W., Williams, I.S., Hand, M., 2007. Testing long-term patterns of basin sedimentation by detrital zircon geochronology, Centralian Superbasin, Australia. Basin Research 19, 335–360.

Maidment, D.W., Hand, M., Williams, I.S., 2013. High grade metamorphism of sedimentary rocks during Palaeozoic rift basin formation in central Australia. Gondwana Research 24, 865–885.

McKillop, M.D., 2013. Adavale Basin. In: Jell, P.A. (Ed.), The Geology of Queensland. Queensland Government, pp. 175–183.

McLean, M.A., Morand, V.J., Cayley, R.A., 2010. Gravity and magnetic modelling of crustal structure in central Victoria: what lies under the Melbourne Zone? Australian Journal of Earth Sciences 57, 153–173.

Meffre, S., Scott, R.J., Glen, R.A., Squire, R.J., 2007. Re-evaluation of contact relationships between Ordovician volcanic belts and the quartz-rich turbidites of the Lachlan Orogen. Australian Journal of Earth Sciences 54, 363–383.

Meinhold, G., Morton, A.C., Avigad, D., 2013. New insights into peri–Gondwana paleogeography and the Gondwana super-fan system from detrital zircon U–Pb ages. Gondwana Research 23, 661–665.

Moore, D.H., Betts, P.G., Hall, M., 2013. Towards understanding the early Gondwanan margin in southeastern Australia. Gondwana Research 23, 1581–1598.

Moore, D.H., Betts, P.G., Hall, M., 2015. Fragmented Tasmania: the transition from Rodinia to Gondwana. Australian Journal of Earth Sciences 62, 1–35.

Morand, V.J., Wohlt, K.E., Cayley, R.A., Taylor, D.H., Kemp, A.I.S., Simons, B.A., Magart, A.P.M., 2003. Glenelg Special Map Area Geological Report. Geological Survey of Victoria Report 123. Geological Survey of Victoria.

Morris, P.A., 1988. A geochemical approach to the characterization of a hidden magmatic arc: the source of the Goonoo Goonoo Mudstone, eastern Australia. Australian Journal of Earth Sciences 35, 81–92.

Murray, C.G., 1986. Metallogeny and tectonic development of the Tasman Fold Belt system in Queensland. Ore Geology Reviews 1, 315–400.

Murray, C.G., 1994. Basement Cores from the Tasman Fold Belt System Beneath the Great Artesian Basin in Queensland. Department of Minerals and Energy. Queensland Geological Record 1994/10, 96 pp.

Murray, C.G., 1997. From geosyncline to fold belt: a personal perspective on the development of ideas regarding the tectonic evolution of the New England Orogen. In: Ashley, P.M., Flood, P.G. (Eds.), Tectonics and Metallogenesis of the New England Orogen. Geological Society of Australia, pp. 1–28. Special Publication 19.

Murray, C.G., Fergusson, C.L., Flood, P.G., Whitaker, W.G., Korsch, R.J., 1987. Plate tectonic model for the Carboniferous evolution of the New England Fold Belt. Australian Journal of Earth Sciences 34, 213–236.

Musgrave, R.J., 2015. Oroclines in the Tasmanides. Journal of Structural Geology 80, 72–98.

Myers, J.S., Shaw, R.D., Tyler, I.M., 1996. Tectonic evolution of Proterozoic Australia. Tectonics 15, 1431–1446.

Myrow, P.M., Hughes, N.C., Goodge, J.W., Fanning, C.M., Williams, I.S., Peng, S., Bhargava, O.M., Parcha, S.K., Pogue, K.P., 2010. Extraordinary transport and mixing of sediment across Himalayan central Gondwana during the Cambrian–Ordovician. Geological Society of America Bulletin 122, 1660–1670.

Neef, G., 2012. Mid Silurian–Carboniferous tectonic and depositional history of the Darling Basin conjugate fault system, western New South Wales: overview. Australian Journal of Earth Sciences 59, 91–117.

Neef, G., Bottrill, R.S., 1991. Early Devonian (Gedinnian) nonmarine strata present in a rapidly subsiding basin in far western New South Wales, Australia. Sedimentary Geology 71, 195–212.

Neef, G., Bottrill, R.S., 2001. Stratigraphy, structure and tectonics of lower Ordovician and Devonian strata of the central part of the Wonominta block, western New South Wales. Australian Journal of Earth Sciences 48, 317–330.

Neef, G., Bottrill, R.S., Ritchie, A., 1995. Phanerozoic stratigraphy and structure of the northern Barrier ranges, western New South Wales. Australian Journal of Earth Sciences 42, 557–570.

Offler, R., Garrad, D., Fardy, J., Seccombe, P.K., 1988. Accretion of N-type ORB basalts in the Nangahrah formation, Woodsreef area, NSW. In: Kleeman, J.D. (Ed.), Proceedings of the New England Orogen Symposium. University of New England, Armidale, New South Wales, pp. 99–104.

Offler, R., Murray, C.G., 2011. Devonian volcanics in the New England Orogen: tectonic setting and polarity. Gondwana Research 19, 706–715.

Offler, R., McKnight, S., Morand, V.J., 1998. Tectonothermal history of the western Lachlan Fold Belt: insights from white mica studies. Journal of Metamorphic Geology 16, 531–540.

Pahl, J.K., Sikorska, M., 2004. Cathodoluminescence study of carbonate cements in the upper Cambrian conglomerates from the Wonominta block, northwestern New South Wales. Australian Journal of Earth Sciences 51, 247–259.

Packham, G.H., Keene, J.B., Barron, L.M., 2003. Middle to late Ordovician hydrothermal veining in the Molong Volcanic Belt, northeastern Lachlan Fold Belt: sedimentological evidence. Australian Journal of Earth Sciences 50, 257–269.

Percival, I.G., 2012. Biotic characteristics of Ordovician deep-water cherts from Eastern Australia. Palaeogeography, Palaeoclimatology. Palaeoecology 367–368, 63–72.

Percival, I.G., Glen, R.A., 2007. Ordovician to earliest Silurian history of the Macquarie Arc, Lachlan Orogen, New South Wales. Australian Journal of Earth Sciences 54, 143–165.

Percival, I.G., Quinn, C.D., Glen, R.A., 2011. Review of Cambrian and Ordovician stratigraphy in New South Wales. Quarterly Notes of the Geological Survey of New South Wales 137, 1–39.

Pickering, K.T., Underwood, M.B., Saito, S., Naruse, H., Kutterolf, S., Scudder, R., Park, J.-O., Moore, G.F., Slagle, A., 2013. Depositional architecture, provenance, and tectonic/eustatic modulation of Miocene submarine fans in the Shikoku Basin: results from Nankai Trough Seismogenic zone experiment. Geochemistry, Geophysics, Geosystems 14, 1722–1739.

PIRSA, 2007. Warburton Basin. http://www.pir.sa.gov.au/__data/assets/pdf_file/0007/26926/prospectivity_warburton.pdf.

Powell, C.McA., 1983. Geology of the N.S.W. South Coast and Adjacent Victoria With Emphasis on the Pre-Permian Structural History. Geological Society of Australia, Specialist Group in Tectonics and Structural Geology, Field Guide # 1, 118 pp.

Powell, C.McA., 1984. Ordovician to earliest Silurian: marginal sea and island arc; Silurian to mid Devonian dextral transtensional margin; Late Devonian and Early Carboniferous: continental magmatic arc along the eastern edge of the Lachlan Fold Belt. In: Veevers, J.J. (Ed.), Phanerozoic Earth History of Australia. Oxford University Press, Oxford, pp. 290–340.

Powell, C.McA., Baillie, P.W., Conaghan, P.J., Turner, N.J., 1993. The mid-Palaeozoic turbiditic Mathinna Group, northeast Tasmania. Australian Journal of Earth Sciences 40, 169–196.

Powell, C.McA., Baillie, P.W., VandenBerg, A.H.M., 2003. Silurian to mid-Devonian basin development of the Melbourne zone, Lachlan Fold Belt, southeastern Australia. Tectonophysics 375, 9–36.

Powell, C.McA., Cole, J.P., Cudahy, T.J., 1985. Megakinking in the Lachlan Fold Belt, Australia. Journal of Structural Geology 7, 281–300.

Preiss, W.V., 2000. The Adelaide geosyncline of south Australia and its significance in Neoproterozoic continental reconstruction. Precambrian Research 100, 21–63.

Quinn, C.D., Percival, I.G., Glen, R.A., Xiao, W.J., 2014. Ordovician marginal basin evolution near the palaeo-Pacific East Gondwana margin, Australia. Journal of the Geological Society of London 171, 723–736.

Raymond, O.L., Liu, S., Gallagher, R., Zhang, W., Highet, L.M., 2012. Surface Geology of Australia 1:1 Million Scale Dataset 2012 Edition. Geoscience, Australia. http://www.ga.gov.au/metadata-gateway/metadata/record/gcat_c8856c41-0d5b-2b1d-e044-00144fdd4fa6/Surface+Geology+of+Australia+1%3A1+million+scale+dataset+2012+edition.

Reed, A.R., 2001. Pre-Tabberabberan deformation in eastern Tasmania: a southern extension of the Benambran Orogeny. Australian Journal of Earth Sciences Australia 48, 785–796.

Rino, S., Kon, Y., Sato, W., Maruyama, S., Santosh, M., Zhao, D., 2008. The Grenvillian and Pan-African orogens: world's largest orogenies through geologic time, and their implications on the origin of superplume. Gondwana Research 14, 51–72.

Rösel, D., Boger, S.D., Möller, A., Gaitzsch, B., Barth, M., Oalmann, J., Zack, T., 2014. Indo-Antarctic derived detritus on the northern margin of Gondwana: evidence for continental-scale sediment transport. Terra Nova 26, 64–71.

Rosenbaum, G., 2012. Oroclines of the southern New England Orogen, eastern Australia. Episodes 35, 187–194.

Schmidt, P.W., 2014. A review of Precambrian palaeomagnetism of Australia: palaeogeography, supercontinents, glaciations and true polar wander. Gondwana Research 25, 1164–1185.

Shaanan, U., Rosenbaum, G., Wormald, R., 2015. Provenance of the early Permian Nambucca Block (eastern Australia) and implications for the role of trench retreat in accretionary orogens. Geological Society of America Bulletin 127, 1052–1063.

Simpson, C.J., Scott, R.J., Crawford, A.J., Meffre, S., 2007. Volcanology, geochemistry and structure of the Ordovician Cargo volcanics in the Cargo – Walli region, central New South Wales. Australian Journal of Earth Sciences 54, 315–352.

Sircombe, K.N., 1999. Tracing provenance through the isotope ages of littoral and sedimentary zircon, eastern Australia. Sedimentary Geology 124, 47–67.

Sircombe, K.N., Hazelton, M.L., 2004. Comparison of detrital zircon age distributions by kernel functional estimation. Sedimentary Geology 171, 91–111.

Smyth, H.R., Hamilton, P.J., Hall, R., Kinny, P.D., 2007. The deep crust beneath island arcs: inherited zircons reveal a Gondwana continental fragment beneath East Java, Indonesia. Earth and Planetary Science Letters 258, 269–282.

Spampinato, G.P.T., Betts, P.G., Ailleres, L., Armit, R.J., 2015. Early tectonic evolution of the Thomson Orogen in Queensland inferred from constrained magnetic and gravity data. Tectonophysics 651–652, 99–120.

Squire, R.J., Campbell, I.H., Allen, C.M., Wilson, C.J.L., 2006a. Did the Transgondwanan Supermountain trigger the explosive radiation of animals on Earth? Earth and Planetary Science Letters 250, 116–133.

Squire, R.J., Stewart, I.R., Zang, W.L., 2006b. Acritarchs in polydeformed and highly altered Cambrian rocks in western Victoria. Australian Journal of Earth Sciences 53, 697–705.

Stuart-Smith, P.G., Black, L.P., 1999. Willaura Sheet 7422, Victoria 1:100 000 Map Geological Report. Australian Geological Survey Organisation Record 1999/38, 50 pp.

Taylor, D., Cayley, R., Skladzien, P., Woodhead, J., Corbett, G., 2014. Geochemistry Expands the Exploration Fairway for the Mineralised Copper Porphyries in Western Victoria. Geological Society of Australia, pp. 148–149. Abstracts 110.

Torsvik, T.H., Cocks, L.R.M., 2013. Gondwana from top to base in space and time. Gondwana Research 24, 999–1030.

Turner, N.J., Black, L.P., Kamperman, M., 1998. Dating of Neoproterozoic and Cambrian orogenies in Tasmania. Australian Journal of Earth Sciences 45, 789–806.

Turner, S., Foden, J., Sandiford, M., Bruce, D., 1993. Sm-Nd isotopic evidence for the provenance of sediments from the Adelaide Fold Belt and southeastern Australia with implications for episodic crustal addition. Geocheminica et Cosmochimica Acta 57, 1837–1856.

Turner, S.P., Haines, P.W., Foster, D., Powell, R., Sandiford, M., Offler, R., 2009. Did the Delamerian Orogeny start in the Neoproterozoic? Journal of Geology 117, 575–583.

Turner, S.P., Kelley, S.P., VandenBerg, A.H.M., Foden, J.D., Sandiford, M., Flöttmann, T., 1996. Source of the Lachlan Fold Belt flysch linked to convective removal of the lithospheric mantle and rapid exhumation of the Delamerian-Ross Fold Belt. Geology 24, 941–944.

VandenBerg, A.H.M., 2003. Silurian to early Devonian the Lachlan Fold Belt at its most diverse. In: Birch, G.D. (Ed.), Geology of Victoria. Geological Society of Australia, pp. 117–155. Special Publication 23.

VandenBerg, A.H.M., Stewart, I.R., 1992. Ordovician terranes of the southeastern Lachlan Fold Belt: stratigraphy, structure and palaeogeographic reconstruction. Tectonophysics 214, 159–176.

VandenBerg, A.H.M., Willman, C.E., Maher, S., Simons, B.A., Cayley, R.A., Taylor, D.H., Morand, V.J., Moore, D.H., Radojkovic, A., 2000. The Tasman Fold Belt System in Victoria. Geology and Mineralisation of Proterozoic to Carboniferous rocks. Geological Survey of Victoria Special Publication, 462 pp.

Veevers, J.J. (Ed.), 2000. Billion-Year Earth History of Australia and Neighbours in Gondwanaland. GEMOC Press, Sydney, 400 pp.

Veevers, J.J., 2015. Beach sand of SE Australia traced by zircon ages through Ordovician turbidites and S-type granites of the Lachlan Orogen to Africa/Antarctica: a review. Australian Journal of Earth Sciences 62, 385–408.

Veevers, J.J., Conaghan, P.J., Powell, C.McA., 1994. Eastern Australia. In: Veevers, J.J., Powell, C.McA. (Eds.), Permian–Triassic Pangean Basins and Foldbelts along the Panthalassan Margin of Gondwanalan, 184. Memoir-Geological Society of America, pp. 11–171.

Veevers, J.J., Saeed, A., 2008. Gamburtsev subglacial mountains provenance of Permian–Triassic sandstones in the Prince Charles mountains and offshore Prydz Bay: integrated U–Pb and TDM ages and host-rock affinity from detrital zircons. Gondwana Research 14, 316–342.

Veevers, J.J., Saeed, A., 2011. Age and composition of Antarctic bedrock reflected by detrital zircons, erratics, and recycled microfossils in the Prydz Bay–Wilkes land–Ross sea–Marie Byrd land sector (70°–240°E). Gondwana Research 20, 710–738.

Veevers, J.J., Saeed, A., Pearson, N., Belousova, E., Kinny, P.D., 2008. Zircons and clay from morainal siltstone at Mt. Rymill (73°S, 66°E), Prince Charles mountains, Antarctica, reflect the ancestral Gamburtsev mountains-Vostok subglacial highlands complex. Gondwana Research 14, 343–354.

Voice, P.J., Kowalewski, M., Eriksson, K.A., 2011. Quantifying the timing and rate of crustal evolution: global compilation of radiometrically dated detrital zircon grains. Journal of Geology 119, 109–126.

Weltje, G.J., von Eynatten, H., 2004. Quantitative provenance analysis of sediments: review and outlook. Sedimentary Geology 171, 1–11.

Williams, I.S., 1998. U-Th-Pb geochronology by Ion microprobe. In: McKibben, M.,A., Shanks III, W.C., Ridley, W.I. (Eds.), Applications of Microanalytical Techniques to Understanding Mineralizing Processes. Reviews in Economic Geology, vol. 7, pp. 1–35.

Williams, I.S., 2001. Response of detrital zircon and monazite, and their U-Pb isotopic systems, to regional metamorphism and host-rock partial melting, Cooma Complex, southeastern Australia. Australian Journal of Earth Sciences 48, 557–580.

Williams, I.S., Pulford, A.K., 2008. The contribution of geochronology to understanding the Palaeozoic geological history of Australia. Australian Journal of Earth Sciences 55, 821–848.

Wingate, M.T.D., Campbell, I.H., Compston, W., Gibson, G.M., 1998. Ion microprobe U–Pb ages for Neoproterozoic basaltic magmatism in south-central Australia and implications for the breakup of Rodinia. Precambrian Research 87, 135–159.

Withnall, I.W., Henderson, R.A., 2012. Accretion on the long-lived continental margin of northeastern Australia. Episodes 35, 166–176.

Withnall, I.W., Hutton, L.J., 2013. North Australian craton. In: Jell, P.A. (Ed.), The Geology of Queensland. Queensland Government, pp. 23–112 (Chapter 2).

Zucchetto, R.G., Henderson, R.A., Davis, B.K., Wysoczanski, R., 1999. Age constraints on deformation of the eastern Hodgkinson Province, north Queensland: new perspectives on the evolution of the northern Tasman Orogenic Zone. Australian Journal of Earth Sciences 46, 105–114.

CHAPTER

14

Utility of Detrital Zircon Grains From Upper Amphibolite Facies Rocks of the Grenville Supergroup, Adirondack Lowlands, Northeastern United States

J. Chiarenzelli[1], *D. Kratzmann*[2], *B. Selleck*[3], *W. deLorraine*[4], *M. Lupulescu*[5]

[1]St. Lawrence University, Canton, NY, United States; [2]Santa Rosa Junior College, Petaluma, CA, United States; [3]Colgate University, Hamilton, NY, United States; [4]St. Lawrence Zinc Company, Gouverneur, NY, United States; [5]New York State Museum, Albany, NY, United States

OUTLINE

Sediment Provenance
http://dx.doi.org/10.1016/B978-0-12-803386-9.00014-9

1. INTRODUCTION

Detrital minerals in sedimentary rocks provide important constraints on source regions. Minerals such as chromium spinel and clinopyroxene (Fedortchouk and LeBarge, 2008), garnet (Takeuchi, 2013), rutile (Meinhold et al., 2008), monazite (Hietpas et al., 2011), apatite (Morton and Yaxley, 2007), among many others, have been used to provide information on provenance and to locate mineral resources. However, among these minerals, zircon, because of its well-documented durability, is paramount for its potential to provide an age "bar code" for sediment source terranes. In addition, trace elements, oxygen isotopes, Hf and Nd isotopes, fission tracks, and other analyses may be used in conjunction with U-Th-Pb geochronology of zircon to compliment the temporal constraints obtained (Harley and Kelly, 2007). However on a cautionary note, because of the general dearth of zircon in mafic rocks, a bias toward felsic sources is possible (Fedo et al., 2003).

In this contribution, detrital zircon U-Th-Pb data is utilized, along with additional geological constraints, to interpret the age, source, and basin evolution of metamorphosed sedimentary rocks of the Grenville Supergroup (GSG) in the Adirondack Lowlands. The Adirondack Lowlands are part of the Grenville Province (Fig. 14.1), widely recognized as the roots of an ancient orogenic system that led to the eventual assembly of the Rodinia Supercontinent (Hoffman, 1991). This work provides a test of the applicability of, and constraints on, the use of detrital zircons in high-grade metasedimentary rocks. It expands upon preliminary studies reported elsewhere (Chiarenzelli et al., 2015) by providing data from additional units in the stratigraphy. Rocks in the Adirondack region (Fig. 14.2) range in metamorphic grade from amphibolite to granulite facies. Within the Adirondack Lowlands metamorphic grade ranges from mid-upper amphibolite facies in the Balmat zinc district to granulite facies to the northeast near Colton, New York. Pelitic lithologies in the stratigraphic sequence (Fig. 14.3) often display extensive partial melting and the crystallization of metamorphic and/or anatectic/igneous zircon in both the melanosome and leucosome (Heumann et al., 2006). Ductile deformation is also widespread and deformational fabrics and folds occur throughout the region. With only a few known exceptions (e.g.,

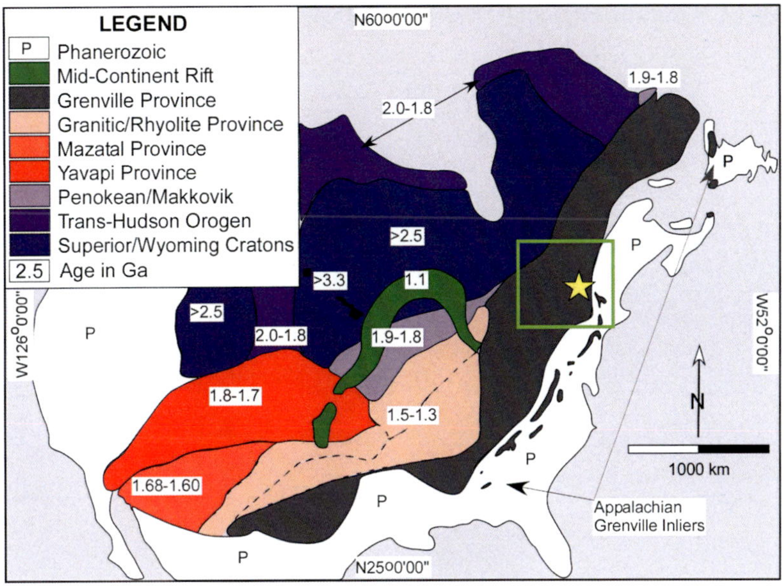

FIGURE 14.1 Location of the study area associated with this chapter. Age provinces of the Canadian Shield in North America are taken from McLelland et al. (2013). *Green outline* shows the location of Fig. 14.2; *yellow star* shows location of Adirondack region within the Grenville Province. Modified from Chiarenzelli et al. (2015).

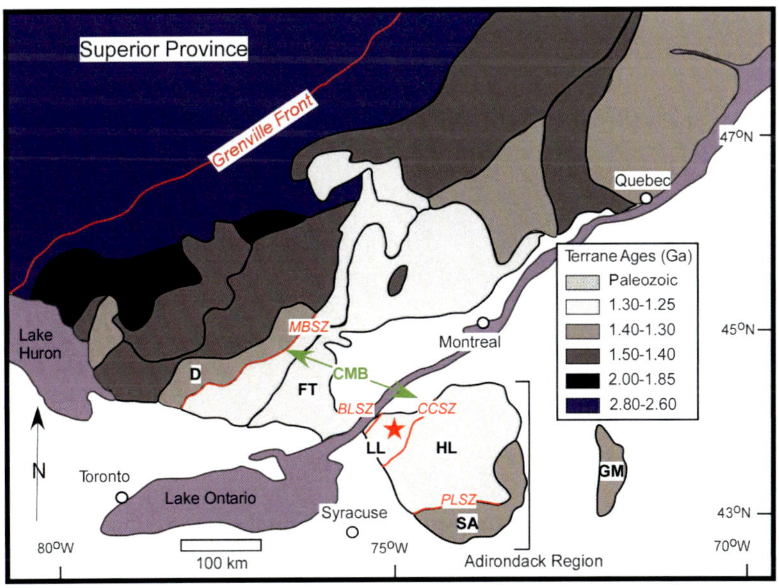

FIGURE 14.2 Map of the southern Grenville Province showing distribution of terranes and ages. Internal shear zones shown in red: BLSZ, CCSZ, MBSZ, and PLSZ, respectively; Black Lake, Carthage–Colton, Maberly, and Piseco Lake shear zones. *CMB*, Central Metasedimentary Belt; *D*, Dysart Suite; *FT*, Frontenac Terrane; *GM*, Green Mountains; *HL*, Adirondack Highlands; *LL*, Adirondack Lowlands; *SA*, Southern Adirondacks. *Star* shows location of sample sites and Fig. 14.4. The diagram is modified from Chiarenzelli et al. (2015).

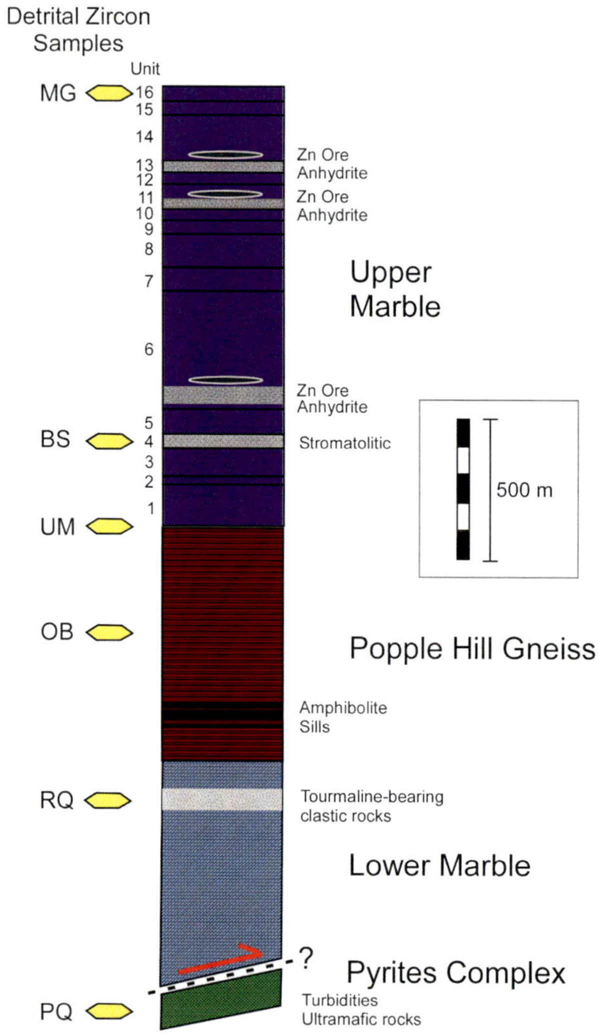

FIGURE 14.3 Simplified stratigraphic summary of the GSG in the Adirondack Lowlands modified from deLorraine and Sangster (1997). Schematic zircon crystals (*yellow with labels inside*) on left denote approximate stratigraphic location of detrital zircon samples investigated in this study. See Table 14.1 for exact locations.

Isachsen and Landing, 1983), primary structures are rarely preserved; however, compositional layering is present, is commonly parallel to foliation, and likely reflects original sedimentary layering (Chiarenzelli et al., 2012, 2015).

Metasedimentary rocks of the GSG (Easton, 1992) are widespread throughout the Grenville Province in the Adirondacks and adjacent Ontario and Quebec (Fig. 14.2). Potentially correlative units also occur in Grenville basement inliers throughout the Appalachians including in Vermont, the New York–New Jersey–Hudson Highlands, and along the spine of the Appalachians (Fig. 14.1). Deposited before the Grenville Orogenic Cycle began they record postdepositional deformation and metamorphic changes associated with orogenesis.

The GSG is best known for a thick (several kilometers) carbonate-dominated sequence now metamorphosed to marble and calc-silicate rock. In the Adirondack Lowlands a tripartite stratigraphy including two thick marble units separated by the siliclastic/volcanic Popple Hill Gneiss has long been recognized (Fig. 14.3; Engel and Engel, 1953; Brown and Engel, 1956; Foose and Carl, 1977; deLorraine and Sangster, 1997). More than a century of exploration for zinc, numerous quadrangle mapping reports, minimal disruption by intrusive igneous rocks, and a host of analytical studies makes the Adirondack Lowlands an ideal geologic terrane to test the efficacy of zircon analyses in provenance studies of highly deformed, high-grade Precambrian rocks.

2. GEOLOGICAL SETTING

Significant tectonic reworking of the hinterland, and the addition of a vast volume of rock occurred during the Grenville Orogenic Cycle (Figs. 14.1 and 14.2; McLelland et al., 1996). The Grenville Orogenic Cycle consists of at least four events over a period more than 250 million years including the Elzevirian (1245–1220 Ma), Shawinigan (1200–1140 Ma), and Grenville orogenies. The Grenville Orogeny consists of the Ottawan (1090–1020) and the Rigolet (1000–980 Ma) pulses (Rivers, 2008). The duration and sequences of tectonic events recorded during the Grenville Orogenic Cycle is similar to those that resulted in the Appalachian Mountains.

The Grenville Front (Fig. 14.2) demarks the northwestward limit of deformation across which older rocks can be traced into the foreland. Archean rocks of the Superior Province and rocks of various Paleoproterozoic to Mesoproterozoic terranes can be found southeast of the Grenville Front. Recent work suggests the Central Metasedimentary Belt (CMB) of the Grenville Province is a failed backarc rift zone/aulacogen (Dickin and McNutt, 2007; Dickin et al., 2015), supplanting earlier interpretations as a tectonic telescoped sequence of arcs (Composite Arc Belt of Carr et al., 2000). This interpretation places the pre-Grenvillian cratonic margin of Laurentia outboard of the volcanic and sedimentary sequence found in the CMB, and disrupted metasedimentary remnants of these found within the Central Granulite Terrane (Wynne-Edwards, 1972).

Prior to the Grenville Orogenic Cycle the margin of Laurentia is postulated to have undergone a prolonged period(s) (1700–1300 Ma) of subduction resulting in development of continental and/or oceanic arcs (Carr et al., 2000; Hanmer et al., 2000; Moretton and Dickin, 2013). The rifted remnants of these arcs occur as basement blocks interspersed within areas underlain by the GSG, as well as semiautochthonous remnants in southern Ontario and Quebec and perhaps beyond (Agustsson et al., 2013; Moretton and Dickin, 2013). In the southern and eastern part of the Adirondack region (McLelland and Chiarenzelli, 1990) and in the Green Mountains of Vermont (Mount Holly Complex; Ratcliffe et al., 1991), 1350–1300 Ma tonalitic gneisses of arc origin are found. Their distribution over a broad area has led many workers to suggest fragmentation of one or more Andean-type arcs by rifting and eventual tectonic reassembly (Carr et al., 2000; Hanmer et al., 2000).

Volcanic and volcaniclastic rocks varying in age from 1290 to 1250 Ma occur in the CMB (Easton, 1992). An ophiolite sequence including rocks of mantle affinity in the Grimsthorpe Domain of the CMB (Smith and Harris, 1996) suggests the opening of a backarc basin (Dickin and McNutt, 2007) of sufficient width to develop oceanic crust within the Ontario segment of the Grenville Province. The extension of this basin tapers northward into Quebec and suggests it is also an aulacogen (Dickin et al., 2015). An investigation of Nd isotopes by Dickin and McNutt (2007) and Moretton and Dickin (2013) has allowed the mapping of plutonic rocks with Nd model ages older and younger than 1350 Ma. In addition to a thick metasedimentary sequence, extensive areas of juvenile (Nd $T_{DM} < 1350$ Ma) plutonic rocks occur within the CMB intruding metasedimentary lithologies (Dickin et al., 2015). A similar and likely contemporaneous basin, the Trans-Adirondack Back-arc Basin (TABB), has been proposed for the Adirondack region by Chiarenzelli et al. (2011a) and also contains a thick sequence of the GSG.

The Adirondack Mountains are a Mesozoic to recent domal uplift, exposing rocks of the Grenville Province (Roden-Tice and Tice, 2005). Rocks of the Adirondack region (Fig. 14.2) are contiguous with the Grenville Province in Canada through exposures along the Frontenac Axis/Thousand Islands region (Isachsen and Fisher, 1970). Three terranes, with distinct geologic histories, are bounded by major structures. The Adirondack Highlands and Lowlands terranes are separated by the Carthage–Colton shear zone (Selleck et al., 2005; Geraghty et al., 1981) with both an early ductile and later brittle history. These two terranes differ in metamorphic grade, proportion of metasedimentary to metaigneous rocks, elevation and relief, and the timing of the terminal metamorphic events. In contrast to the Highlands terrane, the Lowlands generally attained only upper amphibolite facies metamorphic conditions, contain substantially more metasedimentary rocks, are of lesser elevation and relief, and were last deformed during the Shawinigan Orogeny.

The oldest rocks in the Adirondack region, 1350–1300 Ma tonalitic gneisses, occur in the Southern Adirondack Terrane (Fig. 14.2; McLelland and Chiarenzelli, 1990). An east-west trending, left-lateral, strike-slip shear zone, the Piseco Lake shear zone (Gates et al., 2004), separates the Southern Adirondack Terrane from the adjacent portion of the Highlands dominated by massif anorthosite and related rocks (McLelland et al., 2010). Similar, contemporaneous, tonalitic gneisses also occur in the eastern Adirondacks (McLelland and Chiarenzelli, 1990) and in nearby Vermont (Ratcliffe et al., 1991).

3. ANALYTICAL METHODS

Samples were collected from roadside exposures and drill core that could be constrained within the stratigraphy of the Adirondack Lowlands (Figs. 14.3 and 14.4). Zircons were separated from kilogram-size samples by standard methods at the Arizona Laserchron Center. Zircon separates were mounted in epoxy plugs, sectioned approximately half-way through, and imaged in backscattered electron (BSE) mode using a scanning electron microscope. These images were used to navigate and select areas within grain cross-sections for analysis; generally in the center of each grain, and avoiding inclusions, fractures, changes in BSE signature, or other visible heterogeneities, with the exception of oscillatory zoning.

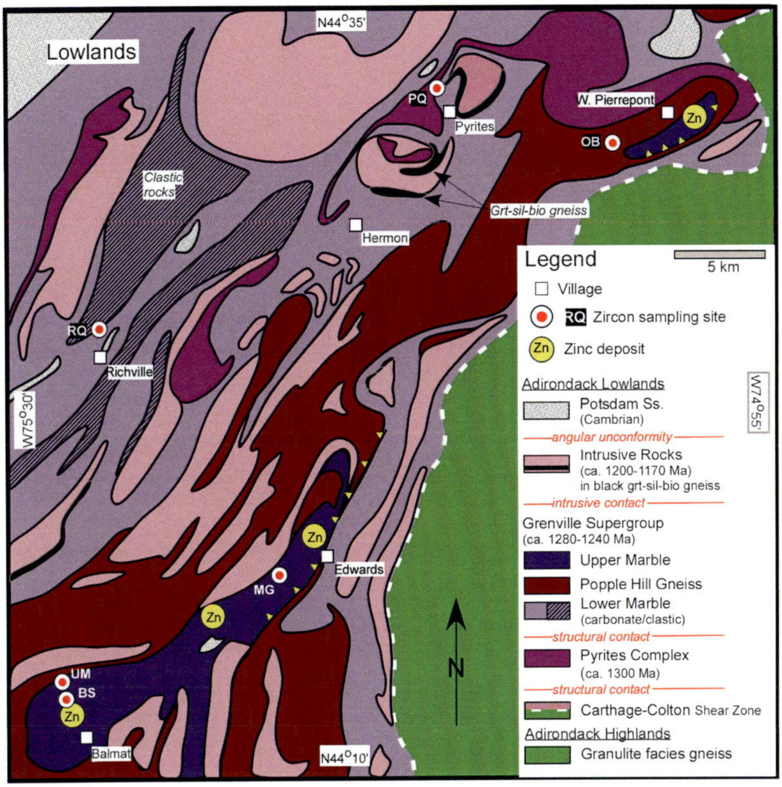

FIGURE 14.4 Simplified geologic map of the Balmat-Pierrepont zinc belt in the Adirondack Lowlands. Detrital zircon sample locations are shown by *red bull's-eye symbols*, with corresponding label, as used in Fig. 14.2 *White dashed line* shows the trace of the Carthage–Colton Shear Zone, boundary between Adirondack Lowlands (NW) and Highlands (SE). Map modified from Isachsen and Fisher (1970) and Chiarenzelli et al. (2015), and initially produced using ESRI Arc geographic information systems.

Although between c. 100 and 300 detrital zircon grains per sample were analyzed, unless otherwise specified, only analyses whose ($^{206}Pb/^{238}U$) age/($^{206}Pb/^{207}Pb$) age was within 3% of Concordia were retained. The details of the analytical procedures and data processing are available from the Arizona Laserchron Center website (https://sites.google.com/a/laserchron.org/laserchron/).

A common approach in detrital zircon studies of sedimentary rocks is use of the youngest age obtained as a maximum age for the time of deposition. This approach is not strictly followed here because of the complicated metamorphic history discussed earlier and the documented growth of metamorphic/anatectic zircon in nearby Popple Hill Gneiss and its correlative units (Heumann et al., 2006). Limitations associated with a two-dimensional view of zircon internal morphology in cross-section and the depth of laser ablation analysis pits warrant caution as unintentional sampling across distinct age domains can result in mixed ages. Other considerations, in addition to imaging and zircon chemical characteristics, such as the geologic history of the region, existing temporal constraints, and previous detrital zircon studies (Peck et al., 2010; Chiarenzelli et al., 2011b) were utilized to assist in our interpretation of the results. The details of the approach are discussed next.

4. RESULTS

The data from this study are shown in Figs. 14.5 and 14.8 and summarized in Table 14.1. Data that has been previously published (Chiarenzelli et al., 2015) is available from the Geological Society of America's data repository; additional datasets can be requested from the senior author. Locations are shown in context to regional geology in Fig. 14.4 and coordinates given in Table 14.1. General characteristics of each of the zircon populations are also shown in Fig. 14.5.

4.1 Pyrites (PQ: Quartz-Rich Layers in Pyritic Turbidities)

Along the Grasse River near the adit to the old pyrite mine at Pyrites, New York, small samples of quartz-rich (~85% SiO_2), cm-scale interlayers within a garnet-sillimanite pelitic gneiss were removed utilizing a chisel and processed for zircon. The rock shows isoclinal folding of interbedded quartzite and metapelitic layers. Centimeter-scale layers are interpreted as the alternation of sand, silt, and mud formed within a turbidite sequence (Chiarenzelli et al., 2015). Three meters structurally below the sample site, a coarse-grained, green, hydrothermally altered peridotite is exposed. These rocks, along with more extensive gabbroic and amphibolitic units, have been named the Pyrites Complex and interpreted as a highly disrupted ophiolite suite (Chiarenzelli et al., 2011a). Continuous exposure, gradation in the composition of the metasedimentary rocks, and the occurrence of chromite (Tiedt and Kelson, 2008) in the pelitic gneiss near the contact suggests the metasedimentary sequence overlies the ultramafic in apparent depositional contact.

A small separate (n = several hundred grains) of zircon was obtained from a kilogram of sample quartzose, sand to silt-sized portions of the aforementioned outcrop. Zircons recovered are relatively small (<100 μm), oscillatory zoned, and have shapes ranging from stubby dipyramids to grains with slightly rounded boundaries (Fig. 14.5F), thought modified by erosion. The zircons from this sample are noteworthy for their homogeneity. The U-content of the zircons analyzed in this sample averaged 243 ± 93 ppm and their U/Th ratio is 1.6 ± 0.4. Most grains are concordant with a range of 104.0–94.7%. Ninety-seven near-concordant grains are plotted (Fig. 14.5) and a group of 86 yielded a weighted average of 1289.7 ± 1.1 Ma. One grain, of smaller size and lacking visible zoning, gave an age of 1176.2 ± 24.1 Ma, in excellent agreement with the timing of Shawinigan orogenesis, and is considered to be metamorphic. Another grain yields an age of 1237.5 ± 24.1 Ma, the timing of Elzevirian orogenesis in the Grenville Province and is also interpreted to be of metamorphic origin. A group of five analyses that are statistically indistinguishable, gave a weighted mean age of 1258.3 ± 7.7 Ma (Mean Square Weighted Distribution (MSWD) = 1.4; Probability (PROB) = 0.22) and are interpreted to be the age of the youngest detrital population. Six other detrital grains range in age between 1372.9 ± 21.2 Ma and 2294.9 ± 21.7 Ma.

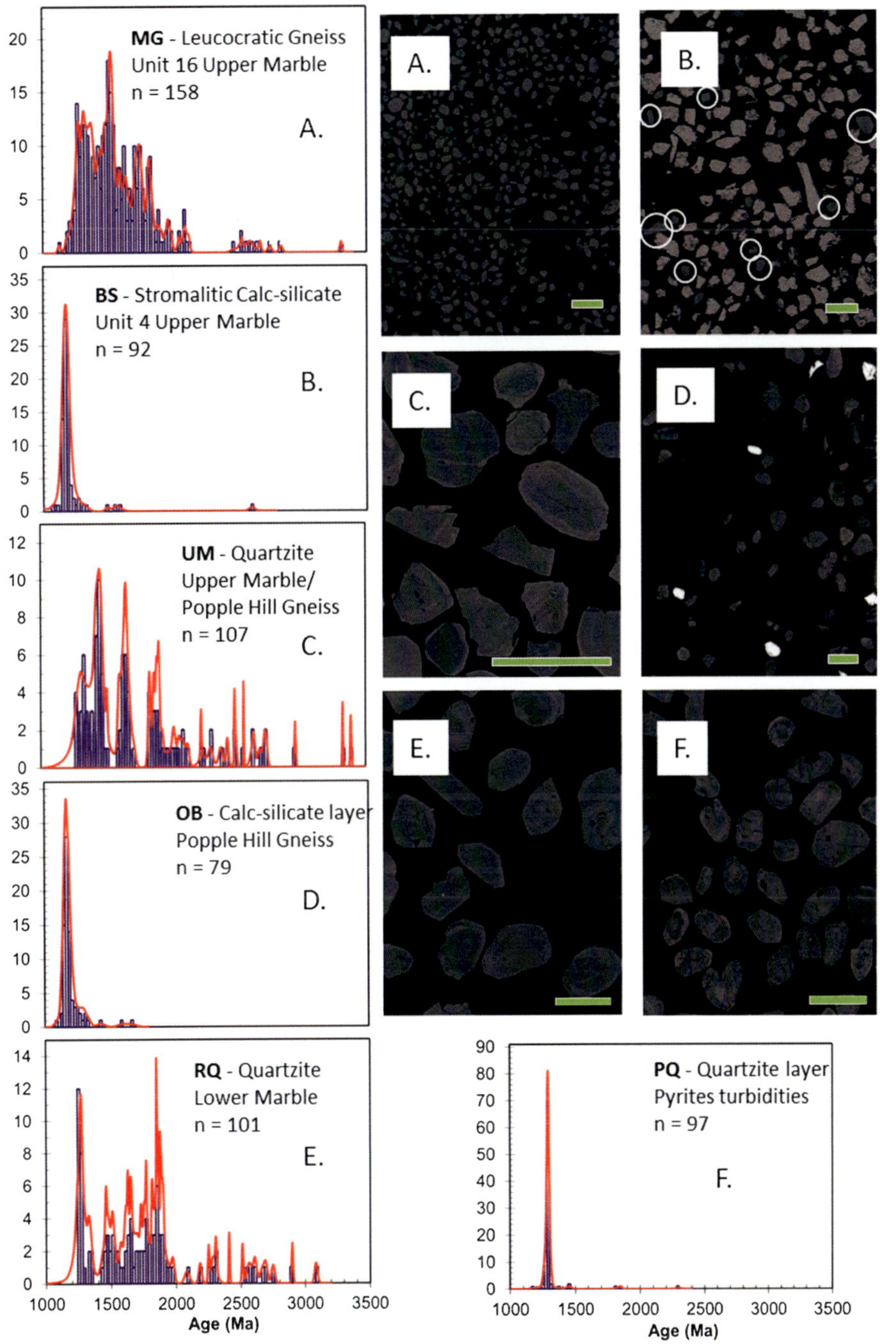

FIGURE 14.5 Probability histograms for detrital zircon samples analyzed in this study (*black letters* A–F) and corresponding scanning electron microscope (back scatter electron mode) photographs (*black letters in white boxes* A–F). The *green bars* in the photographs are 100 microns long. Zircons analyzed in sample BS are *circled* in photograph B. *BS*, Balmat stromatolite bearing calc-silicate; *MG*, median gneiss; *OB*, O'Brien Road calc-silicate layer; *PQ*, pyrites quartzite layer; *RQ*, Richville quartzite layer; *UM*, upper marble quartzite layer.

TABLE 14.1 Location and Summary of Characteristics of Detrital Zircon Samples From Rocks of the Grenville Supergroup, Adirondack Lowlands

Lithology (lat/long)	n (concordant)	Avg. U (ppm) Range (ppm)	U/Th	Largest Mode (Ma)	Youngest Detrital Population[a] (Ma)	Oldest Grain (Ma)
MG: UPPER MARBLE, UNIT 16, MEDIAN GNEISS, TALCVILLE, NY						
Leucogneiss	312	842 ± 548	27.78 to 0.63	1525 Ma	1253 ± 17 (n = 5)	3303.3 ± 8.4
N44°18′56.0″ W075°17′18.7″	(158)	52–2892	3.28 ± 3.23		MSWD = 3.7; PROB = 0.005	
BS: UPPER MARBLE, UNIT 4, BALMAT, NY						
Strom. Marble	93	942 ± 431	11.8 ± 17.6	1173 ± 3.0 Ma	1254.6 ± 21.7 (N = 1)	2607.5 ± 21.3
N44°16′02.8″ W075°24′28.2″	(69)	95–2331	0.5–112.6	MSWD = 1.09 PRB = 0.30		
UM: POPPLE HILL GNEISS (DRILL CORE), TOP OF SECTION, BALMAT, NY						
Quartzite	107	106 ± 75	1.5 ± 1.5	1445.4 Ma	1277.9 ± 13 (n = 5)	3388.3 ± 5
N44°16′46.4″ W075°24′27.2″	(97)	16–380	0.4–13.6		MSWD = 0.09; PROB = 0.994	
OB: POPPLE HILL GNEISS, MIDDLE PART, WEST PIERREPONT, NY						
Rusty calc-silicate	78	145 ± 100	5.0 ± 5.1	1171 ± 4.0 Ma	1260 ± 23 (n = 3)	1667.4 ± 35.2
N44°29′59.6″ W075°04′51.2″	(72)	34–494	0.1–23.3	MSWD = 1.5 PRB = 0.01	MSWD = 0.024; PROB = 0.88	
RQ: LOWER MARBLE, TOURMALINE-BELT, RICHVILLE, NY						
Quartzite	101	163 ± 135	2.0 ± 1.9	1848 Ma	1263.9 ± 4.3 (n = 20)	3082.9 ± 11.8
N44°25′20.0″ W075°23′24.5″	(87)	12–742	0.4–12.4		MSWD = 1.4; PROB = 0.11	
PQ: PYRITES COMPLEX, PYRITES, NY						
Turbidite	101	244 ± 93	1.6 ± 0.4	1289.7 ± 1.1	1258.3 ± 7.7 (n = 5)	2294.9 ± 21.7
N44°31′24.2″ W075°11′24.0″	(97)	24–449	0.6–2.7	MSWD = 1.4 PRB = 0.01	MSWD = 1.4; PROB = 0.22	

MSWD, mean squared weighted distribution; *PROB*, probability.
[a]*See text for how this was determined.*

4.2 Richville (RQ: Tourmaline-Bearing Feldspathic Quartzite/Arkose in Lower Marble)

The Lower Marble includes a number of detrital metasedimentary rocks in addition to the marble and subordinate calc-silicate gneiss (Brown, 1989). Of particular interest is an extensive (50 km) belt of black to reddish-brown tourmaline-rich rocks that are interlayered with dolomitic marble near the top of the sequence (Brown and Ayuso, 1985). Along Rt. 11, just outside of Richville, a meter-thick layer of feldspathic quartzite (arkose) within tourmaline-rich gneiss was sampled for U-Pb zircon geochronology.

A small population (n = 300–400) of zircon grains and a large number of pyrite and tourmaline grains were separated from approximately 1 kg of sample. The zircons ranged in size from 50 to 300 μm and were predominantly rounded to oval in shape, although some are euhedral (Fig. 14.5E). Truncated oscillatory zoning and a few, thin 1–5 μm, partial, euhedral metamorphic overgrowths were observed. Uranium concentrations in zircon range from 12 to 742 ppm and average 163 ± 135 ppm. Ratios of uranium to thorium range from 0.4 to 12.4 and average 2.0 ± 1.9. The vast majority of analyses are concordant.

One hundred and one grains are near concordant (Fig. 14.5) and show a wide range of ages from 1241.6 Ma to 3082.9 Ma. The youngest grain analyzed is 1241.6 ± 41.9 Ma. A cohort of the 20 youngest grains, all within analytical error of one another, gave a weighted mean of 1263.9 ± 4.3 Ma (MSWD = 1.4; PROB = 0.11), and are interpreted as the age of the youngest detrital population. Two large peaks, one at 1260.2 ± 4.7 Ma and the other at 1841 ± 2.1 Ma, are clearly defined on the probability density histogram (Fig. 14.5) and represent the dominant ages found in this sample.

4.3 Popple Hill Gneiss (OB: Sandy, Rusty, Calc-Silicate Interlayer)

A rusty, granular, calc-silicate gneiss was sampled from a 10 cm thick interval in garnet-bearing pelitic portion of the Popple Hill Gneiss near West Pierrepont, New York. The gneiss is interlayered and cut by numerous concordant to discordant leucogranitic gneissic sheets. The Popple Hill Gneiss consists of a thick (several kilometers?) sequence of metamorphosed mud, silt, and sand, that is, at least in part, turbiditic (Chiarenzelli et al., 2012). Samples from partially melted pelitic to psammitic portions of similar gneisses have been investigated for U-Pb zircon geochronology by Heumann et al. (2006) and Bickford et al. (2008) in the Adirondack Lowlands and Highlands and have been shown to have zircons of both detrital and igneous-anatectic origin.

A small population (~500 grains) of zircons was separated from a kilogram-sized sample of rusty calc-silicate gneiss. Zircons range in size from 20 to 80 μm and display a wide range of shapes from nearly rounded and oval to subhedral and faceted (Fig. 14.5D). A number of angular grains also occur.

Seventy-nine grains are near concordant and yielded a range of ages between 1127.0 Ma and 1667.4 Ma (Fig. 14.5). A population of 61 grains yielded an age of 1170.6 ± 4.0 Ma (MSWD = 1.5; PROB = 0.010), which falls in between the range of 1180–1160 Ma interpreted by Heumann et al. (2006) as the time of anataxis in nearby samples of this unit. Two grains, interpreted as the youngest detrital population, give an average age of 1260 ± 23

(MWSD = 0.024; PROB = 0.88). Four grains gave an age of 1223 ± 8 Ma and are interpreted as metamorphic (Elzevirian) or hybrid in origin (MSWD = 0.57; PROB = 0.64).

4.4 Popple Hill Gneiss/Upper Marble (UM: Glassy Quartzite at Lithologic Contact)

A section of drill core from an overturned limb of the Sylvia Lake Syncline near Balmat, New York penetrated the upper portion of the Popple Hill Gneiss and the lower portions of the Upper Marble (Chiarenzelli et al., 2012). The transition is considered to be conformable as the percentage of quartz in Popple Hill Gneiss gradually increases upward into ~30 m of glassy quartzite, then transitions into schist, before the first marble interval is penetrated. Portions of split drill core from several meters of section composed of the glassy quartzite were sampled, crushed, and zircon grains separated.

An excellent yield of several thousand grains was obtained from 1 kg of sample. Grains range from highly rounded to angular (Fig. 14.5C). Their average size is about 100 μm, with large rounded grains up to 300 μm. Smaller angular and euhedral grains are also present. Many zircon grains show truncation of oscillatory zoning (Fig. 14.7), but distinct overgrowths are few, thin, and incomplete. The average uranium content of zircons from the sample was 106 ± 75 ppm and U/Th range from 0.4 to 13.6 and average 1.5 ± 1.5.

One hundred and seven near-concordant grains were plotted and showed a wide range of ages from 1270.8 Ma to 3388.3 Ma (Fig. 14.5). The youngest grain analyzed gave an age of 1270.8 ± 113 Ma (note the large error). A cohort of 5 youngest grains, all within analytical error of one another, gave a weighted mean of 1277.9 ± 13 Ma (MSWD = 0.09; PROB = 0.994) and are interpreted as the age of the youngest detrital population. Two large populations, one at 1446 ± 7.8 Ma and 1650.3 ± 6.2 Ma, occur on the probability density histogram (Fig. 14.5).

4.5 Upper Marble (BS: Unit 4: Balmat Stromatolitic Calc-Silicate Rock)

The outcrop directly across from the entrance to the former Zinc Corporation of America headquarters near Sylvia Lake was sampled for detrital zircon geochronology. The rock sampled is from Unit 4, which exhibits silicified layering interpreted as remnant stromatolites (Isachsen and Landing, 1983). Here the matrix between the sparse, upside-down stromatolite domes (located on the overturned limb of the Sylvia Lake Synform) was sampled and consisted of quartz, dolomite, serpentine, titanite, and gray diopside. A sparse yield of approximately 100 silt-sized (20–50 μm) zircons of rounded to angular shape was obtained (Fig. 14.5B).

The U-content of zircons ranges from 95 to 2331 ppm and averages 947 ± 431 ppm, considerably higher than all other samples. Ratios of U/Th range from 1 to 113 and averages 12 ± 18. Because of the small size of the zircon and high U-content, little in the way of internal features can be discerned as the BSE signal is very homogeneous (Fig. 14.5B). Every grain in the mount larger than the 30 μm analytical spot size was analyzed, yielding 92 data points. Nearly all grains are within ± 3% of Concordia.

The youngest zircon grain analyzed gave an age of 994.7 ± 29.1 Ma, which falls within the time frame noted for the Rigolet pulse of the Grenville Orogeny (Rivers, 2008). A cohort of 66 grains yielded an age of 1172.7 ± 3.0 Ma (MSWD = 1.09; PROB = 0.30). Thirteen grains yield ages ranging from 1214.5 to 2607.5 Ma (Fig. 14.5). Two grains gave a weighted mean of 1224 ± 13 Ma (MSWD = 0.34; PROB = 0.80), interpreted as the timing of Elzevirian metamorphism or analyses that sample across age domain boundaries.

4.6 Upper Marble (MG: Unit 16: Layered Leucogranitic Gneiss)

The Median Gneiss (Unit 16 of the Upper Marble) is the youngest member of the stratigraphic succession of the GSG in the Lowlands. It is primarily a pink, strongly layered, quartzofeldspathic rock with a small percentage of other minerals such as diopside, tourmaline, hornblende, and scapolite. The protolith of the Median Gneiss is not known, although its composition is granitic or arkosic. A 1 kg-sized sample was collected from a small cliff exposure near Talcville, New York and yielded thousands of zircon grains.

The zircon grains range in size from 30 to 100 μm in diameter; most grains are about 50 μm and rounded, but small populations of angular and elongate grains were also noted (Fig. 14.5A). The average U-content is 842 ± 548 ppm and ranges from 52 to 2892 ppm. Ratios of U/Th range from 0.63 to 27.78 and average 3.28 ± 3.23.

Zircon U-Pb analyses yielded ages from 1185.4 to 3303.3 Ma. Over 300 grains were analyzed; however, those falling outside the range of 95–105% concordant were filtered out of the data set leaving 158 analyses to be plotted (Fig. 14.5). Of these, the largest peak on the probability density histogram is 1524.8 Ma. Two grains yield an age of 1234 ± 12 (MSWD = 0.29; PROB = 0.59), falling within the range of timing of Elzevirian orogenesis. The next five oldest grains form a coherent group with an age of 1253 ± 17 (MSWD = 3.7; PROB = 0.005).

5. DISCUSSION

5.1 Age of the Grenville Supergroup in Adirondack Lowlands

In contrast to rocks of the GSG from the CMB of Ontario (1290–1250 Ma; Easton, 1992) and the Franklin Marble (1299 ± 8 Ma to 1240 ± 17 Ma; Volkert et al., 2010) in the NY–NJ Hudson Highlands, rocks of demonstrably volcanic origin have yet to be recognized in the stratigraphic sequence exposed in the Adirondack Lowlands. However, the age of deposition of the Lowlands GSG sequence can be constrained by field relations. The minimum age for the sequence comes from U-Pb zircon dates on Antwerp-Rossie (AR) Suite, which cuts layering, and isoclinal folds within Lower Marble (Fig. 14.6). The AR Suite has been dated several times and yielded zircon U-Pb ages between 1183 and 1207 Ma (1183 ± 7 Ma, McLelland et al., 1992; 1207 + 26/−11, Wasteneys et al., 1999; 1203 ± 13.6, Chiarenzelli et al., 2010). Pegmatites intruding the Lower Marble of the GSG in the Adirondack Lowlands were dated to c. 1195 Ma (U-Pb zircon; Lupulescu et al., 2011). One pegmatite sample also contained xenocrystic zircon ranging in age from 1271 to 1312 Ma; presumably detrital zircon xenocrysts derived from rocks of the GSG it cross-cuts. Other younger suites also cross-cut rocks of the GSG, including plutonic rocks of the Hermon Granite Gneiss, Hyde School Gneiss, and Edwardsville Syenite (Peck et al., 2013). Thus the Lower Marble must have been deposited, buried, and deformed prior to c. 1207 Ma.

FIGURE 14.6 Antwerp-Rossie granitoid intruding and truncating isoclinal fold in lower marble. Geology action figure included for scale.

Where the contact between plutonic rocks and rocks of the GSG is observed it cross-cuts compositional banding and isoclinal folds (Fig. 14.6). This suggests that a strong ductile deformation event, effecting rocks of the GSG, occurred prior to c. 1207 in the Lowlands. By analogy with the nearby CMB this deformation is correlated to the Elzevirian Orogeny (c. 1245–1220 Ma). Thus, from local field relations and correlation with rocks in the CMB, the depositional age of the GSG rocks in the Adirondack region must predate deformation associated with the Elzevirian Orogeny (c. 1245–1220 Ma).

The initiation of sedimentation of the thick, widespread GSG requires the formation of an actively subsiding depositional basin. At ~1300 Ma, rifting occurred with the opening of one or more backarc basins along the southeast margin of Laurentia. Dickin and McNutt (2007), Moretton and Dickin (2013), and Dickin et al. (2015) used Nd model ages to delineate the geometry of this basin in the CMB. The occurrence of ophiolite complexes and mantle rocks in both Ontario (Smith and Harris, 1996) and the Adirondack Lowlands (Chiarenzelli et al., 2011a) requires the eventual formation of oceanic crust flooring at least in part of the basins.

A disrupted sequence of oceanic and mantle rocks identified in the Adirondack Lowlands at Pyrites, New York (Chiarenzelli et al., 2011a) was used to infer the opening of a separate backarc basin in the Adirondacks (TABB). It is separated from the CMB by intervening older rocks of the Frontenac Terrane; however, it could be linked to the west beneath Paleozoic cover. Farther to the southeast, the temporally equivalent Franklin Marble of the NY–NJ Highlands was also formed within a backarc basin (Peck et al., 2009).

Opportunities to directly date the GSG in the Adirondack Lowlands await discovery of the appropriate rocks. The youngest detrital zircon from a basal quartzite in the GSG stratigraphic succession on Wellesley Island in the Frontenac Terrane yielded an age of 1306 ± 16 Ma (Sager-Kinsman and Parrish, 1993) and provides a maximum age for the GSG in the Frontenac and adjacent CMB and Adirondack Lowlands terranes. Apatite from the Lower Marble in the southwestern part of the Adirondack Lowlands has been directly dated to 1274 ± 9 Ma by the Lu-Hf systematics (Barfod et al., 2005) and is interpreted as the timing of diagenetic growth of the apatite and may provide a minimum age constraint for Lower Marble.

The range of maximum depositional ages determined by the zircons analyzed in this study and their position in the stratigraphy is shown in Fig. 14.7. Conservative estimates suggest the entire sequence was deposited between 1306 and 1207 Ma. Other constraints from local (Fig. 14.6) and regional field relations suggest that the backarc basins, which provided the accommodation space for the GSG, began to close at c. 1245 Ma during the Elzevirian Orogeny. Only a very limited number (n = 11) of the U-Pb zircon analyses from the grains analyzed in this study fall into the known range of Elzevirian deformation, plutonism, and metamorphism (Fig. 14.7).

Given the proximity to plutonic rocks of Elzevirian age in the CMB (Fig. 14.2) it would be reasonable to expect a much larger number of local zircons in GSG in the Adirondack Lowlands. Large plutonic bodies of this age occur adjacent to and within the boundary of the backarc failed rift zone (Dickin et al., 2015). In our judgment, the dearth of Elzevirian ages can be explained by deposition of the GSG prior to Elzevirian orogenesis. Because of the problems arising from analyzing distinct age zones in a single grain, the few Elzevirian ages (n = 11) that do occur in the samples analyzed in this study are interpreted as hybrid or metamorphic ages. For these reasons, they are not considered the youngest detrital zircon, as is done in classic detrital zircon studies of sedimentary rocks.

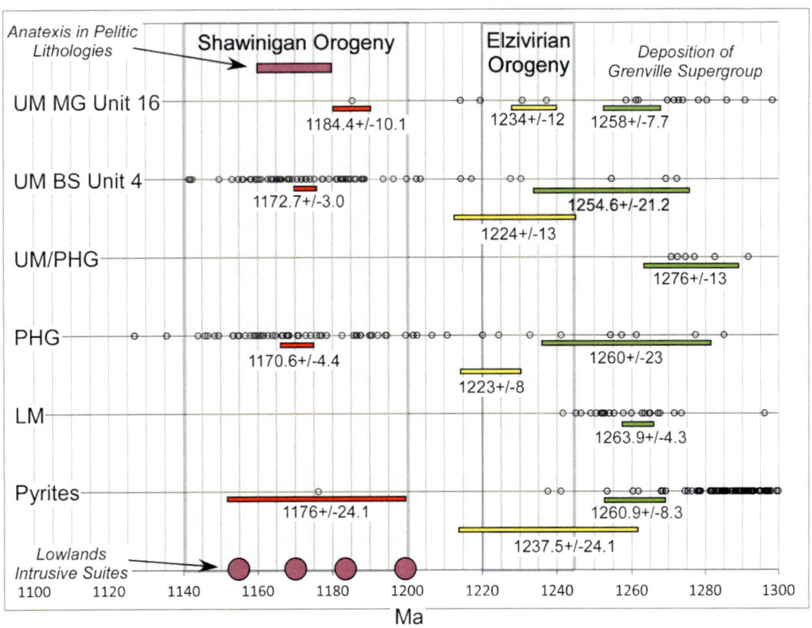

FIGURE 14.7 Schematic timeline showing all zircon U-Pb analyses less than 1300 Ma (*open black circles*). Orogenic events are shown in light gray shading. *Pink symbols* indicate anatectic and intrusive events. *Bars* indicate weighted averages of Shawinigan metamorphic zircon (red), Elzevirian metamorphic zircon (yellow), and youngest detrital zircons (green). See text for detailed explanation.

In contrast to zircon provenance studies in lower-grade rocks, zircons originally deposited in the rocks of the GSG in the Adirondack Lowlands have been through as many as four orogenic pulses representing the total cumulative effects of the Grenville Orogenic Cycle. Many of the grains clearly show metamorphic overgrowths as thin, discontinuous rims and possible areas of recrystallization (Fig. 14.8) and were avoided in this study. However, in the third dimension an ablation pit may sample material of different ages within a zircon crystal. In the study area, extensive anatectic and/or metamorphic zircon growth has been documented in rocks of appropriate composition such as the pelitic and psammitic lithologies of the Popple Hill Gneiss (e.g., Heumann et al., 2006; Bickford et al., 2008) and calc-silicate rocks (this study). These issues point out the difficulties, and emphasize the caution required, when determining the maximum age of deposition zircon analyses in high-grade metasedimentary rocks (also see Chiarenzelli et al., 2011b; Peck et al., 2010).

Our approach to this problem was to assume that the GSG in the Lowlands was deposited before Elzevirian deformation at 1245 Ma for the reasons cited previously. This is consistent with the presence of isoclinal folds that are cross-cut by the c. 1207 Ma Antwerp Rossie suite. These folds must have formed prior to the Shawinigan Orogeny (1200–1140 Ma; Fig. 14.6). Thus, zircon U-Pb ages between c. 1245 and 1220 Ma in all samples (11 out of 527 analyses) are considered metamorphic or mixed ages (Figs. 14.5 and 14.7). Their scarcity also implies a lack of source(s) of this age and infers that the GSG in the Lowlands was deposited prior to nearby magmatism and metamorphism in the Elzevir Terrane and orogenic activity associated with the Elzevirian Orogeny.

Because of the large uncertainty in the data due to analytical methods and isotopic systematics of some zircon grains, the youngest population of zircons, within standard error of one another and interpreted to be detrital, were pooled to calculate a weighted average whenever possible. This yielded a range of maximum deposition ages from 1276 ± 13 to 1254.6 ± 21.2 for the entire sequence (Fig. 14.7). This age range includes the Lu-Hf age determined by Barfod et al. (2005) for apatites from the Lower Marble. This range in ages also falls within the range determined from NY–NJ Hudson Highlands and CMB and is consistent with numerous other geologic constraints noted earlier. However, the data is not precise enough to resolve stratigraphic order in the Lowlands sequence (Figs. 14.5 and 14.7).

5.2 Provenance

The detrital zircon data presented here demonstrate the utility of using detrital zircon ages to track basin evolution in a complexly deformed metasedimentary sequence (Chiarenzelli et al., 2015). The source of detrital material is thought to have shifted during deposition of the GSG in response to tectonic events impacting the TABB. A legitimate question is whether or not these changes represent temporal or geographic differences or both. We argue here that these are indeed temporal shifts and that the detrital zircon data presented are consistent with stratigraphic relations documented in previous studies.

Detrital zircons can only provide a maximum age for deposition and this age is not necessarily close to the actual depositional age. Here we argue that the timing of tectonic events, including opening of the basins that contain rocks of the GSG in Ontario, Quebec, and New York (c. 1300 Ma), and the initiation of Elzevirian tectonism (c. 1245 Ma) resulted in near

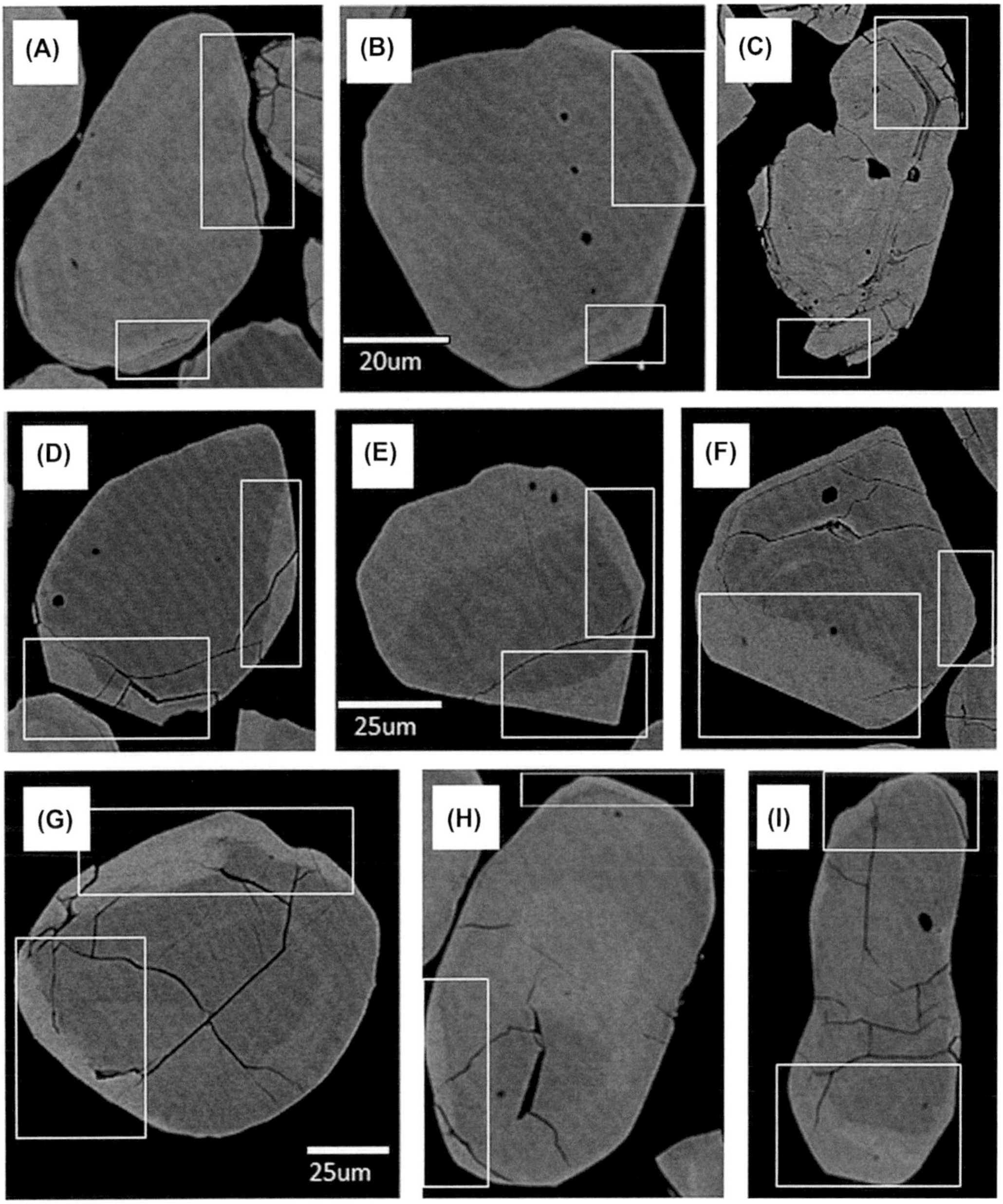

FIGURE 14.8 Possible metamorphic overgrowths (in *white rectangles*) on detrital zircons; thickest examples selected. BSE photographs (A), (B), and (C) from the quartzite (UM) sample at the transition from the Popple Hill Gneiss to the Upper Marble. Scale bar shown for each row. BSE photographs (D), (E), and (F) from the tourmaline-bearing quartzite at Richville (RQ). BSE photographs (G), (H), and (I) from quartzite layers (PQ) in the Pyrites pelitic gneiss. Scale, shown in center photograph, applies to each row of photos.

continuous tectonic activity and magmatism in the Grenville Province (e.g., see summary in Dickin et al., 2015). Thus the maximum ages determined are considered to reasonably represent actual depositional age.

One possible exception to the stratigraphic sequence just outlined is shown in Fig. 14.3, where the relationship between the Pyrites Complex and the Lower Marble is shown as a possible thrust fault. The turbiditic gneisses of the Pyrites Complex are associated with a belt of ultramafic and mafic rocks and were deposited within a deep marine basin, in contrast to the shallow marine depositional environment proposed for the Lower and Upper Marble (Chiarenzelli et al., 2011a, 2012, 2015). This unit is considered allochthonous and is interpreted to have been thrust into its current location along with the ultramafic rocks that structurally underlie it (Chiarenzelli et al., 2011a). Pyrite-rich gneisses also occur at the base of the Popple Hill Gneiss and Lower Marble (Prucha, 1957) and may be correlative to the rocks at Pyrites. If so, they represent contemporaneous distal, deep water, and in part, chemogenic sediments of units deposited on the shelf. The interpreted stages of basin evolution are summarized next.

5.2.1 *Rift*

Quartzite layers from the turbiditic rocks at Pyrites yielded a homogenous population of zircon grains that are euhedral, well-zoned, and geochemically and isotopically similar. Eighty-six out of 97 grains analyzed yielded a unimodal probability density peak at 1289.7 ± 1.1 Ma (Fig. 14.5) and constrain the source of the vast bulk of zircons in the sample. This suggests a source that has either not yet been identified in the Lowlands, was long ago eroded away, or was derived from elsewhere in the Grenville Province. A veneer of felsic volcanic rocks associated with the initial rifting and opening of the TABB is a likely source. In such a scenario the zircons would have been transported to deeper parts of the basin by turbidity currents as the thin volcanic veneer was stripped off the rising rift shoulders.

5.2.2 *Drift*

The GSG in the Lowlands was deposited before Elzevirian deformation at 1245 Ma based on deformation fabrics within the GSG that are cross-cut by c. 1207 Antwerp-Rossie intrusives (Fig. 14.6). The Lower Marble represents carbonate deposition on a subsiding platform, although likely in relatively shallow water. Feldspathic quartzite from a belt of tourmaline-rich rocks sampled at Richville, near the top of the unit, was deposited during a hiatus in carbonate production related to active tectonism and block faulting on the shelf. Alternatively, lowering of sea level may also explain the sudden influence of arkosic clastic material into a carbonate-dominant system. The accumulation of boron, now residing in the tourmaline, may be attributed to evaporitic conditions in a playa lake depositional environment in the putative fault block(s) or accumulation underwater on the shelf (Slack et al., 1984).

The sample of feldspathic quartzite from Richville yielded a small population of zircons with ages ranging from 1241.6 to 3082.9 Ma indicative of detritus from a wide variety of source terranes. The two largest peaks are 1268 Ma and 1848 Ma. The majority (58%) of the zircons yield ages of 1400–2000 Ma, 21% yield ages of 1242–1274 Ma, and 8% of grains were derived from an Archean source. None of the grains yielded ages between 1347 and 1435 Ma, the age of a significant portion of pre-Grenville rocks in Ontario and Quebec to the north (Figs. 14.2, 14.5, and 14.7).

This suggests that the detrital zircon population analyzed was derived from a wide range of sources extending back to the Mesoarchean. However, the source lacked a significant Neoarchean component (i.e., Superior Province) and was composed predominantly of Proterozoic terranes (Fig. 14.1). In addition, a significant population of zircon ages that could be derived from adjacent parts of the Grenville Province is lacking. This is consistent with a source dominated by zircons from the now-buried portions of the Canadian Shield in a broad band across the mid-continent of the United States. Specifically, detrital zircon grains from the Granite-Rhyolite terrane (1.5–1.3 Ga), Mazatal Orogen (1.68–1.60 Ma), Yavapi Orogen (1.8–1.7 Ga), and the Penokean Orogen (1.9–1.8 Ga) are well represented (Figs. 14.1 and 14.5). The presence of this widely dispersed provenance signature in the Richville feldspathic quartzite is consistent with a regional lowering of sea level, during basin evolution, and with terrestrial transport systems (braided streams, aeolian dunes) to the basin.

5.2.3 Foredeep

The Popple Hill Gneiss is predominantly a biotite-quartz-plagioclase gneiss, which in some areas contains large proportions of garnet and/or sillimanite. A variety of protoliths have been ascribed to it ranging from dacitic metavolcanic/volcaniclastic rocks to fine-grained clastic sedimentary rocks (see Chiarenzelli et al., 2012). Chiarenzelli et al. (2012) suggest it is a turbiditic sequence dominated by compositionally and texturally immature sands. Heumann et al. (2006) and Bickford et al. (2008) sampled similar rocks in the Adirondack Lowlands and Highlands and recognized a substantial geographic difference in the age of metamorphic/anatectic zircons. Zircons of Ottawan age (1090–1020 Ma) were restricted largely to the eastern Adirondacks, while samples from most of the Highlands and the Lowlands had only Shawinigan metamorphic zircons (1200–1150 Ma). Most of the detrital zircon ages (78%) from a number of pooled samples from the Adirondack Lowlands are Shawinigan (c. 1180–1160 Ma), consistent with the development of substantial volumes of anatectic melts, represented by leucosome, and zircon growth, at this time. The older, detrital population was made up exclusively of zircon grains yielding ages between 1378 and 1298 Ma (Heumann et al., 2006).

The closet known source of rocks with ages c. 1400–1300 Ma (Fig. 14.2) is located in the southern and eastern Adirondacks (McLelland and Chiarenzelli, 1990). Tonalitic gneisses similar in age and composition to those in the Adirondacks also occur within Grenville inliers in Vermont and nearby areas of the Grenville Province of Ontario and Quebec. Paleogeographic reconstructions suggest the southeastern margin of the TABB was bounded by the Southern Adirondack Terrane. The lack of other older zircon populations could be explained by the northward transport of the detritus derived from tonalitic arc rocks into a developing foreland basin associated with the Elzevirian Orogeny. Sourcing from older portions of the Grenville Province to the north and/or Laurentia to the west would have carried additional, older components into the basin, which are generally lacking. Therefore, the unimodal nature of the detrital population in the Popple Hill Gneiss is consistent with a restricted (c. 1400–1300 Ma) and/or local source.

While the details of foreland basin development remain to be determined, movement of the Southern Adirondack Terrane toward the north prior to Elzevirian orogenesis accompanied by southward directed subduction is envisioned (Fig. 14.9). In such a scenario, a deep basin accompanying flexure of the crust would be produced by thrust/nappe stacking within

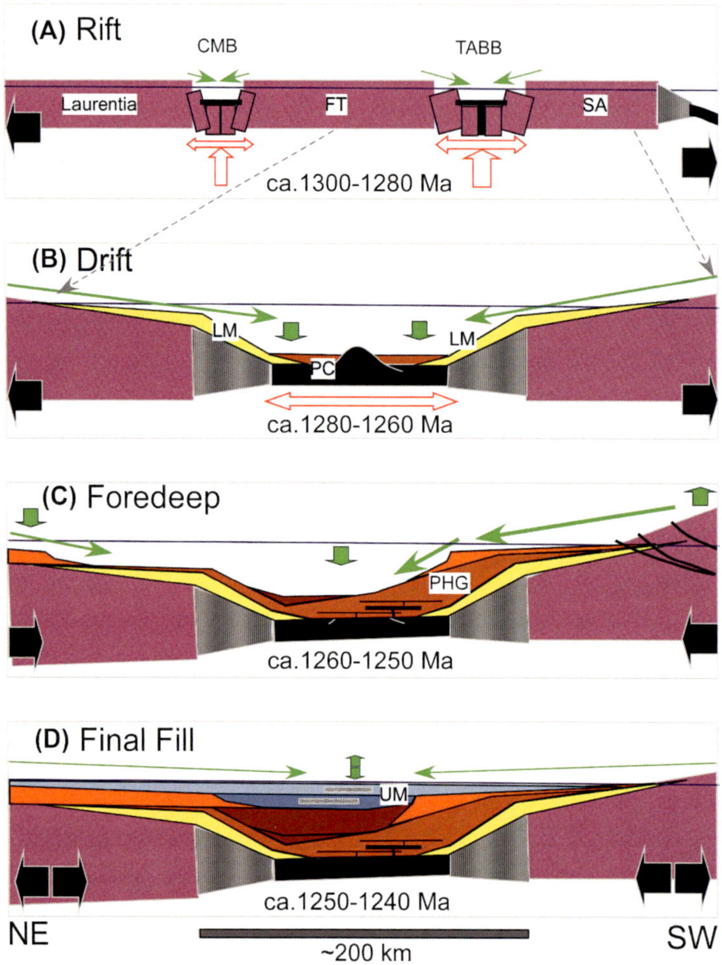

FIGURE 14.9 Schematic carton showing the temporal evolution (1300–1240 Ma) of the Adirondack region during the deposition of the GSG rocks in a series of NW–SE cross-sections. (A) Between 1300 and 1280 Ma ago the margin of Laurentia undergoes extension, block faulting, and rifting, creating a series of back-arc basins (*CMB*, Central Metasedimentary Belt; *TABB*, Trans-Adirondack Back-arc Basin). A continental arc on the margin of Laurentia is split and becomes the Frontenac (FT) and Southern Adirondack (SA) terranes, containing 1350–1300 Ma tonalitic rocks. Subduction dipping to the southeast is shown to the far right of the cross-section. (B) Focusing on the Adirondack region, between c. 1280–1260 Ma sedimentary rocks of the Lower Marble (yellow) are deposited on the rifted edge of and transitional crust (gray) of opposing flanks of the widening TABB. Along the axis of the basin, spreading generates ocean crust (in black) where rocks of the ultramafic Pyrites Complex form (PC). (C) By 1260–1250 Ma, in response to impending collision to the southeast in the Southern Adirondack Terrane a linear foredeep begins to develop centered on the Adirondack region. Turbiditic sediment represented by the Popple Hill Gneiss (PHG), in shades of orange and brown, accumulates in deep water. Note the mafic sill and dike complex in the center of the basin. (D) In the lowermost diagram, between c. 1250–1240 Ma, far-field stresses from the impending Elzevirian Orogeny results in contraction and relaxation of the basin. Now filled with thick accumulations of clastic sedimentary rocks, shallow-water carbonates, evaporites, and sedimentary exhalative zinc deposits (*stippled horizons*) of the Upper Marble (UM, in shades of blue) are deposited in the center of the basin where accommodation space remains. Alternate uplift and subsidence leads to sporadic isolation from the open ocean and evaporite deposition, and migration of hydrothermal, zinc-bearing fluids to the basin interior. Thick stubby *black arrows* represent plate motion; open *red arrows* represent asthenospheric flow; thick blocky *green arrows* show subsidence and tilt in the upper crust; *long narrow arrows* show sediment transport direction. Relative sea level is shown by a thin *blue line* in each cross-section.

or adjacent to the Southern Adirondack Terrane. Such a model also suggests that far-field deformation leading up to the Elzevirian Orogeny impacted on-going sedimentation in the TABB, something we also infer in our later discussion concerning the Upper Marble.

A sample of sandy calc-silicate layer from the Popple Hill Gneiss shows an age distribution of silt-sized zircons with the vast majority (80%) reflecting growth or resetting during Shawinigan orogenesis and anataxis (Fig. 14.7; 1170.6 ± 4 Ma). This interpretation is consistent with field relations that indicate intrusion of rocks ranging in age from 1210 to 1155 Ma into the Popple Hill Gneiss (Chiarenzelli et al., 2010; Peck et al., 2013). Among the zircon analyses whose ages we interpret as detrital (Fig. 14.7), just three are older than 1400 Ma (1432, 1596, and 1667 Ma). Again indicating a dominant source likely derived from the south (i.e., 1400–1300 Ma Southern Adirondack Terrane).

5.2.4 *Transitional Period*

Drill core penetrating from the Sylvia Laker Syncline (Fig. 14.4), through the Popple Hill Gneiss, and continuing into the base of the Upper Marble provides an opportunity to investigate the transition between the two units (Chiarenzelli et al., 2012). Up-section, quartz increases in modal proportions in the upper part of the Popple Hill Gneiss, and eventually thin layers of quartzite amalgamate into over 30 m of pure, glassy quartzite. This gives way to several tens of meters of phyllosilicate and tourmaline-rich schists and the first, decimeter-scale marble interval. A sample of the quartzite at this transition zone yielded a range of zircon U-Pb ages between 1271 and 3383 Ma, all of which are interpreted as detrital. The sample lacked any analyses that could be ascribed to any metamorphic event during the Grenville Orogenic Cycle. The lack of metamorphic zircon in this sample is discussed next.

The probability histogram for the glassy quartzite at the transition between the units (Fig. 14.5) shows peaks at 1445 (Granite Rhyolite Province), 1648 (Mazatal Orogen), and 1901 Ma (Penokean and/or Trans-Hudson Orogens). Theses ages are found within vast regions of the interior of Laurentia (e.g., Fig. 14.1) and are generally lacking or of lesser significance in the nearby Grenville Province (Fig. 14.2). Archean grains comprise only 8.4% of all the analyses. Two of these Archean grains give ages in excess of >3300 Ma (Fig. 14.1), indicating a great age for one of the sources or recycling of very old crustal materials. Rocks of appropriate age exist in the Minnesota River Valley or Winnipeg terranes (shown in Black on Fig. 14.1). Taken together, this information suggests reestablishment of a river system draining eastward from the south-central portion of Laurentia to the TABB. The upward shallowing noted and compositional maturation (quartz-rich) at the top of the Popple Hill Gneiss and base of the Upper Marble argues for the eventual filling of the foredeep basin, reworking of previously deposited sediments, and western expansion of the drainage basin. This expansion would allow for contribution from a much greater number of source terranes and from greater distances than is seen in the samples from lower portions of the Popple Hill Gneiss.

5.2.5 *Basin Closure*

The Upper Marble has been subdivided into 16 lithostratigraphic units (Brown and Engel, 1956; deLorraine and Sangster, 1999), most of them dominated by carbonate and mixed carbonate-siliciclastic rocks. The stratigraphic succession is punctuated by three periods of evaporite deposition, closely followed by the deposition of Zn-rich sedimentary exhalative deposits (deLorraine and Sangster, 1999; Fig. 14.3). Unit 4, consisting of a mixed carbonate-siliclastic rock, contains structures interpreted as remnant stromatolites thought to have been deposited in shallow water along the coast of ancient marine shoreline. Unit 4 yielded a small population of silt-sized zircons compatible with transport by wind. However, 85% of these grains gave a well-constrained age compatible with Shawinigan metamorphism (1173 $\pm$ 3.0 Ma). These zircons cannot be detrital in origin because they are younger than several igneous rock suites that cross-cut Unit 4 (Peck et al., 2013). Based on their lack of internal features (inclusions, zoning, etc.), age, and previous detrital zircon studies in the Lowlands (Heumann et al., 2006), they are interpreted as metamorphic. The remaining zircon grains have ages between 1254 and 2607 Ma, consistent with a detrital origin and a variety of sources derived from within south-central Laurentia.

A sample of the uppermost stratigraphic unit (Unit 16, Median Gneiss) in the Upper Marble, a strongly layered, leucogranitic gneiss, yields zircon whose U-Pb ages ranged between 1258 and 3303 Ma, indicating a detrital, rather than igneous, origin. These ages accounted for over 95% of all the analyses completed and indicate little effect of metamorphism on these grains. Zircon analyses indicate source terranes capable of providing abundance of detritus from 1250 to 2100 Ma in age (Fig. 14.5). However, only 4 grains out of the 159 analyzed yielded Archean ages, again suggesting limited influence of the extensive, west-to-east trending, Abitibi Greenstone Belt (c. 2750–2650 Ma: Corfu, 1993) of the Superior Province to the north (Fig. 14.1) in rocks of the GSG in the Adirondack Lowlands.

This data suggests that the Median Gneiss is indeed of clastic origin, likely composed of arkosic detritus, and that it contains zircon with a wide range of ages extending through most of the Mesoproterozic and early part of the Paleoproterozoic. Yet, it lacks material from nearly all Archean sources, particularly the closest Neoarchean rocks (i.e., Abitibi Greenstone Belt of the Superior Province). This is compatible with sampling and/or reworking of an extensive area to the south and west in Laurentia's interior, but the exclusion of northern sources.

5.3 Use of Zircon in High-Grade Terranes

This study has implications for the use of zircon for provenance analysis in high-grade terranes. Work by Peck et al. (2010) showed that between 62 and 87% of individual zircon grains from the quartzose Irving Pond formation in the Southern Adirondack Terrane have been recrystallized and reset. Thick metamorphic/recrystallized rims predominate, while small cores retain protolith information. Since the Adirondack Highlands are considered to have been metamorphosed to granulite facies, the extent of resetting may be characteristic of the P-T-t conditions.

A similar study on calc-silicate rocks (diopside-rich quartzite) at Chimney Mountain in the central Adirondack Highlands found that detrital grains had been completely recrystallized, gave a well-constrained age related to Ottawan metamorphism, lacked a CL signature suggesting erasure of any original zoning or heterogeneities, and yet retained their size and exterior morphology (Chiarenzelli et al., 2011b). However, zircon grains from a granitic rock sampled approximately 200 m away showed virtually no effect from metamorphism and yielded a Shawinigan intrusive age. In addition, Heumann et al. (2006) and Bickford et al.

(2008) documented large amounts of anatectic or metamorphic zircon in the melanosome and leucosome of pelitic to psammitic gneisses of the Popple Hill Gneiss in both the Adirondack Highlands and Lowlands.

In the present study a clear distinction can be drawn between the metamorphic response of zircons in quartz-rich lithologies and those in rocks with abundant calc-silicate phases. Effects in metamorphosed quartz-rich rocks are limited to sporadically developed rims on detrital zircon grains that are too thin to analyze (Fig. 14.8) and make up a very minor fraction of the total grain volume (<1%). For example, the pure, glassy quartzite from the boundary of the Popple Hill Gneiss and Upper Marble showed few, if any, metamorphic overgrowths and lacked any ages consistent with known periods of Grenvillian metamorphism (Fig. 14.7). However, both calc-silicate bearing rocks (OB, BS) had extensive populations of Shawinigan zircons, as well as older, clearly detrital grains (Figs. 14.5 and 14.7). This suggests that the composition of calc-silicate rocks at upper amphibolite facies allows for the growth of neometamorphic zircon or the resetting of existing grains. In the case of pelitic gneisses, this growth can be attributed to partial melting. In calc-silicate rocks, CO_2-rich fluids or recrystallization of Zr-bearing silicates may play a role in the recrystallization or neometamorphic growth of zircon grains.

The Richville tourmaline-bearing feldspathic quartzite and the Median Gneiss are arkosic in composition but also show relatively little in the way of metamorphic rims and no neometamorphic grains. Thus, it is hypothesized that the more quartz-rich and/or the fewer impurities capable of hosting Zr and/or U a metamorphic rock has, the less the volume of new zircon growth. However, U-Pb isotopic systematics may be partially to completely reset at granulite facies during recrystallization of detrital zircon (Peck et al., 2010; Chiarenzelli et al., 2011b).

5.4 Paleogeographic Constraints

Available evidence suggests the southeast margin of Laurentia (present coordinates) was a long-lived Andean-type continental margin (Hanmer et al., 2000) during the Mesoproterozoic Grenville Orogenic Cycle. Just prior to the deposition of the rocks of the GSG a continental arc, developed on the edge of Laurentia, was rifted apart and a series of small back arc basins developed (e.g., Rivers and Corrigan, 2000; Fig. 14.9). The juvenile crust flooring these basins was separated by older basement blocks as defined by Dickin and McNutt (2007), which are often separated by shear zones or thrusts (Dickin et al., 2008; Moretton and Dickin, 2013). The basins were relatively short-lived (c. 40–50 million years) but were active long enough to produce, and be floored by, oceanic crust as shown by the occurrence of ophiolite sequences and mantle rocks in both the CMB (Smith and Harris, 1996) and the Adirondack Lowlands (Chiarenzelli et al., 2011a). Volcanic rocks associated with rifting and development of the backarc basin have been identified in the CMB but currently only indirect evidence exists in the Adirondack Lowlands, where turbiditic rocks from Pyrites contain a homogeneous and unimodal population of zircons at 1289.7 ± 1.1 Ma (Fig. 14.5), which may have been derived from a local volcanic veneer of this age.

Deposition of sedimentary rocks of the GSG began nearly synchronously in both the CMB and TABB as dry, tropical climatic conditions led to the production of thick sequences of shallow water carbonates rimming the margin of the basins. Nutrient-rich waters allowed for the accumulation of organic matter as shown by graphite deposits, graphitic bands in marble, natural gas discoveries in the Upper Marble in Balmat (Brown, 1932), and the release of hydrogen sulfide gas from fetid marbles in both the Upper and Lower Marble sequences. In the TABB evidence for uplift and episodic clastic sedimentation is provided by quartzites and arkosic tourmaline-rich rocks interlayered with the thick carbonate sequence of the Lower Marble. The development of a northern basin represented by the CMB, which served as a depocenter for detritus from the north, limited delivery of clastic material derived from the adjacent Grenville and Superior provinces to the TABB (Fig. 14.9).

Drainage from the interior of the continent brought detritus from an extensive area in the south-central portion of Laurentia including that of the Mid-Continent, Matatzal, Yavapi, and Penokean-Trans-Hudson Orogens (Fig. 14.10). However, this was interrupted, at least temporarily, by a shift in provenance to a 1400–1300 Ma source terrane. The most likely candidate for this source is the Southern Adirondack Terrane (Fig. 14.2) where calc-alkaline tonalites of the appropriate age have been correlated with the Dysart–Mt. Holly Suite (McLelland and Chiarenzelli, 1990). This detritus occurs in mud-rich turbiditic rocks of the Popple Hill Gneiss in the Lowlands and Highlands equivalents, which infer a significant shift in the nature of sedimentation from shallow water carbonate production. Shallow water deposition occurs both stratigraphically above and below the deep water deposition of the Popple Hill Gneiss. Production of a foreland basin related to a south-dipping subduction zone and far-field stress from an impending collision provides a viable mechanism to account for the shift in provenance and in basin morphology.

5.5 Constraints on the Zinc Ore at Balmat

The GSG hosts several zinc deposits including the world-famous zinc oxide ores at Franklin, New Jersey and the world-class zinc sulfide ore at Balmat, New York. These deposits are two of a handful of economically viable Meosproterozoic zinc deposits in the world. In addition, they occur in high-grade metamorphic terranes. Those at Balmat (Figs. 14.3 and 14.4) are associated with relatively thick evaporite deposits, now mostly anhydrite (deLorraine and Sangster, 1997).

At Balmat, evaporitic units occur beneath each of the three major stratiform zinc deposits and suggest a possible temporal and genetic relationship between the two (Fig. 14.3). Here it is suggested that initial convergence preceding the Elziviran Orogeny resulted in the uplift of the TABB, separating it temporarily from the open ocean. This restricted circulation and hot, dry climatic conditions facilitated evaporite accumulation. Heated chloride-rich basinal brines derived from intercalated evaporite deposits would have been driven by compression of the basin to permeable zones or long-lived fault zones where they vented to the basin floor. Zinc and other metals were scavenged from the sedimentary pile and transported to SEDEX depocenters. Relaxation of tectonic stresses would allow for conditions conducive for continued subsidence and the resumption of dolomitic carbonate, and minor siliclastic deposition. The episodic nature of far-field tectonic forces resulted in at least three preserved periods of evaporitic conditions and subsequent sulfide ore accumulation.

The age of the zinc deposits at Franklin Furnace, New Jersey are bracketed between c. 1290 and 1250 Ma by U-Pb dating of zircons from volcanic units within the section

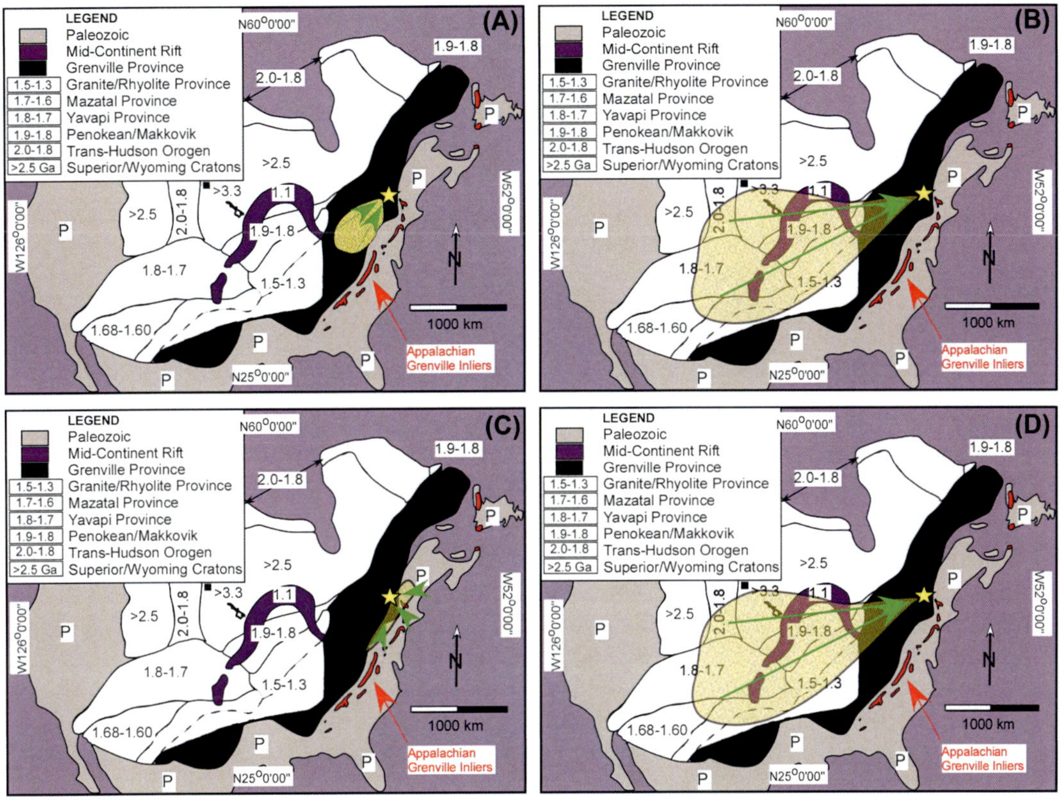

FIGURE 14.10 Series of cartons showing interpreted sediment transport directions to the Trans-Adirondack Back-arc Basin (*yellow star*) during deposition of the GSG superimposed on the Precambrian provinces of the Canadian Shield. Note that much of the southern part of the area shown in the diagrams are now buried beneath younger sedimentary rocks and that the Mid-Continent Rift is shown only for reference, as it did not exist during the time frame in question. Diagrams correlate to schematic cross-sections in Fig. 14.9. (A) Rift phase (c. 1300−1280 Ma). Sediment is derived largely from a local, thin veneer of rift-related volcanic and volcaniclastic rocks approximately 1300 million years old. The basin is long and narrow with a limited lateral extent and axial flow to the northeast. (B) Drift phase (c. 1280−1260). While carbonate sedimentation dominates, clastic detritus is derived from a wide area of the interior of North America with abundant Paleoproterozoic sources from as far away as the southwestern United States. Relatively limited detritus is derived from parts of the Grenville Province and the Archean Superior Province to the north, due to deposition within the intervening Central Metasedimentary Belt basin (see Fig. 14.8A). (C) Foredeep phase c. 1260−1250 Ma. The initial convergence leading up to the Elzevirian Orogeny results in early thrust loading, foundering of the leading edge of the lower plate (to the right of the diagram), and the development of a deep linear trough across the Adirondack region. Sediment is shed into the trough from highlands to the south composed almost exclusively of 1400−1300 Ma arc rocks of the southern Adirondack and adjacent areas; essentially swamping any sediment derived from the west or north. (D) Final Fill (c. 1250−1240 Ma). As stresses from the impending Elzevirian Orogeny propagate from the continental margin to the northwest the crust under the TABB is alternately compressed and relaxed. Little accumulation space remains and sedimentation is dominated by carbonate deposition punctuated by episodic evaporite deposition. What clastic sediment does make it into the basin is derived from western sources as in the drift phase, although likely now much reworking of older sedimentary units occurs. Transparent *yellow overlays* show interpreted drainage basins and narrow *green arrows* show sediment transport.

(Volkert et al., 2010). These ages are in agreement with the range of ages determined for deposition of the GSG in the TABB in the present study and suggest the accumulation of zinc occurred within carbonate rocks of the CMB, Adirondack Lowlands, and Grenville rocks in NY–NJ Hudson Highlands prior to the Elzevirian Orogeny. The backarc basin model also provides the source rocks and elevated heat gradient required to drive hydrothermal circulation systems needed to strip and concentrate the zinc and other metals, while evaporitic rocks in the basin provided a readily available source of sulfur. The interplay between early convergence of the Laurentian margin, heat flow, and appropriate source rocks in the backarc settings led to the widespread production of zinc deposits throughout the GSG (Gauthier and Chartrand, 2005; Peck et al., 2009). In areas without evaporitic rocks as a viable source of sulfur, such as in the Franklin marble (Peck et al., 2009), zinc oxides and silicates may have formed in preference to sulfide ores although other factors may well have been important.

6. CONCLUSIONS

1. Quartzose metasedimentary rocks from the GSG succession in the Adirondack Lowlands yield zircon grains whose detrital morphology and original isotopic systematics are retained. As such, they provide a test of the applicability of detrital zircons in high-grade rocks for provenance, paleogeographic, and basin evolution studies. Quartzite, feldspathic quartzites, and leucogranitic gneisses (arkoses) yield zircon grains with the least proportion of metamorphic rims. Zircon populations from both pelitic and calc-silicate gneisses are dominated by anatectic, metamorphic, and/or recrystallized zircon; some samples contain detrital grains as well.
2. The strategy used to constrain depositional age in a polymetamorphosed sedimentary sequence requires an approach that includes understanding of geologic constraints from field relations, multiple samples from the stratigraphic succession, informed selection of the youngest detrital populations from quartz-rich rocks, and the recognition of possible complications from metamorphic zircons and/or hybrid ages from intersecting age domains in ablation pits. Using this strategy zircons from quartz-rich rocks have been used to constrain the timing of deposition of the GSG in the TABB to the interval between 1276 ± 13 and 1254.6 ± 21.2 Ma. This range of ages is consistent with zircon detrital morphology, the timing of cross-cutting intrusive rocks and the fabric they cut, existing geochronological constraints, and detrital ages reported from other areas where rocks of the GSG occur.
3. In a tectonic context, this time period includes the opening of a series of backarc basins along the southeastern margin of Laurentia (c. 1300 Ma), and convergence related to the collisional Elzevirian Orogeny (c. 1245–1220 Ma). Correspondingly, GSG metasedimentary rocks in the Adirondack Lowlands record the tectonic deformation and metamorphism associated with the Elzevirian and Shawinigan orogenies, as well as the conditions at the earth's surface during their deposition. Detrital zircon populations through the GSG succession record shifts in provenance during basin evolution. Sources change in response to major events during deposition from rifting through initial convergence and eventual collapse of the TABB.

4. The evolution of the TABB can be traced with the use of detrital zircons through the following stages:
   a. Rift: Deep water turbiditic rocks (Pyrites Complex) in conformable contact with mantle peridotites contain a unimodal, homogeneous population of zircon with an age of 1289.7 ± 1.1 Ma from a presently unrecognized syn-rift volcanic source.
   b. Drift: The Lower Marble, the lowermost unit of the GSG of the Adirondack Lowlands, represents establishment of a thick carbonate shelf on the edge of the spreading backarc basin during the drift phase of sedimentation. Carbonate sedimentation was punctuated by faulting, during which clastic sediments were deposited in down-dropped blocks. These arkosic sediments were deposited in alluvial and playa or, more likely, a shallow marine setting in which boron was highly concentrated and now occurs as a 50 km long belt of tourmaline-rich gneisses within the Lower Marble. The belt occurrence near the top of the section may herald foundering of the shelf and transition to deep water deposition of the Popple Hill Gneiss. Detrital zircons have a variable provenance, which suggests derivation from the south central interior of Laurentia and relatively little material from locations in the Grenville and Superior provinces to the north.
   c. Foredeep: A change in provenance and deepening of the depositional environment occurred between deposition of the Lower Marble and Popple Hill Gneiss. Interpreted as deep water turbiditic rocks, the Popple Hill Gneiss contains nearly exclusively detrital zircons that range in age from 1400 to 1300 Ma. The lack of other Proterozoic and Archean components can be readily explained by local sourcing, possibly from the Southern Adirondack Terrane to the south and east (Fig. 14.2), which contains rocks of the appropriate age. The development of the foreland basin occurred in response to thrust loading and flexure of the crust as tectonic forces to the south began to impinge on the Southern Adirondacks Terrane prior to the Elzevirian Orogeny (c. 1245–1220 Ma). The lack of zircons representative of the Canadian Shield to the north may be due to a geographic barrier imposed by the CMB, which would have intercepted sediment from south-flowing rivers.
   e. Basin Closure: Rocks deposited during the conformable transition from the top of the Popple Hill Gneiss to the Upper Marble show upward increase in compositional and textural maturity. A relatively thick (c. 30 m) glassy quartzite near the boundary between the two units yields a wide range of detrital zircon ages extending back to c. 3300 Ma, similar, but not identical to those in the Lower Marble and lacks any zircons thought to be metamorphic in origin. The predominance of ages is compatible with terranes in the southwestern and south-central United States, and is taken to infer reestablishment of east-flowing river systems carrying detritus to the southeast coast of Laurentia.
   f. Convergence, evaporative conditions, and Zn mineralization: The Upper Marble consists of calc-silicate gneisses, dolomitic marble, and siliciclastic rocks punctuated by three periods of evaporite deposition, represented mostly by anhydrite. Each of these evaporate sequences is capped by stratiform sphalerite ± pyrite ± galena deposits interpreted by others as sedimentary exhalatives. This is hypothesized to have been by caused episodic, far-field convergence resulting in basin restriction and the deposition of evaporites. A combination of elevated heat flow from backarc

magmatism, appropriate metal-bearing rocks, sulfur-rich evaporitic rocks in the stratigraphic section, and the migration of heated fluids in response to tectonic stress resulted in the concentration of metals and their eventual break-out on the basin floor. Periodic relaxation of tectonic stress allowed episodic resumption of the carbonate factory.

g. Basin Closure: Leucogranitic rocks of arkosic composition make up the uppermost stratigraphic unit in the Upper Marble and GSG of the Lowlands. These rocks contain a diverse assemblage of zircon populations thought to be derived from the interior of Laurentia and/or reworking of older stratigraphic units in the sequence in response to tectonism and uplift in distal portions of the basin. Additional research is needed to determine the origin and significance of this unit.

h. Zircons yielding ages associated with Shawinigan metamorphism are particularly common in calc-silicate and metamorphosed mud-rich lithologies in the Adirondack Lowlands and Highlands. They differ from detrital grains, generally, in their size (smaller), shape (more angular), U content and U/Th ratios (higher), and lack of internal features (e.g., oscillatory zoning, inclusions, and heterogeneous BSE and CL appearance). None of these characteristics are considered definitive, and evaluation must be done on a sample-by-sample basis. A stromatolitic calc-silicate from Unit 4 of the Upper Marble and a sandy calc-silicate from the Popple Hill Gneiss yield metamorphic ages of 1172.7 ± 3.0 and 1170.6 ± 4.4 Ma, respectively. These ages precisely pin the timing of zircon growth in response to Shawinigan orogenesis in the Lowlands and fall within the time range (c. 1180–1160 Ma) shown by Heumann et al. (2006) for anatexis in pelitic to psammatic gneisses of the Popple Hill Gneiss.

Acknowledgments

This paper was significantly improved by the critical comments and suggestions of Dr. Robert Rainbird and an anonymous reviewer. The first author would like to thank the James S. Street Fund at St. Lawrence University and the Archie F. MacAllaster and Barbara Torrey MacAllaster Professorship in North Country Studies for funding related to this project. Student researchers including Peter Christofferson, Ashley Durham, Evan Koncewicz, and Kathleen Sears are gratefully acknowledged for their help with analysis. We also thank Mark Pecha and the staff at the Arizona Laserchron Center, funded by National Science Foundation grant EAR-1338583, for their help and guidance.

References

Agustsson, K.S., Johnson, S.M., Peck, W.H., Kylander-Clark, A., 2013. Geochemistry and geochronology of the Central Metasedimentary Belt boundary thrust zone thrust sheets in Southern Ontario, Grenville Province. Geosphere 9, 853–863.

Barfod, G.H., Krogstad, E.J., Frei, R., Albarède, F., 2005. Lu-Hf and PbSL geochronology of apatites from Proterozoic terranes: a first look at Lu-Hf isotopic closure in metamorphic apatite. Geochimica et Cosmochimica Acta 69, 1847–1859.

Bickford, M.E., McLelland, J.M., Selleck, B.W., Hill, B.M., Heumann, M.J., 2008. Timing of anatexis in the eastern Adirondack Highlands; implications for tectonic evolution during ca. 1050 Ma Ottawan Orogenesis. Geological Society of America Bulletin 120, 950–961. http://dx.doi.org/10.1130/B26309.1.

Brown, C.E., 1989. Geologic Map of the Beaver Creek Area in the Grenville Lowlands. U.S. Geological Survey Miscellaneous Investigation Series Map I-1725, St. Lawrence County, New York.

Brown, J.S., 1932. Natural gas, salt, and gypsum in Precambrian rocks at Edwards, New York. Bulletin of the American Association of Petroleum Geologists 16 (8), 727–735.

Brown, C.E., Ayuso, R.A., 1985. Significance of tourmaline-rich rocks in the Grenville complex of St. Lawrence County, New York. U.S. Geological Survey Bulletin 1626-C, 1–33.

Brown, J.S., Engel, A.E., 1956. Revision of Grenville stratigraphy and structure in the Balmat-Edwards district, northwest Adirondacks, New York. Geological Society of America Bulletin 67, 1599–1622.

Carr, S.D., Easton, R.M., Jamieson, R.A., Culshaw, N.G., 2000. Geologic transect across the Grenville orogen of Ontario and New York. Canadian Journal of Earth Sciences 37, 193–216. http://dx.doi.org/10.1139/cjes-37-2-3-193.

Chiarenzelli, J., Kratzmann, D., Selleck, B., deLorraine, W., 2015. Age and provenance of Grenville supergroup rocks, Trans-Adirondack Basin, constrained by detrital zircons. Geology 43, 183–186.

Chiarenzelli, J., Hudson, M., Dahl, P., deLorraine, W.D., 2012. Constraints on deposition in the Trans-Adirondack Basin, Northern New York: composition and origin of the Popple Hill Gneiss. Precambrian Research 214–215, 154–171. http://dx.doi.org/10.1016/j.precamres.2011.10.024.

Chiarenzelli, J., Lupulescu, M., Thern, E., Cousens, B., 2011a. Tectonic implications of the discovery of a Shawinigan Ophiolite (Pyrites Complex) in the Adirondack Lowlands. Geosphere 7, 333–356.

Chiarenzelli, J., Valentino, D., Lupulescu, M., Thern, E., Johnston, S., 2011b. Differentiating Shawinigan and Ottawan Orogenesis in the Central Adirondacks. Geosphere 7, 2–22.

Chiarenzelli, J., Regan, S., Peck, W., Selleck, B., Baird, G., Shrady, C., 2010. Shawinigan magmatism in the Adirondack Lowlands as a consequence of closure of the Trans-Adirondack Back-Arc Basin. Geosphere 6, 900–916.

Corfu, F., 1993. The evolution of the southern Abitibi greenstone belt in the light of precise U-Pb geochronology. Economic Geology 88, 1323–1340.

deLorraine, W.F., Sangster, A.L., 1997. Geology of the Balmat mine, New York. In: Field Trip A5, Geological Association of Canada/Mineralogical Association of Canada Joint Annual Meeting, Ottawa, 43 p.

Dickin, A.P., McNutt, R.H., 2007. The Central Metasedimentary Belt (Grenville Province) as a failed back-arc rift zone: Nd isotope evidence. Earth and Planetary Science Letters 259, 97–106. http://dx.doi.org/10.1016/j.epsl.2007.04.031.

Dickin, A., Hynes, E., Strong, J., Wisborg, M., 2015. Testing a Back-arc 'aulacogen' model for the Central Metasedimentary Belt of the Grenville Province. Geology Magazine. http://dx.doi.org/10.1017/S00167568 15000874.

Dickin, A.P., Moretton, K., North, R., 2008. Isotopic mapping of the Allochthon Boundary Thrust in the Grenville Province of Ontario, Canada. Precambrian Res. 167, 260–266.

Easton, R.M., 1992. The Grenville Province and the Proterozoic history of central and southern Ontario, geology of Ontario. In: Thurston, P.C. (Ed.), Ontario Geological Survey, pp. 714–964. Special Volume 4, Part 2.

Engel, A., Engel, C., 1953. Grenville series in the northwest Adirondack Mountains, New York, Part II: origin and metamorphism of the major paragneiss. Geological Society of America Bulletin 64, 1049–1097.

Fedo, C.M., Sircombe, K.N., Rainbird, R.H., 2003. Detrital zircon analysis of the sedimentary record. In: Hanchar, J.M., Hoskin, P.O. (Eds.), Zircon: Experiments, Isotopes, and Trace Element Investigation, Reviews in Mineralogy and Geochemistry, vol. 53. Mineralogical Society of America, pp. 277–303.

Foose, M.P., Carl, J.D., 1977. Setting of alaskite bodies in the northwest Adirondacks, New York. Geology 5, 77–80.

Fedortchouk, Y., LeBarge, W., 2008. Sources of placer platinum in Yukon: provenance study from detrital minerals. Canadian Journal of Earth Sciences 45 (8), 879–896. http://dx.doi.org/10.1139/E08-032.

Gates, A., Valentino, D., Chiarenzelli, J., Solar, G., Hamilton, M., 2004. Exhumed Himalayan-type syntaxis in the Grenville Orogen, northeastern Laurentia. Journal of Geodynamics 37, 337–359.

Gauthier, M., Chartrand, F., 2005. Metallogeny of the Grenville Province revisited. Canadian Journal of Earth Sciences 42, 1719–1734. http://dx.doi.org/10.1139/e05-051.

Geraghty, E.P., Isachsen, Y.W., Wright, S.F., 1981. Extent and Character of the Carthage-Colton Mylonite Zone, Northwest Adirondacks, New York. New York State Geological Survey report to the U.S. Nuclear Regulatory Commission, 83 p.

Hanmer, S., Corrigan, D., Pehrsson, S., Nadeau, L., 2000. SW Grenville Province, Canada: the case against post-1.4 Ga accretionary tectonics. Tectonophysics 319, 33–51.

Harley, S.L., Kelly, N.M., 2007. Zircon Tiny but Timely: Zircon Tiny but Timely. Elements 3, 13–18.

Heumann, M.J., Bickford, M.E., Hill, B.M., McLelland, J.M., Selleck, B.W., Jercinovic, M.J., 2006. Timing of anatexis in metapelites from the Adirondack lowlands and southern highlands: a manifestation of the Shawinigan orogeny and subsequent anorthosite-mangerite-charnockite-granite magmatism. Geological Society of America Bulletin 118, 1283–1298. http://dx.doi.org/10.1130/B25927.1.

Hietpas, J., Sampson, S.D., Moecher, D.P., 2011. A direct comparison of the ages of detrital monazite versus detrital zircon in Appalachian foreland basin sandstones: searching the record of Phanerozoic orogenic events. Earth and Planetary Science Letters 310 (3), 488–497.

Hoffman, P.F., 1991. Did the breakout of Laurentia turn Gondwanaland inside-out? Science 252, 1409–1412.

First Proterozoic stromatolites from the Adirondack massif-Stratigraphic, structural, and depositional implications [abs.]. In: Isachsen, Y.W., Landing (Eds.), Geological Society of America Abstracts with Programs 15 (6), 601.

Isachsen, Y.W., Fisher, D.W., 1970. Geologic Map of New York. Adirondack sheet: New York State Museum, Map and Chart Series 15, scale 1:250000.

Lupulescu, M.V., Chiarenzelli, J.R., Pullen, A.T., Price, J.D., 2011. Using pegmatite geochronology to constrain temporal events in the Adirondack Mountains. Geosphere 7, 23–39.

Meinhold, G., Andres, B., Kostopoulos, D., Reischmann, T., 2008. Rutile chemistry and thermometry as provenance indicator: an example from Chios Island, Greece. Sedimentary Geology 203, 98–111. http://dx.doi.org/10.1016/j.sedgeo.2007.11.004.

McLelland, J., Selleck, B., Bickford, M.E., 2013. Tectonic evolution of the Adirondack Mountains and Grenville Orogen inliers within the USA. Geoscience Canada 40, 1–34.

McLelland, J.M., Daly, J.S., McLelland, J., 1996. The Grenville orogenic Cycle (ca. 1350-1000 Ma): an Adirondack Perspective. Tectonophysics 265, 1–28.

McLelland, J., Chiarenzelli, J., Perham, A., 1992. Age, field, and petrological relationships of the Hyde School Gneiss, Adirondack Lowlands, New York: criteria for an intrusive igneous origin. Journal of Geology 100, 69–90. http://dx.doi.org/10.1086/629572.

McLelland, J.M., Chiarenzelli, J.R., 1990. Geochronological studies of the Adirondack Mountains, and the implications of a Middle Proterozoic tonalite suite. In: Gower, C., Rivers, T., Ryan, C. (Eds.), Mid-Proterozoic Laurentia-Baltica. Geological Association of Canada Special Paper 38, pp. 175–194.

McLelland, J.M., Selleck, B.W., Hamilton, M., Bickford, M.E., 2010. Late- to post-tectonic setting of some major Proterozoic anorthosite-charnockite-mangerite-granite (AMCG) suites. Canadian Mineralogist 48, 1025–1046.

Moretton, K., Dickin, A.P., 2013. Nd isotope mapping of the Dysart gneiss complex: evidence for a rifted block within the Central Metasedimentary Belt of the Grenville Province. Precambrian Research 228, 223–232.

Morton, A., Yaxley, G., 2007. Detrital apatite geochemistry and its application in provenance studies. Geological Society of America Special Papers 420, 319–344.

Peck, W., Selleck, B., Wong, M., Chiarenzelli, J., Harpp, K., Hollocher, K., Lackey, J., Catalano, J., Regan, S., Stocker, A., 2013. Orogenic to postorogenic (1.20–1.15 Ga) magmatism in the Adirondack Lowlands and Frontenac terrane, southern Grenville Province, USA and Canada. Geosphere 9, 1637–1663.

Peck, W.H., Bickford, M.E., McLelland, J.M., Nagle, A.N., Swarr, G.J., 2010. Mechanism of metamorphic zircon growth in a granulite-facies quartzite, Adirondack Highlands, Grenville Province, New York. American Mineralogist 95, 1796–1806.

Peck, W.H., Volkert, R.A., Mansur, A.T., Doverspike, B.A., 2009. Stable isotope and Petrologic evidence for the origin of regional marble-hosted magnetite deposits and the zinc deposits at Franklin and Sterling Hill, New Jersey Highlands, United States. Economic Geology 104, 1037–1054. http://dx.doi.org/10.2113/econgeo.104.7.1037.

Prucha, J., 1957. Pyrite Deposits of St. Lawrence and Jefferson Counties, New York. University of the State of New York, Albany, 87 p.

Ratcliffe, N.M., Aleinikoff, J.N., Burton, W.C., Karabinos, P., 1991. Trondhjemitic, 1.35-1.31 Ga gneisses of the Mount Holly Complex of Vermont: evidence for an Elzevirian event in the Grenville Basement of the United States Appalachians. Canadian Journal of Earth Sciences 28, 77–93.

Rivers, T., 2008. Assembly and preservation of lower, mid, and upper orogenic crust in the Grenville Province – implications for the evolution of large hot long-duration orogens. Precambrian Research 167, 237–259.

Rivers, T., Corrigan, D., 2000. Convergent margin on southeastern Laurentia during the Mesoproterozoic: tectonic implications. Canadian Journal of Earth Sciences 37, 359–383.

Roden-Tice, M., Tice, S., 2005. Regional-scale mid-Jurassic to Late Cretaceous unroofing from the Adirondack Mountains through Central New England based on apatite-fisson track and (U–Th)/He thermochronology. Journal of Geology 113, 535–552.

Sager-Kinsman, A.E., Parrish, R.R., 1993. Geochronology of detrital zircons from the Elzevir and Frontenac terranes, Central Metasedimentary Belt, Grenville Province, Ontario. Canadian Journal of Earth Science 30, 465–473.

Selleck, B., McLelland, J.M., Bickford, M.E., 2005. Granite emplacement during tectonic exhumation: the Adirondack example. Geology 33, 781–784.

Smith, T.E., Harris, M.J., 1996. The Queensborough mafic-ultramafic complex: a fragment of a Meso-Proterozoic ophiolite? Grenville Province, Canada. Tectonophysics 265, 53–82.

Slack, J.F., Herriman, N., Barnes, R.G., Plimer, I.R., 1984. Stratiform tourmalinites in metamorphic terranes and their geologic significance. Geology 12, 713–716.

Takeuchi, M., 2013. Detrital anisotropic grandite garnet as an indicator of denudation level of Permian volcanic arc in the provenance of the South Kitakami Belt, Northeast Japan. Island Arc 22, 477–493. http://dx.doi.org/10.1111/iar.12043.

Tiedt, K., Kelson, C., 2008. Geochemical and mineralogical characterization of ore from the Pyrites and Stella pyrite deposits, St. Lawrence County, NY. In: Geological Society of America Abstracts with Programs, vol. 40, p. 6.

Volkert, R.A., Aleinikoff, J.N., Fanning, C.M., 2010. Tectonic, magmatic, and metamorphic history of the New Jersey Highlands: new insights from SHRIMP U-Pb geochronology. In: Tollo, R.P., Bartholomew, M.J., Hibbard, J.P., Karabinos, P.M. (Eds.), From Rodinia to Pangea: The Lithotectonic Record of the Appalachian Region, Geological Society of America Memoir, vol. 206, pp. 307–346.

Wasteneys, H., McLelland, J.M., Lumbers, S., 1999. Precise zircon geochronology in the Adirondack Lowlands and implications for revising plate-tectonic models of the Central Metasedimentary Belt and Adirondack Mountains, Grenville Province, Ontario and New York. Canadian Journal of Earth Sciences 36, 967–984. http://dx.doi.org/10.1139/cjes-36-6-967.

Wynne-Edwards, H.R., 1972. The Grenville Province. In: Price, R.A., Douglas, R.J.W. (Eds.), Variations in Tectonic Styles in Canada. Geological Association of Canada, pp. 263–334. Special Paper, no. 11.

CHAPTER

15

Detrital Zircon U-Pb Geochronology, Nd Isotope Mapping, and Sediment Geochemistry From the Singhora Group, Central India: Implications Toward Provenance, Its Shift, and Regional Stratigraphic Correlation

K. Das[1], *P.P. Chakraborty*[2], *K. Horie*[3], *Y. Tsutsumi*[4], *S. Saha*[2], *S. Balakrishnan*[5]

[1]Hiroshima University, Higashi-Hiroshima, Japan; [2]University of Delhi, New Delhi, India; [3]National Institute for Polar Research, Tokyo, Japan; [4]National Science Museum, Tsukuba, Japan; [5]Pondicherry University, Pondicherry, India

OUTLINE

Sediment Provenance
http://dx.doi.org/10.1016/B978-0-12-803386-9.00015-0

1. INTRODUCTION

It is increasingly appreciated in studies that understanding geological history of any sedimentary basin remains incomplete unless process-based paleoenvironmental studies of basin fill package(s) are backed with geochemical and geochronological proxies (Chakraborti et al., 2007; Malone et al., 2008; Meert et al., 2010; Meert and Pandit, 2015). Only a comprehensive study involving both physical and chemical aspects of any basin fill can talk about provenance, processes operative therein, and sedimentation history during the initiation and evolution of a basin. Analysis of heavy mineral fractions of clastic sediments, particularly zircon and monazite, is proven as a powerful tool in sediment provenance tracking and drawing stratigraphic correlation (Košler et al., 2002; Machado and Gauthier, 1996; Morton et al., 1996; Whitehouse et al., 1997; Fernandez-Suārez et al., 2000). Application of this technique has offered rich dividends in constraining the depositional age of sedimentary sequences, which otherwise lack biostratigraphic or chronological markers in the form of time-connotative fossil beds, concordant ash fall layers, or cross-cutting intrusions such as dykes and sills. Use of geochemical and isotopic techniques in fine-grained clastics is also a common practice for tracking sediment provenance and tectonic setting thereof (Hayashi et al., 1997; Nagarajan et al., 2007; Ohta, 2008; Lee, 2009; among others). For example, the Nd model age of sedimentary rocks is interpreted to reflect the mean age of mantle extraction of various provenance components and thus helps in assessing the relative proportions of juvenile (mantle-derived) crust and old recycled upper crust in the eroding source region. Indeed, in a heterolithic sedimentary sequence comprising coarse arenaceous (sandy) and argillaceous (shaly) lithologies, both lithic types can contribute in a complementary manner in tracking sediment provenance for the basin—coarse clastic units in terms of heavy mineral geochronology and fine clastics through their geochemical character including Nd isotope signature. Therefore, in Precambrian sedimentary basins with heterolithic sedimentary sequences, application of detrital geochronology and sediment geochemistry can provide significant information on provenance and related tectonic pulses, which otherwise remain very difficult to constrain particularly when the source area is composed dominantly of fine-grained rocks and is a former active continental margin with multiple displaced terrains.

In this backdrop, a number of Mesoproterozoic sedimentary basins (Chhattisgarh, Singhora—the "proto-basin" of Chhattisgarh, Khariar, Ampani, Indravati, and Sukma) hosted within the central Indian Archean Bastar Craton, mostly on its eastern boundary adjoining the Proterozoic orogenic belt of Eastern Ghats Granulite Belt (EGGB; Fig. 15.1A and B), offer a unique opportunity to study many of the aforesaid issues. The origin and evolution of these basins in tandem with adjoining lithosphere dynamics have attracted

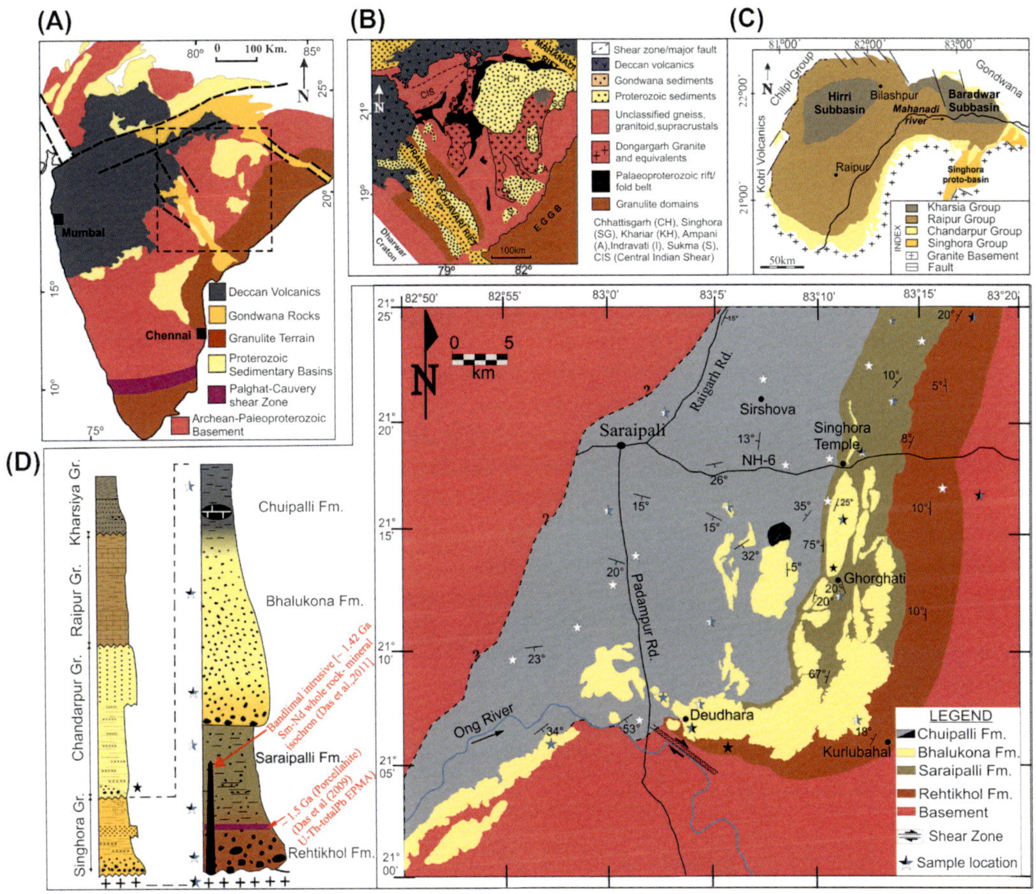

FIGURE 15.1 (A) Broad geological domains in peninsular India. (B) Disposition of the Chhattisgarh Main Basin, Singhora Basin, Khariar Basin, and Ampani Basin at the eastern margin of the Archean Bastar craton adjacent to the Eastern Ghats Granulite Belt (EGGB). (C) Geological map of the Singhora Basin and the main Chhattisgarh Basin. (D) Geological map of the Singhora Basin and its stratigraphic sequence. *Star marks* on the map are the locations of samples collected for the present study. An *open star* represents sample for geochemical study, those filled with black represent samples used for detrital zircon geochronology, and those filled with blue represent samples used for Nd-isotopic study. Stratigraphic succession of Chhattisgarh Supergroup, in general, and of Singhora Group, in particular, is shown in the left. Sample symbols are kept same in the stratigraphic column as are given in the map.

tremendous interest among scientists in recent times. In this study, we aim to track any changes in sediment provenance during the depositional history of the Singhora Basin, a subbasin of the Chhattisgarh Basin (Fig. 15.1B and C), central India, using detrital zircon geochronology, sediment geochemistry, and Nd isotope mapping (stratigraphy-wise variation).

2. GEOLOGICAL AND GEOCHRONOLOGICAL BACKGROUND

On the basis of sedimentation pattern, relative dominance of lithology (viz. clastic-dominated or carbonate-dominated) and contact relationship between litho-stratigraphic successions (viz. unconformable, gradational, etc.), the Chhattisgarh basin is divided into three subbasins: (1) the Singhora proto-basin, (2) the Baradwar subbasin, and (3) the Hirri subbasin (Das et al., 1992, 2003). Its stratigraphic succession is classified into four constituent components, the Singhora, Chandarpur, Raipur, and Kharsiya Groups, in order of superposition (Chakraborty et al., 2015) (Fig. 15.1C). The Singhora Group makes up the Singhora proto-basin, while the overlying units—Chandarpur, Raipur, and Kharsiya—are considered as deposits of the main Chhattisgarh Basin. Proximity with crustal-scale geotectonic features such as rifts (Pranhita-Godavari, Mahanadi), orogenic fold-thrust belt (EGGB), and so on has biased tectonic modeling for the Chhattisgarh basin. Features such as (1) a shrinking, bowl-shape for the basin with a somewhat concentric outcrop pattern, and (2) evaporites in alternation with and overlying the carbonate successions to mark the cessation of basin deposition (Das et al., 1992), point toward a probable desiccated-basin setting for the basin in an intracratonic sag mode. Yet, the two popularly discussed tectonic models for the basin are inclined either toward a rift (Patranabis-Deb, 2004; Chaudhuri et al., 2002; Patranabis-Deb and Chaudhuri, 2007; Dhang and Patranabis-Deb, 2012) or a foreland (Biswal et al., 2002; Chakraborty and Paul, 2008; Gupta, 2012) bound setting.

The fossil-poor Singhora Group represents an arenite-argillite-carbonate sequence (Fig. 15.1D). Rocks of this group are punctuated by magmatic inputs at different stratigraphic levels either as concordant (volcanoclastic tuff) or discordant (dyke) bodies (Chakraborti, 1997; Das et al., 2011). In recent years, detailed process-based sedimentological studies involving different lithostratigraphic intervals of the Singhora Group have documented a wide variety of depositional environment products ranging from a continental alluvial fan/braid plain system to a distal marine shelf beyond stormwave base with a transitional delta in between (Chakraborty et al., 2009, 2011). From sequence stratigraphic appraisal, documentation of spatio-temporal variations in paleocurrent pattern and limited Nd-isotope study across the Saraipalli–Bhalukona transition, Chakraborty et al. (2012a) suggested the presence of a Type-I unconformity at the base of Bhalukona Formation, triggered by tectonically driven forced regression. In addition, a number of studies focused on the basin-bound lithodemic units (Das et al., 2009, 2015; Das, et al., 2011; Bickford et al., 2011a) in order to bracket the age of the Chhattisgarh Basin including the Singhora Basin (c. 1500 Ma (Das et al., 2009) from the bottom and c. 1000 Ma (Patranabis-Deb et al., 2008) from the top). Detrital zircon geochronology was attempted by Bickford et al. (2011a) for the entire Chhattisgarh succession with sampling from selected stratigraphic levels, like basal parts of the Singhora and Chandarpur Group and sandstones of the Raipur Group (i.e., Sarangarh Formation and Sarandih Formation). The results led them to suggest that "regardless of stratigraphic position, there is a unimodal age peak near 2500 Ma" indicating sediment derivation from adjacent granitic and rhyolitic basement rocks. However, this result is potentially over-simplistic in the absence of data from most stratigraphic levels, and did not take into consideration any effect of crustal-scale events in the basin adjoining area in course of the ~500 Ma (~1500 to ~1000 Ma) depositional history of the basin. Nevertheless, all of these studies have helped to decipher the possible tectonic setting at the eastern margin of the East Indian craton (Das et al., 2015). In addition, detailed structural analysis (Saha et al., 2013) revealed the overall deformation pattern in the Singhora basin and its relative timing.

It is noteworthy that despite all these studies there are still ongoing debates on the mechanism and scale of deformation (Deb, 2013; Chakraborty et al., 2013), the exact timing of the basin initiation (Bickford et al., 2011a,b; Das et al., 2011), and the stratigraphic position of the Singhora Group within the overall lithostratigraphic framework of the Chhattisgarh basin (as detailed in Chakraborty et al., 2011; Saha et al., 2013; Basu and Bickford, 2015; Chakraborty et al., 2015). Nevertheless, the Singhora Basin is now considered among one of the better-studied Mesoproterozoic sedimentary basins in India, which still warrants detailed data on the provenance(s) of its sedimentary package and shift in provenance, if any, through its depositional history. Studies related to paleoenvironmental set-ups of different lithic suites and deformation character for the Singhora basin have already been carried out and results are available in literature (Chakraborty et al., 2009, 2012a, 2015; Saha et al., 2013). Presently, attempts are also being made to interpret sedimentary facies and structural data for the adjoining Ampani Basin and to track sediment provenance in the basin using detrital zircon geochronology. The data acquisition from another nearby basin, the Khariar Basin, however, is at a nascent stage and preliminary in character. In order to provide a regional picture, we felt that discussion on provenance analysis data of the Singhora basin should be done in association with data available from other adjoining basins such as Khariar and Ampani (Fig. 15.1B). In the process, an attempt was made to (1) establish relative ages of the basins and (2) build up a history of sedimentary provenance(s) in the region and how this might relate to Meso- to Neoproterozoic tectonic pulses along the margin of East Indian craton.

3. SAMPLE DESCRIPTION

Systematic sampling of all four formations belonging to the Singhora Group (Rehtikhol, Saraipalli, Bhalukona, and Chuipalli) was done in locations spread across the basin (Fig. 15.1D). A total of 21 samples of the Singhora Basin (sample locations as in Fig. 15.1D) were collected for bulk rock (major, trace, rare earth element (REE)), geochemical analysis (1 from arkosic sandstone of Rehtikhol Formation, 7 from heterolithic shale and sandstone/siltstone of Saraipalli Formation, 3 from arenitic sandstone of Bhalukona Formation, and 10 from shale samples of the topmost Chuipalli Formation). Out of 21 samples, 16 samples were used for Sm-Nd isotopic analysis (one sample from the basement granite gneiss of the Bastar Craton, one sample from the Rehtikhol Formation, four samples from the Saraipalli Formation, seven samples from the Bhalukona Formation, and three samples from the Chuipalli Formation). Collection of multiple samples from some of the formations ensured that their well-defined facies variations were represented, which helped us to produce a high-definition Nd-isotope map through the stratigraphic column, in order to establish regional correlatibility. Moreover, zircon geochronology study was carried out from six sandstone/siltstone samples including five samples from Singhora sediment column (two samples from Rehtikhol Formation, one sample from Saraipalli Formation, two samples from the Bhalukona Formation) and one sample from basin-adjacent Bastar Craton at the eastern margin of the basin. We have also analyzed detrital zircon grains separated from sandstones belonging to (1) the lower part of the Chandarpur Group within the Chhattisgarh Supergroup that immediately overlies the Singhora Group and (2) the lowermost part of the basal arenaceous package in the Khariar Group of nearby Khariar Basin.

4. ANALYTICAL METHOD

Zircon grains for SHRIMP analysis were taken from the heavy mineral fraction separated by standard crushing and heavy-liquid techniques. Zircon grains from the samples and the zircon standard AS3 ($^{206}Pb/^{238}U = 0.1859$; Paces and Miller, 1993) were mounted in an epoxy resin and polished until the surface was flat and the centers of the embedded grains were exposed. A combination of backscattered electron (BSE) and cathodoluminescence (CL) images were used to select sites for SHRIMP analysis. The U–Pb dating of these samples were carried out using the SHRIMP II installed at the National Institute of Polar Research, Japan. The experimental conditions and the procedures followed for the measurements were based on Williams (1998). The spot size of the primary ion beam was approximately 20 μm. Obtained U–Pb data were reduced in a manner similar to that described by Williams (1998) using the SQUID Excel macro written by Ludwig (2001). A correction for common Pb was made on the basis of the measured ^{204}Pb and the model for common Pb compositions proposed by Stacey and Kramers (1975).

For Laser Ablation Inductively Coupled Plasma Mass Spectrometry (LA-ICP-MS), zircon U–Pb isotope analysis was performed using a 213 nm Nd-YAG Laser (New Wave Research UP-213) coupled with an Agilent 7500 ICP-MS at the Department of Earth and Planetary Systems Science, Hiroshima University, Japan. The detailed analytical methodology is described by Katsube et al. (2012). The apparatus contains a mixed He-N_2-Ar carrier gas system equipped with a small volume ablation cell, sample aerosol stabilizer (buffering chamber), and charcoal filter attachment. We used a laser spot size of 25 μm. Zircon grains from the two samples dated using this technique were measured in a mixed sequential order to avoid the influences of time-dependent changes in instrumental conditions. Raw data were processed using the data reduction program Pepi-AGE (Dunkl et al., 2008).

The processed data from both analytical procedures were plotted using Isoplot/Ex (Version 4.15; Ludwig, 2008). The isotopic ratios and individual spot ages are quoted at 2σ, and the weighted mean ages are given at the 95% confidence level. The obtained data points are characterized on the basis of their U content and Th/U ratios. For each sample, the probability density plots only show data with <5% discordance to obtain reasonable age populations of analyzed detrital zircon grains.

Selected samples were analyzed for major and trace elements with a WD X-ray fluorescence spectrometry system (Siemens SRS3000) at the Wadia Institute of Himalayan Geology

(WIHG), Dehradun, India. REE concentrations were measured by ICP-MS (Perkin–Elmer) at WIHG, with 0.04 g of the sample subjected to standard acid digestion with HF-HNO_3.

For Sm-Nd isotope analyses, approximately 100 mg of sample was taken for chemical digestion. The sample was precisely weighed (±0.1 mg) and completely digested in acid mixtures. After digestion each sample was split into two halves and an isotope tracer added to one half (isotope dilution, ID fraction). The other half of the solution was used for measuring the sample isotopic composition (IC fraction). The weight of the solutions and added spike were measured accurately during this process. In the next stage, the IC and ID fractions were passed through Bio-Rad AG-X8 cation-exchange resin with HCl calibrated columns. Rb (ID) and Sr (ID and IC) were collected with 2 N HCl and REE fractions were collected with 6 N HCl. The collected solutions were dried. The REE fractions of ID and IC were again diluted with 0.18 N HCl and passed through the calibrated di (2-ethylhexyl) orthophosphoric acid (HDEHP resin) column for Sm and Nd separation. Nd (ID and IC) and Sm (ID) were collected with 0.3 N HCl and 0.4 N HCl, respectively. A few drops of 0.5 M H_3PO_4 were added to the collected solutions and then allowed to dry on a hot plate at a temperature of 120 °C. Subsequently, the dry sample was converted into solution by adding 1 μl HCl for Rb and 1 μl HNO_3 for Sr, Sm, and Nd, and loaded into degassed filaments. For Sm and Nd isotopes, double Re filament was used. The isotope ratios ($^{143/144}$Nd, $^{150/146}$Nd, $^{154/152}$Sm) were determined on a Finnigan-Triton Thermal Ionization Mass Spectrometer (TIMS). The laboratory work was carried out at the National Facility for Isotope Geosciences, Department of Earth Sciences, Pondicherry University, India. Nd isotope ratios were corrected for instrumental mass fractionation by normalizing the measured ratios to ^{144}Nd/^{146}Nd = 0.7219 using an exponential law. The AMES Nd standard was analyzed nine times during the course of this study and yielded a mean ^{143}Nd/^{144}Nd ratio of 0.511968 ± 2, which is similar to the recommended value of 0.511968 ± 4 (Govindaraju, 1994). Further details of sample preparation and Nd isotope analysis are reported in Anand and Balakrishnan (2010).

5. RESULTS

5.1 Bulk Rock Geochemistry of Studied Lithounits

The argillaceous and arenaceous intervals of the basin fill were studied separately to understand the general geochemical character of sediment source rock and its possible variation with time (Table 15.1). Major and trace element data obtained from the argillaceous intervals of the Saraipalli and Chuipalli Formations were compared with respect to PAAS (Post Archean Australian Shale); the best estimate of upper crustal composition. Compared to PAAS, the relative enrichment of large-ion lithophile elements (LILEs) like Rb and Ba (except samples from the Chuipalli Formation), high-field strength elements (HFSEs) like Nb (in some samples), Th, and major oxide K_2O (albeit in a very small proportion) in both the shale formations supports a provenance dominantly of felsic character (Fig. 15.2A and B). Depletion in LILEs like Sr, Pb, and major oxides Na_2O, CaO is possibly resulted from intense chemical weathering in the source region. Selective trace element ratios, which are used in characterizing provenance, like Th/Sc, Th/Co, Th/Cr, and La/Sc, yielded higher values than the reported upper crustal rocks (Condie, 1993) (Table 15.1) and support a

TABLE 15.1 Bulk Rock Geochemical Data for the Representative Studied Samples

																Saraipalli Formation					
	Chuipalli Shale										Bhalukona Sandstone			Shale						Sand-stone	Rehtikhol Sandstone
Sample	TS-1	TS-2	TS-3	TS-4	TS-5	L6/09-	CHIS2	CHIS3	6-10-7	A5	DD-THICK	L18/09	L19/09	STS-1	STS-2	STS-3	STS-4	STS-5	7-22	S1/L	L25/09
SiO_2	60.28	61.88	62.53	60.66	63.59	63.39	67.86	69.29	67.46	65.4	95.9	95.9	83.92	64.05	66.61	68.88	69.06	65.03	56.38	86.41	86.3
TiO_2	0.62	0.82	1.03	0.94	0.59	0.81	0.74	0.71	0.72	0.69	0.07	0.06	0.26	0.83	0.85	0.84	0.86	0.71	0.54	0.64	0.09
Al_2O_3	20.4	18.53	19.72	19.33	18.23	17.49	19.37	19.42	18.95	19.73	1.54	1.48	10.68	21.1	19.48	19.12	18.84	19.45	12.19	7.31	9.82
Fe_2O_3	6.7	7.46	6.04	7.56	6.61	6.54	5.12	3.61	3.28	4.48	0.29	0.06	0.22	2.99	2.78	1.38	1.49	4.703	6.54	1.17	0.29
MnO	0.02	0.02	0.01	0.02	0.01	0.02	0.02	0.012	0.03	0.01	0.03	0.03	0.03	0	0.01	0	0	0.005	0.58	0.03	0.03
MgO	1.2	1.42	0.95	1.22	1.37	1.47	0.88	0.83	0.84	0.78	0.07	0.07	0.11	0.65	0.61	0.59	0.59	0.67	2.9	0.28	0.26
CaO	0.17	0.19	0.11	0.2	0.15	0.45	0.21	0.17	0.35	0.25	0.01	0.01	0.12	0.11	0.09	0.07	0.08	0.06	6.38	0.14	0.06
Na_2O	0.08	0.07	0.07	0.07	0.07	0.05	0.15	0.17	0.1	0.09	0.03	0.01	0.09	0.15	0.14	0.14	0.14	0.1	0.05	0.09	0.11
K_2O	5.62	4.78	5.01	5.01	4.84	4.94	5.01	5.08	4.69	4.97	0.2	0.03	2.21	5.05	5.12	5.22	5.03	5.06	3.55	0.93	3.57
P_2O_5	0.05	0.05	0.05	0.05	0.05	0.04	0.04	0.03	0.03	0.03	0.01	0.01	0.01	0.17	0.04	0.04	0.04	0.03	0.04	0.02	0.01
LOI	4.71	4.51	4.79	5	4.28	6	2	0.18	5.1	4.85	0.31	0.35	1.02	4.36	3.67	3.21	4.01	3.49	12.5	1.59	1.3
Total	99.85	99.73	100.31	100.06	99.79	101.2	101.4	99.502	101.55	101.28	98.46	98.01	98.67	99.46	99.4	99.49	100.14	99.308	101.65	98.61	101.84
Sc	15.59	16.13	15.79	16.25	16.47	15	16.58	16.42	13.5	16	0.7	0.3	2	19	16	15	15	16.44	11.9	5.5	1.1
V	74	91	101	96	90.58	93	103	102.31	97	102	17	17	25	112	95	87	88	99.36	61	55	15
Cr	57	86	94	92	69.7	65	86	78.84	68	68	109	33	12	85	79	97	85	72.87	63	41	17
Co	4	18	4	7	13.37	11	8	7.24	8	9	5	167	72	3	3	2	2	2.18	15	92	117
Ni	20	37	23	29	30.17	26	18	12.96	16	19	17	6	9	6	6	6	6	10.74	30	12	7
Cu	13	14	28	17	8.3	39	17	5.28	14	14	10	4	7	25	24	18	20	11.1	8	15	5
Zn	66	82	58	57	68.09	70	50	32.8	23	25	5	5	5	35	35	19	24	42.52	67	7	5
Ga	23	24	25	22	24.75	23	28	25	21	22	5	5	6	26	24	22	20	25.67	15	7	6
Rb	295	265	268	262	269.33	259	227	237.54	196	240	8	5	59	238	212	209	205	218.32	166	35	95
Sr	36	42	43	49	37.68	44	62	59.45	47	31	4	3	9	117	54	50	51	37.04	67	15	20
Y	47	32	30	30	26.88	19	32	32.27	19	17	5	5	10	42	32	31	32	30.42	25	23	11
Zr	259	177	213	191	147	172	184	170.78	165	151	55	54	66	223	255	268	309	163.64	95	500	68
Nb	23	19	22	24	13.74	17	22	19.67	20	15	5	5	5	18	20	20	21	17.72	8	18	5

Cs	8.85	13.37	15.8	9.9	3.7	na	9.72	6.24	na	na	na	na	na	2.77	bdl	bdl	bdl	bdl	na	na	Na
Ba	526	489	474	459	474.95	526	540	553.72	531	465	64	84	94	1580	852	879	885	538.84	766	90	775
Hf	5.51	2.41	4.99	2.75	2.51	na	5.19	2.22	na	na	na	na	na	5.84	6.26	6.44	7.65	3.32	na	na	Na
Pb	6.57	9.79	5.75	4.44	6.27	13	9.05	6.34	12	11	5	5	4	10.88	7.88	7.88	8.68	8.79	10	10	12
Th	36.33	28.69	25.65	43.11	26.78	20	23.78	28.42	23	21	5	3	10	28.52	27.24	29.71	30.03	27.21	10	19	5
U	3.91	3.27	3.83	1.94	2.26	2.6	3.29	3.19	1.88	2.35	0.07	0.01	2.68	2.39	2.47	2.83	3.5	2.28	1.67	0.32	2.57
La	na	na	na	na	na	50.8	55.5778	na	49.6	52.5	4.56	3.57	9.2	na	na	na	na	na	37.04	19.53	10.3
Ce	na	na	na	na	na	95	122.057	na	99.63	105	8.77	6.76	15.9	na	na	na	na	na	75.64	39.93	18.2
Pr	na	na	na	na	na	10.1	12	na	10.2	10.93	0.83	0.56	2	na	na	na	na	na	7.66	3.83	1.67
Nd	na	na	na	na	na	41.1	45.56	na	40.93	44.22	3.16	2.16	6.4	na	na	na	na	na	30.73	15.08	6.27
Sm	na	na	na	na	na	6.8	8	na	6.91	7.15	0.49	0.41	1	na	na	na	na	na	5.64	2.92	0.94
Eu	na	na	na	na	na	1.33	1.59	na	1.277	1.293	0.08	0.08	0.25	na	na	na	na	na	1.13	0.53	0.31
Gd	na	na	na	na	na	5	7	na	5.02	5.13	0.38	0.35	1	na	na	na	na	na	4.28	2.47	0.7
Tb	na	na	na	na	na	1	1	na	0.9	0.96	0.06	0.06	0	na	na	na	na	na	0.8	0.59	0.1
Dy	na	na	na	na	na	5	6	na	4.87	5.21	0.35	0.35	1	na	na	na	na	na	4.38	3.71	0.53
Ho	na	na	na	na	na	1	1	na	0.98	1.05	0.07	0.07	0	na	na	na	na	na	0.86	0.81	0.1
Er	na	na	na	na	na	3	3	na	3	3	0	0	0	na	na	na	na	na	3	3	0
Tm	na	na	na	na	na	0.47	0.54	na	0.46	0.5	0.03	0.03	0.07	na	na	na	na	na	0.408	0.44	0.057
Yb	na	na	na	na	na	3	4	na	3	3	0	0	0	na	na	na	na	na	2	3	0
Lu	na	na	na	na	na	0.4	0.5	na	0.4	0.44	0.03	0.03	0.06	na	na	na	na	na	0.35	0.41	0.056
LREE						210.14	251.96		213.56	226.22	18.27	13.89	35.38						162.12	84.29	38.39
HREE						13.4	16.54		13.19	14.16	0.96	1.018	2.28						11.77	11.18	1.54
TREE						223.54	268.51		226.76	240.39	19.24	14.917	37.66						173.89	95.48	39.94
Eu/Eu*						0.7	0.65		0.66	0.65	0.62	0.717	0.79						0.7	0.61	1.2
Th/U	9.29	8.77	6.69	22.22	11.84	7.66	7.22	8.9	12.23	8.93	71.42	300	3.73	11.93	11.02	10.49	8.58	11.93	5.98	59.37	1.94
Th/Sc	2.33	1.77	1.62	2.65	1.62	1.33	1.43	1.73	1.7	1.31	7.14	10	6.66	1.5	1.7	1.97	2.01	1.65	0.84	3.45	4.54
Th/Co	8.58	1.62	6.89	5.98	2	1.81	3.16	3.92	2.87	2.33	1	0.017	0.13	10.68	9.01	14.56	12.72	12.48	0.66	0.2	0.04
Th/Cr	0.63	0.33	0.27	0.46	0.38	0.3	0.27	0.36	0.33	0.3	0.04	0.09	0.83	0.33	0.34	0.3	0.35	0.37	0.15	0.46	0.29
La/Sc	3.26	3.05	3.31	3.29	2.96	3.38	3.35	2.98	3.67	3.28	6.51	11.9	6.13	3.98	3.33	3.61	3.44	3.076	3.11	3.55	9.36
Zr/Sc	16.61	10.96	13.48	11.74	8.92	11.46	11.07	10.4	12.22	9.43	78.57	180	44	11.82	16.08	17.77	20.68	9.95	7.98	90.9	61.81

bdl, below detection limit; *na*, not analyzed.

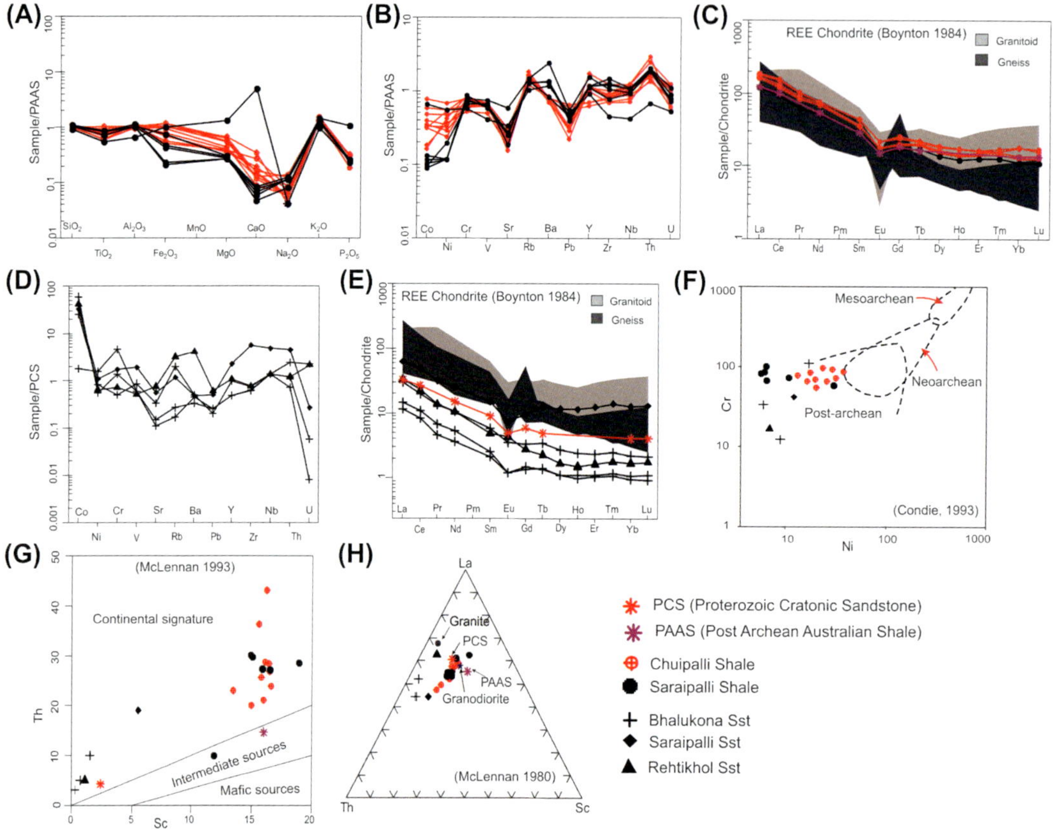

FIGURE 15.2 PAAS (Post Archean Australian Shale) normalized (A) major oxide plot and (B) trace element distribution for the shale units of the Singhora Basin. (C) Chondrite normalized rare-earth element diagram showing overall LREE-enriched pattern with negative Eu-anomaly. Reported data of granitoid and gneiss from the Bastar Craton (Mondal et al., 2006) have been plotted for comparison. (D) Proterozoic Cratonic Sandstone (PCS) normalized trace element plot for sandstone units of the Singhora Basin. (E) Chondrite normalized REE diagram for the sandstone units showing variable REE contents; particularly overall high content of Saraipalli sandstone is to be noted. Reported data of granitoid and gneiss from the Bastar Craton (Mondal et al., 2006) have been plotted for comparison. (F) Bivariate Ni-Cr, (G) Th–Sc, and (H) ternary plot of La-Th-Sc diagrams for all the samples studied showing granite-granodiorite protolith character.

felsic-dominated source. This is further verified by variable depletion in compatible ferromagnesian elements like Co and Ni, a slight depletion in Cr and V, and variable depletions of major oxides Fe_2O_3 and MgO, indicating a very nominal contribution from any mafic source (Fig. 15.2A and B). Chondrite normalized REE plots (Fig. 15.2C) of both the shale units show fractionated REE patterns with enriched light REE (LREE) and flat, heavy REE (HREE) and a negative Eu anomaly ($Eu/Eu^* = 0.65-0.71$). The REE pattern matches well with the geochemical character of Bastar granites and gneisses, which have a Eu/Eu^* value of 0.40 and 0.65, respectively (Mondal et al., 2006).

For sandstone samples, we compared the trace element data with Proterozoic Cratonic sandstone (PCS; Condie, 1993). LILEs, like Rb, and HFSEs, like Nb and Th show variable enrichment or similar values with respect to PCS in both Rehtikhol and Saraipalli sandstones indicating their felsic derivation. LILEs like Ba, Pb, and Sr show near-PCS values or slightly depleted values (except enrichment of Ba in Rehtikhol sandstone). Also, concentrations of these elements within Bhalukona sandstones vary in a range and are comparable with those of PCS (except slight depletions in Sr, Pb, and Ba) (Fig. 15.2D). Trace element ratios, like Th/Sc, Th/Co, Th/Cr, La/Sc in all the three sandstones such as Rehtikhol, Saraipalli, and Bhalukona Formations plot well within the range of values reported from upper crustal felsic rocks (Taylor and McLennan, 1985). Transition elements like Ni, Cr, and V, commonly enriched in mafic rocks, show slight enrichment or depletion in all three sandstones. Hence, we could not ascertain any clear geochemical difference between the sandstones of three different stratigraphic levels. All three sandstones yielded a fractionated REE pattern with enriched LREE and flat HREE (Fig. 15.2E). However, detailed inspection of total REE and some key REE ratios allowed us to draw some distinction between the three sandstones. Though general REE patterns are the same, the three sandstones vary in their total REE content. The Saraipalli sandstone is more enriched in total REE than the Rehtikhol and Bhalukona sandstones. A distinction can also be made between the three sandstones in terms of their Eu/Eu* ratio. Rehtikhol sandstones register a positive Eu anomaly with a Eu/Eu* value of 1.2, whereas the other sandstones record negative anomalies with Eu/Eu* values varying between 0.6 and 0.8. The higher Eu/Eu* value is interpreted as the result of increased plagioclase feldspar abundance in Rehtikhol arkosic sandstone. Support in favor of this contention is provided by the relatively high Sr content in Rehtikhol sandstones.

Further, geochemical data of shale and sandstone samples are plotted in Ni-Cr and Sc-Th bivariant diagrams (Fig. 15.2F and G). All data are found to plot in the post-Archean and continental fields, respectively. In a La-Th-Sc ternary plot, shale and sandstone samples show a mixed source dominated by granite-granodiorites (Fig. 15.2H).

5.2 Nd-Isotope Mapping

In the absence of postdepositional fractionation by surficial processes such as erosion, transportation or diagenesis, the Sm/Nd ratio of sediments, in particular fine-grained sediments, is considered a reliable proxy for Sm/Nd in their source rocks (Armendáriz et al., 2008). In fact, the initial Nd isotopic composition represented by ε^{t}Nd is considered a robust parameter to identify the provenance as well as any sudden shift in the provenance. A total of 16 samples comprising basement and all lithostratigraphic intervals of the Singhora Group (also analyzed for bulk rock geochemistry) were analyzed for their Sm–Nd isotopes. This was attempted to carry out stratigraphic variation in ε^{t}Nd value across the Singhora succession, as an extension of our earlier study (Chakraborty et al., 2012a) in which we generated Nd data only for the transition between Saraipalli and Bhalukona Formations. Table 15.2 presents the dataset obtained from different stratigraphic levels of the Singhora succession. The ε^{t}Nd values for all samples were calculated for the time of formation of the Singhora Group ($t = 1.42$ Ga, age of diabase dyke according to Das et al., 2011) and show consistently negative values, except for one sample of the upper shale, which shows a marginally positive value (Fig. 15.3A). The lowermost Rehtikhol Formation and its underlying granite gneiss

TABLE 15.2 Sm-Nd Isotopic Data for Whole Rock Samples of Singhora Group

Sample	Sm (μg/g)	Nd (μg/g)	$^{147}Sm/^{144}Nd$	$^{143}Nd/^{144}Nd^{a}$	$\varepsilon^{0}Nd$	T_{DM} (Ga)	$\varepsilon^{t}Nd$	$f_{Sm/Nd}$
Tamtora shale	6.423	40.011	0.0850	0.511579 ± 2.3	−20.7	1.86	−0.3	−0.5680
Chuipali shale (A-5)	70.972	401.555	0.0936	0.511604 ± 3.0	−20.2	1.96	−1.4	−0.5244
Chuipalli shale	6.814	37.959	0.0950	0.511655 ± 2.0	−19.2	1.91	−0.7	−0.5169
Bhalukona SSt top most	1.084	5.843	0.1110	0.511253 ± 4.0	−27	2.79	−11.5	−0.4356
Bhalukona SSt top	0.220	1.113	0.1116	0.511369 ± 7.0	−24.7	2.64	−9.3	−0.4326
Bhalukona SSt top	1.114	5.912	0.0998	0.511267 ± 2.8	−26.7	2.51	−9.1	−0.4927
Bhalukona SSt middle	0.358	1.726	0.1100	0.511471 ± 8.4	−22.8	2.46	−7.0	−0.4406
Bhalukona SSt middle	2.881	13.900	0.1250	0.511581 ± 6.0	−20.6	2.68	−7.6	−0.3645
Bhalukona SSt bottom	0.377	1.935	0.1128	0.511300 ± 13	−26.1	2.77	−10.9	−0.4263
Bhalukona SSt contact zone	0.401	2.106	0.1008	0.511731 ± 3.8	−17.7	1.91	−0.2	−0.4877
Saraipalli Sst	3.175	15.791	0.1064	0.511594 ± 5.0	−20.3	2.21	−4.0	−0.4589
Saraipalli Sst	8.837	46.672	0.1146	0.511711 ± 5.0	−18.1	2.20	−3.2	−0.4173
Saraipalli shale	7.883	41.886	0.0996	0.511721 ± 3.0	−17.8	1.90	−0.2	−0.4935
Saraipalli shale	2.684	13.667	0.1185	0.511570 ± 5.0	−20.8	2.51	−6.6	−0.3976
Rehatikhol Fm	0.836	5.500	0.0804	0.510836 ± 2.8	−35.2	2.63	−14.0	−0.5911
Bastar granite	137.376	938.449	0.0775	0.510747 ± 2.4	−36.9	2.67	−15.2	−0.6061

[a]*The errors quoted are 2 × S.E.*

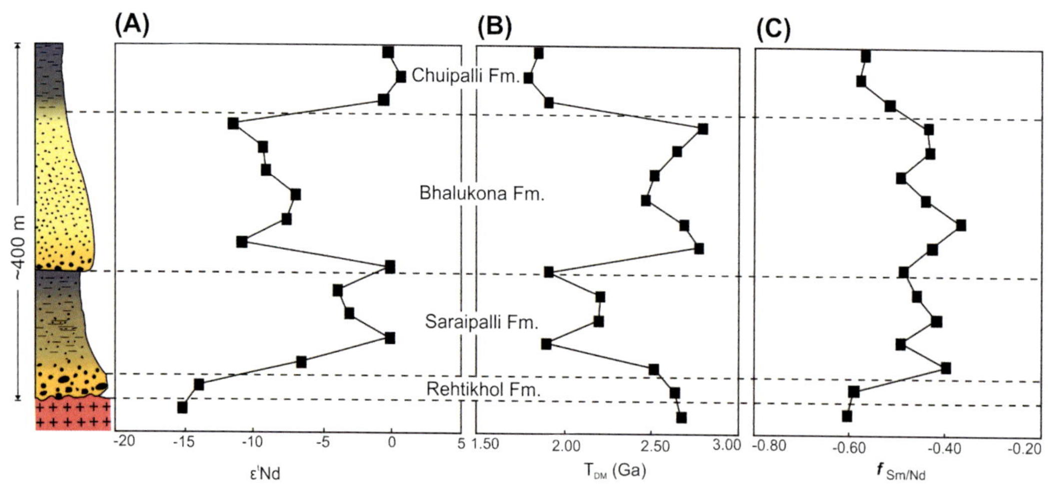

FIGURE 15.3 Nd isotopic composition of rocks from the Singhora Group. The time-corrected Nd isotopic compositions ($\varepsilon^{t}Nd$ = 1.42 Ga, Das, P. et al., 2011), depleted mantle model ages (T_{DM}), and enrichment factors ($f_{Sm/Nd}$) are plotted in (A), (B), and (C), respectively. The Y-axis represents the relative stratigraphic positions of the samples. The maximum error in calculation of $\varepsilon^{t}Nd$ is ±0.2. Note that there is a change of Nd isotopic composition in the early Bhalukona sedimentation history, possibly indicating a sudden provenance change.

yield the strongest negative values of ε^tNd for all samples, which range between −14.03 and −15.23. In the heterolithic Saraipalli Formation, ε^tNd values range between −0.2 and −6.6. On the other hand, six of seven samples from the overlying Bhalukona Formation record a sharp negative shift in values, within the range from −7.0 to −11.5. Only one sample (sample no-10; Table 15.2) that belongs to the boundary between the Bhalukona Formation and its underlying Saraipalli Formation yielded a value of −0.2. From the uppermost Chuipalli Formation, the obtained values are more positive and range between −0.67 and +0.57.

The calculated T_{DM} age data also show systematic changes throughout the stratigraphic sequence (Fig. 15.3B). Basement gneiss and overlying arkosic sandstones of the Rehtikhol Formation reveal a well-constrained age of 2.67–2.63 Ga. Younger and variegated T_{DM} ages are recorded from the overlying Saraipalli Formation ranging between 2.51 and 1.90 Ga. Samples from the Bhalukona Formation, however, show a sharp shift toward an older component, 2.46 to 2.79 Ga, with the exception of the lowermost boundary sample (sample no-10; Table 15.2) that shows a 1.91 Ga T_{DM} age. The uppermost Chuipalli Formation shows a strong presence of younger age values of 1.80–1.91 Ga. Although T_{DM} age data show variations between different stratigraphic levels of the Singhora succession, their enrichment factor ($f_{Sm/Nd}$) varies only in a narrow range between −0.39 and −0.60 (Fig. 15.3C). The most negative values are registered in the basement gneiss, arkosic sandstone of the Rehtikhol Formation, and the uppermost shale samples of the Chuipalli Formation.

5.3 U-Pb Zircon Geochronology

5.3.1 Samples From the Singhora Basin

5.3.1.1 BASEMENT GNEISS (SAMPLE NO-L21/09)

The basement sample analyzed in this study was collected from the gneiss in close proximity to the basement-sediment (arkosic Rehtikhol Formation) contact (also analyzed for Sm-Nd isotopes). The zircon grains from this granitic gneiss sample are mostly euhedral with oscillatory-zoned interiors. Some grains show core-rim structures with dark CL overgrowths on the oscillatory-zoned cores (Fig. 15.4A). Spot analyses on 17 grains (some in core and some near the rim of the grains) are plotted in Concordia diagram, where data show a discordia with upper intercept at 2520 ± 52 Ma and lower intercept at 639 ± 140 Ma (MSWD = 47). The $^{207}Pb/^{206}Pb$ age probability density plot shows two possible peaks (<5% discordance) at 2542 ± 11 Ma and 2475 ± 6 Ma (Table 15.3, Fig. 15.4B). The spots with concordant ages have Th/U ratios between 0.12 and 0.6, while the spots with discordant ages show more variation (0.16–1.20).

5.3.1.2 REHTIKHOL FORMATION (SAMPLE NO-L4/09 AND K-15)

Zircon grains from coarse arkosic sandstones of the Rehtikhol Formation are of different size (~50–300 μm in longer dimension) and shape (euhedral to rounded grains). Many large grains have an inclusion-rich interior and some are spongy in appearance. The grain interiors are mostly fractured, but show oscillatory to lamellar zoning (Fig. 15.5A). The four measured grains with <5% discordance yielded a weighted mean $^{207}Pb/^{206}Pb$ age of 2505 ± 6 Ma (MSWD = 0.58) and the rest of the data form a discordia with upper and lower intercepts

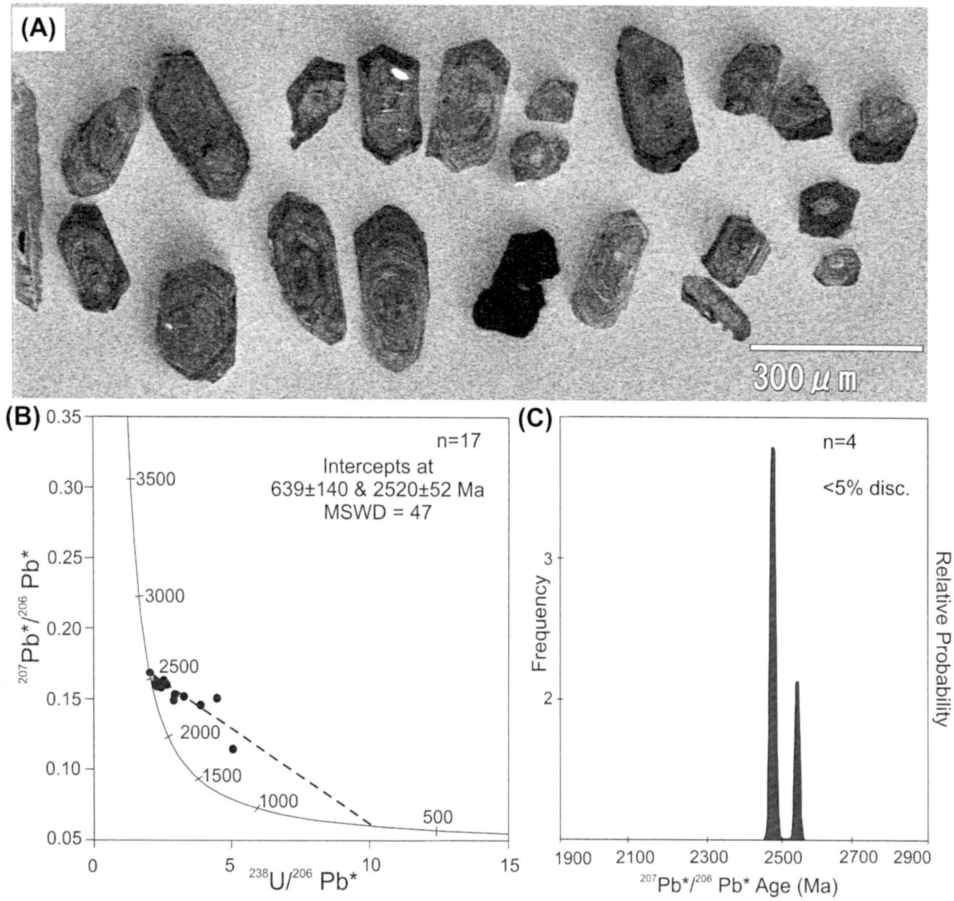

FIGURE 15.4 (A) Scanning electron microscope-Cathodoluminescence (SEM-CL) photomicrographs showing the internal structures of the measured zircon grains of basement gneiss (L21/09). (B) Tera–Wasserburg U–Pb concordia diagram of analyzed zircon grains. ^{207}Pb* and ^{206}Pb* indicate radiometric ^{207}Pb and ^{206}Pb, respectively. (C) Probability distribution diagram for concordant (−5 < Disc (%) < 5; Biao et al., 1996) zircon ^{207}Pb*/^{206}Pb* ages.

at 2495 ± 19 Ma and 562 ± 24 Ma (MSWD = 6.2), respectively (Table 15.3, Fig. 15.5B). Th/U ratios vary in a narrow range between 0.53 and 0.63 for concordant grains and a marginally wide range (0.08–0.63) for discordant grains.

Another sample (K-15) collected from the northeastern corner of the basin yielded several zircon grains. Out of 50 spot analyses from 45 grains of this sample, 36 grains yielded concordant data, and the rest are differentially discordant. The discordia yields an upper intercept age of 2488 ± 13 Ma and lower intercept age of 533 ± 80 Ma (MSWD = 1.03) (Fig. 15.5C). The data show concordant (<5% discordance) ^{207}Pb/^{206}Pb age peaks at 2515 ± 7 Ma, 2471 ± 4 Ma (these two are unmixed from a single broad peak), and 2374 ± 14 Ma (Table 15.4, Fig. 15.5D). The youngest group age (^{207}Pb/^{206}Pb) calculated by Isoplot for this sample is c. 2376 Ma.

TABLE 15.3 U–Pb Zircon Data of the Studied Samples Using SHRIMPIIe

									Apparent Age (Ma)				
									^{204}Pb Corrected				
	$\%^{206}Pb_c$	U (ppm)	Th (ppm)	$^{232}Th/^{238}U$	$^{238}U/^{206}Pb^*$	±%	$^{207}Pb^*/^{206}Pb^*$	±%	$^{206}Pb^*/^{238}U$ Age		$^{207}Pb^*/^{206}Pb^*$ Age		Disc. %
SINGHORA BASIN SAMPLES													
L2109-01.1	0.05	206	48	0.24	2.189	0.01	0.1618	0.001	2426	±10	2475	±6	2.0
L2109-02.1	0.27	532	174	0.34	2.397	0.01	0.1595	0.001	2248	±6	2450	±8	9.0
L2109-03.1	0.69	610	204	0.35	2.958	0.01	0.1490	0.001	1877	±5	2335	±11	24.3
L2109-04.1	0.62	430	139	0.33	2.445	0.01	0.1612	0.002	2210	±7	2468	±22	11.7
L2109-05.1	0.25	1330	202	0.16	5.076	0.01	0.1140	0.001	1159	±2	1864	±10	60.8
L2109-06.1	0.01	267	139	0.54	2.094	0.01	0.1685	0.001	2516	±9	2542	±5	1.0
L2109-07.1	0.07	360	201	0.58	2.371	0.01	0.1586	0.000	2268	±7	2441	±5	7.6
L2109-09.1	0.21	264	136	0.53	2.701	0.01	0.1603	0.001	2030	±7	2458	±7	21.1
L2109-10.1	0.05	261	152	0.60	2.313	0.01	0.1620	0.001	2316	±9	2476	±6	6.9
L2109-11.1	0.02	330	162	0.51	2.208	0.01	0.1621	0.001	2408	±8	2478	±6	2.9
L2109-12.1	0.63	643	204	0.33	3.321	0.01	0.1517	0.001	1697	±4	2366	±10	39.4
L2109-13.1	0.15	393	93	0.25	2.218	0.01	0.1615	0.001	2399	±7	2472	±5	3.0
L2109-14.1	0.89	662	772	1.20	4.505	0.01	0.1504	0.001	1292	±4	2351	±14	81.9
L2109-15.1	0.14	205	127	0.64	2.594	0.02	0.1630	0.001	2102	±13	2487	±8	18.3
L2109-16.1	0.11	692	78	0.12	2.505	0.01	0.1579	0.000	2165	±5	2434	±4	12.4
L2109-17.1	0.17	793	174	0.23	3.000	0.09	0.1531	0.001	1854	±47	2381	±16	28.4
L2109-18.1	0.31	415	166	0.41	3.926	0.01	0.1455	0.001	1463	±5	2294	±9	56.9
L409-01.1	0.01	174	89	0.53	2.110	0.01	0.1646	0.001	2501	±11	2504	±7	0.1
L409-22.1	0.12	545	308	0.58	2.790	0.01	0.1538	0.000	1975	±6	2388	±6	21.0
L409-26.1	0.00	1792	473	0.27	8.322	0.02	0.09137	0.000	731	±2	1454	±6	98.8

(Continued)

TABLE 15.3 U–Pb Zircon Data of the Studied Samples Using SHRIMPIIe—cont'd

									Apparent Age (Ma)				
									^{204}Pb Corrected				
	%$^{206}Pb_c$	U (ppm)	Th (ppm)	$^{232}Th/^{238}U$	$^{238}U/^{206}Pb^*$	±%	$^{207}Pb^*/^{206}Pb^*$	±%	$^{206}Pb^*/^{238}U$ Age		$^{207}Pb^*/^{206}Pb^*$ Age		Disc. %
L409-27.1	0.01	259	139	0.55	2.100	0.01	0.1653	0.001	2511	±9	2511	±6	0.0
L409-28.1	0.92	612	170	0.29	3.595	0.02	0.1469	0.002	1582	±6	2311	±18	46.1
L409-36.1	0.22	677	51	0.078	3.659	0.01	0.1433	0.001	1557	±5	2268	±7	45.6
L409-38.1	0.86	688	308	0.46	3.623	0.01	0.1443	0.001	1571	±5	2280	±15	45.1
L409-41.2	0.16	185	113	0.63	2.168	0.01	0.1642	0.001	2445	±11	2500	±8	2.2
L409-42.1	0.20	318	183	0.59	2.139	0.01	0.1644	0.001	2473	±9	2502	±7	1.2
4L1-03.1	0.02	131	90	0.71	2.131	0.01	0.1636	0.001	2481	±14	2493	±10	0.5
4L1-04.1	0.08	657	288	0.45	2.872	0.02	0.1535	0.000	1926	±9	2385	±5	23.9
4L1-11.1	0.42	78	58	0.77	1.863	0.02	0.1933	0.002	2770	±19	2770	±13	0.0
4L1-14.1	0.21	535	267	0.52	2.122	0.01	0.1634	0.001	2489	±7	2491	±5	0.1
4L1-18.1	0.11	942	1026	1.13	3.724	0.14	0.1668	0.002	1533	±51	2525	±18	64.7
4L1-19.1	0.07	286	268	0.97	2.164	0.01	0.1628	0.001	2448	±10	2485	±6	1.5
4L1-21.1	0.01	323	198	0.63	2.166	0.01	0.1633	0.001	2447	±9	2490	±6	1.8
4L1-29.1	0.17	1824	212	0.12	5.823	0.01	0.09806	0.000	1022	±2	1587	±9	55.4
4L1-30.1	0.03	681	202	0.31	2.184	0.01	0.1633	0.000	2430	±7	2490	±4	2.5
4L1-32.1	0.06	67	64	0.98	1.862	0.02	0.1940	0.001	2771	±20	2776	±10	0.2
4L1-33.1	0.15	97	47	0.51	2.125	0.02	0.1632	0.001	2487	±16	2489	±11	0.1
4L1-34.1	0.26	693	36	0.053	4.416	0.01	0.1235	0.001	1316	±4	2007	±8	52.5
4L1-35.1	0.10	87	139	1.65	1.862	0.02	0.1936	0.001	2771	±18	2773	±10	0.1
4L1-39.1	0.05	542	445	0.85	2.758	0.01	0.1773	0.001	1994	±6	2628	±5	31.8
4L1-39.2	0.31	4806	1957	0.42	7.135	0.02	0.08918	0.001	846	±2	1408	±11	66.5
4L1-40.1	0.06	330	63	0.20	2.120	0.01	0.1634	0.001	2491	±9	2491	±6	0.0

4L1-46.1	0.04	285	98	0.36	1.897	0.01	0.1938	0.001	2729	±11	2774	±5	1.7
4L1-48.1	0.49	697	475	0.70	3.289	0.02	0.1502	0.001	1711	±10	2348	±13	37.2
4L1-51.1	0.03	367	225	0.63	2.169	0.01	0.1634	0.001	2444	±9	2491	±6	1.9
4L1-52.1	0.02	805	276	0.35	3.366	0.01	0.1400	0.000	1677	±5	2227	±5	32.8
4L1-53.1	0.06	459	247	0.56	2.148	0.01	0.1639	0.001	2464	±8	2496	±5	1.3
4L1-55.1	0.13	567	18	0.033	3.154	0.01	0.1460	0.001	1776	±6	2299	±8	29.5
4L1-61.1	0.01	445	208	0.48	2.136	0.01	0.1632	0.000	2476	±8	2489	±5	0.5
4L1-64.1	0.11	686	334	0.50	3.068	0.01	0.1463	0.000	1819	±5	2303	±6	26.6
4L1-74.1	0.10	649	276	0.44	2.683	0.01	0.1538	0.000	2042	±6	2388	±5	17.0
4L1-76.1	0.79	599	706	1.22	2.513	0.01	0.1584	0.001	2159	±7	2438	±12	12.9
4L1-78.1	0.02	298	121	0.42	2.122	0.01	0.1631	0.001	2489	±10	2488	±7	0.0
4L1-81.1	0.29	507	497	1.01	2.423	0.01	0.1555	0.001	2228	±11	2408	±7	8.1
4L1-86.1	0.04	753	170	0.23	3.324	0.01	0.1456	0.000	1696	±5	2295	±5	35.4
L1809-001.1	0.06	113	82	0.76	2.040	0.01	0.1681	0.001	2571	±15	2539	±9	−1.3
L1809-002.1	<0.01	220	151	0.71	2.178	0.01	0.1624	0.001	2436	±12	2480	±6	1.8
L1809-003.1	0.13	54	147	2.80	2.039	0.02	0.1720	0.001	2572	±20	2577	±13	0.2
L1809-004.1	0.02	345	141	0.42	2.026	0.02	0.1728	0.001	2587	±19	2585	±5	−0.1
L1809-005.1	0.01	106	153	1.50	2.109	0.02	0.1684	0.001	2502	±19	2542	±9	1.6
L1809-006.1	0.01	116	90	0.80	2.056	0.02	0.1687	0.001	2555	±16	2545	±9	−0.4
L1809-007.1	0.06	233	129	0.57	2.105	0.02	0.1678	0.001	2506	±15	2536	±7	1.2
L1809−008.1	0.07	100	71	0.73	2.233	0.02	0.1617	0.001	2385	±20	2473	±10	3.7
L1809-009.1	0.02	252	353	1.45	2.099	0.01	0.1688	0.001	2511	±12	2546	±6	1.4
L1809-010.1	0.20	121	156	1.33	2.305	0.02	0.1650	0.001	2323	±14	2507	±10	7.9
L1809-011.1	0.00	71	37	0.53	2.082	0.02	0.1681	0.001	2529	±19	2539	±12	0.4
L1809-012.1	0.27	51	66	1.35	2.071	0.02	0.1676	0.002	2540	±23	2533	±17	−0.2
L1809-013.1	0.22	890	407	0.47	4.140	0.02	0.1394	0.001	1395	±6	2220	±7	59.2

(Continued)

TABLE 15.3 U–Pb Zircon Data of the Studied Samples Using SHRIMPIIe—cont'd

									Apparent Age (Ma)				
									^{204}Pb Corrected				
	%^{206}Pb$_c$	U (ppm)	Th (ppm)	^{232}Th/^{238}U	^{238}U/^{206}Pb*	±%	^{207}Pb*/^{206}Pb*	±%	^{206}Pb*/^{238}U Age		^{207}Pb*/^{206}Pb* Age		Disc. %
L1809-014.1	0.20	50	69	1.41	2.022	0.03	0.1715	0.001	2591	±30	2572	±14	−0.7
L1809-015.1	0.15	295	227	0.79	2.355	0.01	0.1591	0.001	2281	±10	2447	±7	7.2
L1809-016.1	0.41	246	198	0.83	2.548	0.02	0.1644	0.001	2135	±13	2501	±9	17.2
L1809-017.1	0.05	335	149	0.46	1.703	0.02	0.2173	0.001	2978	±25	2961	±8	−0.6
L1809-018.1	0.21	426	143	0.35	2.829	0.01	0.1612	0.001	1951	±8	2468	±7	26.5
L1809-019.1	0.03	213	200	0.97	2.041	0.01	0.1729	0.001	2570	±13	2586	±6	0.6
L1809-020.1	0.11	122	142	1.21	2.113	0.01	0.1679	0.001	2498	±14	2537	±8	1.6
L1809-021.1	0.21	386	256	0.68	2.725	0.02	0.1556	0.001	2015	±12	2409	±6	19.5
L1809-022.1	0.02	43	36	0.87	1.973	0.02	0.1779	0.001	2643	±23	2633	±13	−0.4
L1809-023.1	0.03	133	96	0.75	2.031	0.01	0.1727	0.001	2580	±14	2584	±7	0.1
L1809-024.1	0.12	509	93	0.19	2.270	0.01	0.1674	0.000	2353	±9	2532	±4	7.6
L1809-025.1	0.04	117	83	0.73	2.142	0.01	0.1608	0.001	2469	±14	2464	±8	−0.2
L1809-026.1	0.04	109	105	1.00	2.058	0.02	0.1683	0.001	2553	±21	2541	±8	−0.5
L1809-027.1	0.10	71	60	0.87	2.067	0.02	0.1678	0.001	2543	±18	2536	±10	−0.3
L1809-028.1	0.08	65	29	0.47	1.956	0.02	0.1773	0.001	2661	±19	2628	±10	−1.3
L1809-029.1	0.09	39	35	0.91	1.971	0.02	0.1771	0.002	2646	±25	2626	±16	−0.8
L1809-030.1	0.06	242	130	0.56	1.438	0.01	0.3088	0.001	3403	±14	3515	±4	3.3
L1809-031.1	0.03	42	46	1.14	2.030	0.02	0.1727	0.001	2582	±23	2584	±13	0.1
L1809–032.1	0.02	321	195	0.63	2.123	0.01	0.1685	0.000	2488	±10	2543	±5	2.2
L1809-033.1	0.21	479	120	0.26	2.848	0.01	0.1540	0.001	1940	±8	2391	±6	23.2
L1809-034.1	0.08	110	103	0.96	2.020	0.01	0.1678	0.001	2592	±15	2536	±8	−2.2
L1809-035.1	0.01	157	337	2.22	2.025	0.01	0.1729	0.001	2588	±13	2586	±7	−0.1

L1809-036.1	0.01	136	57	0.43	2.100	0.04	0.1629	0.001	2511	±40	2486	±8	−1.0
L1809-037.1	0.00	349	189	0.56	3.418	0.02	0.1011	0.000	1654	±7	1644	±8	−0.6
L1809-038.1	0.37	400	385	0.99	2.561	0.01	0.1830	0.001	2125	±9	2681	±7	26.1
L1809-039.1	0.07	128	72	0.58	2.023	0.01	0.1723	0.001	2589	±14	2580	±8	−0.3
L1809-040.1	0.02	215	132	0.63	2.050	0.01	0.1691	0.001	2561	±12	2548	±7	−0.5
L1809-041.1	0.34	448	347	0.80	3.049	0.01	0.1604	0.001	1828	±8	2460	±8	34.6
L1809-042.1	0.33	138	102	0.76	3.008	0.02	0.1652	0.001	1850	±11	2509	±12	35.6
L1809-043.1	0.07	51	49	0.99	2.135	0.03	0.1632	0.001	2476	±27	2489	±13	0.5
L1809-044.1	0.02	210	185	0.91	2.015	0.01	0.1727	0.001	2598	±15	2584	±6	−0.5
L1809-045.1	0.17	55	40	0.75	2.015	0.02	0.1717	0.001	2597	±21	2574	±13	−0.9
L1809-046.1	0.07	336	187	0.58	2.189	0.01	0.1635	0.000	2426	±10	2492	±5	2.7
L1809-047.1	0.10	389	262	0.70	2.092	0.01	0.1676	0.000	2518	±10	2534	±5	0.6
L1809-048.1	0.44	640	58	0.094	4.030	0.02	0.1437	0.001	1429	±6	2272	±9	59.0
L1809-049.1	0.02	63	51	0.83	2.099	0.02	0.1619	0.001	2511	±24	2476	±11	−1.4
L1809-050.1	0.00	140	204	1.51	2.036	0.02	0.1731	0.001	2575	±19	2588	±7	0.5
L1809-051.1	0.06	135	139	1.06	2.128	0.01	0.1618	0.001	2484	±14	2475	±8	−0.4
L1809-052.1	0.07	214	167	0.81	2.034	0.01	0.1682	0.001	2578	±12	2540	±6	−1.5
L1809-053.1	0.02	66	84	1.31	1.995	0.02	0.1780	0.001	2620	±19	2634	±10	0.6
L1809-054.1	0.04	40	34	0.87	2.000	0.03	0.1779	0.001	2614	±36	2633	±13	0.7
L1809-055.1	0.00	181	153	0.87	1.996	0.01	0.1783	0.001	2618	±13	2637	±6	0.7
L1809-056.1	0.04	304	139	0.47	2.118	0.01	0.1682	0.001	2493	±11	2540	±5	1.9
L1809-057.1	0.08	33	25	0.77	2.143	0.03	0.1684	0.002	2469	±24	2541	±18	2.9
L1809-058.1	0.18	120	71	0.61	3.448	0.03	0.1005	0.001	1642	±11	1634	±17	−0.5
L1809-059.1	0.04	132	178	1.40	2.068	0.02	0.1682	0.001	2543	±16	2540	±8	−0.1
L1809-060.1	0.24	194	120	0.64	2.386	0.07	0.1751	0.005	2256	±52	2607	±46	15.6
L1809-061.1	0.24	425	258	0.63	2.689	0.01	0.1594	0.001	2038	±9	2449	±7	20.1

(Continued)

TABLE 15.3 U–Pb Zircon Data of the Studied Samples Using SHRIMPIIe—cont'd

									Apparent Age (Ma)				
									^{204}Pb Corrected				
	$\%^{206}Pb_c$	U (ppm)	Th (ppm)	$^{232}Th/^{238}U$	$^{238}U/^{206}Pb^*$	±%	$^{207}Pb^*/^{206}Pb^*$	±%	$^{206}Pb^*/^{238}U$ Age		$^{207}Pb^*/^{206}Pb^*$ Age		Disc. %
L1809-062.1	0.04	190	287	1.56	2.022	0.02	0.1725	0.001	2591	±16	2583	±6	−0.3
L1809-063.1	0.03	118	190	1.66	2.029	0.01	0.1726	0.001	2583	±15	2583	±8	0.0
L1809-064.1	0.01	121	50	0.43	1.985	0.01	0.1782	0.001	2630	±15	2636	±8	0.3
L1809-065.1	0.03	320	83	0.27	2.189	0.01	0.1621	0.001	2425	±13	2478	±5	2.2
L1809-066.1	0.06	111	69	0.64	2.077	0.04	0.1701	0.001	2533	±38	2558	±9	1.0
L1809-067.1	0.09	156	72	0.48	2.030	0.01	0.1687	0.001	2582	±14	2545	±7	−1.4
L1809-068.1	0.05	41	11	0.26	1.903	0.02	0.1885	0.001	2722	±24	2729	±13	0.3
L1809-069.1	0.25	167	79	0.49	2.755	0.02	0.1685	0.001	1997	±11	2543	±9	27.4
L1809-070.1	0.09	75	49	0.67	2.007	0.02	0.1774	0.001	2606	±18	2629	±10	0.9
L1809-072.1	0.06	138	100	0.75	2.138	0.01	0.1681	0.001	2474	±14	2539	±8	2.6
L1809-073.1	0.05	72	60	0.87	2.046	0.02	0.1697	0.001	2565	±18	2555	±12	−0.4
L1809-074.1	0.03	289	165	0.59	2.301	0.02	0.1493	0.001	2326	±15	2338	±8	0.5
L1809-075.1	0.00	257	157	0.63	2.112	0.01	0.1623	0.001	2499	±11	2479	±6	−0.8
L1809-076.1	0.10	108	47	0.45	2.032	0.02	0.1677	0.001	2580	±16	2535	±9	−1.7
L1809-077.1	0.02	247	99	0.41	2.068	0.01	0.1684	0.001	2543	±12	2542	±6	0.0
L1809-078.1	0.16	58	41	0.73	2.058	0.02	0.1672	0.001	2553	±20	2530	±12	−0.9
L1809-079.1	0.13	72	45	0.65	2.144	0.09	0.1633	0.002	2468	±91	2490	±22	0.9
L1809-080.1	0.06	56	68	1.27	2.087	0.02	0.1683	0.001	2524	±20	2540	±12	0.7
L1809-081.1	0.02	206	173	0.87	2.025	0.01	0.1686	0.001	2588	±13	2544	±6	−1.7
L1809-082.1	0.08	107	42	0.41	2.106	0.02	0.1621	0.001	2504	±15	2477	±9	−1.1
L1809-083.1	0.18	97	66	0.70	3.411	0.03	0.1004	0.001	1657	±11	1631	±18	−1.6
L1809-084.1	<0.01	171	100	0.61	1.981	0.01	0.1780	0.001	2635	±13	2634	±6	0.0

L1809-085.1	0.08	87	87	1.03	2.095	0.02	0.1680	0.001	2515	±17	2538	±10	0.9
L1809-086.1	0.08	333	341	1.06	2.099	0.01	0.1683	0.001	2511	±11	2541	±5	1.2
L1809-087.1	0.19	42	37	0.92	2.115	0.02	0.1669	0.002	2496	±23	2527	±16	1.3
L1809-088.1	0.42	271	119	0.45	2.742	0.01	0.1613	0.001	2005	±9	2469	±11	23.2
L1809-089.1	0.01	193	85	0.46	1.992	0.01	0.1783	0.001	2622	±13	2637	±6	0.6
L1809-090.1	0.06	60	36	0.63	3.543	0.03	0.1007	0.001	1603	±14	1637	±21	2.2
L1809-091.1	0.01	467	176	0.39	2.251	0.04	0.1616	0.000	2370	±31	2473	±5	4.3
L1809-091.2	0.02	391	147	0.39	2.207	0.01	0.1623	0.000	2409	±10	2479	±5	2.9
L1809-093.1	0.27	194	185	0.98	2.508	0.02	0.1643	0.001	2163	±11	2501	±10	15.6
L1809-094.1	0.05	66	37	0.59	1.724	0.02	0.2181	0.001	2949	±27	2967	±9	0.6
L1809-095.1	0.00	138	120	0.90	2.079	0.01	0.1686	0.001	2532	±14	2544	±8	0.5
L1809-096.1	0.07	91	109	1.24	2.089	0.02	0.1680	0.001	2522	±17	2538	±10	0.6
L1809-097.1	0.17	38	43	1.17	2.031	0.02	0.1691	0.002	2581	±26	2548	±20	−1.3
L1809-098.1	0.21	625	352	0.58	3.301	0.02	0.1423	0.001	1706	±7	2256	±7	32.2
L1809-099.1	0.16	71	71	1.04	2.094	0.02	0.1619	0.001	2517	±19	2476	±12	−1.6
L1809-100.1	0.01	97	112	1.19	2.047	0.02	0.1730	0.001	2565	±17	2587	±9	0.9
L1809-101.1	0.13	89	95	1.10	2.121	0.03	0.1675	0.001	2490	±27	2532	±10	1.7
L1809-102.1	0.03	219	132	0.62	2.190	0.01	0.1620	0.001	2424	±12	2476	±6	2.1
L1809-103.1	0.06	148	99	0.69	1.989	0.01	0.1780	0.001	2626	±15	2634	±8	0.3
L1809-104.1	0.04	134	94	0.72	3.483	0.03	0.1013	0.001	1627	±10	1648	±15	1.3
L1809-106.1	0.06	408	237	0.60	2.229	0.01	0.1620	0.000	2389	±10	2477	±5	3.7
L1809-107.1	0.02	88	111	1.30	2.082	0.02	0.1689	0.001	2529	±18	2547	±10	0.7
L1809-108.1	0.21	366	156	0.44	2.648	0.02	0.1606	0.001	2065	±11	2462	±7	19.2
L1809-109.1	0.22	190	154	0.84	2.390	0.01	0.1653	0.001	2253	±12	2510	±8	11.4
L1809-110.1	0.03	136	126	0.96	2.043	0.01	0.1728	0.001	2568	±15	2585	±8	0.7
L1809-111.1	0.14	77	84	1.13	2.049	0.02	0.1669	0.001	2562	±19	2527	±11	−1.4

(Continued)

TABLE 15.3 U–Pb Zircon Data of the Studied Samples Using SHRIMPIIe—cont'd

									Apparent Age (Ma)				
									^{204}Pb Corrected				
	$\%^{206}Pb_c$	U (ppm)	Th (ppm)	$^{232}Th/^{238}U$	$^{238}U/^{206}Pb^*$	±%	$^{207}Pb^*/^{206}Pb^*$	±%	$^{206}Pb^*/^{238}U$ Age		$^{207}Pb^*/^{206}Pb^*$ Age		Disc. %
L1809-112.1	0.11	81	64	0.82	2.081	0.02	0.1677	0.001	2529	±24	2534	±11	0.2
L1809-113.1	0.14	66	86	1.34	1.998	0.02	0.1685	0.001	2616	±20	2543	±13	−2.8
L1809-114.1	0.01	277	284	1.06	2.068	0.02	0.1681	0.001	2543	±16	2539	±6	−0.1
L1809-115.1	0.04	345	379	1.14	2.169	0.01	0.1672	0.001	2444	±11	2530	±7	3.5
L1809-116.1	0.02	150	89	0.61	2.032	0.01	0.1728	0.001	2579	±15	2585	±8	0.2
L1809-117.1	0.04	177	58	0.34	3.390	0.02	0.1010	0.001	1666	±10	1642	±12	−1.5
L1809-118.1	0.04	92	123	1.38	2.072	0.02	0.1724	0.001	2539	±17	2581	±10	1.7
L1809-119.1	0.15	472	257	0.56	2.660	0.02	0.1531	0.001	2057	±11	2381	±6	15.7
L1809-120.1	0.02	82	82	1.03	2.104	0.02	0.1622	0.001	2507	±18	2478	±11	−1.1
L1809-121.1	0.12	104	125	1.24	2.077	0.02	0.1678	0.001	2533	±16	2536	±11	0.1
L1809-122.1	0.00	183	97	0.55	2.046	0.01	0.1673	0.001	2565	±14	2531	±7	−1.3
L1809-123.1	0.03	261	182	0.72	2.328	0.01	0.1598	0.001	2304	±12	2454	±7	6.5
L1809-125.1	0.06	180	141	0.81	2.245	0.01	0.1618	0.001	2375	±13	2475	±8	4.2
L1809-126.1	0.08	288	228	0.82	2.750	0.02	0.1420	0.001	1999	±9	2251	±9	12.6
L1809-127.1	0.11	57	66	1.19	2.128	0.02	0.1613	0.001	2483	±21	2470	±15	−0.5
L1809-128.1	0.05	78	31	0.40	2.032	0.02	0.1721	0.001	2580	±19	2579	±11	−0.1
L1809-129.1	0.08	206	126	0.63	2.113	0.01	0.1684	0.001	2498	±13	2541	±7	1.7
L1809-130.1	0.06	107	75	0.73	2.132	0.03	0.1616	0.001	2479	±30	2472	±10	−0.3
L1809-131.1	0.05	166	201	1.25	2.004	0.01	0.1714	0.001	2610	±15	2571	±10	−1.5
L1809-132.1	0.07	310	115	0.38	2.239	0.01	0.1697	0.001	2380	±11	2555	±6	7.3
L1809-133.1	0.02	96	113	1.21	1.976	0.02	0.1780	0.001	2639	±19	2634	±10	−0.2
L1809-134.1	0.17	26	24	0.96	2.091	0.03	0.1669	0.002	2519	±32	2527	±22	0.3

L1809-135.1	0.04	158	119	0.78	1.374	0.01	0.3100	0.001	3526	±26	3521	±5	−0.1
L1809-136.1	0.16	281	248	0.91	2.426	0.01	0.1646	0.001	2225	±11	2504	±8	12.5
L1809-137.1	0.05	275	49	0.18	3.506	0.02	0.1005	0.001	1617	±9	1633	±13	0.9
L1809-138.1	0.04	138	94	0.71	2.107	0.02	0.1620	0.001	2503	±15	2477	±9	−1.1
L1809-139.1	0.04	74	41	0.57	2.024	0.02	0.1726	0.001	2588	±20	2583	±11	−0.2
L1809-140.1	<0.01	131	63	0.50	2.056	0.03	0.1718	0.001	2555	±28	2575	±8	0.8
L1809-141.1	0.09	77	76	1.01	2.040	0.02	0.1722	0.001	2572	±25	2579	±12	0.3
L1809-142.1	0.08	138	202	1.52	2.010	0.02	0.1730	0.001	2604	±16	2587	±8	−0.6
L1809-143.1	<0.01	239	219	0.95	1.908	0.03	0.1702	0.003	2716	±35	2559	±25	−5.8
L1809-144.1	0.05	246	156	0.66	2.202	0.01	0.1625	0.001	2413	±12	2482	±7	2.8
L1809-145.1	0.08	156	204	1.35	2.052	0.01	0.1720	0.001	2559	±15	2577	±8	0.7
L1809-146.1	0.09	132	142	1.11	2.166	0.02	0.1614	0.001	2447	±15	2470	±9	0.9
L1809-147.1	0.02	104	54	0.54	2.099	0.02	0.1683	0.001	2512	±17	2540	±10	1.1
L1809-148.1	0.01	193	118	0.63	1.976	0.01	0.1777	0.001	2640	±14	2632	±7	−0.3
L1809-149.1	0.12	168	124	0.77	2.373	0.02	0.1633	0.001	2267	±13	2491	±9	9.9
L1809-150.1	0.03	273	183	0.69	2.065	0.01	0.1633	0.001	2546	±12	2491	±6	−2.2
L1809-151.1	0.21	73	94	1.33	2.106	0.03	0.1666	0.001	2505	±32	2524	±13	0.8
L1809-152.1	<0.01	126	85	0.70	2.033	0.02	0.1687	0.001	2579	±16	2545	±9	−1.3
L1809-153.1	0.01	225	193	0.89	2.047	0.01	0.1726	0.001	2565	±13	2583	±7	0.7
L1809-154.1	0.20	506	211	0.43	2.643	0.01	0.1574	0.001	2069	±9	2428	±6	17.4
L1809-155.1	0.17	65	56	0.89	2.058	0.02	0.1713	0.001	2553	±21	2570	±13	0.7
L1809-156.1	0.03	154	114	0.76	2.106	0.01	0.1633	0.001	2504	±15	2490	±8	−0.6
L1809-157.1	0.05	133	144	1.11	2.099	0.02	0.1683	0.001	2511	±16	2541	±10	1.2
L1809-158.1	0.03	146	130	0.92	1.369	0.01	0.3101	0.001	3536	±19	3521	±5	−0.4
L1809-159.1	0.06	150	153	1.05	1.986	0.01	0.1685	0.001	2629	±16	2543	±8	−3.3
L1809-160.1	0.05	51	61	1.22	2.066	0.02	0.1681	0.001	2544	±24	2539	±15	−0.2

(Continued)

TABLE 15.3 U–Pb Zircon Data of the Studied Samples Using SHRIMPIIe—cont'd

									Apparent Age (Ma)				
									^{204}Pb Corrected				
	$\%^{206}Pb_c$	U (ppm)	Th (ppm)	$^{232}Th/^{238}U$	$^{238}U/^{206}Pb^*$	±%	$^{207}Pb^*/^{206}Pb^*$	±%	$^{206}Pb^*/^{238}U$ Age		$^{207}Pb^*/^{206}Pb^*$ Age		Disc. %
L1809-161.1	0.04	38	65	1.78	2.069	0.03	0.1682	0.002	2541	±27	2540	±18	−0.1
L1809-162.1	0.05	157	137	0.90	1.972	0.01	0.1717	0.001	2644	±15	2574	±8	−2.7
L1809-163.1	0.04	97	86	0.92	2.010	0.03	0.1728	0.001	2603	±28	2585	±10	−0.7
L1809-164.1	0.16	52	31	0.61	1.941	0.02	0.1641	0.001	2679	±23	2498	±15	−6.8
L1809-165.1	0.06	517	290	0.58	2.355	0.01	0.1582	0.000	2281	±9	2437	±5	6.8
L1809-166.1	0.07	158	136	0.89	1.863	0.01	0.1928	0.001	2770	±15	2767	±7	−0.1
L1809-167.1	<0.01	180	45	0.26	2.116	0.01	0.1615	0.001	2495	±13	2471	±7	−1.0
L1809-168.1	0.02	113	61	0.55	2.018	0.02	0.1688	0.001	2594	±23	2546	±9	−1.9
L1809-169.1	0.01	169	74	0.45	2.059	0.01	0.1728	0.001	2552	±14	2585	±7	1.3
L1809-170.1	0.17	74	111	1.54	2.063	0.02	0.1673	0.001	2548	±20	2531	±12	−0.7
L1809-171.1	0.02	193	140	0.75	2.081	0.01	0.1686	0.001	2529	±13	2544	±7	0.6
L1809-172.1	0.01	193	288	1.54	2.040	0.01	0.1685	0.001	2572	±13	2543	±7	−1.1
L1809-173.1	0.03	117	66	0.58	2.119	0.02	0.1629	0.001	2492	±16	2486	±10	−0.2
L1809-174.1	0.05	265	177	0.69	2.071	0.01	0.1683	0.001	2539	±12	2541	±6	0.1
L1809-175.1	0.10	53	44	0.86	2.054	0.02	0.1715	0.001	2557	±22	2573	±14	0.6
L1809-176.1	0.07	82	52	0.66	2.086	0.02	0.1679	0.001	2525	±18	2537	±11	0.5
L1809-177.1	0.40	394	325	0.85	3.002	0.02	0.1571	0.001	1853	±8	2424	±9	30.8
L1809-178.1	0.07	91	45	0.51	2.058	0.02	0.1682	0.001	2552	±18	2540	±10	−0.5
L1809-179.1	0.03	147	77	0.54	2.051	0.01	0.1726	0.001	2560	±15	2583	±8	0.9
LOWER CHANDARPUR GROUP													
L309-01.1	0.65	327	362	1.15	3.578	0.01	0.1545	0.001	1589	±5	2396	±12	50.8
L309-02.1	0.03	356	200	0.58	2.163	0.01	0.1625	0.000	2450	±7	2482	±5	1.3

L309-03.1	0.06	163	142	0.90	2.119	0.01	0.1629	0.001	2492	±10	2486	±7	−0.3
L309-04.1	0.05	79	67	0.87	2.175	0.03	0.1638	0.001	2438	±31	2495	±12	2.3
L309-05.1	0.14	48	44	0.96	2.119	0.02	0.1623	0.001	2492	±19	2480	±14	−0.5
L309-06.1	0.01	55	39	0.74	2.093	0.02	0.1635	0.001	2518	±18	2492	±11	−1.0
L309-07.1	0.57	659	819	1.29	4.242	0.01	0.1410	0.001	1365	±4	2240	±10	64.1
L309-08.1	0.07	68	81	1.22	2.135	0.02	0.1622	0.001	2476	±16	2479	±12	0.1
L309-10.1	0.02	208	181	0.90	2.126	0.01	0.1624	0.001	2485	±10	2481	±6	−0.2
L309-12.1	0.05	84	56	0.68	2.121	0.01	0.1631	0.001	2490	±14	2488	±10	−0.1
L309-13.1	0.20	71	61	0.89	2.891	0.03	0.1641	0.002	1915	±17	2498	±16	30.5
L309-14.1	0.01	1052	66	0.065	3.723	0.01	0.0958	0.000	1534	±3	1543	±5	0.6
L309-15.1	0.02	114	67	0.61	2.061	0.01	0.1662	0.001	2550	±13	2520	±10	−1.2
L309-16.1	0.04	147	138	0.97	2.161	0.02	0.1629	0.001	2452	±19	2486	±7	1.4
L309-17.1	0.02	423	282	0.69	2.172	0.01	0.1631	0.000	2442	±11	2488	±5	1.9
L309-18.1	0.04	210	268	1.32	2.085	0.01	0.1664	0.001	2525	±10	2522	±6	−0.1
L309-20.1	0.01	106	85	0.84	2.132	0.01	0.1632	0.001	2479	±13	2489	±9	0.4
L309-21.1	<0.01	75	55	0.75	2.120	0.02	0.1633	0.001	2491	±16	2490	±10	0.0
L309-22.1	0.01	110	124	1.17	2.180	0.01	0.1624	0.001	2434	±13	2480	±9	1.9
L309-23.1	0.08	136	180	1.37	2.282	0.02	0.1612	0.001	2343	±15	2469	±10	5.4
L309-24.1	0.05	172	177	1.06	2.191	0.01	0.1631	0.001	2424	±11	2488	±7	2.6
L309-25.1	0.20	360	339	0.97	2.612	0.01	0.1608	0.001	2089	±7	2464	±7	17.9
L309-26.1	0.50	195	175	0.93	2.648	0.02	0.1619	0.001	2065	±16	2476	±11	19.9
L309-27.1	0.05	82	61	0.77	2.121	0.02	0.1622	0.002	2490	±22	2479	±16	−0.4
L309-28.1	0.09	504	526	1.08	3.735	0.01	0.0959	0.000	1529	±5	1545	±9	1.0
L309-29.1	0.06	76	64	0.88	2.130	0.02	0.1624	0.001	2481	±21	2481	±11	0.0
L309-30.1	0.22	45	27	0.62	2.143	0.02	0.1640	0.001	2468	±20	2497	±15	1.2
L309-31.1	0.35	476	543	1.18	3.477	0.01	0.2558	0.001	1630	±5	3221	±8	97.7

(Continued)

TABLE 15.3 U–Pb Zircon Data of the Studied Samples Using SHRIMPIIe—cont'd

									Apparent Age (Ma)				
									^{204}Pb Corrected				
	$\%^{206}Pb_c$	U (ppm)	Th (ppm)	$^{232}Th/^{238}U$	$^{238}U/^{206}Pb^*$	±%	$^{207}Pb^*/^{206}Pb^*$	±%	$^{206}Pb^*/^{238}U$ Age		$^{207}Pb^*/^{206}Pb^*$ Age		Disc. %
L309-32.1	0.07	80	69	0.90	2.111	0.02	0.1627	0.001	2500	±16	2484	±10	−0.6
L309-33.1	0.01	142	108	0.79	2.168	0.01	0.1634	0.001	2445	±12	2491	±8	1.9
L309-34.1	0.05	139	92	0.68	2.120	0.01	0.1627	0.001	2491	±12	2484	±8	−0.3
LOWER KHARIAR SANDSTONE													
14L5-01.1	0.02	79	91	1.19	2.127	0.02	0.1685	0.001	2485	±19	2543	±10	2.4
14L5-02.1	0.14	49	57	1.20	2.118	0.02	0.1622	0.001	2493	±22	2479	±14	−0.6
14L5-03.1	0.06	160	134	0.87	2.108	0.02	0.1682	0.001	2503	±16	2540	±7	1.5
14L5-04.1	0.13	328	222	0.70	2.326	0.02	0.1635	0.001	2305	±13	2492	±6	8.1
14L5-06.1	0.03	128	220	1.78	2.065	0.02	0.1682	0.001	2546	±17	2540	±8	−0.2
14L5-08.1	0.03	243	188	0.80	2.119	0.02	0.1689	0.001	2492	±15	2547	±5	2.2
14L5-09.1	0.92	546	243	0.46	5.035	0.03	0.1353	0.001	1168	±7	2168	±17	85.7
14L5-10.1	0.09	96	120	1.29	2.358	0.02	0.1645	0.001	2279	±17	2503	±11	9.8
14L5-11.1	0.04	134	121	0.93	2.022	0.02	0.1731	0.001	2591	±17	2588	±7	−0.1
14L5-13.1	0.38	291	491	1.74	2.602	0.02	0.1609	0.001	2097	±13	2466	±9	17.6
14L5-14.1	0.76	559	395	0.73	3.890	0.03	0.1524	0.001	1475	±10	2373	±12	60.9
14L5-15.1	0.33	644	534	0.86	4.857	0.04	0.1227	0.001	1207	±8	1996	±19	65.4
14L5-16.1	0.21	288	154	0.55	2.394	0.02	0.1656	0.001	2250	±13	2514	±6	11.7
14L5-17.1	0.34	393	543	1.43	3.205	0.02	0.1522	0.001	1750	±12	2370	±8	35.4
14L5-19.1	0.13	183	173	0.98	2.492	0.02	0.1648	0.001	2175	±16	2506	±8	15.2
14L5-20.1	0.02	78	119	1.57	2.070	0.02	0.1691	0.001	2541	±20	2549	±10	0.3
14L5-21.1	0.03	248	230	0.96	2.108	0.02	0.1622	0.001	2502	±15	2478	±6	−1.0
14L5-22.1	0.32	669	324	0.50	4.400	0.03	0.1243	0.001	1320	±9	2019	±9	52.9

14L5-23.1	0.21	534	492	0.95	3.713	0.02	0.1383	0.001	1537	±9	2206	±7	43.5
14L5-24.1	0.39	175	275	1.62	2.771	0.02	0.1646	0.001	1986	±13	2503	±10	26.0
14L5-25.1	0.17	449	266	0.61	2.806	0.02	0.1599	0.001	1965	±11	2455	±6	25.0
14L5-26.1	0.05	214	148	0.72	2.089	0.02	0.1660	0.001	2522	±15	2518	±6	−0.1
14L5-27.1	0.70	439	610	1.44	3.899	0.03	0.1494	0.001	1472	±9	2339	±13	58.9
14L5-28.1	0.04	89	26	0.30	1.989	0.02	0.1780	0.001	2626	±20	2634	±10	0.3
14L5-29.1	0.22	449	275	0.63	2.899	0.02	0.1503	0.001	1910	±11	2349	±7	23.0
14L5-31.1	0.16	369	299	0.84	2.757	0.02	0.1559	0.001	1995	±12	2412	±6	20.9
14L5-32.1	0.02	65	77	1.22	2.125	0.02	0.1626	0.001	2486	±21	2483	±12	−0.1
14L5-35.1	0.06	583	317	0.56	2.750	0.02	0.1499	0.000	1999	±11	2345	±5	17.3
14L5-36.1	0.04	185	207	1.15	2.137	0.02	0.1628	0.001	2474	±16	2485	±7	0.4
14L5-38.1	0.16	376	333	0.91	2.432	0.02	0.1655	0.001	2221	±13	2512	±5	13.1
14L5-38.2	0.01	208	96	0.48	2.045	0.02	0.1684	0.001	2567	±16	2542	±6	−1.0
14L5-40.1	0.29	226	561	2.56	3.783	0.04	0.1610	0.001	1512	±14	2466	±10	63.1
14L5-41.2	0.81	161	101	0.65	3.097	0.02	0.1362	0.001	1804	±13	2179	±18	20.8
14L5-42.1	0.02	164	153	0.96	2.187	0.02	0.1629	0.001	2427	±16	2486	±7	2.4
14L5-44.1	0.06	138	124	0.93	1.759	0.01	0.2174	0.001	2902	±19	2962	±6	2.1
14L5-46.1	0.05	90	117	1.35	2.046	0.02	0.1726	0.001	2566	±19	2583	±9	0.7
14L5-47.1	0.06	117	132	1.16	2.094	0.02	0.1683	0.001	2517	±17	2540	±8	0.9
14L5-48.1	0.04	200	248	1.29	2.115	0.02	0.1622	0.001	2496	±15	2478	±6	−0.7
14L5-49.1	0.54	592	395	0.69	3.815	0.02	0.1555	0.001	1501	±9	2407	±9	60.4
14L5-51.1	0.28	189	203	1.11	2.466	0.02	0.1653	0.001	2194	±14	2510	±8	14.4
14L5-52.1	0.09	60	88	1.53	2.021	0.02	0.1722	0.001	2591	±22	2579	±11	−0.5
14L5-53.1	0.06	76	83	1.14	2.024	0.02	0.1682	0.001	2588	±20	2540	±10	−1.9
14L5-54.1	0.21	387	356	0.95	2.859	0.02	0.1602	0.001	1933	±12	2457	±7	27.1
14L5-55.1	0.25	360	560	1.61	2.705	0.03	0.1585	0.001	2028	±20	2440	±7	20.3

(*Continued*)

TABLE 15.3 U–Pb Zircon Data of the Studied Samples Using SHRIMPIIe—cont'd

									Apparent Age (Ma)				
									^{204}Pb Corrected				
	$\%^{206}Pb_c$	U (ppm)	Th (ppm)	$^{232}Th/^{238}U$	$^{238}U/^{206}Pb^*$	±%	$^{207}Pb^*/^{206}Pb^*$	±%	$^{206}Pb^*/^{238}U$ Age		$^{207}Pb^*/^{206}Pb^*$ Age		Disc. %
14L5-57.1	0.07	44	42	0.98	2.115	0.02	0.1623	0.001	2496	±24	2480	±14	−0.6
14L5-58.1	0.04	98	74	0.78	2.032	0.02	0.1686	0.001	2580	±19	2543	±9	−1.4
14L5-59.1	0.51	932	976	1.08	4.631	0.03	0.1273	0.001	1260	±8	2062	±11	63.6
14L5-60.1	0.04	156	161	1.06	2.083	0.02	0.1684	0.001	2527	±16	2541	±7	0.6
14L5-61.1	0.21	600	495	0.85	3.727	0.05	0.1385	0.001	1532	±18	2208	±8	44.1
14L5-62.1	0.29	183	165	0.93	2.557	0.02	0.1659	0.001	2128	±14	2517	±9	18.3
14L5-63.1-2	<0.01	222	201	0.93	2.193	0.02	0.1630	0.001	2422	±14	2487	±6	2.7
14L5-64.1	<0.01	63	46	0.75	2.553	0.03	0.1400	0.001	2131	±18	2228	±15	4.5
14L5-65.1-2	0.08	140	153	1.13	2.179	0.02	0.1619	0.001	2435	±16	2476	±8	1.7
14L5-66.1-2	0.64	382	597	1.61	3.543	0.04	0.1618	0.001	1603	±17	2474	±13	54.4
14L5-67.1-2	0.81	560	943	1.74	4.196	0.03	0.1390	0.001	1378	±9	2215	±15	60.7
14L5-69.1	0.13	95	80	0.87	2.020	0.02	0.1725	0.001	2593	±23	2582	±12	−0.4
14L5-70.1	0.47	453	951	2.17	4.213	0.03	0.1412	0.001	1373	±9	2242	±12	63.3
14L5-71.1	0.40	392	321	0.85	2.537	0.02	0.1633	0.001	2142	±12	2490	±8	16.2
14L5-72.1	0.07	82	125	1.59	2.049	0.02	0.1724	0.001	2562	±20	2581	±10	0.7
14L5-73.1	0.09	72	31	0.45	2.103	0.02	0.1622	0.001	2507	±20	2479	±11	−1.1
14L5-74.1	0.26	395	262	0.69	3.172	0.02	0.1576	0.001	1767	±10	2430	±10	37.6
14L5-75.1	0.45	22	24	1.16	2.016	0.03	0.1717	0.002	2597	±32	2574	±24	−0.9
14L5-77.1	0.01	225	114	0.53	2.112	0.02	0.1620	0.001	2498	±17	2477	±6	−0.9
14L5-78.1	0.06	66	72	1.12	2.074	0.02	0.1683	0.001	2536	±25	2541	±10	0.2
14L5–79.1	0.06	113	97	0.89	2.133	0.02	0.1624	0.001	2478	±17	2481	±9	0.1
14L5-80.1	0.02	237	170	0.74	2.096	0.02	0.1689	0.001	2514	±16	2547	±5	1.3

14L5–81.1	0.27	299	540	1.87	2.917	0.03	0.1597	0.001	1900	±14	2452	±7	29.0
14L5-82.1	0.96	615	391	0.66	3.770	0.02	0.1519	0.001	1517	±9	2368	±14	56.1
14L5-84.1	0.34	649	552	0.88	4.521	0.03	0.1214	0.001	1288	±8	1977	±9	53.5
14L5–85.1	0.65	261	392	1.55	2.768	0.02	0.1650	0.001	1988	±12	2507	±11	26.1
14L5-86.1	0.02	87	49	0.59	1.961	0.02	0.1784	0.001	2656	±19	2638	±8	−0.7
14L5–87.1	0.04	121	64	0.55	2.045	0.02	0.1677	0.001	2566	±17	2535	±7	−1.2
14L5-89.1	0.08	126	101	0.82	1.975	0.02	0.1719	0.001	2641	±17	2576	±8	−2.4
14L5–90.1	0.77	49	49	1.03	2.007	0.02	0.1774	0.003	2606	±24	2628	±25	0.9
14L5-91.1	0.03	60	32	0.55	2.027	0.02	0.1678	0.001	2586	±21	2536	±11	−1.9
14L5-92.1	0.28	524	100	0.20	4.187	0.03	0.1414	0.001	1381	±8	2245	±8	62.6
14L5-93.1	0.24	40	45	1.14	2.037	0.02	0.1725	0.001	2575	±24	2582	±13	0.3
14L5-94.1	0.02	163	139	0.88	2.076	0.02	0.1684	0.001	2535	±16	2542	±6	0.3
14L5-95.1	0.07	281	256	0.94	2.218	0.02	0.1622	0.001	2399	±14	2479	±5	3.3
14L5–96.1	0.06	66	25	0.39	2.040	0.02	0.1723	0.001	2572	±21	2580	±10	0.3
14L5-98.1	0.17	270	305	1.17	2.492	0.03	0.1617	0.001	2175	±21	2474	±6	13.7
14L5A-01.1	<0.01	73	45	0.64	2.033	0.02	0.1678	0.001	2578	±20	2536	±9	−1.7
14L5A-02.1	0.09	293	171	0.60	2.455	0.02	0.1641	0.001	2203	±13	2498	±6	13.4
14L5A-03.1	0.084	150	142	0.983	2.066	0.02	0.1679	0.001	2545	±16	2537	±7	−0.3
14L5A-04.1	0.90	813	474	0.60	4.905	0.04	0.1166	0.002	1196	±9	1905	±29	59.3
14L5A-05.1	0.36	154	168	1.12	2.245	0.02	0.1649	0.001	2375	±17	2506	±12	5.5
14L5A-06.1	0.20	352	224	0.66	2.636	0.02	0.1662	0.001	2073	±12	2519	±6	21.5
14L5A-07.1	0.06	397	218	0.57	2.320	0.02	0.1645	0.001	2310	±14	2503	±6	8.3
14L5A-08.1	0.02	158	153	1.00	2.111	0.02	0.1626	0.001	2500	±16	2483	±7	−0.7
14L5A-09.1	0.13	73	107	1.51	2.122	0.02	0.1626	0.001	2489	±20	2483	±11	−0.2
14L5A-10.1	0.21	285	125	0.45	2.530	0.02	0.1658	0.001	2147	±15	2516	±7	17.2
14L5A-11.1	0.12	77	65	0.88	2.180	0.02	0.1623	0.001	2434	±20	2480	±11	1.9
14L5A-12.1	0.07	51	39	0.78	2.063	0.02	0.1687	0.001	2548	±23	2545	±12	−0.1
14L5A-13.1	0.06	285	292	1.06	2.464	0.02	0.1591	0.001	2196	±13	2446	±6	11.4

(Continued)

TABLE 15.3 U–Pb Zircon Data of the Studied Samples Using SHRIMPIIe—cont'd

									Apparent Age (Ma)				
									^{204}Pb Corrected				
									$^{206}Pb^*/^{238}U$ Age		$^{207}Pb^*/^{206}Pb^*$ Age		
	$\%^{206}Pb_c$	U (ppm)	Th (ppm)	$^{232}Th/^{238}U$	$^{238}U/^{206}Pb^*$	±%	$^{207}Pb^*/^{206}Pb^*$	±%					Disc. %
14L5A-15.1	0.20	372	141	0.39	3.067	0.02	0.1607	0.003	1819	±11	2463	±27	35.4
14L5A–16.1	0.01	139	90	0.67	2.071	0.02	0.1685	0.001	2540	±17	2543	±7	0.1
14L5A-17.1	0.42	529	464	0.90	3.305	0.02	0.1485	0.001	1704	±10	2328	±8	36.6
14L5A-24.1	0.02	164	181	1.14	2.057	0.02	0.1683	0.001	2554	±18	2541	±7	−0.5
14L5A-25.1	0.16	499	382	0.79	2.519	0.02	0.1564	0.001	2155	±12	2418	±6	12.2
14L5A-26.1	0.27	620	123	0.20	4.394	0.03	0.1381	0.001	1322	±8	2203	±8	66.7
14L5A-27.1	0.11	153	237	1.60	2.034	0.02	0.1677	0.001	2578	±17	2535	±8	−1.7
14L5A-28.1	0.03	233	161	0.71	2.212	0.02	0.1622	0.001	2405	±15	2479	±6	3.1
14L5A-30.1	0.10	23	22	1.01	2.048	0.03	0.1718	0.002	2563	±32	2575	±19	0.5
14L5A-31.1	0.15	453	383	0.87	2.511	0.02	0.1571	0.001	2161	±13	2424	±12	12.2
14L5A-32.1	0.03	206	188	0.94	2.162	0.02	0.1626	0.001	2451	±15	2483	±7	1.3
14L5A-35.1	0.09	326	216	0.68	2.349	0.02	0.1634	0.001	2287	±14	2491	±6	8.9
14L5A-37.1	0.19	51	136	2.77	2.126	0.02	0.1623	0.001	2485	±23	2480	±14	−0.2
14L5A-38.1	0.07	147	112	0.79	2.044	0.02	0.1724	0.001	2568	±17	2581	±8	0.5
14L5A-39.1	0.30	418	459	1.14	3.410	0.02	0.1524	0.001	1658	±10	2373	±8	43.2

Isotopic ratios are given with 1σ uncertainties, after the correction for common Pb using the ^{204}Pb correction method. Disc. (%) denotes the percentage of discordancy (e.g., Biao et al., 1996).

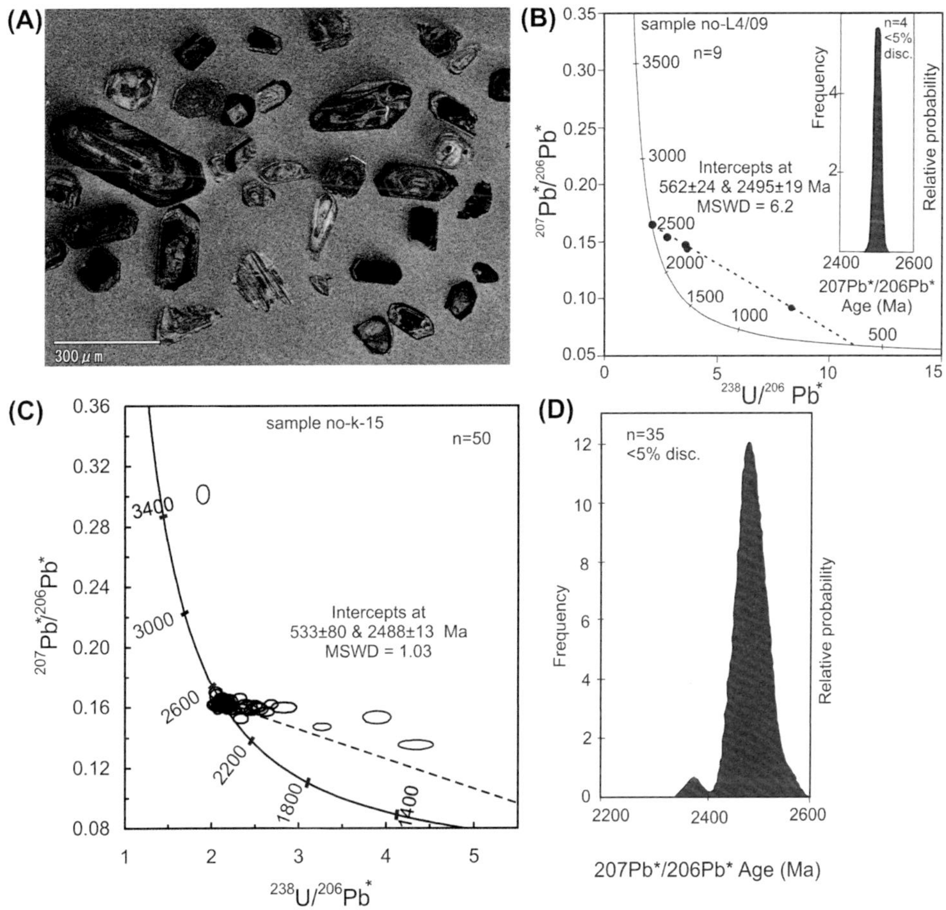

FIGURE 15.5 (A) SEM-CL photomicrographs showing the internal structures of the measured zircon grains of the Rehtikhol Formation (L4/09). (B) Tera–Wasserburg U–Pb concordia diagram and probability distribution diagram for zircon grains measured for sample L4/09. (C) Tera–Wasserburg U–Pb concordia diagram. (D) Probability distribution diagram for zircon grains measured for the sample K-15.

5.3.1.3 SARAIPALLI FORMATION (SAMPLE NO-4/L1)

Fine-grained sandstone of the heterolithic (sandstone-siltstone-shale alternation) Saraipalli Formation yields numerous zircon grains of euhedral to anhedral (rounded) geometry. Most of the grains have an oscillatory zoned interior and many are rich in inclusions. Some grains show a structureless exterior region (Fig. 15.6A). The SHRIMP II data of many of the analyzed grains from this sandstone reveal high values of common Pb. Hence, many data could not be used here. Fifteen of 29 analyses were <5% discordant and yield two strong probability peaks ($^{207}Pb/^{206}Pb$ age) at 2490 ± 5 Ma and 2774 ± 8 Ma. The youngest detrital group age calculated by Isoplot is $^{207}Pb/^{206}Pb$ age of c. 2480 Ma. The Tera-Wasserburg Concordia plot for all the analyzed data shows two apparent discordia lines, one with an upper intercept age of 2596 ± 63 Ma and lower intercept age of 826 ± 110 Ma with MSWD = 175;

TABLE 15.4 U-Pb Isotopic Data for Samples Measured in LA-ICP-MS

Spot Lebel	$^{207}Pb/^{206}Pb^*$	±2σ	$^{238}U/^{206}Pb^*$	±2σ	$^{206}Pb/^{238}U$ Age	±2σ (Ma)	$^{207}Pb/^{235}U$ Age	±2σ (Ma)	$^{207}Pb/^{206}Pb$ Age	±2σ (Ma)	Th/U	Discordance
0010Bh-10	0.165	±0.004	2.121	±0.084	2490	±83	2501	±45	2510	±46	0.819	0%
0013Bh-10	0.204	±0.005	3.493	±0.139	1623	±57	2239	±43	2862	±41	0.421	28%
0014Bh-10	0.101	±0.002	3.641	±0.145	1564	±55	1598	±39	1642	±49	0.600	2%
0015Bh-10	0.170	±0.004	2.157	±0.079	2455	±76	2510	±42	2554	±44	0.542	2%
0016Bh-10	0.170	±0.003	2.245	±0.080	2374	±71	2476	±40	2560	±39	0.674	4%
0017Bh-10	0.163	±0.004	2.271	±0.081	2352	±71	2425	±41	2486	±44	0.622	3%
0018Bh-10	0.154	±0.003	3.818	±0.137	1499	±48	1908	±37	2386	±41	0.552	21%
0019Bh-10	0.169	±0.004	2.133	±0.076	2478	±74	2519	±40	2552	±40	0.848	2%
0020Bh-10	0.168	±0.003	4.307	±0.142	1346	±40	1882	±34	2539	±35	0.241	28%
0023Bh-10	0.167	±0.003	2.103	±0.071	2507	±71	2517	±36	2525	±33	1.057	0%
0024Bh-10	0.150	±0.002	4.703	±0.126	1243	±30	1710	±26	2343	±31	1.539	27%
0025Bh-10	0.163	±0.004	2.196	±0.076	2419	±71	2455	±41	2484	±44	1.078	1%
0026Bh-10	0.150	±0.003	3.775	±0.132	1515	±47	1896	±35	2344	±38	0.894	20%
0028Bh-10	0.170	±0.004	2.228	±0.075	2390	±68	2484	±42	2561	±49	0.898	4%
0029Bh-10	0.166	±0.003	2.731	±0.087	2011	±55	2272	±36	2515	±39	1.100	11%
0033Bh-10	0.191	±0.004	2.454	±0.085	2203	±65	2502	±38	2754	±36	0.348	12%
0034Bh-10	0.165	±0.004	2.010	±0.072	2603	±77	2551	±42	2510	±46	0.623	−2%
0035Bh-10	0.124	±0.002	4.664	±0.125	1252	±30	1564	±26	2014	±34	0.102	20%
0036Bh-10	0.125	±0.002	5.091	±0.402	1156	±84	1502	±65	2031	±35	0.045	23%
0037Bh-10	0.096	±0.002	6.739	±0.208	892	±25	1104	±25	1551	±41	0.772	19%
0038Bh-10	0.161	±0.002	2.224	±0.064	2393	±58	2432	±31	2465	±30	0.674	2%
0039Bh-10	0.097	±0.001	5.415	±0.157	1092	±29	1262	±24	1565	±34	0.933	13%
0040Bh-10	0.149	±0.003	2.424	±0.075	2226	±58	2283	±36	2334	±40	0.380	2%
0043Bh-10	0.301	±0.006	1.396	±0.044	3482	±86	3479	±39	3477	±36	0.490	0%

0044Bh-10	0.171	±0.004	2.077	±0.074	2534	±75	2551	±41	2565	±42	0.338	1%
0045Bh-10	0.169	±0.004	2.158	±0.082	2454	±78	2508	±42	2552	±42	0.397	2%
0046Bh-10	0.161	±0.003	2.131	±0.068	2480	±66	2472	±36	2466	±39	1.011	0%
0047Bh-10	0.166	±0.003	2.132	±0.070	2479	±68	2498	±37	2514	±37	1.140	1%
0048Bh-10	0.173	±0.005	2.230	±0.080	2388	±72	2495	±43	2584	±49	1.421	4%
0049Bh-10	0.160	±0.003	2.431	±0.087	2220	±68	2346	±39	2457	±37	1.189	5%
0050Bh-10	0.167	±0.004	2.665	±0.093	2054	±61	2302	±39	2531	±40	1.061	11%
0053Bh-10	0.168	±0.005	2.056	±0.080	2554	±82	2547	±49	2541	±58	0.897	0%
0054Bh-10	0.163	±0.004	2.235	±0.078	2383	±70	2440	±41	2488	±44	1.006	2%
0055Bh-10	0.167	±0.004	2.219	±0.086	2398	±78	2467	±46	2525	±49	0.502	3%
0056Bh-10	0.161	±0.004	2.204	±0.088	2412	±81	2442	±45	2468	±46	0.716	1%
0057Bh-10	0.167	±0.004	2.122	±0.089	2488	±87	2510	±47	2528	±46	0.548	1%
0058Bh-10	0.169	±0.005	2.278	±0.097	2346	±85	2456	±51	2548	±54	0.387	4%
0059Bh-10	0.154	±0.003	2.698	±0.113	2032	±73	2213	±45	2386	±43	0.468	8%
0005Bh-10	0.096	±0.002	4.166	±0.162	1387	±48	1452	±37	1549	±51	0.622	4%
0060Bh-10	0.164	±0.006	2.144	±0.096	2467	±92	2482	±56	2495	±65	0.718	1%
0061Bh-10	0.169	±0.004	2.075	±0.089	2535	±90	2540	±48	2545	±47	0.494	0%
0062Bh-10	0.189	±0.004	1.888	±0.081	2740	±96	2737	±48	2735	±43	0.630	0%
0063Bh-10	0.168	±0.004	2.093	±0.087	2517	±88	2530	±46	2541	±44	0.475	1%
0064Bh-10	0.160	±0.005	2.145	±0.094	2466	±90	2460	±51	2455	±55	1.311	0%
0006Bh-10	0.155	±0.004	2.682	±0.126	2042	±82	2227	±50	2401	±50	0.395	8%
0007Bh-10	0.172	±0.005	2.047	±0.098	2564	±102	2573	±54	2579	±51	1.231	0%
0008Bh-10	0.163	±0.004	3.290	±0.161	1711	±74	2092	±51	2491	±51	0.727	18%
0009Bh-10	0.164	±0.005	2.020	±0.105	2592	±112	2537	±58	2494	±54	0.910	−2%
005K-15	0.170	±0.002	2.049	±0.059	2563	±61	2560	±32	2557	±28	0.879	0%
006K-15	0.157	±0.002	2.631	±0.071	2077	±48	2257	±28	2426	±27	1.262	9%

(Continued)

TABLE 15.4 U-Pb Isotopic Data for Samples Measured in LA-ICP-MS—cont'd

Spot Lebel	$^{207}Pb/^{206}Pb^*$	$\pm 2\sigma$	$^{238}U/^{206}Pb^*$	$\pm 2\sigma$	$^{206}Pb/^{238}U$ Age	$\pm 2\sigma$ (Ma)	$^{207}Pb/^{235}U$ Age	$\pm 2\sigma$ (Ma)	$^{207}Pb/^{206}Pb$ Age	$\pm 2\sigma$ (Ma)	Th/U	Discordance
007K-15	0.160	±0.002	2.829	±0.118	1951	±71	2207	±42	2454	±30	1.419	13%
008K-15	0.161	±0.003	2.175	±0.063	2438	±59	2451	±35	2462	±41	0.897	1%
009K-15	0.160	±0.002	2.512	±0.080	2160	±59	2316	±33	2457	±29	1.032	7%
010K-15	0.147	±0.001	3.270	±0.085	1720	±39	2004	±25	2312	±22	0.548	17%
011K-15	0.161	±0.002	2.490	±0.074	2176	±55	2332	±31	2471	±27	0.771	7%
012K-15	0.160	±0.002	2.337	±0.067	2297	±56	2379	±29	2451	±23	0.537	4%
013K-15	0.159	±0.002	2.222	±0.059	2396	±54	2421	±29	2443	±29	0.578	1%
014K-15	0.162	±0.003	2.142	±0.064	2470	±61	2471	±34	2472	±35	0.873	0%
018K-15	0.164	±0.003	2.127	±0.055	2485	±53	2493	±32	2501	±37	0.495	0%
019K-15	0.162	±0.003	2.217	±0.053	2400	±48	2440	±30	2474	±35	0.797	2%
020K-15	0.159	±0.003	2.549	±0.079	2134	±56	2296	±33	2443	±32	1.343	8%
021K-15	0.162	±0.002	2.195	±0.061	2420	±56	2448	±31	2472	±29	0.736	1%
022K-15	0.135	±0.002	4.340	±0.164	1337	±46	1691	±35	2164	±33	0.455	27%
023K-15	0.163	±0.003	2.151	±0.062	2461	±59	2474	±34	2484	±39	0.866	1%
024K-15	0.166	±0.002	2.325	±0.062	2306	±52	2418	±29	2514	±30	0.942	5%
025K-15	0.165	±0.003	2.198	±0.050	2418	±46	2464	±28	2503	±34	0.860	2%
026K-15	0.161	±0.002	2.164	±0.056	2449	±53	2459	±29	2467	±27	0.871	0%
027K-15	0.166	±0.002	2.125	±0.053	2487	±51	2505	±28	2520	±27	0.764	1%
031K-15	0.153	±0.002	2.336	±0.070	2297	±58	2338	±31	2374	±29	0.565	2%
032K-15	0.164	±0.003	2.225	±0.060	2393	±54	2449	±32	2495	±37	0.680	2%
033K-15	0.164	±0.002	2.123	±0.046	2488	±45	2491	±25	2493	±27	0.677	0%
034K-15	0.161	±0.003	2.110	±0.059	2501	±58	2483	±33	2469	±37	0.800	−1%
035K-15	0.165	±0.002	2.146	±0.053	2466	±51	2488	±28	2506	±28	0.852	1%
036K-15	0.164	±0.003	2.163	±0.054	2450	±51	2478	±30	2501	±34	1.338	1%

037K-15	0.163	±0.003	2.069	±0.055	2542	±57	2512	±31	2488	±32	1.138	−1%
038K-15	0.162	±0.002	2.388	±0.069	2255	±55	2374	±30	2478	±25	0.242	5%
039K-15	0.301	±0.005	1.898	±0.058	2728	±69	3179	±34	3478	±26	1.007	17%
040K-15	0.153	±0.003	3.894	±0.128	1473	±43	1889	±34	2382	±39	1.143	28%
044K-15	0.162	±0.002	2.041	±0.055	2571	±57	2518	±30	2476	±30	0.885	−2%
045K-15	0.162	±0.004	2.127	±0.065	2484	±64	2479	±37	2474	±44	1.083	0%
046K-15	0.162	±0.002	2.686	±0.061	2040	±40	2266	±24	2477	±25	1.169	11%
047K-15	0.165	±0.002	2.144	±0.053	2468	±51	2487	±29	2503	±30	0.945	1%
048K-15	0.158	±0.002	2.454	±0.063	2204	±48	2326	±28	2435	±29	0.893	6%
049K-15	0.165	±0.003	2.135	±0.055	2476	±53	2496	±31	2512	±35	0.480	1%
050K-15	0.162	±0.002	2.018	±0.054	2594	±57	2527	±29	2474	±27	0.495	−3%
051K-15	0.161	±0.004	2.075	±0.051	2536	±52	2495	±33	2462	±42	1.415	−2%
052K-15	0.161	±0.002	2.059	±0.053	2552	±55	2502	±29	2462	±29	0.705	−2%
053K-15	0.160	±0.002	2.546	±0.068	2136	±49	2304	±28	2457	±25	1.115	8%
057K-15	0.160	±0.003	2.419	±0.084	2230	±66	2352	±38	2459	±36	1.353	5%
058K-15	0.162	±0.003	2.115	±0.052	2496	±52	2488	±31	2481	±35	0.996	0%
059K-15	0.161	±0.003	2.214	±0.062	2402	±56	2438	±32	2469	±35	0.453	1%
060K-15	0.158	±0.002	2.250	±0.060	2370	±53	2406	±28	2436	±24	0.914	2%
061K-15	0.161	±0.002	2.181	±0.067	2433	±63	2452	±32	2468	±29	0.814	1%
062K-15	0.167	±0.003	2.053	±0.057	2558	±59	2544	±34	2532	±37	0.558	−1%
063K-15	0.163	±0.002	2.155	±0.051	2457	±49	2471	±27	2483	±27	0.678	1%
064K-15	0.162	±0.002	2.113	±0.061	2498	±60	2488	±32	2480	±30	0.360	0%
065K-15	0.162	±0.002	2.106	±0.052	2505	±52	2488	±29	2475	±30	0.677	−1%
066K-15	0.164	±0.004	2.100	±0.062	2511	±62	2501	±39	2493	±49	1.286	0%

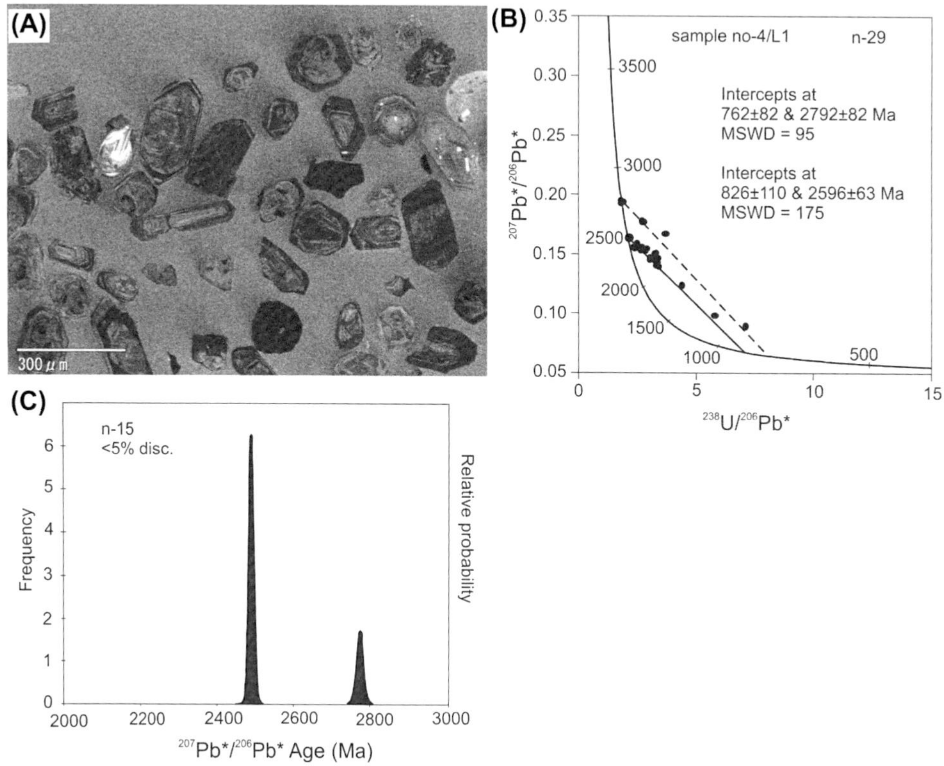

FIGURE 15.6 (A) SEM-CL photomicrographs showing the internal structures of the measured zircon grains of Saraipalli Formation (4/L1). (B) Tera–Wasserburg U–Pb concordia diagram of zircon grains. (C) Probability distribution diagram for concordant zircon $^{207}Pb^{*}/^{206}Pb^{*}$ ages.

the other with an upper intercept age of 2792 ± 82 Ma and lower intercept age of 762 ± 82 Ma with MSWD = 95 (Fig. 15.6C). The probability peak data plots near the upper intercepts of these two concordant age pools (Table 15.3, Fig. 15.6C). Th/U ratios have a wide range (0.20–1.65) within grains with concordant ages, as well as in grains yielding discordant ages (0.05–1.22).

5.3.1.4 BHALUKONA FORMATION (SAMPLE NO-L18/09, BH-10)

Higher up in stratigraphy of the Singhora Group, the sandstone-dominated Bhalukona lithopackage has been sampled at two places for zircon geochronology, one in its lower part and the other from the upper part of the lithopackage (Chakraborty et al., 2012a). The sample from the lower Bhalukona (L18/09) contains mostly rounded zircon grains with occasional irregularly broken grains. The rounded grains are of uniform size and 70–100 μm in diameter, whereas broken grains are relatively small in size (Fig. 15.7A). Unlike sandstone samples from underlying formations, zircon grains from both the Bhalukona samples contain less number of inclusions. Many of the grains exhibit oscillatory and patchy CL interiors. The data analyzed from 176 grains yield both concordant (145 grains <5%

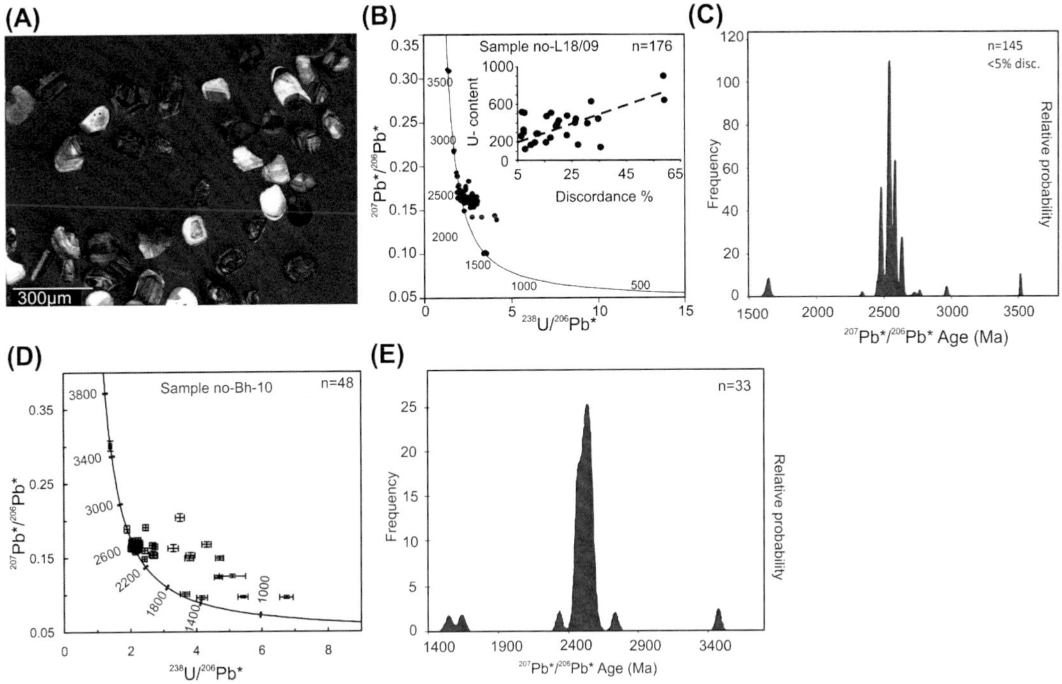

FIGURE 15.7 (A) SEM-CL photomicrographs showing the internal structures of the measured zircon grains from the lower part of Bhalukona Formation (L18/09). (B) Tera–Wasserburg U–Pb concordia diagram of zircon grains measured. The figure in the inset represents discordance versus U-content. (C) Probability distribution diagram for concordant ($-5 < \text{Disc}\ (\%) < 5$) zircon $^{207}Pb^*/^{206}Pb^*$ ages. (D) Tera–Wasserburg U–Pb concordia diagram of zircon grains and age distribution plot of zircon grains of sample Bh-10 (upper part of Bhalukona Formation). $^{207}Pb^*$ and $^{206}Pb^*$ indicate radiometric ^{207}Pb and ^{206}Pb, respectively. Common Pb is corrected using measured ^{204}Pb. Solid curve indicates the concordia curve. (E) Probability distribution diagram for concordant ($-5 < \text{Disc}\ (\%) < 5$) zircon $^{207}Pb^*/^{206}Pb^*$ ages.

discordance) and discordant dates (31 grains) (Fig. 15.7B). The concordant $^{207}Pb/^{206}Pb$ age data (<5% discordance) yield several probability peaks in between 3511 ± 3 Ma and 1639 ± 18 Ma, with strong peaks at 2633 ± 11 Ma, 2582 ± 11 Ma, 2540 ± 12 Ma (being the strongest), and 2478 ± 13 Ma (Table 15.3, Fig. 15.7C). On the Concordia diagram, the discordia line could not be plotted due to the presence of several data groups at the upper intercept region. For the youngest zircon group ($^{207}Pb/^{206}Pb$) age calculated by Isoplot is 1619 (+18/−35) Ma. The values of Th/U ratio are quite variable for spots yielding both concordant and discordant ages (0.09–2.80).

The sample from the upper part of the Bhalukona Formation (Bh-10) yielded hundreds of zircon grains from a small chip (1-inch cube), which were measured after grain mount using the LA-ICPMS technique. A total of 48 grains were analyzed and out of these 33 spots yielded near-concordant (<5% discordance) age values (Table 15.4, Fig. 15.7D). The concordant age data (<5% discordance) reveal a strong probability density peak at 2509 ± 8 Ma, and smaller peaks at 3477 ± 36 Ma, 2735 ± 43 Ma, and 1598 ± 36 Ma (all are 2σ values) (Fig. 15.7E). From the second-youngest zircon group the $^{207}Pb/^{206}Pb$ age calculated by Isoplot is

1643 (+54/−56) Ma at 95% confidence. As there are two adjacent and small peaks in the younger range of the whole profile, we consider the second-youngest one as the upper limit of the age of sedimentation. Also, the spot age near this value, 1642 ± 49 Ma, has lower discordance (<2%).

5.3.2 Sandstones From the Chandarpur Group, Chhattisgarh Main Basin, and the Khariar Group, Khariar Basin

5.3.2.1 CHANDARPUR SANDSTONE (SAMPLE NO-L3/09)

For interbasin comparison and correlation, we carried out zircon age analysis from sandstones in the basal part of the Chandarpur Group of rocks, overlying the Singhora Group. This sandstone contains highly euhedral zircon grains with almost no roundness in their corners. These grains are large (200–300 μm, longer direction), while slightly smaller grains have rounded corners (Fig. 15.8A). A total of 31 grains were analyzed, out of which 25 grains yielded near-concordant data (<5% discordance). The concordia diagram shows that except

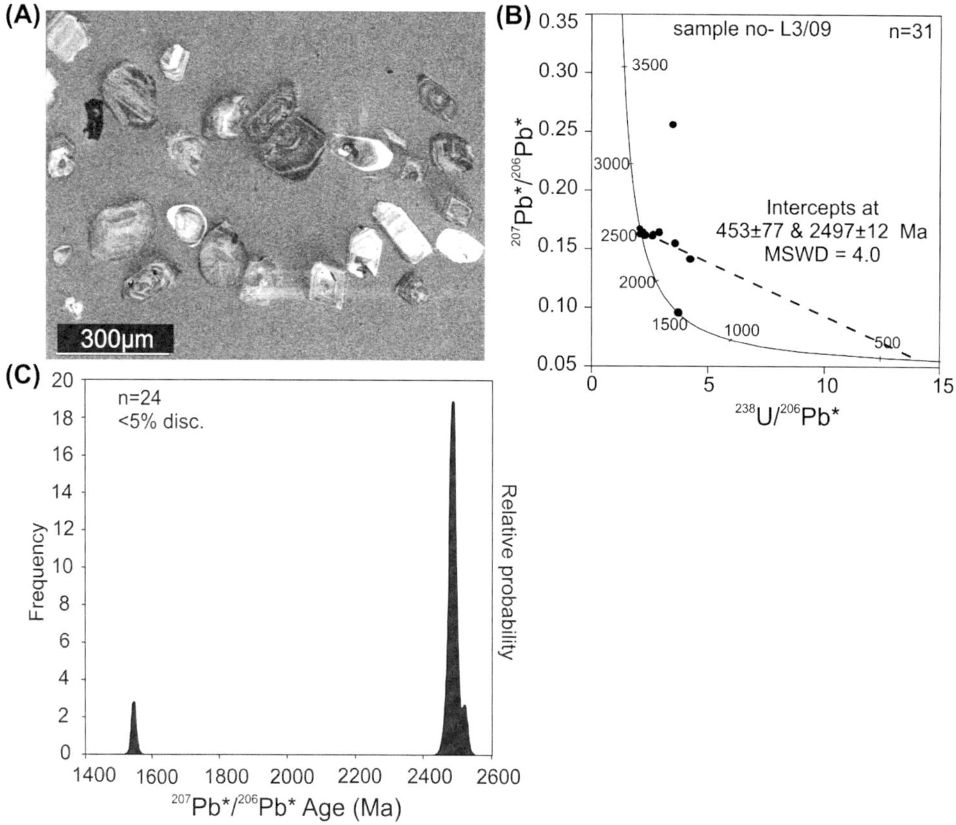

FIGURE 15.8 (A) SEM-CL photomicrographs showing the internal structures of the measured zircon grains from the lower part of Chandarpur Formation (L3/09). (B) Tera–Wasserburg U–Pb concordia diagram of zircon grains measured. (C) Probability distribution diagram for concordant (−5 < Disc (%) < 5) zircon $^{207}Pb^*/^{206}Pb^*$ ages.

for three grains (one having 98% discordance and two other grains having nearly concordant age of c. 1545 Ma), all other analyzed grains plot on a discordia that has its upper and lower intercepts at 2497 ± 12 Ma and at 453 ± 77 Ma (MSWD = 4), respectively (Table 15.3, Fig. 15.8B). The concordant age data (with <5% of discordance) are plotted for their $^{207}Pb/^{206}Pb$ age probability distributions, which yielded two peaks, the strongest one being at 2486 ± 9 Ma and a relatively small peak at 1544 ± 8 Ma (n = 2) (Fig. 15.8C). The youngest zircon group age ($^{207}Pb/^{206}Pb$) calculated by Isoplot is 1541 (+11/−15) Ma.

5.3.2.2 KHARIAR SANDSTONE (BASAL SANDSTONE OF KHARIAR GROUP, SAMPLE NO-14/L5)

Sandstone from the lowermost part of Khariar clastic succession was also analyzed. Zircon grains are of different size and shape; separated grains are rounded and broken (Fig. 15.9A). Though rare, euhedral grains are also present. Grain sizes range from a few tens of μm to 400 μm. A total of 109 grains were analyzed, out of which 57 grains yielded

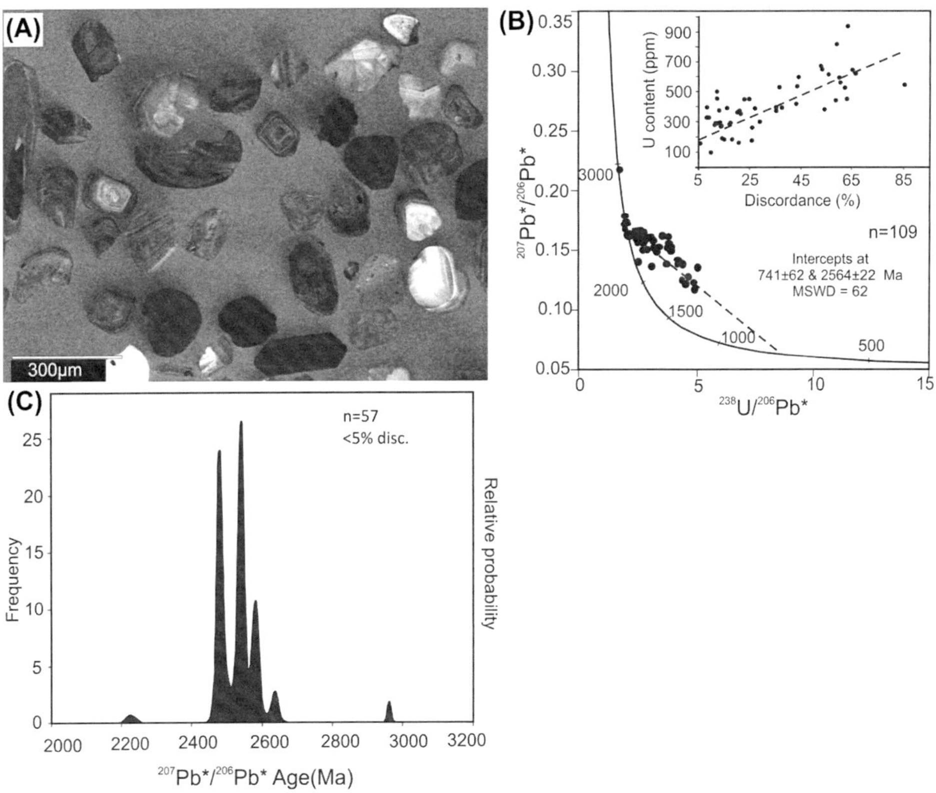

FIGURE 15.9 (A) SEM-CL photomicrographs showing the internal structures of the measured zircon grains from the lower part of Khariar Basin (14/L5). (B) Tera–Wasserburg U–Pb concordia diagram of zircon grains measured. The figure in the inset shows the variation of discordance % with U-content. (C) Probability distribution diagram for concordant (−5 < Disc (%) < 5) zircon $^{207}Pb^*/^{206}Pb^*$ ages.

near-concordant (<5% discordance) age values (Table 15.3, Fig. 15.9B). The Concordia diagram yields discordia line with upper intercept age at 2564 ± 22 Ma and lower intercept age at 741 ± 62 Ma (MSWD = 62) (Fig. 15.9B). The concordant age data (<5% discordance) show major probability peaks of $^{207}Pb/^{206}Pb$ age at 2636 ± 14 Ma, 2581 ± 13 Ma, 2541 ± 11 Ma, and 2481 ± 10 Ma, with minor peaks at c. 2962 and c. 2228 Ma (Fig. 15.9C). Although the 2228 Ma age is obtained from only one analysis of a single grain, it is the youngest concordant age estimated. Irrespective of type of data points, values of Th/U ratio vary widely between 0.20 and 2.77.

6. DISCORDANCE IN U-PB ISOTOPIC VALUES

For all the analyzed samples, irrespective of their stratigraphic positions, we observed variable discordance in the U–Pb isotopic system. Hence, we separately calculated discordia for each of the previous samples. Variable discordance and absence of data in the lower half of these discordia resulted in a large error of lower intercept values. Hence, wide variations in absolute values of lower intercepts were observed. On the other hand, it is also observed that U-contents in zircon grains are widely varying in many of the samples. A plot of U-content against level of discordance in samples shows a moderate-to-strong positive correlation (insets of Figs. 15.7B and 15.9B). Radiation damage of high-U zircon leading to a higher degree of discordance is not uncommon in Precambrian sedimentary environments (Nelson, 2001). Since many measured grains have a high ^{204}Pb problem, we deliberately discarded those data. Moreover, the studied rocks do not show any effect of postdepositional thermal events. Hence, we consider that the observed discordance is not due to any later thermal event of any specific age, but is a secondary artifact.

7. DISCUSSION

7.1 Provenance and Its Change Through Time

Detrital zircon U-Pb data from the entire sedimentary package of the Singhora Basin are presented in a comparative $^{207}Pb/^{206}Pb$ concordant age probability density diagram (Fig. 15.10). The conglomeratic and coarse granular alluvial fan and braid plain system of Rehtikhol Formation is likely to have been sourced from nearby East Indian cratonic rocks. This is supported by the U-Pb zircon data from two samples, which indicate that the provenance remained uniform and appears to be a single detrital age population; resolved statistically into two to three small peaks between c. 2375 and 2515 Ma. The nearby basement rocks yield comparable zircon age spectra. A similar provenance character continued in the Singhora basin through the sedimentation of the heterolithic Saraipalli Formation with the siltstone sample (4/L1) shows two sharp probability peaks at 2490 ± 8 Ma and 2774 ± 10 Ma. However, from the lower part of overlying Bhalukona Formation a greater variation in sources is observed with appearance of several probability peaks dominantly ranging between 2478 ± 13 Ma and 2633 ± 11 Ma (SHRIMP data) in the lower part; and between c. 2317 and c. 2718 Ma in the upper part (LA-ICP-MS data). Besides, peaks of

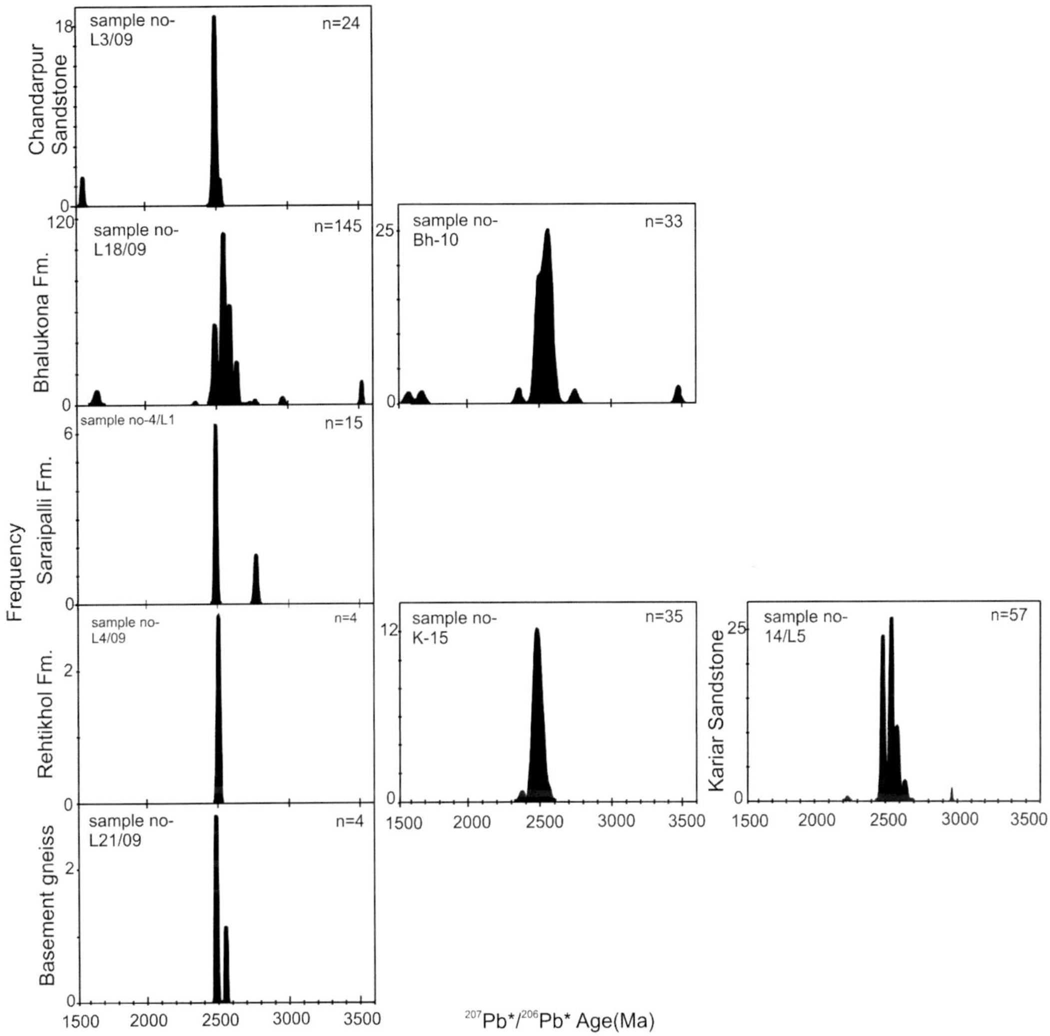

FIGURE 15.10 Probability distribution diagrams of concordant age data ($-5 < \text{Disc}\ (\%) < 5$) for all samples measured in the present study. The plots are arranged according to their stratigraphic positions. Disc (%) means discordancy between ^{238}U-^{206}Pb and ^{207}Pb-^{206}Pb ages (Biao et al., 1996). So, number of data shown is less than in the corresponding concordia plots.

both older and younger components also appeared for the first time in these rocks. The lower part of Bhalukona Formation recorded peaks at 3511 ± 3 Ma and 1639 ± 18 Ma (SHRIMP data), whereas relatively older and younger ages (3477 ± 18 Ma and 1598 ± 17 Ma; LA-ICP-MS data) are recorded from the sample of the upper part. This is contradictory to the idea of Bickford et al. (2011a) that irrespective of stratigraphic level the c. 2500 Ma date remains unimodal in Chhattisgarh sandstones, except for the Sarandih Sandstone present high in the stratigraphic column. Incidences of new ages (both older and younger) within the Bhalukona sandstone indicate the beginning of supply from new provenances in the history of Singhora sedimentation from both a Paleoarchean and from a young Mesoproterozoic provenance. The TTG gneiss of the North Bastar Amagaon Group has reported ages of 3580 ± 14 and 3562 ± 2 Ma (zircon U-Pb age, Ghosh, 2004) and could have supplied the older detrital population. Potential sources for the younger population can be found at the western, northern, and southern peripheries of the Bastar Craton, where there are evidences of tectono–magmatic activities at ~1600 Ma, for example, the Central Indian Tectonic Zone (Bhowmik et al., 2011; Bhandari et al., 2011), its eastern–northeastern extension in the Chhotonagpur Gneissic Complex-North Singhbhum Mobile Belt-Shillong-Meghalaya Gneissic Complex (Maji et al., 2008; Chatterjee et al., 2007; Mahato et al., 2008) at Bhopalpatnam Granulite (Santosh et al., 2004), the southern part of the Eastern Ghats Granulite Belt (Bose et al., 2011; Vijaya Kumar et al., 2010, 2011), and felsic magmatism in the Krishna Province of the Eastern Ghats Belt (Dobmeier et al., 2006). The varied provenance character continued up to the lower part of the Chandarpur Group that overlies the Singhora Group and exposed at the Chhattisgarh Main basin. However, the Paleoarchean contribution is not recorded within the Chandarpur sediment. This observation makes the Chandarpur sandstones distinctive from all stratigraphic intervals of the Singhora Group (including the Rehtikhol Formation with which Bickford and coworkers drew correlation of the Chandarpur Formation) and further, allows consideration of the Singhora Group as a separate lithostratigraphic entity, older than the Chandarpur Group. Also, this understanding reaffirms the four-tier Chhattisgarh stratigraphy—Singhora, Chandarpur, Raipur, and Kharsiya (cf. Chakraborty et al., 2015)—which is debated in some studies considering disputable stratigraphic status for the Singhora Group (Bickford et al., 2011a,b; Patranabis-Deb and Chaudhuri, 2007, 2010; Basu et al., 2013; Basu and Bickford, 2015; Dhang and Patranabis-Deb, 2012).

The stratigraphic variation in Nd-isotope signature also shows a similar variation in corroboration with inference made about the provenance variation through zircon geochronology. The ε^tNd values are comparable for Rehtikhol sandstone and basement gneiss, which together with the similar detrital zircon peak profiles strongly indicate a nearby provenance in the Bastar Craton early in the history of Singhora sedimentation. Sediments of the Saraipalli Formation and lowermost part of the Bhalukona Formation (close to the Saraipalli-Bhalukona boundary) must have been derived from rocks that had less negative ε^tNd values (−0.2 to −6.6). In contrast, major parts of the Bhalukona Formation (i.e., its middle and top stratigraphic levels) were derived from different sources that had more negative ε^tNd values (−7.0 to −11.5). It is important to note that the sediments of both the Saraipalli and the Bhalukona Formations have nearly uniform Sm/Nd ratio (i.e., ~0.1) (McLennan, 2001; Rudnick and Gao, 2003), similar to the average ratio for the continental crust. This ratio shows no correlation with ε^tNd values, indicating that their provenance also had similar Sm/Nd ratios. The enrichment factor, $f_{Sm/Nd}$, is the deviation of $^{147}Sm/^{144}Nd$ in the sample from the chondritic uniform reservoir value, which is similar for these two formations in the range of −0.364 to −0.494 (Fig. 15.3C). Hence, the differences in ε^tNd values can be explained if the provenances are made up of similar rocks but with different ages of juvenile crust formation.

Depleted mantle Nd model ages of a sedimentary rock (T_{DM}) reflect the time of extraction of its source rock from a depleted mantle source. Clastic sediments could have been derived from a uniform source or more likely from multiple sources. In the latter case, Nd model ages would represent a weighted average of the model ages of the different sources. The T_{DM} ages vary between 1900 and 2500 Ma for the Saraipalli Formation, whereas the Bhalukona Formation shows higher ages in the range between 2500 and 2800 Ma except for the sample near the Formation boundary (Table 15.2, Fig. 15.3B). This increase in the T_{DM} ages could result from a shift in sediment provenance to significantly older rocks including input from Mesoarchean granitoid rocks as evidenced from the zircon age spectrum in the lower and upper parts of Bhalukona Formation. Thus, from the calculated ε^tNd values, we conclude that the Saraipalli Formation received more contributions from juvenile continental sources (unfortunately not expressed in the detrital zircon age spectra) than local Bastar Craton basement rocks, and that changed upward in stratigraphy in the Bhalukona Formation with more contribution from older upper continental crustal material. That the contribution from younger aged source rocks was nominal during most parts of the Bhalukona sedimentation as expressed in T_{DM} age is supported by the low intensity of ~1600 Ma probability peak in the detrital zircon age profile. However, within the Saraipalli and Bhalukona Formations variations in the ε^tNd values to the extent of 3 and 2 units can be noticed (Fig. 15.3A; Table 15.2). This could be a result of variations in relative contributions of sediments derived from rocks of slightly different crustal residence ages in the provenance. Hence, the Nd-isotopic composition and the detrital zircon age profiles together show a definite change in source rock from the base of the Bhalukona Formation (and possibly from the Sarapalli Formation according to Nd isotope values). Also, these sudden changes in geochronological and geochemical data has allowed us to put a strong hold on the earlier claim of tectonically forced regression in the Singhora basin during the Bhalukona time (Chakraborty et al., 2012a) on the basis of detailed process-based sedimentological study.

7.2 Timing of Singhora Sedimentation

An attempt has been made to constrain the age of Singhora sedimentation from the age of youngest zircon grain as calculated from the concordant age data. As zircons of comparatively younger age populations are retrieved from the samples of Bhalukona Formation, we relied on the two Bhalukona samples (L18/09, Bh-10) for drawing inference on timing of sedimentation. The youngest zircon group age from the lower part of the Bhalukona sandstone has a $^{207}Pb/^{206}Pb$ age of 1619 (+18/−35) Ma (L18/09; SHRIMP data); while the sample from upper stratigraphic level (Bh-10) yielded the youngest zircon group age as 1643 (+54/−56, 2σ, 95% confidence, 2nd youngest) Ma (LA-ICP-MS data). Hence, we consider that the overall age of sedimentation in the Singhora Basin was <1600 Ma. It is noteworthy that the crystallization age of tuffaceous units in the basal parts of stratigraphic successions in Singhora and Khariar basins are calculated as ~1500 Ma (Das et al., 2009; 1405 ± 9 Ma; Bickford et al., 2011a) and 1455 ± 47 Ma (Das et al., 2009), respectively, which constrains the age of initiation of these basins (Das et al., 2009). Moreover, a diabase intrusive in the Saraipalli Formation has yielded a mineral-whole rock Sm-Nd isochron age of 1421 ± 23 Ma (2σ, MSWD = 1.6; Das et al., 2011). The convergence of all these data strongly support the Mesoproterozoic (<1600 Ma) age of sedimentation in the basin inferred from the detrital zircon age populations observed in the present study.

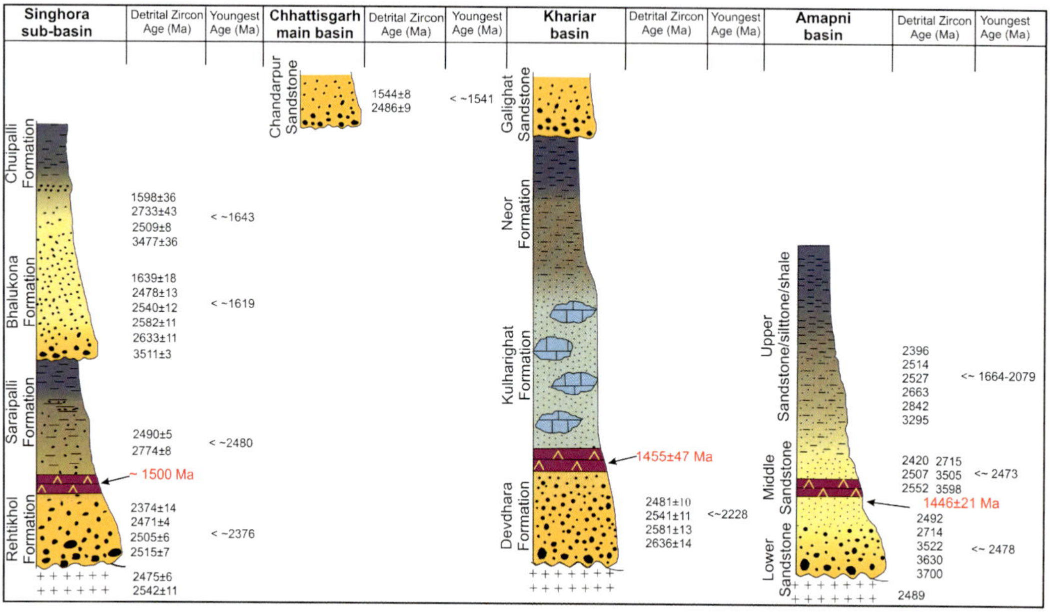

FIGURE 15.11 Interbasinal correlation between Singhora Basin, Khariar Basin, and Ampani Basin on the basis of detrital zircon (near-concordant) ages. The tentative stratigraphic log of the Khariar Basin is after Chakraborty et al. (2012b). That from the Ampani Basin is modified after Das et al. (2015) based on recent structural data from the basin. Detrital zircon data of Ampani Basin is after Saha et al. (2016).

7.3 Regional Correlation

Detrital zircon geochronology from three adjoining but aerially detached basins (Singhora, Ampani, and Khariar; Fig. 15.11) have allowed us to make some comparison between sedimentation history of these basins. While zircon geochronology from the Singhora (present paper) and the Ampani (Saha et al., 2016) basins were undertaken encompassing entire stratigraphic successions of these basins, in the Khariar Basin we could only generate data from sandstones at its basal part. The obtained data (Fig. 15.11) allow a very good correlation of the sedimentation histories of these basins including the character of their provenance(s). In all the three basins, zircon grains from their respective basal sandstone formations have yielded Paleo- to Late Archean ages; while in the Singhora Basin the oldest and youngest peaks recorded are at 2515 and 2376 Ma, the age spread in the Ampani Basin is between c. 3700 and 2500 Ma, and in the Khariar basin between 2636 ± 14 and 2541 ± 11 Ma (with a small peak at c. 2228 Ma), respectively. Further, it is also noted that from the middle stratigraphic levels of both the Singhora (i.e., the Bhalukona Formation) and Ampani Basin (middle shoreface part; Saha et al., 2016) successions a new Mesoproterozoic source (c. 1700 to 1600 Ma) was added with continuation of Paleoproterozoic sources. Addition of a new provenance in the regional scale allows us to suggest a basin-adjacent tectonic activity that would have affected sedimentation in these basins. The youngest zircon populations recorded in the

Singhora and Ampani basins are of 1619 (+18/−35) Ma and 1764 ± 76 Ma age, respectively. This bears a strong indication of coeval Mesoproterozoic sedimentation history in these basins either as a continuous master basin or as contemporary aerially detached basins.

8. CONCLUSION

1. The Mesoproterozoic Singhora basin has been studied for its detrital zircon age populations. For comparison, zircon grains from the basement gneiss, a sandstone of the lower Chandarpur Group and a sandstone from the lower part of the Khariar Basin were also measured using SHRIMP II and LA-ICP-MS.
2. The data from all the studied units of the Singhora basin yield a youngest zircon group age population between c. 1619 and c. 1643 Ma, implying the age of sedimentation is $<$1600 Ma.
3. The lowest stratigraphic unit (i.e., Rehtikhol Formation) yields a strong detrital zircon age peak at ~2500 Ma, showing nearly a single source the same age as the local basement gneisses.
4. A similar provenance continued into the sandstone of the Saraipalli Formation with the additional contribution of an older component at 2774 ± 8 Ma.
5. However, sudden changes appear from the overlying Bhalukona Formation with the addition of oldest and youngest zircon populations, 3511 ± 3 Ma and 1619 (+18/−35) Ma, respectively.
6. This younger age peak continues upward to the lower Chandarpur Group rocks, which distinguish the Chandarpur Group from all stratigraphic intervals belonging to the Singhora Group and fixes the position of the Chandarpur Group above the Singhora Group in the four-tier stratigraphic classification of the Chhattisgarh Supergroup.
7. A similar change (as mentioned earlier from the zircon age populations) in Nd isotopic values is recorded with a sudden inflection from the Bhalukona Formation. This corroborates well with our earlier findings of abrupt paleoenvironmental shift triggered by tectonically guided forced regression at the same interval. Contrasting results appear in Saraipalli Formation where Nd isotopic ratios indicate an addition of younger source but was not reflected from the detrital zircon dataset.
8. Interestingly, almost similar change in detrital zircon age spectra from the middle part of the Ampani Basin further south in the region helps us to carry out an interbasinal correlation.

Acknowledgments

We thank Rajat Mazumder for inviting us to contribute in this volume. Part of this work was done using the financial grant of Department of Science and Technology, India. We thank Ashish Sarkar and Shruti Ranjan Mishra for support during fieldwork, sampling, and Priyabrata Das for generating TIMS data. Yasutaka Hayasaka and Yoshio Takahashi helped us by offering the LA-ICP-MS facility. PPC acknowledges the financial help from University of Delhi in form of extending an R&D project in 2014–15. SS was supported by DST, India in the form of INSPIRE Fellowship during part of this work. I.C.W. Fitzsimons and Sankar Bose have critically assessed the manuscript. Their comments and suggestions (including text modification by I.C.W. Fitzsimons) really helped in improving the quality of this manuscript.

References

Anand, R., Balakrishnan, S., 2010. Pb, Sr and Nd isotope systematics of metavolcanic rocks of the Hutti greenstone belt, Eastern Dharwar craton: constraints on age, duration of volcanism and evolution of mantle sources during Late Archean. Journal of Asian Earth Science 39, 1–11.

Armendáriz, M., López-Guijarro, R., Quesada, C., Pin, C., Bellido, F., 2008. Genesis and evolution of a syn-orogenic basin in transpression: insights from petrography, geochemistry and Sm–Nd systematics in the Variscan Pedroches basin (Mississippian, SW Iberia). Tectonophysics 461, 395–413.

Basu, A., Schieber, J., Patranabis-Deb, S., Dhang, P.C., 2013. Recycled detrital quartz grains are sedimentary rock fragments indicating unconformities: examples from the Chhattisgarh Supergroup, Bastar Craton, India. Journal of Sedimentary Research 83, 368–376.

Basu, A., Bickford, M.E., 2015. An alternate perspective on the opening and closing of the intracratonic Purana basins in peninsular India. Journal of Geological Society of India 85, 5–25.

Bhandari, A., Pant, N.C., Bhowmik, S.K., Goswami, S., 2011. 1.6 Ga ultrahigh- temperature granulite metamorphism in the Central Indian Tectonic Zone: insights from metamorphic reaction history, geothermobarometry and monazite chemical ages. Geological Journal 46, 198–216.

Bhowmik, S.K., Wilde, S.A., Bhandari, A., 2011. Zircon U-Pb/Lu-Hf and monazite chemical dating of the Tirodi biotite gneiss: implication for latest Palaeoproterozoic to Early Mesoproterozoic orogenesis in the Central Indian Tectonic Zone. Geological Journal 46, 574–596.

Biao, S., Nutman, A.P., Dunyi, L., Jiashan, W., 1996. 3800 to 2500 Ma crustal evolution in the Anshan area of Liaoning Province, northeastern China. Precambrian Research 78, 79–94.

Bickford, M.E., Basu, A., Patranabis-Deb, S., Dhang, P.C., Schieber, J., 2011a. Depositional history of the Chhattisgarh basin, central India: constraints from new SHRIMP zircon ages. Journal of Geology 119, 33–50.

Bickford, M.E., Basu, A., Patranabis-Deb, S., Dhang, P.C., Schieber, J., 2011b. Depositional history of the Chhattisgarh basin, central India: constraints from new SHRIMP zircon ages: a reply. Journal of Geology 119, 553–556.

Biswal, T.K., Biswal, B., Mitra, S., Moulik, M.R., 2002. Deformation pattern of the NW Terrane boundary of the Eastern Ghats mobile belt India: a tectonic model and correlation with Antarctica. Gondwana Research 5, 45–52.

Bose, S., Dunkley, D.J., Dasgupta, S., Das, K., Arima, M., 2011. India-Antarctica- Australia-Laurentia connection in the Paleoproterozoic– Mesoproterozoic revisited: evidence from new zircon U-Pb and monazite chemical age data from the Eastern Ghats Belt, India. Geological Society of America Bulletin 123, 2031–2049.

Chakraborti, R., Basu, A.R., Chakraborty, A., 2007. Trace element and Nd-isotopic evidence for sediment sources in the mid-Proterozoic Vindhyan Basin, central India. Precambrian Research 159, 260–274.

Chakraborti, S., 1997. Elucidation of the sedimentary history of the Singhora Group of rocks, Chhattisgarh Supergroup, M.P. Records of Geological Survey of India 130, 184–187.

Chakraborty, P.P., Paul, S., 2008. Depositional pattern across forced regressive- transgressive sea-level history on a Neoproterozoic siliciclastic shelf: Chandarpur Group, central India. Precambrian Research 162, 227–247.

Chakraborty, P.P., Das, K., Sarkar, A., Das, P., 2009. Fan-delta and storm-dominatedshelf sedimentation in the Proterozoic Singhora Group, Chhattisgarh Supergroup,central India. Precambrian Research 70, 88–106.

Chakraborty, P.P., Das, K., Tsutsumi, Y., Horie, K., 2011. Depositional history of the Chhattisgarh basin, central India: constraints from new SHRIMP zircon ages: a discussion. Journal of Geology 119, 549–552.

Chakraborty, P.P., Das, P., Das, K., Saha, S., Balakrishnan, S., 2012a. Regressive depositional architecture on a Mesoproterozoic siliciclastic ramp: sequence stratigraphic and Nd isotopic evidences from Bhalukona Formation, Singhora Group, Chhattisgarh Supergroup, Central India. Precambrian Research 200, 129–148.

Chakraborty, P.P., Das, P., Saha, S., Das, K., Mishra, S., Paul, P., 2012b. Microbial mat related structures (MRS) from Mesoproterozoic Chhattisgarh and Khariar basins, Central India and their bearing on shallow marine sedimentation. Episodes 35, 513–523.

Chakraborty, P.P., Das, K., Saha, S., Das, P., Karmakar, S., Mamtani, M.A., 2013. Reply to the discussion of Deb (2013) on the paper of Saha et al. (2013) entitled 'Tectono- magmatic evolution of the Mesoproterozoic Singhora basin, central India: evidence for compressional tectonics from structural data, AMS study and geochemistry of basic rocks'. Precambrian Research 236, 297–302.

Chakraborty, P.P., Saha, S., Das, P., 2015. Geology of Mesoproterozoic Chhattisgarh Basin, Central India: current status and future goals. Geological Society of London Memoirs 43, 185–205.

Chaudhuri, A.K., Mukhopadhyay, J., Patranabis-Deb, S., Mukherjee, M.K., Ghosh, G., 2002. The Purana basins of southern cratonic province of India- a case for Mesoproterozoic fossil rifts. Gondwana Research 5, 23–33.

Chatterjee, N., Mazumdar, A.K., Bhattacharya, A., Saikia, R.R., 2007. Mesoproterozoic granulites of the Shillong–Meghalaya Plateau: evidence of westward continuation of the Prydz Bay Pan-African suture into northeastern India. Precambrian Research 152, 1–26.

Condie, K.C., 1993. Chemical composition and evolution of the upper continental crust: contrasting results from surface samples and shales. Chemical Geology 104, 1–37.

Das, D.P., Kundu, A., Das, N., Dutta, D.R., Kumaran, K., Ramamurthy, S., Thangavelu, C., Rajaiya, V., 1992. Lithostratigraphy and sedimentation of Chhattisgarh basin. Indian Minerals 46, 271–288.

Das, D.P., Dutta, N.K., Dutta, D.R., Thanavellu, C., Baburao, K., 2003. Singhora Group the oldest Proterozoic litho-package of Eastern Bastar craton and its significance. Indian Minerals 57, 127–138.

Das, K., Yokoyama, K., Chakraborty, P.P., Sarkar, A., 2009. Basal tuffs and contemporaneity of the Chhattisgarh and Khariar Basins based on new dates and geochemistry. Journal of Geology 117, 88–102.

Das, K., Chakraborty, P.P., Hayasaka, Y., Kayama, M., Saha, S., Kimura, K., 2015. ~1450 Ma regional felsic volcanism at the fringe of the East Indian Craton: constraints from geochronology and geochemistry of tuff beds from detached sedimentary basins. Geological Society of London Memoir 43, 207–221.

Das, P., Das, K., Chakraborty, P.P., Balakrishnan, S., 2011. 1420 Ma diabasic intrusives from the Mesoproterozoic Singhora Group, Chhattisgarh Supergroup, India: implications towards non-plume intrusive activity. Journal of Earth System Science 120, 1–14.

Deb, G.K., 2013. Discussion of Saha et al. (2012, Precambrian Research) Tectono-magmatic evolution of the Mesoproterozoic Singhora basin, central India: evidence for compressional tectonics from structural data, AMS study and geochemistry of basic rocks. Precambrian Research 230, 248–257.

Dhang, P.C., Patranabis-Deb, S., 2012. Lithostratigraphy of the basal part of the Chhattisgarh Supergroup around Singhora-Saraipali area and its tectonic implication. In: Tiwari, R.P. (Ed.), Cenozoic Tectonics, Seismology and Palaeobiology of the Eastern Himalayas and Indo-Myanmar Range, Geological Society of India Memoir, vol. 77, pp. 493–515.

Dobmeier, C., Lütke, S., Hammerschmidt, K., Mezger, K., 2006. Emplacement and deformation of the Vinukonda meta-granite (Eastern Ghats, India) – implications for the geological evolution of peninsular India and for Rodinia reconstructions. Precambrian Research 146, 165–178.

Dunkl, I., Mikes, T., Simon, K., Von Eynatten, H., 2008. Brief introduction to the Windows program Pepita: data visualization, and reduction, outlier rejection, calculation of trace element ratios and concentration from LA-ICP-MS data. In: Sylvester, P. (Ed.), Laser Ablation ICP-MS in the Earth Sciences: Current Practices and Outstanding Issues, Mineralogical Association of Canada Short Course, vol. 40, pp. 334–340.

Fernandez-Suãrez, J., Guttiérez-Alfonso, G., Jenner, G.A., Tubrett, M.N., 2000. New ideas on the Proterozoic-Early Palaeozoic evolution of NW Iberia: insights from U–Pb detrital zircon ages. Precambrian Research 102, 185–206.

Govindaraju, K., 1994. Computation of working values and sample description for 383 geostandards. Geostandards Newsletter 18, 1–158.

Ghosh, J.G., 2004. 3.56 Ga tonalite in the central part of the Bastar Craton, India: oldest Indian date. Journal of Asian Earth Science 23, 359–364.

Gupta, S., 2012. Strain localization, granulite formation and geodynamic setting of 'hot orogens': a case study from the Eastern Ghats Province. Indian Geological Journal 47, 334–351.

Hayashi, K.I., Fujisawa, H., Holland, H.D., Ohmoto, H., 1997. Geochemistry of ~1.9 Ga sedimentary rocks from northeastern Labrador, Canada. Geochimica et Cosmochimica Acta 61, 4115–4137.

Katsube, A., Hayasaka, Y., Sakaguchi, A., Takahashi, Y., 2012. U-Pb zircon dating using Nd-YAG (213 nm) Laser ablation-ICP-MS, and evaluating the consistency with SHRIMP dating. Journal of Geological Society of Japan 118, 762–767.

Košler, J., Fonneland, H., Sylvester, P., Tubrett, M., Pedersen, R.-B., 2002. U–Pb dating of detrital zircons for sediment provenance studies—a comparison of laser ablation ICPMS and SIMS techniques. Chemical Geology 182, 605–618.

Lee, Y.I., 2009. Geochemistry of shales of the Upper Cretaceous Hayang Group, SE Korea: implications for provenance and source weathering at an active continental margin. Sedimentary Geology 215, 1–12.

Ludwig, K.R., 2001. SQUID Version: 1.03 – A User's Manual. Berkeley Geochronology Center Special Publication No. 2. Berkeley Geochronology Center, Berkeley, CA, USA, p. 19.

Ludwig, K.R., 2008. Isoplot 3.70. Berkeley Geochronology Center Special Publication No. 4, p. 76.

Machado, N., Gauthier, G., 1996. Determination of $^{207}Pb/^{206}Pb$ ages on zircon and monazite by laser ablation ICPMS and application to a study of sedimentary provenance and metamorphism in southeastern Brazil. Geochimica et Cosmochimica Acta 60, 5063–5073.
Mahato, S., Goon, S., Bhattacharya, A., Mishra, B., Bernhardt, H.J., 2008. Thermo-tectonic evolution of the North Singhbhum Mobile Belt: a view from the western part of the belt. Precambrian Research 162, 102–107.
Maji, A.K., Goon, S., Bhattacharya, A., Mishra, B., Mahato, S., Bernhardt, H.J., 2008. Proterozoic polyphase metamorphism in the Chhotanagpur Gneissic Complex (India), and implication for trans-continental Gondwanaland correlation. Precambrian Research 162, 385–402.
Malone, S.J., Meert, J.G., Banerjee, D.M., Pandit, M.K., Tamrat, E., Kamenov, G.D., Pradhan, V.R., Sohl, L.E., 2008. Paleomagnetism and detrital zircon geochronology of the Upper Vindhyan Sequence, Son Valley and Rajasthan, India: a ca. 1000 Ma closure age for the Purana Basins? Precambrian Research 164, 137–159.
McLennan, S., 2001. Relationships between trace element compositions of sedimentary rocks and upper continental crust. Geochemistry, Geophysics and Geosystems 2, 1525–2027.
Meert, J.G., Pandit, M.K., Vimal, P.R., Banks, J., Sirianni, R., Stroud, M., Newstead, B., Gifford, J., 2010. The Precambrian tectonic evolution of India: a 3.0 billion year odyssey. Journal of Asian Earth Science 39, 483–515.
Meert, J.G., Pandit, M.K., 2015. The Archean and Proterozoic history of Peninsular India: tectonic framework for Precambrian sedimentary basins in India. Geological Society of London 43, 29–54.
Mondal, M.E.A., Hussain, M.F., Ahmad, T., 2006. Continental growth of Bastar craton, central Indian shield during Precambrian via multiphase subduction and lithospheric extension/rifting: evidence from geochemistry of gneisses, granitoids and mafic dykes. Journal of Geosciences 49, 137–151.
Morton, A.C., Claoué-Long, J.C., Berge, C., 1996. SHRIMP constraints on sediment provenance and transport history in the Mesozoic Statfjord formation, North Sea. Journal of Geological Society of London 153, 915–929.
Nagarajan, R., Armstrong-Altrin, J.S., Nagendra, R., Madhavaraju, J., Moutte, J., 2007. Petrography and geochemistry of terrigenous sedimentary rocks in the Neoproterozoic Rabanpalli formation, Bhima basin, southern India: implications for paleoweathering conditions, provenance and source rock composition. Journal of Geological Society of India 70, 297–312.
Nelson, D.R., 2001. Determination of depositional age and provenance of clastic sedimentary rocks. Sedimentary Geology 141, 37–60.
Ohta, T., 2008. Measuring and adjusting the weathering and hydraulic sorting effects for rigorous provenance analyses of sedimentary rocks: a case study from the Jurassic Ashikita Group south-west Japan. Sedimentology 55, 1687–1701.
Paces, J.B., Miller, J.D., 1993. Precise U-Pb ages of Duluth Complex and related mafic intrusions, northeastern Minnesota: geochronological insights to physical, petrogenetic, paleomagnetic, and tectonomagmatic processes associated with the 1.1 Ga midcontinent rift system. Journal of Geophysical Research 98, 13997–14013.
Patranabis-Deb, S., 2004. Proterozoic stratigraphy of the area around Sarangarh in the eastern part of the Chhattisgarh Basin, Madhya Pradesh, India. Gondwana Research 7, 323–337.
Patranabis-Deb, S., Chaudhuri, A.K., 2007. A retreating fan delta system in the Neoproterozoic Chhattisgarh rift basin, central India: major controls on its evolution. AAPG Bulletin 91, 785–808.
Patranabis-Deb, S., Chaudhuri, A.K., 2010. Sedimentological products and processes in the Mesoproterozoic Chhattisgarh basin and contemporary tectonics in Central India. Indian Journal of Geology 80, 139–155.
Patranabis-Deb, S., Bickford, M.E., Hill, B., Chaudhuri, A.K., Basu, A., 2008. SHRIMP ages of zircon in the uppermost tuff in Chhattisgarh Basin in Central India require ~500-Ma adjustment in Indian Proterozoic Stratigraphy: a reply. Journal of Geology 116, 540–542.
Rudnick, R.L., Gao, S., 2003. Composition of the continental crust. Treatise on Geochemistry 3, 1–64.
Saha, S., Das, K., Chakraborty, P.P., Das, P., Karmakar, S., Mamtani, M.A., 2013. Tectono-magmatic evolution of the Mesoproterozoic Singhora basin, central India: evidence for compressional tectonics from structural data, AMS study and geochemistry of basic rocks. Precambrian Research 227, 276–294.
Saha, S., Das, K., Hidaka, H., Kimura, K., Chakraborty, P.P., Hayasaka, Y., 2016. Detrital zircon geochronology (U-Pb SHRIMP and LA-ICPMS) from the Ampani Basin, Central India: implication for provenance and Mesoproterozoic tectonics at East Indian cratonic margin. Precambrian Research 281, 363–383.
Santosh, M., Yokoyama, K., Acharyya, S.K., 2004. Geochronology and Tectonic Evolution of Karimnagar and Bhopalpatnam Granulite Belts, Central India. Gondwana Research 7, 501–518.

Stacey, J.S., Kramers, J.D., 1975. Approximation of terrestrial lead isotope evolution by a two-stage model. Earth and Planetary Science Letter 26, 207–221.

Taylor, S.R., McLennan, S.M., 1985. The Continental Crust: Its Composition and Evolution: An Examination of the Geochemical Record Preserved in Sedimentary Rocks. Blackwell Sciences, Oxford, p. 312.

Vijaya Kumar, K., Ernst, W.G., Leelanandam, C., Wooden, J.L., Grove, M.J., 2010. First Paleoproterozoic ophiolite from Gondwana. Geochronologic–geochemical documentation of ancient oceanic crust from Kandra, SE India. Tectonophysics 487, 22–32.

Vijaya Kumar, K., Leelanandam, C., Ernst, W.G., 2011. Formation and fragmentation of the Palaeoproterozoic supercontinent Columbia: evidence from the Eastern Ghats Granulite Belt, southeast India. International Geology Review 53, 1297–1311.

Whitehouse, M.J., Bridgewater, D., Park, R.G., 1997. Detrital zircon ages from the Loch Maree Group, Lewisian Complex, NW Scotland: confirmation of Palaeoproterozoic Laurentia–Fennoscandia connection. Terra Nova 9, 260–263.

Williams, I.S., 1998. U-Th-Pb geochronology by ion microprobe. In: McKibben, M.A., Shanks, W.C.P., Ridley, W.I. (Eds.), Applications of Microanalytical Techniques to Understanding Mineralizing Processes, Reviews in Economic Geology, vol. 7. Society of Economic Geologists, Littleton, CO., USA, pp. 1–35.

CHAPTER

16

Deciphering Sedimentary Provenance and Timing of Sedimentation From a Suite of Metapelites From the Chotanagpur Granite Gneissic Complex, India: Implications for Proterozoic Tectonics in the East-Central Part of the Indian Shield

A. Dey[1], *S. Mukherjee*[1], *S. Sanyal*[1], *M. Ibanez-Mejia*[2,3], *P. Sengupta*[1]

[1]Jadavpur University, Kolkata, India; [2]Massachusetts Institute of Technology, Cambridge, MA, United States; [3]University of Rochester, Rochester, NY, United States

OUTLINE

Sediment Provenance
http://dx.doi.org/10.1016/B978-0-12-803386-9.00016-2

1. INTRODUCTION

Precambrian metamorphic complexes of the Indian shield and the slices of such rocks in the Cenozoic metamorphic belt of the lesser Himalaya often expose a sequence of quartzite-metapelites-metacarbonate rocks that can be best interpreted as the metamorphic equivalent of a stable shelf depositional sequence, including sandstone-shale and limestone assemblages (Malone et al., 2008; Basu et al., 2008; Basu and Bickford, 2014; Sengupta et al., 2015). Timing of sedimentation and provenance of the precursors of this metasedimentary litho-ensemble as well as their postdepositional thermal history play a crucial role in unraveling the tectonic history of the Indian shield over the Precambrian time. Unfortunately, polyphase deformation, partial melting, and metasomatism virtually obliterate the sedimentary attributes of these supracrustal rocks; hence, no meaningful sedimentary parameters can be recovered from these high grade rocks. However, owing to its refractory nature over a wide range of physico-chemical conditions, U-Pb isotopic compositions of zircon, a common accessory mineral in many metasedimentary rocks, serve as a powerful tool in deciphering their sedimentary provenance, bracketing their time of sedimentation and placing robust constraints on the postdepositional thermal history of these metasediments (Corfu et al., 2003; Harley et al., 2007; Upadhyay et al., 2009; Kooijman et al., 2011; Sengupta et al., 2015). A few attempts, however, have been made to unravel the sedimentary and superimposed metamorphic history of the metasedimentary rocks from the Indian shield through interpretation of internal morphology and U-Pb isotopic compositions of the inherited zircon grains (Collins et al., 2003; Buick et al., 2006; Upadhyay et al., 2009; Bose et al., 2011; Kooijman et al., 2011; Kaur et al., 2013; Sengupta et al., 2015).

In this chapter we present internal morphology and U-Pb isotopic compositions of zircon populations from three metapelitic samples collected from the north eastern part of the Chotanagpur Granite Gneissic Complex (CGGC). The internal structures of zircon and $^{207}Pb/^{206}Pb$ apparent ages of different textural domains of the zircon populations place tight constraints on the likely provenance from which the precursor sediments of these metapelites might have been derived and their timing of deposition. $^{207}Pb/^{206}Pb$ apparent dates of the rims that developed in situ around the detrital cores of zircon suggests the presence of at least three major pulses of high grade metamorphism over the time span between ~1680 Ma and ~900 Ma.

Combining the published geological and geochronological information with the observation made in this study it is argued that extensive sedimentary basins were formed during the late Paleoproterozoic time on a unified basement of the Indian shield consisting of the cratonic blocks of Singhbhum, Bastar, Aravalli, Bundelkhand, and Dharwar. Disintegration of the Columbia Supercontinent is deemed responsible for the formation of these sedimentary basins due to breakup of the Archaean-early Paleoproterozoic basement of the Indian shield where stable continent shelf sediments (quartzite-shale-lime stone) were deposited at the dusk of Paleoproterozoic Era.

2. REGIONAL GEOLOGY

The CGGC of the east Indian shield is separated from the Meso Archaean Singhbhum Craton by the Late Paleoproterozoic North Singhbhum Fold Belt (NSFB), and from the Paleoproterozoic rocks that are interleaved with or formed the basement of the Phanerozoic rocks of the Lesser Himalaya (hereafter, lesser Himalaya) by Indo-Gangetic alluvium (Fig. 16.1A). Phanerozoic Rajmahal Traps and Paleo- to Neoproterozoic rocks of the Central Indian Tectonic Zone (CITZ) lie outside the eastern and the western boundaries of the CGGC, respectively (Fig. 16.1A and B). Because of younger Gondwana cover sediments, the direct relation between CITZ and CGGC cannot be established; however, geophysical data suggest that the CITZ is a western extension of the latter (Verma, 1985). Furthermore, a number of studies have proposed that the ~1600 Ma old rocks of the CGGC continue further northeast to the gneissic belt of Sillong-Meghalaya (SMGC; Fig. 16.1A) (Chatterjee et al., 2007; Hossain et al., 2007; Ameen et al., 2007; Chatterjee and Ghose, 2011). It is the opinion of a group of workers that displacement along the east–west (E–W) trending Dauki Fault separated the CGGC from the gneissic basement of the Shillong Plateau (Evans, 1964). Migmatitic and non-migmatitic felsic orthogneisses that dominate the lithology of the CGGC contain enclaves of metasedimentary and metaigneous rocks of variable dimensions (reviewed in Sanyal and Sengupta, 2012). Field features and limited petrological studies are consistent with the fact that the metasedimentary and metaigneous enclaves underwent high to ultra-high temperature metamorphism before being caught up in the protolith of the felsic orthogneisses (reviewed in Sanyal and Sengupta, 2012). Subsequent polyphase deformation and high temperature metamorphism produced the migmatitic banding in the felsic orthogneisses, which defines a roughly E–W trending regional planar structure. A body of anorthosite was emplaced at ~1550 Ma (Chatterjee et al., 2008) and was also subsequently deformed along with the enclosing felsic orthogneisses under granulite facies condition (Chatterjee et al., 2008; Maji et al., 2008). Swarms of mafic dykes and pegmatite veins that cut the E–W trending foliation of the felsic orthogneiss, which now appear defining N–S trending open to tight folds, mark the last major deformational and metamorphic (under upper amphibolite facies) event in the heartland of the CGGC (Sanyal and Sengupta, 2012). A prominent N–S trending shear zone is reported to have run through the eastern fringe of the CGGC and the adjoining NSFB (Chatterjee et al., 2010). The lithological map of the studied areas that are located in the NE part of the CGGC is presented in Fig. 16.1C. Metasedimentary rocks that occur as enclaves within felsic orthogneisses are composed of metapelite (garnet-sillimanite migmatitic gneiss, commonly referred to as khondalite), minor calc-silicate granulite, and rare quartzite.

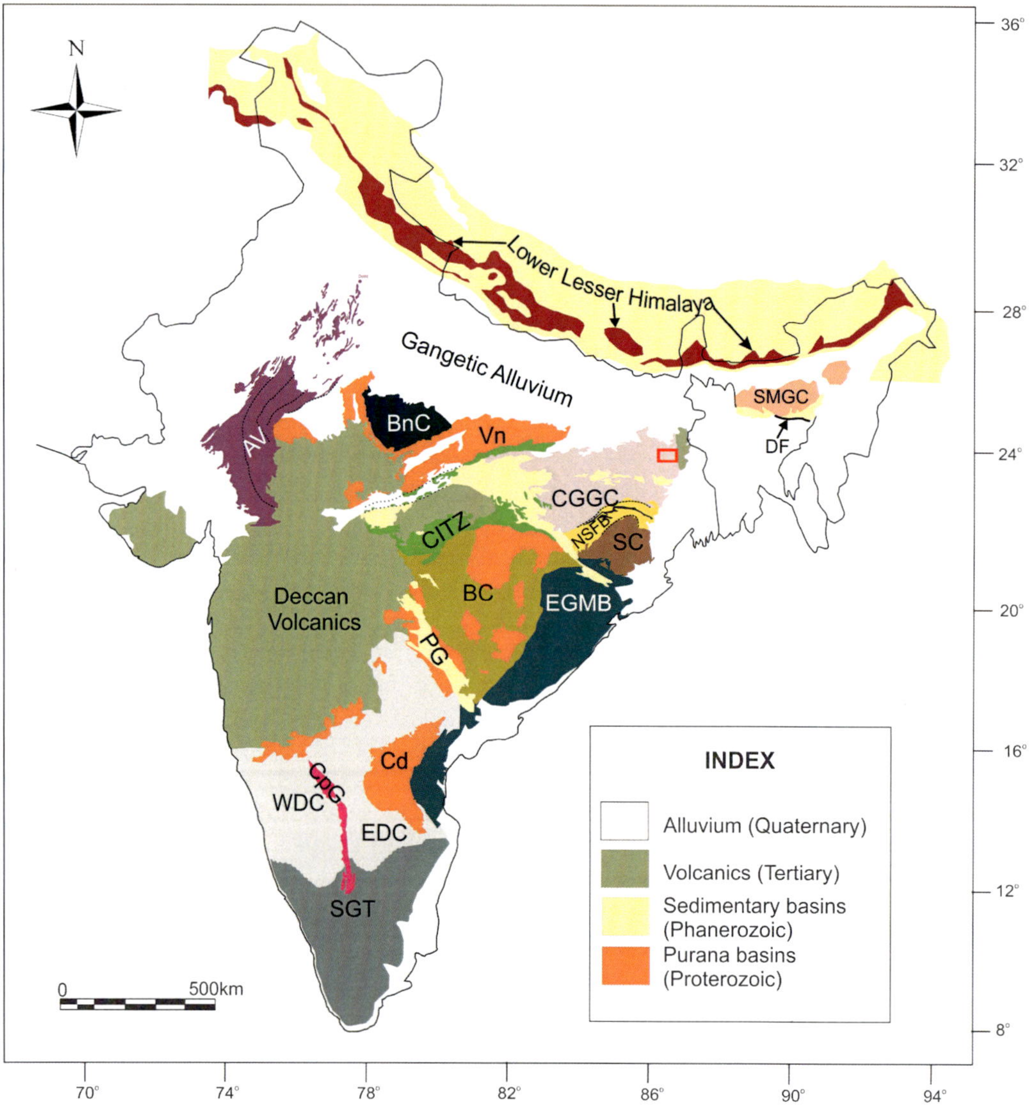

FIGURE 16.1A : Geological map representing the Indian crystalline basement and sedimentary cover. The *red box* at the eastern part represents our study area. *AV*, Aravalli craton; *BC*, Bastar craton; *BnC*, Bundelkhand craton; *CGGC*, Chotanagpur Granite Gneissic Complex; *CITZ*, Central Indian Tectonic Zone; *CpG*, Closepet Granite; *Cd*, Cudappah basin; *DF*, Dauki Fault; *EGMB*, Eastern Ghats mobile belt; *EDC*, Eastern Dharwar craton; *NSFB*, North Singhbhum Fold Belt; *PG*, Pranhita Godavari basin; *SC*, Singhbhum craton; *SGT*, Southern granulitic terrane; *SMGC*, Shillong Meghalaya Gneissic Complex; *Vn*, Vindhyan basin; *WDC*, Western Dharwar craton.

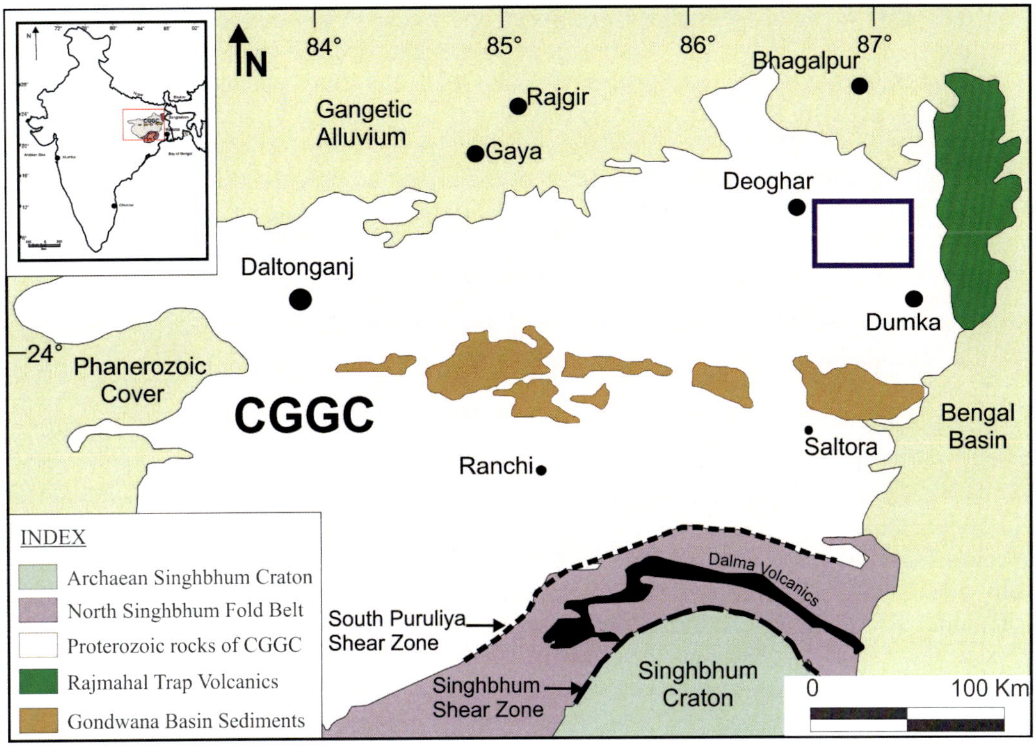

FIGURE 16.1B Geological map of the Chotanagpur Granite Gneissic Complex (CGGC) and part of the North Singhbhum Fold Belt (NSFB) (modified after Acharyya, 2003). *Blue square box* indicates the study area located in the northeastern part of the terrane.

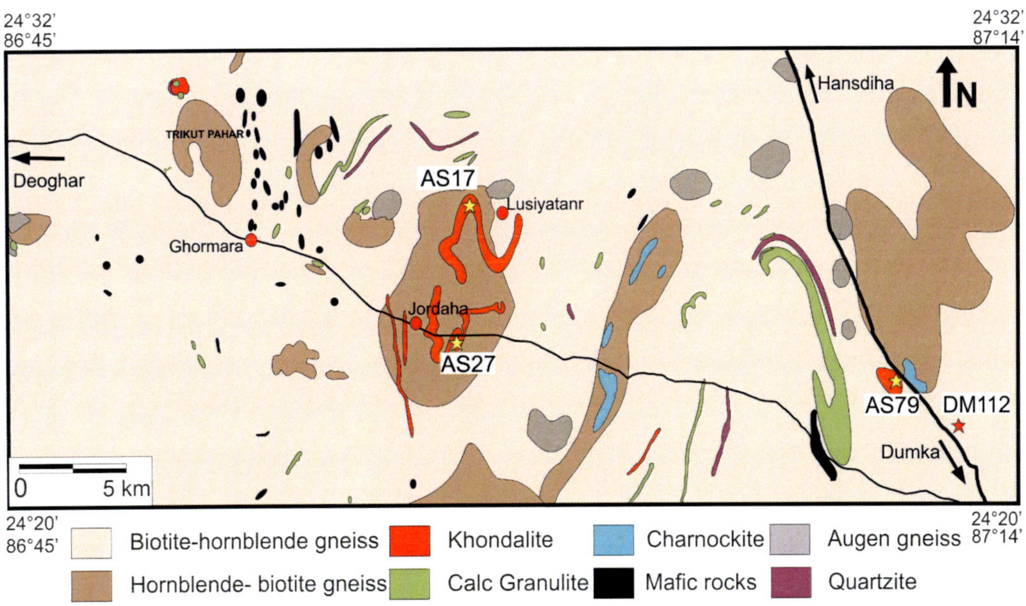

FIGURE 16.1C Map showing major lithological distributions of the study area (modified after Bhattacharya, 1976). *Yellow stars* indicate sample locations of this study and the *red star* indicates location of the metapelite analyzed by Rekha et al. (2011).

Elongated rafts of metasedimentary rocks and the mafic dykes define map scale open to tight folds (Fig. 16.1C). Migmatitic banding of the metapelites and the enclosing felsic orthogneiss also appear deformed and define the N–S trending folds. Published chemical dates of monazite (Chatterjee et al., 2010; Rekha et al., 2011; Sanyal and Sengupta, 2012) and U-Pb date of zircon from metapelite as well as form felsic gneiss (Rekha et al., 2011; Mukherjee et al., 2015) suggest that the last major high-grade metamorphism and accompanied regional deformation in the CGGC occurred in the time span of ~1100–900 Ma. Detailed description of the three metapelites samples chosen for this study and their relations with the enclosing felsic orthogneisses are described next.

3. SAMPLE DESCRIPTION

Three metapelite samples were collected for zircon geochronology. These samples are located between Deoghar and Dumka, in the northeastern segment of the CGGC (Fig. 16.1B and C). All three metapelite samples are collected from pods that are enclosed within the felsic orthogneisses. The attitudes of gneissic banding of the enclosing felsic orthogneiss are at discordance with the attitudes of the gneissic banding of metapelites. In a few places the gneissic banding of the host felsic orthogneiss swerves around the pods of metapelite. The migmatitic banding in the metapelites veers and becomes parallel with the gneissic banding of the host felsic gneisses. All these features suggest that the felsic magma, the protolith of the felsic orthogneiss, captured the pods of migmatitic metapelites during its emplacement and share a similar history of younger deformation and metamorphic overprinting. Detailed petrological study of metapelites and the adjoining rocks are beyond the purview of this chapter. Nevertheless, to put the observations in perspective some salient petrological features of the three representative metapelite samples that were chosen for zircon chronology are presented here.

3.1 AS79 (24°23.427′, 87°10.192′)

This sample was taken from a small outcrop located west of Dumka-Hansdiha road, at about 20 km northwest of Dumka town (Fig. 16.1C). This garnet-sillimanite-biotite gneiss shows well-developed melanosomes and leucosomes. The melanosome is dominated by garnet of variable size (up to 3 cm diameter), sillimanite and biotite. Quartz and feldspars in the melanosome never exceed 30 vol%. The leucosomes, on the other hand, are dominated by quartzofeldspathic minerals (>90 vol%) with minor garnet (<10 vol%). These leucosomes that are presumed to be quenched melt occur as irregular patches or as discontinuous bands within the melanosomes (Fig. 16.2A). Locally, laterally continuous bands made up of quartz, garnet, and biotite with minor feldspar are present in the metapelites (Fig. 16.2B). These quartzofeldspathic bands seem to represent arenaceous layers in the protolith of the pelitic rock. Microscopic study reveals that alkali-feldspar grains of leucosome and melanosome are perthitic. Rarely hercynite occurs as an included phase within garnet porphyroblasts that are present in the melanosome. No primary biotite is present in the studied metapelite. However, an aggregate of biotite and sillimanite replaces the porphyroblastic garnet,

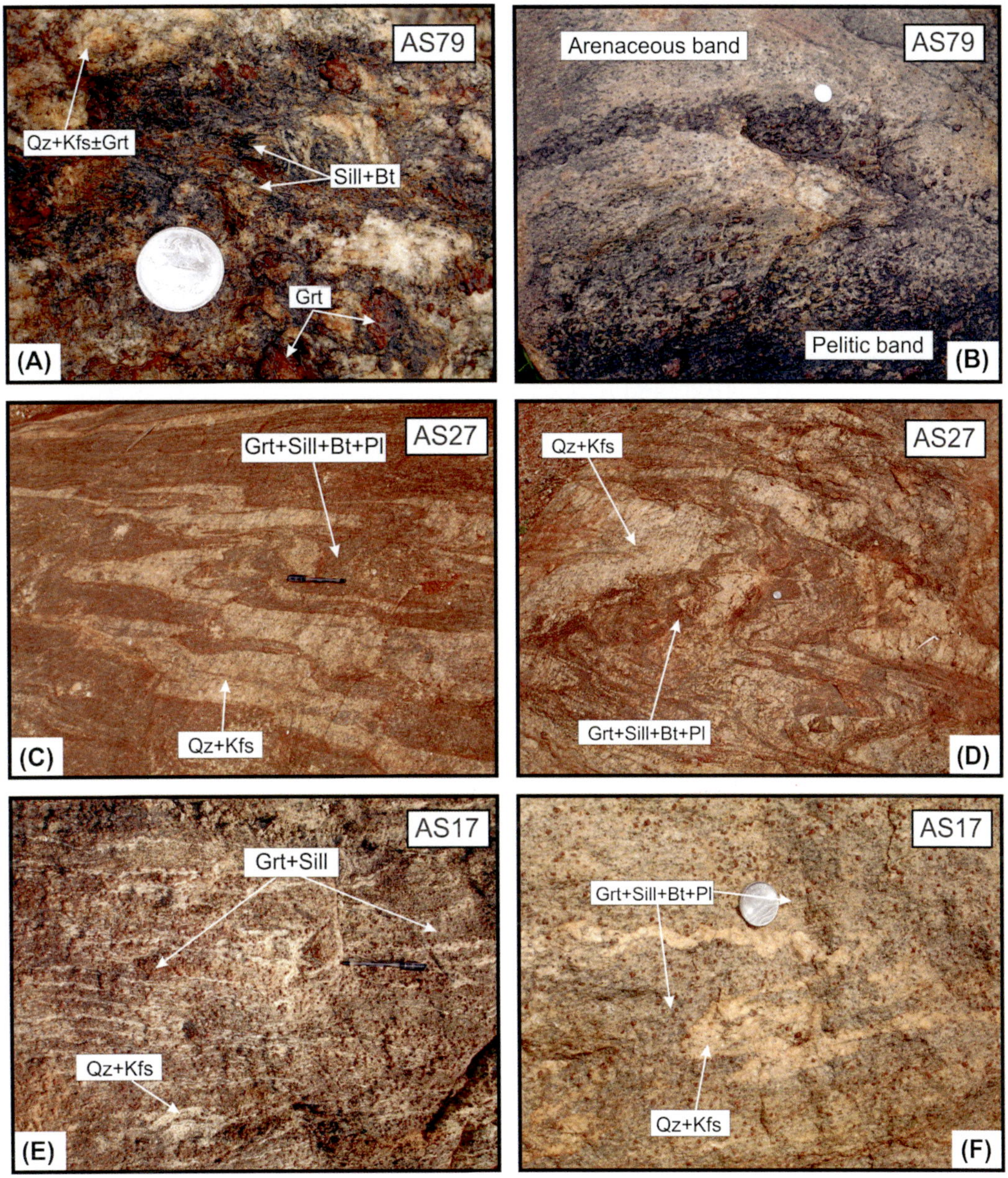

FIGURE 16.2 Field feature of the studied samples. (A) Irregular patches or discontinuous bands of leucosomes within the melanosomes (AS79). (B) Massive meta-arenaceous bands present in association with metapelitic bands (AS79). (C, D) Thick leucosomes of AS27 with alternate melanosomes defining folds with N–S closure. (E) Migmatitic banding in AS17 with garnet clots in melanosome. (F) Garnet-free thick leucosomes exhibiting N–S folds in AS 17. Mark the development of axial planar foliation defined by sillimanite -biotite and very fine leucosomes. *Grt*, Garnet; *Sill*, Sillimanite; *Bt*, Biotite; *Qz*, Quartz; *Kfs*, K-feldspar; *Pl*, Plagioclase.

suggesting retrogression from the peak metamorphic assemblage. These observations and the results of quantitative geothermobarometry suggest an ultrahigh temperature metamorphism (>900°C) in the metapelitic enclaves (Sanyal and Sengupta, 2012; Dey et al., 2013). Zircon grains observed in the thin sections occur mostly as inclusions within garnet porphyroblasts or as dispersed grains within quartzofeldspathic groundmass.

3.2 AS27 (24°24.510′, 86°57.179′)

This sample was collected from an outcrop close to a roadside mound near Jordaha, about 10 km southeast of Ghormara along the Deoghar-Dumka road (Fig. 16.1C). This rock exhibits prominent migmatitic banding defined by alternation of bands of leucosome (quartz + Kfeldspar ± plagioclase) and melanosome (garnet + sillimanite + biotite + plagioclase + Kfeldspar + quartz) (Fig. 16.2C and D). Compared to the melanosomes, leucosomes are thick and represent the dominant component of the migmatite. The migmatitic bandings define open to tight folds having closure toward the north or south (Fig. 16.2C). The gneissic banding of the host felsic orthogneiss also defines the same N−S orientation for the alignment of the fold-axes. Similar to AS79, thin and laterally continuous aranaceous bands are present in this sample. Certain petrographic features, namely the preponderance of mesoperthite and the absence of primary biotite, suggest partial melting and formation of migmatitic banding due to high-temperature dehydration melting reactions (>900°C; Sanyal and Sengupta, 2012). The zircons that were observed in the thin sections occur as inclusion within garnet porphyroblasts as well as within the quartzofeldspathic matrix.

3.3 AS17 (24°26.426′, 86°57.580′)

This metapelite sample was collected from an outcrop near Lusiyatanr, about 10 km east of Ghormara (Fig. 16.1C). Compared to AS79 and AS27, AS17 is dominated by sillimanite (>25 vol%). The arenaceous bands that are present in the previously described locations are absent in this rock. This feature suggests formation of the protolith of AS17 in deeper water conditions with a more limited supply of coarse-grained detrital components. Alternation of leucosome (quartz + kfeldspar ± plagioclase ± garnet) and melanosome (garnet + sillimanite + plagioclase + ilmenite + kfeldspar + quartz) define the prominent migmatitic banding of this rock (Fig. 16.2E). The thick leucosomes are folded in N−S closing asymmetric folds with development of axial-planar foliation defined by the alignment of sillimanite-biotite and very fine leucosomes (Fig. 16.2F). Garnet-rich pods (>50 vol% garnet) are common in the melanosomes but rare in leucosomes (Fig. 16.2E). A detailed microscopic study revealed hercynite occurring as inclusion within garnet porphyroblasts. No primary biotite is observed, indicating extensive dehydration melting of the biotite-rich assemblage at high temperature (>900°C). Secondary biotite and acicular sillimanite were also observed to replace the porphyroblastic garnet and feldspar grains. Under the microscope, the majority of the observed zircons occur within recrystallized leucosomal segregations.

4. ANALYTICAL METHODS, ZIRCON MORPHOLOGY, AND RESULTS

4.1 Analytical Methods for Zircon U-Pb Geochronology

Zircons from the metasedimentary samples described earlier were separated using standard magnetic and gravimetric techniques in Presidency University, India. After separation, grains were picked under a binocular microscope, mounted in epoxy plugs, and carefully polished in order to expose an internal surface for geochronologic analysis. Given the possible complexity and polycyclic growth history of zircons recovered from the studied samples, all grains were imaged by cathodoluminescence using a secondary electron microscope (SEM-CL) prior to isotopic analysis; all imaging was conducted at the Arizona Laserchron SEM facilities, University of Arizona. These images were used as a guide to locate the spot analyses, and the observed growth textures are used here in an interpretive fashion in order to discuss the petrologic and tectonic significance of the obtained dates (see discussion in the following section).

After imaging, U-Pb geochronologic analyses were conducted by laser ablation-multicollector-inductively coupled plasma-mass spectrometry (LA-MC-ICP-MS) methods at the Arizona LaserChron facilities, University of Arizona, using a Photon Machines *Analyte-G2* laser ablation system coupled with a Nu instrument *Plasma I* MC-ICP-MS. The collector block of the Nu plasma has the advantage that it allows for different measurement configurations depending on the required ablation-spot diameter and the amount of Pb present in a particular sample (Ibanez-Mejia et al., 2014). For large ablation spot sizes (i.e., 25–40 μm diameter), all Pb masses with the exception of 204 were measured using faraday cups; this configuration is referred to in the text and tables as FAR. For complexly zoned or very small zircon crystals/overgrowths, where a spot size $\leq$12 μm was required, all Pb masses were measured using ion multipliers in order to allow reliable measurements of small ablation volumes; this configuration is referred to as IC. Details on the data acquisition, processing, and uncertainty propagation protocols for both measurement modes are described in detail in Ibanez-Mejia et al. (2014). Instrumental bias, drift, and interelement fractionation corrections were performed by the standard-sample bracketing approach, using an in-house Sri Lanka zircon crystal with a known ID-TIMS age of 563.5 ± 3.2 Ma (Gehrels et al., 2008) as primary reference material.

In order to independently assess the accuracy of our isotopic measurements, zircon crystals from a Duluth complex anorthosite (FC-1 locality; Paces and Miller, 1993) were used as secondary reference material and analyzed alongside the unknowns. From 46 analyses of FC-1 conducted during the course of this study, we obtain a weighted-mean $^{207}Pb/^{206}Pb$ age of 1098.8 ± 3.3/9.4 Ma (2σ, MSWD = 0.34, n = 46), which is in agreement with the known ID-TIMS age of this locality (i.e., 1099.0 ± 0.6 Ma) and further validates the accuracy of our results.

For sample AS79, where the observed detrital zircon cores were relatively large, most analyses were conducted using a 25 μm spot diameter and FAR collector configuration. Subsequently, some of the smaller-scale complexities and thin metamorphic overgrowths were analyzed using the IC configuration and an 8 μm ablation-spot diameter (Figs. 16.3 and16.4). Given the on-average smaller zircon grain sizes found in samples AS17 and AS27, all spots conducted in these two samples utilized the IC collector configuration and 8 μm ablation spot diameters. In both cases, each ablation spot consisted of a 70-pulse laser burst, firing at 7 Hz and with a constant surface energy density of c. 7 J cm^{-2}. Using the Analyte-G2, this configuration results in craters of ~4 μm in depth during the ablation of zircon (Ibanez-Mejia et al., 2015).

4.2 Zircon Morphology and Results

In this section we describe the morphology, internal structure and geochronologic results from zircon grains of the three studied metapelite samples. The program Isoplot-4.15 (Ludwig, 2003) was used for statistical analysis of the age data and construction of concordia plots. For each sample, the moderately reverse-discordant data that overlap concordia within uncertainty are considered as concordant ages. Details of the zircon U–Pb LA-MC-ICP-MS analyses of samples are presented in Table 16.1.

4.2.1 AS79

Zircon grains recovered from this sample are of variable size, shapes, and morphology. The length of the zircon grains varies between 100 and 450 μm. A few large zircon grains with principal axes >200 μm have elongated aspect ratios (>2:1), and display oscillatory-zoned growth structures holding euhedral shape with resorbed boundaries (Fig. 16.3A). These features indicate a magmatic origin for these zircons that were later affected by the high-grade metamorphic event (e.g., Corfu et al., 2003). The medium-sized (150–300 μm) grains have mostly elliptical shapes and CL imaging reveals they generally consist of a sub-rounded core enclosed within homogenous overgrowths (Fig. 16.3B–F). The xenocrystic cores are either dark, metamictized (Fig. 16.3C), or oscillatory zoned (Fig. 16.3D). These cores have corrugated boundaries (Fig. 16.3C), indicative of abrasion presumably caused by erosion and sedimentation process and hence implying a detrital history (Corfu et al., 2003). The cores are overgrown by thin low-luminescence mantles and high-luminescence rims (Fig. 16.3D–F), respectively, interpreted as formed in situ during metamorphism. The other variety of zircon grains are medium-sized (100–200 μm), equant, and subrounded, with large homogeneous featureless domains (Fig. 16.4A). Their morphology suggests that these zircons are also metamorphically grown (Corfu et al., 2003).

A total of 28 spots were analyzed on 25 grains, out of which 17 analyses show <2% discordance. The homogeneous, featureless, subrounded zircon grains that we interpret were neoformed during metamorphism, have an overall low Th/U ratio (0.01–0.4, with ~80% data having Th/U $\leq$ 0.2), which further supports their metamorphic origin (Rubatto, 2002; Harley et al., 2007). Dates (with $\leq$2% discordance) obtained from these grains yield a weighted mean age of 941 ± 13 Ma (MSWD = 0.75). This Neoproterozoic age is considered to date the last major metamorphic event suffered by the rock.

Data (with <2% discordance) from oscillatory-zoned euhedral grains as well as from xenocrystic cores fall into two distinct age groups. The majority of them yield $^{207}Pb/^{206}Pb$ dates between 1864 and 1814 Ma with a smaller population ranging from ~2700 to 2100 Ma. On a

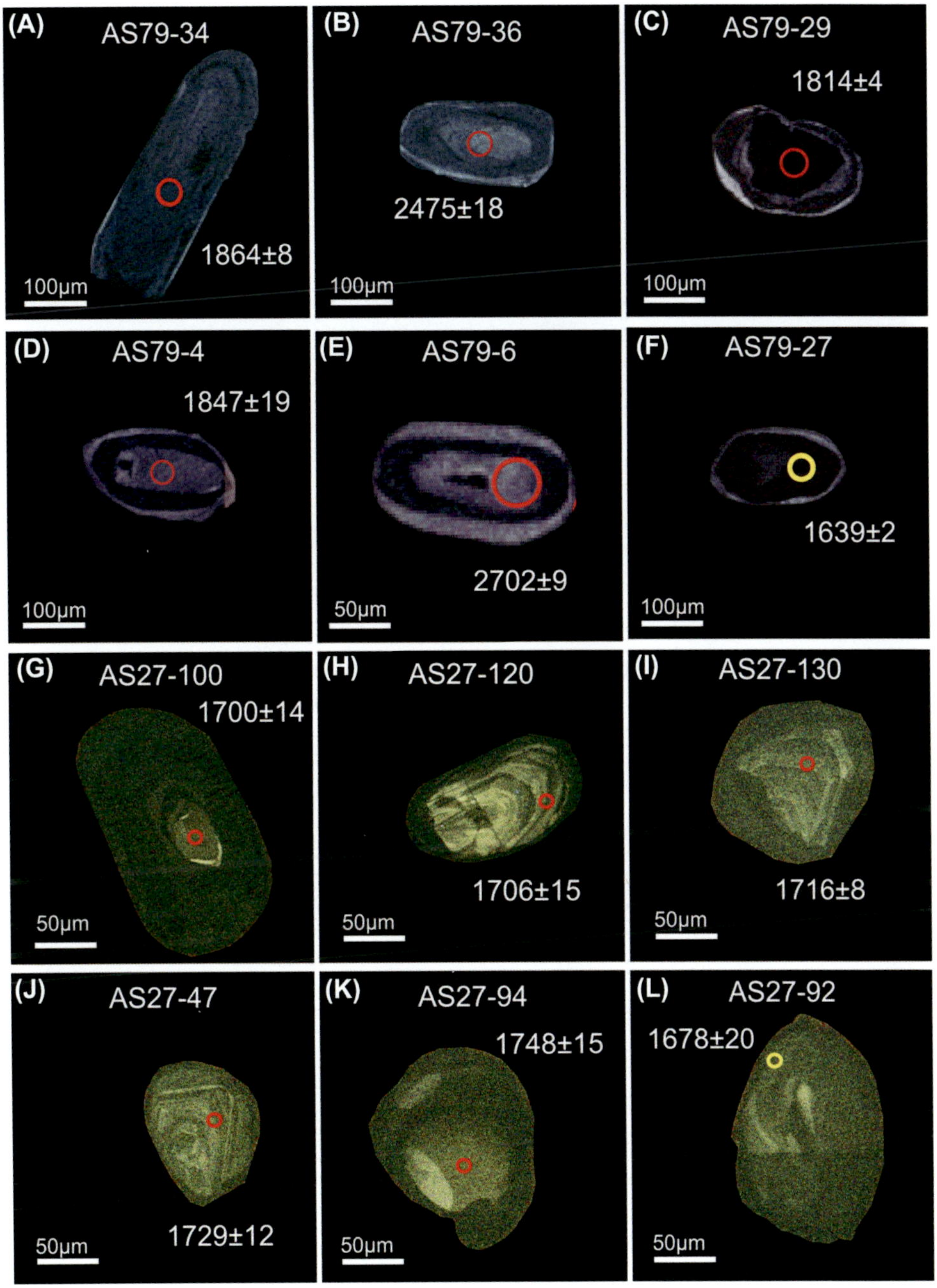

FIGURE 16.3 Cathodo-luminescence (CL) images of zircon grains from the studied samples: AS79: (A)–(F); AS27: (G)–(L). Numbers on the top of the grains indicate spot numbers and the dates indicate $^{207}Pb/^{206}Pb$ ages. The *red spots* indicate detrital grains whereas the *yellow ones* indicate metamorphic domains.

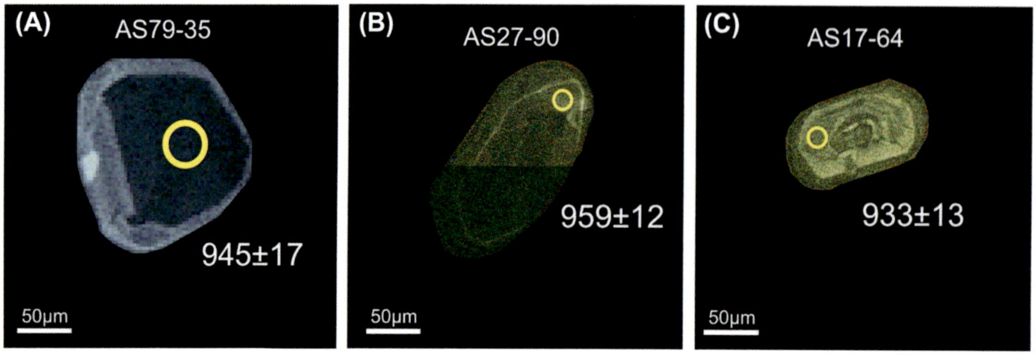

FIGURE 16.4 Cathodo-luminescence (CL) images of zircon grains from the studied samples: AS79: (A); AS27: (B); AS17: (C). Numbers on top of the grains indicate spot numbers and the dates refer to the $^{207}Pb/^{206}Pb$ ages of the youngest metamorphic event.

probability density curve the first group exhibits a peak at 1856 ± 7 Ma (Fig. 16.5A). The youngest concordant detrital age of this cluster is 1814 ± 4 Ma. Data from the older group has peaks at 2693 Ma, 2441 Ma, and 2139 Ma (Fig. 16.5A). Both of these groups have Th/U ratio >0.2 with ~80% of them >0.5, pointing toward a magmatic origin of the detrital grains (Scherer et al., 2007).

Two concordant analyses from dark homogeneous overgrowths define two distinct single-data peaks at 1639 ± 2 Ma and 1468 ± 10 Ma with Th/U ratio <0.2 (Table 16.1). These last two overgrowth dates are possibly reflecting older metamorphic or thermal events experienced by the rock.

4.2.2 AS27

The zircon grains obtained from this sample are subhedral to anhedral, with prismatic or ovoid shapes. Their length varies between 100 and 300 μm. The prismatic grains are elongated to tabular (aspect ratio >2:1) with subrounded crystal face terminations indicating metamorphic resorption (Fig. 16.3G). Other grains are equant to subequant (aspect ratio from 2:1 to 1:1) with rounded grain boundaries. The majority of the grains have xenocrystic cores with single or multiple homogenous (Fig. 16.3I–K) or oscillatory-zoned overgrowths (Fig. 16.3G–H). The cores are subrounded (Fig. 16.3G and K) to sub-angular (Fig. 16.3I) and exhibit a variety of internal structures like strong oscillatory-zoning (Fig. 16.3H–J), patchy sector zoning (Fig. 16.3G), or completely homogenous weakly luminous to dark portions (Fig. 16.3K). The sharp truncation of zoning against the overgrowth or the corrugation of the core boundaries (Fig. 16.3H–J) are features indicative of a detrital origin (Corfu et al., 2003). The zircon grains devoid of any distinct cores are either oscillatory-zoned prismatic grains (Fig. 16.4B) or are equant subrounded grains with large homogenous featureless domains.

A total of 69 spots were analyzed, which include both core and overgrowth domains. Out of these 69 analyses, 21 exhibit <2% discordance or are concordant within uncertainty. The oscillatory-zoned prismatic or homogeneous subrounded grains without evident detrital cores yield a weighted mean age of 962 ± 7 Ma (MSWD = 0.64). Morphological features

TABLE 16.1 U-Pb (Zircon) Geochronologic Analyses of Chotanagpur Metapelites

Analysis	U (ppm)	$^{206}Pb/^{204}Pb$	Isotope Ratios $^{206}Pb^*/^{207}Pb^*$	± (%)	$^{207}Pb^*/^{235}U^*$	± (%)	$^{206}Pb^*/^{238}U$	± (%)	Error Corr.	Apparent Ages (Ma) $^{206}Pb^*/^{238}U^*$	± (Ma)	$^{207}Pb^*/^{235}U$	± (Ma)	$^{206}Pb^*/^{207}Pb^*$	± (Ma)	Best Age (Ma)	± (Ma)	Conc (%)	Th/U	Morphology of the Zircon Grain
SAMPLE AS 79(24°23.427′, 87°10.192′)																				
Concordant Data <2% Discordance																				
AS79_6b	62	33,603,735	5.39	0.57	12.97	2.10	0.51	2.0	0.96	2647.0	43.9	2677.7	19.8	2701.7	9.5	2701.7	9.5	98	0.70	Detrital
AS79_36	59	56,412	6.18	1.08	10.20	1.92	0.46	1.6	0.83	2426.4	32.0	2453.2	17.7	2475.4	18.2	2475.4	18.2	98	0.99	Detrital
AS79_10	324	74,216	6.30	0.29	9.87	1.56	0.45	1.5	0.98	2401.7	30.7	2423.1	14.4	2441.2	5.0	2441.2	5.0	98	1.08	Detrital
AS79_20	98	68,933	6.51	0.58	9.31	3.33	0.44	3.3	0.98	2348.1	64.6	2369.2	30.6	2387.5	9.9	2387.5	9.9	98	0.81	Detrital
AS79_08	258	114,241	7.51	0.26	7.08	2.02	0.39	2.0	0.99	2104.1	36.0	2121.8	18.0	2138.9	4.6	2138.9	4.6	98	0.54	Detrital
AS79_34	215	53,525	8.77	0.44	5.28	1.42	0.34	1.3	0.95	1867.8	21.9	1866.3	12.1	1864.5	8.0	1864.5	8.0	100	0.82	Detrital
AS79_17	323	206,851	8.82	0.34	5.09	3.16	0.33	3.1	0.99	1817.6	49.7	1835.2	26.8	1855.2	6.1	1855.2	6.1	98	0.34	Detrital
AS79_23	256	83,746	8.82	0.32	5.16	1.87	0.33	1.8	0.99	1839.3	29.5	1846.3	15.9	1854.2	5.8	1854.2	5.8	99	1.53	Detrital
AS79_04	80	93,619	8.85	1.07	5.09	1.55	0.33	1.1	0.72	1822.6	17.8	1834.2	13.1	1847.4	19.4	1847.4	19.4	99	0.69	Detrital
AS79_24	117	84,068	8.83	0.56	5.23	1.14	0.33	1.0	0.87	1861.7	16.1	1857.4	9.7	1852.7	10.1	1852.7	10.1	100	0.58	Detrital
AS79_40	382	142,224	8.97	0.28	4.90	1.60	0.32	1.6	0.98	1784.2	24.5	1802.9	13.5	1824.6	5.1	1824.6	5.1	98	0.27	Detrital
AS79_29	654	98,525	9.02	0.22	4.95	0.97	0.32	0.9	0.97	1806.6	15.0	1810.2	8.2	1814.4	4.0	1814.4	4.0	100	1.21	Detrital
AS79_27	1115	246,578	9.92	0.13	4.06	0.87	0.29	0.9	0.99	1652.6	12.5	1646.8	7.1	1639.3	2.4	1639.3	2.4	101	0.11	Metamorphic
AS79_30b	231	20,298,471	10.86	0.51	3.18	1.90	0.25	1.8	0.96	1441.0	23.6	1451.7	14.7	1468.2	9.7	1468.2	9.7	98	0.15	Metamorphic
AS79_28	192	72,801	14.29	1.26	1.49	1.72	0.15	1.2	0.68	925.3	10.1	925.9	10.4	927.5	25.8	927.5	25.8	100	0.10	Metamorphic
AS79_35	321	132,432	14.17	0.85	1.56	1.86	0.16	1.7	0.89	960.1	14.7	955.6	11.5	945.2	17.3	945.2	17.3	102	0.04	Metamorphic

(Continued)

TABLE 16.1 U-Pb (Zircon) Geochronologic Analyses of Chotanagpur Metapelites—cont'd

			Isotope Ratios							Apparent Ages (Ma)										
Analysis	U (ppm)	^{206}Pb/^{204}Pb	$^{206}Pb^*$/$^{207}Pb^*$	± (%)	$^{207}Pb^*$/$^{235}U^*$	± (%)	$^{206}Pb^*$/^{238}U	± (%)	Error Corr.	$^{206}Pb^*$/$^{238}U^*$	± (Ma)	$^{207}Pb^*$/^{235}U	± (Ma)	$^{206}Pb^*$/$^{207}Pb^*$	± (Ma)	Best Age (Ma)	± (Ma)	Conc (%)	Th/U	Morphology of the Zircon Grain
AS79_13	126	33,637	14.16	1.40	1.51	1.56	0.15	0.7	0.44	927.3	6.0	933.1	9.5	947.0	28.6	947.0	28.6	98	0.13	Metamorphic
Discordant Data >2% Discordance																				
AS79_06	72	54,475	5.45	0.57	12.25	2.33	0.48	2.3	0.97	2545.0	47.4	2623.5	21.8	2684.6	9.5	2684.6	9.5	95	0.69	Detrital
AS79_19	198	148,305	6.74	0.82	7.28	3.17	0.36	3.1	0.97	1962.5	51.9	2146.9	28.3	2328.2	14.0	2328.2	14.0	84	0.19	Detrital
AS79_30	419	58,100	8.51	0.46	5.44	2.62	0.34	2.6	0.98	1868.1	41.9	1891.8	22.5	1918.0	8.2	1918.0	8.2	97	0.46	Detrital
AS79_32	612	140,472	9.09	0.18	4.73	2.09	0.31	2.1	1.00	1749.5	32.0	1772.8	17.6	1800.3	3.3	1800.3	3.3	97	0.83	Detrital
AS79_21	464	82,897	9.17	0.45	4.46	2.96	0.30	2.9	0.99	1674.2	43.1	1723.4	24.5	1783.7	8.1	1783.7	8.1	94	0.15	Detrital
AS79_14	549	14,243	9.26	3.42	4.35	4.45	0.29	2.9	0.64	1652.9	41.6	1703.2	36.8	1765.7	62.4	1765.7	62.4	94	0.21	Detrital
AS79_15	428	13,090	9.51	2.32	3.65	6.25	0.25	5.8	0.93	1447.6	75.2	1560.7	49.9	1717.4	42.7	1717.4	42.7	84	0.49	Detrital
AS79_06	249	307,131	9.64	1.10	3.76	2.24	0.26	1.9	0.87	1505.7	26.1	1584.8	18.0	1691.7	20.3	1691.7	20.3	89	0.56	Detrital
AS79_26	260	59,142	9.72	0.53	4.08	2.44	0.29	2.4	0.98	1630.0	34.3	1650.7	19.9	1677.2	9.8	1677.2	9.8	97	0.05	Metamorphic
AS79_25B	732	13,388	10.26	0.27	3.47	1.22	0.26	1.2	0.98	1480.1	15.8	1520.3	9.6	1576.6	5.0	1576.6	5.0	94	0.03	Metamorphic
AS79_16	74	22,235	14.18	3.07	1.46	3.34	0.15	1.3	0.40	900.0	11.1	912.7	20.1	943.5	62.8	943.5	62.8	95	0.20	Metamorphic
SAMPLE AS27 (24°24.510′, 86°57.179′)																				
Concordant Data <2% Discordance																				
AS27_141	151	8,995,987	9.34	1.32	4.48	4.33	0.30	4.1	0.95	1709.7	61.9	1727.8	36.0	1750.6	24.1	1750.6	24.1	98	0.60	Detrital
AS27_94	94	180,731,623	9.35	0.82	4.75	2.40	0.32	2.3	0.94	1798.9	35.5	1775.3	20.1	1748.4	14.9	1748.4	14.9	103	1.18	Detrital
AS27_144	232	7,429,255	9.40	0.64	4.57	2.31	0.31	2.2	0.96	1749.9	34.0	1744.2	19.2	1738.2	11.8	1738.2	11.8	101	0.67	Detrital
AS27_174	457	73,865,073	9.40	0.48	4.50	2.21	0.31	2.2	0.98	1727.1	32.6	1731.5	18.3	1737.7	8.8	1737.7	8.8	99	0.75	Detrital
AS27_66	241	23,946,415	9.42	0.56	4.44	2.14	0.30	2.1	0.96	1708.2	30.9	1719.7	17.7	1734.5	10.3	1734.5	10.3	98	0.88	Detrital

AS27_110	406	25,503,870	9.44	0.65	4.49	2.89	0.31	2.8	0.97	1729.5	42.6	1729.5	24.0	1730.3	12.0	1730.3	12.0	100	0.63	Detrital
AS27_47	462	78,164,601	9.45	0.68	4.37	2.09	0.30	2.0	0.95	1688.7	29.4	1706.7	17.3	1729.5	12.4	1729.5	12.4	98	0.57	Detrital
AS27_126	414	23,421,616	9.46	0.37	4.64	2.39	0.32	2.4	0.99	1781.1	36.8	1756.2	20.0	1727.5	6.7	1727.5	6.7	103	0.69	Detrital
AS27_160	132	8,794,843	9.51	0.80	4.45	2.37	0.31	2.2	0.94	1726.5	33.7	1721.7	19.6	1716.7	14.7	1716.7	14.7	101	0.58	Detrital
AS27_130	241	116,360,411	9.51	0.44	4.36	2.38	0.30	2.3	0.98	1694.8	34.8	1704.1	19.6	1716.5	8.2	1716.5	8.2	99	0.69	Detrital
AS27_31	145	11,250,333	9.52	0.75	4.30	2.29	0.30	2.2	0.95	1675.7	32.0	1692.7	18.9	1714.7	13.8	1714.7	13.8	98	1.05	Detrital
AS27_118	151	11,188,477	9.54	0.65	4.46	2.18	0.31	2.1	0.96	1736.3	31.7	1724.3	18.1	1710.7	11.9	1710.7	11.9	101	0.78	Detrital
AS27_120	78	2,852,060	9.57	0.84	4.33	2.44	0.30	2.3	0.94	1695.4	34.2	1699.9	20.2	1706.3	15.4	1706.3	15.4	99	1.05	Detrital
AS27_100	276	34,961,390	9.60	0.80	4.34	2.57	0.30	2.4	0.95	1703.3	36.5	1701.4	21.2	1700.0	14.7	1700.0	14.7	100	0.74	Detrital
AS27_52	90	4,355,187	9.62	0.84	4.48	2.45	0.31	2.3	0.94	1753.2	35.3	1727.2	20.3	1696.5	15.5	1696.5	15.5	103	0.92	Detrital
AS27_92	190	7,347,748	9.71	1.11	4.39	2.88	0.31	2.7	0.92	1735.9	40.4	1709.6	23.8	1678.4	20.4	1678.4	20.4	103	0.72	Metamorphic
AS27_178	244	13,204,851	10.09	1.23	3.80	3.32	0.28	3.1	0.93	1581.9	43.3	1592.3	26.7	1606.8	22.9	1606.8	22.9	98	0.38	Metamorphic
AS27_116	134	21,244,832	10.24	1.00	3.69	2.85	0.27	2.7	0.94	1562.9	37.1	1570.0	22.8	1580.3	18.7	1580.3	13.7	99	0.67	Metamorphic
AS27_127	65	32,078,958	11.05	0.95	3.12	2.27	0.25	2.1	0.91	1437.1	26.5	1436.5	17.4	1436.5	18.1	1436.5	18.1	100	1.54	Metamorphic
AS27_61	312	7,701,556	11.24	0.78	2.91	1.87	0.24	1.7	0.91	1373.1	21.0	1384.6	14.1	1403.2	14.9	1403.2	14.9	98	1.25	Metamorphic
AS27_46	593	11,822,224	14.01	0.61	1.56	2.35	0.16	2.3	0.97	949.0	20.0	954.5	14.5	968.0	12.5	968.0	12.5	98	0.01	Metamorphic
AS27_119	602	19,242,067	14.07	0.60	1.55	2.42	0.16	2.3	0.97	945.0	20.6	949.1	14.9	959.5	12.2	959.5	12 2	98	0.00	Metamorphic
AS27_90	699	29,266,317	14.07	0.60	1.54	1.97	0.16	1.9	0.95	939.0	16.3	944.8	12.1	959.3	12.3	959.3	12.3	98	0.02	Metamorphic
Discordant Data >2% Discordance																				
AS27_12	412	19,081,071	9.08	1.44	4.63	3.35	0.30	3.0	0.90	1715.3	45.5	1754.0	27.9	1801.3	26.1	1801.3	26.1	95	0.09	Detrital
AS27_108	201	1,170,942	9.11	1.85	3.24	4.72	0.21	4.3	0.92	1250.3	49.4	1466.0	36.7	1794.8	33.7	1794.8	33.7	70	1.07	Detrital
AS27_112	171	7,563,494	9.12	0.77	4.49	3.76	0.30	3.7	0.98	1676.7	54.3	1728.5	31.2	1792.7	14.1	1792.7	14.1	94	0.43	Detrital
AS27_122	114	9,324,217	9.36	2.32	3.86	6.56	0.26	6.1	0.94	1502.7	82.2	1606.2	52.9	1745.5	42.4	1745.5	42.4	86	0.96	Detrital

(Continued)

TABLE 16.1 U-Pb (Zircon) Geochronologic Analyses of Chotanagpur Metapelites—cont'd

Analysis	U (ppm)	$^{206}Pb/^{204}Pb$	Isotope Ratios $^{206}Pb^*/^{207}Pb^*$	± (%)	$^{207}Pb^*/^{235}U^*$	± (%)	$^{206}Pb^*/^{238}U$	± (%)	Error Corr.	Apparent Ages (Ma) $^{206}Pb^*/^{238}U^*$	± (Ma)	$^{207}Pb^*/^{235}U$	± (Ma)	$^{206}Pb^*/^{207}Pb^*$	± (Ma)	Best Age (Ma)	± (Ma)	Conc (%)	Th/U	Morphology of the Zircon Grain
AS27_43	186	177,856,082	9.44	0.56	4.33	2.57	0.30	2.5	0.98	1676.2	37.0	1699.8	21.2	1729.8	10.4	1729.8	10.4	97	0.68	Detrital
AS27_84	293	48,253,355	9.51	0.41	4.25	2.01	0.29	2.0	0.98	1659.2	28.7	1684.1	16.5	1716.0	7.5	1716.0	7.5	97	0.70	Detrital
AS27_72	581	30,354,727	9.52	0.60	3.96	2.57	0.27	2.5	0.97	1559.7	34.7	1627.0	20.9	1716.0	11.0	1716.0	11.0	91	0.43	Detrital
AS27_99	176	12,944,327	9.55	0.61	4.24	2.16	0.29	2.1	0.96	1662.1	30.3	1682.5	17.7	1708.9	11.2	1708.9	11.2	97	0.61	Detrital
AS27_109	60	3,726,199	9.56	1.22	4.17	4.12	0.29	3.9	0.96	1638.8	57.0	1668.4	33.8	1706.7	22.5	1706.7	22.5	96	0.64	Detrital
AS27_102	389	30,802,583	9.56	0.74	4.00	2.80	0.28	2.7	0.96	1580.6	37.9	1635.1	22.8	1706.6	13.6	1706.6	13.6	93	0.76	Detrital
AS27_189	278	11,030,103	9.57	0.58	4.05	2.45	0.28	2.4	0.97	1597.2	33.7	1644.1	20.0	1705.4	10.7	1705.4	10.7	94	0.80	Detrital
AS27_186	211	7,331,900	9.58	0.59	4.19	2.30	0.29	2.2	0.97	1646.9	32.3	1671.8	18.8	1704.0	10.9	1704.0	10.9	97	0.64	Detrital
AS27_16	346	91,744,914	9.59	0.44	4.18	2.12	0.29	2.1	0.98	1645.6	30.1	1670.1	17.3	1702.0	8.1	1702.0	8.1	97	0.67	Detrital
AS27_114	312	15,159,446	9.62	0.50	4.05	2.22	0.28	2.2	0.97	1606.0	30.8	1644.8	18.1	1695.7	9.2	1695.7	9.2	95	1.93	Detrital
AS27_39	765	16,920,3379	9.63	0.62	3.94	1.86	0.27	1.8	0.94	1565.7	24.4	1621.0	15.1	1694.4	11.4	1694.4	11.4	92	1.03	Detrital
AS27_129	203	16,528,994	9.63	0.59	3.93	3.11	0.27	3.1	0.98	1563.4	42.4	1619.4	25.2	1693.7	10.8	1693.7	10.8	92	0.75	Detrital
AS27_36	221	34,318,067	9.66	0.87	4.13	2.54	0.29	2.4	0.94	1639.5	34.5	1660.6	20.7	1688.2	16.0	1688.2	16.0	97	0.75	Detrital
AS27_137	370	43,492,942	9.81	0.88	3.65	3.39	0.26	3.3	0.97	1487.6	43.4	1559.6	27.0	1659.2	16.3	1659.2	16.3	90	0.43	Detrital
AS27_138	509	39,524,390	9.87	0.78	3.76	3.24	0.27	3.1	0.97	1538.1	43.1	1585.1	26.0	1649.0	14.5	1649.0	14.5	93	0.07	Detrital
AS27_07	224	12,361,212	9.91	0.76	3.46	3.37	0.25	3.3	0.97	1431.6	42.2	1517.3	26.6	1639.9	14.1	1639.9	14.1	87	0.69	Detrital
AS27_146	163	13,433,609	9.98	1.06	3.75	2.96	0.27	2.8	0.93	1550.1	38.1	1583.1	23.7	1628.1	19.6	1628.1	19.6	95	0.72	Detrital
AS27_51	205	13,509,879	10.06	0.66	3.70	2.64	0.27	2.6	0.97	1542.2	35.1	1572.1	21.1	1613.4	12.2	1613.4	12.2	96	0.75	Metamorphic
AS27_71	633	94,064,535	10.10	2.19	2.57	3.87	0.19	3.2	0.82	1112.8	32.6	1292.5	28.3	1605.7	40.9	1605.7	40.9	69	0.60	Detrital
AS27_133	460	23,322,288	10.12	1.68	3.46	3.26	0.25	2.8	0.86	1459.0	36.4	1518.3	25.7	1602.7	31.4	1602.7	31.4	91	0.15	Detrital

AS27_134	149	25,135,232	10.12	0.79	3.60	2.70	0.26	2.6	0.96	1513.7	34.8	1550.2	21.4	1601.0	14.8	1601.0	14.8	95	0.56	Detrital
AS27_170	465	26,133,622	10.14	1.01	3.46	2.76	0.25	2.6	0.93	1461.5	33.6	1517.5	21.8	1597.4	18.9	1597.4	18.9	91	0.41	Detrital
AS27_91	696	64,496,070	10.23	0.50	3.23	2.02	0.24	2.0	0.97	1384.4	24.4	1463.9	15.7	1582.1	9.3	1582.1	9.3	87	0.68	Detrital
AS27_75	122	6,806,098	10.33	1.22	3.26	3.31	0.24	3.1	0.93	1409.1	38.9	1471.8	25.7	1564.3	22.9	1564.3	22.9	90	0.67	Detrital
AS27_45	651	26,359,632	10.36	0.57	2.99	2.36	0.22	2.3	0.97	1307.8	27.1	1405.6	18.0	1557.9	10.6	1557.9	10.6	84	0.12	Detrital
AS27_101	274	8,207,874	10.45	0.58	3.37	1.80	0.26	1.7	0.95	1466.1	22.3	1497.0	14.1	1541.8	10.9	1541.8	10.9	95	1.06	Metamorphic
AS27_190	827	12,417,147	10.46	1.12	3.25	3.32	0.25	3.1	0.94	1421.6	39.8	1469.7	25.8	1540.8	21.1	1540.8	21.1	92	0.28	Metamorphic
AS27_166	308	11,543,526	10.46	1.05	2.91	3.54	0.22	3.4	0.95	1287.4	39.5	1385.0	26.8	1539.7	19.8	1539.7	19.8	84	0.56	Detrital
AS27_175	237	214,789,137	10.49	1.02	3.06	3.52	0.23	3.4	0.96	1350.8	41.0	1423.2	26.9	1534.0	19.2	1534.0	19.2	88	0.97	Detrital
AS27_81	221	14,699,219	10.56	0.59	3.06	2.16	0.23	2.1	0.96	1357.1	25.4	1422.3	16.5	1522.2	11.0	1522.2	11.0	89	0.83	Detrital
AS27_183	425	28,241,954	10.73	0.65	2.83	3.30	0.22	3.2	0.98	1281.4	37.6	1362.2	24.7	1492.2	12.4	1492.2	12.4	86	0.41	Detrital
AS27_64	679	53,609,167	10.85	0.45	2.72	1.93	0.21	1.9	0.97	1251.4	21.3	1334.6	14.3	1471.7	8.6	1471.7	8.6	85	0.31	Detrital
AS27_177	284	16,910,462	10.94	0.69	2.98	2.54	0.24	2.4	0.96	1368.4	30.1	1402.3	19.3	1455.1	13.2	1455.1	13.2	94	0.51	Metamorphic
AS27_89	374	18,257,059	10.99	1.63	2.43	3.41	0.19	3.0	0.88	1142.4	31.3	1252.0	24.5	1446.5	31.1	1446.5	31.1	79	0.74	Detrital
AS27_155	452	23,246,570	11.05	1.63	2.28	4.02	0.18	3.7	0.91	1080.0	36.5	1204.7	28.4	1436.5	31.0	1436.5	31.0	75	0.38	Detrital
AS27_154	542	13,938,905	11.10	2.03	2.33	3.85	0.19	3.3	0.85	1109.8	33.3	1222.3	27.4	1427.7	38.8	1427.7	38.8	78	0.12	Detrital
AS27_17	738	10,715,677	11.62	1.18	2.33	3.26	0.20	3.0	0.93	1157.7	32.2	1222.4	23.2	1339.4	22.8	1339.4	22.8	86	0.34	Detrital
AS27_193	888	37,058,460	11.74	1.45	2.08	2.42	0.18	1.9	0.80	1051.9	18.8	1142.2	16.6	1319.0	28.0	1319.0	28.0	80	0.17	Detrital
AS27_135	803	63,771,299	11.86	0.98	2.22	2.66	0.19	2.5	0.93	1126.0	25.5	1186.6	18.6	1299.6	19.1	1299.6	19.1	87	0.04	Detrital
AS27_179	528	27,732,573	11.98	0.76	2.20	2.36	0.19	2.2	0.95	1127.5	23.1	1180.7	16.5	1280.4	14.7	1280.4	14.7	88	0.08	Detrital
AS27_167	231	9,590,698	12.72	1.24	1.91	2.76	0.18	2.5	0.89	1048.7	23.8	1085.9	18.4	1162.1	24.7	1162.1	24.7	90	0.13	Detrital
AS27_143	268	14,210,765	12.77	1.54	1.92	3.29	0.18	2.9	0.88	1056.0	28.4	1088.5	22.0	1155.2	30.5	1155.2	30.5	91	0.37	Metamorphic
AS27_113	47	4,423,513	13.06	2.09	1.82	4.64	0.17	4.1	0.89	1026.8	39.3	1053.4	30.4	1109.8	41.8	1109.8	41.8	93	0.83	Metamorphic
AS27_56	663	290,408,792	13.21	1.39	1.78	4.00	0.17	3.8	0.94	1015.9	35.3	1038.4	26.0	1086.8	27.9	1086.8	27.9	93	0.05	Metamorphic

(*Continued*)

TABLE 16.1 U-Pb (Zircon) Geochronologic Analyses of Chotanagpur Metapelites—cont'd

Analysis	U (ppm)	$^{206}Pb/^{204}Pb$	Isotope Ratios $^{206}Pb^*/^{207}Pb^*$	± (%)	$^{207}Pb^*/^{235}U^*$	± (%)	$^{206}Pb^*/^{238}U$	± (%)	Error Corr.	Apparent Ages (Ma) $^{206}Pb^*/^{238}U^*$	± (Ma)	$^{207}Pb^*/^{235}U$	± (Ma)	$^{206}Pb^*/^{207}Pb^*$	± (Ma)	Best Age (Ma)	± (Ma)	Conc (%)	Th/U	Morphology of the Zircon Grain
AS17 (24°26.426′, 86°57.580′)																				
Concordant Data <2% Discordance																				
AS17_66	137	3,364,032	14.05	0.97	1.55	2.13	0.16	1.9	0.89	945.5	16.7	950.2	13.1	962.1	19.9	962.1	19.9	98	0.05	Metamorphic
AS17_7	912	46,875,317	14.12	0.55	1.53	1.95	0.16	1.9	0.96	936.6	16.3	941.2	12.0	952.9	11.2	952.9	11.2	98	0.03	Metamorphic
AS17_27	561	25,336,683	14.28	0.77	1.54	2.23	0.16	2.1	0.94	951.6	18.5	944.7	13.7	929.6	15.8	929.6	15.8	102	0.03	Metamorphic
AS17_12	186	3,627,216	14.29	1.08	1.54	2.53	0.16	2.3	0.90	956.0	20.3	947.2	15.6	927.7	22.2	927.7	22.2	103	0.05	Metamorphic
AS17_55	224	13,608,347	14.19	0.96	1.58	2.77	0.16	2.6	0.94	970.1	23.4	961.3	17.2	942.3	19.7	942.3	19.7	103	0.00	Metamorphic
Discordant Data >2% Discordance																				
AS17_34	1011	16,717,157	14.12	0.70	1.37	2.03	0.14	1.9	0.94	849.3	15.2	878.3	12.0	953.0	14.3	953.0	14.3	89	0.03	Metamorphic
AS17_5	874	17,818,015	14.15	0.59	1.50	1.98	0.15	1.9	0.95	922.1	16.2	929.5	12.0	947.9	12.1	947.9	12.1	97	0.03	Metamorphic
AS17_32	375	11,146,737	14.16	0.59	1.48	2.35	0.15	2.3	0.97	910.7	19.3	921.1	14.2	947.0	12.2	947.0	12.2	96	0.01	Metamorphic
AS17_69	165	5,145,234	14.16	0.88	1.49	2.55	0.15	2.4	0.94	919.4	20.5	927.2	15.5	946.8	18.1	946.8	18.1	97	0.00	Metamorphic
AS17_54	411	22,809,168	14.20	0.63	1.41	2.56	0.14	2.5	0.97	872.3	20.2	891.4	15.2	940.0	12.9	940.0	12.9	93	0.02	Metamorphic
AS17_51	126	18,397,514	14.21	1.03	1.44	2.49	0.15	2.3	0.91	892.0	18.8	905.3	14.9	938.7	21.2	938.7	21.2	95	0.15	Metamorphic
AS17_64	253	11,852,785	14.25	0.66	1.43	2.73	0.15	2.6	0.97	891.4	22.0	903.4	16.3	933.6	13.5	933.6	13.5	95	0.02	Metamorphic
AS17_28	103	3,633,546	14.25	0.92	1.45	2.87	0.15	2.7	0.95	901.4	22.9	910.5	17.3	933.4	18.9	933.4	18.9	97	0.03	Metamorphic
AS17_50	613	11,898,203	14.28	0.92	1.40	2.50	0.14	2.3	0.93	871.3	19.0	887.6	14.8	929.1	19.0	929.1	19.0	94	0.02	Metamorphic
AS17_36	225	10,223,422	14.29	0.65	1.41	2.07	0.15	2.0	0.95	880.9	16.1	894.2	12.3	928.2	13.4	928.2	13.4	95	0.14	Metamorphic

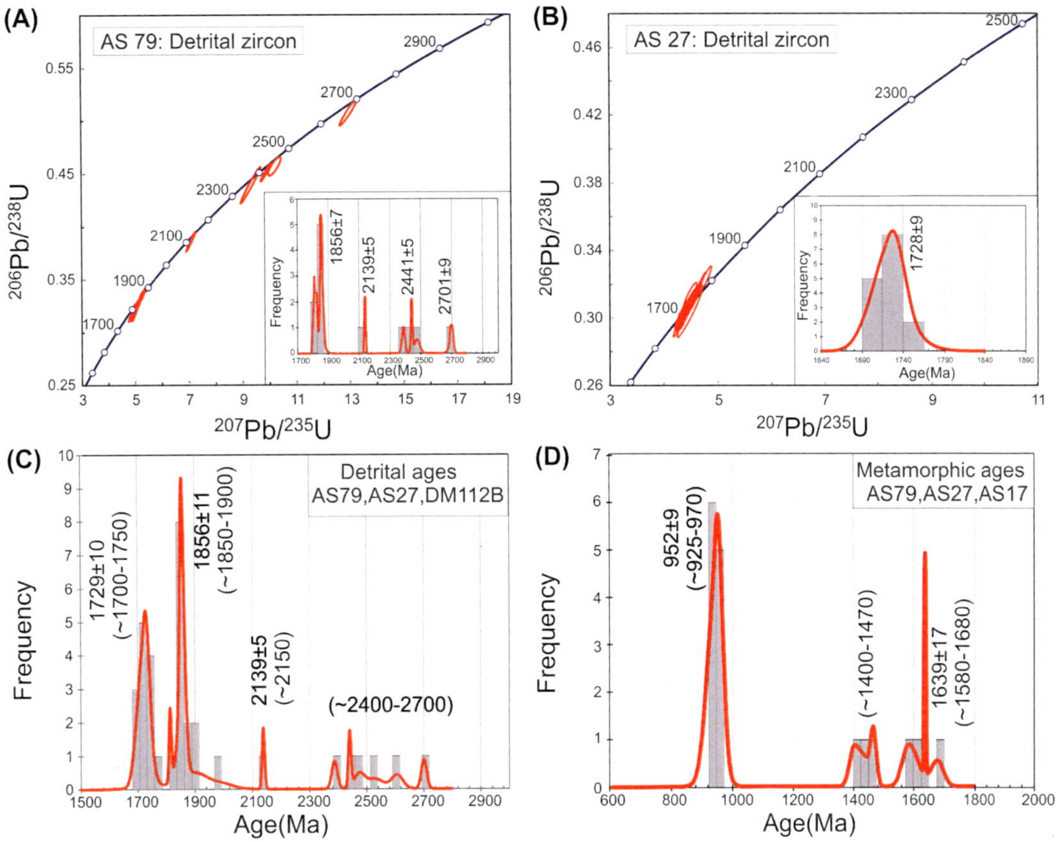

FIGURE 16.5 Concordia diagrams showing the results of U–Pb analyses of zircon grains from metasedimentary rocks. Histograms plus probable peaks of the age cluster of the detrital zircon grains with <2% discordance are shown in the inset: (A) sample AS79; (B) sample AS27. (C) The age histogram plus probability curve for all detrital zircon from our samples AS79, AS27 and sample DM112B from Rekha et al. (2011). (D) The age histograms plus probability curve for all metamorphic ages from all the three samples AS79, AS27, and AS17.

suggest that these grains are crystallized from partial melt and/or metamorphically grown (Corfu et al., 2003), implying that a high-grade metamorphic event occurred at that age. All these grains have low Th/U (<0.1) typical of metamorphic zircons under certain conditions (Rubatto, 2002; Harley et al., 2007). Spot analysis from the xenocrystic detrital cores yield both concordant (<2% discordance) and discordant data. On a probability density plot, the concordant data defines a bell-shaped curve with a peak at 1728 ± 9 Ma ($^{207}Pb/^{206}Pb$ data; Fig. 16.5B). The youngest concordant $^{207}Pb/^{206}Pb$ detrital age is 1696 ± 15 Ma. More than 80% of these detrital grains have Th/U > 0.2 with the vast majority having >0.6, thus suggesting a magmatic origin. A few concordant analyses with $^{207}Pb/^{206}Pb$ apparent dates between 1436(±18)–1403(±15) Ma and 1678(±20)–1580(±19) Ma are obtained from overgrowths on the detrital zircon core and discrete featureless homogeneous zircon grains. We interpret these ages to date the older metamorphic or thermal events recorded by the studied rocks. Among them, the oldest metamorphic age is provided by a concordant analysis with a $^{207}Pb/^{206}Pb$ apparent date of 1678 ± 20 Ma. It is intriguing that the Th/U ratio of these dated zircon is high (>0.7), a feature that is likely to develop in magmatic zircon (Scherer et al., 2007). Harley et al. (2007) showed that the Th/U ratio of zircon alone cannot be used as a fingerprint of its genesis (i.e., magmatic vs. metamorphic) and that morphology and internal texture of zircon is a more reliable factor to determine the conditions under which this mineral grows in metamorphic rocks.

4.2.3 AS17

This sample contains a lesser amount of zircon. They are medium-sized (100–250 μm), and euhedral to anhedral in shape. The euhedral grains are prismatic, elongated to tabular (aspect ratio ≥2:1) and exhibit distinct oscillatory zoning (Fig. 16.4C). All imaged crystals have a thin, dark, homogenous outer rim, which makes the crystal face boundary blunt and resorbed. The anhedral grains are elliptical to ovoid in shape with bright subrounded to multifaceted luminous cores mantled by dark rims. The luminous cores are homogenous, sharing smooth boundaries with the overgrowth. Some zircon grains are smaller (≤100 μm), equant in shape (aspect ratio 1:1), and have less luminous to dark featureless internal structure. Incidentally no detrital zircons have been found from this sample.

Fifteen spot analyses have been done from 15 grains. Whether it be the oscillatory zoned euhedral grains, luminous homogeneous cores enclosed in dark mantles, or equant featureless small grains, all concordant (<2% discordance) analyses obtained from these crystals fall within a restricted range from 928 to 962 Ma and define a weighted mean age of 945 ± 17 Ma (MSWD = 2.8). All of these data have Th/U ratio <0.2 (90% of data have <0.1) supporting their metamorphic origin. The presence of oscillatory-zoned grains and typical metamorphic ones, defining age populations that overlap within uncertainty, are a clear indication for the presence of partial melts during metamorphism.

5. DISCUSSION

5.1 Age of the Sediment Deposition

Ascertaining the age of sedimentation (basin formation age) of metamorphosed supracrustal rocks is a difficult task owing to absence of datable authigenic minerals. Nevertheless, the age of the youngest detrital zircon population and the oldest metamorphic age can be used to bracket the time interval of sedimentation, even in strongly deformed, high-grade granulitic terranes (Collins et al., 2003; Upadhyay et al., 2009; Kooijman et al., 2011; Ibanez-Mejia et al., 2011, 2015; Sato et al., 2011a, b; 2012, Bhowmik et al., 2014; Sarkar et al., 2014; Sengupta et al., 2015). Fig. 16.5A and B show the results of concordant to near concordant (<2% discordance) U-Pb dates of detrital zircons for two of the samples (AS27, AS79) analyzed for this study. Lack of detrital component of zircon makes the other sample (AS17) unsuitable for provenance analyses or to bracket the age of sedimentation. Using the criteria mentioned earlier to constrain a permissible age range for sedimentary-protolith accumulation, we observe that detrital and metamorphic zircons analyzed from samples AS79 and AS27 constrain the age of sedimentation in the range from 1814 to 1639 Ma and 1696 to 1678 Ma, respectively.

In a previous study, Rekha et al. (2011) dated two paragneiss samples (DM-60 and DM112B) that are close to our location AS79. Of these, DM-60 is devoid of any detrital zircon and contains only metamorphic ones, whereas DM-112B contains both. The U-Pb ages of detrital zircon and its metamorphic overgrowth bracket the age of sedimentation between 1764 Ma and ~1650 Ma. Within the uncertainties of U-Pb zircon ages, the age of basin formation constrained from data given by Rekha et al. (2011) are in good agreement with the sedimentation age range we constrain from sample AS79. Compared to AS79, zircon populations of sample AS27, which was collected approximately 30 km west of AS79, suggest a more tightly bracketed age of sedimentation (i.e., in the range from 1696 to 1678 Ma). Lack of young detrital zircon (~1700 Ma) in AS79 can be explained by sample bias or complex hydrodynamic regimes of the feeder river systems.

5.2 Provenance Interpretations and Potential Sediment Source Areas

In order to trace the footprint (provenance) of the detrital zircons present in the metapelites, U-Pb ages of sample DM112B (Rekha et al., 2011) are combined with our results from samples AS79 and AS27 because of the close proximity of these sampling locations. As stated earlier, sample AS17 does not preserve any detrital zircon populations and therefore will not be considered in the following discussion. Concordant detrital zircon populations from samples AS27, AS79, and DM112B cover a wide time span ranging from 1696 to 2701 Ma (Fig. 16.5C). However, the apparent $^{207}Pb/^{206}Pb$ dates of these detrital zircon grains in the three samples define relatively narrow clusters within the following age ranges: 2700–2400 Ma, 2150 Ma, 1900–1850 Ma, and 1750–1700 Ma (Fig. 16.5C). Among these four age clusters, zircon populations that fall in the time spans of 1900–1850 Ma and 1750–1700 Ma dominate in terms of their frequencies (Fig. 16.5C). It seems very likely, therefore, that the protoliths of these rocks were sourced from a hinterland domain where rocks having ages ~1900–1700 Ma were abundant and data from the metamorphically grown zircon grains ascertain that after deposition these pelitic sediments suffered three tectonothermal events at c. 1640 Ma, c. 1450 Ma, and c. 950 Ma (Fig. 16.5D).

Tracing the source of the sedimentary protoliths of the studied metapelites from the age spectrum of the detrital zircon components is not straightforward. Some of the critical issues in these regards are as follows:

1. It is difficult to ascertain if the detrital zircons are directly derived from the rocks in which they crystallized (i.e., first cycle; directly derived from their primary source) or if they are polycyclic and derived from sedimentary sequences where they previously occurred as detrital components (secondary sources; Sengupta et al., 2015).
2. The rocks from which the detrital zircons were originally sourced may not have been dated yet (e.g., Collins et al., 2003; Moecher and Samson, 2006).
3. The actual source rocks might by hidden beneath younger supracrustal rocks or active sediments (e.g., beneath the cover of the Deccan Traps, Indo Gangetic Alluvium, and/or upper Vindhyan sedimentary sequences).
4. Their sources might have already been completely eroded away.
5. There exists a controversy regarding the time since when the Indian craton behaved like a unified land mass. Bhowmik et al. (2010, 2012) opined that the southern and northern

parts of the Indian shield got finally amalgamated at ~1000 Ma along the CITZ. This is in contradiction with the proposals of Meert et al. (2010), Chakraborty et al. (2012), and Meert and Pandit (2015), who considered widespread development of Paleo- to Neoproterozoic platformal deposits on a unified Indian landmass. The ~1700–1600 Ma old sedimentary basin formation ages constrained from parts of the Cuddappah Supergroup, Pakhal Group (both lying south of CITZ), and Lower Vindhyan rocks (lying north of CITZ) (Fig. 16.6) are consistent with the view of Chakraborti et al. (2012) and Meert et al. (2010). It is, however, possible that the Paleoproterzoic landmass of the Indian shield got separated in the subsequent times and got finally amalgamated along the CITZ at ~1000 Ma. In view of the existing information we favor the interpretation that the Indian shield behaved as a unified landmass at least from ~1700 Ma.

In spite of these complexities, an attempt has been made to identify the potential areas from where the detrital zircon grains of the studied rocks were sourced, by comparing them with the exposed rocks of the Indian shield and the Paleoproterozoic rocks now exposed in the lesser Himalaya. Fig. 16.7 shows the different magmatic and metamorphic events recorded in different parts of the Indian shield during the age range from Archean to early Mesoproterozoic times. Figs. 16.6 and 16.8 show the dated detrital zircon populations from sedimentary/metasedimentary sequences whose inferred depositional ages are older than the basin formation age constrained in this study. By combining the different tectonothermal events and their assigned ages as recorded in different parts of the Indian shield, the following inferences regarding the provenance of the studied metapelites are made:

1. Detrital zircon populations of the age group 2700–2450 Ma could have been derived from Neoarchaean rocks of the Aravalli, the Singhbhum, the Bastar, and/or the eastern and western Dharwar craton (Fig. 16.7 and the references therein). Some of the unmetamorphosed and/or metamorphosed sedimentary sequences from the lesser Himalaya, CITZ, Aravalli craton, Cuddapah Group, Eastern Ghat, and the Southern Granulite Terrane (SGT) contain detrital components of this age (Fig. 16.6) and can also serve as a potential source for some of the detrital zircons found in the studied CGGC metapelites. However, because of large distance with respect to the CGGC, rocks from the SGT, the Cuddapah, the Dharwar craton, and the southern Eastern Ghats do not seem to be the likely sources from which the Neoarchaean to early Paleoproterozoic detrital zircons found in our studied metapelites could have been derived. Therefore, we favor the interpretation that zircon grains in this age range were derived from the lesser Himalayan Proterozoic enclaves, the CITZ, Singhbhum, Aravalli, and/or Bastar craton.
2. Paleoproterozoic rocks with ages ~2100 Ma have been reported from a few areas including a suite of mafic dykes of Bastar craton (French et al., 2008; Srivastava et al., 2011) and Bundelkhand craton (MallikharjunaRao et al., 2005). Some metapelites from the southern part of CITZ (Bhowmik et al., 2005) as well as from the present lesser Himalayan region also contain ~2100 Ma detrital zircon (Fig. 16.6). In view of less fertile nature of mafic rocks as a potential source of zircons (Moecher and Samson, 2006; Dickinson, 2008), detrital components of zircons in the metapelites from the CITZ and Proterozoic lesser Himalayan sequence seems to be the likely source for the ~2100 Ma detrital zircon grains in the studied metapelite.

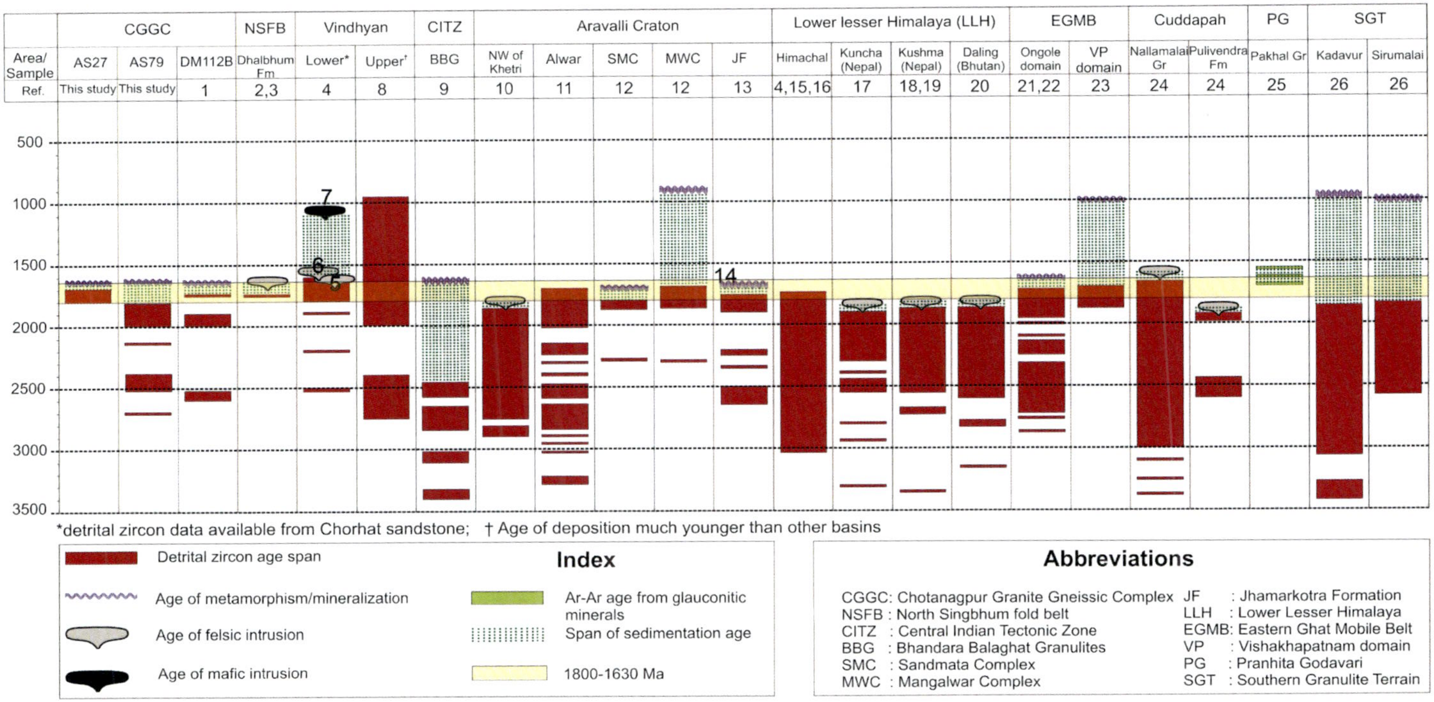

FIGURE 16.6 U-Pb ages of detrital zircon and age of oldest metamorphism/age of intrusion in Paleoproterozoic (meta)-sedimentary rocks from CGGC, NSFB, Vindhyan, CITZ, Aravalli, Himalaya, EGMB, Cudappah, PG valley, and SGT. The *yellow box* exhibits broadest span of basin formation age for CGGC metapelites. The numbers in figure represent references: (1) Rekha et al. (2011); (2) Nelson et al. (2007); (3) Reddy et al. (2009); (4) McKenzie et al. (2011); (5) Ray et al. (2002); (6) Rasmussen et al. (2002); (7) Gregory et al. (2006); (8) Malone et al. (2008); (9) Bhowmik et al. (2014); (10) Kaur et al. (2013); (11) Kaur et al. (2011a); (12) Buick et al. (2006); (13) McKenzie et al. (2013); (14) Deb et al. (1989); (15) Webb et al. (2011); (16) Richards et al. (2005); (17) Martin et al. (2011); (18) DeCelles (2000); (19) Gehrels et al. (2011); (20) Long et al. (2011); (21) Sarkar et al. (2014); (22) Henderson et al. (2014); (23) Bose et al. (2011); (24) Collins et al. (2014); (25) Conrad et al. (2011); (26) Kooijman et al. (2011).

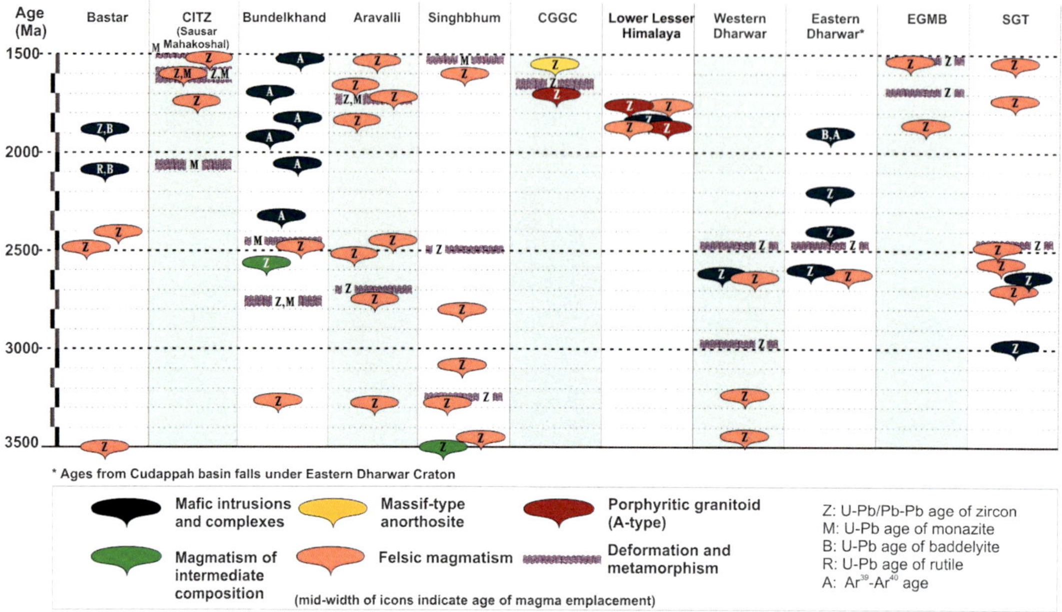

FIGURE 16.7 Timing of magmatic and metamorphic events in different cratons and mobile belts of India. References: Bastar Craton: Sarkar et al. (1993), Ghosh (2004), Bickford et al. (2014), Panigrahi et al. (2004), Panigrahi et al. (2002), Ahmad et al. (2009), Ahmad et al. (2008), Srivastava et al. (2011), French (2007), French et al. (2008); CITZ: Ahmad et al. (2009), Bhowmik et al. (2005), Bhandari et al. (2011), Bhowmik et al. (2011), Bora et al. (2013); Bundelkhand craton: Mondal et al. (2002), Saha et al. (2011), Mallikharjuna Rao et al. (2005); Aravalli Craton: Buick et al. (2006), Kaur et al. (2013), Wiedenbeck et al. (1996), Dharma Rao et al. (2011), Deb et al. (2002), Mukhopadhya et al. (2000), Kaur et al. (2009), Gupta et al. (1998), Biju-Sekhar et al. (2003), Kaur et al. (2011b), Kaur et al. (2007), Sarkar et al. (1989), Bhowmik et al. (2010), Sivaraman and Raval (1995); Singhbhum Craton: Chatterjee et al. (2010), Rekha et al. (2011), Nelson et al. (2007), Reddy et al. (2009), Mukhopadhyay et al. (2008), Acharyya et al. (2010), Mishra et al. (1999), Tait et al. (2011), Basu et al. (1996), Bandyopadhyay and Chakrabarti (2001), Misra et al. (2000), Mahato et al. (2008), Chatterjee et al. (2013); CGGC: Chatterjee et al. (2008), Rekha et al. (2011), Chatterjee and Ghose (2011); Lower Lesser Himalaya: Kohn et al. (2010); Western Dharwar Craton: Ramakrishnan and Vaidyanathan (2008), Rekha et al. (2013); Eastern Dharwar Craton: French et al. (2008), Ramakrishnan and Vaidyanathan (2008), Halls et al. (2007), Chatterjee and Bhattacharji (2001); EGMB: Bose et al. (2011), Dobmeier and Raith (2003); SGT: Ghosh et al. (2004), Shaji et al. (2014), Plavsa et al. (2012), Peucat et al. (2013), Brandt et al. (2014), Plavsa et al. (2014). Abbreviations are same as in Fig. 16.6.

3. The extant geochronological data shows that igneous/metamorphic rocks having ages between ~1900 and 1700 Ma are abundant in the rocks of Aravalli craton and in the Precambrian rocks of the Lesser Himalaya (Fig. 16.7). Some of the available data from the Eastern Ghats Belt also fall in this range (Fig. 16.7). Sedimentary sequences or their metamorphic equivalents from the Aravalli craton, Lesser Himalaya, Eastern Ghats Belt, and the SGT record detrital zircons from ~1900 to 1700 Ma (Fig. 16.6). In principle, rocks from some or all of these areas can act as the potential source domains for 1900–1700 Ma old detrital zircons present in the studied metapelites. The geochronologic comparison presented above provides several possibilities for the source domains from which the detrital populations of the CGGC metapelites could be derived, but in

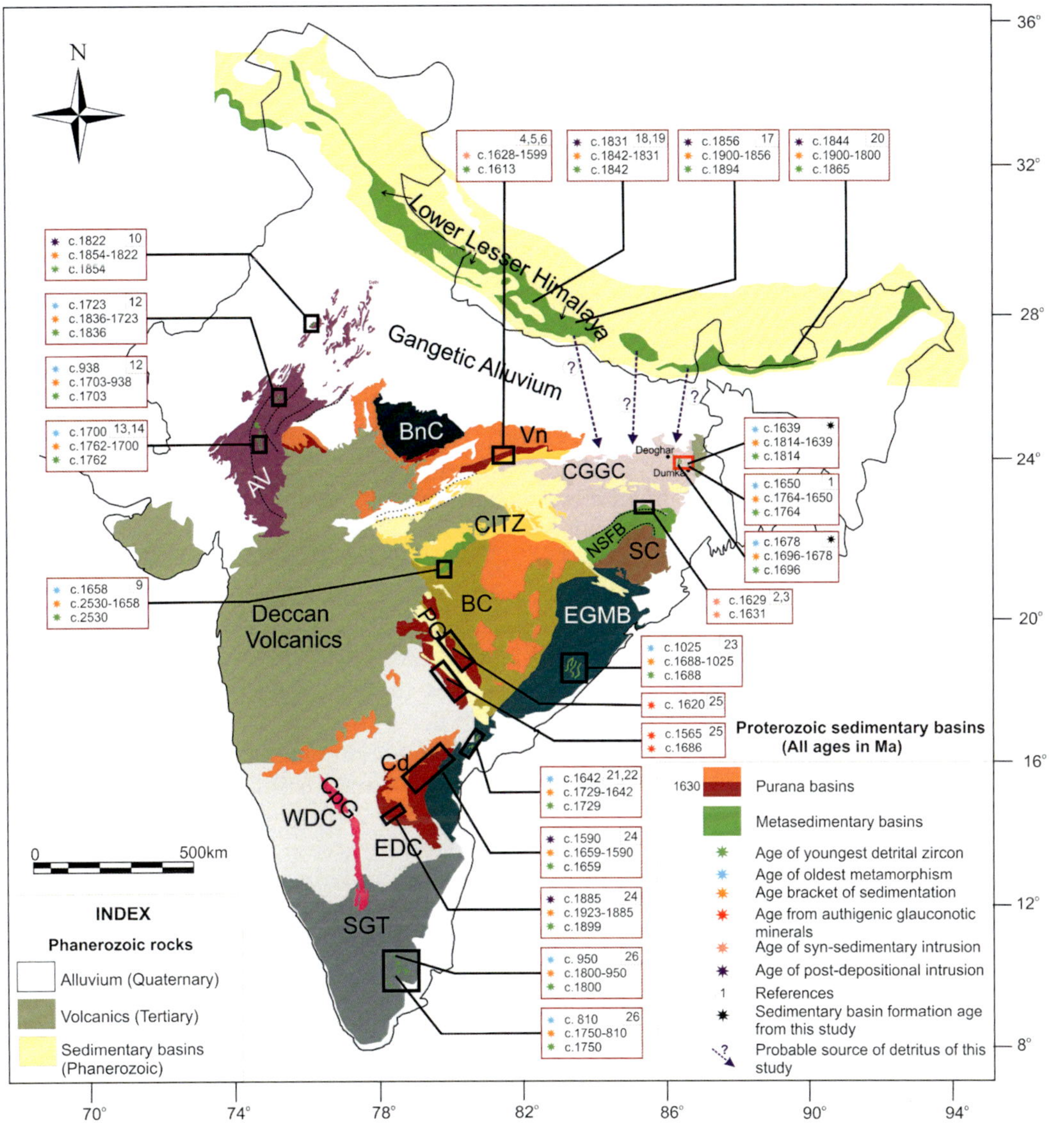

FIGURE 16.8 Geological and geochronological attributes of the crystalline basement and sedimentary cover in the Indian landmass. The purana basins in which sediment deposition started before 1630 Ma are shown in dark brown color. In Proterozoic metasedimentary basins also deposition started before 1630 Ma. Reference numbers at the corner of the boxes are same as in Fig. 16.6. The *blue dashed lines* with *arrow* indicate schematic sediment transport paths for protolith of CGGC metapelites from probable source of Lower Lesser Himalaya. Abbreviations are same as Fig. 16.1A.

view of their proximity to the CGGC, rocks from the Aravalli craton, the Bundelkhand craton, the Singhbhum Craton, the CITZ, and the Lesser Himalaya seem to be the most plausible sources. Based on the currently available geochronological data, and keeping the geographic position of CGGC in mind, it is proposed that the Paleoproterozoic rocks now occurring within Lesser Himalaya are the most likely candidates from which the dominant detrital zircon age populations (1900 Ma−1700 Ma) of our studied samples could have been sourced.

5.3 Tectonic Significance of the Paleoproterozoic Sedimentary Basins of India

In order to place the results obtained from this study in a broader tectonic context, Fig. 16.8 depicts some Paleoproterozoic basins (traditionally referred to as Purana basins) developed in the peninsular India and several places of the Lesser Himalaya. The timing of formation of these sedimentary basins is constrained from three different procedures, as follows: (1) ages of most of the metamorphosed sedimentary basins are constrained from dates obtained from detrital and metamorphic zircon grains (e.g., metasedimentary rocks from the CGGC presented here, the CITZ, Aravalli Fold Belt, Eastern Ghats, and Madurai Block of the SGT); (2) some sequences have been dated using ages from syn-sedimentary acid volcanics exposed within the sequence, such as those used to constrain the age of the Chandil Formation of the NSFB, metasedimentary rocks from Lesser Himalaya, and unmetamorphosed sedimentary sequences of the Cuddapah and lower Vindhyan basins; lastly, (3) some dates from authigenic glauconite have been used to constrain the age of the Pranhita-Godavari Basin. Because of absence of young age components in the detrital zircon spectra, only a broad time bracket could be obtained for some of the other metasedimentary basins (e.g., metasedimentary basins of CITZ, parts of EGB and SGT, Figs. 16.6 and 16.8). Notwithstanding these limitations, it is evident from Fig. 16.8 that the vestiges of Late Paleoproterozoic sedimentary sequences and their metamorphosed equivalent scatter over a vast domain of the Archaean-early Proterozoic basement of the Indian shield, covering an area wider than several tens of thousands of km^2. Several factors such as younger orogenic phases and consequent uplift and erosion, formation of younger volcanic and sedimentary cover sequences, and the limited geochronological dataset available to date, implies that the present extent of the Proterozoic sedimentary sequences reflects only a conservative minimum estimate of their original distribution (Meert and Pandit, 2015). Leaving aside the areas where relatively broad sedimentary basin formation age estimates have been obtained, Fig. 16.6 suggests that in many parts of the Indian shield Paleoproterozoic sedimentation occurred in two narrow time spans, namely ~1930−1830 Ma and ~1800−1630 Ma. Barring the rocks of the Pulivendla formation, metasedimentary units of the Khetri and Lesser Himalayan belt that belongs to the older age cluster, sedimentation estimates in all other areas including the CGGC are included within the younger age cluster of ~1800−1630 Ma. Several studies have documented that the massive Supercontinent Columbia, formed during ~1900−1800 Ma due to the welding of Laurentia, Baltica, Siberia, Amazonia, West Africa, South Africa, India, Australia, Antarctica, and North China, eventually broke down during the time period from 1600 to 1400 Ma (Rogers and Santosh, 2002, 2003, 2009; Condie, 2002; Zhao et al., 2011; Meert, 2012). In some places, an earlier period of relaxation (c. 1900−1700 Ma) immediately following the maximum packing of Columbia has also been reported (Goldberg, 2010; Kusky et al., 2007). One of the primary evidences of supercontinent break down is the formation of large-scale sedimentary basins associated with intracontinental rifting, and eventually by the development of stable continental shelf sediments associated to the postbreakup drift phase (i.e., quartzite + shale with or without limestone) (Dickinson, 1974). The Paleoproterozoic sedimentary basins that are depicted in Fig. 16.8 show a preponderance of quartzite + shale ± limestone sequence (or their metamorphic equivalents quartzite + quartzofeldspathic to metapelitic rocks + calc-silicates and marbles). Recent studies on these Purana sedimentary sequences have supported the idea that these basins were initiated in a rift-related setting (Bose et al., 2001; Conrad et al., 2011; Patranabis-Deb et al., 2012). This information and the constrained ages of sedimentation from several unmetamorphosed and metamorphosed Purana basins (Fig. 16.8) suggest large-scale fragmentation of the Archean and early Paleoproterozoic basement domains of Indian shield, possibly in association with the breakup of the Columbia supercontinent. It is not clear, however, from the extant geological information if this regional extension eventually led to the formation of true ocean basins, or only formed intracratonic basins floored by extended continental crust (like the Purana basins) in which the late Paleoproterozoic sedimentation took place. Nevertheless, in either of these cases the idea that the Indian shield behaved as a coherent landmass onto which late Paleoproterozoic sedimentation took place soon after the assembly of the Columbia supercontinent remains a distinct possibility. This contention is in good agreement with the study of Sengupta et al. (2015).

Acknowledgments

A.D. and S.M. acknowledge the financial support from the Council of Scientific and Industrial Research, New Delhi and University Grant Commission, New Delhi, respectively. P.S. and S.S. acknowledge the grants received from the programs University Potential for Excellence (Phase II), Jadavpur University and Center of Advance Studies (phase VI), Department of Geological Sciences, Jadavpur University. Staff members of the ALC are thanked for their kind assistance while obtaining the zircon CL images and geochronologic results presented in this study. We thank Prof. Nilanjan Chatterjee and Prof. Somnath Dasgupta for their critical but very constructive reviews that have significantly improved the clarity of the text. We thank Prof. Rajat Mazumder for inviting us to contribute with this chapter in the special volume on sediment provenance.

References

Acharyya, S.K., 2003. The nature of Mesoproterozoic Central Indian Tectonic Zone with exhumed and reworked older granulites. Gondwana Research 6 (2), 197–214.

Ameen, S.M.M., Wilde, S.A., Kabir, Z., Akon, E., Chowdhury, K.R., Khan, S.H., 2007. Palaeoproterozoic granitoids in the basement of Bangladesh: a piece of the Indian shield or an exotic fragment of the Gondwana jigsaw? Gondwana Research 12, 380–387.

Ahmad, T., Kaulina, T.V., Wanjari, N., Mishra, M.K., Nitkina, E.A., 2009. U-Pb zircon chronology and Sm-Nd isotopic characteristics of the Amgaon and Tirodi Gneissic Complex, Central Indian Shield: constraints on Precambrian crustal evolution. In: Singh, V.K., Chandra, R. (Eds.), Second International Conference on "Precambrian Continental Growth and Tectonism": Book of Abstracts, Bundelkhand University, Jhansi. Excel India Publishers, New Delhi, pp. 137–138.

Ahmad, T., Wanjari, N., Kaulina, T.V., Mishra, M.K., Nitkina, E.A., Geochemical and isotopic characteristics of the Amgaon gneissic complex, Central Indian Shield: constraints on Precambrian crustal evolution. In: Proceedings International Symposium 'Geoscience Resources and Environments of Asian Terranes (GREAT-2008), 4thIGCP516, and 5APSEG;' November 24–26, 2008, Bangkok, Thailand.

Acharyya, S.K., Gupta, A., Orihashi, Y., 2010. New U-Pb zircon ages from Palaeo- Mesoarchean TTG gneisses of the Singhbhum Craton, eastern India. Geochemical Journal 44, 81–88.

Basu, A., Patranabis-Deb, S., Schieber, J., Dhang, P.C., 2008. Stratigraphic position of the ~1000 Ma Sukhda Tuff (Chhattisgarh Supergroup, India) and the 500 Ma question. Precambrian Research 167, 383–388.

Basu, A., Bickford, M.E., 2014. Contributions of zircon U–Pb geochronology to understanding the volcanic and sedimentary history of some Purana basins, India. Journal of Asian Earth Sciences 91, 252–262.

Buick, I.S., Allen, C., Pandit, M., Rubatto, D., Hermann, J., 2006. The Proterozoic magmatic and metamorphic history of the Banded Gneiss Complex, central Rajasthan, India: LA-ICP-MS U–Pb zircon constraints. Precambrian Research 151, 119–142. http://dx.doi.org/10.1016/j.precamres.2006.08.006.

Bose, S., Dunkley, D.J., Dasgupta, S., Das, K., Arima, M., 2011. India-Antarctica-Australia-Laurentia connection in the Paleoproterozoic-Mesoproterozoic revisited: evidence from new zircon U-Pb and monazite chemical age data from the Eastern Ghats Belt, India. Geological Society of America Bulletin 123, 2031–2049. http://dx.doi.org/10.1130/B30336.1.

Bhattacharya, B.P., 1976. Metamorphism of the Precambrian rocks of the central part of SanthalParganas district, Bihar. Quaternary Journal of the Geology, Mining and Metallurgical Society of India 48, 183–196.

Bhowmik, S., Wilde, S.A., Bhandari, A., BasuSarbadhikari, A., 2014. Zoned Monazite and Zircon as Monitors for the Thermal History of Granulite Terranes: an example from the Central Indian Tectonic Zone. Journal of Petroleum 55 (3), 585–621.

Bhowmik, S.K., Bernhardt, H.-J., Dasgupta, S., 2010. Grenvillian age high-pressure upper amphibolite-granulite metamorphism in the Aravalli-Delhi Mobile Belt, Northwestern India: new evidence from monazite chemical age and its implication. Precambrian Research 178, 168–184. http://dx.doi.org/10.1016/j.precamres.2010.02.015.

Bhowmik, S.K., Wilde, S.A., Bhandari, A., Pal, T., Pant, N.C., 2012. Growth of the greater Indian Landmass and its assembly in Rodinia: geochronological evidence from the Central Indian Tectonic Zone. Gondwana Research 22, 54–72. http://dx.doi.org/10.1016/j.gr.2011.09.008.

Bickford, M.E., Basu, A., Kamenov, G.D., Mueller, P.A., Patranabis-Deb, S., Mukherjee, A., 2014. Petrogenesis of 1000 Ma Felsic Tuffs, Chhattisgarh and Indravati Basins, Bastar Craton, India: geochemical and Hf isotope constraints. Journal of Geology 122, 43–54. http://dx.doi.org/10.1086/674802.

Bhowmik, S.K., Basu Sarbadhikari, A., Spiering, B., Raith, M.M., 2005. Mesoproterozoic reworking of Palaeoproterozoic ultrahigh-temperature granulites in the Central Indian Tectonic Zone and its implications. Journal of Petroleum 46, 1085–1119. http://dx.doi.org/10.1093/petrology/egi011.

Bhandari, A., Chandra Pant, N., Bhowmik, S.K., Goswami, S., 2011. ~1.6 Ga ultrahigh-temperature granulite metamorphism in the Central Indian Tectonic Zone: insights from metamorphic reaction history, geothermobarometry and monazite chemical ages. Geological Journal 46, 198–216. http://dx.doi.org/10.1002/gj.1221.

Bhowmik, S.K., Wilde, S.A., Bhandari, A., 2011. Zircon U-Pb/Lu-Hf and monazite chemical dating of the Tirodi biotite gneiss: implication for latest Palaeoproterozoic to Early Mesoproterozoic orogenesis in the Central Indian Tectonic Zone. Geological Journal 46, 574–596. http://dx.doi.org/10.1002/gj.1299.

Bora, S., Kumar, S., Yi, K., Kim, N., Lee, T.H., 2013. Geochemistry and U-Pb SHRIMP zircon chronology of granitoids and microgranular enclaves from Jhirgadandi Pluton of Mahakoshal Belt, Central India Tectonic Zone, India. Journal of Asian Earth Sciences 70–71, 99–114. http://dx.doi.org/10.1016/j.jseaes.2013.03.006.

Biju-Sekhar, S., Yokoyama, K., Pandit, M.K., Okudaira, T., Yoshida, M., Santosh, M., 2003. Late Paleoproterozoic magmatism in Delhi Fold Belt, NW India and its implication: evidence from EPMA chemical ages of zircons. Journal of Asian Earth Sciences 22, 189–207. http://dx.doi.org/10.1016/S1367-9120(02)00188-8.

Basu, A.R., Sharma, M., Premo, W.R., 1996. U-Pb age of an older Metamorphic Group mica schist: earliest terrain of the eastern Indian Craton. In: Saha, A.K. (Ed.), Recent Researches in Geology and Geophysics of the Precambrians, Recent Researches in Geology, vol. 16. Hindustan Publishing Corporation, New Delhi, pp. 93–102.

Bandyopadhyay, P.K., Chakrabarti, A.K., Deomurari, M.P., Misra, S., 2001. 2.8 Ga old anorogenic granite-acid volcanic association from western margin of Singhbhum-Orissa craton, eastern India. Gondwana Research 4, 465–475.

Brandt, S., Raith, M.M., Schenk, V., Sengupta, P., Srikantappa, C., Gerdes, A., 2014. Crustal evolution of the Southern Granulite Terrane, south India: new geochronological and geochemical data for felsic orthogneisses and granites. Precambrian Research 246, 91–122. http://dx.doi.org/10.1016/j.precamres.2014.01.007.

Bose, P.K., Sarkar, S., Chakraborty, S., Banerjee, S., 2001. Overview of the Meso- to Neoproterozoic evolution of the Vindhyan basin, central India. Sedimentary Geology 141 (2), p.395–419.

Corfu, F., Hanchar, J.M., Hoskin, P.W.O., Kinny, P.D., 2003. Atlas of zircon textures. Reviews in Mineralogy and Geochemistry 53, 469–500.

Collins, A.S., Kröner, A., Fitzsimons, I.C.W., Razakamanana, T., 2003. Detrital foot-print of the Mozambique Ocean: U/Pb SHRIMP and Pb evaporation zircon geochronology of metasedimentary gneisses in Eastern Madagascar. Tectonophysics 375, 77–99.

Chatterjee, N., Mazumdar, A.C., Bhattacharya, A., Saikia, R.R., 2007. Mesoproterozoic granulites of the Shillong-Meghalaya Plateau: evidence of westward continuation of the Prydz Bay Pan-African suture into Northeastern India. Precambrian Research 152, 1–26.

Chatterjee, N., Ghose, N.C., 2011. Extensive Early Neoproterozoic high grade metamorphism in North Chotanagpur Gneissic Complex of the Central Indian Tectonic Zone. Gondwana Research 20, 362–379. http://dx.doi.org/10.1016/j.gr.2010.12.003.

Chatterjee, N., Crowley, J.L., Ghose, N.C., 2008. Geochronology of the 1.55Ga Bengal anorthosite and Grenvillian metamorphism in the Chotanagpur gneissic complex, eastern India. Precambrian Research 161, 303–316. http://dx.doi.org/10.1016/j.precamres.2007.09.005.

Chatterjee, N., Banerjee, M., Bhattacharya, A., Maji, A.K., 2010. Monazite chronology, metamorphism-anatexis and tectonic relevance of the Eastern Indian Tectonic Zone. Precambrian Research 179, 99–120.

Collins, A.S., Patranabis-Deb, S., Alexander, E., Bertram, C.N., Falster, G.M., Gore, R.J., et al., 2014. Detrital mineral age, radiogenic isotopic stratigraphy and tectonic significance of the Cuddapah Basin, India. Gondwana Research. http://dx.doi.org/10.1016/j.gr.2014.10.013.

Conrad, J.E., Hein, J.R., Chaudhuri, A.K., Patranabis-Deb, S., Mukhopadhyay, J., Deb, G.K., et al., 2011. Constraints on the development of Proterozoic basins in central India from 40Ar/39Ar analysis of authigenic glauconitic minerals. Geological Society of America Bulletin 123, 158–167. http://dx.doi.org/10.1130/B30083.1.

Chakraborty, P.P., Sarkar, S., Patranabis-Deb, S., 2012. Tectonics and sedimentation of Proterozoic Basins of Peninsular India. In: Proceedings of the Indian National Science Academy 78 No. 3 September, pp. 393–400.

Chatterjee, P., De, S., Ranaivoson, M., Mazumder, R., Arima, M., 2013. Geoscience Frontiers A review of the ~ 1600 Ma sedimentation, volcanism, and tectono-thermal events in the Singhbhum craton, Eastern India. Geoscience Frontiers 4, 277–287. http://dx.doi.org/10.1016/j.gsf.2012.11.006.

Chatterjee, N., Bhattacharji, S., 2001. Petrology, geochemistry and tectonic settings of the mafic dikes and sills associated with the evolution of the Proterozoic Cuddapah Basin of South India. Proceedings of the Indian Academy of Sciences (Earth and Planetary Science) 110, 433–453.

Condie, K.C., 2002. Breakup of a Paleoproterozoic supercontinent. Gondwana Research 5, 41–43.

Dey, A., Mukherjee, S., Sanyal, S., Sengupta, P., 2013. Petrology and chemical dating of monazite from khondalite from a part of Chhotanagpur Granite Gneiss Complex, Jharkhand, India: evidence for Neoproterozoic reworking of Mesoproterozoic high to ultra-high temperature lower crust. In: Abstract Volume of Precambrian Continental Growth and Tectonism, Jhansi, IAGR Conference Series No. 16, pp. 50–51.

Deb, M., Thorpe, R.I., Cumming, G.L., Wagner, P.A., 1989. Age, source and Strati-graphic implications of Pb isotope data for conformable, sediment-hosted, base-metal deposits in the Proterozoic Aravalli–Delhi Orogenic Belt, Northwestern India. Precambrian Research 43, 1–22.

DeCelles, P.G., 2000. Tectonic implications of U-Pb Zircon Ages of the Himalayan Orogenic Belt in Nepal. Science (80-.) 288, 497–499. http://dx.doi.org/10.1126/science.288.5465.497.

Dharma Rao, C.V., Santosh, M., Purohit, R., Wang, J., Jiang, X., Kusky, T., 2011. LA-ICP-MS U–Pb zircon age constraints on the Paleoproterozoic and Neoarchean history of the Sandmata Complex in Rajasthan within the NW Indian Plate. Journal of Asian Earth Sciences 42, 286–305. http://dx.doi.org/10.1016/j.jseaes.2011.01.018.

Deb, M., Thorpe, R., Krstic, D., 2002. Hindoli Group of Rocks in the Eastern Fringe of the Aravalli-Delhi Orogenic Belt-Archean Secondary Greenstone Belt or Proterozoic Supracrustals? Gondwana Research 5, 879–883. http://dx.doi.org/10.1016/S1342-937X(05)70922-9.

Dobmeier, C., Raith, M.M., 2003. Crustal architecture and evolution of the Eastern Ghats Belt and adjacent regions of India. In: Yoshida, M., Windley, B.F., Dasgupta, S. (Eds.), Proterozoic East Gondwana: Supercontinent Assembly and Breakup. Geological Society, London, Special Publications, 206, 169–202.

Dickinson, W.R., 2008. Impact of differential zircon fertility of granitoid basement rocks in North America on age populations of detrital zircons and implications for granite petrogenesis. Earth and Planetary Science Letters 275 (1–2), 80–92.

Dickinson, W.R., 1974. Plate tectonics and sedimentation. In: Dickinson, W.R. (Ed.), Tectonics and Sedimentation; Society of Economic Paleontologists and Mineralogists Special Publication. 22, 1–27.

Evans, P., 1964. The tectonic framework of Assam. Journal of the Geological Society of India 5, 80–96.

French, J.E., 2007. U-Pb Dating of the Paleoproterozoic Mafic Dyke Swarms of the South Indian Shield: Implications for Paleocontinental Reconstructions and Identifying Ancient Mantle Plume Events (Ph.D. thesis). University of Alberta, Edmonton, Alberta, 373 pp.

French, J.E., Heaman, L.M., Chacko, T., Srivastava, R.K., 2008. 1891–1883 Ma Southern Bastar–Cuddapah mafic igneous events, India: a newly recognized large igneous province. Precambrian Research 160, 308–322. http://dx.doi.org/10.1016/j.precamres.2007.08.005.

Gehrels, G., Valencia, V.A., Ruiz, J., 2008. Enhanced precision, accuracy, efficiency, and spatial resolution of U–Pb ages by laser ablation-multicollector-inductively coupled plasma-mass spectrometry. Geochemistry, Geophysics, Geosystems 9, 3.

Gregory, L.C., Meert, J.G., Pradhan, V., Pandit, M.K., Tamrat, E., Malone, S.J., 2006. A paleomagnetic and geochronologic study of the MajhgawanKimberlite, India: implications for the age of the VindhyanSuperGroup. Precambrian Research 149, 65–75.

Gehrels, G., Kapp, P., DeCelles, P., Pullen, A., Blakey, R., Weislogel, A., et al., 2011. Detrital zircon geochronology of pre-Tertiary strata in the Tibetan-Himalayan orogen. Tectonics 30. http://dx.doi.org/10.1029/2011TC002868 n/a–n/a.

Ghosh, J.G., 2004. 3.56 Ga tonalite in the central part of the Bastar Craton, India: oldest Indian date. Journal of Asian Earth Sciences 23, 359–364. http://dx.doi.org/10.1016/S1367-9120(03)00136-6.

Gupta, P., Guha, D.B., Chattopadhyay, B., 1998. Basement-cover relationship in the Khetri copper belt and the emplacement mechanism of the granite massifs, Rajasthan, India. Journal of the Geological Society of India 52, 417–432.

Ghosh, J.G., Dewit, M.J., Zartman, R.E., 2004. Age and tectonic evolution of Neoproterozoic ductile shear zones in the Southern Granulite Terrane of India, with implications for Gondwana studies. Tectonics 23. http://dx.doi.org/10.1029/2002TC001444. TC3006.

Goldberg, A.S., 2010. Dyke swarms as indicators of major extensional events in the 1.9–1.2 Ga Columbia supercontinent. Journal of Geodynamics 50, 176–190. http://dx.doi.org/10.1016/j.jog.2010.01.017.

Harley, S.L., Kelly, N.M., Möller, A., 2007. Zircon behaviour and the thermal histories of mountain chains. Elements 3, 25–30.

Hossain, I., Tsunogae, T., Rajesh, H.M., Chen, B., Arakawa, Y., 2007. Palaeoproterozoic U-Pb SHRIMP zircon age from basement rocks in Bangladesh: a possible remnant of the Columbia supercontinent. Comptes Rendus Geoscience 339, 979–986.

Henderson, B., Collins, A.S., Payne, J., Forbes, C., Saha, D., 2014. Geologically constraining India in Columbia: the age, isotopic provenance and geochemistry of the protoliths of the Ongole Domain, Southern Eastern Ghats, India. Gondwana Research 26, 888–906. http://dx.doi.org/10.1016/j.gr.2013.09.002.

Halls, H.C., Kumar, A., Srinivasan, R., Hamilton, M.A., 2007. Paleomagnetism and U–Pb geochronology of easterly trending dykes in the Dharwar craton, India: feldspar clouding, radiating dyke swarms and the position of India at 2.37 Ga. Precambrian Research 155, 47–68.

Ibanez-Mejia, M., Gehrels, G.E., Ruiz, J., Vervoort, J.D., Eddy, M.P., Li, C., 2014. Small-volume baddeleyite (ZrO2) U–Pb geochronology and Lu–Hf isotope geochemistry by LA-ICP-MS. Techniques and applications. Chemical Geology 384, 149–167.

Ibanez-Mejia, M., Pullen, A., Arenstein, J., Gehrels, G.E., Valley, J., Ducea, M.N., 2015. Unraveling crustal growth and reworking processes in complex zircons from orogenic lower-crust: the Proterozoic Putumayo Orogen of Amazonia. Precambrian Research 267, 285–310.

Ibanez-Mejia, M., Ruiz, J., Valencia, V.A., Cardona, A., Gehrels, G., Mora, A.R., 2011. The Putumayo Orogen of Amazonia and its implications for Rodinia reconstructions: new U–Pb geochronological insights into the Proterozoic tectonic evolution of northwestern South America. Precambrian Research 191, 58–77.

Kooijman, E., Upadhyay, D., Mezger, K., Raith, M.M., Berndt, J., Srikantappa, C., 2011. Response of the U–Pb chronometer and trace elements in zircon to ultrahigh-temperature metamorphism: the Kadavur anorthosite complex, southern India. Chemical Geology 290, 177–188. http://dx.doi.org/10.1016/j.chemgeo.2011.09.013.

Kaur, P., Zeh, A., Chaudhri, N., Gerdes, A., Okrusch, M., 2013. Nature of magmatism and sedimentation at a Columbia active margin: insights from combined U–Pb and Lu–Hf isotope data of detrital zircons from NW India. Gondwana Research 23, 1040–1052. http://dx.doi.org/10.1016/j.gr.2012.07.008.

Kaur, P., Zeh, A., Chaudhri, N., Gerdes, A., Okrusch, M., 2011a. Archaean to Palaeoproterozoic crustal evolution of the Aravalli mountain range, NW India, and its hinterland: the U–Pb and Hf isotope record of detrital zircon. Precambrian Research 187, 155–164. http://dx.doi.org/10.1016/j.precamres.2011.03.005.

Kaur, P., Chaudhri, N., Raczek, I., Kröner, A., Hofmann, A.W., Okrusch, M., 2011b. Zircon ages of late Palaeoproterozoic (ca. 1.72–1.70Ga) extension-related granitoids in NE Rajasthan, India: regional and tectonic significance. Gondwana Research 19, 1040–1053. http://dx.doi.org/10.1016/j.gr.2010.09.009.

Kaur, P., Chaudhri, N., Raczek, I., Kröner, A., Hofmann, A.W., 2009. Record of 1.82 Ga Andean-type continental arc magmatism in NE Rajasthan, India: insights from zircon and Sm–Nd ages, combined with Nd–Sr isotope geochemistry. Gondwana Research 16, 56–71. http://dx.doi.org/10.1016/j.gr.2009.03.009.

Kaur, P., Chaudhri, N., Raczek, I., Kröner, A., Hofmann, A.W., 2007. Geochemistry, zircon ages and whole-rock Nd isotopic systematics for Palaeoproterozoic A-type granitoids in the northern part of the Delhi belt, Rajasthan, NW India: implications for late Palaeoproterozoic crustal evolution of the Aravalli craton. Geological Magazine 144, 361. http://dx.doi.org/10.1017/S0016756806002950.

Kohn, M.J., Paul, S.K., Corrie, S.L., 2010. The lower Lesser Himalayan sequence: a Paleoproterozoic arc on the northern margin of the Indian plate. Geological Society of America Bulletin 122, 323–335. http://dx.doi.org/10.1130/B26587.1.

Kusky, T., Li, J., Santosh, M., 2007. The Paleoproterozoic North Hebei Orogen: North China craton's collisional suture with the Columbia supercontinent. Gondwana Research. http://dx.doi.org/10.1016/j.gr.2006.11.012.

Ludwig, K.R., 2003. Isoplot/Ex Version 3.00: A Geochronological Toolkit for Microsoft Excel. Berkeley Geochronology Center, Berkeley, CA.

Long, S., McQuarrie, N., Tobgay, T., Rose, C., Gehrels, G., Grujic, D., 2011. Tectonostratigraphy of the Lesser Himalaya of Bhutan: implications for the along-strike stratigraphic continuity of the northern Indian margin. Geological Society of America Bulletin 123, 1406–1426. http://dx.doi.org/10.1130/B30202.1.

Malone, S.J., Meert, J.G., Banerjee, D.M., Pandit, M.K., Tamrat, E., Kamenov, G.D., Pradhan, V.R., Sohl, L.E., 2008. Palaeomagnetism and detrital zircon geochronology of the Upper Vindhyan Sequence, Son Valley and Rajasthan, India: a c. 1000 Ma closure age for the Purana basins? Precambrian Research 164, 137–159.

Maji, K., Goon, S., Bhattacharya, A., Mishra, B., Mahato, S., Bernhardt, H.-J., 2008. Proterozoic polyphasemetamorphism in the Chhotanagpur Gneissic Complex (India), and implication for trans-continental Gondwanaland correlation. Precambrian Research 162, 385–402. http://dx.doi.org/10.1016/j.precamres.2007.10.002.

Mukherjee, S., Dey, A., Sanyal, S., Ibanez-Mejia, M., Sengupta, P., 2015. Petrology and U-Pb Geochronology of Zircon in a Suite of Charnockitic Gneisses from parts of the Chotanagpur Granite Gneiss Complex (CGGC). In: East Indian Shield: Evidence for Reworking of a Mesoproterozoic Basement during the Formation of Rodinia Supercontinent. Abstract Volume of the International Symposium on Antarctic Earth Sciences, Goa, pp. 110–111.

McKenzie, N.R., Hughes, N.C., Myrow, P.M., Xiao, S., Sharma, M., 2011. Correlation of Precambrian–Cambrian sedimentary successions across northern India and the utility of isotopic signatures of Himalayan lithotectonic zones. Earth and Planetary Science Letters 312, 471–483. http://dx.doi.org/10.1016/j.epsl.2011.10.027.

McKenzie, N.R., Hughes, N.C., Myrow, P.M., Banerjee, D.M., Deb, M., Planavsky, N.J., 2013. New age constraints for the Proterozoic Aravalli–Delhi successions of India and their implications. Precambrian Research 238, 120–128. http://dx.doi.org/10.1016/j.precamres.2013.10.006.

Martin, A.J., Burgy, K.D., Kaufman, A.J., Gehrels, G.E., 2011. Stratigraphic and tectonic implications of field and isotopic constraints on depositional ages of Proterozoic Lesser Himalayan rocks in central Nepal. Precambrian Research 185, 1–17. http://dx.doi.org/10.1016/j.precamres.2010.11.003.

Moecher, D.P., Samson, S.D., 2006. Differential zircon fertility of source terranes and natural bias in the detrital zircon record: implications for sedimentary provenance analysis. Earth and Planetary Science Letters 247, 252–266.

Meert, J.G., Pandit, M.K., Pradhan, V.R., Banks, J., Sirianni, R., Stroud, M., Newstead, B., Gifford, J., 2010. Precambrian crustal evolution of Peninsular India: a 3.0 billion year odyssey. Journal of Asian Earth Sciences 39, 483–515. http://dx.doi.org/10.1016/j.jseaes.2010.04.026.

Meert, J.G., Pandit, M.K., 2015. The Archaean and Proterozoic history of Peninsular India: tectonic framework for Precambrian sedimentary basins in India. Geological Society, London, Memoirs 43, 29–54.

Mondal, M.E., Goswami, J., Deomurari, M., Sharma, K., 2002. Ion microprobe $^{207}Pb/^{206}Pb$ ages of zircons from the Bundelkhand massif, northern India: implications for crustal evolution of the Bundelkhand–Aravalliprotocontinent. Precambrian Research 117, 85–100. http://dx.doi.org/10.1016/S0301-9268(02)00078-5.

MallikharjunaRao, J., PoornachandraRao, G.V.S., Widdowson, M., Kelley, S.P., 2005. Evolution of Proterozoic mafic dyke swarms of the Bundelkhand Granite Massif, Central India. Current Science 88, 502–506.

Mukhopadhya, D., Bhattacharyya, T., Chattopadhyay, N., Lopez, R., Tobisch, O.T., 2000. Anasagar gneiss: a folded granitoid pluton in the Proterozoic South Delhi Fold Belt central Rajasthan. Proceedings of the Indian Academy of Sciences (Earth and Planetary Science) 109, 21–37.

Mukhopadhyay, J., Beukes, N.J., Armstrong, R.A., Zimmermann, U., Ghosh, G., Medda, R.A., 2008. Dating the oldest Greenstone in India: a 3.51 Ga precise U-Pb SHRIMP zircon age for dacitic lava of the Southern Iron Ore Group, Singhbhum Craton. Journal of Geology 116, 449–461.

Mishra, S., Deomurari, M.P., Wiedenbeck, M., Goswami, J.N., Ray, S., Saha, A.K., 1999. $^{207}Pb/^{206}Pb$ zircon ages and the evolution of the Singhbhum Craton, eastern India: an ion microprobe study. Precambrian Research 93, 139–151.

Misra, S., Moitra, S., Bhattacharya, S., Sivaraman, T.V., 2000. Archean granitoids at the Contact of Eastern Ghat Granulite Belt and Singhbhum-Orissa Craton, in Bhuban-Rengali Sector, Orissa, India. Gondwana Research 3, 205–213. http://dx.doi.org/10.1016/S1342-937X(05)70097-6.

Mahato, S., Goon, S., Bhattacharya, A., Mishra, B., Bernhardt, H., 2008. Thermo-tectonic evolution of the North Singhbhum Mobile Belt (eastern India): a view from the western part of the belt. Precambrian Research 162, 102–127. http://dx.doi.org/10.1016/j.precamres.2007.07.015.

Meert, J.G., 2012. What's in a name? The Columbia (Paleopangaea/Nuna) supercontinent. Gondwana Research 21, 987–993. http://dx.doi.org/10.1016/j.gr.2011.12.002.

Nelson, D.R., Bhattacharya, H.N., Misra, S., Dasgupta, N., Altermann, W., 2007. New SHRIMP U–Pb zircon dates from the Singhbhum craton, Jharkhand– Orissa region, India. In: Banerjee, S. (Ed.), Abstract Volume of the International Conference on Precambrian Sedimentation and Tectonics and Second GPSS Meeting. Indian Institute of Technology, Bombay, p. 47.

Paces, J.B., Miller, J.D.J., 1993. Precise U–Pb ages of Duluth Complex and related mafic intrusions, northeastern Minnesota: geochronological insights to physical, petrogenetic, paleomagnetic, and tectonomagmatic processes associated with the 1.1 Ga Midcontinent Rift System. Journal of Geophysical Research 98, 13997–14013.

Panigrahi, M.K., Bream, B.R., Misra, K.C., Naik, R.K., 2004. Age of granitic activity associated with copper/molybdenum mineralization at Malanjkhand, Central India. Mineral Deposition 39, 670–677. http://dx.doi.org/10.1007/s00126-004-0441-0.

Panigrahi, M.K., Misra, K.C., Bream, B., Naik, R.K., 2002. Genesis of the granitoid affiliated copper-molybdenum mineralization at Malanjkhand central India: facts and problems. In: Proceedings of the 11th Quadrennial IAGOD Symposium and Geocongress, Windhoek, Namibia (Electronic Extended Abstract).

Plavsa, D., Collins, A.S., Foden, J.F., Kropinski, L., Santosh, M., Chetty, T.R.K., et al., 2012. Delineating crustal domains in Peninsular India : age and chemistry of orthopyroxene-bearing felsic gneisses in the Madurai Block. Precambrian Research 198–199, 77–93. http://dx.doi.org/10.1016/j.precamres.2011.12.013.

Peucat, J.J., Jayananda, M., Chardon, D., Capdevila, R., Fanning, C.M., Paquette, J.L., 2013. The lower crust of the Dharwar Craton, Southern India: Patchwork of Archean granulitic domains. Precambrian Research 227, 4–28. http://dx.doi.org/10.1016/j.precamres.2012.06.009.

Plavsa, D., Collins, A.S., Payne, J.L., Foden, J.D., Clark, C., Santosh, M., 2014. Detrital zircons in basement metasedimentary protoliths unveil the origins of southern India. Geological Society of America Bulletin 126, 791–811. http://dx.doi.org/10.1130/B30977.1.

Patranabis-Deb, S., Saha, D., Tripathy, V., 2012. Basin stratigraphy, sea-level fluctuations and their global tectonic connections—evidence from the Proterozoic Cuddapah Basin. Geological Journal. http://dx.doi.org/10.1002/gj.1347.

Rekha, S., Upadhyay, D., Bhattacharya, A., Kooijman, E., Goon, S., Mahato, S., Pant, N.C., 2011. Lithostructural and chronological constraints for tectonic restoration of Proterozoic accretion in the Eastern Indian Precambrian shield. Precambrian Research 187, 313–333.

Rubatto, D., 2002. Zircon trace element geochemistry: partitioning with garnet and the link between U–Pb ages and metamorphism. Chemical Geology 184, 123–138.

Reddy, S.M., Clarke, C., Mazumder, R., 2009. Temporal constraints on the evolution of the Singhbhum Crustal Province from U-Pb SHRIMP data. In: Saha, D., Mazumder, R. (Eds.), Paleoproterozoic Supercontinents and Global Evolution. International Association for Gondwana Research Conference Series, Abstract Volume, vol. 9, pp. 17–18.

Ray, J.S., Martin, M.W., Veizer, J., Bowring, S.A., 2002. U-Pb zircon dating and Sr isotope systematics of the VindhyanSupergroup, India. Geology 30, 131–134. http://dx.doi.org/10.1130/0091-7613(2002)030<0131:upzdas>2.0.co;2.

Rasmussen, S., Bose, P.K., Sarkar, S., Banerjee, S., Fletcher, I.R., McNaughton, N.J., 2002. 1.6 Ga U-Pb zircon age for the Chorhat Sandstone, lower Vindhyan, India: possible implications for early evolution of animals. Geological Society of America 30, 103. http://dx.doi.org/10.1130/0091-7613(2002)030<0103:GUPZAF>2.0.CO;2.

Richards, A., Argles, T., Harris, N., Parrish, R., Ahmad, T., Darbyshire, F., et al., 2005. Himalayan architecture constrained by isotopic tracers from clastic sediments. Earth and Planetary Science Letters 236, 773–796. http://dx.doi.org/10.1016/j.epsl.2005.05.034.

Ramakrishnan, M., Vaidyanathan, R., 2008. Geology of India. Geological Society of India, Bangalore, vol. 1.

Rekha, S., Viswanath, T.A., Bhattacharya, A., Prabhakar, N., 2013. Meso/Neoarchean crustal domains along the north Konkan coast, western India: the Western Dharwar Craton and the Antongil- Masoara Block (NE Madagascar) connection. Precambrian Research 233, 316–336.

Rogers, J.J.W., Santosh, M., 2002. Configuration of Columbia, a Mesoproterozoic supercontinent. Gondwana Research 5, 5–22.

Rogers, J.J.W., Santosh, M., 2003. Supercontinents in Earth history. Gondwana Research 6, 357–368.

Rogers, J.J.W., Santosh, M., 2009. Tectonics and surface effects of the supercontinent Columbia. Gondwana Research 15, 373–380.

Sengupta, P., Raith, M.M., Kooijman, E., Talukdar, M., Chowdhury, P., Sanyal, S., Mezger, K., Mukhopadhyay, D., 2015. Provenance, timing of sedimentation and metamorphism of metasedimentary rock suites from the Southern Granulite Terrane, India. Geological Society, London, Memoirs 43, 297–308.

Sanyal, S., Sengupta, P., 2012. Metamorphic evolution of the Chhotanagpur Granite Gneiss Complex of the East Indian Shield: current status. Geological Society, London, Special Publications 365, 117–145. http://dx.doi.org/10.1144/SP365.7.

Scherer, E.E., Whitehouse, M.J., Münker, C., 2007. Zircon as a monitor of crustal growth. Elements 3, 19–24.

Sato, K., Santosh, M., Tsunogae, T., Chetty, T.R.K., Hirata, T., 2011a. Subduction–accretion–collision history of the Gondwana suture in southern India: a laser ablation ICP-MS study of zircon chronology. Journal of Asian Earth Sciences 40, 162–171.

Sato, K., Santosh, M., Tsunogae, T., Chetty, T.R.K., Hirata, T., 2011b. Laser ablation ICP mass spectrometry for zircon U–Pb geochronology of metamorphosed granite from the Salem Block: implication for Neoarchaean crustal evolution in southern India. Journal of Mineralogical and Petrological Sciences 106, 1–12.

Sato, K., Santosh, M., Chetty, T.R.K., Hirata, T., 2012. U–Pb zircon geochronology of granites and charnockite from southern India: implications for magmatic pulses associated with plate tectonic cycles within a Precambrian suture zone. Geological Journal 47, 237–252.

Sarkar, T., Schenk, V., Appel, P., Berndt, J., Sengupta, P., 2014. Two-stage granulite formation in a Proterozoic magmatic arc (Ongole domain of the Eastern Ghats Belt, India): Part 2. LA-ICP-MS zircon dating and texturally controlled in-situ monazite dating. Precambrian Research 255, 467–484.

Sarkar, G., Corfu, F., Paul, D.K., McNaughton, N.J., Gupta, S.N., Bishui, P.K., 1993. Early Archaean crust in Bastar craton, central India: a geochemical and isotopic study. Precambrian Research 62, 127–137.

Srivastava, R.K., Heaman, L.M., French, J.E., Filho, C.F.F., 2011. Evidence for a Paleoproterozic event of metamorphism in the Bastar craton, Central India: P-T-t constraints from mineral chemistry and U-Pb geochronology of maficdykes. Episodes 34 (1), 113–124.

Saha, L., Pant, N.C., Pati, J.K., Upadhyay, D., Berndt, J., Bhattacharya, A., et al., 2011. Neoarchean high-pressure margarite–phengitic muscovite–chlorite corona mantled corundum in quartz-free high-Mg, Al phlogopite–chlorite schists from the Bundelkhand craton, north central India. Contributions to Mineralogy and Petrology 161, 511–530. http://dx.doi.org/10.1007/s00410-010-0546-7.

Sarkar, G., Barma, T.R., Corfu, F., 1989. Timing of continental arc magmatism in northwest India: evidence from U–Pb geochronology. Journal of Geology 97, 607–612.

Sivaraman, T.V., Raval, U., 1995. U–Pb isotopic study of zircons from a few granitoids of Delhi–Aravalli Belt. Journal of the Geological Society of India 46, 461–475.

Shaji, E., Santosh, M., He, X.-F., Fan, H.-R., Dev, S.G.D., Yang, K.-F., et al., 2014. Convergent margin processes during Archean–Proterozoic transition in southern India: geochemistry and zircon U–Pb geochronology of gold-bearing amphibolites, associated metagabbros, and TTG gneisses from Nilambur. Precambrian Research 250, 68–96. http://dx.doi.org/10.1016/j.precamres.2014.05.021.

Tait, J., Zimmermann, U., Miyazaki, T., Presnyakov, S., Chang, Q., Mukhopadhyay, J., Sergeev, S., 2011. Possible juvenile Palaeoarchaean TTG magmatism in eastern India and its constraints for the evolution of the Singhbhum craton. Geological Magazine 148, 340–347.

Upadhyay, D., Gerdes, A., Raith, M.M., 2009. Unraveling sedimentary provenance and tectonothermal history of high to ultra-high temperature metapelites using zircon and monazite chemistry: a case study from the Eastern Ghats Belt, India. Journal of Geology 117, 665–683.

Verma, R.K., 1985. Gravity Field, Seismicity and Tectonics of the Indian Peninsula and the Himalayas. Allied Publishers Private Limited, New Delhi, p. 213.

Webb, A.G., Yin, A., Harrison, T.M., Célérier, J., Gehrels, G.E., Manning, C.E., Grove, M., 2011. Cenozoic tectonic history of the Himachal Himalaya (northwestern India) and its constraints on the formation mechanism of the Himalayan orogen. Geosphere 7 (4), 1013–1061. http://dx.doi.org/10.1130/GES00627.1.

Wiedenbeck, M., Goswami, J.N., Roy, A.B., 1996. 1996. An ion microprobestudy of single zircons from the Amet granite, Rajasthan. Journal of the Geological Society of India 48, 127–137.

Zhao, S., Li, M., Sun, S.A., 2011. Wilde assembly, accretion, and break-up of the Palaeo-Mesoproterozoic Columbia supercontinent: record in the North China Craton revisited. International Geology Review 53 (11–12), 1331–1356.

CHAPTER

17

SEM-CL Fabric Analysis of Quartz Framework Population From the Mesoarchean Keonjhar Quartzite From Singhbhum Craton, Eastern India: Implications for the Understanding of the Upper Continental Crust

S. De[1], *J. Mukhopadhyay*[1,3], *C.A. Rosiere*[2], *S. Ghosh*[1]

[1]Presidency University, Kolkata, India; [2]Federal University of Minas Gerais, Belo Horizonte, Brazil; [3]University of Johannesburg, Auckland Park, South Africa

OUTLINE

Sediment Provenance
http://dx.doi.org/10.1016/B978-0-12-803386-9.00017-4

1. INTRODUCTION

Provenance study of siliciclastic successions is based primarily on petrographic and chemical compositional analyses (Crook, 1974; Schwab, 1975; Dickinson and Suczek, 1979). Standard petrographic procedures for the determination of arenite modal composition have been fruitfully used to distinguish different tectonic settings of the source area (e.g., Dickinson and Suczek, 1979; Dickinson, 1985; Zuffa, 1985). Geochemical proxies using several major, trace, and rare earth element (REE) compositional trends also provide reasonable discrimination among different tectonic settings of the source terrane (Bhatia, 1983; Bhatia and Crook, 1986; for an alternative view, see also Basu et al., 2016). Besides the bulk rock petrographic and chemical compositional data, the nature of the major mineral component, i.e., the characteristics of the quartz grains in sandstones, has been employed by some researchers to obtain additional information about the proportion of different constituents such as shallow or deep-crustal, deformed or undeformed components in the source terrane (Johnsson and Basu, 1993). Framework quartz being the most resistant component in arenites has an advantage over petrographic and geochemical proxies from bulk rocks that are variably altered during diagenesis. Cathodoluminescence (CL) analysis of quartz population is an important tool for discriminating grains of diverse origin to characterize the source terrane. Fabric investigations using a CL detector attached to a scanning electron microscope (SEM) is a relatively new tool for differentiating quartz from different kinds of source rocks in provenance studies (e.g., Zinkernagel, 1978; Matter and Ramseyer, 1985; Marshall, 1988). Quartz grains under SEM-CL display a variety of internal features such as zoning, healed fractures, dark CL streaks, and patches, which do not appear under the petrographic microscope. The variation in the luminescence intensity (differential CL) from dark (weak or no CL) to bright (strong CL) is related to the crystallization and cooling history of minerals, as well as later metamorphic changes. Seyedolali et al. (1997) proposed the application of variable luminosity patterns and related microstructures to recognize igneous, metamorphic, and shock-metamorphic features, and Boggs and Krinsley (2006) used it as an indication of different conditions of their origin under varying composition of parent fluids.

Zinkernagel (1978) and Augustsson et al. (2011) applied color CL for the study of quartz and detrital zircons from sandstone in interpreting changes in sediment source. SEM-CL fabric analysis of the quartz grains reveals additional details of internal structure when used in conjunction with optical petrographic studies. Kwon and Boggs (2002) used both petrographic and SEM-CL techniques to evaluate the provenance of Tertiary sandstone samples in the Cheju Basin, northeast East China Sea, southwest of the Korean Peninsula. Bernet and Bassett (2005) examined the optical properties of quartz grains in quartz-rich sedimentary rocks of New Zealand and compared these properties to the features revealed by SEM-CL imaging of the same grains.

Here we applied SEM-CL to the study of framework quartz grains with the purpose to examine the nature of the source terrane for the Mesoarchean quartzarenites of the Keonjhar Quartzite (Ghosh et al., 2016) from the Singhbhum craton, eastern India (Fig. 17.1). Ghosh et al. (2016) based on petrographic and geochemical proxies suggest a craton interior to recycled orogen source terrane settings for these arenites. We demonstrate how the SEM-CL technique, when applied in conjunction with petrographic analyses, makes it possible to distinguish the relative proportions of different rock types in order to characterize the upper crustal constitution during the Paleo-Mesoarchean.

2. GEOLOGIC BACKGROUND

The cratonic block of the Singhbhum–Odisha region in eastern India is mainly constituted by Precambrian rocks ranging in age from Paleoarchean to Mesoproterozoic (Saha, 1994) (Fig. 17.1). The core of the craton is mainly comprised by the multicomponent Singhbhum Granite batholith with ages ranging between 3.4 and 3.1 Ga (Mishra et al., 1999; Mukhopadhyay, 2001; Acharyya et al., 2010; Tait et al., 2011; Nelson et al., 2014; Upadhyay et al., 2014). The batholith includes enclaves of Paleo-Mesoarchean (3.51 to 3.39 Ga) (BIF) banded iron formation-bearing greenstone belts (Acharyya, 1993) collectively known as the Iron Ore Group (IOG) (Saha, 1994). The granitoid–greenstone core of the craton is unconformably overlain by Meso- to Neoarchean siliciclastic rocks exposed in a several detached outcrop belts bordering the core that include the Phuljhari Formation in the lower parts of the Dhanjori succession in the north, and in the Keonjhar, Daitari, and Pallahara areas in the south and west of the block (Mukhopadhyay et al., 2014). The Keonjhar Quartzite, which we have studied for the SEM-CL analysis, belongs to the Mesoarchean siliciclastic sequence that is exposed west of Keonjhar (Fig. 17.1). This unit is primarily made up of aggradational parasequences of shallow marine wave-tide-reworked shelf arenites (Mukhopadhyay et al., 2012; Ghosh et al., 2016), and unconformably overlies the Singhbhum Granite and the IOG locally with the Mesoarchean Keonjhar Paleosol along the unconformity (Mukhopadhyay et al., 2014).

The Archean cratonic core is bordered in the north by the Singhbhum Shear Zone, which traces the limits to the Neoarchean to Mesoproterozoic North Singhbhum Mobile Belt (NSMB). The NSMB includes volcaniclastic and siliciclastic formations of diverse depositional settings ranging from aeolian to fluvial and shallow marine (see review of Mazumder et al., 2012), and comprises three major stratigraphic units, namely the Dhanjori Group, the Singhbhum Group, and the Dalma Volcanics (Mazumder et al., 2012). The siliciclastic-carbonate succession of the late Mesoproterozoic Kolhan Group (Mukhopoadhyay et al., 2006) records the youngest stratigraphic unit on the Singhbhum craton. The Kolhan Group unconformably overlies the granite-greenstone basement in the northwestern part of the granitoid core. The Chottanagpur Granite Gneissic Complex (CGGC) lies in north of the NSMB. High-grade metamorphic rocks of the Rengali Province and the Eastern Ghats mobile belt are tectonically juxtaposed with the granite-greenstone craton along its southern margin.

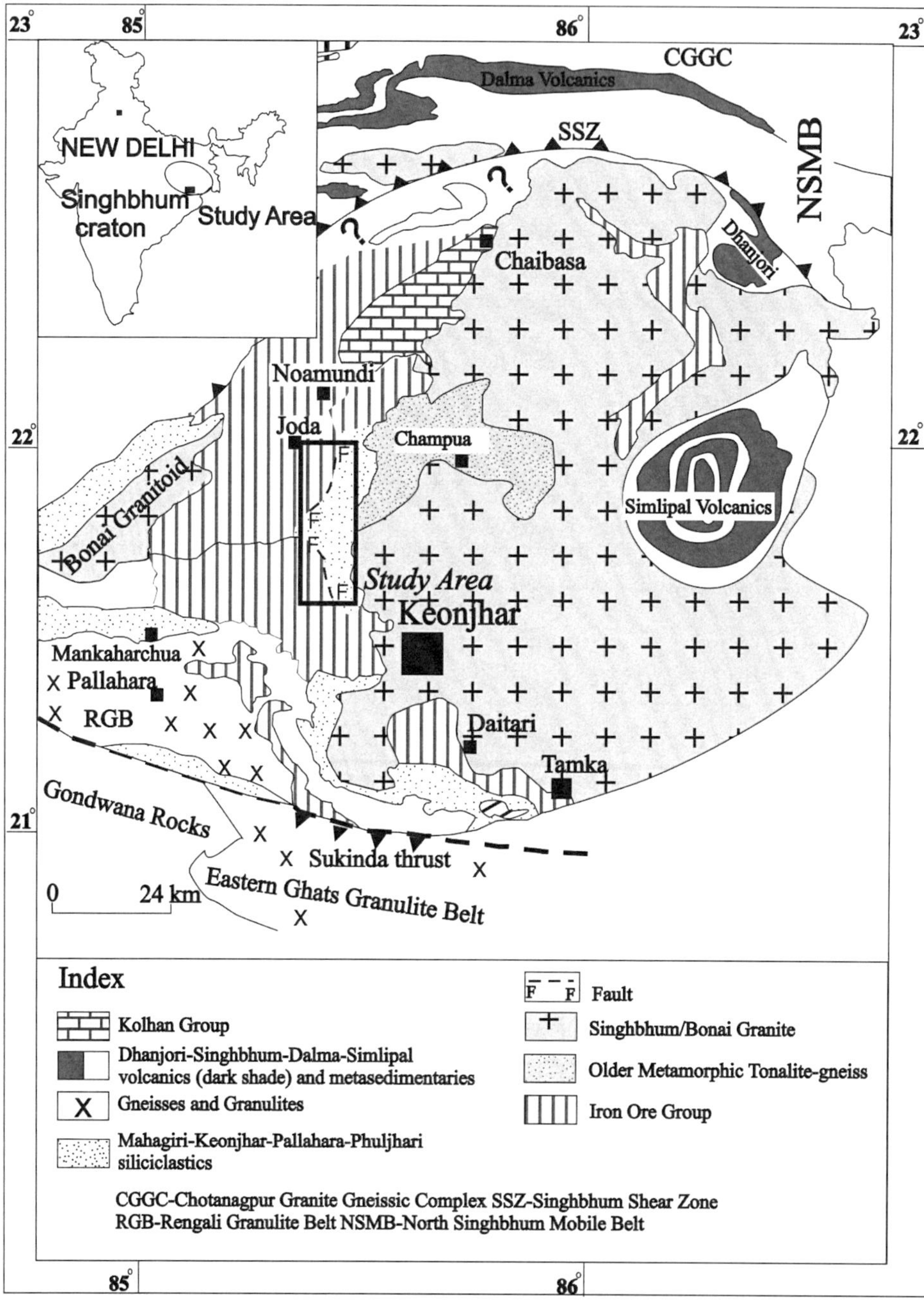

FIGURE 17.1 Generalized geological map of the Singhbhum craton. Modified after Saha (1994).

3. MATERIALS AND METHODS

Fresh medium-grained quartz arenite samples were collected from the study area covering the range of the stratigraphy. Polished thin-sections were prepared for petrographic and SEM- CL studies. The petrographic study was carried out with a Nikon POl200 petrological microscope at Presidency University. Modal compositional analyses of arenites were carried out on framework components such as quartz (Q), feldspar (F), and lithic fragments (L) for classifying the arenites according to the ternary discrimination plots after Dickinson (1985). A minimum of 750 representative points were counted from each thin section. The notable feature of these arenites is the near absence of feldspar in them. Petrographic studies reveal that though the matrix proportion is very low in these arenites, the matrix is now represented by epi- or pseudomatrix. The absence of feldspar points to the possibility that given the Mesoarchean age of the arenite, feldspars might have undergone burial diagenetic transformation into clay matrix (Ghosh et al., 2016). To account for the presumable diagenetic changes in a closed system for the siliciclastic sediments (cf. Ali et al., 2014), we used the chemical composition of the arenites following the method of Cox and Lowe (1996) to calculate the QtFLt (total quartz-feldspar-lithic fragments, Fig. 17.3A), Qm-F-Lt (monocrystalline quartz-feldspar and lithic fragments, Fig. 17.3B), and Qm-P-K (monocrystalline quartz-plagioclase feldspar and K-feldspar, Fig. 17.3C) ternary plots (Dickinson, 1985), using both chemical (Table 17.1) and petrographic analyses to evaluate the detrital constituents of the matrix.

TABLE 17.1 Major Element Composition of the Arenites

Sample Number	SiO_2 (wt%)	Al_2O_3 (wt%)	Fe_2O_3 (wt%)	MnO (wt%)	MgO (wt%)	CaO (wt%)	Na_2O (wt%)	K_2O (wt%)	TiO_2 (wt%)	P_2O_5 (wt%)	LOI (wt%)	Total (wt%)
KNJ-48A	92.76	4.21	0.37	0.01	0.10	0.01	0.02	1.31	0.06	0.02	0.65	99.52
KNJ-48J	94.00	3.62	0.24	0.01	0.10	0.01	0.02	1.15	0.11	0.02	0.10	99.38
KNJ-49A	90.32	5.68	0.39	0.01	0.11	0.01	0.03	1.65	0.14	0.02	1.32	99.68
KNJ-49B	90.30	5.67	0.46	0.01	0.11	0.01	0.02	1.65	0.12	0.02	0.87	99.24
KNJ-46C	97.23	1	0.07	0.02	<0.01	0.02	0.03	0.14	0.02	0.01	1.21	99.75
KNJ-45J	96.79	0.96	0.04	0.02	<0.01	0.01	0.02	0.17	0.02	0.01	1.98	100.02
KNJ-45C	95.96	0.84	0.02	0.02	<0.01	0.01	0.02	0.13	0.02	0.01	1.83	98.86
KNJ-45I	96.32	1.27	0.05	0.02	<0.01	0.01	0.02	0.27	0.05	0.01	0.97	98.99
KNJ-45H	96.83	0.92	0.04	0.02	<0.01	0.01	0.02	0.16	0.03	0.01	1.22	99.26
KNJ-45L	96.32	1.27	0.05	0.02	<0.01	0.01	0.02	0.27	0.03	0.01	0.58	98.59
KNJ-46B	96.86	0.95	0.03	0.02	<0.01	0.01	0.03	0.12	0.02	0.01	1.25	99.31
KNJ-45K	96.42	1.25	0.06	0.02	<0.01	0.01	0.02	0.25	0.04	0.01	0.93	99.02
KNJ-46A	96.91	1.04	0.04	0.02	<0.01	0.01	0.03	0.15	0.03	0.01	0.84	99.09
KNJ-48E	94.5	3.04	0.16	0.02	0.06	0.01	0.02	0.86	0.07	0.01	0.76	99.51
KNJ-48C	94.27	2.94	0.15	0.02	0.06	0.01	0.02	0.85	0.1	0.01	1.1	99.53

The samples were powdered to −200 mesh using a chromium-free tool steel mill (Pulverisette 9 Fritz-GMBH). Major element analyses were carried out by using X-ray fluorescence spectrometry (XRF Model Panalytical Axios WD-XRF) at the Department of Geology, Presidency University. The pressed powder pellets were analyzed on the XRF with matrix correction procedure at 60 kV, 170 mA with nominal analysis time of 300 s for all major oxides. Overall precision for major and minor oxides is <5%. The precision and accuracy are based on multiple analyses of international rock standards [JG-2 (granite), NIM G (granite)]. Results of analyses of the standard samples are given in Appendix A.

Fabric analyses of more than 1500 quartz grains from different locations and mounted in polished, carbon-coated thin sections were carried out by a JSM 6510 scanning electron microscope with an attached cathodoluminescence detector (SEM-CL) at the Institute of Geoscience, Federal University of Minas Gerais, Belo Horizonte, Brazil. This instrument was commonly operated at 30 kv and a beam current of 10 nA. CL images were recorded in 18−19 working distance.

4. RESULTS

4.1 Petrography

The arenites are medium-grained and are moderately to well sorted (Fig. 17.2A). The sandstone consists of monocrystalline quartz (80−90%), polycrystalline quartz (2−8%), feldspar (<4%), and rock fragments (<6%), which are mainly sedimentary and metasedimentary. Compositional maturity of the sandstone is also reflected in the prevalence of monocrystalline quartz in the framework. Undulose (>5-degree extinction) monocrystalline quartz (Fig. 17.2B) is more common than clear grains (<5-degree extinction). The arenites are further distinguished using the modal compositional (Table 17.2) data in the standard tectonic discrimination plots (Dickinson, 1985).

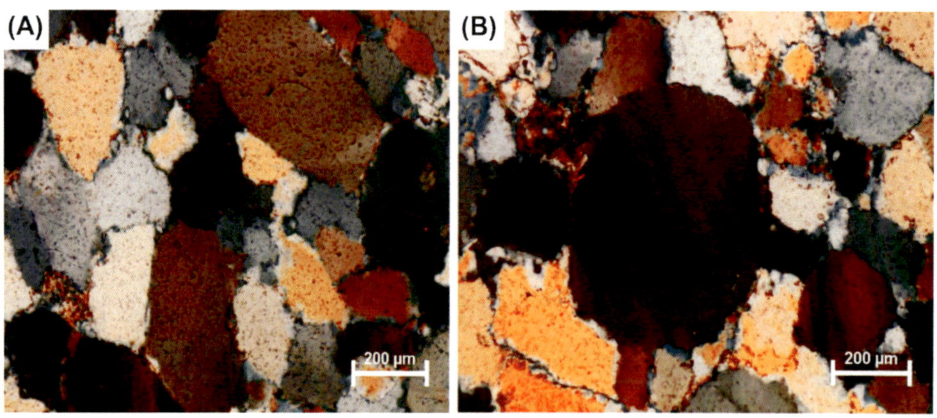

FIGURE 17.2 Transmitted light photomicrograph of arenites from the Keonjhar Quartzite. (A) Quartzarenite with well-sorted quartz framework population that represents the bulk of the formation. (B) Well-rounded quartz famework grain with very low or nonundulose extinction.

TABLE 17.2 Modal Compositions of Arenites From Point-Count Method

Sample Number	Quartz		K-Feldspar	Feldspar		Lithic fragment		Opaque Minerals	Musco Vite	Matrix	Total
	Qm	Qp		Plagioclase Feldspar	Altered Feldspar	Sedimentary Lithic Grains	Unidentified Lithic Grains				
KNJ-48A	82.20	3.01	0.00	0.00	1.80	0.00	0.00	0.00	0.00	9.00	96.01
KNJ-48J	86.04	1.90	0.00	0.00	2.50	0.70	0.10	0.00	0.00	8.01	99.25
KNJ-49A	79.30	4.80	0.00	0.00	1.02	5.80	0.40	0.00	0.00	9.05	100.37
KNJ-49B	80.01	5.30	0.00	0.00	3.50	0.00	1.40	0.00	0.00	10.00	100.21
KNJ-46C	90.12	5.89	0.00	0.00	0.00	0.00	0.00	0.00	0.00	4.10	100.11
KNJ-45J	84.23	7.80	0.00	0.00	0.00	0.00	0.00	0.00	0.00	7.91	99.94
KNJ-45C	85.70	6.50	0.00	0.00	1.50	0.00	0.00	0.00	0.00	5.56	99.26
KNJ-45I	83.33	5.10	0.00	0.00	0.00	0.00	0.00	0.00	0.00	10.01	98.44
KNJ-45H	87.91	3.40	0.00	0.00	0.00	0.00	0.00	0.00	0.00	7.89	99.20
KNJ-45L	87.77	3.91	0.00	0.00	0.00	0.00	0.00	0.00	0.00	8.98	100.66
KNJ-46B	91.05	5.25	0.00	0.00	0.00	0.00	0.10	0.00	0.00	2.81	99.21
KNJ-45K	86.78	6.50	0.00	0.00	0.00	0.00	0.00	0.00	0.00	6.97	100.25
KNJ-46A	92.35	4.12	0.00	0.00	0.00	0.00	1.10	0.00	0.00	2.10	99.67
KNJ-48E	80.04	4.50	1.50	0.00	0.00	0.45	0.02	0.00	0.00	12.01	98.52
KNJ-48C	83.51	3.80	0.85	0.00	0.00	0.72	0.03	0.00	0.00	11.03	99.94

All abbreviations are after Dickinson and Suczek (1979).

4.1.1 *QtFLt Tectonic Discrimination Diagram*

In this triangular diagram, Qt represents total quartzose grains, including monocrystalline quartz and polycrystalline quartz, monocrystalline feldspar grains (F), and total lithic fragments (Lt) including igneous, sedimentary, and metamorphic varieties. After recalculation using the Cox and Lowe (1996) spreadsheet for accommodating possible diagenetic transformation of feldspars to secondary matrix, the compositional ranges show total quartz proportion of more than 80–96%, 1–11% for feldspar, and 1–8% for the lithic fragments. All the samples fall within the range of craton interior to recycled orogen (Fig. 17.3A).

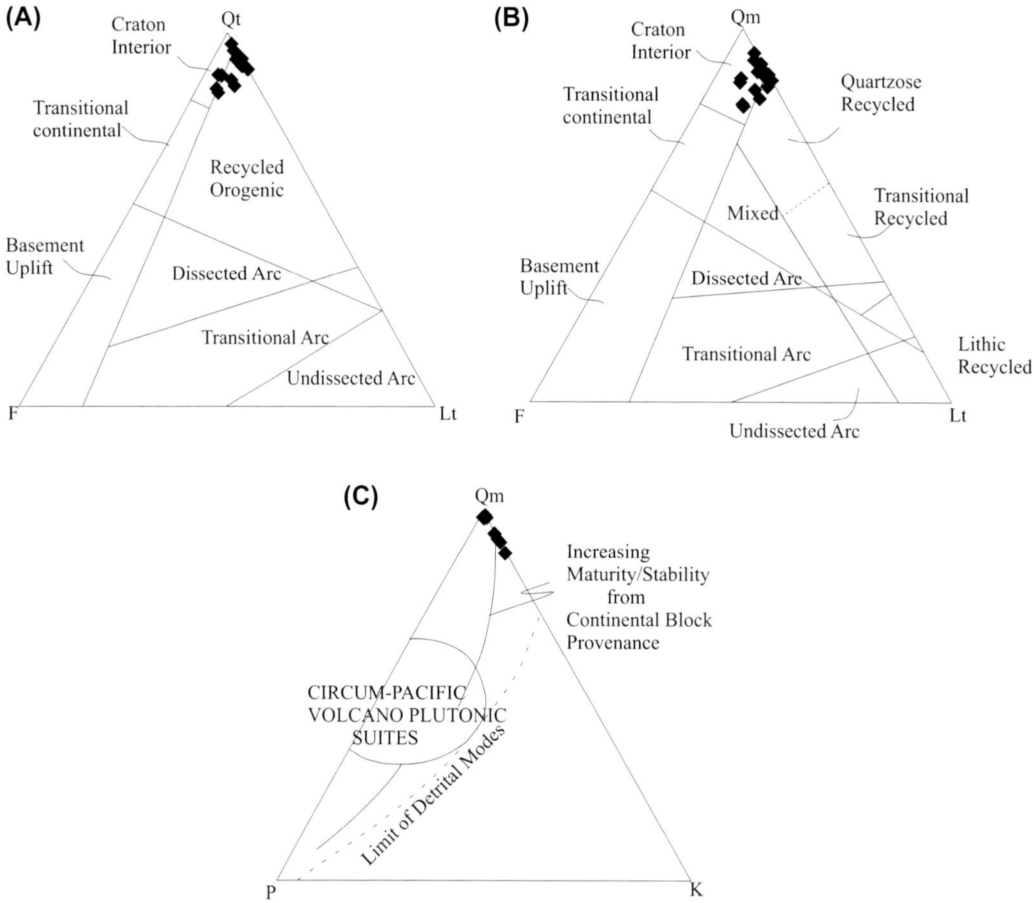

FIGURE 17.3 Modal compositional plots of arenites. (A) Qt-F-Lt compositional plots (after Dickinson and Suczek, 1979) of arenites. Note samples cluster in cratonic interior to recycled orogen fields. (B) Qm-F-Lt compositional plots (after Dickinson, 1985) of arenites. Note the entire samples cluster in the quartzose recycled to cratonic interior origin. (C) Qm-P-K compositional plots (after Dickinson and Suczek, 1979) of arenites. Note the entire samples cluster in the stable continental block provenance.

4.1.2 *Qm-F-Lt Tectonic Discrimination Diagram*

This diagram further discriminates the monocrystalline quartz grains (Qm), feldspar grains (F), and lithic fragments (Lt) including igneous, sedimentary, and metamorphic varieties. In the QmFLt plot, all the samples of the study area fall in the craton interior to quartzose recycled orogen fields (Fig. 17.3B). The near absence or negligible proportion of polycrystalline quartz grains and metamorphic lithic fragments suggests that the source region may have been uplifted in a noncompressional (extensional or transform) setting (see Dickinson, 1985, for discussion). In contrast, the predominant monocrystalline quartz composition indicates compositional maturity.

4.1.3 *Qm-P-K Tectonic Discrimination Diagram*

In the diagram only monocrystalline quartz grains (Qm) are plotted against variable proportions of plagioclase (P) and K-feldspar grains (K). The points cluster in the Qm field, which clearly shows that they are sourced from a highly mature and stable continental block provenance (Fig. 17.3C), presumably a largely quartz-rich passive margin setting with sands derived from plate interior, or stable continental margins.

4.2 Scanning Electron Microscope and Cathodoluminescence Analysis

Krinsley and Tovey (1978) identified several parameters for SEM-CL: absence of luminosity, variation of the contrast of CL, presence of narrow band structure (simple to complex in nature), mottled CL texture, and morphology of dark areas boundaries (sharp or diffuse). Based on these features, Seyedolali et al. (1997) further summarized genetic classes of quartz grains (Table 17.4). Five types of quartz grains have been identified in the Keonjhar Quartzite based on their SEM-CL signatures (Table 17.3).

4.2.1 *Type A*

Type A grains are low luminescent and are characterized by thin ($<10\ \mu m$) distinct, black hairline structures (CL weak/absent) that represent healed fractures (Fig. 17.4A–D) (after Seyedolali et al., 1997). The fractures are sparsely distributed and widely spaced as several sets of lines, which intersect each other at nearly constant angles (Fig. 17.4B). The closed fractures appear to have undergone healing or possibly were filled up by precipitation of some nonluminescent material (may be SiO_2 or any other foreign material) (Fig. 17.4D). The healed fractures locally occur in more than two sets with different orientations that resemble a cobweb pattern often referred to as spider CL-texture (cf. Rusk and Reed, 2002; Boggs and Krinsley, 2006, p. 69). The proportion of this CL type ranges in samples from 20% to 80%.

4.2.2 *Type B*

The second class of quartz grains occurs in the proportion of less than 2% to about 60%. They display homogeneous and patchy or mottled CL intensity (Fig. 17.5A–D) and are nearly devoid of healed fractures.

TABLE 17.3 SEM-CL Fabric Analysis of the Quartz Framework From Keonjhar Quartzite

		Type A	Type B	Type C	Type D	Type E
Sample Number	Healed Quartz	Dark Streak and Patchy Quartz	Medium-Grade Metamorphic Quartz	Low-Grade Quartz	Complex Shearing/ Brittle Deformation Quartz/Metamorphic Quartz	High-Grade Metamorphic Quartz
KNJ-48A_1	50.27	10.2	28.61	7.11	0	0
KNJ-48A_2	49.58	14.58	35.83	0	0	0
KNJ-48A_3	65.1	0	19.21	15.69	0	0
KNJ-48A_4	64.83	13.09	25.17	0	0	0
KNJ-48A_5	45.9	18.99	22.56	0	4.45	1.22
KNJ-48A_6	47.14	28.57	24.29	0	0	0
KNJ-48A_7	55.78	31.67	15.83	0	0	0
KNJ-48A_8	52.86	15.72	31.43	0	0	0
KNJ-48A_9	44.44	0	24.44	0	6.11	0
KNJ-48A_10	39.59	17.65	32.94	0	5.07	3.75
KNJ-48A_11	54.09	9.98	23.18	0	7.69	5.04
KNJ-48A_12	47.69	21.78	26.67	0	5.64	0
KNJ-48A_13	40.45	18.18	25.45	0	12.91	6.74
KNJ-49B_1	48.21	21.4	23.21	0	4.94	2.2
KNJ-49B_2	33.33	27.78	25.93	0	6.78	6.18
KNJ-49B_3	21.52	20.12	40.01	0	12.97	7.15
KNJ-49B_4	39.09	30	13.33	0	12.02	6.07
KNJ-49B_5	38.3	6.06	45.45	0	10.33	7.85
KNJ-49B_6	22.58	0	58.06	0	11.13	8.22
KNJ-49B_7	66.67	0	15.28	0	11.11	7.17
KNJ-49B_8	53.85	13.85	10.77	0	12.67	8.86
KNJ-49B_9	51.22	6.25	25.32	0	13.79	4.96
KNJ-49B_10	58.76	11.07	13.42	0.01	10.14	6.87
KNJ-45A_1	38.78	28.02	26.52	0	4.4	5.72
KNJ-45A_2	48.01	10.52	15.57	0	5.76	0
KNJ-45A_3	56.71	15.17	35.75	0.5	0.5	0
KNJ-45A_4	25.01	53.4	15.45	0	6.99	0
KNJ-45A_5	41.75	35.75	19.15	0	2.86	2.29
KNJ-45A_6	87.5	5.01	1.2	0	4.32	2.85
KNJ-45A_7	21.02	39.19	35.15	0	4.11	0
KNJ-45A_8	20.11	55.75	21.74	0	4.12	0

TABLE 17.3 SEM-CL Fabric Analysis of the Quartz Framework From Keonjhar Quartzite—cont'd

		Type A	Type B	Type C	Type D	Type E
Sample Number	Healed Quartz	Dark Streak and Patchy Quartz	Medium-Grade Metamorphic Quartz	Low-Grade Quartz	Complex Shearing/ Brittle Deformation Quartz/Metamorphic Quartz	High-Grade Metamorphic Quartz
KNJ-45A_9	30.75	45.72	15.01	0	4.85	4.87
KNJ-45A_10	89.01	5.05	3.89	0	2.42	0
KNJ-45A_11	59.75	10.17	20.52	0	5.91	3.51
KNJ-45A_12	21.55	56.75	22.01	0	0.52	0
KNJ-45A_13	41.11	37.79	15.71	0	4.24	0
KNJ-45A 14	48.68	20.01	26.77	0	5.72	0

All abbreviations are after Seyedolali et al. (1997) and Laubach (1997).

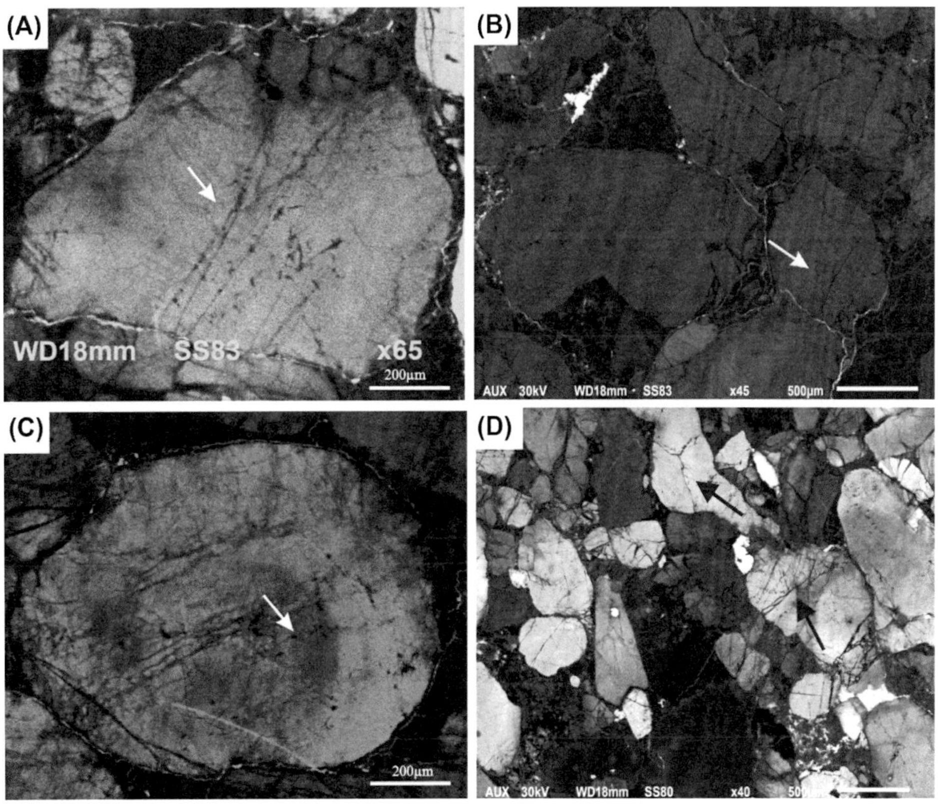

FIGURE 17.4 SEM-CL images of type A (plutonic) quartz from the Keonjhar Quartzite. (A) Type A quartz with low luminescent and thin hairline healed fracture. Note healed fracture (*arrow*). (B) Type A quartz where healed fracture filled by secondary fluid inclusion (foreign material or trace element like Li, Al, Ti) that produce dark streak and patches. Note healed fractures are filled by foreign material and produce dark patches (*arrow*). (C) Dark streak and patches (*arrow*) in Type A quartz grain. (D) Sparsely distributed fractures in more than one set within low luminescent plutonic quartz (*arrow*). Note the fractures intersect at constant angles (*arrow*).

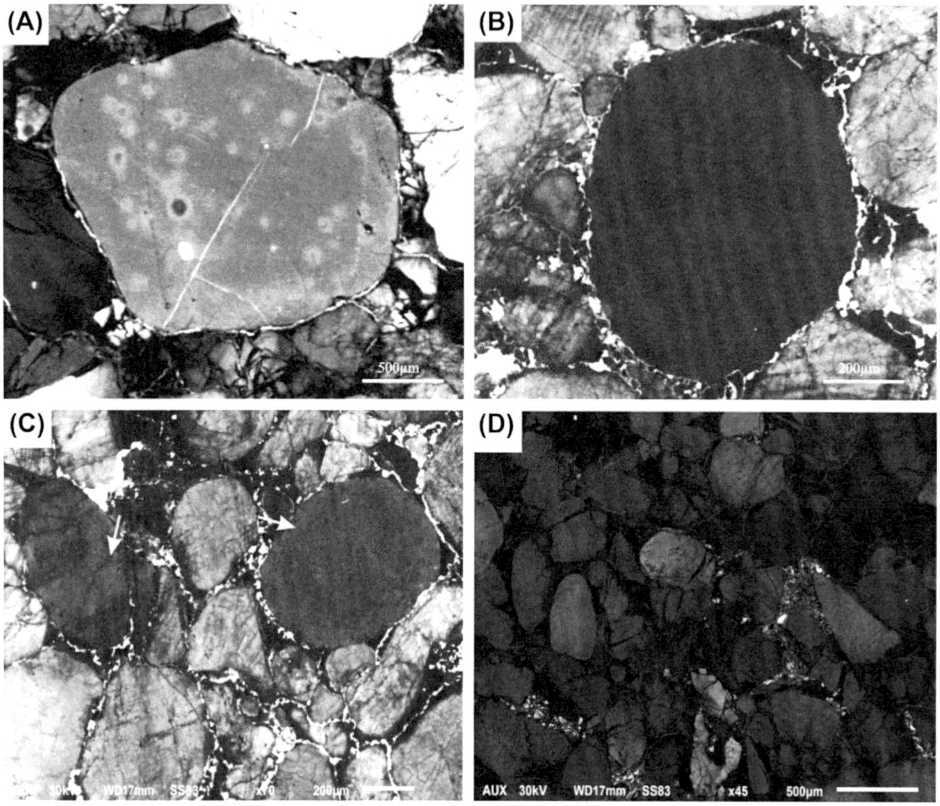

FIGURE 17.5 SEM-CL images of type B and type C (metamorphic quartz) quartz from the Keonjhar Quartzite. (A) Inhomogeneous and patchy mottled luminosity in type B (medium-grade metamorphic) quartz. (B) Dark non-differentiated CL character of low-grade metamorphic quartz type C (central dark grain). (C) Dark CL patterns of image of type C (low-grade metamorphic) dominated detrital quartz population. Note partly healed fractures in the grain on the left side of the image (*arrow*). (D) CL pattern of framework quartz population with both type B and C (metamorphic) grains.

4.2.3 *Type C*

Type C grains are characterized by very weak to dark CL intensity with patchy distribution (Fig. 17.5B). They have indistinct and partly preserved healed fractures. The relative proportion of this grain type is very low, up to 18%.

4.2.4 *Type D*

Quartz grains of type D have internal fractures of variable extent and intensity (Fig. 17.6A–C). The fractures are mainly confined within grains and show variably curved geometry (Fig. 17.6A) typical of the category II fractures of Laubach (1997). Locally, fractures render a polycrystalline character to the grains (Fig. 17.6B). Fractures filled up with foreign materials with thick and dark appearance are comparable with the category III inherited fractures of Laubach (1997). This type of CL signature represents up to 20% of grain population in a sample.

FIGURE 17.6 SEM-CL images of type D (fractured) quartz from the Keonjhar Quartzite. (A) Medium-to-high grade deformed quartz grain (moderate to high luminosity) produce category II (see Table 17.4 for description) microfractures that have curved to straight traces arranged in simple to complex intersecting patterns. (B) Grain (center) with category II microfractures giving rise to polycrystallinity of the quartz grain (*arrow*). (C) Grains of category III microfractures are inherited in nature; those fractures are secondarily filled up. (D) Cathodoluminescence (CL) image of type E high-grade (bright luminosity) quartz.

4.2.5 *Type E*

In type E grains, negligible proportions of quartz grains show bright luminescence (Fig. 17.6D). The bright grains are homogenous and smooth in appearance (Fig. 17.6D). Such bright homogenous luminescence has been ascribed to high-grade metamorphic quartz (Boggs and Krinsley, 2006, p. 74).

5. DISCUSSION

5.1 Quartz Genetic Types

The tectonic discrimination plots from modal framework composition suggest that the arenites are sourced from recycled orogenic to craton interior provenance. Blatt (1967)

TABLE 17.4 Characteristic Features of Quartz Grains From Different Source Rocks Revealed by SEM-CL Fabric Analysis.

Quartz Type and Quartz Source	Features	Description
Type A (plutonic)	Healed fracture Dark CL streak and Patches (spider-like fabrics)	Black hairline structures (CL weak/absent), typically sparsely distributed and widely spaced in more than one set of lines intersecting at nearly constant angles Low-intensity CL in the form of broad irregularly shaped black patches, black streak, and irregular patches of dark CL (weak or nonluminescence) associated with fractures
Types B and C (low-to-medium grade metamorphic)	Healed fractures Indistinct mottling Nondifferential CL	Healed fractures partially preserved, which disappear with increasing metamorphic grade Very weak to dark CL Homogenous CL character
Type B (medium-grade metamorphic quartz)	Mottled inhomogeneous fabric	Bright CL (high-grade metamorphic quartz), inhomogeneous, patchy or mottled
Type D (brittle deformed quartz/high-grade metamorphic quartz)	Category I Category II Category III	Microfractures cut across grain boundaries without having a consistent relation to grain centers or grain contacts Microfractures with curved to straight traces arranged in simple to complex intersecting patterns Microfractures with inhomogeneous dark fillings
Type E	Homogeneous fabric	Very bright homogeneous CL intensity

Compiled After Seyedolali et al. (1997) and Laubach (1997).

distinguishes igneous and metamorphic parentage of framework quartz grains based mainly on the degree of undulosity in the extinction pattern when viewed under transmitted light. The SEM-CL signatures of the five classified types may be related to different genetic types (criteria summarized in Table 17.4 after Seyedolali et al., 1997; Laubach, 1997). The type A grains with healed fractures and dark patchy streaks are comparable to plutonic quartz. Dark CL patches in plutonic quartz are believed to be due to the presence of inclusions (Muller, 2000; Boggs and Krinsley, 2006, p. 69). Petrographic study of the nonundulose quartz did not reveal the presence of inclusions, but in SEM-CL imaging small amounts of the scattered vacuoles are observed that also locally display a linear trend. More common randomly scattered inclusions are likely to be due to fluid trapped during crystallization. The dark streaks and patches (>15%) indicate the effect of lattice defect within the grains during the emplacement stage of the pluton (cf. Passchier and Trouw, 1996; Stunitz, 1998). The dark CL patches also represent micropores filled with foreign material (probably amorphous silica), that are likely to have been created initially by dissolution (Boggs and Krinsley, 2006). Moreover, the dark CL patches may also represent areas of differential strain within the grains. Spider texture (Boggs and Krinsley, 2006, p. 69) in quartz only appears in the CL images. Muller (2000) interprets this texture in plutonic quartz as a secondary feature formed during the cooling magmatic stage. Decrepitation commonly generates microcracks around the fluid inclusion that move outward and produce further defects in the grain structure.

Dynamic recrystallization also produces variation of the intensity of luminescence. High-grade metamorphic quartz that has undergone secondary recrystallization and grain growth at high temperature shows very bright homogeneous CL intensity due to the presence of large numbers of trapped inclusions and impurities, whereas medium-to-low grade metamorphic quartz displays very weak to dark CL intensity due to the development of relative clean recovered and recrystallized grains. Type B and C quartz grains (Fig. 17.5D) with homogeneous weak to moderate CL intensity suggest their low-to medium-grade metamorphic source (Seyedolali et al., 1997). The partial preservation or the absence of healed fractures also distinguishes them from plutonic quartz.

Mottled texture under SEM-CL analysis has also been ascribed to the irregular distribution of point defects due to activator ions, i.e., accumulation of some trace element in some sites such as Li, Al, K, Ti, etc. (Zinkernagel, 1978; Matter and Ramseyer, 1985; Ramseyer et al., 1988; Walker and Burley, 1991) during crystallization from a melt. Some of the quartz grains of type B showing patchy or mottled texture suggest incomplete recrystallization or accompanying annealing (Boggs and Krinsley, 2006). The origin of dark quartz grains or grains with nondifferential and internal fractures is unknown, but it may reflect irregularly distributed internal strain in the grains (Seyedolali et al., 1997). Tectonically brittle deformed quartz may display a complex internal fracturing (Dickinson and Milliken, 1995; Laubach, 1997; Milliken and Laubach, 2000; Makowitz and Milliken, 2003).

The last type of quartz grains (type D) display internal confined fractures. This signature is related to cataclastic deformed terranes (cf. Laubach, 1997), and they closely resemble fractured grains of categories II and III as described by Laubach (1997) and Seyedolali et al. (1997). The bright luminescent grains (type E, Fig. 17.6D) are likely to be from sources of higher metamorphic grade.

5.2 Statistical Significance of the Determination of Quartz Source

Relative proportions of different genetic types of quartz vary from sample to sample (Table 17.3). Though plutonic terranes represent the main source of the quartz framework (type A), other quartz types in combination constitute a variable proportion in many samples. To arrive at a statistically significant inference about the dominant genetic quartz type we have employed the techniques of Sign test and Wilcoxon signed-rank test (described in Appendix B) for testing the null hypothesis that plutonic quartz is not the dominant quartz population. The statistical investigation clearly suggests that there is a large predominance of plutonic over metamorphic quartz in the Keonjhar Quartzite.

5.3 Source and Tectonic Distribution

Quartz-rich rocks are typically associated with passive continental margin (Crook, 1974; Schwab, 1975). Quartzose sands are derived from source rocks of stable cratonic interior and from recycled orogen tectonic terranes (Dickinson, 1985). Ghosh et al. (2016) describe the tectonic setting based on petrography and geochemical proxies for Keonjhar Quartzite. From their petrographic analysis they suggest that the presence of the quartz grain with abraded overgrowth, abundant illitic pseudomatrix, and phyllite and mica schist fragment supports the hypothesis of a recycled orogenic source component.

Based on immobile trace element ratios (Th/Sc, La/Sc, and La/Th) and REE pattern (negative Eu-anamoly i.eEu/Eu* = 0.81−1.56) these authors also suggest a cratonic to recycled orogen provenance (Ghosh et al., 2016). Recycling of metamorphic and brittle sheared metamorphic quartz (categories I and II in Table 17.4) typically involves deformation and uplift of foreland basin successions (cf. Dickinson and Suczek, 1979; Dickinson, 1985). The quartz-rich nature of the framework and the presence of metasedimentary rock fragments are an indication of a recycled sedimentary source. The presence of low-grade metamorphic quartz further supports that OMG and IOG metasediments acted as a minor source for the sandstone.

5.4 SEM-CL Study of Quartz Framework and Implications for the Mesoarchean Upper Crust

The SEM-CL study indicates the prevalence of plutonic quartz over other genetic types in Mesoarchean the Keonjhar Quartzite. The next dominant population comes from the low-to-medium grade metamorphic terrane and tectonically deformed quartz sources that represent a subordinate proportion of the population. These observations indicate that the sediments were sourced mainly from igneous batholiths, but also support the presence of recycled metasedimentary components. The nearly total absence of quartz grains with brighter and homogeneous CL pattern, characteristic of high-grade metamorphic source, is noteworthy, and presumably indicates that high-grade metamorphic rocks were not significant components in the source terrane. This may have a further implication that collisional accretion for the development of upper crust might not have been an important geodynamic process during the formation of the Paleo-Mesoarchean Singhbhum craton.

6. CONCLUSIONS

The modal framework compositions of the arenites from the Keonjhar Quartzite suggest recycled orogen to craton interior provenance. The tectonic discrimination diagrams also indicate the supermature nature of the sandstone derived from the interior of a stable continental craton. SEM-CL fabric analysis of the quartz grains identifies five types of quartz: plutonic quartz with healed fractures and mottled CL, medium-grade metamorphic quartz with mottled CL intensity, low-grade metamorphic quartz with low or patchy luminescence, quartz with internal brittle fractures, and higher metamorphic-grade bright luminescent quartz. The predominance of plutonic quartz over the other types suggests that high-grade metamorphic components from collisional geodynamics were not significant in the Paleo-Mesoarchean upper continental crust in the Singhbhum craton.

Acknowledgments

We acknowledge support for field work in the study area from the UGC-CAS programme of the Department of Geology, Presidency University, Kolkata. JM and CR received support for this work from the DST India grant under India–Brazil Inter Governmental programme, Pr. 401020/2013-4. We thank the Institute of Geoscience, Federal University of Minas Gerais, Belo Horizonte, Brazil for giving us the opportunity to operate the scanning electron microscope and accomplish cathodoluminescence analyses supported by the project. This is a part of the PhD programme of SD. SD acknowledges Prof. N.V. Krishnachaitanya Yerroju of Presidency University Department of Statistics for introducing her to the applied statistical analysis. The authors are thankful to A. Basu and P. G. Eriksson for very constructive reviews of the earlier versions of the manuscript.

References

Acharyya, S.K., 1993. Greenstones from Singhbhum craton, their Archaean character, oceanic crustal affinity and tectonics. Proceeding National Academy of Sciences India 63 (A), 211–222.

Acharyya, S.K., Gupta, A., Orihashi, Y., 2010. New U-Pb zircon ages from Paleo- Mesoarchean TTG gneisses of the Singhbhum craton, eastern India. Geochemical Journal 44, 81–88.

Ali, S., Stattegger, K., Schönberg, D.G., Frank, M., Kraft, S., Kuhnt, W., 2014. The provenance of Cretaceous to Quaternary sediments in the Tarfaya basin, SW Morocco: evidence from trace element geochemistry and radiogenic Nd-Sr isotopes. Journal of African Earth Sciences 90, 64–76.

Augustsson, C., Rusing, T., Adams, C.J., Chmiel, H., Kcabayoglu, M., Buld, M., Zimmermann, U., Brendt, J., Kooijman, E., 2011. Detrital quartz and zircon combined: the production of mature sand with short transportation paths along the Cambrian West Gondwana margin, Northwestern Argentina. Journal of Sedimentary Research 81, 284–298.

Basu, A., Bickford, M.E., Deasy, R., 2016. Inferring Tectonic Provenance of Siliciclastic Rocks From Their Chemical Compositions: A Dissent. http://dx.doi.org/10.1016/j.sedgeo.2015.11.013.

Bernet, M., Bassett, K., 2005. Provenance analysis by single-quartz grain SEM-CL/optical microscopy. Journal of Sedimentary Research 75, 492–500.

Bhatia, M.R., 1983. Plate tectonics and geochemical composition of sandstones. Journal of Geology 91, 611–627.

Bhatia, M.R., Crook, K.A.W., 1986. Trace element characteristics of graywackes and tectonic setting discrimination of sedimentary basins. Contributions to Mineralogy and Petrology 92, 181–193.

Blatt, H., 1967. Provenance determinations and recycling of sediments. Journal of Sedimentary Petrology 37, 1031–1040.

Boggs, S., Krinsley, D., 2006. Application of Cathodoluminescence Imaging to the Study of Sedimentary Rocks, 165 pp. Springer, Cambridge.

Cox, R., Lowe, D.R., 1996. Quantification of the effects of secondary matrix on the analysis of sandstone composition, and a petrographic-chemical technique for retrieving original framework grain modes of altered sandstones. Journal of Sedimentary Research 66, 548–558.

Crook, K.A.W., 1974. Lithogenesis and geotectonics: the significance of compositional variation in flysch arenites (graywackes). Society of Economic Paleontologists and Mineralogists 19, 304–310 special publication.

Dickinson, W.W., Milliken, K.L., 1995. The diagenetic role of brittle deformation in compaction and pressure solution, Etjo Sandstone, Nambia. Journal of Geology 103, 339–347.

Dickinson, W.R., 1985. Interpreting provenance relations from detrital modes of sandstones. In: Zuffa, G.G. (Ed.), Provenance of Arenites. Reidel, Dordrecht, pp. 333–361.

Dickinson, W.R., Suczek, C.A., 1979. Plate tectonics and sandstone compositions. AAPG Bull 63, 2164–2182.

Ghosh, S., De, S., Mukhopadhyay, J., 2016. Provenance of >2.8 Ga Keonjhar quartzite, Singhbhum craton, Eastern India: implications for the nature of Mesoarchean upper crust and geodynamics. Journal of Geology 124, 331–351.

Johnsson, M.J., Basu, A., 1993. Processes controlling the composition of clastic sediments. Geological Society of America Special Paper 284, 342 pp.

Krinsley, D.H., Tovey, N.K., 1978. Cathodoluminescence in quartz sand grains. In: Johari, O. (Ed.), Scanning Electron Microscopy. SEM Inc., AMF O'Hare, Illinois, pp. 887–894.

Kwon, Y.-N., Boggs Jr., S., 2002. Provenance interpretation of Tertiary sandstones from the Cheju Basin (NE East China Sea): a comparison of conventional petrographic and scanning cathodoluminescence techniques. Sedimentary Geology 152, 29–43.

Laubach, S.E., 1997. A method to detect natural fracture strike in sandstones. American Association of Petroleum Geologists Bulletin 81, 604–623.

Makowitz, A., Milliken, K.L., 2003. Quantification of brittle deformation in burial compaction, Frio and Mount Simon Formation sandstones. Journal of Sedimentary Research 73, 1007–1021.

Marshall, D.J., 1988. Cathodoluminescence of Geologic Materials, 146 pp. Massachusetts, Unwin Hyman, Boston.

Matter, A., Ramseyer, K., 1985. Cathodoluminescence petrography as a tool for sedimentary provenance studies of sandstones. In: Zuffa, G.G. (Ed.), Provenance of Arenites. Reidel, Dordrecht, pp. 191–211.

Mazumder, R., van Loon, A.J., Mallik, L., Reddy, S.M., Arima, M., Altermann, W., Eriksson, P.G., De, S., 2012. Mesoarchaean-Palaeoproterozoic stratigraphic record of the Singhbhum crustal province, eastern India: a synthesis. In: Mazumder, R., Saha, D. (Eds.), Palaeoproterozoic of India. Geological Society of London Special Publication 365, pp. 29–47.

Milliken, K.L., Laubach, S.E., 2000. Brittle deformation in sandstone diagenesis revealed by scanned cathodoluminescence imaging with application to characterization of fractured reservoirs. In: Pagel, M., Barbin, V., Blanc, P., Ohnenstetter, D. (Eds.), Cathodoluminescence in Geosciences. Springer Verlag, Berlin, pp. 225–243.

Mishra, S., Deomurari, M.P., Wiedenbeck, M., Goswami, J.N., Ray, S., Saha, A.K., 1999. $^{207}Pb/^{206}Pb$ zircon ages and the evolution of the Singhbhum craton, eastern India: an ion microprobe study. Precambrian Research 93, 139–151.

Mukhopadhyay, D., 2001. The Archean nucleus of Singhbhum: the present state of knowledge. Gondwana Research 4, 307–318.

Mukhopadhyay, J., Ghosh, G., Nandi, A.K., Chaudhuri, A.K., 2006. Depositional setting of the Kolhan Group: its implications for the development of a Meso to Neoproterozoic deep-water basin on the South Indian craton. South African Journal of Geology 109, 183–192.

Mukhopadhyay, J., Ghosh, G., Zimmermann, U., Guha, S., Mukherjee, T., 2012. A 3.51 Ga bimodal volcanics-BIF-ultramafic succession from Singhbhum craton: implications for Palaeoarchaean geodynamic processes from the oldest greenstone succession of the Indian subcontinent. Geological Journal 47, 284–311.

Mukhopadhyay, J., Crowley, Q., Ghosh, S., Ghosh, G., Chakrabarti, K., Misra, B., Bose, S., 2014. Oxygenation of the Archean atmosphere: new paleosol constraints from eastern India. Geology 42, 923–926.

Muller, A., 2000. Cathodoluminescence and Characterisation of Defect Structures in Quartz With Applications to the Study of Granitic Rocks. Doctoral Dissertation, Georg-August- Universitatzu Gottingen, 229 pp.

Nelson, D.R., Bhattacharya, H.N., Thern, E.R., Altermann, W., 2014. Geochemical and ion-microprobe U-Pb zircon constraints on the Archaean evolution of Singhbhum craton, eastern India. Precambrian Research 255, 412–432.

Passchier, C.W., Trouw, R.A.J., 1996. Microtectonics. Springer, Berlin, 325 pp.

Ramseyer, K., Baumann, J., Matter, A., Mullis, J., 1988. Cathodoluminescence colours of alpha-quartz. Mineralogical Magazine 52, 669–677.

Rusk, B., Reed, M., 2002. Scanning electron microscope-cathodoluminescence analysis of quartz reveals complex growth histories in veins from the Butte porphyry copper deposit, Montana. Geology 30, 727–730.

Saha, A.K., 1994. Crustal evolution of Singhbhum-North Orissa, eastern India. Geological Society of India. Memoir 27, 341 pp.

Schwab, F.L., 1975. Framework mineralogy and chemical composition of continental margin-type sandstone. Geology 3, 487–490.

Seyedolali, A., Krinsley, D.H., Boggs Jr., S., O'Hara, P.F., Dypvik, H., Goles, G.G., 1997. Provenance interpretation of quartz by scanning electron microscope- cathodoluminescence fabric analysis. Geology 25, 787–790.

Stunitz, H., 1998. Syndeformational recrystallization – dynamic or compositionally induced? Contributions to Mineralogy and Petrology 131, 219–236.

Tait, J., Zimmermann, U., Miyasaki, T., Presnyakov, S., Chang, Q., Mukhopadhyay, J., Sergeev, S., 2011. Possible juvenile Palaeoarchaean TTG magmatism in eastern India and its constraints for the evolution of the Singhbhum craton. Geological Magazine 148, 340–347.

Upadhyay, D., Chattopadhyay, S., Kooijman, E., Mezger, K., Berndt, J., 2014. Magmatic and metamorphic history of Paleoarcheantonalite-trondhjemite-granodiorite (TTG) suite from the Singhbhum craton, eastern India. Precambrian Research 252, 180–190.

Walker, G., Burley, S., 1991. Luminescence petrography and spectroscopic studies of diagenetic minerals. In: Barker, C.E., Kopp, O.C. (Eds.), Luminescence Microscopy and Spectroscopy: Quantitative and Qualitative Applications. SEPM Short Course, vol. 25, pp. 83–96.

Zinkernagel, U., 1978. Cathodolumninescence of quartz and its application to sandstone petrology. In: Füchtbauer, H., Lisitzyn, A.P., Milliman, J.D., Seibold, K. (Eds.), Contributions to Sedimentology, vol. 8, pp. 1–67.

Zuffa, G.G., 1985. Provenance of Arenites. Reidel, Dordrecht, 408 pp.

APPENDICES

Appendix A

Analyses of Rock Standards

Major Oxides	Certified Value (JG-2)	Analyzed Value (JG-2)	Certified Value (NIM G)	Analyzed Value (NIM G)
SiO_2	76.95	76.45	75.70	75.56
Al_2O_3	12.41	12.22	12.08	11.87
Fe_2O_3	0.92	0.78	2.02	1.98
MnO	0.015	0.011	0.02	0.01
MgO	0.04	0.04	0.06	0.57
CaO	0.8	0.79	0.78	0.73
Na_2O	3.55	3.41	3.36	3.29
K_2O	4.72	4.59	4.99	4.78
TiO_2	0.04	0.03	0.09	0.077
P_2O_5	0.002	0.001	0.01	0.01

Appendix B

Two statistical methods are performed to evaluate consistent differences between pairs of observations. For the given pair (plutonic quartz and metamorphic quartz) of the observations for each sample of our study area, the Sign test and Wilcoxon signed-rank test help to determine if one member of the pair tends to be significantly greater than the other member of the pair.

Sign test

Define:

X_i, concentration level of plutonic quartz in the ith sandstone sample
Y_i, concentration level of metamorphic quartz in the ith sandstone sample

In order to check the proper source region of the sandstone, we performed the Sign Test with a 37-sample data set, where:

$$D_i = X_i - Y_i.$$

A hypothesis of interest here is that D has a median of 0, i.e., $M_D = 0$.
So:

$$H_0:\ M_D = 0$$

$$H_1:\ M_D > 0$$

Define:

$$Z_i = \begin{cases} 1 \text{ if } & D_i > 0 \\ 0 \text{ if } & D_i < 0 \end{cases}$$

Ignore the sample observation for which $D_i = 0$ and reduce the sample size accordingly.
Under H_0:
$P(Z_i = 1) = P(D_i > 0) = 0.5.$
Define K as number of times $Z_i = 1$ (where i = 1,...37).
So in this case, the P- value will be:

$$\sum_{K_0}^{n} \binom{n}{i} (0.5)^n$$

where K_0 is the observed value of K.

Here $n = 37$, $K_0 = 34$.

If the P value is <.5, there is strong evidence against the null hypothesis. The P value from the Sign test of the unmetamorphosed siliciclastic rocks from our study area is 6.1649×10^{-8}, which is <0.5. This suggests that there is strong evidence against the null hypothesis, and suggests that the proportion of the plutonic quartz is predominant over metamorphic quartz.

The Sign Test does not make use of information available in the sample in its entirety, since it does not take into consideration the magnitude of the difference between observations. In order to make use of the information on magnitude of differences, we performed the Wilcoxon signed-rank test.

Wilcoxon signed-rank test

Define:

X_i, concentration level of plutonic quartz in the i-th sandstone
Y_i, concentration level of metamorphic quartz in the i-th sandstone

Where:

$$D_i = X_i - Y_i \text{ (where i } = 1, 2, \ldots n)$$

To test H_0: $M_D = 0$, we first arrange |D1|, |D2|... |Dn| in increasing order of magnitude and assign ranks 1,2,...n, where M_D is the median difference.

Let:

$$T^+ = \text{the sum of the ranks of positive 'Di's.}$$

$$T^- = \text{the sum of the ranks of negative 'Di's.}$$

Then, under H_0, we expect T^+ and T^- to be the same. Note that:

$$T^+ + T^- = \sum_{i=1}^{n} i = \frac{n(n+1)}{2}$$

So that, T^+ and T^- are linearly related and offer equivalent criteria.

The statistic T^+ (or T^-) is known as the **Wilcoxon** statistic. A large volume of T^+ (or equivalently, a small value of T^-) means that most of the large deviations from 0 are positive, and therefore we reject H_0 in favor of the alternative, H_1: $M_D > 0$.

So here the rejection criterion is as follows:

Reject H_0 if $T^+ > C_\alpha$

where C_α is a value such that P $(T^+ > C_\alpha) = \alpha$, where α = level of significance.
In our study area, n = 37. Since n > 30, we can use the Z-test:

$$Z = \frac{T^+ - E_{H_0}(T^+)}{\sqrt{Van_{H_0}(T^+)}}$$

where:

$$E_{H_0}(T^+) = \frac{n(n+1)}{4}$$

$$Van_{H_0}(T^+) = \frac{n(n+1)(2n+1)}{24}$$

Here we reject the H_0 at α level of significance, if the observed value of $Z > Z_\alpha$, where Z_α is the value such that P $(Z > Z_\alpha) = \alpha$. In our case, n = 37, and let us consider $\alpha = 0.05$. From the normal distribution data table we can find out the value of the Z_α. If the calculated value of Z is greater than the table value, we conclude that plutonic quartz is predominant over metamorphic quartz. The Z value of our data set from this test is 4.50, which is well above the Z_α 1.645 at $\alpha = 0.05$. This rejects the H_0 at α level of significance, because our observed value of $Z > Z_\alpha$, where Z_α is the value such that P $(Z > Z_\alpha) = \alpha$.

CHAPTER

18

Provenance of Detrital Pyrite in Archean Sedimentary Rocks: Examples From the Witwatersrand Basin

G. *da Costa*[1], A. *Hofmann*[1], A. *Agangi*[2]

[1]University of Johannesburg, Auckland Park, South Africa; [2]Curtin University, Bentley, WA, Australia

OUTLINE

Sediment Provenance
http://dx.doi.org/10.1016/B978-0-12-803386-9.00018-6

1. INTRODUCTION

Detrital pyrite is a redox-sensitive mineral that occurs as an accessory phase in clastic sedimentary rocks deposited prior to the initial rise of atmospheric oxygen in the early Paleoproterozoic. Its preservation in sedimentary rocks older than c. 2.4 Ga and its disappearance after that time indicates a rapid rise in atmospheric oxygen to levels >0.01 atm (Grandstaff, 1980; Holland, 1984; Rasmussen and Buick, 1999; Johnson et al., 2014). Only rarely is detrital pyrite found in younger sedimentary successions, related to specific conditions such as rapid burial in water-saturated and reduced sediments, for example, cretaceous to recent sedimentary sequence in southern New Zealand (Youngson et al., 2006). Detrital pyrite is abundant in some conglomerates enriched in gold, constituting the famous placer gold quartz pebble conglomerates of the Witwatersrand-type (Fig. 18.1). In this contribution we discuss the potential sources of detrital pyrite from different geological environments and the methodologies used to study detrital pyrite in order to evaluate their provenance. We use the Witwatersrand Basin in South Africa as an example of a sedimentary basin known for its gold and uranium mineralization associated with detrital pyrite.

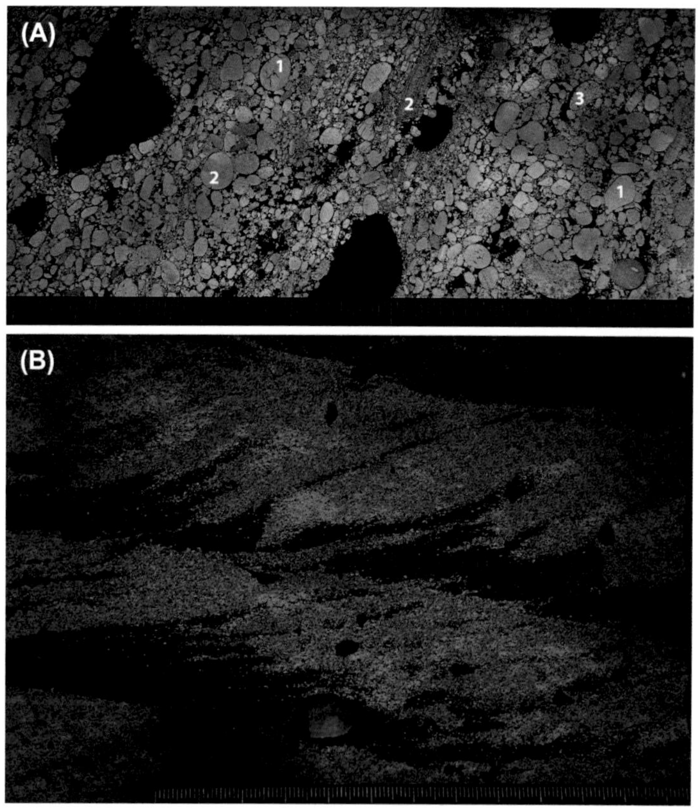

FIGURE 18.1 Photographs of detrital pyrite-rich sedimentary rocks. (A) Polished slab of Pyrite Leader Reef, Black Reef Formation, Modder East Mine, South Africa. Detrital pyrite types: (1) type M; (2) type IP; (3) type IR. (B) Climbing ripple cross-laminated pyritic sandstone, Main Reef Leader, Main Formation, Geldenhuis Deep Mine, South Africa.

2. PROVENANCE OF DETRITAL PYRITE

Detrital pyrite can be derived from erosion of any pyrite-bearing lithology, as long as conditions during weathering, transport, deposition, diagenesis, and metamorphism will allow pyrite to be preserved in the sedimentary rocks. Potential sources for detrital pyrite may include (1) sediments and sedimentary rock successions (Fig. 18.2A), (2) volcanogenic massive sulfide (VMS) deposits (Fig 18.2B), (3) sedimentary exhalative (SEDEX) deposits, (4) magmatic, magmatic-hydrothermal, or metamorphism-related hydrothermal sulfide deposits (Fig. 18.2C), and (5) igneous rocks (Fig. 18.2D). Because pyrite can be stable under a large range of metamorphic grades (Craig and Vokes, 1993), any potential source rock subjected to metamorphism may equally yield pyrite upon erosion. However, pyrite does start to recrystallize at greenschist facies grade and may be converted to pyrrhotite during metamorphism at lower amphibolite facies grade associated with host rock dehydration reactions (Craig and Vokes, 1993; Tomkins, 2010).

FIGURE 18.2 Examples of potential source rocks of detrital pyrite. (A) Neoarchean black shale with diagenetic pyrite, Fife Scott Formation, Gweru greenstone belt, Zimbabwe. (B) Sample of deformed and metamorphosed VMS consisting of euhedral pyrite crystals with minor chalcopyrite interstratified with quartz, Otjihasi Mine, Namibia. (C) Quartz vein with pyrite mineralization. Redwing Gold Mine, Mutare greenstone belt, Zimbabwe. *(Photo courtesy of Roelf le Roux).* (D) Coarse-grained pyroxenite with pyrrhotite interstitial to pyroxene and plagioclase, Merensky Reef, Bushveld Complex, South Africa.

2.1 Sedimentary Pyrite Formation

Pyrite is a common component of marine organic-rich muds and derived sedimentary rocks deposited along continental margins. Most of the world's sediment carried by rivers to the oceans is deposited today along the continental margins, where organic matter, derived from terrestrial plant debris delivered by rivers and marine organisms, is also being deposited (Berner, 1985). In the modern marine environment, anoxic conditions arise from the local consumption of dissolved oxygen by bacteria and other microorganisms living near the sediment–water interface, preventing migration of O_2 to deeper levels in the sediment (Berner, 1985). As a result, decomposition of organic matter at depth takes place only by anoxic processes in the continental margin sediment through bacterial sulfate reduction. Sulfate-reducing bacteria in association with fermentative microorganisms obtain energy by oxidizing organic matter to CO_2 while reducing dissolved interstitial sulfate to H_2S, as shown by the following reaction (Berner, 1985): $2CH_2O + SO_4^{2-} = H_2S + 2HCO_3$.

The hydrogen sulfide produced by this reaction migrates upward or laterally to oxic portions of sediments to be converted back to sulfate or reacts with detrital iron minerals to form iron sulfide phases, such as mackinawite (FeS) or greigite (Fe_3S_4) among others (e.g., Rickard, 2012), that may convert into pyrite (FeS_2). This process of organic matter decomposition in continental margin sediments is intertwined with both processes of sulfate reduction and pyrite formation (Berner, 1985).

The factors controlling sedimentary pyrite formation is diffusion of sulfate as well as the availability of organic matter and reactive iron (Fe^{2+}). In anoxic marine environments, pyrite formation is limited by the abundance and reactivity of detrital iron minerals and, less so, by the amount of locally deposited organic matter (Berner, 1984, 1985).

Organic decomposition, sulfate reduction, and pyrite formation also take place in oxygenated bottom waters, deep-sea sediments, swamp and lake sediments, and in euxinic environments. Euxinic environments arise from tectonic, physiographic, climatic, and oceanic factors resulting in isolation of stagnant deep water bodies from contact with atmospheric oxygen, allowing the build-up of H_2S (Berner, 1985).

Once H_2S is formed in the sediment it reacts with various iron minerals to form iron sulfide. The first phase, as shown under laboratory conditions, is generally noncrystalline FeS, mackinawite ($Fe_{1+x}S$), and greigite (Fe_3S_4), all being thermodynamically unstable under sedimentary conditions (Berner, 1970). Such instability is observed by the transformation of FeS to pyrite in modern sediments during early diagenesis and the rarity of mackinawite and greigite in ancient sedimentary rocks (Berner, 1970). The last phase of pyrite formation is the transformation of FeS to pyrite, according to the overall reaction proposed by Berner (1970): $FeS + S^0 = FeS_2$, where FeS corresponds to iron monosulfide (i.e., mackinawite or greigite).

The reaction pathway is the FeS conversion via sulfur addition, with sulfur acting as electron acceptor (Schoonen, 2004). Two other reaction pathways are also possible with FeS conversion via sulfur addition in the presence of a nonsulfur electron acceptor, and FeS conversion via iron loss and an electron acceptor (Schoonen, 2004).

The formation of pyrite in organic-rich sediments can be traced back well into the Archean. Evidence for microbial sulfate reduction has been described from rocks as old as 3.5 Ga (e.g., Ohmoto et al., 1993; Shen et al., 2001). Especially prior to the great oxidation event (GOE) in the Paleoproterozoic, dissolved Fe^{2+} was present in high concentrations in marine waters (Holland, 1984). Total organic carbon contents of Archean shales are similar to those of the Neogene (Lyons et al., 2014). Therefore, the main limiting factor for pyrite formation by microbial sulfate reduction may have been low contents of dissolved sulfate in the oceans prior to the GOE (Crowe et al., 2014).

Pyrite in sedimentary rocks occurs in different morphologies, which may be related to the type of precursor phases (Wang and Morse, 1996; Goldhaber, 2003; Rickard, 2012). The most common forms of pyrite in sedimentary rocks are: (1) framboids, ranging from 5 to 25 μm (Wang and Morse, 1996), (2) aggregates of small pyrite crystals, with irregular size and shape (Wang and Morse, 1996), and (3) spheroidal and nodular pyrite typically 1–2 cm but up to 10 cm in diameter (Rickard, 2012, Fig. 18.2A).

Marcasite, the orthorhombic dimorph of pyrite (FeS_{2m}), is common in hydrothermal deposits, such as deep-ocean hydrothermal-vent mineralization (Wilkin and Barnes, 1996; Rickard, 2012), but has also been reported from marine shales (Schieber, 2011). Marcasite forms crystals with a radiating, acicular habit and it is often mistaken as pyrite crystals, but can be reliably identified by X-ray diffraction analysis (Rickard, 2012).

2.2 Pyrite in Igneous Rocks

Iron sulfides are commonly found as minor and accessory minerals in igneous rocks. Pyrrhotite is more reduced than pyrite and is the stable iron sulfide formed from magmas (Fig. 18.2B; Hall, 1986). Pyrite does occur in igneous rocks, however, including both basaltic and felsic igneous rocks, where its origin may be related to late crystallization (Hall, 1986; Naldrett, 1989). Under constant pressure, five factors are important for the solubility of iron sulfides in a silicate magma: (1) oxygen fugacity (fO_2), (2) temperature, (3) FeO and TiO_2 contents, (4) oxidation state, and (5) mafic versus felsic components (Naldrett, 2004). As magmas ascend to environments of lower pressure and temperature, their ability to dissolve sulfide increases. Even though the magma may have left the source region saturated in sulfide, it progressively becomes less saturated during its ascent (Naldrett, 2004). The addition of silica to a magma decreases its capability to dissolve sulfide, possibly resulting in segregation of immiscible sulfide (Naldrett, 2004). Contamination of a magma by sulfur-bearing phases may have the same result. Oxidation of a melt is a process also capable for precipitating sulfide without silicate formation, as has been reported from the Bushveld Complex, South Africa (Naldrett, 2004). Sulfide solubility in silicate melts is affected by the sulfate/sulfide ratio and reaches a minimum for fO_2 at the SO_4^{2-}/S^{2-} redox boundary (around nickel–nickel oxide values) (Carroll and Rutherford, 1987; Baker and Moretti, 2011; Jugo et al., 2005). Another mechanism capable of saturating a magma in sulfide is reduction by crystallization of magnetite according to the reaction $8FeO + SO_4^{2-} = 4Fe_2O_3 + S^{2-}$ (Carmichael and Ghiorso, 1986), which causes a decrease of Fe^{3+}/Fe^{2+} by removal of Fe^{3+} (Sun et al., 2004).

Pyrite occurs as a common minor or accessory mineral in norite, oxide-rich gabbro, hydrothermally altered andesite, dolerite, and tuff, and is also a common phase in dolerite and diorite (Naldrett et al., 1970; Obiora and Charan, 2010; Tabelin et al., 2012). Pyrite commonly occurs as cubic crystals and aggregates, typically around 0.1 mm in diameter, and also as pyritohedra, varying in size from 100 μm to 1.0 mm (Tabelin et al., 2012; Voroshin et al., 2014).

2.3 Volcanogenic Massive Sulfide and Sedimentary Exhalative Deposits

VMS deposits are known from c. 3.5 Ga up to the present and have formed on or near the seafloor by hydrothermal fluid circulation associated with volcanic activity (Huston et al., 2010; Hannington, 2014; Tornos et al., 2015). They are an important source of Pb, Zn, Cu, and other metals (Co, Sn, Ga, Bi, etc.), including Au and Ag (Zheng et al., 2013; Tornos et al., 2015). VMS deposits are stratiform, lenticular, and/or replacive orebodies of massive (>60%) to semimassive sulfide accumulations (Huston et al., 2010; Tornos et al., 2015). Many VMS deposits have formed in backarc basin and intracontinental rifts/protoocean basins associated with rapid extension and subsidence, as well as submarine volcanism under deep marine conditions (Zheng et al., 2013; Hannington, 2014).

VMS deposits are found worldwide and the age distribution is not uniform, with the Ordovician and Carboniferous deposits being economically more important than others (Ohmoto, 1996). World-class Archean VMS deposits are restricted to the Neoarchean and include occurrences in the Abitibi greenstone belt in Canada (Mueller et al., 2009) and in the Yilgarn Craton, western Australia (Hollis et al., 2015).

Pyrite represents the most common ore mineral in Phanerozoic VMS deposits, followed by chalcopyrite, sphalerite, and pyrrhotite; pyrrhotite becomes more abundant further back in time (Hannington, 2014). The sulfide minerals are fine-grained, showing colloform, concretionary, brecciated, and framboidal textures; coarser grained euhedral crystals, crystal aggregates, and granoblastic textures are common in deposits affected by hydrothermal recrystallization, diagenesis, burial metamorphism, or deformation, ranging from 0.5 to 2 mm in size (Marcoux et al., 1996; Hannington, 2014).

SEDEX deposits are stratiform, sediment-hosted Pb–Zn sulfide mineralizations that formed by submarine venting of hydrothermal fluids within fault-controlled basins in intracratonic or continental margin rift zones. They date as far back as 2.0 Ga (Lyons et al., 2006; Leach et al., 2010). SEDEX deposits contain abundant S-bearing phases, mainly including pyrite, pyrrhotite, sphalerite, galena, and barite. Ore fluids were principally hot metalliferous basinal brines showing no direct link to volcanic activity, while reduced S was largely derived from seawater sulfate reduction (Lyons et al., 2006; Leach et al., 2010).

2.4 Magmatic-Hydrothermal or Metamorphism-Related Hydrothermal Sulfide Deposits

Many hydrothermal ore deposits are linked in space and time to magmatic systems that provide the heat that drives hydrothermal circulation and metal redistribution (Hedenquist and Lowenstern, 1994). Dehydration reactions during the prograde path of regional metamorphism may equally drive hydrothermal systems, water–rock interaction, and mineralization. Of specific relevance for this article are hydrothermal ore deposits, specifically lode-gold deposits of Archean granitoid-greenstone terranes. They typically contain pyrite (together with pyrrhotite, arsenopyrite, chalcopyrite) associated with quartz-carbonate vein sets, shear zones, and wall rock alteration envelopes (Fig. 18.2D; Ho et al., 1994; Ridley et al., 1996; Cassidy et al., 1998). Host rocks can be very variable, including plutonic and volcanic igneous rocks as well as sedimentary rocks, typically at greenschist to lower amphibolite facies grade. In the lode-gold deposit of Kambalda, Western Australia, Ho et al. (1994) described a variety of pyrite shapes, which were attributed to mineralization followed by corrosion or dissolution. Shapes include single or interpenetrating cubes, irregular or parallel aggregates, and sub- to anhedral grains with conchoidal or irregular pitted surface. Pyrite may be massive or disseminated (Ho et al., 1994). The dominant pyrite size ranges from 0.2 mm for fine grains to 1 mm for very coarse grains (Ho et al., 1994).

3. STABILITY OF DETRITAL PYRITE AND ITS PRESERVATION POTENTIAL

Pyrite and its preservation on the Earth's surface is a factor of atmospheric, hydrospheric, and pedospheric conditions and mechanical stability (Rickard, 1997, 2012). Pyrite forms in most Earth surface environments, as it has an extensive stability over pH = 2–10. Pyrite has stability even in more oxic environments, above the boundary between oxic and anoxic (or suboxic) systems, which is usually taken to be where $\{SO_4^{2-}\} = \{S(\text{-II})\}$ (Rickard, 2012). Another factor of pyrite preservation is its resistance to mechanical abrasion. The cleavage of pyrite is poor along {100} and imperfect; that is, it breaks with conchoidal fracture (Tauson et al., 2015). Together with a hardness of 6.5, we can assume that pyrite can be transported for long distances before deposition. As has been shown for the Southland placers of New Zealand, pyrite and marcasite can form in a diagenetic environment and can survive erosion and transport in streams and be reburied in younger sediments (Falconer et al., 2006; Youngson et al., 2006). Pyrite (and arsenopyrite) can be transported for tens of kilometers in fluvial systems, while marcasite may survive at least short distance transport under oxic and acidic conditions in streams (Craw et al., 2003).

The redox sensitivity of pyrite has been used as an indicator of low oxygen levels at the time of detrital pyrite deposition during the Archean, when the atmosphere was largely anoxic (Rasmussen and Buick, 1999; England et al., 2002). The rare presence of detrital pyrite in sedimentary rocks younger than c. 2.4 Ga at a time of atmospheric oxygen levels sufficiently high for pyrite oxidation may be linked to aridity or rapid erosion. The maximum O_2 constraints (i.e., redox-sensitivity), obtained from pyrite preservation in the sedimentary succession, are related to physical transport scenarios and deposition (Holland, 1984; Rasmussen and Buick, 1999; Johnson et al., 2014). This can explain the reason for detrital pyrite to be found in areas with rapid erosion and transport or high aridity, as otherwise oxidative weathering during transport would not preserve pyrite grains (Johnson et al., 2014).

Once pyrite gets buried, it may remain stable during burial diagenesis, as exemplified by its abundance in pre-2.4 Ga sedimentary deposits, unless subjected to the influx of oxidizing fluids at a later stage in the history of the sedimentary successions. Even high-grade metamorphic conditions are not enough to destroy the evidence of the origin and history of pyrite grains, as, for example, evidenced from the retention of crystal shape, mineral inclusions, early fabrics, or chemical zoning (Craig and Vokes, 1998).

4. TOOLS TO STUDY DETRITAL PYRITE IN SEDIMENTARY ROCKS

Research on detrital pyrite provenance interfaces several geological disciplines, such as mineralogy, geochemistry, sedimentology, and geochronology. The study of detrital pyrite includes aspects of grain size and morphology, internal texture, nature of inclusions, chemical composition, and age, and a large range of analytical techniques can be employed. Conventional petrographic and mineral-geochemical methods using the petrographic reflected light microscope, scanning electron microscope (SEM), and electron probe microanalyzer (EPMA) are increasingly coupled to various mass spectrometric methods that allow bulk or in situ analysis of the (isotope) geochemical composition of pyrite. Much progress has been made during the last decades to better constrain the origin of pyrite (detrital, authigenic) in sedimentary rocks and the environmental conditions under which it formed, helping to understand not only the geological past, but to also help understand mechanisms of the formation of ore deposits frequently associated with it.

4.1 Sample Preparation

Detrital pyrite is investigated in situ from polished rock slabs, thin sections, and block mounts. In addition, pyrite may also be liberated from the rock matrix. The hydrofluoric acid liberation technique was used by Neuerburg (1975) to analyze the morphology of pyrite. This technique results in dissolution of silicate rock matrix, but does not attack the pyrite particles, which can be investigated for their morphology and their single grain chemical and isotopic composition (Hallbauer, 1977; Utter, 1978; Hallbauer and von Gehlen, 1983; Barton and Hallbauer, 1996; Hofmann et al., 2009). Other techniques include liberation of individual pyrite grains with a dentist's drill, crushing a sample followed by heavy-liquid separation, and hand picking (Saager, 1976; Barton and Hallbauer, 1996; Hofmann et al., 2009). Microdrilling to generate pyrite powder for chemical analysis is also frequently employed.

4.2 Mineralogical Characterization

Reflected light microscopy of polished thin sections with or without oil immersion is the classic and normally the first step for detrital pyrite studies (Ramdohr, 1958; Saager and Mihálik, 1967; Saager, 1970, 1976; England et al., 2002; Guy et al., 2014). Etching of the thin section with concentrated nitric acid may reveal some additional textural features (Saager and Mihálik, 1967). Except where recrystallized, compositional and textural features of pyrite, such as the presence of inclusions or crystal growth habits, can be readily identified and are generally used for a differentiation into different genetic types. More detailed petrographic work is done with the SEM. This technique allows imaging of the surface features of individual grains using secondary electron mode, while investigating intergrowths with other minerals on the microscale using back-scattered electron mode (Hallbauer and von Gehlen, 1983; Utter, 1978; England et al., 2002; Guy et al., 2012; Agangi et al., 2015a).

Some researchers investigated pyrite from polished sections subjected to color staining with $KMnO_4$ dissolved in sulfuric acid. This staining can determine relative variations of As concentration in pyrite, which may be a result of changing fluid composition during pyrite growth (Fleet et al., 1989, 1993). Oxidation of arsenian pyrite can be determined by X-ray photoelectron spectroscopy and laser-ionization mass spectrometry (Fleet et al., 1993).

4.3 Trace Elements

Trace elements in pyrite include a large range of siderophile and chalcophile elements, the concentrations and ratios of which can provide information on the setting of pyrite formation.

The concentration of major (Fe, S) and some trace elements of pyrite in thin sections can be determined qualitatively with energy dispersion spectrometry attached to the SEM or quantitatively by EPMA (Guy et al., 2010; Agangi et al., 2013, 2015a), although only As, Co, Ni, and Pb may be present at concentrations above detection limits for quantitative analysis.

Absolute concentrations of a large range of trace elements of pyrite can be obtained by inductively coupled plasma mass spectrometry (ICP-MS) from solutions of separate pyrite grains (Hofmann et al., 2009) or in situ using a laser ablation (LA) system (Agangi et al., 2013, 2015a).

Older techniques of trace element analysis of pyrite used atomic absorption spectrophotometry of dissolved pyrite grains (Saager, 1976; Barton and Hallbauer, 1996). The distribution of major and trace elements can be determined in situ using X-ray compositional mapping with EPMA (Agangi et al., 2013), proton-induced X-ray emission (Przybylowicz et al., 2001; Agangi et al., 2015b), or element mapping using LA-ICP-MS (Large et al., 2013; Agangi et al., 2015a). These techniques allow for the evaluation of chemical growth zoning recording the evolution of mineralizing fluids, but also secondary events of brecciation, veining, resorption, and recrystallization (Fig. 18.3).

4.4 Stable Isotope Analysis

The sulfur isotope composition ($\delta^{33}S$, $\delta^{34}S$, and $\delta^{36}S$) is generally determined with an isotope ratio mass spectrometer from pyrite or Ag_2S obtained by S extraction following the Cr reduction method of Canfield et al. (1986).

Fe isotope ($\delta^{54}Fe$ and $\delta^{56}Fe$) analysis is performed by multicollector ICP-MS following dissolution of pyrite sample in a $HNO_3{-}HCl$ acid mixture (Hofmann et al., 2009). The combination of Fe and S isotope analysis using in situ secondary ion mass spectrometry is a method that is increasingly being used (Agangi et al., 2015a), having the advantage to

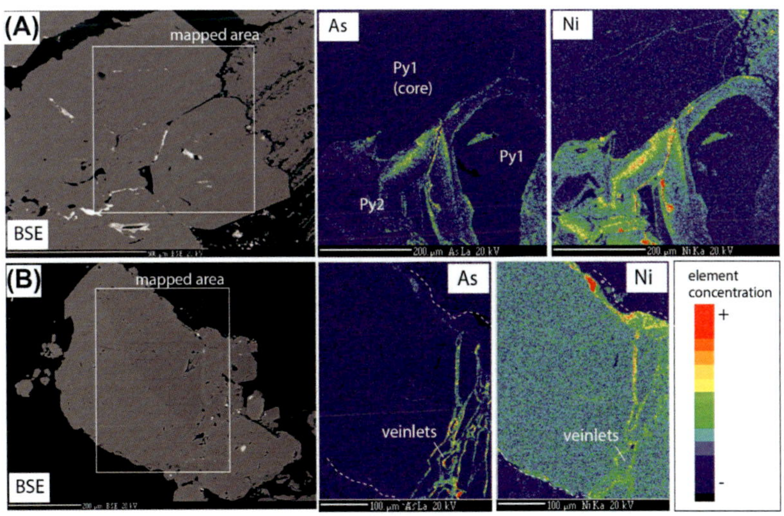

FIGURE 18.3 BSE images and X-ray compositional maps of euhedral compact pyrite from the Ventersdorp Contact Reef (A) and Carbon Leader Reef (B), Witwatersrand Basin. Brighter colors in the X-ray maps indicate higher element concentration. Elemental maps allow the identification of overgrowths (Py2 in A) and veinlets (B).

evaluate isotopic variation on the microscale. Multiple S and Fe isotope analysis of Archean detrital pyrite grains provides information on their provenance, specifically if pyrite was derived from a magmatic or sedimentary source (Hofmann et al., 2009).

Sulfur isotopes can fractionate by both mass-dependent, biotic, and abiotic processes. Thermochemical sulfate reduction coupled to the oxidation of organic matter or ferruginous iron-bearing minerals in hydrothermal systems can result in large fractionation of $\delta^{34}S$ between sulfate and sulfide (Ohmoto and Goldhaber, 1997). At lower temperatures during early diagenesis, microorganisms can impart fractionations during their sulfur metabolism, particularly during the process of sulfate reduction (Canfield, 2001).

Mass-independent fractionation (MIF) of S isotopes is caused by SO_2 photolysis under ultraviolet radiation (Farquhar et al., 2001). MIF-S has been attributed to the absence of an ozone shield in the largely anoxic Archean atmosphere (Farquhar et al., 2000; Pavlov and Kasting, 2002). Multiple S isotope analysis thus can distinguish between S that passed through the Archean atmosphere ($\Delta^{33}S \neq 0‰$) and magmatic S incorporated into hydrothermal fluids ($\Delta^{33}S = 0‰$). Sulfides present in VMS deposits that formed by seawater circulation through hydrothermal systems and seawater sulfate reduction commonly carry near to zero or relatively small negative $\Delta^{33}S$ values (Farquhar and Wing, 2005; Jamieson et al., 2006; Ueno et al., 2008). In contrast, disseminated sulfides in Archean black shales have mostly positive $\Delta^{33}S$ values inherited from photochemically produced reduced S species such as elemental S (Farquhar and Wing, 2003, 2005; Ono et al., 2009). Pyrite nodules in Archean organic matter–rich shales with negative $\Delta^{33}S$ values might reflect seawater sulfate reduction in sediments during diagenesis.

Fe isotope fractionation occurs during redox changes, fluid–mineral interactions, and biological processes (see Dauphas and Rouxel, 2006; Anbar and Rouxel, 2007; Johnson et al., 2008 for reviews). While igneous rocks and siliciclastic sedimentary rocks with low carbon and sulfur contents have limited Fe isotope variations, diagenetic pyrites in Archean black shales show highly variable and mostly negative values ranging from −3.5 to +0.5‰ (Rouxel et al., 2005; Yamaguchi et al., 2005; Archer and Vance, 2006; Marin-Carbonne et al., 2014). In contrast, Fe-oxides in Archean BIFs are frequently characterized by positive $\delta^{56}Fe$ values up to 1.6‰ (Johnson et al., 2003; Dauphas et al., 2004; Rouxel et al., 2005; Whitehouse and Fedo, 2007). Modern seafloor hydrothermal pyrite deposits have variable but mostly negative $\delta^{56}Fe$ values ranging from −2.1 to −0.1‰, a wider range than that of seafloor-hydrothermal fluids defined at around $-0.4 \pm 0.3‰$ (Rouxel et al., 2008). Precambrian VMS deposits are characterized by pyrite having a similar range of $\delta^{56}Fe$ values.

4.5 Radiogenic Isotope Analysis

The isotope ratios of lead ($^{206}Pb/^{204}Pb$, $^{207}Pb/^{204}Pb$) of detrital pyrites can be analyzed with thermal ionization mass spectrometry (TIMS) following dissolution of pyrite or in situ using LA-ICP-MS (Barton and Hallbauer, 1996; Poujol et al., 1999; Meffre et al., 2008). Pb isotopes are mainly used to provide constraints on pyrite formation age, based on a model Pb growth curve, and on the Pb source, which may provide hints to pyrite provenance or post-depositional processes.

The Rhenium–Osmium isotope system is based on the radioactive decay of ^{187}Re to ^{187}Os with a half-life of 41.6 Ga and is used as a process tracer and geochronometer. Rh and Os are chalcophile and siderophile and, in sedimentary environments, become fixed under reducing conditions in organic-rich shales and sulfides (Stein and Hannah, 2015). The Re–Os isotopic data ($^{187}Re/^{188}Os$, $^{187}Os/^{188}Os$) is determined by negative TIMS (Creaser et al., 1991). Single pyrite samples can be dated accurately if they have low common Os concentrations. If not, then the isochron approach requires analyses of multiple, contemporaneously formed samples (Stein and Hannah, 2015).

5. CASE STUDY: THE WITWATERSRAND BASIN

Pyrite is the most abundant ore mineral in gold-bearing conglomerates (reefs) of the Witwatersrand Basin (Fig. 18.4). Much of it is detrital in origin and thus survived weathering and transport to the site of deposition as clasts, although authigenic types, formed in situ as a result of postdepositional processes are also common. The location of the detrital source and the nature of it are largely unknown.

5.1 Setting

The Witwatersrand Basin is located in the central portion of the Archean Kaapvaal Craton, South Africa, and hosts the richest and the single most import gold province worldwide (Saager et al., 1982a; Frimmel et al., 2005; Frimmel, 2008; Phillips, 2013). In the Witwatersrand Basin, the total resource of known gold is 96 703 t Au at 2–10 g/t, representing the biggest

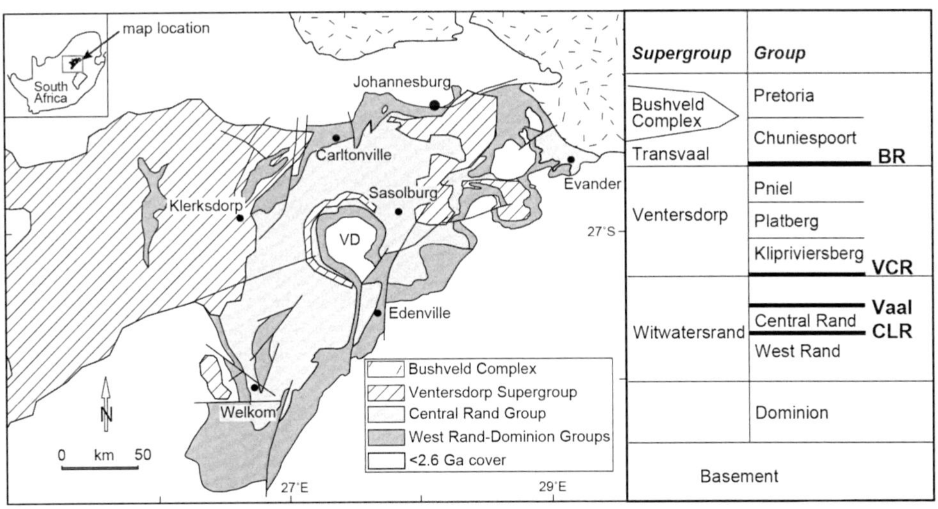

FIGURE 18.4 Witwatersrand basin location and simplified stratigraphic column of the Witwatersrand Basin and Transvaal Supergroup (modified from Frimmel, 2005; Agangi et al., 2015a). *BR*, Black Reef; *CLR*, Carbon Leader Reef; *Vaal*, Vaal Reef; *VCR*, Ventersdorp Contact Reef; *VD*, Vredefort Dome.

gold paleoplacer deposit by far (91% of all paleoplacer gold; Frimmel, 2008). The Witwatersrand Basin-fill consists of the predominantly sedimentary Witwatersrand Supergroup that is subdivided into the West Rand Group and the conformably to locally unconformably overlying Central Rand Group (Fig. 18.4). The West Rand Group is a sedimentary succession deposited between 2985 ± 14 and 2914 ± 8 Ma (Armstrong et al., 1990; Kositcin and Krapez, 2004) and nonconformably overlies Paleoarchean granitoid-greenstone basement and the volcano-sedimentary succession of the c. 3.07 Ga Dominion Group (Armstrong et al., 1990; Jolley et al., 2004). The Central Rand Group is overlain by the volcano-sedimentary succession of the Ventersdorp Supergroup dated at 2714 ± 8 and 2709 ± 4 Ma (Armstrong et al., 1990). A minimum age for deposition of the Central Rand Group was provided by the maximum age of authigenic xenotime (YPO_4) estimated at ~2780 to ~2765 Ma (England et al., 2001; Kositcin et al., 2003). Different xenotime phases were reported by Kositcin et al. (2003) and interpreted, based on different ages, as igneous-detrital (>2800 Ma), diagenetic (~2780 Ma), and hydrothermal (<2780 Ma).

Apart from gold, uranium and platinum group elements are also being extracted from the reefs. Au/U-bearing reefs of the Witwatersrand Basin frequently contain abundant detrital pyrite and are concentrated in the Central Rand Group (e.g., Main Reef, Vaal Reef). Economic to subeconomic gold mineralization associated with detrital pyrite is also found in conglomerates of the Dominion and West Rand groups as well as at the base of the Ventersdorp and Transvaal Supergroups (Ventersdorp Contact Reef and Black Reef, respectively; Fig. 18.4). There are different hypotheses on the genesis of the gold mineralization of the Witwatersrand reefs, which has been a controversy for many decades. Most researchers concur with the hydrothermally modified placer deposit model, although the syngenetic model has gained some momentum in recent years (Agangi et al., 2015a; Frimmel and Hennigh, 2015; Heinrich, 2015).

A large variety of detrital accessory minerals have been described from the gold-bearing reefs (Ramdohr, 1958; Feather and Koen, 1975; Hallbauer, 1986; Janisch, 1986; Phillips and Law, 2000). The most common are perhaps chromite, zircon, monazite, rutile, and uraninite, while others may indicate more specific source rocks, such as columbite, corundum, cassiterite, garnet, molybdenite, cobaltite, and diamond. According to Hallbauer (1986), pyrite makes up 3–5% of Witwatersrand conglomerates, of which between 50% and 75% represents rounded (i.e., detrital) grains. As detrital pyrite is frequently associated with gold (Feather and Koen, 1975), many authors have developed different pyrite classification schemes in order to better understand the nature and origin of the reefs and the environmental conditions of their formation.

5.2 Detrital Pyrite Types

Two main classes of pyrite occur in the Witwatersrand basin: (1) detrital pyrite and (2) epigenetic pyrite. In this study we focus on detrital pyrite, which can generally be differentiated from epigenetic pyrite by evidence of abrasion. While this evidence may be found in the truncation of internal structures (e.g., zonation, growth banding, microveining) at the outer margins of rounded grains in pyrite (e.g., MacLean and Fleet, 1989; England et al., 2002), the more direct approach is to simply regard rounding as evidence for abrasion. Rounded pyrite representing sulfidized detrital rock or mineral fragments do occur (Hirdes and Saager, 1983; Myers et al., 1993), although sulfidization is frequently not complete, thus allowing distinction between detrital and pseudomorphic pyrite. A large literature base exists on the description of detrital pyrite (Ramdohr, 1958; Saager and Mihálik, 1967; Saager, 1970; Utter, 1978; Hallbauer and von Gehlen, 1983; Barton and Hallbauer, 1996; England et al., 2002; Guy et al., 2012, 2014). In this contribution we use a simple, but comprehensive scheme of detrital pyrite classification.

Among the different detrital pyrites (Fig. 18.5) from the various gold-bearing conglomerates of the Witwatersrand Basin, two main groups of pyrite are present, based on composition: massive and inclusion-bearing. Both pyrite types are commonly present in many reefs together, and may even form different portions of the same grain (e.g., Hofmann et al., 2009). The most common inclusions are sulfide minerals (chalcopyrite, galena, sphalerite, and pyrrhotite) and a variety of silicates, most notably biotite, chlorite, and quartz (Hallbauer, 1986). Massive pyrite (type M) may contain some, but frequently contains no inclusions at all (Figs. 18.1 and 18.5A,B). In inclusion-bearing pyrite (type I), mineral inclusions typically represent 10–50 vol% of pyrite grains. Depending on the arrangement of the inclusions within a grain, three main subtypes can be differentiated. In random inclusion-bearing pyrite (type IR) (Figs. 18.1 and 18.5E,F), the inclusions do not show any preferred orientation.

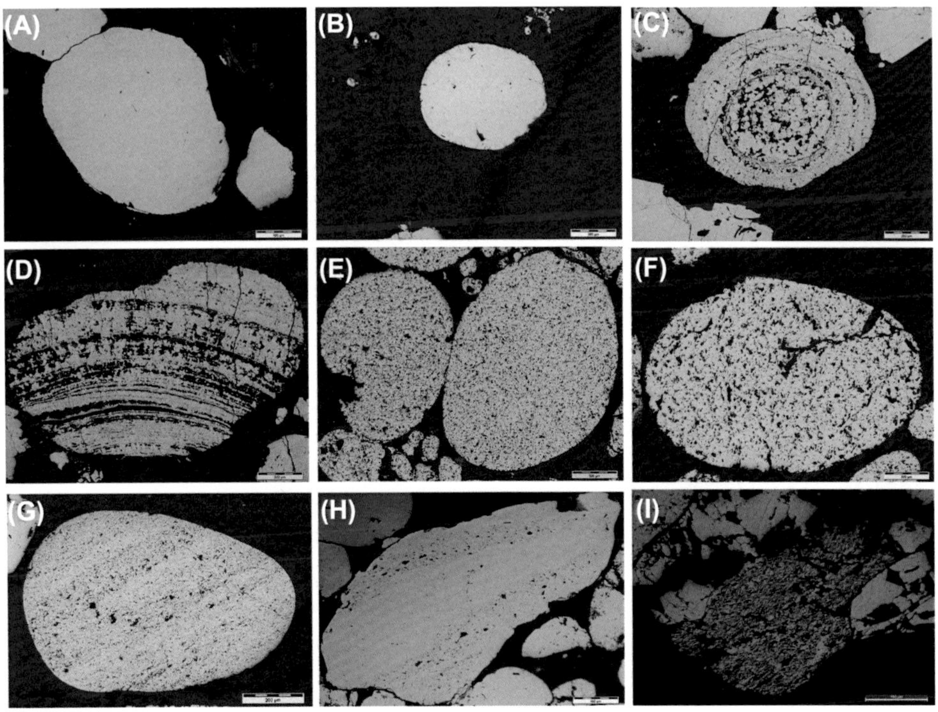

FIGURE 18.5 Detrital pyrite types. (A and B) Massive pyrite (type M) from Black Reef Leader, Modderfontein. (C and D) Concentrically laminated pyrite (type IC) from the Ventersdorp Contact Reef. (E and F) Rounded pyrite with random inclusions (type IR) from pyrite pebble conglomerate, Black Reef Leader, Modderfontein. (G and H) Planar laminated pyrite (type IP) from Black Reef Leader, Modderfontein. (I) Crystalline pyrite (type C) showing skeletal textures, Black Reef Leader, Modderfontein.

In planar laminated pyrite (type IP) (Figs. 18.1 and 18.5G,H), inclusions are concentrated in parallel layers, whereas in concentrically laminated pyrite (type IC) (Fig. 18.5C,D) inclusion-rich and inclusion-poor layers are concentrically arranged. Other pyrite types include crystalline grains, for example, with skeletal texture (Fig. 18.5I). Detrital pyrite size is variable, with rare grains being as large as 5 cm in diameter in some reefs, although most pyrite is found as sand-sized grains. A decrease in grain size of detrital pyrite has been observed in the direction of the paleoslope in some reefs (e.g., Minter et al., 1986).

5.3 Pyrite Provenance

Much sedimentological work has been done on Witwatersrand basin strata, most notably in the 1970s (e.g., Minter, 1976, 1978; Minter and Loen, 1991), which indicated sediment transport toward the center of the basin, with the former source situated mainly to the north and west of the present basin margin. Depositional environments have varied through time from continental to shallow-marine subjected to base level changes, with many of the reefs attributed to have formed in braided fluvial environments (Els, 1998).

Provenance of the siliciclastic sediment has been evaluated from many different angles. The chemical composition of shale in the Witwatersrand basin has been used as an indicator for the average composition of the source rocks. Trace element modeling of shale has indicated a granitoid-greenstone provenance, with proportions of granite, tonalite, basalt, and komatiite having varied with time during deposition of the Witwatersrand Supergroup (Wronkiewicz and Condie, 1987). Oxygen isotope analyses of sand- and pebble-sized quartz supported a provenance of granitoids, pegmatites, as well as hydrothermal veins for the quartz (Vennemann et al., 1996). Indices of chemical alteration recorded in Witwatersrand strata have been used to indicate intense, acidic chemical weathering at times (Sutton et al., 1990; Maynard et al., 1991; Frimmel, 2005). Detrital zircon and xenotime age spectra (Kositcin and Krapez, 2004) have assisted in provenance, age of deposition, as well as tectonic discrimination, by recognizing systematic changes in provenance with time.

The systematic description of detrital pyrite and evaluation of its provenance started with the detailed ore petrological studies of Ramdohr (1958). Since then, many new observations have been made, which include some of the following: (1) further detailed petrological descriptions (Saager, 1970; Hallbauer, 1986; England et al., 2002; Guy et al., 2010), (2) trace element analysis (Meyer et al., 1990; Barton and Hallbauer, 1996; Koglin et al., 2010; Large et al., 2013; Agangi et al., 2013; Gregory et al., 2015), (3) radiogenic isotope studies (Köppel and Saager, 1974; Kirk et al., 2001), and (4) stable isotope analyses (England et al., 2002; Hofmann et al., 2009; Guy et al., 2012, 2014; Agangi et al., 2015a).

The morphology and texture of pyrite grains, the nature of mineral inclusions, trace element geochemistry, and isotopic composition as well as age are the main variables that have been used to evaluate their source. A number of potential sources for detrital pyrite have been described in the literature and these include sedimentary sources (e.g., diagenetic pyrite in fine-grained sedimentary rocks; Dimroth, 1979; Guy et al., 2010; Hofmann et al., 2009), syngenetic sedimentary sources (early diagenetic pyrite; Hallbauer, 1986; Agangi et al., 2015a), exhalative and syngenetic sulfide deposits (Hutchinson and Viljoen, 1988), hydrothermally altered granites (Klemd and Hallbauer, 1987; Robb and Meyer, 1990), as well as volcanic rocks, vein mineralization, and/or iron formations of greenstone belts (e.g., Saager et al., 1982b).

Textures of certain types of pyrite clearly indicate a sedimentary provenance. The texture of concentrically laminated pyrite suggests growth of these grains lamina by lamina in a muddy substrate and eventual incorporation into the conglomerates after only limited transport and reworking (Agangi et al., 2015a). As such, they formed more or less in situ, close to the site of final deposition. The same applies to reworked crystalline nodules that have preserved a radial marcasite texture, although marcasite is no longer preserved due to its conversion to pyrite at elevated temperatures. Crystalline skeletal pyrites with internal growth structure that imply unidirectional growth in open-pores (England et al., 2002) probably also represents syndepositional pyrite growth possibly in the conglomerates themselves. Pyrite-rich intraclasts have the same origin. In addition, many inclusion-rich pyrites contain inclusions pointing to a sedimentary environment, and this is especially the case for inclusion-rich pyrite containing carbonaceous matter (sooty pyrite; Large et al., 2013). In fact, probably most of the detrital pyrites types show textures that can be found in diagenetic sedimentary pyrite. Nevertheless, some overlap exists with textures found in SEDEX and VMS deposits.

This does not mean, however, that all the pyrite was sourced from sedimentary or exhalative deposits. Any pyrite-bearing rock could potentially have contributed pyrite detritus. However, many of the sources proposed do not necessarily have pyrite in abundances required to contribute the large amounts of detrital pyrite found in many reefs. For example, altered granites contain euhedral to subhedral disseminated pyrite and anhedral pyrite fillings associated with quartz-carbonate veins (Klemd and Hallbauer, 1987). Such rocks could contribute silt-to sand-size massive pyrite clasts, but not pebble-sized clasts or inclusion-bearing pyrites.

Trace element composition of detrital pyrite from the Witwatersrand Basin has been reported in several studies (Meyer et al., 1990; Barton and Hallbauer, 1996; Hofmann et al., 2009; Koglin et al., 2010; Large et al., 2013; Agangi et al., 2013, 2015a). Table 18.1 shows a summary of the trace element composition of different pyrite types found in the reefs of the Witwatersrand Basin. Massive pyrite exhibits lower concentrations in trace elements and gold than inclusion-rich pyrites (types IC and IR) (Agangi et al., 2013, 2015a; Large et al., 2013). The content of gold also differs from pyrite type M and the other two pyrite types: the higher content of gold is present in pyrite types IR and IC, due to inclusion-rich texture possibly allowing detrital gold to be trapped during pyrite growth.

The abundance of redox sensitive trace elements (Ni, Cu, Cr, As, Ag, Hg, Te, V, U) of pyrite type IR is similar to the composition of diagenetic pyrites commonly developed in mudstone or black shales from suboxic to anoxic environments (Large et al., 2013; Agangi et al., 2013). The trace element composition of pyrite type IC is reminiscent of pyrite formed in anoxic sedimentary environments (black shales); the inclusions of quartz, chlorite, and muscovite and the different trace element composition of the laminae indicates a greenschist facies metamorphism of fine-grained sediment (silt–clay) (Agangi et al., 2013, 2015a). This conclusion contradicts the origin of concentrically laminated pyrite derived from sulfidization of ooids (England et al., 2002), which was not observed by Agangi et al. (2013, 2015a).

Co and Ni substitute for Fe in pyrite and their absolute concentrations and ratios have been used, with a limited success, in the past as a provenance indicator (Meyer et al., 1990).

TABLE 18.1 Comparison of Trace Element Contents (in ppm) of Different Detrital Pyrite Types From the Witwatersrand Basin

	Type M (Massive)		Type IC (Concentrically Laminated)		Type IR (Random Inclusion)	
References	Gold	Others	Gold	Others	Gold	Others
Large et al. (2013)	0.002 to 0.22	Poor in trace elements. As, Co, Ni and Cu excess of 100 ppm	–	–	0.02 to 1.14	Elevated trace element contents: As (2047), Ni (1092), Co (656), Cu (185), Cr (105), and Pb (115)
Agangi et al. (2013)	<1	Poor in trace elements. Some grains rich in Co, Ni, Ag, Sb, Pb, Mn, Cu, Tl, Hg	6.4 (average)	High values of Pb, Sb, Au, Ag, Tl, Sn, Cu, Mn, Mo	–	Rich in Co, Ni, Ag, Sb, Pb, Mn, Cu, Tl, Hg
Agangi et al. (2015a)	–	–	Up to 40,000[a]	Significant amounts of Au, Ag, As, Ni, Co, Sb, Mo, Cu, Tl	–	High values of Sb, Au, Ag, Tl, Cu, Mn, Mo

[a]*Measured in grains containing visible gold.*

Co/Ni ratios less than one are often associated with a sedimentary (diagenetic) origin (Loftus-Hills and Solomon, 1967). Hofmann et al. (2009) noted that single grain bulk pyrite analysis of trace elements can be problematic for provenance analysis because of the common occurrence of inclusions and evidence for remobilization of elements susceptible to hydrothermal alteration, such as Au, Ni, and As (Meyer et al., 1990; England et al., 2002). Spatially resolved in situ analysis on the other hand is better suited to investigate complex pyrite grains as it allows targeting domains of inferred primary sulfide composition. In their LA-ICP-MS study, Koglin et al. (2010) argued for an unspecified hydrothermal source for massive detrital pyrite on the basis of high Co/Ni ratios and low Au concentrations.

Several radiogenic isotope studies have been undertaken on Witwatersrand pyrite. Lead isotopic composition of detrital pyrite concentrates were compared with sulfides from Archean greenstone belt-hosted gold deposits of South Africa by Köppel and Saager (1974). A similar isotopic composition was argued to indicate a greenstone belt provenance for the detrital pyrite. These inferences, however, may need to be treated with care, as the analysis used concentrates of detrital pyrite of potentially very different types. Pb–Pb dating of single detrital pyrite grains by Barton and Hallbauer (1996) indicated derivation from a c. 3000 Ma old source region, although hydrothermal fluid flow was reported to have had a strong influence on the Pb–Pb system.

Four pyrite particles from a whole rock sample of Vaal Reef were subjected to Re–Os analysis (Kirk et al., 2001) and defined an isochron with an age of 2.99 ± 0.11 Ga. While the type of pyrite was not specified by Kirk et al. (2001), a detrital rather than epigenetic origin was inferred, as the age either predates or overlaps the age of deposition of the Central Rand Group in which the Vaal Reef is situated.

Mathur et al. (2013) analyzed Re and Os contents of Au-rich sulfides from the Carbon Leader Reef, the Ventersdorp Contact Reef, and the Black Reef. Sulfides included unspecified types of mainly pyrite and likely included mixtures of detrital and epigenetic pyrite. Isochron ages for individual reefs were found to be younger than Witwatersrand sediment deposition, suggesting hydrothermal remobilization of sulfides.

Coupled isotope analyses of $\delta^{56}Fe$, $\delta^{34}S$ and $\delta^{33}S$ of detrital pyrite are a powerful tool for provenance analysis (Hofmann et al., 2009). Detrital pyrite from the lowermost Dominion Group yielded relatively unfractionated values, thus suggesting derivation from a crustal source, which may include sulfides of magmatic, magmatic-hydrothermal, or metamorphism-related hydrothermal origin, hosted in granitoid-greenstone terrains. This interpretation is supported by (1) the nature of the detrital pyrite grains that are predominantly massive and not coarser than sand-size and (2) the setting of the Dominion Group resting directly on granitoid basement (Hofmann et al., 2009).

In contrast, reefs from the Witwatersrand Supergroup and overlying successions are characterized by MIF-S ($\Delta^{33}S$ as negative as $-1.35‰$) and large range of Fe isotope values ($\delta^{56}Fe$ between $-1.1‰$ and $1.2‰$; Hofmann et al., 2009), suggesting a supracrustal source, potentially including diagenetic sulfides, SEDEX, and VMS deposits.

6. CONCLUSIONS

Research on detrital pyrite in Archean sedimentary rocks is a fascinating subject due to the large variation in pyrite morphology and economic implications. A growing number of different tools is available to determine the provenance of detrital pyrite in clastic sedimentary successions. Whether hydrothermal, paleoplacer, synsedimentary processes, or their combination caused the gold mineralization in the Witwatersrand Basin, detrital pyrite holds clues to this debate and will continue to fascinate researchers and motivate them for further research on sediment and gold provenance.

Acknowledgments

This research was supported by the DST-NRF Centre of Excellence for Integrated Mineral and Energy Resource Analysis (CIMERA) via a Doctoral Fellowship to Giuliana Costa. Patrick Eriksson kindly provided a review of the article. We thank Evan Cook (Gold1), Simone Hartmann (AngloGold Ashanti), and Sunet Dercksen (Sibanye Gold) for access to drill core and arranging underground visits.

References

Agangi, A., Hofmann, A., Wohlgemuth-Ueberwasser, C.C., 2013. Pyrite zoning as a record of mineralization in the Ventersdorp contact reef, Witwatersrand, South Africa. Economic Geology 108, 1243–1272.

Agangi, A., Hofmann, A., Rollion-Bard, C., Marin-Carbonne, J., Cavalazzi, B., Large, R., Meffre, S., 2015a. Gold accumulation in the Archean Witwatersrand basin, South Africa—evidence from concentrically laminated pyrite. Earth-Science Reviews 10, 27–53.

Agangi, A., Przybyłowicz, W., Hofmann, A., 2015b. Trace element mapping of pyrite from Archean gold deposits – a comparison between PIXE and EPMA. Nuclear Instruments and Methods in Physics Research Section B: Beam Interactions with Materials and Atoms 348, 302–306.

Anbar, A.D., Rouxel, O., 2007. Metal stable isotopes in paleoceanography. Annual Review of Earth and Planetary Sciences 35, 717–746.

Archer, C., Vance, D., 2006. Coupled Fe and S isotope evidence for Archean microbial Fe (III) and sulfate reduction. Geology 34, 153–156.

Armstrong, R.A., Compston, W., de Wit, M.J., Williams, I.S., 1990. The stratigraphy of the 3.5–3.2 Ga Barberton greenstone belt revisited: a single zircon ion microprobe study. Earth and Planetary Science Letter 101 (1), 90–106.

Baker, D.R., Moretti, R., 2011. Modeling the solubility of sulfur in magmas: a 50-year old geochemical challenge. Reviews in Mineralogy and Geochemistry 73 (1), 167–213.

Barton, E.S., Hallbauer, D.K., 1996. Trace-element and U–Pb isotope compositions of pyrite types in the Proterozoic Black Reef, Transvaal Sequence, South Africa: implication on genesis and age. Chemical Geology 133, 173–199.

Berner, R.A., 1970. Sedimentary pyrite formation. American Journal of Science 268, 1–23.

Berner, R.A., 1984. Sedimentary pyrite formation: an update. Geochimica et Cosmochimica Acta 48 (4), 605–615.

Berner, R.A., 1985. Sulfate reduction, organic matter decomposition and pyrite formation [and discussion]. Philosophical Transactions of the Royal Society of London. Series A, Mathematical and Physical Sciences 315 (1531), 25–38.

Canfield, D.E., Raiswell, R., Westrich, J.T., Reaves, C.M., Berner, R.A., 1986. The use of chromium reduction in the analysis of reduced inorganic sulfur in sediments and shales. Chemical Geology 54 (1–2), 149–155.

Canfield, D.E., 2001. Biogeochemistry of sulfur isotopes. Reviews in Mineralogy and Geochemistry 43 (1), 607–636.

Carmichael, I.S.E., Ghiorso, M.S., 1986. Oxidation-reduction relations in basic magma: a case for homogeneous equilibria. Earth and Planetary Science Letters 78 (1–2), 200–210.

Carroll, M.R., Rutherford, M.J., 1987. The stability of igneous anhydrite: experimental results and implications for sulfur behavior in the 1982 El Chichon Trachyandesite and other evolved magmas. Journal of Petrology 28 (5), 781–801.

Cassidy, K.F., Groves, D.I., McNaughton, N.J., 1998. Late-Archean granitoid-hosted lode-gold deposits, Yilgarn Craton, Western Australia: deposit characteristics, crustal architecture and implication for ore genesis. Ore Geology Reviews 13, 65–102.

Craig, J.R., Vokes, F.M., 1993. The metamorphism of pyrite and pyritic ores: an overview. Mineralogical Magazine 57, 3–18.

Craig, J.R., Vokes, F.M., 1998. Pyrite: physical and chemical textures. Mineralium Deposita 34, 82–101.

Craw, D., Falconer, D., Youngson, J.H., 2003. Environmental arsenopyrite stability and dissolution: theory, experiment, and field observations. Chemical Geology 199 (1), 71–82.

Creaser, R.A., Papanastassiou, D.A., Wasserburg, G., 1991. Negative thermal ion mass spectrometry of osmium, rhenium and iridium. Geochimica et Cosmochimica Acta 55 (1), 397–401.

Crowe, S.A., Paris, G., Katsev, S., Jones, C., Kim, S.T., Zerkle, A.L., Canfield, D.E., 2014. Sulfate was a trace constituent of Archean seawater. Science 346 (6210), 735–739.

Dauphas, N., van Zuilen, M., Wadhwa, M., Davis, A.M., Marty, B., Janney, P.E., 2004. Clues from Fe isotope variations on the origin of early Archean BIFs from Greenland. Science 306, 2077–2080.

Dauphas, N., Rouxel, O.J., 2006. Mass spectrometry and natural variations of iron isotopes. Mass Spectrometry Reviews 25, 515–550.

Dimroth, E., 1979. Significance of diagenesis for the origin of Witwatersrand-type uraniferous conglomerates. Philosophical Transactions of the Royal Society of London A: Mathematical, Physical and Engineering Sciences 291 (1381), 277–285.

Els, B.G., 1998. The auriferous Late Archean sedimentation systems of South Africa: unique palaeo-environmental conditions? Sedimentary Geology 120 (1), 205–224.

England, G.L., Rasmussen, B., McNaughton, N.J., Fletcher, I.R., 2001. SHRIMP U–Pb ages of diagenetic and hydrothermal xenotime from the Archean Witwatersrand Supergroup of South Africa. Terra Nova 13 (5), 360–367.

England, G.L., Rasmussen, B., Krapez, B., Groves, D.I., 2002. Paleoenvironmental significance of rounded pyrite in siliciclastic sequences of the Late Archean Witwatersrand Basin: oxygen-deficient atmosphere or hydrothermal alteration? Sedimentology 49, 1133–1156.

Falconer, D.M., Craw, D., Youngson, J.H., Faure, K., 2006. Gold and sulfide minerals in Tertiary quartz pebble conglomerate gold placers, Southland, New Zealand. Ore Geology Reviews 28, 525–545.

Farquhar, J., Bao, H., Thiemens, M., 2000. Atmospheric influence of Earth's earliest sulfur cycle. Science 289, 756–758.

Farquhar, J., Savarino, J., Airieau, S., Thiemens, M.H., 2001. Observation of wavelength sensitive mass-independent sulfur isotope effects during SO_2 photolysis: implications for the early atmosphere. Journal of Geophysical Research 106, 1–11.

Farquhar, J., Wing, B.A., 2003. Multiple sulfur isotopes and the evolution of the atmosphere. Earth and Planetary Science Letters 213, 1–13.

Farquhar, J., Wing, B.A., 2005. The terrestrial record of stable sulfur isotopes: a review of the implications for evolution of Earth's sulfur cycle. In: McDonald, I., Boyce, A.J., Butler, I.B., Herrington, R.J., Polya, D.A. (Eds.), Mineral Deposits and Earth Evolution, vol. 248. Geological Society of London. Special Publication, pp. 167–177.

Feather, C.E., Koen, G.M., 1975. The mineralogy of the Witwatersrand reefs. Mineral Science Engineering 7 (3), 189–224.

Fleet, M.E., MacLean, P.J., Barbier, J., 1989. Oscillatory-zoned as-bearing pyrite from strata-bound and stratiform gold deposits: an indicator of ore fluid evolution. Economic Geology Monograph 6, 356–362.

Fleet, M.E., Chryssoulis, S.L., MacLean, P.J., Davidson, R., Weisener, C.G., 1993. Arsenian pyrite from gold deposits; Au and As distribution investigated by SIMS and EMP, and color staining and surface oxidation by XPS and LIMS. The Canadian Mineralogist 31 (1), 1–17.

Frimmel, H.E., 2005. Archean atmospheric evolution: evidence from the Witwatersrand gold fields, South Africa. Earth-Science Reviews 70, 1–46.

Frimmel, H.E., 2008. Earth's continental crustal gold endowment. Earth and Planetary Science Letters 267, 45–55.

Frimmel, H.E., Hennigh, Q., 2015. First whiffs of atmospheric oxygen triggered onset of crustal gold cycle. Mineralium Deposita 50 (1), 5–23.

Frimmel, H.E., Groves, D.I., Kirk, J., Ruiz, J., Chesley, J., Minter, W.E.L., 2005. The Formation and Preservation of the Witwatersrand Goldfields, the World's Largest Gold Province. Economic Geology 100th Anniversary Volume, pp. 769–797.

Goldhaber, M.B., 2003. Sulfur-rich sediments. In: Treatise on Geochemistry. Sediments, Diagenesis, and Sedimentary Rocks, vol. 7. Elsevier, Amsterdam, pp. 257–288.

Grandstaff, D.E., 1980. Origin of uraniferous conglomerates at Elliot lake, Canada and Witwatersrand, South Africa: implications for oxygen in the Precambrian atmosphere. Precambrian Research 13, 1–26.

Gregory, D.D., Large, R.R., Halpin, J.A., Baturina, E.L., Lyons, T.W., Wu, S., Danyushevsky, L., Sack, P.J., Chappaz, A., Maslennikov, V.V., Bull, S.W., 2015. Trace element content of sedimentary pyrite in black shales. Economic Geology 110, 1389–1410.

Guy, B.M., Beukes, N.J., Gutzmer, J., 2010. Paleoenvironmental controls on the texture and chemical composition of pyrite from non-conglomeratic sedimentary rocks of the Mesoarchean Witwatersrand Supergroup, South Africa. South African Journal of Geology 113 (2), 195–228.

Guy, B.M., Ono, S., Gutzmer, J., Kaufman, A.J., Lin, Y., Fogel, M.L., Beukes, N.J., 2012. A multiple sulfur and organic carbon isotope record from non-conglomeratic sedimentary rocks of the Mesoarchean Witwatersrand Supergroup, South Africa. Precambrian Research 216–219, 208–231.

Guy, B.M., Ono, S., Gutzmer, J., Lin, Y., Beukes, N.J., 2014. Sulfur sources of sedimentary "buckshot" pyrite in the auriferous conglomerates of the Mesoarchean Witwatersrand and Ventersdorp Supergroups, Kaapvaal Craton, South Africa. Mineralium Deposita 49, 751–775.

Hall, A.J., 1986. Pyrite-pyrrhotite redox reactions in nature. Mineralogical Magazine 50 (6), 223–229.

Hallbauer, D.K., 1977. Morphology and microtexture of minerals from fossil gold and uranium placers. Chamber of Mines of South Africa, Research Review 36–50.

Hallbauer, D.K., 1986. The mineralogy and geochemistry of Witwatersrand pyrite, gold, uranium and carbonaceous matter. Mineral Deposits of Southern Africa I & II, 731–752.

Hallbauer, D.K., von Gehlen, K., 1983. The Witwatersrand pyrites and metamorphism. Mineralogical Magazine 47, 473–479.

Hannington, M.D., 2014. Volcanogenic massive sulfide deposits. In: Treatise on Geochemistry, second ed., vol. 13. Elsevier Ltd., pp. 463–488

Hedenquist, J.W., Lowenstern, J.B., 1994. The role of magmas in the formation of hydrothermal ore deposits. Nature 370 (6490), 519–527.

Heinrich, C.A., 2015. Witwatersrand gold deposits formed by volcanic rain, anoxic rivers and Archaean life. Nature Geoscience 8 (3), 206–209.

Hirdes, W., Saager, R., 1983. The Proterozoic Kimberley Reef Placer in the Evander Goldfield. Borntraeger, Witwatersrand, South Africa (No. 20).

Ho, S.E., McNaughton, N.J., Groves, D.I., 1994. Criteria for determining initial lead isotopic compositions of pyrite in Archean lode-gold deposits: a case study at Victory, Kambalda, Western Australia. Chemical Geology 111, 57–84.

Hofmann, A., Bekker, A., Rouxel, O., Rumble, D., Master, S., 2009. Multiple sulfur and iron isotope composition of detrital pyrite in Archean sedimentary rocks: a new tool for provenance analysis. Earth and Planetary Science Letters 286, 436–445.

Holland, H.D., 1984. The Chemical Evolution of the Atmosphere and Oceans. Princeton University Press, Princeton, NJ, 582 p.

Hollis, S.P., Yeats, C.J., Wyche, S., Barnes, S.J., Ivanic, T.J., Beldord, S.M., Davidson, G.J., Roache, A.J., Wingate, M.T.D., 2015. A review of volcanic-hosted massive sulfide (VHMS) mineralization in the Archean Yilgarn Craton, Western Australia: tectonic, stratigraphic and geochemical associations. Precambrian Research 260, 113–135.

Huston, D.L., Pehrsson, S., Eglington, B.M., Zaw, K., 2010. The geology and metallogeny of volcanic-hosted massive sulfide deposits: variations through geologic time and with tectonic setting. Economic Geology 105 (3), 571–591.

Hutchinson, R.W., Viljoen, R.P., 1988. Re-evaluation of gold source in Witwatersrand ores. South African Journal of Geology 91 (2), 157–173.

Jamieson, J.W., Wing, B.A., Hannington, M.D., Farquhar, J., 2006. Evaluating isotopic equilibrium among sulfide mineral pairs in Archean ore deposits: case study from the Kidd Creek VMS deposit, Ontario, Canada. Economic Geology 101, 1055–1061.

Janisch, P.R., 1986. Gold in South Africa. Journal of the South African Institute of Mining and Metallurgy 86 (8), 273–316.

Johnson, C.M., Beard, B.L., Beukes, N.J., Klein, C., O'Leary, J.M., 2003. Ancient geochemical cycling in the Earth as inferred from Fe isotope studies of banded iron formations from the Transvaal Craton. Contributions to Mineralogy and Petrology 144, 523–547.

Johnson, C.M., Beard, B.L., Roden, E.E., 2008. The iron isotope fingerprints of redox and biogeochemical cycling in modern and ancient Earth. Annual Review of Earth and Planetary Science 36, 457–493.

Johnson, J.E., Gerpheide, A., Lamb, M.P., Fischer, W.W., 2014. O_2 constraints from Paleoproterozoic detrital pyrite and uraninite. Geological Society of America Bulletin 125 (5–6), 813–830.

Jolley, S.J., Freeman, S.R., Barnicoat, A.C., Phillips, G.M., Knipe, R.J., Pather, A., Fox, N.P.C., Strydom, D., Birch, M.T.G., Henderson, I.H.C., Rowland, T.W., 2004. Structural controls on Witwatersrand gold mineralization. Journal of Structural Geology 26, 1067–1086.

Jugo, P.J., Luth, R.W., Richards, J.P., 2005. An experimental study of the sulfur content in basaltic melt saturated with immiscible sulfide or sulfate liquids at 1300°C and 1.0 GPa. Journal of Petrology 46 (4), 783–798.

Kirk, J., Ruiz, J., Chesley, J., Titley, S., Walshe, J., 2001. A detrital model for the origin of gold and sulfides in the Witwatersrand basin based on Re–Os isotopes. Geochimica et Cosmochimica Acta 65 (13), 2149–2159.

Klemd, R., Hallbauer, D.K., 1987. Hydrothermally altered peraluminous Archean granites as a provenance model for Witwatersrand sediments. Mineralium Deposita 22 (3), 227–235.

Koglin, N., Frimmel, H.E., Minter, W.E.L., Brätz, H., 2010. Trace-element characteristics of different pyrite types in Mesoarchean to Paleoproterozoic placer deposits. Mineralium Deposita 45, 159–180.

Köppel, V.H., Saager, R., 1974. Lead isotope evidence on the detrital origin of Witwatersrand pyrites and its bearing on the provenance of the Witwatersrand gold. Economic Geology 69 (3), 318–331.

Kositcin, N., McNaughton, N.J., Griffin, B.J., Fletcher, I.R., Groves, D.I., Rasmussen, B., 2003. Textural and geochemical discrimination between xenotime of different origin in the Archean Witwatersrand Basin, South Africa. Geochimica et Cosmochimica Acta 67 (4), 709–731.

Kositcin, N., Krapez, B., 2004. Relationship between detrital zircon age-spectra and the tectonic evolution of the Late Archean Witwatersrand basin, South Africa. Precambrian Research 129, 141–168.

Large, R.R., Meffre, S., Burnett, R., Guy, B., Bull, S., Gilbert, S., Goemann, K., Danyushevsky, L., 2013. Evidence for an intrabasinal source and multiple concentration processes in the formation of the Carbon Leader Reef, Witwatersrand Supergroup, South Africa. Economic Geology 108 (6), 1215–1241.

Leach, D.L., Bradley, D.C., Huston, D., Pisarevsky, S.A., Taylor, R.D., Gardoll, S.J., 2010. Sediment-hosted lead-zinc deposits in Earth history. Economic Geology 105 (3), 593–625.

Loftus-Hills, G., Solomon, M., 1967. Cobalt, nickel and selenium in sulfides as indicators of ore genesis. Mineralium Deposita 2, 228–242.

Lyons, T.W., Gellatly, A.M., McGoldrick, P.J., Kah, L.C., 2006. Proterozoic sedimentary exhalative (SEDEX) deposits and links to evolving global ocean chemistry. In: Kesler, S.E., Ohmoto, H. (Eds.), Evolution of Early Earth's Atmosphere, Hydrosphere, and Biosphere—Constraints From Ore Deposits, vol. 198. Geological Society of America Memoir, pp. 169–184.

Lyons, T.W., Reinhard, C.T., Planavsky, N.J., 2014. The rise of oxygen in Earth's early ocean and atmosphere. Nature 506 (7488), 307–315.

MacLean, P.J., Fleet, M.E., 1989. Detrital pyrite in the Witwatersrand gold fields of South Africa: evidence from truncated growth banding. Economic Geology 84, 2008–2011.

Marcoux, E., Moëlo, Y., Leistel, J.M., 1996. Bismuth and cobalt minerals as indicators of stringer zones to massive sulfide deposits, Iberian Pyrite Belt. Mineralium Deposita 31, 1–26.

Marin-Carbonne, J., Rollion-Bard, C., Bekker, A., Rouxel, O., Agangi, A., Cavalazzi, B., Wohlgemuth-Ueberwasser, C., Hofmann, A., McKeegan, K., 2014. Coupled Fe and S isotope variations in pyrite nodules from Archean shales. Earth and Planetary Science Letters 392, 67–79.

Mathur, R., Gauert, C., Ruiz, J., Linton, P., 2013. Evidence for mixing of Re–Os isotopes at <2.7 Ga and support of a remobilized placer model in Witwatersrand sulfides and native Au. Lithos 164–167, 65–73.

Maynard, J.B., Ritger, S.D., Sutton, S.J., 1991. Chemistry of sands from the modern Indus river and the Archean Witwatersrand basin: implications for the composition of the Archean atmosphere. Geology 19 (3), 265–298.

Meffre, S., Large, R.R., Scott, R., Woodhead, J., Chang, Z., Gilbert, S.E., Danyushevsky, L.V., Maslennikov, V., Hergt, J.M., 2008. Age and pyrite Pb-isotopic composition of the giant Sukhoi Log sediment-hosted gold deposit, Russia. Geochimica et Cosmochimica Acta 72, 2377–2391.

Meyer, F.M., Oberthür, T., Robb, L.J., Saager, R., Stupp, H.D., 1990. Cobalt, nickel and gold in pyrite from primary gold deposits and Witwatersrand reefs. South African Journal of Geology 93, 70–82.

Minter, W.E.L., 1976. Detrital gold, uranium, and pyrite concentrations related to sedimentology in the Precambrian Vaal Reef Placer, Witwatersrand, South Africa. Economic Geology 71 (1), 157–176.

Minter, W.E.L., 1978. A sedimentological synthesis of placer gold, uranium and pyrite concentrations in Proterozoic Witwatersrand sediments. In: Miall, A.D. (Ed.), Fluvial Sedimentology, vol. 5. Canadian Soc. Petrology Geology Mem, pp. 801–829.

Minter, W.E.L., Hill, W.C.N., Kidger, R.J., Kingsley, C.S., Snowden, P.A., 1986. The Welkom goldfield. In: Annhaeusser, C.R., Maske, S. (Eds.), Mineral Deposits of Southern Africa: Marshalltown, Geological Society of South Africa, vol. 1, pp. 497–539.

Minter, W.E.L., Loen, J.S., 1991. Palaeocurrent dispersal patterns of Witwatersrand gold placers. South African Journal of Geology 94 (1), 70–85.

Mueller, W.U., Stix, J., Corcoran, P.L., Daigneault, R., 2009. Subaqueous calderas in the Archean Abitibi greenstone belt: an overview and new ideas. Ore Geology Reviews 35, 4–46.

Myers, R.E., Zhou, T., Phillips, G.N., 1993. Sulphidation in the Witwatersrand Goldfields: evidence from the Middelvlei reef. Mineralogical Magazine 57, 395.

Naldrett, A.J., Bray, J.G., Gasparrini, E.L., Podolsky, T., Rucklidge, J.C., 1970. Cryptic variation and the petrology of the Sudbury nickel irruptive. Economic Geology 65 (2), 122–155.

Naldrett, A.J., 1989. Magmatic Sulfide Deposits. Oxford Monographs on Geology and Geophysics No. 14. Oxford University Press, 186 p.

Naldrett, A.J., 2004. Magmatic Sulfide Deposits: Geology, Geochemistry and Exploration. Springer-Verlag, 727 p.

Neuerburg, G.J., 1975. A procedure using hydrofluoric acid for quantitative mineral separations from silicate rocks. Journal of Research of the U.S. Geological Survey 3 (3), 377–378.

Obiora, S.C., Charan, S.N., 2010. Geochemical constraints on the origin of some intrusive igneous rocks from the Lower Benue rift, southeastern Nigeria. Journal of African Earth Sciences 58, 197–210.

Ohmoto, H., Kakegawa, T., Lowe, D.R., 1993. 3.4-Billion-year-old biogenic pyrites from Barberton, South Africa: sulfur isotope evidence. Science 262 (5133), 555–557.

Ohmoto, H., 1996. Formation of volcanogenic massive sulfide deposits: the Kuroko perspective. Ore Geology Reviews 10, 135–177.

Ohmoto, H., Goldhaber, M.B., 1997. Sulfur and carbon isotopes. Geochemistry of Hydrothermal Ore Deposits 3, 517–611.

Ono, S., Beukes, N.J., Rumble, D., 2009. Origin of two distinct multiple-sulfur isotope compositions of pyrite in the 2.5 Ga Klein Naute formation, Griqualand west basin, South Africa. Precambrian Research 169, 48–57.

Pavlov, A.A., Kasting, J.F., 2002. Mass-independent fractionation of sulfur isotopes in Archean sediments: strong evidence for an anoxic Archean atmosphere. Astrobiology 2, 27–41.

Phillips, G.N., Law, J.D.M., 2000. Witwatersrand gold fields; geology, genesis, and exploration. Reviews in Economic Geology 13, 439–500.

Phillips, G.N., 2013. Witwatersrand gold: discovery matters. Applied Earth Science 122 (2), 122–127.
Poujol, M., Robb, L.J., Respaut, J.P., 1999. U–Pb and Pb–Pb isotopic studies relating to the origin of gold mineralization in the Evander Goldfield, Witwatersrand Basin, South Africa. Precambrian Research 95, 167–185.
Przybylowicz, W.J., Mesjasz-Przybylowicz, J., Pineda, C.A., Churms, C.L., Ryan, C.G., Prozesky, V.M., Frei, R., Slabbert, J.P., Padayachee, J., Reimold, W.U., 2001. X-Ray Spectrometry 30 (3), 156–163.
Ramdohr, P., 1958. New observations on the ores of Witwatersrand in South Africa and their genetic significance. Transactions of the Geological Society of South Africa 61, 1–50.
Rasmussen, B., Buick, R., 1999. Redox state of the Archean atmosphere: evidence form detrital heavy minerals in ca. 3250–2750 Ma sandstones from the Pilbara Craton, Australia. Geology 27 (2), 115–118.
Rickard, D., 1997. Kinetics of pyrite formation by the H_2S oxidation of iron (II) monosulfide in aqueous solutions between 25 and 125°C: the rate equation. Geochimica et Cosmochimica Acta 61 (1), 115–134.
Rickard, D., 2012. Sulfidic sediments and sedimentary rocks. In: Developments in Sedimentology, vol. 65. Elsevier, 801 p.
Ridley, J., Mikucki, E.J., Groves, D.I., 1996. Archean lode-gold deposits: fluid flow and chemical evolution in vertically extensive hydrothermal systems. Ore Geology Reviews 10, 249–293.
Robb, L.J., Meyer, F.M., 1990. The nature of the Witwatersrand hinterland; conjectures on the source area problem. Economic Geology 85 (3), 511–536.
Rouxel, O., Bekker, A., Edwards, K., 2005. Iron isotope constraints on the Archean and Paleoproterozoic ocean redox state. Science 307, 1088–1091.
Rouxel, O., Shanks, W.C., Bach, W., Edwards, K., 2008. Integrated Fe and S isotope study of seafloor hydrothermal vents at East Pacific Rise 9–10°N. Chemical Geology 252, 214–227.
Saager, R., Mihálik, P., 1967. Two varieties of pyrites from the basal reef of the Witwatersrand system. Economic Geology 62, 719–731.
Saager, R., 1970. Structures in pyrite from the basal reef in the Orange free state gold-field. Transactions of the Geological Society of South Africa 73 (Part 1), 29–46.
Saager, R., 1976. Geochemical Aspects of the Origin of Detrital Pyrite in Witwatersrand Conglomerates. Economic Geology Research Unit, University of the Witwatersrand. Information circular n 105: 21 pp.
Saager, R., Utter, T., Meyer, M., 1982a. Pre-Witwatersrand and Witwatersrand conglomerates in South Africa: a mineralogical comparison and bearings on the genesis of gold-uranium placers. In: Ore Genesis. Springer, Berlin Heidelberg, pp. 38–56.
Saager, R., Meyer, M., Muff, R., 1982b. Gold distribution in supracrustal rocks from Archean greenstone belts of southern Africa and from Paleozoic ultramafic complexes of the European Alps; metallogenic and geochemical implications. Economic Geology 77 (1), 1–24.
Shen, Y., Buick, R., Canfield, D.E., 2001. Isotopic evidence for microbial sulphate reduction in the early Archean era. Nature 410, 77–81.
Schieber, J., 2011. Marcasite in black shales—a mineral proxy for oxygenated bottom waters and intermittent oxidation of carbonaceous muds. Journal of Sedimentary Research 81 (7), 447–458.
Schoonen, M.A.A., 2004. Mechanisms of sedimentary pyrite formation. In: Amend, J.P., Edwards, K.J., Lyons, T.W. (Eds.), Sulfur biogeochemistry-past and present, Geological Society of America, Special Paper 379, pp. 117–134.
Stein, H., Hannah, J., 2015. Rhenium–Osmium geochronology: sulfides, shales, oils, and mantle. Encyclopedia of Scientific Dating Methods 707–723.
Sun, W., Arculus, R.J., Kamenetsky, V.S., Binns, R.A., 2004. Release of gold-bearing fluids in convergent margin magmas prompted by magnetite crystallization. Nature 431, 975–978.
Sutton, S.J., Ritger, S.D., Maynard, J.B., 1990. Stratigraphic control of chemistry and mineralogy in metamorphosed Witwatersrand quartzites. Journal of Geology 98, 329–341.
Tabelin, C.B., Igarashi, T., Tamoto, S., Takahashi, R., 2012. The roles of pyrite and calcite in the mobilization of arsenic and lead from hydrothermally altered rocks excavated in Hokkaido, Japan. Journal of Geochemical Exploration 19, 17–31.
Tauson, V.L., Akimov, V.V., Lipko, S.V., Spiridonov, A.M., Budyak, E.E., Belozerova, O.Y., Smagunov, N.V., 2015. Typomorphism of pyrite of the Sukhoi Log deposit (East Siberia). Russian Geology and Geophysics 56, 1394–1413.
Tomkins, A.G., 2010. Windows of metamorphic sulfur liberation in the crust: implications for gold deposit genesis. Geochimica et Cosmochimica Acta 74 (11), 3246–3259.

Tornos, F., Peter, J.M., Allen, R., Conde, C., 2015. Controls on the siting and style of volcanogenic massive sulfide deposits. Ore Geology Reviews 68, 142–163.

Ueno, Y., Ono, S., Rumble, D., Maruyama, S., 2008. Quadruple sulfur isotope analysis of ca. 3.5 Ga Dresser Formation: new evidence for microbial sulfate reduction in the early Archean. Geochimica et Cosmochimica Acta 72, 5675–5691.

Utter, T., 1978. Morphology and geochemistry of different pyrite types from the Upper Witwatersrand system of the Klerksdorp Goldfield, South Africa. Geologische Rundschau 67 (2), 774–804.

Vennemann, T.W., Kesler, S.E., Frederickson, G.C., Minter, W.E.L., Heine, R.R., 1996. Oxygen isotope sedimentology of gold- and uranium-bearing Witwatersrand and Huronian Supergroup quartz-pebble conglomerates. Economic Geology 91 (2), 322–342.

Voroshin, S.V., Tyukova, E.E., Newberry, R.J., Layer, P.W., 2014. Orogenic gold and rare metal deposits of the Upper Kolyma District, Northeastern Russia: relation to igneous rocks, timing, and metal assemblages. Ore Geology Reviews 62, 1–24.

Wang, Q., Morse, J.W., 1996. Pyrite formation under conditions approximating those in anoxic sediments. Marine Chemistry 52, 99–121.

Whitehouse, M.J., Fedo, C.M., 2007. Microscale heterogeneity of Fe isotopes in N3.71 Ga banded iron formation from the Isua Greenstone Belt, southwest Greenland. Geology 35, 719–722.

Wilkin, R.T., Barnes, H.L., 1996. Pyrite formation by reactions of iron monosulfides with dissolved inorganic and organic sulfur species. Geochimica et Cosmochimica Acta 60 (21), 4167–4179.

Wronkiewicz, D.J., Condie, K.C., 1987. Geochemistry of Archean shales from the Witwatersrand Supergroup, South Africa: source-area weathering and provenance. Geochimica et Cosmochimica Acta 51, 2401–2416.

Yamaguchi, K.E., Johnson, C.M., Beard, B.L., Ohmoto, H., 2005. Biogeochemical cycling of iron in the Archean–Paleoproterozoic Earth: constraints from iron isotope variations in sedimentary rocks from the Kaapvaal and Pilbara Cratons. Chemical Geology 218, 135–169.

Youngson, J.H., Craw, D., Falconer, D.M., 2006. Evolution of Cretaceous–Cenozoic quartz pebble conglomerate gold placers during basin formation and inversion, southern New Zealand. Ore Geology Reviews 28 (4), 451–474.

Zheng, Y., Zhang, L., Chen, Y.J., Hollings, P., Chen, H.Y., 2013. Metamorphosed Pb–Zn–(Ag) ores of the Keketale VMS deposit, NW China: evidence from ore textures, fluid inclusions, geochronology and pyrite compositions. Ore Geology Reviews 54, 167–180.

CHAPTER

19

Ice Ages in Earth History: Puzzling Paleolatitudes and Regional Provenance of Ice Sheets on an Evolving Planet

G.M. *Young*

University of Western Ontario, London, ON, Canada

OUTLINE

Sediment Provenance
http://dx.doi.org/10.1016/B978-0-12-803386-9.00019-8

1. INTRODUCTION

The record preserved in sedimentary rocks suggests that throughout most of Earth's 4.6-Ga history the surface of the planet was free of large ice-covered areas (continental glaciers or ice sheets). It is therefore fortuitous that humans evolved during one of the planet's brief glacial episodes so that direct observation of glaciers and their processes and products was possible. Without evidence from observations of modern glaciers, interpretation of ancient glacial deposits would have been a much greater intellectual challenge. As it was, interpretation of the widespread cover of unconsolidated materials in northern Europe and North America was to become the focus of one of the greatest controversies in the history of geology. By the beginning of the 19th century these deposits were well known. Conventional wisdom had them explained as the products of floods in which boulders of foreign rocks (erratics) drifted long distances in icebergs. Thus the British Geological Survey adopted the practice of publishing "drift" and "solid" editions of geological maps. Despite the publication of a number of papers refuting the "deluvialist" theory for these surficial deposits, it took the energy and enthusiasm of the Swiss paleontologist, Louis Agassiz (1842), to bring the geological community to believe that much of the northern hemisphere had recently been under an extensive and thick cover of ice. Thus the concept of an Ice Age entered the mainstream of geological thought and thereafter geologists began identifying similar deposits in the record provided by sedimentary rocks, particularly in Gondwanan Permo-Carboniferous deposits, where they played a pivotal role in the equally hard-fought battle over continental drift. The first recognized Precambrian glacial deposits are in Neoproterozoic rocks of the Dalradian Supergroup on the island of Islay on the west coast of Scotland (Thomson, 1871), but some of the Cryogenian glaciations have subsequently been shown to be the world's most widespread (Harland, 1964; Hambrey and Harland, 1981; Arnaud et al., 2011).

The main requirements for formation of glaciers are simple: elevated continental lithosphere; an adequate supply of moisture in the atmosphere; and temperatures that, over a significant number of years, remain sufficiently low during the warmest part of the year to permit net accumulation of snow (ultimately compacted into ice) so that a thick ice cover is developed. Large land-based ice masses are generally considered as being either Alpine (mountain) glaciers or more extensive continental glaciers (ice caps or ice sheets). Mountain glaciers form under similar conditions to those that give rise to continental ice sheets, but the latter are able, under certain circumstances, to descend to low altitudes and enter the world oceans. Mountain glaciers may develop at any latitude (provided the ground surface is high enough) but at lower latitudes the ice is destroyed by rising temperatures as it descends to lower altitudes. The deposits of such glaciers are therefore susceptible to reworking by nonglacial processes and are unlikely to be recognized if they are preserved in the geological record. Large portions of continental glaciers may also terminate in areas above sea level. For example, the Pleistocene ice sheets of North America have draped much of the continent in a layer of fertile crushed rock stripped largely from the Canadian shield to the north. Because of subsequent erosion, such deposits are rarely found in the geological record (Deynoux, 1982), whereas marine glacial deposits are preserved in a variety of basinal settings such as rifts, continental margins, and foreland basins.

2. DIAGNOSTIC CHARACTERISTICS OF GLACIAL DEPOSITS

Diamictites (unconsolidated equivalent, diamictons) are conglomerates that consist of rock fragments (generally of varied lithology and size) "floating" in a matrix of fine material that is commonly unweathered (Fig. 19.1A). Although diamictites are one of the most important indicators of ancient glacial activity, such rocks may form in many environments, so additional criteria are necessary for identification of glacial activity during deposition of such rocks. Additional important diagnostic features include striated and faceted rock fragments that have been scratched and planed by rubbing against each other or a hard substrate as they were carried along in a glacier. In like fashion, the substrate to a glacier may also bear striations where hard stones were forced across its surface. Similar more deeply furrowed surfaces may be produced where ice moved across a semiconsolidated substrate of contemporary sediment (Young et al., 2004; Fig. 19.3).

Another indicator of the former existence of glaciers is the presence of ice-rafted debris in the form of dropstones (Fig. 19.1B), large isolated clasts that are present in fine-grained, stratified sediments, and commonly show penetration of underlying layers. Such dropstones are interpreted as having been released from melting icebergs calved from glaciers that entered a body of water (sea or lake). Mud pellets, considered to have been rendered cohesive by compression while being transported in ice, are also preserved in such deposits

FIGURE 19.1 Photographs of some features of glaciogenic rocks. (A) Typical diamictite of the Paleoproterozoic Gowganda Formation, Ontario, Canada, showing a variety of clasts "floating" in a matrix of rock flour (ground rock). Large granitic clast is about 20 cm across. (B) Large rock fragment in bedded sedimentary rocks of the Gowganda Formation. Such unusually large fragments are interpreted as "dropstones" emplaced by melting of glacial ice and are collectively known as ice-rafted debris. Hammer for scale.

(Ovenshine, 1970). Both dropstones and mud pellets are common in laminated sediments known as varved clays. Such sediments typically display alternating fine and coarse laminae (generally of silt- and clay-size material) that are thought to have formed in proglacial lakes, the coarser lamina representing summer melt-out and the fine material being mainly the result of settling from suspension during the winter season. Some varves contain variable amounts of carbonate minerals, planktonic microorganisms, and other organic matter. Early investigation of varves from the Pleistocene glaciations was carried out in Sweden at the end of the 19th and beginning of the 20th centuries (de Geer, 1911), and such deposits have subsequently been studied in many glaciogenic rock successions extending as far back as the Early Peoterozoic Huronian Supergroup (Pettijohn and Bastron, 1959; Young, 1969, 2001).

These are among the main physical criteria used to identify ancient glacial deposits, but chemical criteria such as the Chemical Index of Alteration have also been employed (e.g., Nesbitt and Young, 1982; Young and Nesbitt, 1999). Low values of the CIA have been used to identify rocks that have undergone little chemical weathering, but even under glacial conditions there can be subtle chemical differences among different size grades because of different susceptibilities of minerals to comminution and sorting. In some cases, the CIA has been used to show that there were warmer intervals within a thick succession of glaciogenic rocks (Visser and Young, 1990; Bahlburg and Dobrzinski, 2011).

3. CHANGES BETWEEN SOURCE AND SINK

Glacial deposits are among the least exciting when it comes to studying changes that can occur between source and sink, for the low temperatures at which they typically form and the general dearth of chemical reactions involving water ensure that the influence of chemical weathering is minimal. On the other hand, such materials can undergo a high degree of physical breakdown during transport on and in moving ice masses, so that extremely fine, physically pulverized material (rock flour) is a common product. Such materials are extremely useful as chemical indicators of provenance.

4. GLACIATIONS AND CONTINENTAL DRIFT

To utilize the uniformitarian principle, the starting point must be the present distribution of glaciers on the planet. Most of the world's glacial ice is in circumpolar locations, with the majority in Antarctica and Greenland. Because of the tilt of the Earth's spin axis (obliquity of the ecliptic), circumpolar regions receive less solar energy during the year than areas situated at other latitudes. This distribution of climatic zones is generally assumed for the past, but as more information was gathered about ancient glacial deposits some puzzling aspects began to emerge. Particularly problematic was the wide distribution of evidence of Permo-Carboniferous glaciations, especially the fact that some were discovered in tropical latitudes. This problem was only resolved when pioneers of continental drift, like the German astronomer and meteorologist Alfred Wegener (in translations, 1966, 2001), suggested that the continents had "wandered" across the surface of the planet so that ancient glacial deposits found in warm climatic zones today could have formed in, or near to, polar regions. In fact the distribution of the Permo-Carboniferous glaciers played an important role in converting skeptics to the theory of continental drift, but it was the birth of a new field in geophysics, known as rock magnetism, that provided a firm foundation for the continental drift theory and paved the way for discovery of sea-floor spreading and what is now known as plate tectonics. Thus the riddle of Permo-Carboniferous glacial deposits preserved in equatorial latitudes was resolved and the remarkable vision of the early "continental drifters" was belatedly recognized. Our view of the planet was forever changed.

5. PALEOMAGNETIC PUZZLES

Paleomagnetic results, complemented by equally important advances in geochronological techniques, were subsequently applied to glacial deposits of greater and greater antiquity, but many problems remain, particularly in regard to older Precambrian deposits. Glacial deposits, in common with most sedimentary rocks, are notoriously difficult to date, especially in the Precambrian (most of Earth history), which has yielded few diagnostic correlatable fossils. Although direct dating is not possible, ages may be obtained from underlying, contemporary, and intrusive igneous rocks, and many clues may be garnered from the burgeoning data bank of U-Pb dates from detrital zircons. Likewise, in spite of additional layers of sophistication, it can still be difficult to obtain reliable latitudes from ancient glacial deposits, and this difficulty appears to increase with the age of the deposits (Evans and Raub, 2011, and references therein). Although initially these new techniques brought great success in the interpretation of latitudes of Phanerozoic glacial deposits and resolved longstanding problems regarding former positioning of continents, it gradually became apparent that they also opened a Pandora's box of perplexing problems for those investigating Precambrian glaciogenic rocks. The crux of the matter is that magnetic results from most Precambrian glacial deposits appear to indicate that they were deposited at, or near to, sea level in low latitudes, in clear contradistinction to those of the Phanerozoic. Among the many proposed explanations of these extraordinary results are compactional flattening, giving rise to less steep magnetization (indicating lower latitudes), different (multipolar) magnetic fields, superimposed magnetizations related to repeated thermo-tectonic processes, and rapid movement of lithospheric plates, or indeed the entire silicate shell of the planet (true polar wander), so that a sedimentary deposit could be formed at one latitude and receive a magnetic signature in a quite different setting. None of these proposals, however, appears to be satisfactory, so that even the most carefully scrutinized data appear to indicate that glaciations, particularly those of the Cryogenian Period (850–630 Ma), appear to have formed at latitudes now typified by the Earth's highest surface temperatures.

6. UNUSUAL ROCK ASSOCIATIONS

Many glaciogenic rocks occur in an unusual and perplexing stratigraphic context. Many Cryogenian diamictites are associated with unexpected rock types such as limestones and dolostones, which are usually associated with warmer climatic settings at the present day and throughout most of the Phanerozoic Eon. Some diamictites, especially those of the second great Cryogenian glaciations (Marinoan), are surprisingly and immediately overlain by peculiar "cap carbonates," with an unusual array of structures, including columnar stromatolites, vertical pencil-sized pipe structures, and tepee-like structures that have been interpreted by some (Allen and Hoffman, 2005) as unusually large-scale wave ripples. The older of the two widespread Cryogenian glaciations (Sturtian) is commonly associated with iron formation—a rock type that virtually disappeared from the geologic record about a billion years earlier.

7. THE CRYOGENIAN CONUNDRUM

Precambrian glacial deposits are widespread (Mawson, 1949; Harland, 1964; Young, 1970; Hambrey and Harland, 1981; Ojakangas, 1988) and have consistently indicated deposition at low latitudes. Two hypotheses have emerged from attempts to explain these data. One is the snowball Earth hypothesis (SEH), a name coined by Kirschvink (1992) in a short paper in a large volume (Schopf and Klein, 1992) and expanded in a paper by Hoffman et al. (1998), which contributed to a resurgence of interest in paleoclimatology (Hoffman, 2011a; Fig. 19.2). The idea that the Earth's surface was encased in an ice cover (both continents and oceans) at least twice during the Cryogenian period remains controversial, but definitive answers to complex questions are sometimes only reached long after the dust raised by the controversy has settled.

The unusual juxtaposition of glacial deposits and carbonate rocks in late Proterozoic successions was noted much earlier by people such as Spencer (1971), Williams (1975), and Roberts (1976). Although there were earlier attempts to determine the latitudes at which Proterozoic glacial deposits accumulated, the low latitude obtained by Embleton and Williams (1986) for what they considered at the time to be periglacial "varvites," turned out to be the first reliable evidence for low latitude deposition of Neoproterozoic glaciogenic rocks. The association of carbonate rocks with glacial deposits was taken by Hoffman (2011b) as evidence of descent of glaciers to sea level at low latitudes. It was assumed that the disposition of climatic zones across the Earth's surface resembled that of today (Fig. 19.2A), so that discovery of glacial deposits at sea level in tropical latitudes was taken to indicate that continents at higher latitudes would also have been glaciated, leading to the snowball Earth hypothesis. It is ironic that Williams (1975, 2008) interpreted the same data in a quite different way that became the main alternative to the SEH. The low latitudes obtained from Neoproterozoic glacial deposits were interpreted by Williams (1975) to indicate that the distribution of climatic zones on the early Earth was quite different from that of today (and the Phanerozoic). He suggested that this could be explained by the planet having a greatly increased tilt of its spin axis relative to its orbital plane. If the tilt becomes greater than 54 degrees, it has been calculated that solar radiation in a given year would be less in equatorial than in polar regions (Fig. 19.2B). Thus if the Earth entered a glacial phase, areas at low latitudes would be glaciated first. Implicit in this theory is the idea that glaciations would have occurred *preferentially* at low latitudes so that occurrence of glacial deposits at sea level in tropical latitudes would not a priori imply that the entire

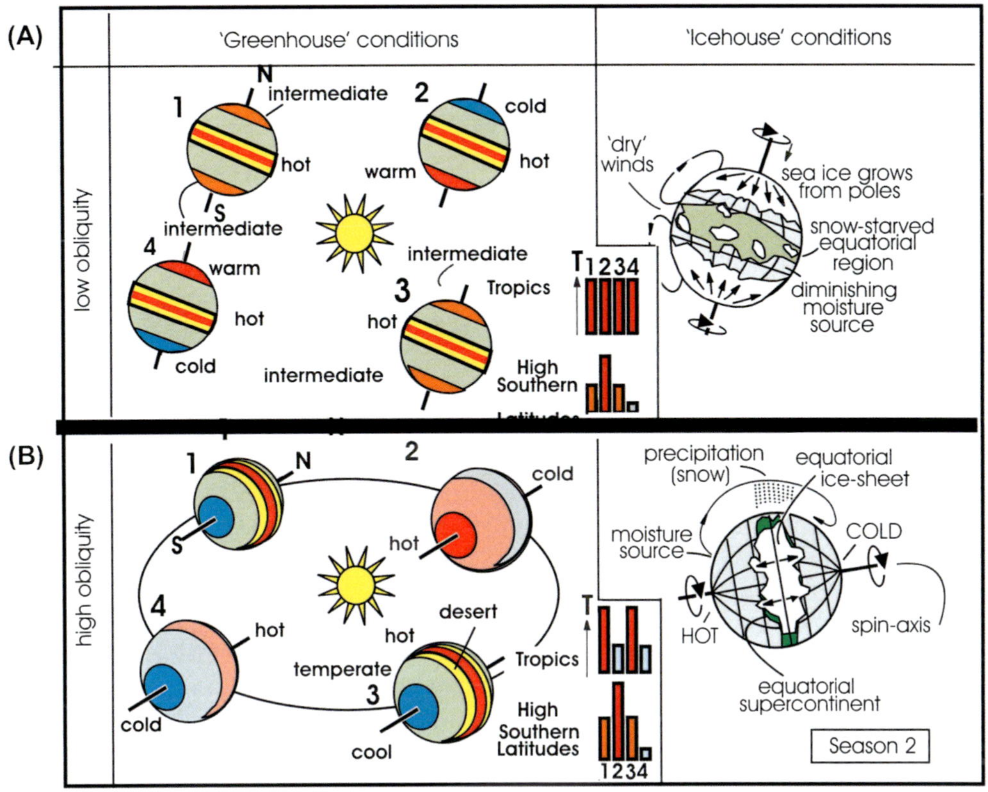

FIGURE 19.2 Schematic representation of climatic zones on Earth during a year (A) under the present obliquity and (B) under a much higher obliquity (>54 degrees), modified from Williams (1975). In Part A, tropical zones are shown in red, desert areas are yellow, temperate zones are green, and circumpolar latitudes are shown in various colors depending on the season. Histograms show that the tropical temperatures remain relatively constant throughout the year, whereas circumpolar regions experience considerable fluctuations. Note that under these conditions, during a glacial episode ice sheets should first develop in circumpolar regions and spread toward lower latitudes (right side). Part B illustrates the high obliquity theory of Williams (1975, 2008). With a high obliquity, the distribution of climatic zones on the surface of the planet would be quite different with extreme seasonality in polar regions but with seasonality also at low latitudes (see histograms). If the Earth descended into an ice age under these conditions, ice sheets would develop preferentially at low latitudes because, in a given year, such areas would receive less solar energy than polar regions. In this case ice sheets would initiate in low latitudes and spread toward the poles. Supporters of SEH favor condition A, with ice expanding to cover most of the planet, whereas the HOT favors the situation depicted in part B. See text for discussion.

planet was encased in ice. Because, like the SEH, the high obliquity theory (HOT) involves a radical departure from what, in the short time-span of human observations, is regarded as the norm, it too has been greeted with a degree of skepticism. Perhaps the most pressing problem with the HOT is that of bringing about radical changes in the orientation of the planet's spin axis. It was suggested by Williams that the Moon-forming collision could have caused the early Earth to adopt a much greater obliquity, but the question remains regarding how and when the inferred high-obliquity planet adopted its present moderately tilted attitude. As pointed out next, a combination of geological and geophysical data may offer some clues.

7.1 Cryogenian Global Sources and Sinks

On the Earth today there is little doubt about the geography of major ice caps (continental glaciers). Large glaciated areas exist where there are continental masses at high latitudes and the ice caps move, in general, from higher to lower latitudes. Now that most of the northern ice cap has vanished, ice movement directions are much more variable because of the increased importance of local topography. Regional transport in the northern hemisphere would have been more obvious at the height of the Pleistocene glacial advances when much of North America and Europe was under a thick carapace of ice that, on a regional scale, was advancing to the south. Thus the source in North America was largely the Canadian shield, and most glacial deposits were either scattered across the surface of the continent or delivered to the sea. Most of the continental Pleistocene deposits are probably ephemeral, whereas those in the marine realm (on continental shelves) have a much greater preservation potential. With increasing ice thickness, underlying topography becomes less important in controlling major ice movements, which should become directed from continent to ocean and during periods of ice sheet expansion, toward lower latitudes.

It is precisely this scenario that is envisaged by adherents of the snowball Earth hypothesis. The process is seen as being enhanced and accelerated by runaway albedo due to a positive feedback loop caused by global heat loss due to the highly reflective nature of snow and ice cover (Budyko, 1969) so that eventually the entire surface of the planet would have been ice covered. Thus the SEH is seen as a more extreme version of the present day situation.

According to the HOT, glaciations should be initiated in equatorial latitudes and should spread toward (but never achieve) the highest latitudes. This is because of excessively high temperatures at polar latitudes during those parts of the year when a polar region was pointed toward the sun throughout the entire diurnal rotation (Fig. 19.2B). The source of glacial sediments under the HOT would lie in low-latitude continents, and ice should grow from near the equator toward the poles, the opposite of the situation at the present day and envisaged in the SEH.

7.2 Challenging the Snowball Earth

Many arguments have been raised against the SEH, including evidence of open water during deposition of glacial deposits (Arnaud, 2004), evidence of deposition of weathered materials, indicating warm, humid conditions (Leather et al., 2002; Bahlburg and Dobrzinski, 2011), survival of organisms through supposed snowball Earth conditions (Xiao, 2004; Corsetti et al., 2006; Moczydlowska, 2008), and alternative explanations for cap carbonates and iron formations as hydrothermal deposits related to contemporary volcanism in rift settings (Yeo, 1981; Cox et al., 2013). It was also pointed out by Schmidt and Williams (1995) that large-scale periglacial sand wedges and polygons associated with Cryogenian glacial deposits that formed near the equator in South Australia require large annual temperature fluctuations, which appear to be at odds with their low depositional latitude. Such temperature fluctuations are unknown at low latitudes today but would be expected under the HOT (Fig. 19.2B). Much of the sedimentological evidence of open water was countered by Hoffman (2011b) when he pointed out that most deposition under snowball Earth conditions would have occurred during the destruction of the ice, which he regarded as having been very rapid and related to the buildup of atmospheric CO_2 to a critical level.

Several studies have shown that interglacials/interstadials, some with open water, existed during Neoproterozoic glaciations, so that it is necessary to explain several rapid global glaciations/deglaciations during a single glacial epoch. It is also difficult to explain how as much as 5000 m of glacial sediments (Preiss, 2000) could be delivered during rapid melting of glaciers that had existed for millions of years under an essentially nonexistent or inactive hydrological regime, when most oceans were frozen. Some authors have pointed out that the Sturtian deposits represent more than one glacial advance–retreat cycle (Young and Gostin, 1990, 1991; Busfield and Le Heron, 2014; Lan et al., 2015).

8. THE BARREN OR BORING BILLION

The long Proterozoic Eon, from 2.5 Ga to 542 Ma, can be thought of in terms of two very glacially active periods near the beginning and end, with a long nonglacial period between about 1.8 and 1.7 Ga and 800–700 Ma (the barren or boring billion). The billion-year-long period is usually designated by negative connotations. For example, there is little or no definitive evidence of glaciations or the deposition of iron formations. Investigations of stable isotopes and other geochemical parameters reveal little evidence of major changes in the composition of seawater. There are various other parameters related to igneous, metamorphic, and economic geology (Bradley, 2011) that reveal unusual attributes of the barren billion. The relative stability of conditions during the barren billion has been attributed by some (Roberts, 2013) to relative tectonic quiescence, with core parts of the supercontinent Columbia remaining intact through to the formation of the younger supercontinent, Rodinia. On the other hand the amalgamation of Rodinia during the Grenville orogeny was one of the greatest mountain-building episodes in Earth's history. Perhaps the key factor was the absence of a great breakup episode between formation of the two supercontinents, for the fragmentation of supercontinents may have played an important role in the onset of glacial episodes and associated atmospheric changes during several supercontinental cycles (Nance et al., 2013; Young, 2013a).

Among all these negative signals there are three small indications that there may have been some glacial activity near the beginning and end of the barren billion years (Fig. 19.3). In the Kimberley Basin in Western Australia, Williams (2005) reported the presence of subglacial meltwater channels and glacio-fluvial deposits, considered to have formed at about 1.8 Ga in low latitudes. A low latitude occurrence of c. 1.9 Ga periglacial phenomena was reported by Kuipers et al. (2013) from central Sweden. Tentatively dated at about 1.3 Ga, Geboy et al. (2013) reported glacial deposits from eastern Brazil. These minor occurrences are dated at near the introduction and end of the barren billion, and indicate that the Earth came under some glacial influence (some at low latitudes). These minor occurrences stand in marked contrast to the relatively abundant occurrences of glaciogenic rocks near the beginning and end of the Proterozoic Eon.

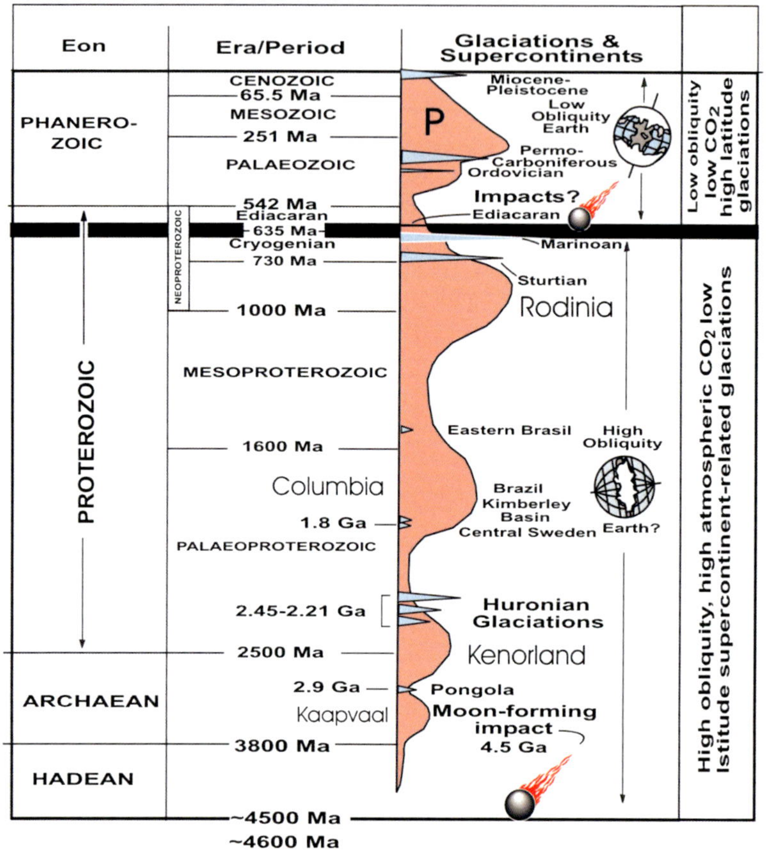

FIGURE 19.3 Relationship between supercontinents (after Bleeker, 2004) and glaciations throughout geological time. Width of the pink ornament signifies the relative size and degree of amalgamation of continental lithosphere. Note that there is little evidence of Archean glaciations. The great Proterozoic glaciations appear to have taken place during breakup of supercontinents whereas the Permo-Carboniferous glaciation occurred during assembly of Pangaea. There is little evidence of glaciation on the supercontinent Columbia. The small globes at right show possible changes in the tilt of the Earth's spin axis (obliquity of the ecliptic) through time. A change in the Earth's obliquity may have occurred during the Ediacaran Period. The inferred time of the transition from high to low obliquity is shown by the heavy black line. The alternative theory to explain the great Proterozoic glaciations is the snowball Earth hypothesis (Kirschvink, 1992; Hoffman et al., 1998). See text for explanation and discussion.

9. EARLY PROTEROZOIC GLACIATIONS: THE HURONIAN GLACIAL EVENT

Although widely dispersed, glacial deposits of Early Proterozoic age are few enough that they may be briefly considered individually. Glacial deposits that are now known to be about half the age of the Earth were first described by Coleman (1907, 1908) from the north shore of Lake Huron in Ontario, Canada. Their early discovery, wide extent, and magnificent exposures have justifiably led to these glacial deposits being recognized as the "type" example of early Proterozoic glacial activity and used for comparisons and correlations with similar units around the world.

9.1 North America and NW Europe

These two great regions, are considered together because of their presumed proximity prior to opening of the North Atlantic Ocean.

The Huronian Supergroup is essentially a southerly thickening wedge of mostly siliciclastic sedimentary rocks that achieves a maximum thickness of about 12,000 m in the southern part of the Huronian outcrop belt. As shown in (Fig. 19.4), the supergroup is considered to represent a rift-to-drift transition with the thickest succession preserved in a rift basin which was overstepped at the time of deposition of the Gowganda Formation and succeeding marine-dominated Huronian formations (Fig. 19.5). Following the discovery of rich placer uranium

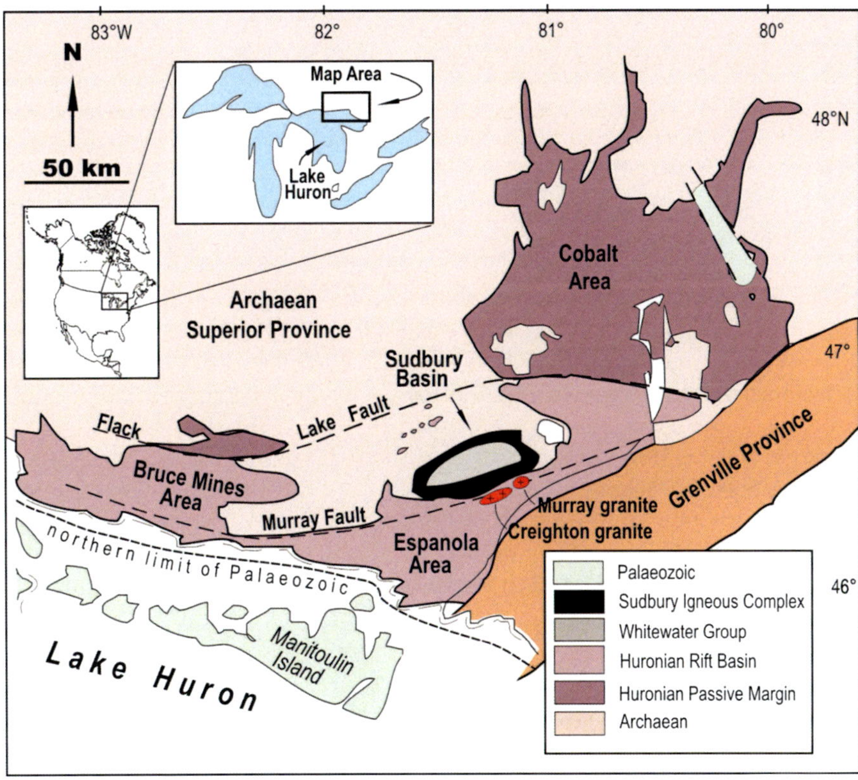

FIGURE 19.4 Sketch map to show the distribution of the Huronian Supergroup (2.45–2.21 Ga) on the north shore of Lake Huron. The lower part of the Huronian Supergroup (below the Gowganda Formation) is present only in the area south of the Flack Lake Fault (see cross-section in Fig. 19.5), which is interpreted as a rift basin. In the area to the north, the Gowganda and succeeding marine formations lie on Archean basement rocks and are thought to represent a passive margin succession. The Huronian Supergroup contains three glaciogenic formations but only the youngest (Gowganda Formation) is present beyond the confines of the rift basin.

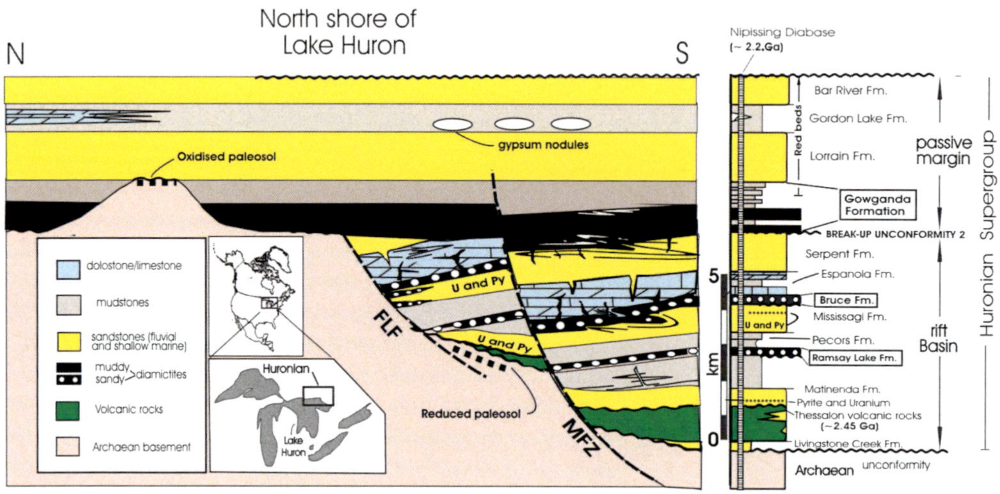

FIGURE 19.5 Cross-section and stratigraphic log to show the succession of formations in the Huronian Supergroup. There are three glaciogenic formations, the Ramsay Lake, Bruce, and Gowganda, but note that the Gowganda and succeeding formations overstep the Murray and Flack Lake faults (see Fig. 19.4 for location of faults) to lie directly on Archean rocks. The Huronian Supergroup is thought to have formed during fragmentation of the supercontinent, Kenorland, with the sub-Gowganda unconformity and overstep representing the breakup unconformity. The widespread and largely marine character of the Gowganda and succeeding formations means that they have much greater potential for regional and international correlation than the underlying rift-filling lower Huronian succession.

deposits near the base of the supergroup, the area was the focus of intensive mapping programs by both federal and provincial geological surveys, with attendant erection of stratigraphic schemes and nomenclature (Robertson et al., 1969; Roscoe, 1969; Frarey, 1977; Card, 1978). The tectonic setting of the Huronian Supergroup has been and remains controversial, having been considered as an aulacogen (Young, 1983), a strike-slip basin (Long, 2004), and a passive margin (Dietz, 1964; Young and Nesbitt, 1985; Bennett et al., 1991; Young, 2015). It was recently argued by Young (2014) that breakup of the Archean supercontinent, Kenorland, took place at the time of deposition of the widespread glacial unit, the Gowganda Formation, for it is the first Huronian formation to display evidence of marine influence (Lindsey, 1971) and is much more widespread than the underlying Huronian formations, which are considered to have been deposited in a restricted rift basin (Young, 2014). The Huronian Supergroup was deposited between about 2450 Ma (Krogh et al., 1984; Ketchum et al., 2013) and 2217 ± 9 Ma (Corfu and Andrews, 1986). These dates provide temporal limits for accumulation of the thick sedimentary pile, but they do not provide precise ages for any of the sedimentary formations. The older date comes from volcanic rocks near the base of the succession and the younger is from the intrusive Nipissing diabase suite. There are three glaciogenic formations in the Huronian Supergroup, but none of them has been directly or precisely dated, so that attempts to correlate these formations to Proterozoic glacial deposits in other parts of the world are uncertain (Young, 2014; Fig. 19.4). Geochronological studies by Vallini et al. (2006) provided a zircon age of 2317 ± 6 Ma from the Enchantment Lake Formation in the Animikie basin, immediately west of the Huronian rocks. This glacial unit, which is thought to be correlative to the Gowganda Formation, is overlain by the Sunday Quartzite, which was found to contain zircons with an age of 2306 ± 9 Ma. Thus these formations are presumably older than, or equivalent in age to, these zircon crystals. A zircon date of 2308 ± 8 and ±10 Ma, obtained from the Gordon Lake Formation in the upper part of the Huronian Supergroup (Fig. 19.5), was interpreted by Rasmussen et al. (2013) as being from a thin tuff and was therefore thought to represent the depositional age of the formation. This interpretation is, however, thrown into question by the younger age obtained by Vallini et al. (2006) from zircon in what is probably an older formation—the Sunday Qurtzite, which is probably equivalent to the Lorrain Formation. Thus the age of the Gowganda Formation remains uncertain. The age of 2.31 Ga (Enchantment Lake Formation) could be close to the depositional age of its putative correlative, the Gowganda Formation, or it could indicate that both are younger than that date. In spite of claims to the contrary, none of the Huronian sedimentary formations is precisely dated.

Within and beyond North America, early attempts at using glacial deposits as time-markers to correlate Proterozoic supracrustal successions were made by Pettijohn (1943), Young (1970), and Ojakangas (1988), but even in the immediately adjacent Animikie basin, there has been much controversy concerning correlation of such rocks (see Young, 2013b, for a summary). Part of the problem was that the lower half of the Huronian Supergroup is missing in the Animikie basin, a situation that was rationalized recently (Young, 2015) by invoking deposition of the Proterozoic rocks on lower and upper plates of a separating continent (Lister et al., 1986). Glacial deposits resembling the Gowganda Formation have been found in Michigan, at Chibougamau in northern Quebec, in Wyoming, about 2000 km to the SW, and in Nunavut on the west side of Hudson Bay (Fig. 19.6). Among these occurrences, the only one containing rocks that closely resemble the lower Huronian succession (with its two included glaciogenic formations) is the Snowy Pass Supergroup in SE Wyoming (Karlstrom and Houston, 1984). The distribution of the Gowganda Formation and correlatives in North America was interpreted by Young (1970) as representing the marginal deposits of a continental ice sheet that covered much of the supercontinent, Kenorland, at the time of its breakup. This interpretation involves centrifugal ice dispersal and assumes that the areas were in a spatial relationship that approaches that of the present day. A more "mobilistic" interpretation was offered by Roscoe and Card (1993) and Ernst and Bleeker (2010), who suggested juxtaposing all of these now widely separated areas, including the Karelian region of NW Europe, against the southern margin of the Huronian basin. These two interpretations infer quite different areal distributions of ice-cover (Fig. 19.7), for the "fixist" interpretation would mean that the sediments of the Gowganda Formation and correlatives were widely distributed (in a centrifugal pattern) and that most were derived from the western part of what is now the Superior province. Based on thickness and facies changes, paleocurrent directions and clast orientations, ice movements for the deposits on the south side of the inferred ice sheet were in a southerly direction. On the other hand the "mobilist" interpretation requires an unusual situation involving centripetal ice movement in contrast to modern and Phanerozoic continental glaciations, which seem to have involved radial dispersal of ice (e.g., Antarctica or Greenland). On the other hand, we live in a world on which continents are widely distributed or are in the act of amalgamating, whereas the Huronian Supergroup and equivalents were formed at a time of continental breakup.

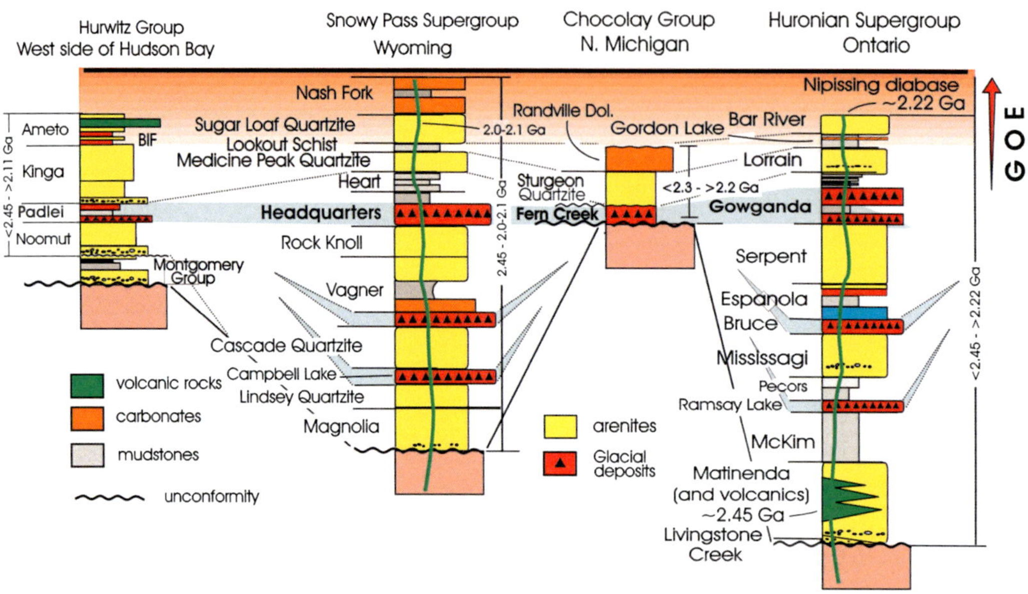

FIGURE 19.6 Lithostratigraphic correlation among widely separated areas of North America (for locations see Fig. 19.7). Note that the correlations are based on available geochronological data and on the glaciogenic Gowganda Formation and overlying aluminous quartz arenites. The near-identical succession of the Snowy Pass Supergroup in Wyoming (Karlstrom and Houston, 1984) has been interpreted to indicate a similar tectonic and climatic history along parts of the same developing continental margin (see Fig. 19.7), or it was proposed that the Wyoming rocks might be the displaced and rotated "other side" of the Huronian basin (Roscoe and Card, 1993; Ernst and Bleeker, 2010). The first red beds in the Huronian succession are present in the Lorrain Formation, which overlies the Gowganda Formation. As noted by Roscoe (1973), these beds are considered to indicate that the atmosphere was oxygenated—this important turning point in Earth history has since become widely accepted (Holland, 2006) as the Great Oxidation Event (GOE and *arrow* at top right). It has been proposed by Shields-Zhou and Och (2011) that atmospheric oxygenation might be related to great glaciations because melting of ice sheets flooded the oceans with nutrient-rich sediment that stimulated blooms of photosynthetic microorganisms.

The Huronian contains some interesting stratigraphical juxtapositions; for example, the glaciogenic Bruce Formation is overlain by carbonate rocks (both limestones and dolostones) and the extensive glacial deposits of the Gowganda Formation are succeeded by thick quartz arenites that contain aluminous minerals (kaolinite, diaspore, and aluminosilicates derived therefrom), suggesting strong chemical weathering (Young, 1973). There is also clear evidence of development of deep paleosols both near the beginning and end of deposition of the Huronian succession (Gay and Grandstaffe, 1980; Rainbird et al., 1990). The carbonates of the Espanola Formation have been compared by Bekker et al. (2005) to cap carbonates that commonly occur above much younger (Cryogenian—especially Marinoan) glacial deposits attributed by some to a snowball Earth condition (Hoffman, 2011b). These carbonate rocks have taken on a special significance in the sense that they have been considered as indicators of global glaciations (snowball Earth events) even in cases where there is only very local evidence of such rocks. A cautionary note was sounded by Young (2013a) that, apart from stable isotope analyses, very little geochemical work has been carried out on these

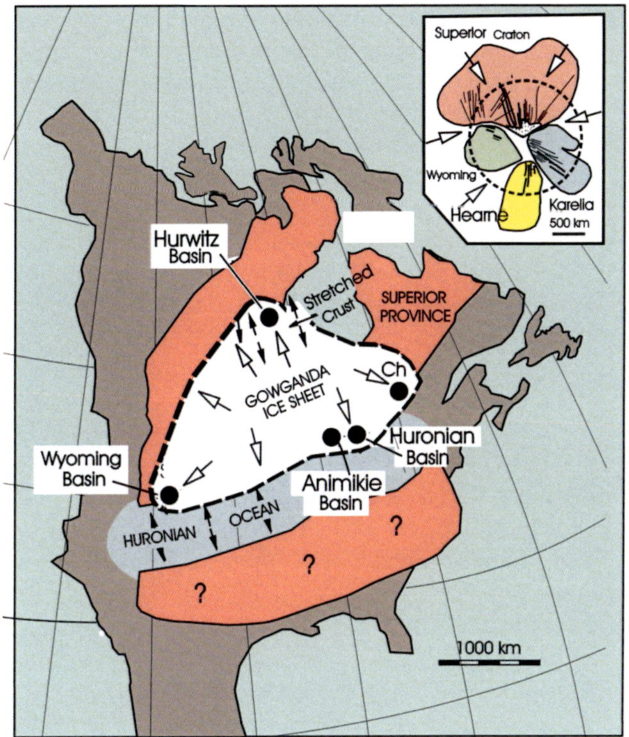

FIGURE 19.7 Sketch map to show the possible location of a large ice sheet in North America during deposition of the Gowganda Formation and correlatives (see Fig. 19.6). A passive margin and small ocean basin are thought to have formed on the south side of the Gowganda ice sheet but at the northern limit of the ice sheet the crust may have undergone stretching to produce an intracratonic basin (Aspler et al., 2001). *Arrows* on ice sheet show the inferred centrifugal ice drainage pattern. An alternative paleogeographic reconstruction by Ernst and Bleeker (2010) is shown in the insert at top right. Note the centripetal ice sheet that is inferred from this reconstruction. *Ch*, Chibougamau.

carbonates. Restricted distribution of the Espanola Formation, together with stable isotopic data and possible evidence of hydrothermal influence, suggest deposition in a restricted lacustrine environment in a rift setting. In a little-cited but visionary paper, Roscoe (1973) pointed out important indicators of atmospheric oxidation during deposition of the Huronian Supergroup.

In NW Europe, glacial deposits of Paleoproterozoic age are present in northern Finland and in the northwestern part of Russia. The "type area" for these deposits is in Finland, where Marmo and Ojakangas (1984, 1985) found definitive evidence of glacial conditions on the Fennoscandian Shield in the Urkaavaara Formation. Similar glaciogenic rocks are known from several greenstone belts in Karelia and in the Kola Peninsula area of Russia. Following their discovery there have been numerous attempts to correlate these glacial deposits to the Huronian Supergroup in Canada and elsewhere (Melezhik et al., 2013, their Fig. 7.29, and references therein). According to Gärtner et al. (2014), glacial deposits forming part of the

Seidorechka and Polisarkal Sedimentary formations in the Imandra-Varzuga Greenstone Belt of the Kola Province of the Fennoscandian shield have yielded ages indicating that they were deposited between about 2430 and 2411 Ma. Based on these dates, they suggested that the glacial deposits in the Fennoscandian area may be correlated to the two older glacial formations of the Huronian Supergroup. It must, however, be kept in mind that the younger date (<2411 Ma) is from detrital zircon, so that the depositional age could be considerably younger. Like the Huronian Supergroup the Fennoscandian glacial rocks have been interpreted as rift deposits, formed in an extensional setting that accompanied breakup of the Archean supercontinent, Kenorland. In Fennoscandia, however, the glacial deposits are much more intimately involved with volcanic rocks and they lack the three clearly defined glacial formations of the Huronian Supergroup and their tripartite cyclicity (Fig. 19.4). It is clear that in the Huronian Supergroup, the two older glacial formations are only locally developed, in contrast to the widespread nature of the youngest (Gowganda Formation and equivalents) (Young, 2014). These relationships perhaps suggest that some of the Fennoscandian glacial deposits may have formed at the same time as the Gowganda Formation or, as suggested by Gärtner et al. (2014), their common association with early Paleoproterozoic volcanic rocks may point to an earlier, syn-rift origin. More detailed geochronological data are needed, but the Fennoscandian occurrences appear to be part of a widespread Huronian Glacial Event.

9.2 The Gondwanan Continents

Paleoproterozoic glacial deposits are also known from other continents, notably from South Africa, Australia, and India.

9.2.1 South Africa

In South Africa there are several glaciogenic units. Stratigraphy of the Proterozoic rocks of this area has been reviewed by Bekker et al. (2001, 2005), Eriksson and Alternann (2013), Hoffman (2013), Tang and Chen (2013), and Young (2014), and is only summarized here.

Unlike the Huronian Supergroup, which is mainly siliciclastic and which formed in an extensional setting during breakup of an Archean supercontinent, the South African glacial deposits have been attributed to deposition in a compressive tectonic setting (Eriksson and Altermann, 2013) and are commonly associated with chemical sedimentary rocks such as iron formations and carbonates (Fig. 19.8). Glaciogenic rocks are disposed in three separate basins among which there are stratigraphic differences so that correlation is difficult. It was proposed by Visser (1971) that two of the South African basins, the Griqualand and West Transvaal basins, may have been separated by a high-standing region known as the Vryburg Ridge, which he suggested might have been the source of glaciogenic materials in the two basins. There is evidence of at least three glacial episodes, and it was suggested by Kirschvink et al. (2000), Bekker et al. (2005), and Kopp et al. (2005) that glacial deposits of the Makganyene Formation in the Transvaal basin of South Africa formed during a worldwide Paleoproterozoic glacial episode, like the "snowball" events postulated for the Cryogenian glaciations (Kirschvink, 1992; Hoffman et al., 1998).

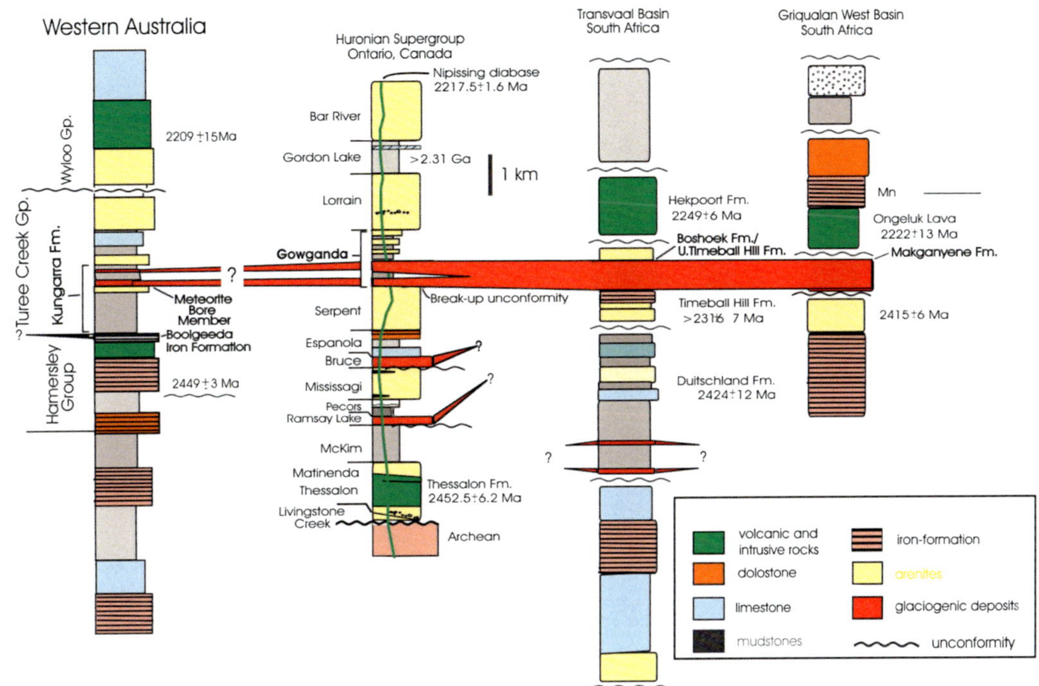

FIGURE 19.8 Comparison of Paleoproterozoic stratigraphic columns from Western Australia, Ontario, Canada, and two basins in South Africa. The Gowganda Formation is tentatively correlated with glacial deposits in the other two areas but these are ultimately dependent on reliable and unequivocal geochronological data. Two glacial advance–retreat cycles in the Gowganda Formation (see Fig. 19.9) may correspond to those recently recognized by Van Kranendonk and Mazumder (2015) in Western Australia.

From the results of geochronological investigations by Rasmussen et al. (2013) and Beukes et al. (2013), it was concluded that both the Makganyene and Upper Timeball Hill formations may be younger than all of the Huronian glacial formations, so that there could have been four Paleoproterozoic glaciations in total with the oldest Huronian one (Ramsay Lake Formation) not represented in South Africa and the youngest one in South Africa (upper Timeball Hill and Makganyene) not present in North America. It must, however, be recognized that the Huronian glaciogenic formations remain undated and that it may be imprudent to invoke a global glaciation on the basis of a single occurrence, in contrast to the widespread nature of the Cryogenian glaciations, to which they have been compared.

Attempts to establish correlations with the Huronian Supergroup are rendered difficult by the general paucity of precise dates and the local nature of the two older Huronian glacial units.

9.2.2 *Western Australia*

In Western Australia, the Paleoproterozoic Meteorite Bore Member of the Turee Creek Group was recognized by Trendall (1976, 1979) as having a glacial component, but it was later shown by Krapez (1996) and Martin (1999) to include mass flow deposits and turbidites thought to have been deposited in a compressive retro-arc or foreland basin related to orogenesis on the south side of the Hamersley province. The Meteorite Bore Member is loosely constrained between about 2.45 and 2.21 Ga (Lepland, 2013 and references therein). The upper age limit is from the Cheela Springs Basalt in the unconformably overlying Wyloo Group (Martin, 1999). These rocks, and possibly the Makganyene Formation in South Africa, have been interpreted by some as having formed in a foreland basin setting. Recent work on these rocks in Western Australia by Mazumder et al. (2015) and Van Kranendonk and Mazumder (2015) supports the idea of two separate glacial units within the Kungarra Formation, which they related to global sea-level changes. Because the Meteorite Bore Member falls within the same broad time frame as the Huronian glacial deposits and the Gowganda Formation contains evidence of two glacial advance–retreat cycles (Fig. 19.9), it is interesting to consider the possibility that they may have formed at the same time and reflect the same changes in sea level. Such correlations are tentative and must await definitive geochronological results.

9.2.3 *India*

Paleoproterozoic metasedimentary rocks including diamictites and laminites with dropstones have been described from the basal part of the Sausar Group in central India. These rocks have been interpreted as glacial deposits (Mohanty, 2006). They lie unconformably above Archean gneiss (Tirodi gneiss) and are overlain by carbonate rocks and manganiferous fine siliciclastic rocks. Available age dates suggest that these rocks were deposited as part of the Huronian Glacial Event. The association with carbonates and Mn-rich sedimentary rocks is reminiscent of the Paleoproterozoic glacial rocks of South Africa, but more detailed paleomagnetic and geochronological studies are needed.

9.2.4 *Summary of Paleoproterozoic Glacial Occurrences*

Evidence of Paleoproterozoic glaciations is most abundant in North America, but is also known from NW Europe, India, South Africa, and Western Australia. Glacial deposits of Paleoproterozoic age in North America and Fennoscandia were laid down during the breakup of the early supercontinent, Kenorland. Of three Huronian glaciogenic formations, the most widespread is the Gowganda Formation, which is thought to have been deposited on a newly formed passive margin, at a time of basin subsidence and sea level rise. Although the Huronian Supergroup includes two older glacial formations, they were probably restricted in distribution because they formed in a rift setting. The older glacial formations may be approximately equivalent to those of the Fennoscandian shield but more geochronological data are needed. Whereas the Huronian Supergroup is known to have been deposited between about 2.45 and 2.2 Ga, none of the glacial formations has been precisely dated. In spite of the restricted distribution of the two lower Huronian glacial units and the lack of precise dates, the Huronian succession has been used as a template for correlation with Paleoproterozoic glacial deposits elsewhere, with highly contradictory results (Young, 2014). There are several glaciogenic units in South Africa. There have been various attempts at correlation of these rocks to the Huronian formations and recently it has been suggested that the Makganyene Formation may be entirely younger than the Huronian Supergroup and may have been deposited during a "snowball" glaciation, although its occurrence is quite local

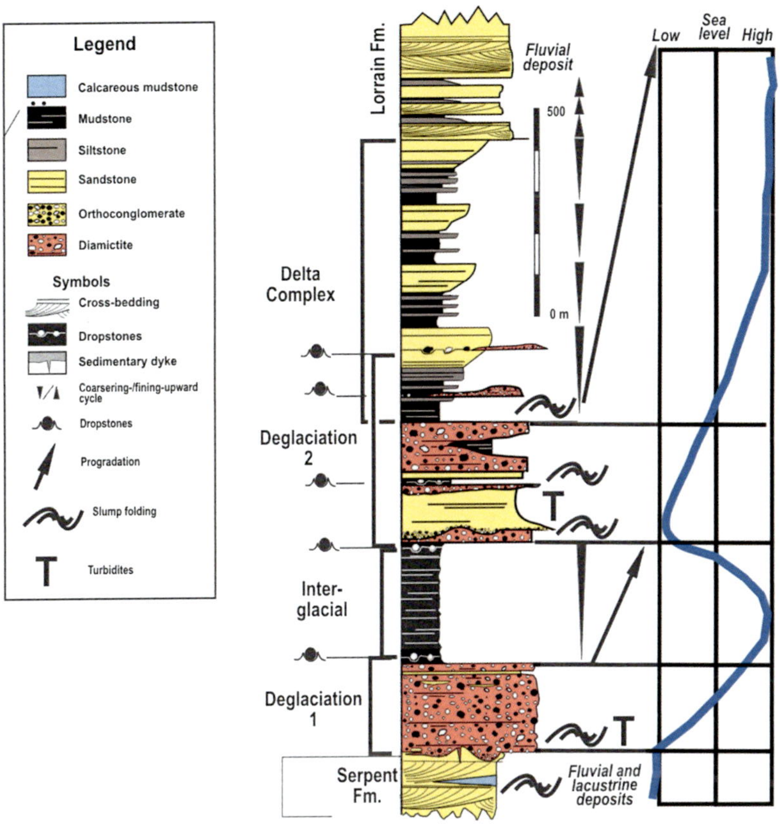

FIGURE 19.9 Stratigraphic log to represent the Gowganda Formation in the southern part of the Huronian outcrop belt. There are two diamictite-rich glacial successions, each overlain by siliciclastic rocks that bear no physical evidence of glacial activity and are interpreted as interglacial and postglacial deposits formed mainly during periods of elevated sea level. Both form coarsening-upward sequences that suggest progradation. The upper half of the Gowganda Formation consists of several postglacial coarsening-upwards deltaic sequences. Because these sediments accumulated during a rift–drift transition they contain abundant evidence of contemporary tectonic activity in the form of slumped beds, coarse mass flow deposits, turbidites (T), and syn-sedimentary faults (Young and Nesbitt, 1985). Thus the inferred sea level rise due to melting of the ice sheet was accompanied by basin subsidence.

(Rasmussen et al., 2013; Beukes et al., 2013). None of these correlations is definitively constrained by reliably interpreted geochronological data, so that for the time being, it is probably prudent to denote only a broadly defined Huronian Glacial Event (Tang and Chen, 2013; Young, 2014) in anticipation of more definitive criteria for detailed correlations.

A low latitude was obtained (Williams and Schmidt, 1997) by paleomagnetic investigation of the Gowganda and Lorrain formations and from hematite-rich breccias and a paleosol near Ville Marie (Schmidt and Williams, 1999) that are considered to be about the same age as the Lorrain Formation (Rainbird et al., 1990). Some of these paleomagnetic results were questioned by Hilburn et al. (2005), but the low paleolatitude is supported by the several investigations of Williams and Schmidt.

10. ARCHEAN GLACIAL DEPOSITS

There is little definitive evidence of glacial conditions in rocks deposited during the first half of geological history—the Archean Eon. It has been claimed by some such as Du Toit (1954) and Crowell (1983) that there is evidence of glacial influence on deposition of part of the Witwatersrand Supergoup in South Africa and in Archean rocks elsewhere (Hambrey and Harland, 1981), but the most persuasive evidence probably comes from the Pongola Supergroup (von Brunn and Gold, 1993; Young et al., 1998), which is a thick succession of Archean (~2.9 Ga) supracrustal rocks in the eastern part of the Kaapvaal Craton of southern Africa. The glacial succession includes diamictites with clasts of different types, some of which are striated and faceted. Associated silty laminites contain isolated clasts (dropstones) with splash-up structures, interpreted as evidence of the presence of floating icebergs derived from glaciers. A glacial origin is also supported by geochemical data from matrix material of the diamictites and from argillites. On the basis of well-constrained paleomagnetic investigations (Nhleko, 2004), the Pongola glacial deposits are thought to have formed at a latitude of about 43–48 degrees.

Diamictites interpreted as glaciomarine laminites with ice-rafted clasts were described by Ojakangas et al. (2014) from a ~2.7-Ga metasedimentary succession in the Dharwar Craton in central India. Similar glaciogenic rocks of about the same age have been described from the footwall of the Stillwater Complex in Montana in the USA (Page, 1981).

11. WHY DID SPORADIC GLACIATIONS OCCUR?

The short answer is that we do not know. Perhaps an even more immediate question is why early Earth history was not one of continuous frigidity, because it is believed that the Sun was much less radiant (about 30% less according to some estimates) during the early part of our planet's history. The geological record of abundant Archean sedimentary rocks, both siliciclastic and chemical precipitates, puts lie to that possibility, for these rocks provide clear evidence of the existence of abundant water on the young Earth's surface. Resolution of this problem, which came to be known as the "faint young Sun paradox" (Sagan and Mullen, 1972), lies in the discovery of the importance of greenhouse gases, including water vapor, carbon dioxide, and methane. A corollary of this interpretation is that the partial pressure of such greenhouse gases must have decreased with time, to accommodate the rising radiative power of the Sun. Whereas this interpretation would allow the Earth's surface to remain within the temperature range of liquid water (as indicated by the presence of water-lain sedimentary rocks throughout Earth history), it fails to explain why the Earth was gripped sporadically by ice ages (Fig. 19.3). There have been numerous attempts to link glaciations to atmospheric changes. It has been suggested that methane was an important greenhouse gas in the early atmosphere but it was sporadically diminished by rising oxygen levels, leading to cool periods. A potential problem with this theory is that periods of increased oxygenation of the atmosphere appear to have occurred following glaciations rather than before them (Shields-Zhou and Och, 2011; Young, 2013a). It has been proposed that oxygen levels were raised by fertilization of the oceans as a result of increased postglacial or tectonically-induced sediment influx. It was proposed by Eyles and Januszczak (2004) that the onset of glaciations was due to uplift of rift shoulders during the breakup of large continental masses (the "zipper-rift" model), which would involve diachronous development of glaciations as rifting extended laterally. With the development of ideas concerning the supercontinental cycle (Nance et al., 2013) came the suggestion (Young, 2013a) that the great Proterozoic glaciations might have been triggered by weathering-related drawdown of CO_2 during periods of supercontinentality. Because weathering rates are controlled, among other factors, by temperature, it was noted by Walker et al. (1981) that there is a negative feedback loop that could have caused global temperatures to fluctuate, at certain times, between glacial and nonglacial conditions, i.e., as CO_2 is withdrawn from the atmosphere by weathering it gets colder but low temperatures reduce weathering rates so that CO_2 (mainly from mantle emanations) increases in the atmosphere, bringing the ice age to an end. A similar scheme was illustrated by Raub and Kirschvink (2008, Fig. 19.3), who showed that the glaciations at the beginning and end of the Proterozoic Eon appear to have occurred at the end of periods of supercontinentality. That relationship appears to have ended in the latest Proterozoic (late Ediacaran Period) when glaciations occurred exclusively in circumpolar latitudes and their occurrence was predicated, among other things, on the presence of a continental mass, or part thereof, at high latitudes. It was suggested by Santosh et al. (2014) that some of the latest Proterozoic glaciations were initiated by continental uplift associated with the Pan-African orogeny, just as the Cenozoic glaciation has been linked by some to the Alpine-Himalayan orogenic episode. Thus there appears to be a radical change in the mode of glaciations in the last 570 Ma or so compared to that which prevailed during the earlier part of geological history (Williams, 1975; Evans, 2003; Young, 2013c).

12. SNOWBALL EARTH VERSUS HIGH OBLIQUITY

The differences between most Precambrian and later glacial deposits are quite profound, because glaciations up to about the middle of the Ediacaran Period (those from which reliable paleomagnetic results have been obtained) appear to have formed at low latitudes in climatic zones that today are mostly described as tropical. This enigmatic relationship, together with sedimentological indicators of deposition at or near to sea level, was taken by some (Kirschvink, 1992; Hoffman et al., 1998) as a priori evidence that areas of the planet which were closer to the poles must also have been glaciated. Thus it was suggested that, as a result of runaway albedo (Budyko, 1969), the surface of the entire planet was encased in snow and ice—the SEH. There are many questions regarding the validity of the she, but perhaps the most serious is that no high-latitude glacial deposits have been definitively identified in the early part of the Earth's history whereas all Phanerozoic glaciations occurred in circumpolar latitudes. It has been argued that there were no continent-bearing plates located in high latitudes during the Precambrian, but this beggars belief when it is considered that the time involved (from the oldest glacial deposits to the time when they assumed their "modern" high latitude distribution) is over 2 billion years.

The extraordinary low-latitude distribution of the earlier glacial deposits, together with compelling evidence of strong, presumably seasonal, temperature variations, led

Williams (1975, 2008) to suggest that glacial deposits at low latitudes did not require the whole planet to be frozen but rather, that the Earth's climatic regime was radically different during the early (largest) part of Earth history. He explained the lack of high-latitude glacial deposits as being due to the Earth's spin axis having a quite different orientation. It has been calculated that if the obliquity of the ecliptic were to exceed 54 degrees it would cause today's tropical zones to receive less annual solar radiation than the polar regions, thus providing a possible explanation for the distribution of ancient glacial deposits and the evidence of strong seasonal temperature variations. It has been argued that the distribution of evaporites (Evans, 2006) shows that the distribution of climatic zones was the same on the Cryogenian Earth as it is today, but in addition to temperature, world distribution of such deposits is controlled by atmospheric circulation, which would have been similar to the present one during the two "summer" seasons depicted in Fig. 19.2B. The planet may have obtained a high obliquity early in its history as a result of the huge impact commonly invoked to explain the origin of the Moon. On the other hand, the mechanism that brought it to its present low obliquity (~23 degrees) is unknown, but the change may have occurred during the Ediacaran Period and may be related to development of the world's deepest $\partial^{13}C$ excursion, known as the Shuram anomaly. It was proposed (Young, 2013c) that the Shuram anomaly might be related to a large marine impact. Although it is unlikely that a large change in the orientation of the Earth's spin axis could be the direct result of such an impact without causing destruction of the planet, a smaller oceanic impact could have caused avalanching of oceanic lithosphere (Condie, 1998) and a major readjustment of mass in the planet's interior, resulting in inertial interchange true polar wander (Kirschvink et al., 1997, 2005) and altering the orientation of the Earth's spin axis.

13. GLACIATIONS, IMPACTS, AND HIGHER LIFE FORMS

It is widely recognized (McLaren, 1983) that environmental changes, especially if they are rapid and severe, can contribute not only to mass extinctions but also to innovative evolutionary radiations. When the widespread nature of many Precambrian glaciations was recognized, it was suggested (Hoffman et al., 1998) that the evolution of higher life forms (metazoans) was influenced greatly by environmental changes associated with the great Cryogenian glaciations. Many dramatic biological changes also occurred in the Ediacaran Period, long after the ice from these glacial episodes had melted. In addition, some of the most dramatic biological events appear to be related to impacts. For example, it was shown by Gray et al. (2003) that there was an important extinction-radiation event involving acritarchs following the Acraman impact in South Australia. It was suggested by Young (2013c) that the Shuram anomaly was related to a large marine impact. The isotopic anomaly, regardless of its origin, was also accompanied by an extinction event (McFadden et al., 2008), which saw the extirpation of most acanthomorphic acritarchs and was followed by expansion of the enigmatic Ediacaran biota. The origin of a negative $\partial^{13}C$ excursion, immediately below the Precambrian–Cambrian boundary, is unknown, but it appears to have coincided with glaciations in a number of locations (Caby and Fabre, 1981; Bertrand-Sarfati et al., 1995; Deynoux et al., 2006; Vernheta et al., 2012). Both the ages and nature of most of these Ediacaran events remain uncertain, but it appears that the dramatic evolutionary events that culminated in the Cambrian explosion of metazoan life forms were influenced by Ediacaran impacts as well as glaciations.

14. SUMMARY AND CONCLUSIONS

Ice ages are relatively rare events in the long history of our planet. About 2700 million years elapsed before the first known definitive evidence of glaciation was preserved in the rock record—the Pongola glaciation at about 2.9 Ga. Although compelling, the evidence of Archean glaciations is only locally preserved. There followed a hiatus of about 450 Ma until the next glacial "event," which has become synonymous with the supergroup in which it was discovered—the Huronian Supergroup. The first Huronian glacial deposits occurred shortly after extrusion of volcanic rocks of the Thessalon Formation and equivalents at about 2.45 Ga. There are three formations in the Huronian Supergroup that are considered to be glaciogenic, but the older, Ramsay Lake and Bruce formations probably formed in a rift setting, whereas the Gowganda Formation, which is much more widespread, was probably deposited during and following continental fragmentation. The Gowganda Formation is overlain by aluminous orthoquartzites, and this unusual rock association has been used to effect correlations with similar rocks as far away as SE Wyoming and in Nunavut, on the west side of Hudson Bay. In other parts of the world (NW Europe, India, South Africa, and W Australia), glacial rocks of approximately the same age as the Huronian Supergroup have been identified, and many preliminary correlations, mostly to the Huronian, have been made. Most of these correlations are tentative and must await precise geochronological data. In the meantime it is probably prudent to include these glaciogenic formations in a single Huronian Glacial Event.

The barren billion, from about 1.8 to 0.8 Ga, contains little evidence of glacial influence, but recently fluvio–glacial activity was documented in Australia, and periglacial structures were described from central Sweden. Both are thought to have formed just prior to the beginning of the barren billion. Low latitudes have been suggested for both. Glacial deposits in eastern Brazil may have been deposited at about 1.3 Ga. There is no compelling evidence of widespread glaciations during this long period of time.

During the Cryogenian Period the Earth underwent its most widespread glaciations (Sturtian and Marinoan). The wide distribution of these glacial deposits and their near-ubiquitous low latitudes led to the snowball Earth hypothesis. An alternative explanation, the high obliquity theory, invokes an early Earth in which glaciations would have been initiated at low latitudes. The jury is still out on the explanation for the Earth's great glaciations, and only time will tell. The great thickness of some Cryogenian glacial sequences (up to 5000 m) is difficult to explain under the SEH, for even if most of the sediments were deposited during recession of the supposed world-enclosing ice sheet, destruction of the ice and reactivation of the hydrological cycle are thought to have been extremely rapid, so that it is difficult to envisage the means of accumulating such impressive thicknesses of glaciogenic sediment, and their stratigraphical complexity. On the other hand the high obliquity theory would have retained open seas at the poles (at least seasonally), so that there would have been a permanent moisture source and a vigorous hydrologic cycle throughout the glacial events. In addition the HOT provides an explanation for strong seasonal temperature variations at low latitudes.

Under the SEH the regional source of glacial sediments would have been in the polar regions, with the main sink in lower latitudes, as is the case today and as it was throughout the Phanerozoic and probably latest Ediacaran. If the continents were mainly located in a band at low latitudes and freezing initiated at the poles and migrated toward equatorial regions as postulated in the SEH, there would have been a limited source of moisture to produce the thick Cryogenian glacial successions reported from continental areas. By contrast, the HOT would have involved a radically different distribution of climatic zones with accumulation of ice near the equator and spreading toward, but never reaching, the poles because of elevated temperatures at high latitudes during parts of the year when the polar regions received continuous solar radiation.

The Ediacaran Period was also a time of sporadic glaciations but none seems to have been of magnitude comparable to those of the previous period. The Earth was also subjected to the large Acraman impact in this period and (more speculatively) a second large marine impact, that may have caused the world's deepest negative $\partial^{13}C$ excursion, the Shuram anomaly. The Ediacaran Period ended with terminal Precambrian glaciations, extremely elevated $^{87}Sr/^{86}Sr$ ratios in carbonate rocks, and the Cambrian explosion of animals that include most extant phyla. As more detailed investigations of the biota of the Ediacaran Period emerge, it appears that this very turbulent period of Earth history, involving multiple glaciations and possibly more than one large impact event, may have caused a series of extinction–radiation cycles that culminated in the emergence of the shelly fauna that gives the false impression of extremely rapid evolution of a myriad of complex organisms. The latest Ediacaran and all subsequent glacial deposits have yielded paleomagnetic evidence of circumpolar distribution.

Acknowledgments

I would like to thank the Natural Sciences and Engineering Research Council of Canada (NSERC) for supporting studies of ancient glacial deposits for many years. I also acknowledge discussions, help, and encouragement from colleagues and graduate students at the University of Western Ontario and in particular from H.W. Nesbitt and C.M. Fedo, who provided valuable insights into the use of geochemistry in investigation and interpretation of Huronian rocks and modern analogues. I am grateful for thoughtful and careful reviews by R. W. Ojakangas and G. E. Williams.

References

Agassiz, L., 1842. The glacial theory and its recent progress. New Edinburgh Philosophical Journal 33, 217–263.

Allen, P.A., Hoffman, P.F., 2005. Extreme winds and waves in the aftermath of a Neoproterozoic glaciation. Nature 433, 123–127.

Arnaud, E., Halverson, G.P., Shields-Zhou, G. (Eds.), 2011. The Geological Record of Neoproterozoic Glaciations. Geological Society, London, Memoir, vol. 36, 752 p.

Arnaud, E., 2004. Giant cross-beds in the Neoproterozoic Port Askaig formation, Scotland: implications for snowball earth. Sedimentary Geology 165, 155–174.

Aspler, L.B., Wisotzek, I.E., Chiarenzelli, J.R., Losonczy, M.F., Cousens, B.L., McNicoll, V.J., Davis, W.J., 2001. Paleoproterozoic intracratonic basin processes, from breakup of Kenorland to assembly of Laurentia: Hurwitz basin, Nunavut, Canada. Sedimentary Geology 141/142, 287–318.

Bahlburg, H., Dobrzinski, N., 2011. A review of the Chemical Index of Alteration (CIA) and its application to the study of Neoproterozoic glacial deposits and climate transitions. In: Arnaud, E., Halverson, G.P., Shields-Zhou, G. (Eds.), The Geological Record of Neoproterozoic Glaciations, Geological Society, London, Memoir, vol. 36, pp. 81–92.

Bekker, A., Kaufman, A.J., Karhu, J.A., Beukes, N.J., Swart, Q.D., Coetzee, L.L., Eriksson, K.A., 2001. Chemostratigraphy of the proterozoic Duitschland Formation, South Africa: implications for coupled climate change and carbon cycling. American Journal of Science 301, 261–285.

Bekker, A., Kaufman, A.J., Karhu, J.A., Eriksson, K.A., 2005. Evidence for proterozoic cap carbonates in North America. Precambrian Research 137, 167–206.

Bennett, G., Dressler, B.O., Robertson, J.A., 1991. The Huronian Supergroup and associated intrusive rocks. In: Thurston, P.C., Williams, H.R., Sutcliffe, R.H., Scott, G.M. (Eds.), Geology of Ontario, Part 1. Ontario Geological Survey, Toronto, pp. 549–591.

Bertrand-Sarfati, J., Moussine-Pouchkine, A., Amard, B., Ait Kaci Ahmed, A., 1995. First Ediacaran fauna found in Western Africa and evidence for an early Cambrian glaciation. Geology 23, 133–136.

Beukes, N.J., Frei, D., Vorster, C., 2013. Detrital Zircon Age Constraint of the Makganyene Snowball Earth Event (South Africa) at 2.25-2.22 Ga: The Youngest of Four Huronian Glaciations? [abst.] Geological Society of America, p. 340. Abstracts with Programs 45, No 7.

Bleeker, W., 2004. Taking the pulse of planet earth: a proposal for a new multi-disciplinary Flagship Project in Canadian solid earth sciences. Geoscience Canada 31, 179–190.

Bradley, D.C., 2011. Secular trends in the geologic record and the supercontinent cycle. Earth-Science Reviews 108, 16–33.

Budyko, M.I., 1969. The effect of solar radiation variations on the climate of the Earth. Tellus 21, 611–619.

Busfield, M.E., Le Heron, D.P., 2014. Sequencing the Sturtian icehouse: dynamic ice behaviour in South Australia. Journal of the Geological Society 171, 443–456.

Caby, R., Fabre, J., 1981. Late Proterozoic to early Palaeozoic diamictites, tillites and associated glacigenic sediments in the Série Pourprée of western Hoggar, Algeria. In: Harland, W.B., Hambrey, M.J. (Eds.), Earth's Pre-Pleistocene Glacial Record. Cambridge University Press, pp. 140–145.

Card, K.D., 1978. Geology of the Sudbury-Manitoulin Area, Districts of Sudbury and Manitoulin. Ontario Geological Survey Report, vol. 166, 239 p.

Coleman, A.P., 1907. A lower Huronian ice age. American Journal of Science 23, 187–192.

Coleman, A.P., 1908. The lower Huronian ice age. Journal of Geology 16, 149–158.

Condie, K.C., 1998. Episodic continental growth and supercontinents: a mantle avalanche connection? Earth and Planetary Science Letters 163, 97–108.

Corfu, F., Andrews, A.J., 1986. A U–Pb age for mineralized Nipissing diabase, Gowganda, Ontario. Canadian Journal of Earth Sciences 23, 107–109.

Corsetti, F.A., Olcott, A.N., Bakermans, C., 2006. The biotic response to Neoproterozoic snowball Earth. Palaeogeography, Palaeoclimatology, Palaeoecology 232, 114–130.

Cox, G.M., Halverson, G.P., Minarik, W.G., Le Heron, D.P., Macdonald, F.A., Bellefroid, E.J., Strauss, J.V., 2013. Neoproterozoic iron formation: an evaluation of its temporal, environmental and tectonic significance. Chemical Geology 362, 232–249.

Crowell, J.C., 1983. Ice ages recorded on Gondwanan continents. Transactions of the Geological Society of South Africa 86, 237–262.

De Geer, G., 1911. A geochronology of the last 12,000 years. In: Proceedings of the International Geological Congress Stockholm (1910), vol. 1, pp. 241–257.

Deynoux, M., Affaton, P., Trompette, R., Villeneuve, M., 2006. Pan-African tectonic evolution and glacial events registered in Neoproterozoic to Cambrian cratonic and foreland basins of West Africa. Journal of African Earth Sciences 46, 397–426.

Deynoux, M., 1982. Periglacial polygonal structures and sand wedges in the late Precambrian glacial formations of the Taoudeni Basin in Adrar of Mauritania (West Africa). Palaeogeography, Palaeoclimatology, Palaeoecology 39, 55–70.

Dietz, R.S., 1964. Sudbury structure as an astrobleme. The Journal of Geology 73, 412–421.

Du Toit, A.L., 1954. The Geology of South Africa. Oliver and Boyd, London, 611 p.

Embleton, B.J.J., Williams, G.E., 1986. Low latitude of deposition for late Precambrian periglacial varvites in South Australia: implications for palaeoclimatology. Earth and Planetary Science Letters 79, 419–430.

Eriksson, P.H., Altermann, W., 2013. Paleoproterozoic glacial deposits of South Africa (Chapter 7.2.2). In: Melezhik, V.A., Kump, L.R., Fallick, A.E., Strauss, H., Hanski, E.J., Prave, A., Lepland, A. (Eds.), Reading the Archive of Earth's Oxygenation, Global Events and the Fennoscandian Arctic Russia – Drilling Early Earth Project, vol. 3. Springer, Berlin, pp. 1083–1096.

Ernst, R.E., Bleeker, W., 2010. Large igneous provinces (LIPs), giant dike swarms, and mantle plumes: significance for breakup events within Canada and adjacent regions from 2.5 Ga to the present. Canadian Journal of Earth Sciences 47, 695–739.

Evans, D.A.D., Raub, T.D., 2011. Neoproterozoic glacial palaeolatitudes: a global update. In: Arnaud, E., Halverson, G.P., Shields-Zhou, G. (Eds.), The Geological Record of Neoproterozoic Glaciations, Geological Society, London, Memoir, vol. 36, pp. 93–112.

Evans, D.A.D., 2003. A fundamental Precambrian–Phanerozoic shift in earth's glacial style? Tectonophysics 375, 353–385.

Evans, D.A.D., 2006. Proterozoic low orbital obliquity and axial-dipolar geomagnetic field from evaporite palaeolatitudes. Nature 444, 51–55.

Eyles, N., Januszczak, N., 2004. 'Zipper-rift': a tectonic model for Neoproterozoic glaciations during the breakup of Rodinia after 750 Ma. Earth-Science Reviews 65, 1–73.

Frarey, M.J., 1977. Geology of the Huronian belt between Sault Ste. Marie and Blind river, Ontario. Geological Survey of Canada Memoir 383, 87.

Gärtner, C., Bahlburg, H., Melezhik, V.A., Berndt, J., 2014. Dating Paleoproterozoic glacial deposits of the Fennoscandian shield using detrital zircons from the Kola Peninsula, Russia. Precambrian Research 246, 281–295.

Gay, A.I., Grandstaffe, D.E., 1980. Chemistry and mineralogy of Precambrian paleosols at Elliot Lake, Ontario, Canada. Precambrian Research 12, 349–373.

Geboy, N.J., Kaufman, A.J., Walker, R.J., Misi, A., Oliviera, T.F., Miller, K.E., Azmy, K., Kendall, B., Poulton, S.W., 2013. Re–Os age constraints and new observations of Proterozoic glacial deposits in the Vazante Group, Brazil. Precambrian Research 238, 200–213.

Grey, K., Walter, M.R., Calver, C.R., 2003. Neoproterozoic biotic diversification: snowball Earth or aftermath of the Acraman impact? Geology 31, 459–462.

Hambrey, M.J., Harland, W.B., 1981. Earth's Pre-Pleistocene Glacial Record. Cambridge University Press, Cambridge, 1003 p.

Harland, W.B., 1964. Critical evidence for a great infra-Cambrian glaciation. Geologische Rundschau 54, 45–61.

Hilburn, I.A., Joseph L. Kirschvink, J.L., Tajikab, E., Tadab, R., Hamanob, Y., Yamamotob, S., 2005. A negative fold test on the Lorrain Formation of the Huronian Supergroup: uncertainty on the latitude of the Proterozoic Gowganda glaciation and implications for the great oxygenation event. Earth and Planetary Science Letters 232, 315–332.

Hoffman, P.F., Kaufman, A.J., Halverson, G.P., Schrag, D.P., 1998. A Neoproterozoic snowball earth. Science 281, 1342–1346.

Hoffman, P.F., 2011a. A history of Neoproterozoic glacial geology, 1871–1997. In: Arnaud, E., Halverson, G.P., Shields-Zhou, G. (Eds.), The Geological Record of Neoproterozoic Glaciations, Geological Society, London, Memoir, vol. 36, pp. 17–37.

Hoffman, P.F., 2011b. Strange bedfellows: glacial diamictite and cap carbonate from the Marinioan (635 Ma) glaciations in Namibia. Sedimentology 58, 57–119.

Hoffman, P.F., 2013. The Great Oxidation and a Siderian snowball Earth: MIF-S based correlation of Paleoproterozoic glacial epochs. Chemical Geology 362, 143–156. http://dx.doi.org/10.1016/j.chemgeo.2013.04.018.

Holland, H.D., 2006. The oxygenation of the atmosphere and oceans. Philosophical Transactions of the Royal Society B361, 903–915.

Karlstrom, K.E., Houston, R.S., 1984. The Cheyenne belt: analysis of a Proterozoic suture in southern Wyoming. Precambrian Research 25, 415–446.

Ketchum, K.Y., Heaman, L.M., Bennett, G., Hughes, D.J., 2013. Age, petrogenesis and tectonic setting of the Thessalon volcanic rocks, Huronian Supergroup, Canada. Precambrian Research 233, 144–172. http://dx.doi.org/10.1016/j.precamres.2013.04.009.

Kirschvink, J.L., Ripperdan, R.L., Evans, D.A.D., 1997. Evidence for a large-scale reorganization of Early Cambrian continental masses by inertial interchange true polar wander. Science 277, 541–545.

Kirschvink, J.L., Gaidos, E.J., Bertani, L.E., Beukes, N.J., Gutzmer, J., Maepa, J.N., Steinberger, R.E., 2000. Paleoproterozoic snowball Earth: extreme climatic and geochemical global change and its biological consequences. Proceedings of the National Academy of Science of the United States of America 97 (4), 1400–1405.

Kirschvink, J.L., Mound, J., Mitrovitca, J.X., Raub, T.D., Evans, D.A.D., Kopp, R.E., November 2005. Rapid inertial-interchange true polar wander during Ediacaran time induced by intrusion of the Sept-Îles Intrusive Suite, Quebec, Canada: a possible trigger for massive sea-level and carbon isotopic excursions. In: Proceedings of the Gondwana XII Conference, Mendoza, Argentina. http://cig.museo.unlp.edu.ar/gondwana/.

Kirschvink, J.L., 1992. Late Proterozoic low-latitude glaciation: the snowball Earth. In: Schopf, J.W., Klein, C. (Eds.), The Proterozoic Biosphere. Cambridge University Press, Cambridge, pp. 51–52.

Kopp, R.E., Kirschvink, J.L., Hilburn, I.A., Nash, C.D., 2005. The Paleoproterozoic snowball Earth: a climate disaster triggered by the evolution of oxygenic photosynthesis. Proceedings of the National Academy of Science of the United States of America 102 (32), 11,131–11,136.

Krapez, B., 1996. Sequence stratigraphic concepts applied to the identification of basin-filling rhythms in Precambrian successions. Australian Journal of Earth Sciences 43, 355–380.

Krogh, T.E., Davis, D.W., Corfu, F., 1984. Precise U-Pb zircon and baddeleyîte ages for the Sudbury Structure. In: Pye, E.G., Naldrett, A.J., Giblin, P.E. (Eds.), Geology and Ore Deposits of the Sudbury Structure, Ontario Geological Survey, Special volume 1, pp. 431–446.

Kuipers, G., Beunk, F.F., van der Wateren, F.M., 2013. Periglacial evidence for a 1.91–1.89 Ga old glacial period at low latitude, Central Sweden. Geology Today 29 (6), 218–221.

Lan, Z., Li, X.-H., Zhang, Q., Li, Q.-L., 2015. Global synchronous initiation of the 2nd episode of Sturtian glaciation: SIMS zircon U–Pb and O isotope evidence from the Jiangkou Group, South China. Precambrian Research 267, 28–38.

Leather, J., Allen, P.A., Brasier, M.D., Cozzi, A., 2002. Neoproterozoic snowball Earth under scrutiny: evidence from the Fiq glaciations of Oman. Geology 30, 891–894.

Lepland, A., 2013. Paleoproterozoic glacial deposits of Australia (Chapter 7.2.4). In: Melezhik, V.A., Kump, L.R., Fallick, A.E., Strauss, H., Hanski, E.J., Prave, R., Lepland, A. (Eds.), Reading the Archive of Earth's Oxygenation, Global Events and the Fennoscandian Arctic Russia – Drilling Early Earth Project, vol. 3. Springer, Berlin, pp. 1090–1091.

Lindsey, D.A., 1971. Glacial marine sediments in the Precambrian Gowganda Formation at Whitefish falls, Ontario (Canada). Palaeogeography, Palaeoclimatology, Palaeoecology 9, 7–25.

Lister, G.S., Etheridge, M.A., Symonds, P.A., 1986. Detachment faulting and the evolution of passive continental margins. Geology 14, 246–250.

Long, D.G.F., 2004. The tectonostratigraphic evolution of the Huronian basement and subsequent basin fill: geological constraints on impact models of the Sudbury event. Precambrian Research 129, 203–223.

Marmo, J.S., Ojakangas, R.W., 1984. Lower Proterozoic glaciogenic deposits, eastern Finland. Geological Society of America Bulletin 98, 1055–1062.

Martin, D.M.B., 1999. Depositional setting and implications of Paleoproterozoic glaciomarine sedimentation in the Hamersley province, western Australia. Geological Society of America Bulletin 111, 189–203.

Mawson, D., 1949. The Late Precambrian ice age and glacial record of the Bibliando dome. Journal and Proceedings of the Royal Society of New South Wales 82, 150–174.

Mazumder, R., Van Kranendonk, M.J., Altermann, W., 2015. A marine to fluvial transition in the Paleoproterozoic Koolbye formation, turee Creek Group, western Australia. Precambrian Research 258, 161–170.

McFadden, K.A., Huang, J., Chu, X., Jiang, G., Kaufman, A.J., Zhou, C., Yua, X., Shuhai Xiao, S., 2008. Pulsed oxidation and biological evolution in the Ediacaran Doushantuo formation. Proceedings of the National Academy of Sciences of the United States of America 105, 3197–3202.

McLaren, D.J., 1983. Bolides and biostratigraphy: address as Retiring President of the Geological Society of America, October 1982. Geological Society of America Bulletin 94, 313–324.

Melezhik, V.A., Young, G.M., Eriksson, P.G., Altermann, W., Kump, L.R., Lepland, A., 2013. Huronian-age glaciation (Chapter 7.2). In: Melezhik, V.A., Kump, L.R., Fallick, A.E., Strauss, H., Hanski, E.J., Prave, R., Lepland, A. (Eds.), Reading the Archive of Earth's Oxygenation, Global Events and the Fennoscandian Arctic Russia – Drilling Early Earth Project, vol. 3. Springer, Berlin, pp. 1059–1109.

Moczydlowska, M., 2008. The Ediacaran microbiota and the survival of snowball Earth conditions. Precambrian Research 167, 1–15.

Mohanty, S., 2006. Evidence of volcanism and glaciation from the Sausar Group, Central India. Journal of the Geological Society of India 68, 764–768.

Nance, R.D., Murphy, B., Santosh, M., 2013. The supercontinent cycle: a retrospective essay. Gondwana Research 24, 4–29.

Nesbitt, H.W., Young, G.M., 1982. Early Proterozoic climates and plate motions inferred from major element chemistry of lutites. Nature 299, 715–717.

Nhleko, N., 2004. The Pongola Supergroup in Swaziland (D.Phil. thesis). Rand Afrikaans University, Johannesburg, 285 p.

Ojakangas, R.W., Srinivasan, R., Hegde, V.S., Chandrakant, S.M., Srikantia, S.V., 2014. The Talya conglomerate: an Archaean (~2.7 Ga) glaciomarine formation, western Dharwar Craton, southern India. Current Science 106, 387–396.

Ojakangas, R.W., 1985. Early Proterozoic glaciation: the dropstone unit-diamictite association. In: Laajoki, K., Paakkola, J. (Eds.), Proterozoic Exogenic Processes and Related Metallogeny, Geological Survey of Finland Bulletin, vol. 331, pp. 51–72.

Ojakangas, R.W., 1988. Glaciation: an uncommon mega-event as a key to intracontinental and intercontinental correlation of early Proterozoic basin fill, North American and Baltic cratons. In: Kleinspehn, K.L., Paola, C. (Eds.), New Perspectives in Basin Analysis. Springer, Berlin, pp. 431–444.

Ovenshine, A.T., 1970. Observations of iceberg rafting in Glacier Bay, Alaska, and the identification of ancient ice-rafted deposits. Geological Society of America Bulletin 81, 891–894.

Page, N.J., 1981. The Precambrian diamictite below the base of the Stillwater Complex, Montana. In: Hambrey, M.J., Harland, W.B. (Eds.), Earth's Pre-Pleistocene Glacial Record. Cambridge University Press, Cambridge, pp. 821–823.

Pettijohn, F.J., Bastron, H., 1959. Chemical composition of argillites of the Cobalt Series (Precambrian) and the problem of soda-rich sediments. Geological Society of America Bulletin 70, 593–600.

Pettijohn, F.J., 1943. Basal Huronian conglomerates of Menominee and Calumet districts, Michigan. Journal of Geology 51, 387–397.

Preiss, W.V., 2000. The Adelaide Geosyncline of south Australia and its significance in Neoproterozoic continental reconstruction. Precambrian Research 100, 21–36.

Rainbird, R.H., Nesbitt, H.W., Donaldson, J.A., 1990. Formation and diagenesis of a sub-Huronian saprolith, comparison with a modern soil profile. Journal of Geology 98, 801–822.

Rasmussen, B., Bekker, A., Fletcher, I.R., 2013. Correlation of proterozoic glaciations based on U–Pb zircon ages for tuff beds in the Transvaal and Huronian Supergroups. Earth and Planetary Research Letters 382, 173–180.

Raub, T.D., Kirschvink, J.L., 2008. A pan-Precambrian link between deglaciation and environmental oxidation. In: Cooper, A.K., Barrett, P.J., Stagg, B., Storey, B., Stump, E., Wise, W., the 10th ISAES editorial team (Eds.), Antarctica: A Keystone in a Changing World. Proceedings of the 10th International Symposium on Antarctic Earth Sciences. National Academies Press, Washington, DC, USA.

Roberts, J.D., 1976. Late Precambrian dolomites, Vendian glaciation, and synchroneity of Vendian glaciations. Journal of Geology 84, 47–63.

Roberts, N.M.W., 2013. The boring billion? Lid tectonics, continental growth and environmental change associated with the Columbia supercontinent. Geoscience Frontiers 4, 681–691.

Robertson, J.A., Frarey, M.J., Card, K.D., 1969. The Federal-Provincial Committee on Huronian Stratigraphy: progress report. Canadian Journal of Earth Sciences 6, 335–336.

Roscoe, S.M., Card, K.D., 1993. The reappearance of the Huronian in Wyoming: rifting and drifting of ancient continents. Canadian Journal of Earth Sciences 30, 2475–2480.

Roscoe, S.M., 1969. Huronian Rocks and Uraniferous Conglomerates in the Canadian Shield. Geological Survey of Canada. Paper 68–40, 205 p.

Roscoe, S.M., 1973. The Huronian Supergroup, an Aphebian succession showing evidence of atmospheric evolution. In: Young, G.M. (Ed.), Huronian Stratigraphy and Sedimentation. The Geological Association of Canada, pp. 31–47. Special Paper 12.

Sagan, C., Mullen, G., 1972. Earth and Mars, evolution of atmospheres and surface temperatures. Science 177, 52–56.

Santosh, M., Maruyama, S., Sawaki, Y., Meert, J.G., 2014. The Cambrian explosion: Plume-driven birth of the second ecosystem on earth. Gondwana Research 25, 945–965.

Schmidt, P.W., Williams, G.E., 1995. The Neoproterozoic climatic paradox: equatorial palaeolatitude for Marinoan glaciation near sea level in South Australia. Earth and Planetary Science Letters 134, 107–124.
Schmidt, P.W., Williams, G.E., 1999. Paleomagnetism of the Paleoproterozoic hematitic breccia and paleosol at Ville-Marie, Quebec: further evidence for the low latitude of Huronian glaciations. Earth and Planetary Science Letters 172, 273–285.
Schopf, J.W., Klein, C. (Eds.), 1992. The Proterozoic Biosphere. Cambridge University Press, Cambridge, 1353 p.
Shields-Zhou, G.A., Och, L.M., 2011. The case for a Neoproterozoic Oxygenation Event: geochemical evidence and biological consequences. GSA Today 21, 4–11.
Spencer, A.M., 1971. Late pre-Cambrian glaciation in Scotland. Geological Society of London Memoir 6, 99.
Tang, H., Chen, Y., 2013. Global glaciations and atmospheric change at ca. 2.3 Ga. Geoscience Frontiers 4, 583–596. http://dx.doi.org/10.1016/j.gsf.2013.02.003. 374.
Thomson, J., 1871. On the Stratified Rocks of Islay. Report of the 41st Meeting of the British Association for the Advancement of Science. John Murray, London, Edinburgh, pp. 110–111.
Trendall, A.F., 1976. Striated and Faceted Boulders from the Turee Creek Formation – Evidence for a Possible Huronian Glaciation on the Australian Continent. Geological Survey of Western Australia, Annual Report, 1975, pp. 88–92.
Trendall, A.F., 1979. A Revision of the Mount Bruce Supergroup. Geological Survey of Western Australia, Annual Report, 1978, pp. 63–71.
Vallini, D.A., Cannon, W.F., Schulz, K.J., 2006. Age constraints for Paleoproterozoic glaciation in the lake superior region: detrital zircon and hydrothermal xenotime ages for the Chocolay Group, Marquette Range Supergroup. Canadian Journal of Earth Sciences 43, 571–591.
Van Kranendonk, M.J., Mazumder, R., 2015. Two glacio-eustatic cycles in the Paleoproterozoic Turee Creek Group, western Australia. Geological Society of America Bulletin 127, 596–607. http://dx.doi.org/10.1130/B31025.1.
Vernheta, E., Youbia, N., Chellaib, E.H., Villeneuvee, M., El Archif, A., 2012. The Bou-Azzer glaciation: evidence for an Ediacaran glaciation on the West African Craton (Anti-Atlas, Morocco). Precambrian Research 196/197, 106–112.
Visser, J.N.J., Young, G.M., 1990. Major element geochemistry and climatology of the Permo-Carboniferous glacigene Dwyka Formation and post-glacial mudrocks in Southern Africa. Palaeogeography, Palaeoclimatology, Palaeoecology 81, 49–57.
Visser, J.N.J., 1971. The deposition of the Griquatown glacial member in the Transvaal Supergroup. Transactions of the Geological Society of South Africa 74, 187–199.
Von Brunn, V., Gold, D.J.C., 1993. Diamictite in the Archaean Pongola sequence of southern Africa. Journal of African Earth Sciences 16, 367–374.
Walker, J.C.G., Hays, P.B., Kasting, J.F., 1981. A negative feedback mechanism for the long-term stabilization of Earth's surface temperature. Journal of Geophysical Research 86 (C10), 9776–9782.
Wegener, A., 1966. The Origin of Continents and Oceans (translation from German version of 1929). Dover Publications, New York, 246 p.
Wegener, A., 2001. The origins of the continents (translation from German version of 1912). Journal of Geodynamics 32, 29–63.
Williams, G.E., Schmidt, P.W., 1997. Paleomagnetism of Paleoproterozoic Gowganda and Lorrain Formations, Ontario: low latitude for Huronian glaciation. Earth and Planetary Science Letters 153, 157–169.
Williams, G.E., 1975. Late Precambrian glacial climate and the Earth's obliquity. Geological Magazine 112, 441–465.
Williams, G.E., 2005. Subglacial meltwater channels and glaciofluvial deposits in the Kimberley Basin, Western Australia: 1.8 Ga low-latitude glaciations coeval with continental assembly. Journal of the Geological Society, London 162, 111–124.
Williams, G.E., 2008. Proterozoic (pre-Ediacaran) glaciation and the high obliquity, low-latitude ice, strong seasonality (HOLIST) hypothesis: principles and tests. Earth-Science Reviews 87, 61–93.
Xiao, S., 2004. Neoproterozoic glaciations and the fossil record. In: Jenkins, G.S., McMenamin, M.A.S., McKey, C.P., Sohl, L. (Eds.), The Extreme Proterozoic: Geology, Geochemistry, and Climate, Geophysical Monograph, vol. 146. American Geophysical Union, Washington, DC, pp. 199–214.
Yeo, G.M., 1981. The late Proterozoic Rapitan glaciation in the northern Cordillera. In: Campbell, F.H.A. (Ed.), Proterozoic Basins of Canada. Geological Survey of Canada, pp. 25–46. Paper 81–10.

Young, G.M., Gostin, V.A., 1990. Sturtian glacial deposition in the vicinity of the Yankaninna anticline, North Flinders basin, south Australia. Australian Journal of Earth Sciences 37, 447–458.

Young, G.M., Gostin, V.A., 1991. Late Proterozoic (Sturtian) succession of the North Flinders Basin, south Australia; an example of temperate glaciation in an active rift setting. In: Anderson, J.R., Ashley, G.M. (Eds.), Glacial Marine Sedimentation: Palaeoclimatic Significance. Geological Society of America, pp. 207–222. Special Paper 261.

Young, G.M., Nesbitt, H.W., 1985. The lower Gowganda Formation in the southern part of the Huronian outcrop belt, Ontario, Canada: stratigraphy, depositional environments and tectonic setting. Precambrian Research 29, 265–301.

Young, G.M., Nesbitt, H.W., 1999. Palaeoclimatology and provenance of the glaciogenic Gowganda Formation (proterozoic), Ontario, Canada: a chemostratigraphic approach. Geological Society of America Bulletin 111, 264–274.

Young, G.M., von Brunn, V., Gold, D.J.C., Minter, W.E.L., 1998. Earth's oldest reported glaciation: physical and chemical evidence from the Archaean Mozaan Group (~2.9 Ga) of South Africa. The Journal of Geology 106, 523–538.

Young, G.M., Minter, W.E.L., Theron, J.N., 2004. Geochemistry and palaeogeography of late Ordovician glaciogenic sedimentary rocks in the Table Mountain Group, South Africa. Palaeogeography, Palaeoclimatology, Palaeoecology 214, 323–345.

Young, G.M., 1969. Geochemistry of early Proterozoic tillites and argillites of the Gowganda Formation, Ontario, Canada. Geochimica et Cosmochimica Acta 33, 483–492.

Young, G.M., 1970. An extensive early Proterozoic glaciation in North America. Palaeogeography, Palaeoclimatology, Palaeoecology 7, 85–100.

Young, G.M., 1973. Tillites and aluminous quartzites as possible time markers for Middle Precambrian (Aphebian) rocks of North America. In: Young, G.M. (Ed.), Huronian Stratigraphy and Sedimentation. The Geological Association of Canada, pp. 97–127. Special Paper 12.

Young, G.M., 1983. Tectono-sedimentary history of early Proterozoic rocks of the northern Great Lakes region. In: Medaris Jr., L.G. (Ed.), Early Proterozoic Geology of the Great Lakes Region, Geological Society of America Memoir, vol. 160, pp. 15–32.

Young, G.M., 2001. Comparative geochemistry of Pleistocene and Paleoproterozoic (Huronian) glaciogenic laminated deposits: relevance to crustal and atmospheric composition. The Journal of Geology 109, 463–477.

Young, G.M., 2013a. Precambrian supercontinents, glaciations, atmospheric oxygenation, metazoan evolution and an impact that may have changed the second half of Earth history. Geoscience Frontiers 4, 247–261.

Young, G.M., 2013b. Paleoproterozoic glacial deposits of North America (Chapter 7.2.2). In: Melezhik, V.A., Kump, L.R., Fallick, A.E., Strauss, H., Hanski, E.J., Prave, A., Lepland, A. (Eds.), Reading the Archive of Earth's Oxygenation, Global Events and the Fennoscandian Arctic Russia – Drilling Early Earth Project, vol. 3. Springer, Berlin, pp. 1067–1082.

Young, G.M., 2013c. Evolution of Earth's climatic system: evidence from ice ages, isotopes and impacts. GSA Today 23, 4–10.

Young, G.M., 2014. Contradictory correlations of Paleoproterozoic glacial deposits: local, regional or global controls? Precambrian Research 247, 33–44.

Young, G.M., 2015. Did prolonged two-stage fragmentation of the supercontinent Kenorland lead to arrested orogenesis on the southern margin of the superior province? Geoscience Frontiers 6, 419–435. http://dx.doi.org/10.1016/j.gsf.2014.04.003.

CHAPTER

20

The Isua Supracrustal Belt of the North Atlantic Craton (Greenland): Spotlight on Sedimentary Systems With the Oldest Preserved Sedimentary Structures (~3.7, ~3.75, and ~3.8 Ga)

A.P. Nutman[1,2], V.C. Bennett[3], C.R.L. Friend[4], A.R. Chivas[1]

[1]University of Wollongong, Wollongong, NSW, Australia; [2]Chinese Academy of Geological Sciences, Beijing, China; [3]The Australian National University, Canberra, ACT, Australia; [4]Glendale, Oxon, United Kingdom

OUTLINE

Sediment Provenance
http://dx.doi.org/10.1016/B978-0-12-803386-9.00020-4

1. INTRODUCTION

The Isua supracrustal belt of southern West Greenland (Fig. 20.1) is the world's only locale in Eoarchean rocks where there are lacunae of sufficiently low strain and low metamorphic grade (i.e., below the migmatization level), such that any details of the origin and evolution of the most ancient sedimentary sequences can be established. As such, across the past five decades, Isua has been a key locality to understand the earliest preserved surficial environments, sediment provenance, the early atmosphere-hydrosphere system, and the interaction of this system with the lithosphere. Starting from Keto and Kirki (1967), similarities between the Isua supracrustal belt rocks and ophiolitic assemblages have been noted, which is in accord with the suprasubduction setting interpretations proposed in this millennium (e.g., Polat and Hoffman, 2003; Furnes et al., 2007; Dilek and Polat, 2008; Nutman et al., 2009; Friend and Nutman, 2010). Suprasubduction zone environments are dynamic in terms of both tectonic evolution on a 1-million-year timescale and also the juxtaposition of rocks that formed in deep-to-shallow water and emergent environments (see overview by Cawood et al., 2009). Therefore, despite its age and the rarity of preservation of sedimentary structures due to superimposed ductile deformation (Nutman et al., 1984), Isua affords the unique opportunity to investigate a variety of environments (detrital and chemical sedimentary—shallow-to-deep water) near the start of the rock record (Fig. 20.2).

A major issue in extracting information from Isua rocks is to discriminate protolith features (both sedimentary structures and chemical composition) from those created by strong deformation and metasomatism in superimposed tectonothermal events. Due to this, structures of potential sedimentary origin in Isua rocks have always been hotly debated: sedimentary versus tectonic ? sedimentary versus metasomatic? This contribution explores the extent of our knowledge of Isua sedimentary rocks—partly by review of localities known about since the 1970s–80s, and partly from new information from low-strain domains discovered in the 2000s–10s.

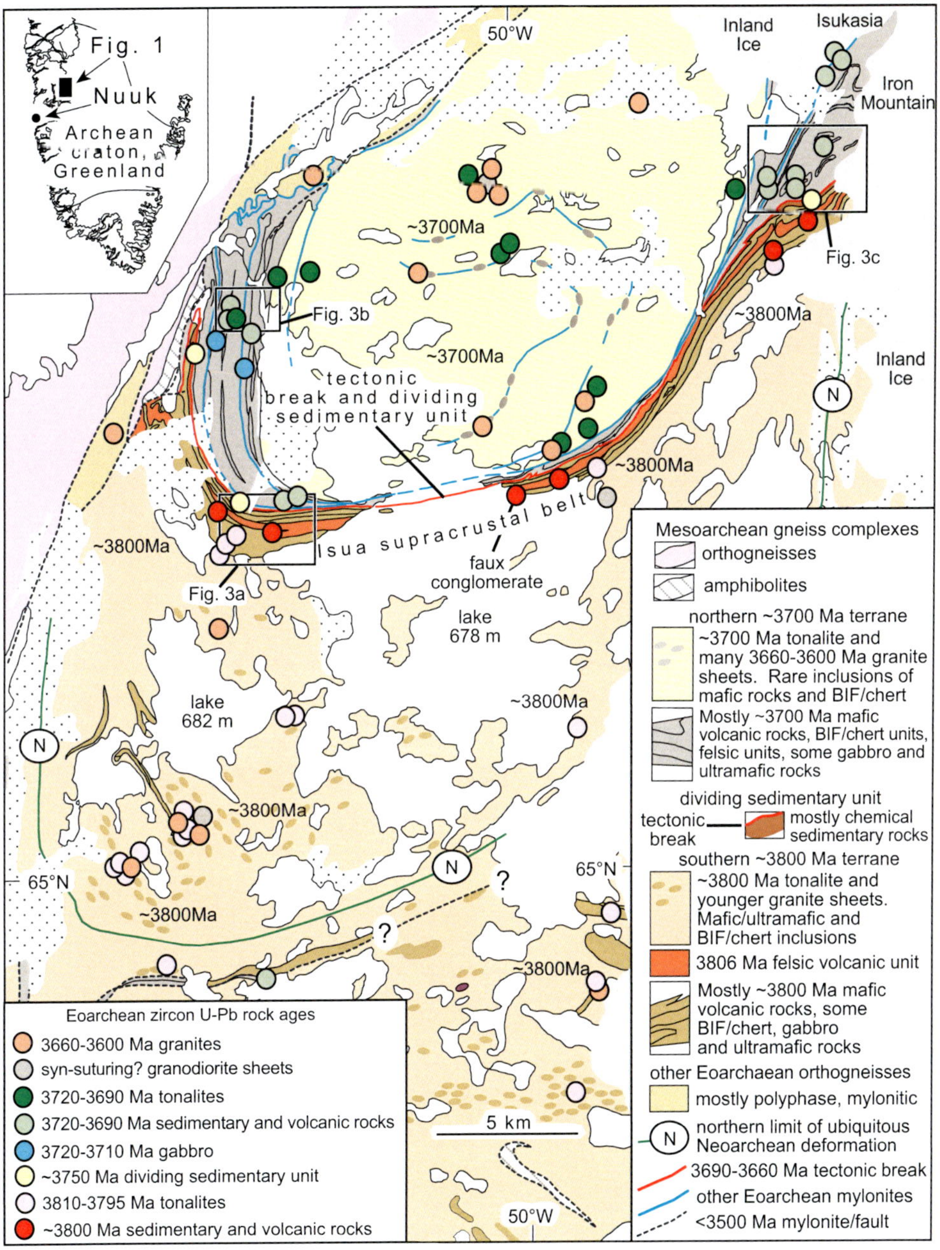

FIGURE 20.1 Summary geological map of the Isua supracrustal belt and adjacent gneisses.

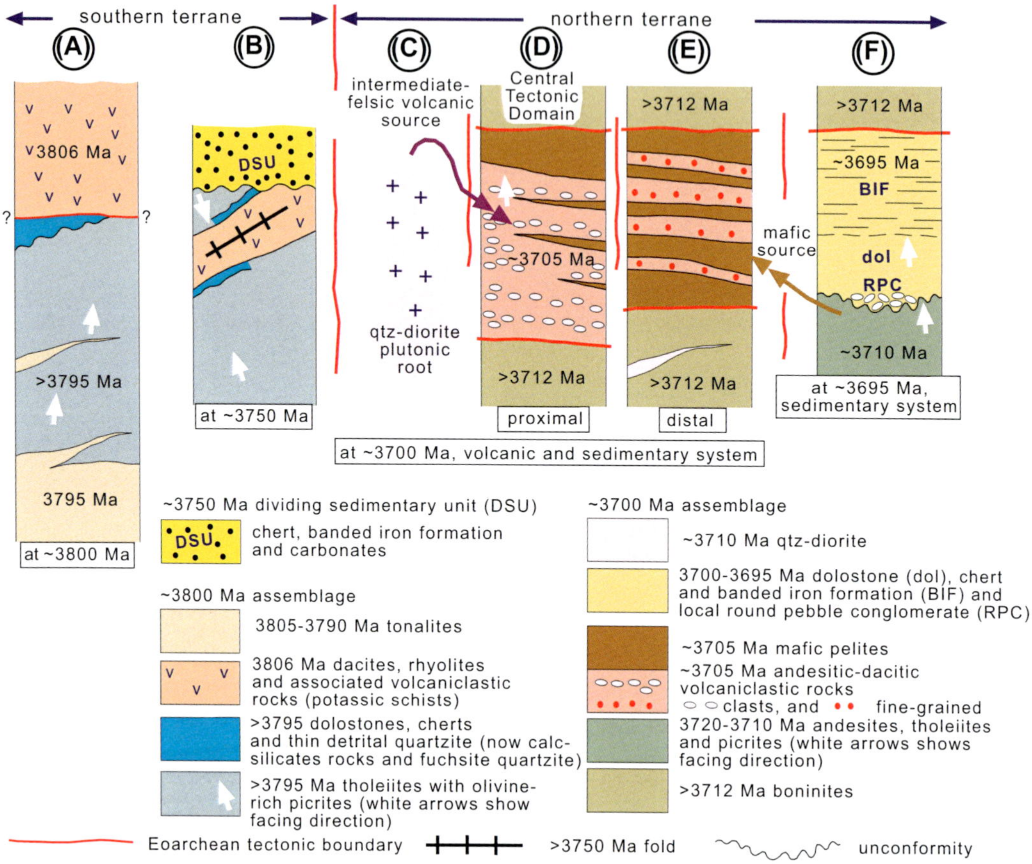

FIGURE 20.2 Reconstructions of different volcanic-sedimentary sequences within the Isua supracrustal belt. (A) ~3800 Ma dolostones and quartzites within a volcanic sequence. (B) The dividing sedimentary unit (DSU) interpreted to lie unconformably on ~3800 Ma rocks, but lie in tectonic contact with ~3700 Ma rocks, (C–E) ~3710 Ma volcanogenic sequence; (C) the magmatic source, (D) proximal facies, and (E) distal facies. (F) 3700–3695 Ma dolostones, cherts, and banded iron formation deposited as a transgressive sequence over ≥3710 Ma volcanic rocks.

2. GEOLOGY OF THE ISUA SUPRACRUSTAL BELT

The 35 km long Isua supracrustal belt is the largest supracrustal rock unit within the Eoarchean ~3000 km^2 Itsaq Gneiss Complex of the North Atlantic Craton in southern West Greenland (Allaart, 1976; Nutman et al., 1996; Nutman and Friend, 2009; Fig. 20.1). The belt is bounded by metaplutonic complexes, with the one to the south dominated by 3820–3795 Ma tonalite protoliths and that to the north by 3710–3685 Ma tonalite protoliths cut by 3650–3640 Ma granite sheets (Nutman et al., 1993, 1996, 1997, 2000; Crowley et al., 2002; Crowley, 2003; Nutman and Friend, 2009). The southern edge of the belt is intruded by ~3800 Ma tonalite sheets, which were emplaced into already deformed supracrustal rocks (Nutman et al., 1996, 1997, 1999; Crowley, 2003). To the west, the belt is in tectonic contact with Mesoarchean rocks (Nutman and Friend, 2009), while eastward it is obscured by the inland ice.

U-Pb zircon dating with uncertainties of ≤ ± 10 Ma demonstrate that the Isua supracrustal belt contains two tectonically juxtaposed slices of unrelated supracrustal rocks ~100 million years different in age (Nutman et al., 1996, 1997, 2009; Crowley et al., 2002; Crowley, 2003; Kamber et al., 2005). The southern portion of the belt consists of a ~3800 Ma assemblage, whereas the northern and median portions of the belt comprise a ~3700 Ma assemblage (Fig. 20.1; Nutman et al., 1996, 1997; Nutman and Friend, 2009, and references therein). The ~3750 Ma dividing sedimentary unit (DSU) of metachert, banded iron formation (BIF), and dolomitic rocks is associated with the ~3800 Ma portion, and lies between the ~3700 and ~3800 portions (Nutman and Friend, 2009; Nutman et al., 2009). The northern contact of the DSU with the ~3700 Ma assemblage is an Eoarchean mylonite. The southern contact of the DSU, although deformed, is not mylonitic in character and is interpreted as a (tectonized) unconformity over ~3800 Ma rocks (Fig. 20.2; Nutman and Friend, 2009; Nutman et al., 2009). Therefore, it has been proposed that the ~3800 terrane with the DSU and the 3700 Ma terrane evolved separately and were tectonically juxtaposed before 3660 Ma—the age of the first igneous intrusions common to both terranes (Nutman et al., 1997, 2002, 2009; Crowley, 2003).

The dominant lithologies within both the ~3700 and ~3800 Ma assemblages are similar, comprising amphibolites derived from basaltic, picritic, and boninitic (~3700 Ma only for the latter) volcanic rocks, felsic volcanic rocks, and felsic volcaniclastic and chemical sedimentary rocks, particularly chert and BIF (e.g., Allaart, 1976; Dymek and Klien, 1988; Nutman et al., 1984, 2007, 2009, 2010, 2013; Komiya et al., 1999; Polat et al., 2002; Polat and Hofmann, 2003; Bolhar et al., 2004, 2005; Furnes et al., 2007; Jenner et al., 2009). Gabbros and ultramafic rocks are volumetrically minor compared to volcanic and sedimentary rocks. The ultramafic rocks were derived from both cumulates in gabbro intrusions (e.g., Dymek et al., 1988) and from tectonic slices of the upper mantle (Friend and Nutman, 2011; Kaczmarek et al., 2016).

Although Neoarchean strain in the Isua supracrustal belt is low to moderate, most of the belt was strongly deformed in the Eoarchean (e.g., Nutman et al., 1984), as demonstrated by using the ~3500 Ma Ameralik dykes (a suite of metamorphosed diabase dykes) as strain markers (e.g., Bridgwater and McGregor, 1974; Nutman, 1986; Nutman et al., 2004). Thus, primary volcanic and sedimentary structures have survived in only a few places, and the compositional layering in the volcanic and sedimentary rocks is mostly of transposed tectonic origin (Nutman et al., 1984, 1996, 2002, 2007; Myers, 2001). The most common relict structures are pillows within mafic volcanic rocks, and more rarely, bedding in BIF and in felsic volcaniclastic rocks (e.g., Nutman et al., 1984, 1997, 2007; Komiya et al., 1999; Rosing, 1999; Solvang, 1999; Furnes et al., 2007). All Isua supracrustal belt rocks have been polymetamorphosed in events from ~3650 to 2690 Ma, with the maximum grade ranging from lower amphibolite (500–550°C) to middle amphibolite facies (up to 650°C) in different mylonite-bounded panels (Rollinson, 2003; Nutman et al., 2013).

3. ~3800 Ma METASEDIMENTARY ROCKS

The southern part of the Isua supracrustal belt is dominated by mafic metavolcanic rocks that flank a unit of potassic metavolcanic felsic schists extending throughout the length of the belt (Fig. 20.1). The ~3800 Ma age of this package is demonstrated by U-Pb zircon ages of

3800–3795 Ma for tonalites that intrude the metavolcanic rocks and 3807–3804 Ma eruptive/depositional ages for the potassic felsic schists (Compston et al., 1986; Nutman et al., 1996, 1997, 2002, 2015; Crowley, 2003; Hiess et al., 2009). The potassic felsic schists have mostly very low Na_2O/K_2O ratios (<0.2) yet high K_2O (up to 11 wt%), which are characters attributed to early alteration, probably weathering (Nutman et al., 1984, 2015). The potassic felsic schists occupy the core of an early isoclinal fold (see Fig. 20.3; Nutman et al., 1984; Nutman and Friend, 2009). This is why locally preserved pillow structures in the amphibolites south of the potassic felsic schists face north, whereas pillows in amphibolites north of the potassic felsic schists face south (Fig. 20.3a). This further indicates that the potassic felsic schist unit originally rests above the amphibolites, and could be stratigraphically younger (Fig. 20.2a; Nutman et al., 2002). The metavolcanic mafic rocks have major and trace element chemistry resembling that of island arc tholeiites (Polat and Hoffman, 2003; Jenner et al., 2009). High-Mg schists form discontinuous units along the belt (Fig. 20.3a) and on most Isua geological maps are shown as ultramafic rocks. Chemically they are neither like upper mantle rocks nor basal parts of tholeiitic layered intrusions. Instead they chemically resemble olivine-rich picrites (Polat and Hofmann, 2003).

3.1 Intrabasalt Cherts

The pillow basalts that dominate the ~3800 Ma sequence contain ferruginous chert layers (Fig. 20.3a). These are <5 m thick with the most continuous followed along strike for ~5 km. However, others extend for <1 km along strike. In all cases they are strongly deformed and devoid of relict sedimentary features. Given the high strain of these units, it is uncertain if they are discontinuous because they mark cryptic layer-parallel tectonic décollements within the volcanic sequence, or if this is an original feature. Regardless, the recurring chert layers are best interpreted as marking hiatii in ~3800 Ma volcanic activity allowing deposition of almost pure chemical sediments.

3.2 Dolostones, Cherts, and Detrital Sandstones

A heterogeneous unit of dolomitic marble, calc-silicate rocks, and rare fuchsitic quartzite crops out between the metavolcanic mafic rocks and the potassic felsic schist unit (Fig. 20.3a). Because of the early isoclinal fold, the dolomite and calc-silicate unit flanks both sides of the schist unit (Fig. 20.3a). The calc-silicates are more continuous throughout the belt, but the fuchsitic quartzite is rarer, and always crops out between calc-silicate rocks and the ~3806 Ma potassic felsic schist unit (Fig. 20.3a). Due to the high abundance of carbonates in the dolomite and calc-silicate unit, it has been strongly modified during several amphibolite facies tectonothermal events (Nutman et al., 1984). Consequently, the main lithology is clinopyroxene ± amphibole calc-silicate ± quartz or carbonate. There is abundant evidence of reaction between CO_2-rich fluids and the adjacent rocks, particularly those of mafic-ultramafic composition. This occurred during metamorphic decarbonatization reactions. This has led some previous investigations to regard these calc-silicate rocks to be entirely intracrustal metasomatic in origin (e.g., Rosing et al., 1996). However, Allaart (1976) reported pure dolomite marble within these calc-silicate rocks (analysis 170728; Table 20.1). During recent fieldwork we have relocated Allaart's dolomite marble locality

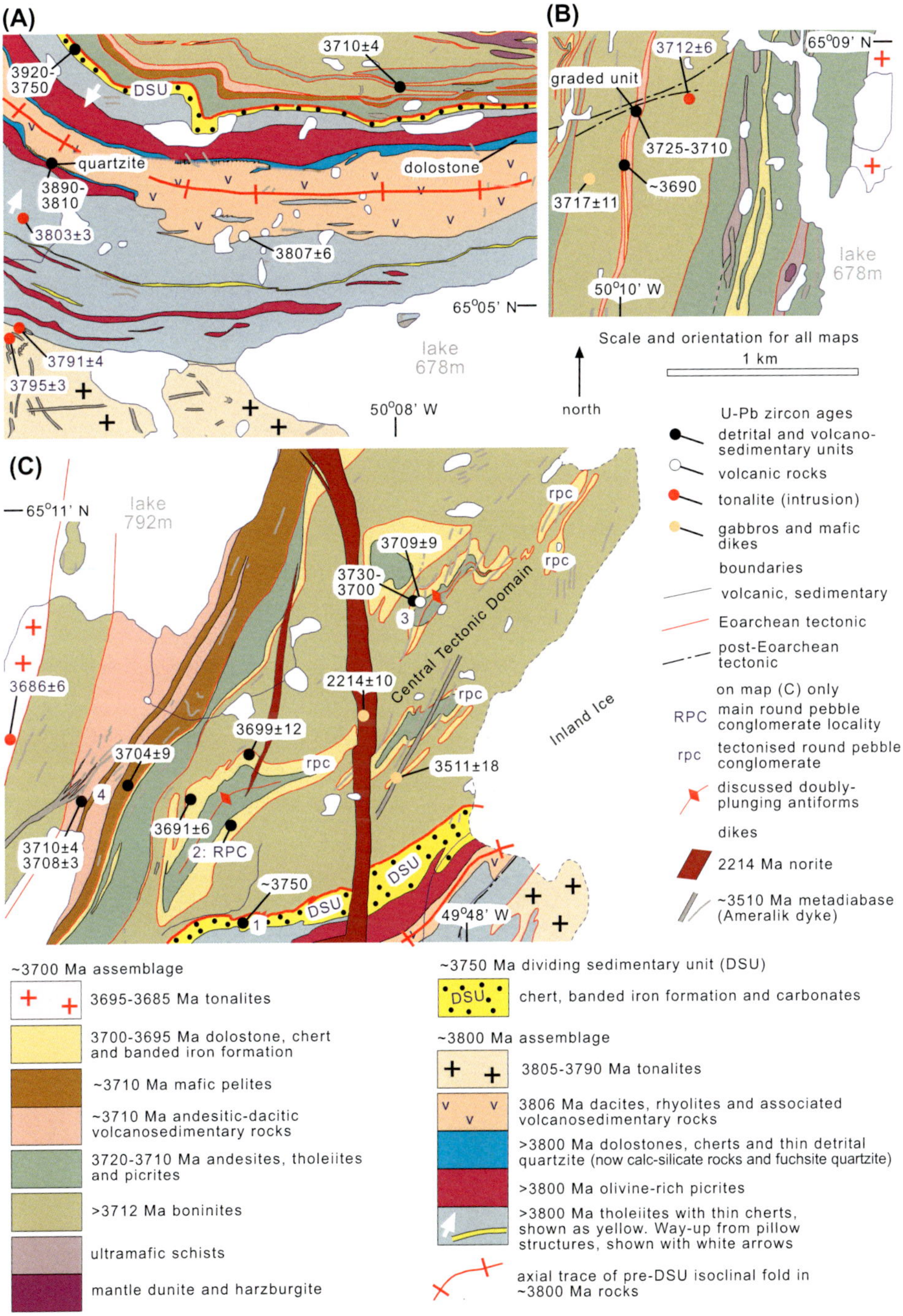

FIGURE 20.3 Detailed geological maps (A, B, C) of parts of the Isua supracrustal belt (see Fig. 20.1 for locations; after maps of Nutman and Friend, 2009).

TABLE 20.1 Representative Whole Rock Geochemistry

	~3800 Ma Terrane				Dividing Sediment Unit			≥3710 Ma Volcanic and Volcano-Sedimentary Rocks							
Sample No	G05/07	465401	JHA170728	248203	G93/28	G07/22	G05/17	G05/32	G05/22	G12/99	G91/73	292,285	G91/76	G05/28	G05/29
Protolith	Picrite	Basalt	Chemical Sediment	Rhyolite	Chemical Sediment	Chemical Sediment	BIF	Basalt (pillow core)	Andesite Weathered	Andesite Weathered	Andesitic Sedimentary Rock	Andesitic Sedimentary Rock	Dacitic Sedimentary Rock	Andesitic Sedimentary Rock	Andesitic Sedimentary Rock
Latitude	g65°09.8′N		g65°05.4′N	m65°05.8′N	m65°05.63′N	65°08.059′N	65°09.8689′N	g65°07.9′N	g65°10.8′	Below RPC	m65°10.3′N	m65°11.9′N	g65°10.2′N	g65°08.9′N	g65°08.9′N
Longitude	g49°48.9′W		g50°06.5′W	m50°00.5′W	m50°10.45′W	50°11.464′W	49°48.936′W	g50°09.8′W	g49°48.2′		m49°50.5′W	m49°48.4′W	g49°50.4′W	g50°10.0′W	g50°10.0′W
Zircon Age (Ma)	~3800	≥3803 ± 3	~3800	3806 ± 2	~3750	~3750	~3750	≥3718	3710 ± 9	~3710	3710 ± 4	3710 ± 4	3710 ± 4	≤3712	≤3712
SiO_2 (wt%)	39.68	53.11	3.19	68.92	71.86	71.28	66.26	55.53	63.04	60.54	59.92	60.04	77.33	59.16	59.46
TiO_2	0.10	0.87	0.02	0.44	0.01	0.01	0.01	0.81	0.84	0.40	0.65	0.59	0.20	0.59	0.54
Al_2O_3	5.97	10.74	0.04	15.30	0.31	0.16	0.19	14.26	13.36	11.50	15.65	14.99	12.50	16.18	15.13
Fe_2O_3	10.28	14.00	0.06	0.69	12.42	20.91	28.41	11.77	7.67	10.79	6.11	13.97	1.89	7.00	7.48
FeO			3.47			n.d.	n.d.								
MnO	0.17	0.42	0.24	0.02	0.42	0.20	0.20	0.22	0.25	0.23	0.12	0.22	0.04	0.37	0.35
MgO	34.34	8.86	22.84	0.69	3.08	2.21	1.48	5.18	2.85	2.24	4.64	2.11	0.71	2.88	3.61
CaO	1.77	8.92	23.51	0.84	6.21	2.46	1.64	7.40	3.17	3.99	5.08	0.76	1.64	3.80	4.30
Na_2O	0.09	2.87	Bld	0.58	0.03	0.04	0.06	4.07	0.65	0.13	3.44	0.23	3.65	0.56	1.56
K_2O	0.01	0.13	Bld	11.09	0.04	0.01	0.02	0.25	5.25	4.40	2.21	5.65	2.04	3.85	2.44
P_2O_5	0.01	0.06	n.d.	0.18	0.02	0.01	0.02	0.02	0.13	0.01	0.12	0.04	0.02	0.11	0.10
LOI	6.94	1.29	44.95	1.03		2.73	1.74	0.29	2.54	5.54			1.07	5.41	5.03
Total	99.36	101.27	98.32	99.78	94.4	100.01	100.04	99.79	99.77	99.77	97.94	98.6	101.09	99.91	100.00
							Trace Elements (ppm)								
Cr	3753	565	24	29	16.0	13	29	73	529				152	150	36
Ni	1731	121			43.0			73	89.9				78	106	32

Co	113	61	2.6		6.80	4.2	3.6	45			9.7	16	19
Rb	0.42	0.6	0.7	116	1.56	0.3	0.8	3.16	118	40	130	94	172
Sr	48.7	44	141	18	8.70	3.4	4.3	106	30.4	60	37	116	131
Ba	3.94	10.1	14	1230	7.00	1.9	45	25	425	164	160	108	198
Y	4.10	20	11.6	5.5	4.60	4.0	4.4	19	20.6	20.9	11	14	23
Zr	4.84	65.3	0.8	138	2.20	2.7	2.8	69	126	217	131	120	205
Hf	0.16		Bld	3.5	0.06	0.06	0.08	1.98	3.58		3.7	3.4	5.4
Nb	0.08	2.04	0.21	4.7	0.08	0.04	0.04	2.73	4.65	8	4.3	3.9	7.0
Ta	0.01	0.19	Bld	1.3	0.01	<0.019	0.02	0.18	0.35		0.31	0.29	0.50
Th	0.02	0.78	0.07	4.8	0.08	0.02	0.05	0.85	4.41		3.4	3.41	5.2
U	0.01	0.12	Bld	0.66	0.03	0.071	<0.013	0.17	0.73		0.71	0.72	0.82
La	0.30	3.58	2	28.6	2.167	1.432	1.192	6.006	15.5	25.2	14.853	20.151	28.39
Ce	0.20	10.02	2.84	63.3	3.314	2.144	1.810	14.10	34.2	52.1	32.055	42.850	62.44
Pr	0.03	1.45	0.46	7.48	0.362	0.254	0.210	1.950	4.25	6.80	3.876	5.092	7.80
Nd	0.16	7.13	1.97	26.8	1.513	1.020	0.902	9.263	17.7	26.6	15.349	19.958	30.68
Sm	0.08	2.45	0.67	4.23	0.335	0.292	0.250	2.616	4.03	5.03	3.113	4.065	5.91
Eu	0.04	0.9	0.56	0.77	0.261	0.190	0.175	0.811	1.13	0.66	0.819	1.062	1.42
Gd	0.26	3.24	0.81	2.24	0.467	0.357	0.464	3.133	0.70	4.21	2.720	3.409	5.17
Tb	0.07	0.51	0.11	0.286	0.075	0.046	0.072	0.567	4.05	0.66	0.410	0.524	0.75
Dy	0.56	3.34	0.89	1.20	0.444	0.406	0.511	3.396	3.96	3.59	2.163	2.775	4.28
Ho	0.14	0.72	0.21	0.195	0.107	0.097	0.115	0.690	0.77	0.74	0.427	0.522	0.86
Er	0.50	2.02	0.58	0.56	0.323	0.283	0.357	2.026	2.15	2.12	1.273	1.467	2.35
Tm													
Yb	0.56	1.96	0.56	0.521	0.265	0.273	0.332	1.811	1.84	2.20	1.248	1.350	2.19
Lu	0.09	0.3	0.09	0.076	0.043	0.040	0.058	0.267	0.27		0.182	0.193	0.32

		~3700 Ma Sedimentary sequence										
SM/GR/97/22	**SM/GR/93/57**	**173067**	**G11/63**	**G11/74**[a]	**G11/74**[a]	**G11/72**[a]	**G11/72**[a]	**G11/72**	**G04/54**	**G05/85**	**IF-G#**	**G91/75**
Mafic Pelite	**Mafic Pelite**	**RPC Matrix**	**Chemical Sediment**	**Chemical Sediment**	**Laminated Sandstone**	**Evaporitic Nodule**	**Evaporitic Matrix**	**Evaporitic Bulk**	**Chemical Sedimentary**	**BIF**	**BIF**	**Chert**
m65°11.9′N	**m65°10.4′N**		**g65°10.8′N**	**g65°10.8′N**	**g65°10.8′N**	**g65°10.8′N**	**g65°10.8′N**	**G65°10.8′N**	**g65°10.8′N**	**g65°10.3′N**		**m65°12.3′N**
m49°48.5′W	**m49°50.3′W**		**g49°48.2′W**	**g49°48.2′W**	**g49°48.2′W**	**g49°48.2′W**	**g49°48.2′W**	**g49°48.2′W**	**g49°48.2′W**	**g49°49.2′W**		**m49°47.9′W**
~3710	**~3710**	**~3700**	**~3700**	**~3700**	**~3700**	**~3700**	**~3700**	**~3700**	**~3700**	**~3700**	**~3700**	**~3700**
52.22	61.47	57.26		42.23	60.26	86.22	93.50		54.04	53.78	41.20	97.07
0.75	0.72	0.92		0.07	0.12	0.01	0.02		0.07	0.01	0.01	<0.01
14.98	16.41	14.02		3.45	2.87	0.26	1.20		1.72	0.19	0.02	0.16
7.61		1.64		7.60	6.30	0.28	4.92		12.35	46.10	55.85	2.62
		10.32										
0.09	0.09	0.19		0.54	0.30	0.03	0.16		0.79	0.06	0.04	0.08
3.03	2.01	3.17		3.72	1.58				7.16	2.63	1.89	0.67
4.57	3.00	2.03		21.21	14.25	0.20	0.23		13.49	0.23	1.55	0.94
2.78	3.51	0.25							0.03	0.07	0.03	0.01
3.02	2.23	5.57		0.75	1.05	0.08	0.65		0.53	0.01	0.01	Bld
0.19	0.14	0.07							0.01	0.05	0.06	0.01
		3.98								−3.14	−1.58	
89.24	89.58	99.42							90.19			101.56
113	1922	37	140	106	124	n.d.	87	210	6	4	13	13
50	560	23		103	85	31	107		18	23	8	8
17	68	3.1							2.9	29.0	0.7	0.7
116	1.9	17	17	4	11	1	13	22	0.13	0.04	0.39	0.39
121	40	18	18	30	16	1	15	63	0.4	3	1.6	1.6
134	6	349	9640	760	1736	1400	12,300	1295	1	1.5	1	1.0

14	9	3.1	8					6.6	3.2	9	0.9	0.9
201	16.5	14	14	6	8	Bld	15	9	0.6	1	0.5	0.5
5.3	0.54	0.41	0.50					0.30	0.02	0.04	0.02	0.024
6.7	0.24	0.46	0.60			1.00	2.00	0.50	0.09	0.1	0.03	0.027
0.47		0.03	0.10					Bld	Bld	0.2	0.01	0.006
4.4	0.14	0.30	0.25				1.00	0.13	0.03	0.1	0.03	0.026
0.83		0.06	0.10		2.00		5.00	0.06	0.02	0.02	0.01	0.012
15.14	1.032	2.041	2.4					2.2	1.959	2.8	0.461	0.461
35.56	2.476	3.111	3.5					3.4	3.026	4	0.610	0.610
4.59	0.349	0.351	0.39					0.41	0.305	0.4	0.063	0.063
18.59	1.664	1.378	1.4					1.6	1.231	1.8	0.251	0.251
3.87	0.615	0.307	0.29					0.4	0.242	0.4	0.066	0.066
0.85	0.196	0.263	0.15					0.28	0.175	0.39	0.056	0.056
3.34	0.879	0.362	0.67					0.66	0.304	0.74	0.083	0.083
0.48	0.177	0.059	0.12					0.11	0.05		0.014	0.014
2.68	1.358	0.353	0.73					0.66	0.329	0.8	0.087	0.087
0.53	0.349	0.077	0.19					0.17	0.077	0.2	0.021	0.021
1.44	1.303	0.230	0.530					0.460	0.263	0.63	0.059	0.059
			0.080					0.070				
1.36	1.447	0.211	0.49					0.46	0.305	0.6	0.060	0.060
0.21	0.217	0.033	0.08					0.09	0.053	0.09	0.008	0.008

Positions: *m*, read off topographic map (Nutman, 1986); *g*, WGS84 GPS reading.

[a]*Analysis by hand-held-XRF on heterogeneous rocks. X-ray energy insufficient to measure* Na_2O, *ages in italics are age constraints from other samples;* RPC, *round pebble conglomerate.*

(dolostone in Fig. 20.3a). The marble lens is < 5 m thick and <50 m long and is situated in the nose of a minor fold that is parasitic on the early isoclinal closure in the potassic felsic schist unit. It is bounded by heterogeneous calc-silicate ± quartz ± calcite rocks, which we interpret as the metamorphic product of interlayered siliceous rocks (metachert) and dolomitic carbonate. We regard the dolomite marble lacuna as the center of a thick carbonate layer, which was protected from destruction by the metamorphic reactions that formed the calc-silicates along its fringes.

The dolomite marble is massive, with no layering visible from the microscopic to mesoscopic scale (Fig. 20.4a), has a grain size of 1−0.25 mm, and a granoblastic texture. It is devoid of sedimentary or tectonic structures. Accessory quartz and a few small pyrite grains are present. The rock is dominated by ferroan-dolomite, with $SiO_2 \sim 3$ wt%, $TiO_2 \sim 0.02$ wt% and $Al_2O_3 \sim 0.04$ wt% (Table 20.1; Fig. 20.5a). In terms of trace elements, it displays a low Rb/Sr ratio of 0.005, Sr = 141 ppm, Ba = 14 ppm, and Zr < 1 ppm, and has a well-defined seawater-like rare earth element + yttrium (REE + Y) signature (Fig. 20.5b; Nutman et al., 2010). This REE + Y signature is quite distinct from that expected for metasomatic carbonate, or even other sedimentary environments such as seafloor hydrothermal or lacustrine (Kamber et al., 2014). We interpret the marble to be of sedimentary origin, and that its protolith was a dolostone deposited in a marine setting, largely devoid of contamination by terrigenous sources. The associated calc-silicate rocks indicate that the dolostones were interlayered with siliceous rocks, probably cherts.

The rare fuchsitic quartzite is ≤ 1 m thick. The small occurrences of these rocks are all strongly deformed and thereby devoid of any original (sedimentary) structures (geochronology locality is marked as quartzite in Fig. 20.3a). The quartzite carries slightly-rounded, small (mostly <100 μm) oscillatory-zoned zircons with a non-Gaussian age distribution of ~3890, 3850 (most frequent) and 3810 Ma (Nutman et al., 1997, 2009). These zircons are interpreted to be detrital, indicating that the protolith of the quartzite was a clastic sedimentary rock. Its mature character (almost pure quartz) and the 80 million years spread of detrital zircon ages indicate that its provenance was a strongly weathered, somewhat older, quartzo-feldspathic crust. The prominence of ~3850 Ma zircons is in accord with recognition of ~3850 Ma quartzo-feldspathic gneisses south of the Isua supracrustal belt (Nutman et al., 2009).

3.3 ~3800 Ma Felsic Conglomerates

A 3806 Ma potassic felsic schist unit is continuous for almost 30 km in the Isua supracrustal belt (Fig. 20.1); the mineralogy is quartz + K-feldspar + biotite ± muscovite ± plagioclase ± tremolite ± carbonate ± garnet. High strain is prevalent, such that in most places its layering is entirely tectonic (Nutman et al., 2015). Much debated are features found in the core of a steeply plunging parasitic fold in the central segment of the unit where it features felsic prolate nodular structures up to 1 m long (Fig. 20.4b), set in a more strongly eroded carbonate-bearing matrix (locality indicated in Fig. 20.1 as faux-conglomerate). When discovered in the 1970s, these structures were interpreted as acid volcanic structures (Allaart, 1976), formed either directly by volcanic air-fall (bombs) or as a volcanogenic mass-flow sedimentary deposit. Based on this volcanic interpretation, the nodular structures were the target of the first U-Pb zircon geochronology on Isua

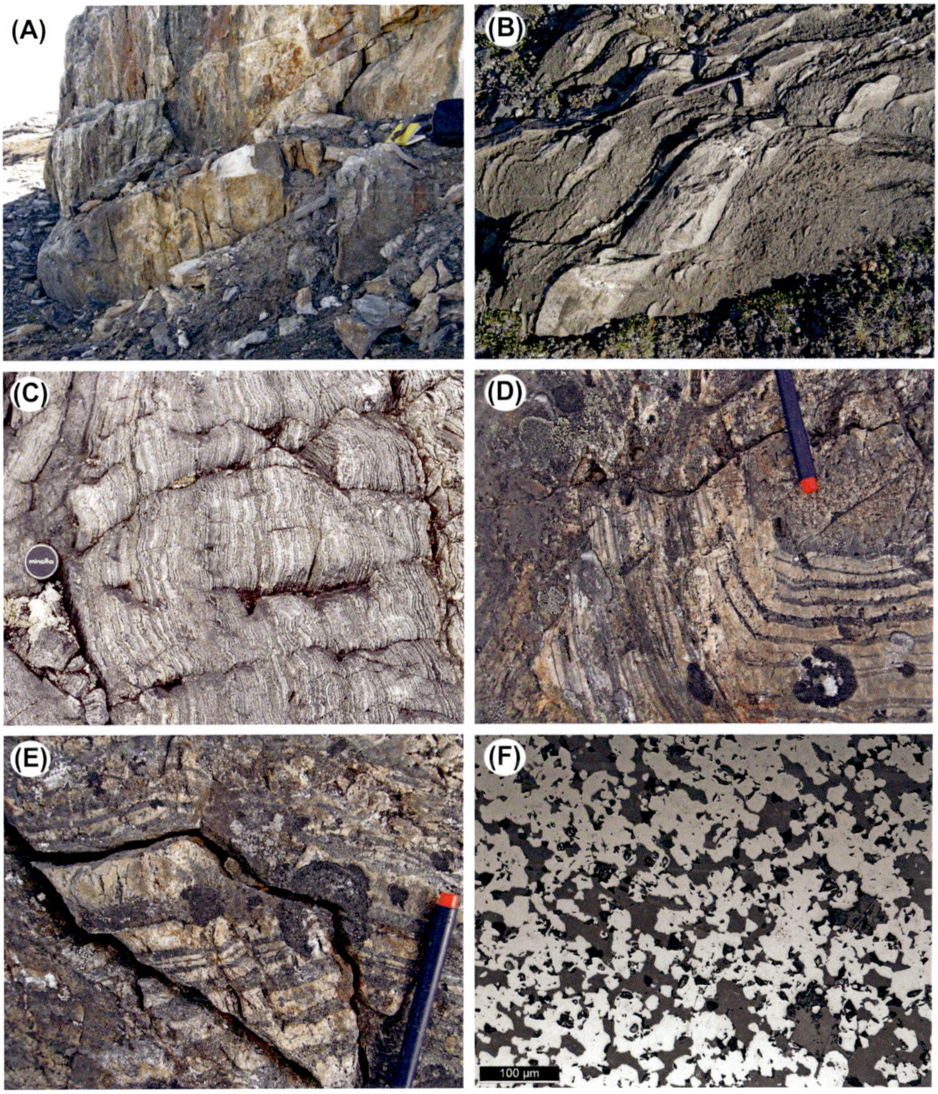

FIGURE 20.4 (A) Homogeneous, well-preserved (nonmetasomatized) ~3800 Ma dolostone, locality dolostone in Fig. 20.3a, showing lack of metasomatic veining. (B) Nodular structures (lighter gray) in altered rhyolitic rocks (faux-conglomerate locality in Fig. 20.1). This is not a genuine conglomerate, with the nodular structures resulting from a combination of variable degrees of early alteration and tectonic disruption (Nutman et al., 2015). (C) Typical, strongly deformed banded iron formation (quartz + magnetite + amphibole) of the dividing sedimentary unit (DSU). (D) Edge of low-strain lacuna within the DSU in which the original sedimentary layering is preserved (right-hand side of image). Locality is marked as 1 in Fig. 20.3c. (E) Detail of original sedimentary/diagenetic layering in DSU banded iron formation. (F) Reflected light photomicrograph of original sedimentary/diagenetic layering in DSU banded iron formation, showing 50–100 μm intergrowth of quartz (dark gray) + magnetite (light gray) ± minor tremolite. Note that the Fe-rich layers actually consist of fine-grained granoblastic aggregates of metamorphic quartz + magnetite. This demonstrates these layers were not deposited as pure oxides/hydroxides.

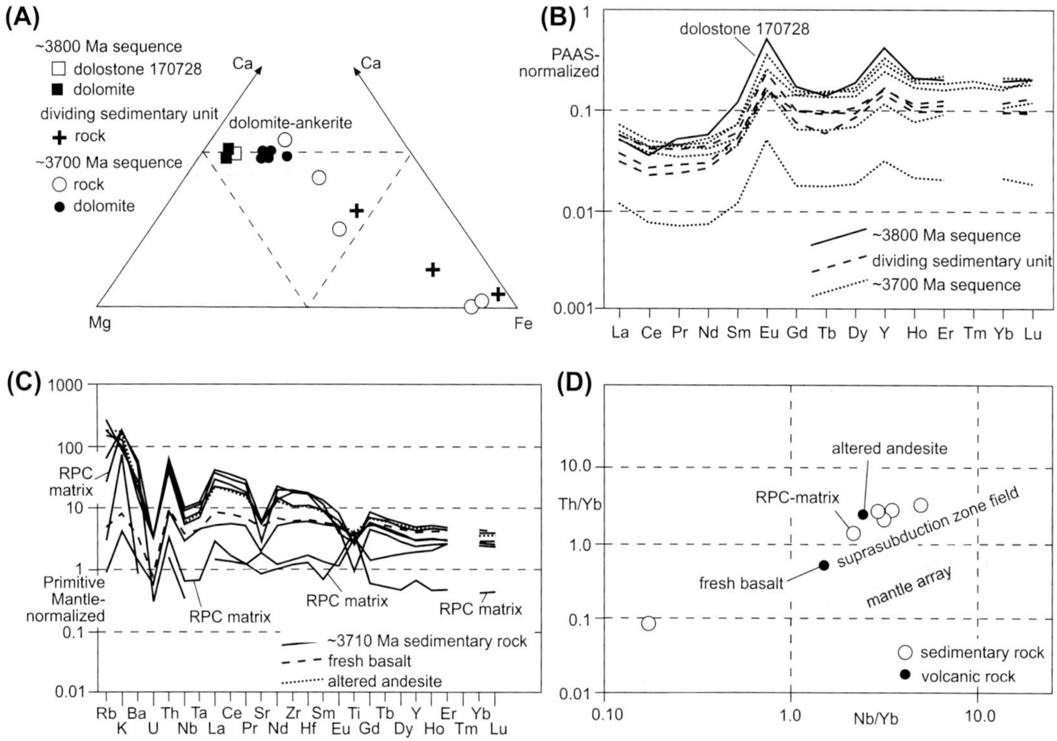

FIGURE 20.5 (A) Ca-Mg-Fe molecular proportions ternary plot for Isua supracrustal belt chemical sedimentary rocks (carbonates and BIF), (B) PAAS-normalized (McClennan, 1989) rare earth element + yttrium plot for Isua supracrustal belt chemical sedimentary rocks (carbonates and BIF). (C) Primitive mantle normalized (McDonough and Sun, 1995) trace element plot for ~3710 Ma felsic and mafic schists and reference unaltered igneous rocks. (D) Th/Yb–Nb/Th plot (after Pearce, 2008) for ~3710 Ma felsic and mafic schists and reference unaltered igneous rocks. The fresh basalt refers to unaltered tholeiitic compositions from this part of the belt (Table 20.1). A sedimentary rock is plotted to demonstrate the unmodified arc-like signature of such rocks.

supracrustal belt rocks, with all early and subsequent studies indicating an age of ~3800 Ma (Michard-Vitrac et al., 1977; Baadsgaard et al., 1984; Compston et al., 1986; Nutman et al., 1996, 2002, 2015; Hiess et al., 2009).

Nutman et al. (1984) concurred with the 1970s acid volcanic and/or volcano-sedimentary origin. Lines of evidence for this presented in that paper were the (1) rare preservation of graded layering, and (2) survival of microtextures resembling fiamme within the putative volcanic bombs. Nutman et al. (1984) also documented the deviation from igneous compositions, particularly high K_2O correlated with low Na_2O, which they interpreted as alteration soon after deposition. However, Nutman et al. (1984) also noted that the nodular structures occur only in the core of a fold, and they reinterpreted them as tectonic boudins of more competent massive felsic rocks between layers containing carbonate. The unit was re-evaluated again by Rosing et al. (1996), who concluded that it represents a strongly metasomatized, carbonated tonalite sheet intruded into the belt (i.e., a plutonic and not volcanic or sedimentary rock). Myers (2001) reinterpreted the unit as a strongly metasomatized series of *mafic* volcanic rocks. Given that the potassic schists contain a large quantity of oscillatory-zoned magmatic zircon, this latter interpretation is the least likely for the unit. Nutman et al. (2015) undertook a further investigation of the potassic felsic schist, involving detailed intrazircon petrography and geochemistry integrated with detailed whole rock petrography of the nodular structures and integration of zircon and whole rock stable isotope geochemistry. Overall the magmatic zircons are small (mostly ≤100 μm in length), suggesting crystallization in a volcanic or hypabyssal environment. This latest study suggests that its protolith was a package of layered dacitic-rhyolitic rocks that were weathered at the time of eruption. Thus the putative bomb/mass flow structures should be removed from the inventory of Isua supracrustal belt sedimentary structures. Nonetheless, the rocks of the potassic schist unit as a whole comprise ~3800 Ma volcanic and volcaniclastic protoliths (Nutman et al., 2015).

4. ~3750 Ma DIVIDING SEDIMENTARY UNIT

4.1 Relationship with ~3800 and ~3700 Ma Assemblages

A unit of metachert, BIF, and calc-silicate rocks runs the length of the Isua supracrustal belt (Fig. 20.1). It has the singular distinction of separating all the rocks with ~3700 Ma zircon age determinations in the north of the belt from all those with ~3800 Ma age determinations to the south and has thus been named the DSU (Nutman and Friend, 2009). The DSU is highly tectonized, and in places, particularly in the east, it is clearly seen to be in Eoarchean tectonic contact with the rocks to the north (e.g., Myers, 2001; Nutman and Friend, 2009; Nutman et al., 2002, 2007). In the central reaches of the belt, the DSU is considerably thinned on the limbs of early asymmetric folds (Fig. 20.4c), to the degree that it is excised locally (see map of Nutman and Friend, 2009). However, despite the strongly tectonized state, the DSU does form an essentially continuous unit throughout the belt.

4.2 Time of Deposition

Zircon geochronology was undertaken on highly deformed magnetite and carbonate-bearing lithologies, and a magnetite BIF from a low-strain zone in which sedimentary layering is preserved (locality 1 in Fig. 20.3c; Nutman et al., 2009). The carbonate-bearing samples yielded <20 small zircons with ages ranging between 3920 and ~3750 Ma, whereas the pure magnetite BIF in the east yielded only two tiny zircons (≤50 μm), both with ages of ~3750 Ma (Nutman et al., 2009). Thus all three samples contain a component of ~3750 Ma zircons, which from CL images are oscillatory-zoned and devoid of abrasion, and are overall better preserved than older zircons in these samples (Nutman et al., 2009). A weakly deformed 3658 ± 6 Ma granite dyke cuts already deformed DSU rocks (Nutman et al., 2009). Our preferred interpretation is that the ~3750 Ma zircons are a volcanogenic component, and thereby give a proxy age for the deposition of the DSU, whereas the granite sheet indicates that by 3658 ± 6 Ma the DSU was already strongly deformed.

4.3 Lithological Character

Because of the strongly deformed state of the DSU, at most places layering in it is entirely tectonic, without relicts of sedimentary structure (Fig. 20.4c). The only known low-strain lacunae are at one locality in the east of the belt (locality 1 in Fig. 20.3c). These small domains (maximum dimension ~5 m each) reside in a part of the DSU dominated by magnetite-bearing BIF. The strongly deformed portions have discontinuous banding of magnetite ± Fe-rich amphibole, which displays asymmetric folds verging toward a large, steeply plunging synformal closure south of the Isua supracrustal belt. The margins of the low-strain lacunae are abrupt, and show that the magnetite banding in the strongly deformed varieties comprises not only a severely transposed and disrupted sedimentary layering, but also a magnetite foliation (Fig. 20.4d). Within the lacunae the layering is on a scale of up to several cm thick (as opposed to the mm-scale in strongly deformed varieties), and can in some instances show subtle graded layering given by the variation of quartz versus magnetite content (Fig. 20.4e). Individual layers can be nodular or discontinuous. Given the low strain, these variations in the layering are regarded as depositional and/or diagenetic. In the low-strain domains the magnetite layers consist of aggregates of magnetite + quartz ± tremolite with a grain size typically <100 μm (Fig. 20.4f). In the more typical, strongly deformed Isua BIF, this texture has been erased, because quartz and magnetite have segregated into alternating bands, to give a pseudo-sedimentary layering. Two westerly carbonate-bearing samples and the easterly BIF (Samples G93/28, G07/22, and G05/17, respectively in Table 20.1) have a well-developed seawater-like REE + Y signature (Fig. 20.5b) and low Zr and Al_2O_3. As such, these rocks are interpreted as a chemical sedimentary rock with only minor contamination from volcanic or detrital material (Friend et al., 2008; Nutman et al., 2009).

5. ~3710 Ma CLASTIC AND 3700–3695 Ma CHEMICAL SEDIMENTARY ROCKS

5.1 Tectonic Setting and Strain Partitioning

The ~3700 Ma portion of the Isua supracrustal belt is a tectonic collage of different lithologies, that in the Eoarchean were assembled into their present configuration; albeit different workers may not agree on the details of this process (e.g., Komiya et al., 1999; Nutman et al., 2002; Furnes et al., 2007; Nutman and Friend, 2009; Friend and Nutman, 2010). This tectonic collage is dominated by arc-related rocks (e.g., Polat and Hoffman, 2003) that evolved from ≥3720 Ma to ~3695 Ma (Friend and Nutman, 2010; Nutman et al., 2014). The >25 million years evolution of this package saw a shift of lithology, from early boninite and basalt-picrite of island arc geochemical affinity (Polat and Hoffman, 2003), through andesites-dacites, and finally to chemical sedimentary rocks with input of a minor felsic volcanogenic component (Nutman and Friend, 2009; Friend and Nutman, 2010; Nutman et al., 2014). As such, this ancient assemblage follows a path very similar to that of Phanerozoic island arcs evolving from a juvenile to mature stage (e.g., Shervais, 2001). The 25 million year time span was punctuated by tectonic events, and by 3660 Ma was followed by juxtaposition of the ~3700 and ~3800 Ma terranes (Nutman et al., 2002, 2014; Crowley, 2003; Friend and Nutman, 2010). The early tectonic imbrication is illustrated in Fig. 20.3b, which shows a portion of the northwestern end of the belt, where a thin BIF unit carrying rare ~3690 Ma volcanogenic zircon (providing a proxy of its depositional age) was tectonically intercalated with juvenile felsic sedimentary rocks carrying ≥3710 Ma zircons, prior to being tectonically intercalated between units of >3712 Ma boninites. This early tectonic activity destroyed much of the original sedimentary and volcanic structures, such that they only occur as relicts in rare, small, low-strain lacunae distributed within the more prevalent L-S tectonites.

5.2 ~3710 Ma Felsic Volcano-Sedimentary Rocks and Mafic Pelites

A package of predominantly felsic rocks and mafic mica schists crops out at the northeastern end of the Isua supracrustal belt (Fig. 20.3c). Its contacts are mylonites, separating it from the Central Tectonic Domain to the east and from boninitic amphibolites to the west (Fig. 20.3c). Felsic rocks of this package contain several domains of low strain. The facing directions of sedimentary structures surviving in the low-strain zones suggest that the package becomes consistently younger to the southeast. However, this does not preclude that it is tectonically imbricated. Felsic rocks in the western part of the panel are predominantly andesitic in composition (sample G91/73; Table 20.1) and massive to layered. They contain rare examples of graded layering, defined by variation in the abundance of coarse plagioclase grains (Fig. 4b in Nutman et al., 1984). To the southeast, the felsic units are dacitic (e.g., sample G91/76; Table 20.1) and carry ample magmatic zircon with a unimodal age population of ~3705 Ma in several samples (Nutman et al., 1996, 1997, 2009; Kamber et al., 2005). These layers are graded, with the volume and size of felsic volcanic clasts (Fig. 20.6a) consistently decreasing eastward. Each graded felsic layer is succeeded by a layer of homogeneous, fine-grained mafic schist (Nutman et al., 1997). Up sequence (to the east) the mafic schists become prevalent with rare, thin felsic schist layers (Nutman et al., 1984).

Panels of felsic and mafic schists are found as tectonic intercalations with boninitic amphibolites in the western part of the Isua supracrustal belt. Felsic schists from two localities have given small yields of ~3710 Ma zircons (Fig. 20.3a, b; Nutman et al., 1997, 2009). In all

FIGURE 20.6 Structures in ~3710 Ma felsic and mafic schist assemblage. (A) Deformed volcanic clasts in unit with overall dacitic composition (locality 4 in Fig. 20.3c). (B) Graded felsic schists (graded unit locality in Fig. 20.3b).

but one locality, these rocks are so strongly deformed that all original structures have been obliterated. The approximately 10 m across low-strain locality (Fig. 20.3b) was discovered in the early 1970s by the late Vic McGregor, and was described by Nutman et al. (1984) and Fedo et al. (2001). It has been the focus of considerable study, due to the discovery of ^{13}C-depleted graphite flakes within it, which has been interpreted as an Eoarchean biogenic signature (Rosing, 1999).

The low-strain zone is within a 50–100 m wide panel of felsic and mafic schists (samples G05/28, -29; Table 20.1) that can be followed north–south for >1 km. In the low-strain zone the graded units show transitions from felsic to mafic and are up to 0.5 m thick (Fig. 20.6b). The bases of the graded units are sharp. Packets of several graded felsic layers with little or no interlayered mafic schist are common. The upper parts of complete individual units are dark gray to black mafic metapelites of andesitic composition (Nutman and Friend, 2009). Nutman et al. (1984) suggested that these graded units formed as turbiditic Bouma sequences. Although Fedo et al. (2001) concurred with deposition below the wave base, they suggested that alternative depositional mechanisms should be contemplated such as debris flows and sediment sculpted by bottom currents.

The composition of the felsic rocks described earlier is broadly andesitic trending to dacitic, with trace element and whole rock ^{143}Nd isotopic signatures indicating that the protoliths are juvenile intermediate to acid calc-alkaline rocks of arc affinity (Fig. 20.5c,d; e.g., Jacobsen and Dymek, 1988; Bolhar et al., 2005; Nutman et al., 2014). The protoliths of the mafic schists are interpreted to be pelites derived from weathered mafic (volcanic) rocks (Nutman et al., 1984; Bolhar et al., 2005). The geochemical variation of both the felsic and mafic schists indicate the sources were variably weathered, which is supported by their whole rock $\delta^{18}O_{(VSMOW)}$ values of $> +11$ (Table 20.2). We propose that the ~3710 Ma felsic schists were sourced from a calc-alkaline volcanic center, with the clast-rich layers in the east of the belt representing proximal mass-flow deposits, and the finer-grained graded units in the west of the belt representing distal deposits—deposited by mass-flow or turbiditic currents below the wave base. The interlayered mafic schists were derived from weathered mafic crust. Kamber et al. (2005) suggested their source could be Hadean, whereas we consider it more feasible that ≥3710 Ma Eoarchean basalts were the source, particularly as such rocks are actually found in the same ~3700 Ma tectonic panels of the Isua supracrustal belt.

5.3 Setting of Sedimentary Rocks in the Central Tectonic Domain

In the east of the Isua supracrustal belt, the central tectonic domain of Appel et al. (1998) contains several refolded isoclines (examples shown in Fig. 20.3c) that because they are cut by a metadiabase Ameralik dyke with a U-Pb protolith zircon age of 3511 ± 18 Ma (Fig. 20.3c; Nutman and Friend, 2009), must have formed pre-3500 Ma. The isoclines are strongly noncylindrical, such that they are doubly plunging and closed at both ends. Low-strain domains in sedimentary rocks are found in the axial regions of these folds. A southwesterly plunging mineral lineation and rodding is pervasive throughout all rocks. Even with this L-fabric, looking down-lineation the total strain can appear minimal, whereas deformation is more apparent when looking orthogonal to the fabric (Nutman, 1986; Fedo, 2000).

TABLE 20.2 Oxygen Isotope Data

Sample	Type	Age	$\delta^{18}O_{(VSMOW)}$	Source
G05/21	Weathered basalt/andesite	~3710 Ma	15.7	University of Wollongong
G05/22	Weathered basalt/andesite	3709 ± 10 Ma	16.8	University of Wollongong
G12/99	Weathered basalt/andesite	~3710 Ma	16.6	University of Wollongong
G12/101	Mafic schist	~3710 Ma	11.8	University of Wollongong
G12/113	Mafic schist	~3710 Ma	12.6	University of Wollongong
G04/46	Dacitic volcano-sedimentary	3710 ± 4 Ma	13.0	University of Wollongong
229403	Diorite	3715 ± 10 Ma	7.3	Baadsgaard et al. (1986)
03–3.5	Basalt pillow	≥3718 Ma	9.9	Furnes et al. (2007)
03–3.13	Basalt pillow	≥3718 Ma	6.8	Furnes et al. (2007)
03–3.22	Basalt pillow	≥3718 Ma	7.3	Furnes et al. (2007)
03–3.23	Basalt pillow	≥3718 Ma	6.9	Furnes et al. (2007)
03–3.31	Basalt pillow	≥3718 Ma	6.7	Furnes et al. (2007)
03–3.39	Basalt pillow	≥3718 Ma	6.5	Furnes et al. (2007)
03–3.40	Basalt pillow	≥3718 Ma	6.6	Furnes et al. (2007)
1A2-IG06	Basalt pillow	≥3718 Ma	6.8	Furnes et al. (2007)
1B4-IG06	Basalt pillow	≥3718 Ma	6.8	Furnes et al. (2007)
2A-IG06	Basalt pillow	≥3718 Ma	6.9	Furnes et al. (2007)
2B2A-IG06	Basalt pillow	≥3718 Ma	7.1	Furnes et al. (2007)
3A2A-IG06	Basalt pillow	≥3718 Ma	6.8	Furnes et al. (2007)
5B1-IG06	Basalt pillow	≥3718 Ma	7.9	Furnes et al. (2007)
5C-IG06	Basalt pillow	≥3718 Ma	7.6	Furnes et al. (2007)
6B-IG06	Basalt pillow	≥3718 Ma	6.8	Furnes et al. (2007)

The cores of doubly plunging antiforms consist of amphibolites of basaltic and andesitic affinity, around which are complete or partial rims of dolomitic rocks, chert, and BIF. The intervening (synformal) closures contain magnesian amphibolites of boninitic chemical affinity (Fig. 20.3c; Polat et al., 2002), in which pillows are locally preserved (Komiya et al., 1999). Thus, these boninitic amphibolites structurally overlie the other amphibolites and sedimentary rocks (Fig. 20.2c). The base of the hanging wall boninitic amphibolites is a chloritic metamylonite, developed from the intense deformation of the pillowed boninite protolith. This mylonite is interpreted as a folded thrust (Nutman et al., 2002, 2010). It excises variable amounts of the underlying lithological package, so that in some places the hanging wall boninitic amphibolites are in contact with BIF and quartzites (metacherts?), whereas elsewhere they are in contact with footwall amphibolites (Fig. 20.3c). This tectonic contact is definitely pre-3511 Ma, based on the age of a cross-cutting Ameralik dyke (Fig. 20.3c).

Metabasalts and andesites with arc-like trace element signatures (Nutman et al., 2014) are the structurally lowest unit, occurring in the cores of the pre-3510 Ma antiforms (Fig. 20.3c). Most probably represent massive flows. In addition to the massive varieties, vesicular (Appel et al., 1998) and pillowed metabasalts are also present (Solvang, 1999; Nutman et al., 2014). The vesicular varieties have quartz amygdales up to a few millimeters across, and were first brecciated by quartz veins and then cut by superimposed carbonate veins. The intensity of carbonate veining increases toward the adjacent dolomite + quartz sedimentary rocks. At one locality, pillows are preserved of (altered) andesitic composition, with the pillows' younging direction toward the adjacent sedimentary rocks (Nutman et al., 2014; analysis G05/22, Table 20.1). These have yielded rare 3709 ± 10 Ma magmatic zircon (Nutman et al., 2010), giving the maximum depositional age for the overlying sedimentary rocks.

5.4 Round Pebble Conglomerate: Unconformity and Overlying Dolomitic Rocks

The round pebble conglomerate (RPC; Allaart, 1976; Nutman et al., 1984; Fedo, 2000) has been long regarded as an enigmatic unit. In Fig. 20.3c the locality where it has been studied in most detail is capitalized as RPC, whereas other known localities are shown as lowercase rpc (Nutman et al., 1984; Fedo, 2000). Some workers have interpreted the RPC as entirely tectonic in origin (Myers, 2001). However, in most investigations it has been regarded as an L-tectonite metaconglomerate, with the most detailed analysis and description by Fedo (2000). Previous workers have documented the RPC's polymict nature (Fig. 20.7a,b) with biotite-amphibolite, coarse-grained polycrystalline quartz, and rare layered chert clasts in a biotite-rich matrix. The biotite-amphibolite clasts are regarded as altered (weathered) mafic volcanic rocks (Nutman et al., 1984; Fedo, 2000). The monomineralic quartz clasts may have been derived from a combination of fragments of vein quartz and possibly amygdules from coarsely vesicular basaltic rocks. Evidence for the latter might be demonstrated in Fig. 20.7b, where a clast consists of a large, rounded, coarse-grained polycrystalline quartz nodule, on one side of which seems to be part of the original wall of mafic volcanic rocks. This would be strong support for the observation first made by Fedo (2000) that the clasts include vesicular volcanic rocks.

The RPC at the most studied locality is least deformed, and original relationships with the adjacent lithologies are displayed. At one side the RPC has a sharp contact with fine-grained massive biotite-rich garnetiferous amphibolites (Fig. 20.7c). Given their altered chemical composition and elevated $\delta^{18}O_{(VSMOW)}$ value of + 16.6, these rocks are interpreted to be metabasalt/andesite affected by low temperature alteration (sample G12/99; Tables 20.1, 20.2). On the other side, the RPC is in sharp contact with fine-grained layered dolomite and dolomite + quartz rocks with accessory phlogopite (Fig. 20.7d). At this locality, there is no shearing on both contacts of the RPC and no change in the aspect ratio of the conglomerate's quartz clasts. This indicates that here, the boundaries of the RPC are not tectonic breaks, and it appears there is not even a significant strain gradient across them. Therefore we conclude that both contacts are original; an unconformity with underlying altered

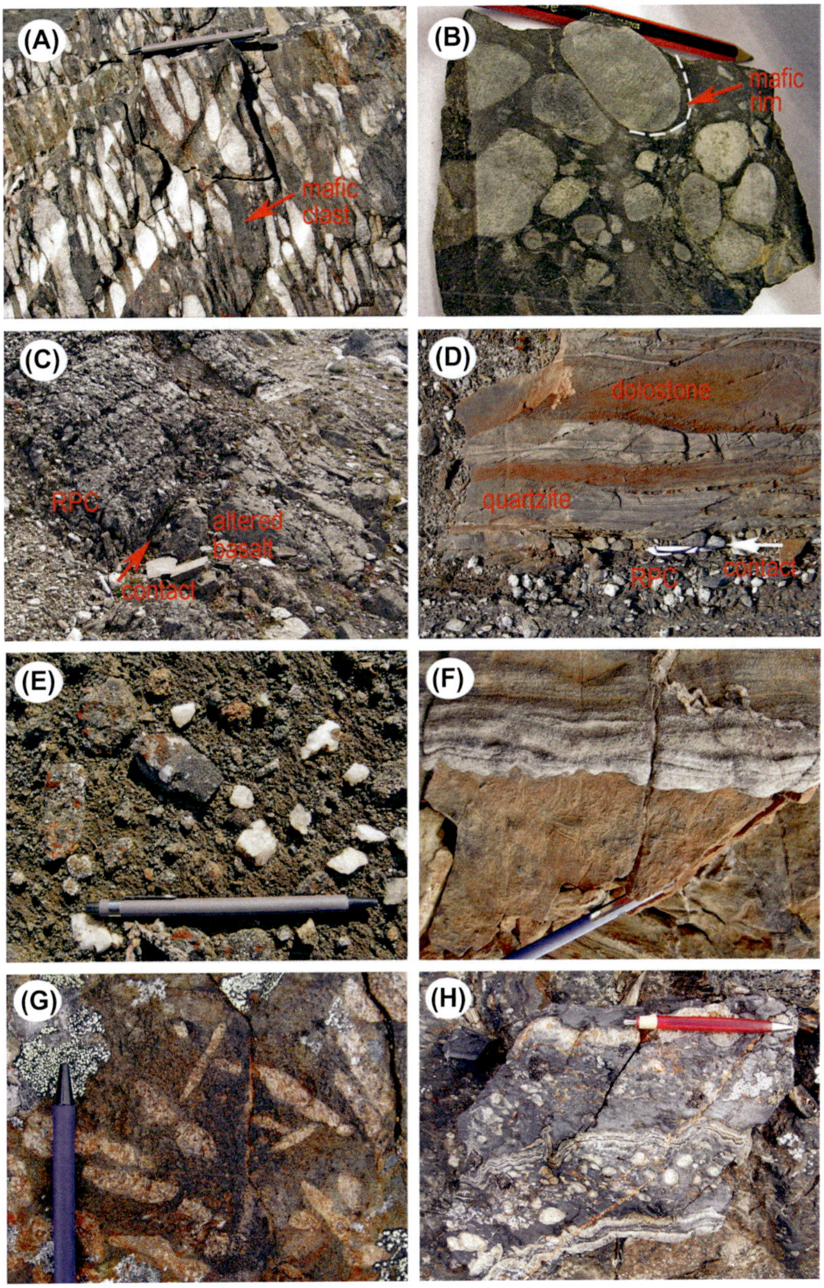

FIGURE 20.7 Images of the ~3700 Ma sedimentary sequence. (A) General overview of the round pebble conglomerate (locality 2: RPC in Fig. 20.3c). (B) Sawn slab of the RPC showing its polymict nature with quartz (qtz) and altered volcanic (volc) clasts. The dotted boundary shows the margin a clast interpreted to consist of mostly a quartz amygdale with a fringe of (altered) basalt. (C) Nontectonized contact of the RPC with underlying altered basaltic-andesitic volcanic rocks, indicated by red arrow. Pen in foreground for scale. (D) Interlayered dolostone (brown) and quartz + dolomite rocks (off-white) at the RPC locality. At bottom of image (partly obscured by debris) is a nontectonized contact between the RPC and the dolomite-quartz rocks (red arrow = where pen for scale rests). (E) Modern lag deposit of mafic schist fragments, clay and quartz fragments developed on top of an altered ultramafic body, West Greenland. (G–H) Low-strain lacuna in dolostone unit, where sedimentary structures are preserved, indicated as 3 on Fig. 20.3c. (F) Cross-lamination in quartz + dolomite sedimentary rock, with laminae containing higher concentration of phlogopite and dolomite. The laminated unit overlies a massive dolostone at bottom of image. (G) Quartz + dolomite rock with randomly orientated lenses/blades rich in quartz. Possible replaced barite in evaporitic sedimentary rock. (H) Layer with quartz nodules in BIF. The nodules are interpreted to be diagenetic in origin.

metabasalts/andesites and a conformable one with overlying dolomite and dolomite + quartz rocks (Fig. 20.2g). However, elsewhere at the other localities (indicated as rpc in Fig. 20.3c), the RPC and its contacts are strongly sheared.

In the quartz + dolomite layers, a fine-scale lamination is preserved. Based on the low-strain domain in the antiformal closure to the north where sedimentary structures are preserved in dolostones (see later), these dolomite + quartz rocks are interpreted to have a sedimentary protolith.

Accepting the protolith of the RPC as a conglomerate has led to debate concerning the depositional environment, particularly given that it appears as an anomaly in an assemblage of mainly pillow basalts, pillow breccias, cherts, and BIF (Fedo, 2000). Dimroth (1982) suggested that it formed in a tidal channel, whereas Fedo (2000) suggested that it represents a mass flow deposit, proximal to exposed, weathered mafic volcanic sources. Although we concur with the provenance of the clasts suggested by Fedo, we suggest instead that the RPC could represent a lag deposit, formed by reworking of surficial altered debris on an erosional surface. To illustrate this, Fig. 20.7e shows a modern lag deposit of altered mafic material of variable grain size together with clasts of vein quartz, which developed on top of an Archaean ultramafic body on coastal exposures in Western Greenland.

5.5 Depositional Structures in ~3700 Ma Banded Iron Formations, Dolomitic and Quartz-Rich Sedimentary Rocks of the Central Tectonic Domain

In the Central Tectonic Domain (Fig. 20.3c), ~3710 Ma basaltic and andesitic volcanic rocks, with the local lenses of the RPC discussed earlier, are overlain by dolomite + quartz rocks followed by metacherts and BIF (Nutman et al., 2014). The latter contain a few ≤50 μm oscillatory zoned zircons with a uniform age of 3700–3695 Ma, that are interpreted as derived from small amounts of volcanic ash contaminating the BIF and cherts (Nutman et al., 2002, 2009). Therefore these zircons are a proxy for a ~3700 Ma age of deposition for dolomite + quartz rocks followed by metacherts and BIF (Nutman et al., 2009). There are rare low-strain domains in these rocks, mostly in the cores of the noncylindrical folds indicated in Fig. 20.3c. In dolomitic rocks at locality 3 in that figure, low strain has assisted in maintaining high XCO_2 during 500–550°C peak metamorphism, preventing the growth of tremolite by quartz + dolomite reaction. This has helped to preserve fine-scale sedimentary structures in these rocks. On one outcrop sedimentary cross-lamination is well preserved in quartz + dolomite rocks overlying massive dolostones (Fig. 20.7f). The cross-lamination is hummocky, with clearly developed erosional surfaces. This style of cross-lamination is most widely developed in shallow water settings, particularly where there is repeated change of current direction, such as controlled by tides (Van Kranendonk, 2006 and references therein).

Also at this locality, massive to weakly layered quartz-rich rocks with minor carbonate contain a layer of randomly orientated elongate lenses (Fig. 20.7g). The lenses are predominantly quartz, whereas the matrix is dominated by dolomite + quartz + hyalophane. A bulk sample of the rock has ~ 1 wt% BaO (Sample G11/72; Table 20.1). The favored interpretation of these rocks is that they represent evaporites, where large crystals (originally barite?) grew in a soft matrix. Barite is the preferred phase, given the high barium content of these rocks. These structures resemble ones described from modern sabkha environments, and in evaporitic intervals in Paleoarchean sequences from the eastern Pilbara terrane, Western Australia (Van Kranendonk, 2006). In this interpretation, the barite was replaced during diagenesis or superimposed metamorphism.

Succeeding the dolomite-rich rocks are massive units of quartzite with thin calc-silicate layers (Fig. 20.3b of Nutman et al., 2010). The protoliths of these are interpreted as chert, with thin layers of dolomitic carbonate. These rocks locally preserve breccias of jumbled laminated siliceous and carbonate clasts that have been interpreted as tempestites (Nutman et al., 2012). The cherts with dolomitic layers are succeeded by BIF and massive chert, which are continuous with the large exposures of these rocks in fold closures at the eastern end of the Isua supracrustal belt (Moorbath et al., 1973; Allaart, 1976; Nutman et al., 1984; sample IFG, Table 20.1). Most of the BIF has a tectono-metamorphic quartz + magnetite ± grunerite layering on the scale of a few mm, but as with the 60 million year older DSU, sedimentary structures are preserved in low-strain lacuna, particularly in the far-eastern end of the belt. Thus Frei and Polat (2007) describe pretectonic layering of the same magnitude as found in the DSU, with quartz layers alternating with layers containing quartz + magnetite. At another locality in BIF at the far-eastern end of the belt, layers of quartz-rich nodules are preserved in a quartz + magnetite matrix, between alternating more continuous layers (Fig. 20.7h). These rocks resemble in morphology younger Western Australian BIF in which nodular structures are diagenetic silica concretions developed over original chert + Fe-clay ± carbonate layering, with the magnetite having developed at a later stage in diagenesis (Rasmussen et al., 2013). The Isua BIF nodules have a sparse distribution of Fe-silicate minerals ± carbonate within them, but with abundant magnetite in the surrounding matrix. Like the examples from Western Australia, a diagenetic origin is attributed to the silica nodules in the Isua ~3700 Ma BIFs. All the chemical sedimentary rocks described previously display seawater-like REE + Y signatures (Fig. 20.5a; Table 20.1; Bolhar et al., 2004; Friend et al., 2008), which using the logic of Kamber et al. (2014) favor a marine rather than a hydrothermal, estuarine, or lacustrine depositional environment.

6. ISUA DEPOSITIONAL ENVIRONMENTS

6.1 Depositional Environment of ~3800 Ma Sedimentary Rocks

The discontinuous ferruginous cherts within the ~3800 Ma pillow basalt sequences represent hiatii within arc-related mafic magmatism. Given the strong deformation and recrystallization of these units, a more detailed interpretation is not possible. The sequences of calc-silicate rocks that cap the mafic volcanic rocks were derived from interlayered cherts and dolostones. In places they are in contact with basaltic rocks, whereas elsewhere the contact is with higher-Mg rocks interpreted to be picritic basalts (Fig. 20.3a). Approaching the contact with the (overlying) calc-silicates, the mafic volcanic rocks start to contain biotite—a phase that is minor or lacking in the mafic rocks away from the contact. This might indicate that the contact represents a deformed and altered unconformity, whereby laterally the dolostone and cherts rest upon basalt or picrite (Figs. 20.2a, 20.3a). The major and trace element compositions of relict well-preserved dolostone supports a marine origin and shows minimal influx of terrigenous material (e.g., very low TiO_2, Al_2O_3, Rb, Zr, Cr, and well-resolved seawater-like REE + Y signature; sample 170728; Table 20.1 and Fig. 20.5). This lithology was succeeded by a discontinuous fuchsitic quartzite, with a non-Gaussian age distribution of zircons (~3890, 3850, and 3810 Ma), indicating influx of mature sediment derived from quartzo-feldspathic crust. These were succeeded by the unit of 3806 Ma altered dacites and rhyolites. We interpret the whole sequence as marking a change in the dynamics of an evolving arc system at ~3800 Ma, from a primitive basalt and picrite assemblage including pillow basalts, with a hiatus, prior to eruption of more felsic volcanic rocks as represented by the 3806 Ma dacites and rhyolites. This hiatus seemed to involve first the deposition of chemical sedimentary rocks (shallow marine?) on the mafic volcanic package, and then immediately prior to eruption of the 3806 Ma dacites and rhyolites, an influx of older detrital material, represented by the fuchsite quartzite. We speculate that this might be related to a change in arc dynamics, from an early entirely intraoceanic assemblage, to the arrival at the convergent plate boundary of a fragment of quartzo-feldspathic crust formed in a slightly older (3890–3850 Ma) arc.

6.2 Depositional Environment of ~3750 Ma Sedimentary Rocks

Although the contact of the DSU and the ~3800 Ma rocks is strongly deformed, in no place has a mylonite been detected that would indicate early dislocation between the two. Secondly, the ~3800 Ma mafic rocks tend to become more biotite-rich toward the contact, suggesting that it might be the site of early alteration. This is interpreted to be due to weathering reactions, because neither the fresh mafic rocks nor the overlying chemical sedimentary rocks are rich in potassium, and thus metasomatic alteration is unlikely. Thirdly, the southern edge of the DSU is in contact with a diversity of lithologies in the ~3800 Ma package. Our interpretation is that prior to deposition of the DSU, the ~3800 Ma package had already been isoclinally folded (exemplified by the closure in the 3806 Ma unit of dacites and rhyolites) and then uplifted and eroded (Fig. 20.2b). Thus there are several lines of evidence to support an unconformity between the ~3800 Ma volcanic rocks and the DSU.

The DSU is dominated by BIF, cherts, and dolomitic rocks, all now very strongly deformed apart from a single locality in the east of the belt with small low-strain lacunae. Major and trace element analyses and zircon dating studies of DSU lithologies indicate chemical sedimentary protoliths, with minor influx of terrigenous material. They all display seawater-like REE + Y signatures, with low contents of Al_2O_3, TiO_2, Rb, Zr, and Cr (Fig. 20.5a; Friend et al., 2008; Nutman et al., 2009). The presence of rare older zircons with ages between 3920 and 3800 Ma indicate a minor contribution from quartzo-feldspathic crust. Currently, orthogneisses with ages of up to 3890 Ma have been detected in association with ~3800 Ma rocks in the Itsaq Gneiss Complex (Nutman et al., 2007; Hiess et al., 2009). The ~3920 Ma detrital zircons in the Isua supracrustal belt indicate that such rocks might occur in the Itsaq Gneiss Complex but have yet to be detected. Again, like the ~3800 Ma dolomite and chert dominated unit within the southern ~3800 Ma terrane, the DSU appears to have been deposited as a sequence of predominantly chemical sedimentary rocks on top of volcanic rocks of arc-affinity. Thus it represents either a hiatus in arc evolution, or maybe even cessation of a multistage complex arc development initiated from ~3800 Ma, and perhaps as early as ~3900 Ma.

6.3 Depositional Environments of ~3700 Ma Sedimentary Rocks

The ~3700 Ma assemblage of the Isua supracrustal belt shows two distinct sedimentary environments (Fig. 20.2c, d). The first, best-preserved in the Central Tectonic Domain (Fig. 20.3c) is illustrated by a sequence of 3700–3695 Ma dolostones, cherts, and then BIF deposited unconformably on altered ≥3709 Ma basaltic-andesitic volcanic rocks with arc-like geochemical affinities. Locally the unconformity is marked by a polymict conglomerate known as the RPC, which contains clasts of variably weathered mafic arc rocks, rare chert, quartz derived from veins, or possibly large amygdules. Thus provenance is from the underlying mafic volcanic sequence, either as a proximally sourced conglomerate (Fedo, 2000) or as an essentially in situ lag deposit (this book chapter).

The dolostones, associated cherty rocks, and BIF all show seawater-like REE + Y signatures, quite distinct from signatures seen in other environments, such as hydrothermal systems or lacustrine environments (Kamber et al., 2014). Where sedimentary structures are preserved in low-strain lacunae, structures such as tempestite breccias and cross-lamination support shallow water deposition. This is in keeping with these rocks having been deposited on a weathered surface of volcanic rocks, marked by RPC lenses. The elevated barium contents are suggestive of, but do not prove, restricted basin evaporative conditions. Therefore these rocks are interpreted as a ~3700 Ma transgressive sequence over a mafic volcanic basement, with first shallow water dolomitic rocks, and then with deepening water, the deposition of BIF.

The other sedimentary environment is represented by felsic and mafic schists, whose protoliths were deposited at ~3710 Ma. The felsic rocks were probably derived from an arc-related calc-alkaline volcanic center. In the east of the Isua supracrustal belt the felsic schists are overall coarser-grained, with some units rich in variegated large clasts of dacitic composition. These are interpreted as mass flow deposits proximal to the volcanic center. In the west of the belt, at a single low-strain locality, finer-grained graded felsic rocks are preserved. These are interpreted as distal mass flow deposits (Fedo et al., 2001) or as turbidites (Nutman et al., 1984) sourced from the volcanic center. The associated mafic schists are interpreted as having pelite protoliths, derived from distal weathered mafic volcanic sources (Nutman et al., 1984; Bolhar et al., 2005). These are interlayered with the felsic schists. One possible scenario is that deposition of the pelites, derived from weathering of uplifted weathered mafic arc rocks, was interrupted by episodic eruptions from a ~3710 Ma calc-alkaline volcanic center that shed debris flows and perhaps related turbidite deposits into a basin. Elevated $\delta^{18}O_{(VSMOW)}$ of the felsic volcanic rocks and major and trace element indications of their moderate to strong weathering suggests that this volcanic center was subareal.

6.4 Samples of Complex Eoarchean Arc Systems

All the preserved sedimentary environments extracted from rocks of the Isua supracrustal belt are regarded as arc-related, and formed during punctuated events over a ~100 million year period. Sedimentary environments were either entirely free of influx of older terrigenous material, or it was minimal. This is indicated by the clean chemistry of the chemical sedimentary rocks of all ages and also the scarcity of old predepositional detrital zircons (Nutman et al., 2009). At ~3700 Ma, there is robust evidence for deposition of chemical sedimentary rocks during drowning of a weathered surface formed on marginally older volcanic rocks, with first above wave base deposition in shallow water of rocks rich in dolomite, followed by deeper water deposition of BIF.

This unconformity scenario can also be invoked for deposition of the ~3750 Ma DSU, and of ~3800 Ma dolomitic rocks in the southern terrane, albeit because of higher strain, the evidence is more tenuous. This is a pattern that is very familiar in the development of (largely intraoceanic) arcs throughout geological history, whereby volcanic rocks formed during one phase of arc activity are tectonically uplifted, eroded, and then submerged, with transgressive chemical sedimentary sequences deposited over the top of them. In the early Precambrian such sequences were dominated by largely dolomitic carbonate followed by chert and BIF (as is proposed here for Isua), whereas in the Phanerozoic, the sedimentary rocks in this cycle are largely limestones (±dolomites). The zircon U-Pb chronology of the Isua rocks show that cessation of volcanic activity, uplift and erosion followed by the transgression with drowning has a short time scale, $\leq$ 10 million years. However, this time constraint is governed by the uncertainties on the U-Pb zircon ages, and in reality the time scale could have been much shorter, as seen in modern environments.

The felsic rocks are derived from coeval calc-alkaline volcanic centers whereas mafic pelites are derived from weathering of mafic volcanic rocks, with probable proximal and distal environments being preserved in Isua. In one locality in the west of the belt (Fig. 20.3b), these rocks have been tectonically intercalated with BIF carrying ~3695 Ma volcanogenic zircon. This suggests that the ~3710 Ma volcanogenic sedimentary sequences and the 3700–3695 Ma chemical sedimentary sequences were not interrelated but were tectonically juxtaposed after they had formed.

One important feature of the Isua supracrustal rocks is the scarcity of older (>3800 Ma) detrital zircons derived from evolved (zircon-saturated) crust. Despite intensive searches only rare older (up to 3890 Ma) zircons from the ~3800 Ma quartzite in the southern terrane have been identified, but this is volumetrically a very minor component. Likewise, the DSU contains sparse detrital zircons back in age to ~3920 Ma. The chemical sedimentary rocks of the northern ~3700 Ma package carry mostly volcanogenic zircons with only rare pre-3700 Ma zircons with the oldest being ~3740 Ma (Nutman et al., 2009). This suggests that these older components are mostly local contaminants coming from oldest components of arc systems that were evolving in an intraoceanic setting, and not from large amounts of older continental crust.

6.5 Climate Control

The Isua sedimentary rocks provide an important insight into the Eoarchean climate and environment, that despite the faintness of the young sun (Walker et al., 1981), sedimentary structures such as the tempestites indicate that marine environments were open to the air, rather than ice-capped (Nutman et al., 2012). Two scenarios are possible to maintain an equable climate under the faint young sun. The first is that the content of greenhouse gases (carbon dioxide and/or methane) was higher (e.g., Kasting, 1993), or that on an early Earth without extensive continental crust or icecaps the planetary albedo was much lower (Rosing et al., 2010). The evidence from Isua, with shallow water dolomite-rich sequences of three ages certainly favors a higher carbon dioxide content of the atmosphere, enhancing warming.

Thus we consider that a more greenhouse atmosphere was an important feature to maintain Eoarchean temperatures above freezing. However, the albedo effect cannot be discounted, and could well be a contributing factor. The reason for this is that the model that best fits the radiogenic isotope (Sr, Nd, Hf) database for the world's oldest rock record is that in the Eoarchean and Hadean, continental crustal volumes were small, and grew progressively throughout Earth's history (e.g., Moorbath, 1975; Bennett et al., 1993, 2007). Thus in the Eoarchean, 30% less bright continental crust areas and more ocean would have definitely decreased planetary albedo.

7. CONCLUSIONS

1. In the Isua supracrustal belt, there are sedimentary rocks deposited at ~3800, 3750, and 3700 Ma.
2. At all three times there are sequences of ferroan dolostones and siliceous rocks, which are chemical sedimentary rocks deposited on marginally older volcanic sequences of arc-like geochemical affinity. In the case of the ~3700 Ma sequence, evidence of deepening of the basin is preserved, leading to the deposition of voluminous BIF.
3. In the ~3700 Ma sequence, andesitic to dacitic sedimentary rocks are interpreted as a series of proximal to distal mass flow deposits and possible turbidites sourced from a calc-alkaline volcanic center. These were emplaced into a sequence of mafic mudstones sourced from weathered, marginally older mafic volcanic rocks.
4. All Isua supracrustal belt sedimentary rocks formed in convergent plate boundary environments that were the ancient analogs of island arcs. Influx of rare older detrital material comes from the older components of the evolving arcs, rather than from extensive areas of ancient (Hadean?) continental crust.
5. Preserved sedimentary structures provide evidence of open-to-air rather than ice-covered marine environments, despite the faint early sun. Increased greenhouse gas content (particularly carbon dioxide) was an important factor maintaining an equable climate. However, lower albedo on an early Earth with a small extent of continental crust may have been a contributing factor.

Acknowledgments

Current work on the Isua supracrustal belt is supported by Australian Research Council Discovery Grant DP120100273.

References

Allaart, J.H., 1976. The pre-3760 m. y. old supracrustal rocks of the Isua area, central West Greenland, and the associated occurrence of quartz-banded ironstone. In: Windley, B.F. (Ed.), The Early History of the Earth. Wiley, London, pp. 177–189.

Appel, P.W.U., Fedo, C.M., Moorbath, S., Myers, J.S., 1998. Recognizable primary volcanic and sedimentary features in a low-strain domain of the highly deformed, oldest known (≈ 3.7–3.8 Gyr) Greenstone Belt, Isua, West Greenland. Terra Nova 10, 57–62.

Baadsgaard, H., Nutman, A.P., Bridgwater, D., McGregor, V.R., Rosing, M., Allaart, J.H., 1984. The zircon geochronology of the Akilia association and the Isua supracrustal belt, West Greenland. Earth Planetary Science Letters 68, 221–228.

Baadsgaard, H., Nutman, A.P., Rosing, M., Bridgwater, D., Longstaffe, F.J., 1986. Alteration and metamorphism of Amîtsoq gneisses from the Isukasia area, West Greenland: Recommendations for isotope studies of the early crust. Geochimica et Cosmochimica Acta 50, 2165–2172.

Bennett, V.C., Nutman, A.P., McCulloch, M.T., 1993. Nd isotopic evidence for transient, highly depleted mantle reservoirs in the early history of the Earth. Earth and Planetary Science Letters 119, 299–317.

Bennett, V.C., Brandon, A., Nutman, A.P., 2007. Coupled ^{142}Nd-^{143}Nd isotopic evidence for Hadean mantle dynamics. Science 318, 1907–1910.

Bolhar, R., Kamber, B.S., Moorbath, S., Fedo, C.M., Whitehouse, M.J., 2004. Characterisation of early Archaean chemical sediments by trace element signatures. Earth Planetary Science Letters 222, 43–60.

Bolhar, R., Kamber, B.S., Moorbath, S., Whitehouse, M.J., Collerson, K.D., 2005. Chemical characterization of earth's most ancient clastic metasediments from the Isua Greenstone Belt, southern West Greenland. Geochimica et Cosmochimica Acta 69, 1553–1573.

Bridgwater, D., McGregor, V.R., 1974. Field work on the very early Precambrian rocks of the Isua area, southern West Greenland. Rapport Grønlands Geologiske Undersøgelse 65, 49–54.

Cawood, P.A., Kröner, A., Collins, W.J., Kusky, T.M., Mooney, W.D., Windley, B.F., 2009. Accretionary orogens through Earth history. In: Cawood, P.A., Kröner, A. (Eds.), Earth Accretionary Systems in Space and Time, Special Publications of the Geological Society, vol. 318, pp. 1–36.

Compston, W., Kinny, P.D., Williams, I.S., Foster, J.J., 1986. The age and lead loss behaviour of zircons from the Isua supracrustal belt as determined by ion microprobe. Earth and Planetary Science Letters 80, 71–81.

Crowley, J.L., 2003. U-Pb geochronology of 3810–3630 Ma granitoid rocks south of the Isua greenstone belt, southern West Greenland. Precambrian Research 126, 235–257.

Crowley, J.L., Myers, J.S., Dunning, G.R., 2002. Timing and nature of multiple 3700–3600 Ma tectonic events in intrusive rocks north of the Isua greenstone belt, southern West Greenland. Geological Society of America Bulletin 114, 1311–1325.

Dilek, Y., Polat, A., 2008. Suprasubduction zone ophiolites and Archean tectonics. Geology 36, 431–432. http://dx.doi.org/10.1130/Focus052008.1.

Dimroth, E., 1982. The oldest rocks on Earth: stratigraphy and sedimentology of the 3.8 billion years old Isua supracrustal sequence. In: Sedimentary Geology of Highly Metamorphosed Precambrian Complexes. "Nauka" Publishing House, Moscow, pp. 16–27.

Dymek, R.F., Klien, C., 1988. Chemistry, petrology and origin of banded iron-formation lithologies from the 3800 Ma Isua supracrustal belt, West Greenland. Precambrian Research 39, 247–302.

Dymek, R.F., Brothers, S.C., Schiffries, C.M., 1988. Petrogenesis of ultramafic metamorphic rocks from the 3800 Ma Isua supracrustal belt, West Greenland. Journal of Petrology 29, 1353–1397.

Fedo, C.M., 2000. Setting and origin of problematic rocks from the >3.7 Ga Isua greenstone belt, southern West Greenland: earth's oldest coarse clastic sediments. Precambrian Research 101, 69–78.

Fedo, C.M., Myers, J.S., Appel, P.W.U., 2001. Depositional setting and paleogeographic implications of Earth's oldest supracrustal rocks, the >3.7 Ga Isua Greenstone Belt, West Greenland. Sedimentary Geology 141–142, 61–78.

Frei, R., Polat, A., 2007. Source heterogeneity for the major components of ~3.7 Ga banded iron formations (Isua greenstone belt, western Greenland): tracing the nature of interacting water masses in BIF formation. Earth and Planetary Science Letters 253, 266–281.

Friend, C.R.L., Nutman, A.P., 2010. Eoarchean ophiolites? New evidence for the debate on the Isua supracrustal belt, southern West Greenland. American Journal of Science 310, 826–861.

Friend, C.R.L., Nutman, A.P., 2011. Dunites from Isua, Greenland: a ca. 3720 Ma window into subcrustal metasomatism of depleted mantle. Geology 39, 663–669.

Friend, C.R.L., Bennett, V.C., Nutman, A.P., Norman, M.D., 2008. Seawater-like trace element signatures (REE + Y) of Eoarchaean chemical sedimentary rocks from southern West Greenland, and their corruption during high-grade metamorphism. Contributions Mineralogy and Petrology 155, 229–246.

Furnes, H., de Wit, M., Staudigel, H., Rosing, M., Muehlenbachs, K., 2007. Vestige of Earth's oldest ophiolite. Science 215, 1704–1707.

Hiess, J., Bennett, V.C., Nutman, A.P., Williams, I.S., 2009. In situ U–Pb, O and Hf isotopic compositions of zircon and olivine from Eoarchaean rocks, West Greenland: new insights to making old crust. Geochimica et Cosmochimica Acta 73, 4489–4516.

Jacobsen, S.B., Dymek, R.F., 1988. Nd and Sr isotope systematics of clastic metasediments from Isua, West Greenland: identification of pre-3.8 Ga differentiated crustal components. Journal of Geophysical Research 93, 338–354.

Jenner, F.E., Bennett, V.C., Nutman, A.P., Friend, C.R.L., Norman, M.D., Yaxley, G., 2009. Evidence for subduction at 3.8 Ga: geochemistry of arc-like metabasalts from the southern edge of the Isua Supracrustal Belt. Chemical Geology 261, 83–98.

Kaczmarek, M.A., Reddy, S., Friend, C.R.L., Nutman, A.P., Bennett, V.C., 2016. Earth's oldest mantle fabrics indicate Eoarchaean subduction. Nature Communications 7. http://dx.doi.org/10.1038/ncomms10665.

Kamber, B.S., Whitehouse, M.J., Bolhar, R., Moorbath, S., 2005. Volcanic resurfacing and the early terrestrial crust: zircon U-Pb and REE constraints from the Isua Greenstone Belt, southern West Greenland. Earth and Planetary Science Letters 240, 276–290.

Kamber, B.S., Webb, G.E., Gallagher, M., 2014. The rare earth element signal in Arcaean microbial carbonate: information on ocean redox and biogenicity. Journal of the Geological Society, London 171, 745–763.

Kasting, J.F., 1993. Earth's early atmosphere. Science 259, 920–926.

Keto, L., Kurki, J., 1967. Report on the Exploration Activity at Isua 1967. Kryolitselskabet Øresund A/S Prospektering Report Lodged as Report 20024 at the Geological Survey of Denmark and Greenland.

Komiya, T., Maruyama, S., Masuda, T., Appel, P.W.U., Nohda, S., 1999. The 3.8–3.7 Ga plate tectonics on the Earth; Field evidence from the Isua accretionary complex, West Greenland. Journal Geology 107, 515–554.

McClennan, S.M., 1989. Rare earth elements in sedimentary rocks: influence of provenance and sedimentary processes. In: Lipin, B.R., McKay, G.A. (Eds.), Geochemistry and Mineralogy of Rare Earth Elements. Mineralogical Society of America, Washington, DC, USA, pp. 169–200.

McDonough, W.F., Sun, S.S., 1995. The composition of the Earth. Chemical Geology 120, 223–253.

Michard-Vitrac, A., Lancelot, J., Allegre, C.J., Moorbath, S., 1977. U-Pb ages on single zircons from early Precambrian rocks of West Greenland and the Minnesota River Valley. Earth and Planetary Sciences Letters 35, 449–453.

Moorbath, S., 1975. Evolution of Precambrian crust from strontium isotopic evidence. Nature 254, 395–398.

Moorbath, S., O'Nions, R.K., Pankhurst, R.J., 1973. Early Archaean age for the Isua iron formation, West Greenland. Nature 245, 138–139.

Myers, J.S., 2001. Protoliths of the 3.8–3.7 Ga Isua greenstone belt, West Greenland. Precambrian Research 105, 129–141.

Nutman, A.P., 1986. The geology of the Isukasia region, southern West Greenland. Grønlands geologiske Undersøgelse Bulletin 154, 80 pp.

Nutman, A.P., Friend, C.R.L., 2009. New 1:20000 geological maps, synthesis and history of the Isua supracrustal belt and adjacent gneisses, Nuuk region, southern West Greenland: a glimpse of Eoarchaean crust formation and orogeny. Precambrian Research 172, 189–211.

Nutman, A.P., Allaart, J.H., Bridgwater, D., Dimroth, E., Rosing, M.T., 1984. Stratigraphic and geochemical evidence for the depositional environment of the early Archaean Isua supracrustal belt, southern West Greenland. Precambrian Research 25, 365–396.

Nutman, A.P., Friend, C.R.L., Kinny, P.D., McGregor, V.R., 1993. Anatomy of an early Archaean gneiss complex: 3900 to 3600 Ma crustal evolution in southern West Greenland. Geology 21, 415–418.

Nutman, A.P., McGregor, V.R., Friend, C.R.L., Bennett, V.C., Kinny, P.D., 1996. The Itsaq Gneiss Complex of southern West Greenland; the world's most extensive record of early crustal evolution (3900–3600 Ma). Precambrian Research 78, 1–39.

Nutman, A.P., Bennett, V.C., Friend, C.R.L., Rosing, M.T., 1997. $\sim$3710 and $\geq$3790 Ma volcanic sequences in the Isua (Greenland) supracrustal belt; structural and Nd isotope implications. Chemical Geology 141, 271–287.

Nutman, A.P., Bennett, V.C., Friend, C.R.L., Norman, M.D., 1999. Meta-igneous (non-gneissic) tonalites and quartz-diorites from an extensive ca. 3800 Ma terrain south of the Isua supracrustal belt, southern West Greenland: constraints on early crust formation. Contributions Mineralogy Petrology 137, 364–388.

Nutman, A.P., Friend, C.R.L., Bennett, V.C., McGregor, V.R., 2000. The early Archaean Itsaq Gneiss Complex of southern West Greenland: the importance of field observations in interpreting dates and isotopic data constraining early terrestrial evolution. Geochimica et Cosmochimica Acta 64, 3035–3060.

Nutman, A.P., Friend, C.R.L., Bennett, V.C., 2002. Evidence for 3650–3600 Ma assembly of the northern end of the Itsaq gneiss complex, Greenland: implication for early Archean tectonics. Tectonics 21 article 5.

Nutman, A.P., Friend, C.R.L., Bennett, V., 2004. Dating of the Ameralik dyke swarms of the Nuuk district, southern West Greenland: mafic intrusion events starting from c. 3510 Ma. Journal of the Geological Society 161, 421–430. London.

Nutman, A.P., Friend, C.R.L., Horie, H., Hidaka, H., 2007. Construction of pre-3600 Ma crust at convergent plate boundaries, exemplified by the Itsaq Gneiss Complex of southern West Greenland. In: van Kranendonk, M.J., Smithies, R.H., Bennett, V.C. (Eds.), Earth's Oldest Rocks. Elsevier, pp. 187–218.

Nutman, A.P., Bennett, V.C., Friend, C.R.L., 2007. c. 3850 Ma tonalites in the Nuuk region, Greenland: geochemistry and their reworking within an Eoarchaean gneiss complex. Contributions to Mineralogy and Petrology 154, 385–408. http://dx.doi.org/10.1007/s00410-007-0199-3.

Nutman, A.P., Friend, C.R.L., Paxton, S., 2009. Detrital zircon sedimentary provenance ages for the Eoarchaean Isua supracrustal belt, southern West Greenland: juxtaposition of an imbricated ca. 3700 Ma juvenile arc assemblage against an older complex with 3920–3800 Ma components. Precambrian Research 172, 212–233.

Nutman, A.P., Friend, C.R.L., Bennett, V.C., Wright, D., Norman, M.D., 2010. ≥3700 Ma pre-metamorphic dolomite formed by microbial mediation in the Isua supracrustal belt (W. Greenland): simple evidence for early life? Precambrian Research 183, 725–737.

Nutman, A.P., Bennett, V.C., Friend, C.R.L., 2012. Waves and weathering at 3.7 Ga: geologic evidence for an equable climate under the faint early Sun. Australian Journal of Earth Sciences 59, 167–176.

Nutman, A.P., Bennett, V.C., Friend, C.R.L., Hidaka, H., Yi, K., Lee, S.R., Kamiichi, T., 2013. The Itsaq gneiss complex of Greenland: episodic 3900 to 3660 Ma juvenile crust formation and recycling in the 3660 to 3600 Ma Isukasian orogeny. American Journal of Science 313, 877–911.

Nutman, A.P., Bennett, V.C., Friend, C.R.L., 2014. The emergence of the Eoarchaean proto-arc: evolution of a c. 3700 Ma convergent plate boundary at Isua, southern West Greenland. Special Publications of the Geological Society 389. http://dx.doi.org/10.1144/SP389.5.

Nutman, A.P., Bennett, V.C., Chivas, A.R., Friend, C.R.L., Liu, X.M., Dux, F.W., 2015. 3806 Ma Isua rhyolites and dacites affected by low temperature Eoarchaean surficial alteration: Earth's earliest weathering. Precambrian Research 268, 323–338.

Pearce, J.A., 2008. Geochemical fingerprinting of oceanic basalts with applications to ophiolite classification and the search for Archean oceanic crust. Lithos 100, 14–48.

Polat, A., Hofmann, A.W., 2003. Alteration and geochemical patterns in the 3.7–3.8 Ga Isua greenstone belt, West Greenland. Precambrian Research 126, 197–218.

Polat, A., Hofmann, A.W., Rosing, M.T., 2002. Boninite-like volcanic rocks in the 3.7–3.8 Ga Isua greenstone belt, West Greenland: geochemical evidence for intra-oceanic subduction zone processes in the early Earth. Chemical Geology 184, 231–254.

Rasmussen, B., Meier, D.B., Krapez, B., Muhling, J.R., 2013. Iron silicate microgranules as precursor sediments to 2.5-billion-year-old banded iron formations. Geology 41, 435–438.

Rollinson, H., 2003. Metamorphic history suggested by garnet-growth chronologies in the Isua greenstone belt, West Greenland. Precambrian Research 126, 181–196.

Rosing, M.T., 1999. ^{13}C-depleted carbon microparticles in >3700 sea-floor sedimentary rocks from West Greenland. Science 283, 674–676.

Rosing, M.T., Rose, N.M., Bridgwater, D., Thomsen, H.S., 1996. Earliest part of the Earth's stratigraphic record: a reappraisal of the >3.7 Ga Isua (Greenland) supracrustal sequence. Geology 24, 43–46.

Rosing, M.T., Bird, D.K., Sleep, N.H., Bjerrum, C.J., 2010. No climate paradox under the faint early Sun. Nature 464, 744–747.

Shervais, J.W., 2001. Birth, death, and resurrection: the life cycle of suprasubduction zone ophiolites. Geochemistry Geophysics Geosystems 2. Paper Number 2000GC000080.

Solvang, M., 1999. An Investigation of Metavolcanic Rocks from the Eastern Part of the Isua Greenstone Belt, West Greenland. Geological Survey of Denmark and Greenland (GEUS) Internal Report, Copenhagen, Denmark, p. 62.

Van Kranendonk, M.J., 2006. Volcanic degassing, hydrothermal circulation and the flourishing of early life on Earth: new evidence from the Warrawoona Group, Pilbara Craton, Western Australia. Earth-Science Reviews 74, 197–240.

Walker, J.C.G., Hays, P.B., Kasting, J.F., 1981. A negative feedback mechanism for the long-term stabilization of the earth's surface temperature. Journal of Geophysical Research 86, 9776–9782.

Index

'*Note:* Page numbers followed by "f" indicate figures and "t" indicate tables.'

S

T

U

V

W

X

Z

FSC
www.fsc.org
MIX
Papier | Fördert
gute Waldnutzung
FSC® C083411

Kolibri 360 GmbH
Ferdinand-Jühlke-Straße 7
99095 Erfurt, Deutschland
produktsicherheit@kolibri360.de